Scientific Machine Learning

Bikram Pratim Bhuyan · Amylia Ait Saadi ·
Yassine Meraihi · Amar Ramdane-Cherif

Scientific Machine Learning

Foundations, Algorithms, and Applications

Springer

Bikram Pratim Bhuyan
LISV Laboratory
Université Paris Saclay
Vélizy-Villacoublay, France

Yassine Meraihi
LIST Laboratory
University of M'Hamed Bougara Boumerdes
Boumerdès, Algeria

Amylia Ait Saadi
LISV Laboratory, ESME Sudria
Université Paris-Saclay
Vélizy-Villacoublay, France

Amar Ramdane-Cherif
LISV Laboratory
Université Paris-Saclay
Vélizy-Villacoublay, France

ISBN 978-981-95-7855-9 ISBN 978-981-95-7856-6 (eBook)
https://doi.org/10.1007/978-981-95-7856-6

This Springer imprint is published by the registered company Springer Nature Singapore Pte Ltd.
The registered company address is: 152 Beach Road, #21-01/04 Gateway East, Singapore 189721, Singapore

Contents

Part I
Foundations

Chapter 1
Introduction to Scientific Machine Learning

Any sufficiently advanced technology is indistinguishable from magic.

Arthur C. Clarke, Profiles of the Future

Abstract Scientific Machine Learning (SciML) represents a paradigm shift in computational science: the synthesis of physics-based modeling with data-driven machine learning. This chapter introduces the field by examining its emergence from the convergence of scientific computing and deep learning, establishing a working definition that emphasizes the synergistic combination of domain knowledge and learned representations. We trace the historical arc from early neural PDE solvers through the physics-informed neural network revolution to today's operational deployments in weather forecasting, fusion energy, drug discovery, and materials science. The chapter presents a taxonomy of SciML problems (forward, inverse, control, and discovery) and methods (physics-informed networks, neural operators, hybrid models, structure-preserving architectures, and neuro-symbolic approaches), providing readers with a conceptual map for the detailed treatments that follow. We conclude with an assessment of open challenges that define the field's research frontier.

1.1 Introduction

On February 25, 2025, the European Centre for Medium-Range Weather Forecasts (ECMWF) made history: for the first time, an artificial intelligence system began issuing operational weather forecasts to national meteorological services worldwide. The AI model (running on graph neural networks at 28 km resolution) matched or exceeded the accuracy of physics-based numerical weather prediction on most metrics, while consuming approximately one thousand times less computational energy. A fifteen-day ensemble forecast that once required hours on a supercomputer now completes in eight minutes on a single chip.

B. P. Bhuyan et al., *Scientific Machine Learning*,
https://doi.org/10.1007/978-981-95-7856-6_1

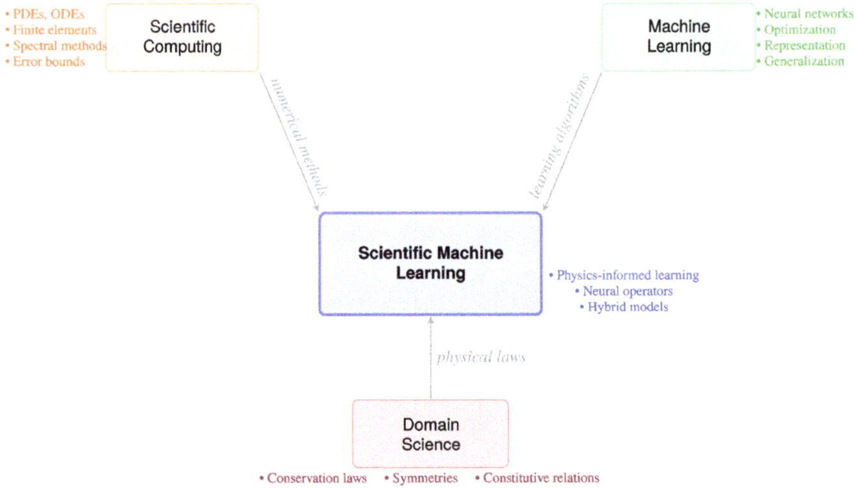

Fig. 1.1 Scientific machine learning emerges at the intersection of three disciplines: scientific computing provides numerical methods and error analysis; machine learning contributes optimization and representation learning; domain science supplies the physical laws that constrain solution spaces. The synthesis enables capabilities that none of these fields can achieve alone

Halfway around the world, at the Swiss Plasma Center, a deep reinforcement learning agent controls the magnetic coils of a tokamak fusion reactor. The agent processes ninety sensor measurements and adjusts nineteen coil voltages ten thousand times per second, shaping plasma configurations that human operators could never achieve manually. Similar AI controllers are being integrated into the SPARC tokamak under construction in Massachusetts, bringing commercial fusion energy closer to reality.

In pharmaceutical laboratories, drugs designed entirely by artificial intelligence have entered Phase 3 clinical trials; the final stage before regulatory approval. What traditionally required four to six years of development was accomplished in eighteen months. The 2024 Nobel Prize in Chemistry recognized the transformation: Demis Hassabis, John Jumper, and David Baker received the award for computational methods that have fundamentally altered structural biology and drug discovery.

These are not isolated curiosities. They represent the operational emergence of **Scientific Machine Learning**; a synthesis of physics-based modeling and data-driven learning that is reshaping computational science across every domain.

The story of Scientific Machine Learning is fundamentally about a recognition: that physical knowledge encoded in equations and empirical knowledge encoded in data are not competing sources of truth, but complementary constraints on physical reality.

For centuries, scientists have developed mathematical models of natural phenomena; Newton's laws of motion, Maxwell's equations of electromagnetism, Navier-Stokes equations for fluid flow. These models encode deep structural understanding: conservation of mass, energy, and momentum; symmetries under rotation and translation; causality and locality. When we know the governing equations, we can simulate physical systems with extraordinary fidelity. Modern weather forecasting, aircraft design, and nuclear reactor analysis all depend on this capability.

Yet physics-based simulation faces fundamental limitations. The governing equations for turbulent flow, quantum many-body systems, or protein folding are often intractable; either because analytical solutions do not exist or because numerical solutions require prohibitive computational resources. The curse of dimensionality defeats grid-based methods in high-dimensional spaces. Mesh generation for complex geometries consumes enormous human effort. And crucially, physics-based models cannot easily incorporate experimental observations; they are simulation tools, not learning systems.

Machine learning offers a different paradigm. Neural networks can approximate arbitrarily complex functions from data. They scale to high dimensions. They learn representations automatically, without manual feature engineering. And they seamlessly incorporate observational data; indeed, data is their lifeblood. The successes of deep learning in computer vision, natural language processing, and game playing demonstrate remarkable pattern recognition capabilities.

But pure data-driven learning has its own limitations. Neural networks are interpolation engines: they excel within the distribution of their training data but extrapolate poorly to new conditions. A model trained on subsonic flow may produce nonsense at supersonic speeds. Without physical constraints, predictions can violate conservation laws, produce negative concentrations, or generate non-physical solutions. And the data requirements are enormous; collecting sufficient training data for scientific applications is often impractical or impossible.

Definition: Scientific Machine Learning

Scientific Machine Learning (SciML) is the field concerned with the integration of physics-based models and machine learning methods to solve scientific and engineering problems. SciML encompasses:

- Embedding physical constraints (differential equations, conservation laws, symmetries) into machine learning architectures and training procedures
- Using machine learning to accelerate, augment, or replace components of physics-based simulations
- Learning unknown physics from data while respecting known physical principles
- Solving inverse problems that infer parameters or fields from observations.

The defining characteristic of SciML is the *synergistic combination* of domain knowledge and data-driven learning, where physics constrains the hypothesis space and data calibrates the model.

Scientific Machine Learning provides the framework for honoring all sources of knowledge (Fig. 1.1). Physical constraints act as regularizers that guide learning toward solutions consistent with governing equations, conservation laws, and symmetries. Data provides the observations that calibrate models, correct for incomplete physics, and enable inference of unknown parameters. The result is a class of methods that can:

- Solve partial differential equations without meshes, using neural networks as continuous function approximators
- Extrapolate beyond training distributions by leveraging physical structure
- Learn from sparse, noisy data by using physics as an implicit regularizer
- Solve inverse problems and forward problems within a unified framework
- Discover unknown governing equations from observational data
- Achieve orders-of-magnitude speedups over classical solvers while maintaining physical fidelity.

The stakes extend far beyond academic curiosity. Climate modeling requires simulating coupled Earth systems across scales from cloud microphysics to global circulation. Drug discovery demands predicting molecular interactions across a chemical space of 10^{60} possible compounds. Fusion energy control must operate in real-time within plasma environments too hostile for direct measurement. Materials discovery aims to search billions of candidate structures for those with desired properties. In each domain, the synthesis of physics and learning offers capabilities that neither approach can achieve alone.

1.2 What Is Scientific Machine Learning?

To understand what Scientific Machine Learning is, it helps to understand what it is *not*; and what gap it fills between existing approaches.

1.2.1 The Third Pillar, Extended

The history of science can be understood through its modes of inquiry. For millennia, **theory**; mathematical reasoning about physical principles; and **experiment**; systematic observation and measurement of nature; defined the scientific enterprise. The late twentieth century saw the emergence of a third pillar: **computational simulation**. When equations became too complex for analytical solution and experiments too expensive or dangerous, numerical simulation on computers provided a new way to investigate nature.

Scientific Machine Learning represents an extension of this third pillar into a fourth mode: **learning-augmented computation**. Where traditional simulation

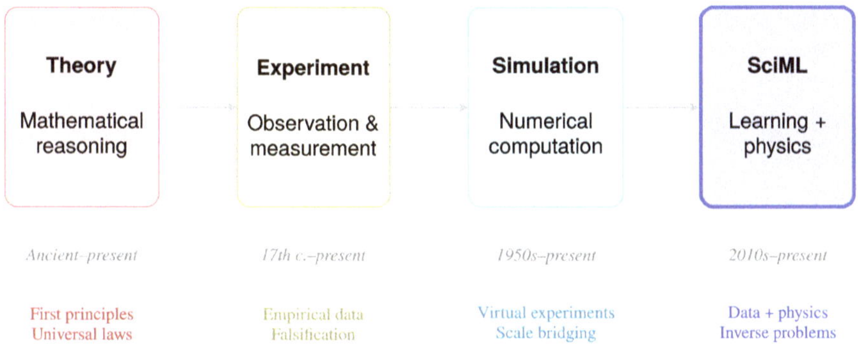

Fig. 1.2 The four pillars of scientific inquiry. Theory provides mathematical frameworks; experiment yields empirical observations; simulation enables virtual experiments; scientific machine learning extends simulation with the ability to learn from data while respecting physical constraints

solves known equations with known parameters, SciML enables computation that learns from data while respecting physical constraints. The distinction is not merely technical; it represents a different relationship between the scientist and the model (Fig. 1.2).

1.2.2 Contrasting Paradigms

To appreciate the distinctive character of SciML, consider how different approaches handle a canonical problem: predicting the temperature distribution in a material given boundary conditions and heat sources.

Classical Numerical Methods.
The traditional approach discretizes the heat equation on a mesh, yielding a system of algebraic equations solved by matrix methods. Finite difference, finite element, or spectral methods each offer trade-offs between accuracy, efficiency, and geometric flexibility. The approach is rigorous: convergence theory guarantees that as the mesh is refined, the numerical solution approaches the true solution. Error bounds can be computed a priori.

But classical methods require mesh generation; a labor-intensive process for complex geometries. They scale poorly to high dimensions (the curse of dimensionality makes three-dimensional problems expensive and higher dimensions intractable). They cannot easily incorporate observational data; if sensors measure temperature at scattered locations, this information cannot naturally inform the solution. And when physical parameters are unknown, solving the inverse problem requires expensive iterative procedures.

Pure Machine Learning.
A data-driven approach would collect input-output pairs; boundary conditions and heat sources as inputs, temperature fields as outputs; and train a neural network

	Classical Methods	Pure ML	SciML
Physical consistency	✓✓	×	✓
Data incorporation	×	✓✓	✓✓
High dimensions	×	✓	✓
Inverse problems	~	~	✓✓
Mesh requirement	Required	None	None

Fig. 1.3 Comparative strengths of classical numerical methods, pure machine learning, and Scientific Machine Learning. SciML inherits the physical consistency of classical methods while gaining the data incorporation and meshless computation of machine learning, with particularly strong capabilities for inverse problems

to approximate the mapping. Given sufficient data, the network learns to predict temperature fields for new inputs.

But where does the training data come from? Either from expensive simulations (in which case why not use the simulator directly?) or from experiments (rarely available at sufficient scale). The trained model has no knowledge of the heat equation; it may produce solutions with negative temperatures, violate energy conservation, or fail catastrophically on boundary conditions outside the training distribution. There is no error bound, no convergence guarantee, no physical interpretability.

Scientific Machine Learning.
The SciML approach trains a neural network to represent the temperature field, but incorporates the heat equation directly into the training loss. At randomly sampled points in the domain, the network must satisfy:

$$\frac{\partial u}{\partial t} = \alpha \nabla^2 u + f(x, t) \tag{1.1}$$

Boundary conditions and initial conditions are enforced similarly. The network learns a solution that is consistent with the governing physics, even without labeled data from simulations or experiments.

If sensor measurements are available, they can be incorporated as additional constraints. If the thermal diffusivity α is unknown, it can be learned jointly with the temperature field. The same framework handles forward problems (simulating given known parameters) and inverse problems (inferring parameters from observations).

1.2.3 *The Hybrid Philosophy*

The essential insight of Scientific Machine Learning is that physics and data play complementary roles (Fig. 1.3).

- **Physics as constraint**: Governing equations, conservation laws, and symmetries restrict the space of possible solutions. Rather than learning arbitrary functions, the model learns within the manifold of physically plausible solutions.
- **Data as calibration**: Observations provide information that physics alone cannot supply; parameter values, forcing functions, initial conditions, or corrections for model inadequacy.

This complementarity yields several practical advantages:

Key Advantages of the Hybrid Approach

1. **Data efficiency**: Physics constraints reduce the effective dimensionality of the learning problem. Models can learn from sparse observations that would be insufficient for pure data-driven approaches.
2. **Physical plausibility**: Solutions respect conservation laws, symmetries, and other physical constraints by construction, not by hoping the network learns them from data.
3. **Extrapolation**: Physics provides structure for generalization beyond the training distribution. A model constrained to satisfy conservation of energy cannot predict energy-violating solutions, even in novel conditions.
4. **Unified framework**: Forward simulation, inverse problems, and uncertainty quantification can be addressed within the same architecture, simply by adjusting which quantities are known and which are to be inferred.
5. **Meshless computation**: Neural networks represent continuous functions, avoiding the mesh generation bottleneck of classical methods.

1.2.4 *A Taxonomy of Problems*

Scientific Machine Learning addresses four major classes of problems, distinguished by what is known and what must be inferred (Fig. 1.3).

Forward Problems.
Given the governing equations, parameters, initial conditions, and boundary conditions, compute the solution. This is the classical simulation problem, but SciML approaches offer advantages in speed (neural surrogates), geometry (meshless methods), and dimensionality (escaping the curse of dimensionality).
Inverse Problems.

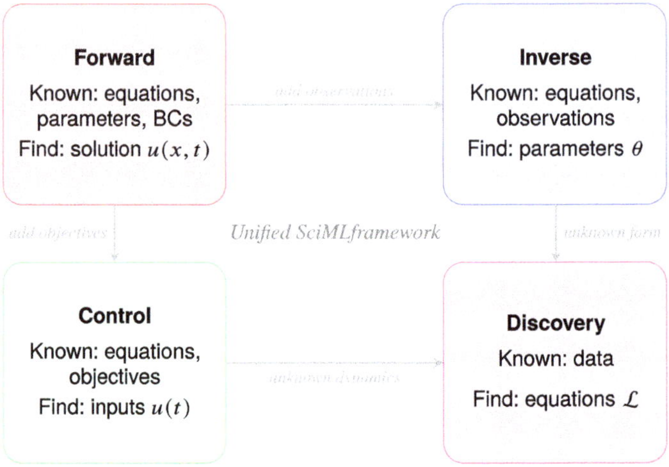

Fig. 1.4 The four classes of problems addressed by Scientific Machine Learning. Forward problems compute solutions given equations and parameters; inverse problems infer parameters from observations; control problems find optimal inputs; discovery problems learn the governing equations themselves. SciML provides a unified framework for all four

Given observations of the solution (possibly sparse, noisy, and indirect), infer the parameters, initial conditions, or forcing functions that produced it. Inverse problems are ubiquitous in science: inferring material properties from measured displacements, estimating turbulent viscosity from flow observations, reconstructing subsurface structure from seismic waves.

Control and Optimization.
Given a physical system, find the control inputs that drive it toward a desired state while respecting constraints. This encompasses optimal design (shape optimization, material selection), process control (manufacturing, chemical engineering), and active control (robotics, plasma physics).

Discovery.
Given observational data, infer the governing equations themselves. This is the most ambitious goal: automated scientific discovery, where the form of physical laws is learned from measurements (Fig. 1.4).

1.3 A Brief History: From Neural PDEs to Operational Systems

The intellectual roots of Scientific Machine Learning extend deeper than the recent explosion of deep learning might suggest. Understanding this history illuminates both the conceptual foundations and the technological catalysts that enabled the field's current capabilities.

1.3.1 Early Precursors (1990s–2000s)

The idea of using neural networks to solve differential equations appeared in the literature as early as 1998, when Lagaris, Likas, and Fotiadis demonstrated that feedforward networks could approximate solutions to both ordinary and partial differential equations [1]. Their approach; training networks to satisfy differential equations at collocation points; contained the conceptual kernel of what would later become physics-informed neural networks.

But the approach languished for nearly two decades. The neural networks of the 1990s were shallow, training was slow, and the available optimization methods struggled with the complex loss landscapes induced by differential equation constraints. More fundamentally, the scientific computing community had mature, well-understood methods; finite elements, spectral methods, multigrid solvers; that worked reliably. Why trade proven convergence theory for the black-box behavior of neural networks?

During this period, machine learning developed largely independently. The focus was on pattern recognition: image classification, speech recognition, natural language processing. The connection to physical simulation seemed distant.

1.3.2 The Deep Learning Catalyst (2012–2017)

The ImageNet revolution of 2012 transformed machine learning. Deep convolutional networks, trained on GPUs with large datasets, achieved dramatic improvements in image classification. The subsequent years saw deep learning penetrate domain after domain: object detection, machine translation, game playing.

Several technical advances proved crucial for Scientific Machine Learning:

- **GPU acceleration**: Training neural networks became thousands of times faster, making previously intractable optimization problems feasible.
- **Automatic differentiation**: Frameworks like TensorFlow (2015) and PyTorch (2016) made computing gradients of arbitrary computational graphs routine. This capability is essential for physics-informed learning, where loss functions involve derivatives of the network with respect to its inputs.
- **Modern architectures**: Residual connections, normalization techniques, and attention mechanisms produced networks that could be trained to greater depth and expressivity.
- **Software ecosystems**: The democratization of deep learning through open-source frameworks lowered barriers to entry for researchers in other fields.

These advances set the stage for a convergence with scientific computing.

1.3.3 The PINN Revolution (2017–2019)

The watershed moment arrived in 2019 with the publication of "Physics-informed neural networks" [2]. The paper introduced a unified framework that embedded differential equations directly into the neural network training loss:

$$\mathcal{L} = \mathcal{L}_{\text{data}} + \lambda \mathcal{L}_{\text{physics}} \tag{1.2}$$

where $\mathcal{L}_{\text{data}}$ penalizes mismatch with observations and $\mathcal{L}_{\text{physics}}$ penalizes violation of the governing equations at collocation points. The same architecture could solve forward problems (simulating with known parameters), inverse problems (inferring parameters from data), and data assimilation problems (combining sparse observations with physical models). The impact was immediate and substantial. The paper has accumulated over 19,500 citations as of late 2025, spawning an entire ecosystem of extensions, variations, and applications. The term "physics-informed neural networks" entered the vocabulary of computational scientists worldwide.

1.3.4 Methodological Expansion (2019–2023)

The years following the PINN paper saw rapid methodological diversification:

Neural ODEs (2018).
Chen et al. [3] introduced neural ordinary differential equations, receiving the NeurIPS Best Paper Award. By treating the layers of a neural network as a continuous transformation governed by a differential equation, neural ODEs established deep connections between neural networks and dynamical systems theory.

DeepONet (2021).
Lu et al. [4] introduced the Deep Operator Network, which learns mappings between function spaces rather than between finite-dimensional vectors. Grounded in the universal approximation theorem for operators, DeepONet enabled neural networks to learn solution operators; maps from input functions (boundary conditions, forcing terms) to output functions (solutions).

Fourier Neural Operators (2021).
Li et al. [5], from Anima Anandkumar's group at Caltech, introduced neural operators that learn in Fourier space. By parameterizing convolutional kernels in the frequency domain, Fourier Neural Operators achieved resolution invariance and demonstrated speedups of 1000× or more over traditional PDE solvers for turbulent flow problems.

Universal Differential Equations (2021).
Rackauckas et al. [6] introduced a framework for embedding neural networks within classical differential equation solvers. Rather than replacing physics with learning, Universal Differential Equations use neural networks to represent unknown components (constitutive relations, closure terms) while retaining classical numerical methods for time integration and known physics.

Structure-Preserving Architectures.
Hamiltonian Neural Networks, Lagrangian Neural Networks, and symplectic integrators embedded physical structure directly into network architectures, guaranteeing conservation laws by construction rather than hoping networks learn them from data.

The Method Landscape

By 2023, the SciML landscape included multiple complementary approaches:

- **PINNs**: Physics as soft constraints in the loss function
- **Neural Operators**: Learning solution operators between function spaces
- **Hybrid Models**: Neural components within classical solvers
- **Structure-Preserving Networks**: Physics encoded in architecture
- **Neuro-Symbolic Methods**: Combining neural learning with symbolic reasoning.

Each approach offers distinct advantages depending on the problem structure, available data, and computational requirements.

1.3.5 The Operational Era (2024–Present)

The period from 2023 to 2025 marked the transition from research demonstrations to operational deployment. Scientific Machine Learning moved from laboratory prototypes to production systems handling real-world decisions.

The most visible milestone came in weather forecasting. Google DeepMind's GraphCast [7], published in 2023, demonstrated that graph neural networks could outperform ECMWF's operational physics-based model on most metrics. GenCast [8], extended this to probabilistic ensemble forecasting, matching the physics-based ensemble on 97.2% of 1320 forecast metrics. When ECMWF launched its own AIFS system operationally in February 2025, Scientific Machine Learning officially entered the infrastructure of global weather prediction.

Similar transitions occurred across domains. DeepMind's AlphaFold had already transformed structural biology; the 2024 Nobel Prize in Chemistry recognized its impact. Materials discovery accelerated through GNoME, which identified 2.2 million new crystal structures. Pharmaceutical companies invested billions in AI-driven drug discovery, with the first fully AI-designed drugs reaching advanced clinical trials.

The field had matured from "interesting research direction" to "proven technology for high-stakes applications."

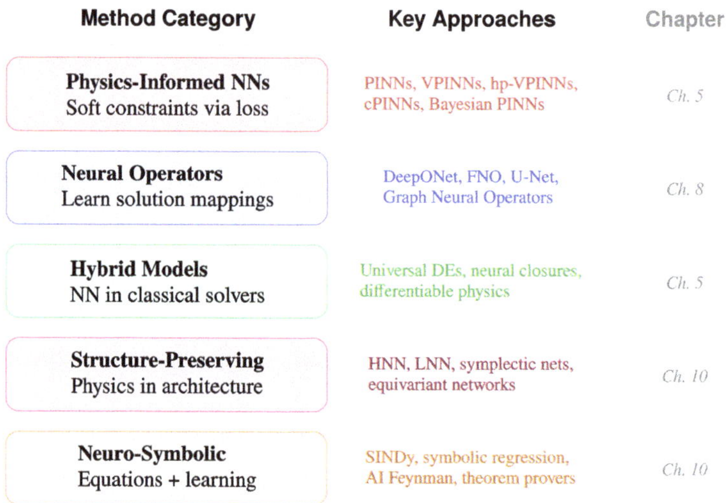

Fig. 1.5 Taxonomy of scientific machine learning methods, organized by how physical knowledge enters the learning framework. Each category is covered in detail in the indicated chapters

1.4 The SciML Landscape: Methods and Applications

Having traced the historical development of Scientific Machine Learning, we now provide a systematic overview of the methods and applications that constitute the field. This section serves as a map to the terrain explored in subsequent chapters.

1.4.1 A Taxonomy of Methods

Scientific Machine Learning encompasses multiple methodological approaches, distinguished by how physical knowledge is incorporated into the learning framework (see Fig. 1.5).

Physics-Informed Neural Networks (PINNs).

The most widely adopted approach embeds physical constraints as penalty terms in the training loss. A neural network $u_\theta(x, t)$ approximates the solution, and the loss includes terms measuring PDE residual, boundary condition violations, and (optionally) data mismatch:

$$\mathcal{L}(\theta) = \underbrace{\frac{1}{N_d}\sum_{i=1}^{N_d} |u_\theta(x_i) - u_i^{\text{obs}}|^2}_{\text{data loss}} + \lambda_{\text{pde}} \underbrace{\frac{1}{N_r}\sum_{j=1}^{N_r} |\mathcal{N}[u_\theta](x_j)|^2}_{\text{physics loss}} + \lambda_{\text{bc}}\mathcal{L}_{\text{bc}} \tag{1.3}$$

This "soft constraint" approach is flexible and easy to implement but requires careful tuning of loss weights.

Neural Operators.
Rather than learning a single solution, neural operators learn the mapping from input functions to output functions; the solution operator itself. DeepONet uses a branch-trunk architecture; Fourier Neural Operators parameterize convolutions in spectral space. Once trained, a neural operator can evaluate solutions for new inputs without retraining, enabling real-time inference.

Hybrid Models.
Universal Differential Equations and related approaches embed neural networks within classical numerical solvers. Known physics is handled by established methods (Runge-Kutta integrators, finite element discretizations); neural networks represent only the unknown components (constitutive relations, subgrid closures). This preserves the stability and interpretability of classical methods while gaining the flexibility of learning.

Structure-Preserving Architectures.
Hamiltonian and Lagrangian neural networks encode energy-based physics directly into the network architecture. Symplectic integrators preserve phase-space volume. Equivariant networks respect symmetries (rotation, translation, permutation). These "hard constraint" approaches guarantee physical properties by construction, at the cost of reduced architectural flexibility.

Neuro-Symbolic Methods.
Symbolic regression (SINDy, PySR, AI Feynman) seeks to discover interpretable equations from data. Logic-augmented approaches combine neural networks with formal reasoning. The goal is not just prediction, but understanding; producing symbolic expressions that scientists can analyze and verify.

1.4.2 Application Domains

Scientific Machine Learning finds applications across the full spectrum of science and engineering. We highlight several domains where the impact has been most substantial (sneak peek at Table 1.1).

1.5 Open Challenges and Research Frontiers

Despite remarkable operational successes, fundamental challenges remain that limit broader deployment and define active research priorities. Honest assessment of these limitations is essential both for setting realistic expectations and for identifying the most important directions for future work.

Table 1.1 Representative SciML applications across domains, with key methods and exemplar systems

Domain	Applications	Primary methods	Exemplar systems
Fluid dynamics	Turbulence modeling, flow prediction, aerodynamics	Neural operators, PINNs, hybrid closures	FourCastNet, CFD surrogates
Weather and Climate	Forecasting, downscaling, extreme events	Graph NNs, diffusion models	GraphCast, GenCast, AIFS
Materials science	Property prediction, structure discovery	Equivariant GNNs, potentials	GNoME, MACE, NequIP
Drug discovery	Binding prediction, molecule generation	Diffusion models, GNNs	AlphaFold, DiffDock
Plasma physics	Tokamak control, disruption prediction	RL, PINNs, surrogates	DeepMind tokamak
Structural Eng.	Stress analysis, health monitoring	PINNs, neural operators	Digital twins
Robotics	Dynamics learning, control	Hamiltonian NNs, RL	Sim-to-real transfer

Generalization Beyond Training Distributions

The most significant limitation of current SciML methods is their difficulty generalizing beyond training distributions. Neural operators and physics-informed networks perform well within the regime of their training data, but accuracy degrades; sometimes catastrophically; for out-of-distribution inputs.

This limitation is particularly severe because *extrapolation is often precisely what scientific applications require*. We want to predict behavior at higher Reynolds numbers than any in the training set, under operating conditions never previously encountered, or for geometries that differ substantially from training examples. Physics constraints help; they prevent certain classes of unphysical predictions; but do not guarantee accurate extrapolation.

Long-Time Integration and Stability

Autoregressive neural operators accumulate errors over time. Unlike classical numerical methods, which have rigorous stability theory (CFL conditions, A-stability, symplectic integration), neural networks lack formal stability certificates. Small errors compound exponentially, and trajectories diverge from the true solution.

Most published evaluations assess performance only within training time horizons, carefully avoiding the regime where methods fail. Emerging approaches; embedding classical time integrators within neural architectures, enforcing PDE consistency at each step, using symplectic or energy-conserving networks; offer partial solutions, but comprehensive stability guarantees remain elusive.

Uncertainty Quantification

Mission-critical applications in aerospace, nuclear engineering, and medicine require not just predictions but confidence intervals. Current approaches to uncertainty quantification in SciML include Bayesian PINNs that place distributions over network weights, ensemble methods that aggregate predictions from multiple models, and deep Gaussian processes that provide uncertainty estimates.

Each approach has limitations: Bayesian methods are computationally expensive; ensemble methods require training multiple models; calibration across different physical regimes is challenging.

Additional Challenges

Several other challenges warrant mention:

- **Spectral bias**: Neural networks preferentially learn low-frequency components, struggling with multi-scale phenomena where sharp gradients or fine features are important.
- **Curse of dimensionality**: While neural networks scale better than grid-based methods, very high-dimensional PDEs remain challenging.
- **Interpretability**: The representations learned by neural networks often lack clear physical interpretation, making it difficult to extract scientific insight.
- **Computational cost**: While inference is fast, training physics-informed models can be expensive, sometimes requiring weeks of GPU time for complex problems.

Warning: Current Limitations

SciML methods are powerful but not panaceas. Users should be aware:

- Extrapolation beyond training distributions remains unreliable
- Long-time integration requires careful attention to stability
- Uncertainty estimates are often poorly calibrated
- Verification and validation frameworks are immature
- Training can be computationally expensive and sensitive to hyperparameters

The field is advancing rapidly, but current methods require informed application.

1.6 About This Book

This textbook provides a comprehensive treatment of Scientific Machine Learning, from mathematical foundations through algorithmic methods to application case studies. Our goal is to equip readers with both the conceptual understanding and practical skills needed to apply SciML methods to real problems.

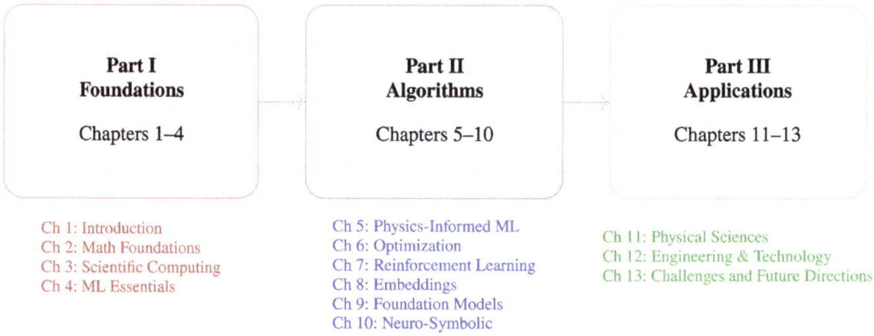

Fig. 1.6 Organization of the book into three parts: foundations establishes the mathematical and computational prerequisites; algorithms covers the core SciML methods; applications demonstrates deployment in physical sciences and engineering

Scope and Philosophy

We adopt a philosophy of **progressive complexity**, presenting each topic through multiple layers of abstraction as shown in Fig. 1.6.

1. **Everyday analogies** that build intuition from familiar concepts
2. **Visual diagrams** that illustrate key ideas geometrically
3. **Conceptual explanations** in accessible prose
4. **Mathematical formulations** with rigorous definitions and theorems
5. **Working implementations** with actual computational results

This layered approach serves multiple audiences: general readers can engage with the intuitive explanations and diagrams; graduate students can work through the mathematical development; practitioners can focus on implementations and code examples.

Structure of the Book

The book is organized into three parts:

Part I: Foundations (Chaps. 1–4).

Part I establishes the mathematical and computational prerequisites for SciML. Chapter 2 covers linear algebra, calculus, function spaces, differential equations, optimization, probability, and numerical stability. Chapter 3 surveys scientific computing paradigms and software frameworks (JAX, PyTorch, Julia/SciML). Chapter 4 introduces machine learning concepts for readers from scientific computing backgrounds.

Part II: Algorithms (Chaps. 5–10).

Part II forms the technical core, presenting the major algorithmic approaches: physics-informed learning, optimization, reinforcement learning, embeddings, foundation models, and neuro-symbolic methods.

Part III: Applications (Chaps. 11 and 12).

Part III demonstrates SciML methods in practice across physical sciences and engineering domains, with working implementations and benchmark comparisons.

How to Read This Book

Different readers will approach this book with different backgrounds and goals:

- **General readers** can focus on intuitive explanations, diagrams, and colored boxes
- **Graduate students** should work through the mathematical development carefully
- **Practitioners** can focus on implementations and application chapters
- **Researchers** may find advanced sections and open problems most valuable.

1.7 The Road Ahead

Scientific Machine Learning is a young field, and its most transformative applications likely lie ahead. We conclude by highlighting three directions that will shape the coming years.

Scientific Foundation Models

Just as large language models have transformed natural language processing, foundation models for science promise to transform computational science. These are large neural networks pre-trained on massive scientific datasets, then fine-tuned for specific tasks. Early examples include Microsoft's Aurora for atmospheric science and the Polymathic AI project building models that transfer knowledge across physics, chemistry, and biology.

Self-Driving Laboratories

The integration of AI planning, robotic synthesis, and automated analysis enables closed-loop experimentation without human intervention. These "self-driving laboratories" can explore experimental spaces far faster than traditional approaches, accessing conditions too dangerous or tedious for human researchers.

Physics-Informed Digital Twins

Digital twins; virtual replicas updated continuously from sensor data; gain new capabilities through SciML. By 2027, analysts predict that 50% of large industrial companies will utilize digital twins with over 21 billion connected sensors. The European Destination Earth initiative is building digital twins of the entire Earth system for climate adaptation.

1.8 Conclusion

The convergence of physics-based modeling and machine learning represents a recognition that theoretical knowledge and empirical knowledge are complementary constraints on physical reality. Neither alone is sufficient; together, they enable capabilities that would be impossible otherwise. For researchers and practitioners entering this field, the opportunity is substantial: to participate in reshaping how computational science operates, from algorithm design through industrial deployment. The chapters that follow provide the foundations needed to engage with this transformation. The remainder of this book develops these ideas systematically. We begin with the mathematical foundations that underpin Scientific Machine Learning.

References

1. Lagaris, I.E., Likas, A., Fotiadis, D.I.: Artificial neural networks for solving ordinary and partial differential equations. IEEE Trans. Neural Netw. **9**(5), 987–1000 (1998)
2. Raissi, M., Perdikaris, P., Karniadakis, G.E.: Physics-informed neural networks: a deep learning framework for solving forward and inverse problems involving nonlinear partial differential equations. J. Comput. Phys. **378**, 686–707 (2019)
3. Chen, R.T., Rubanova, Y., Bettencourt, J., Duvenaud, D.K.: Neural ordinary differential equations. In: Advances in Neural Information Processing Systems, p. 31 (2018)
4. Lu, L., Jin, P., Pang, G., Zhang, Z., Karniadakis, G.E.: Learning nonlinear operators via deeponet based on the universal approximation theorem of operators. Nat. Mach. Intell. **3**(3), 218–229 (2021)
5. Li, Z., Kovachki, N.B., Azizzadenesheli, K., Bhattacharya, K., Stuart, A., Anandkumar, A.: Fourier neural operator for parametric partial differential equations. In: International Conference on Learning Representations
6. Rackauckas, C., Ma, Y., Martensen, J., Warner, C., Zubov, K., Supekar, R., Skinner, D., Ramadhan, A., Edelman, A.: Universal differential equations for scientific machine learning. arXiv preprint arXiv:2001.04385 (2020)
7. Lam, R., Sanchez-Gonzalez, A., Willson, M., Wirnsberger, P., Fortunato, M., Alet, F., Ravuri, S., Ewalds, T., Eaton-Rosen, Z., Hu, W., et al.: Learning skillful medium-range global weather forecasting. Science **382**(6677), 1416–1421 (2023)
8. Price, I., Sanchez-Gonzalez, A., Alet, F., Andersson, T.R., El-Kadi, A., Masters, D., Ewalds, T., Stott, J., Mohamed, S., Battaglia, P.: GenCast: diffusion-based ensemble forecasting for medium-range weather. arXiv preprint arXiv:2312.15796 (2023)

Chapter 2
Mathematical and Computational Foundations

Do not worry about your difficulties in Mathematics. I can assure you mine are still greater.

Albert Einstein, a letter to a struggling student

Abstract This chapter establishes the mathematical language and computational tools that underpin Scientific Machine Learning. We begin with linear algebra foundations that govern both numerical stability and learning dynamics. The treatment of calculus emphasizes automatic differentiation and the adjoint-backpropagation equivalence that makes physics-informed training computationally tractable. Function spaces and variational methods provide the infinite-dimensional perspective essential for understanding PDEs, Galerkin methods, and neural operator theory. We develop the taxonomy of differential equations (ODEs and PDEs) with attention to well-posedness, discretization, and the interplay between continuous and discrete analysis. Optimization fundamentals cover the gradient-based methods, constraints, and nonconvex landscapes characteristic of SciML training. Probability and information theory support uncertainty quantification and Bayesian approaches. Throughout, we emphasize numerical stability, floating-point considerations, and differentiable programming patterns that distinguish practical SciML implementations from textbook formulations.

2.1 Introduction

In the summer of 2029, a graduate student at MIT (let's call her Maya) spent three frustrating weeks debugging what she was certain was a software error. Her physics-informed neural network, designed to solve Burgers' equation for shock wave propagation, produced beautiful results for smooth initial conditions but generated complete nonsense when shocks formed.

She checked her code line by line. She verified her automatic differentiation. She tested on simpler equations. Everything worked perfectly—until it didn't.

B. P. Bhuyan et al., *Scientific Machine Learning*,
https://doi.org/10.1007/978-981-95-7856-6_2

The breakthrough came not from debugging but from her advisor's bookshelf. Buried in a 1983 numerical analysis textbook, she found the answer: her network was working flawlessly. The *mathematics* was the problem.

Burgers' equation develops discontinuous solutions (shocks) from smooth initial data. The classical notion of derivative simply doesn't apply at a shock. Maya's loss function, which penalized the PDE residual computed via automatic differentiation, was asking her network to satisfy an equation that, in the strong sense, had no solution. The mathematics she needed wasn't calculus but the theory of weak solutions, Sobolev spaces, and distributions.

"That textbook chapter," Maya later told her research group, "saved me months. I realized I wasn't fighting my code. I was fighting a century of mathematical physics that I'd never properly learned."

Maya's story is not unique. The researchers pushing Scientific Machine Learning forward are rediscovering a truth that numerical analysts have known for decades: *the mathematics matters*. Not as an abstract exercise, but as a practical tool that determines whether your code works or fails in ways that no amount of hyperparameter tuning can fix.

2.1.1 Why This Chapter Exists

This is not a mathematics textbook. We will not prove every theorem or develop every idea from first principles. But we will establish the mathematical vocabulary and core concepts that make the rest of this book meaningful.

The challenge is that Scientific Machine Learning sits at an unusual intersection. It draws on:

- **Linear algebra**—not just matrix multiplication, but conditioning, sensitivity, and the geometry of high-dimensional spaces
- **Calculus**—not just derivatives, but automatic differentiation, adjoints, and the chain rule at scale
- **Functional analysis**—Sobolev spaces, weak solutions, and the infinite-dimensional perspective essential for PDEs
- **Differential equations**—existence, uniqueness, stability, and the taxonomy of PDEs
- **Optimization**—convexity, constraints, stochastic gradients, and nonconvex landscapes
- **Probability**—Bayesian inference, concentration inequalities, and information geometry
- **Numerical analysis**—floating-point arithmetic, conditioning, and the difference between mathematical and computational reality.

Few researchers arrive with deep expertise in all these areas. Physicists may know PDEs but not optimization. Machine learning researchers may know backpropagation but not functional analysis. Engineers may know numerical methods but not measure theory.

This chapter provides the common foundation. We emphasize the concepts that matter most for SciML, the connections between ideas from different fields, and the practical implications that determine whether implementations succeed or fail.

2.2 Linear Algebra Foundations

2.2.1 *Vector Spaces, Norms, Inner Products*

In Scientific Machine Learning (SciML), most of the objects we work with can be described as vectors. The weights of a neural network, the state of a dynamical system, the columns of a data matrix, or even a function sampled at discrete points all live naturally in a *vector space*. Thinking in vector spaces gives us a common language for talking about linear combinations (superposition), bases (feature sets or modes), and dimension (degrees of freedom).

Definition: Vector Space

A **vector space** over a field $\mathbb{F}$ (typically $\mathbb{R}$ or $\mathbb{C}$) is a set V equipped with two operations: vector addition and scalar multiplication. For all $u, v, w \in V$ and $\alpha, \beta \in \mathbb{F}$:

$$(u + v) + w = u + (v + w), \quad u + v = v + u,$$
$$\exists 0 \in V : u + 0 = u, \quad \forall u \exists -u : u + (-u) = 0,$$

$$\alpha(u + v) = \alpha u + \alpha v, \quad (\alpha + \beta)u = \alpha u + \beta u, \quad (\alpha\beta)u = \alpha(\beta u), \quad 1 \cdot u = u.$$

The elements of V are called *vectors*, and the elements of $\mathbb{F}$ are *scalars*.

To see this in action, let us take a simple running example in $\mathbb{R}^3$. Consider (a 2D projection is shown in Fig. 2.1)

$$\mathbf{v} = (2, -1, 3), \qquad \mathbf{w} = (1, 4, -2).$$

If we scale $\mathbf{v}$ by 2, we obtain $2\mathbf{v} = (4, -2, 6)$. If we add $\mathbf{v}$ and $\mathbf{w}$, we obtain $\mathbf{v} + \mathbf{w} = (3, 3, 1)$. Both results remain in $\mathbb{R}^3$, demonstrating closure under addition and scalar multiplication. These simple operations illustrate the heart of vector spaces: we can add and scale, and the results always stay inside.

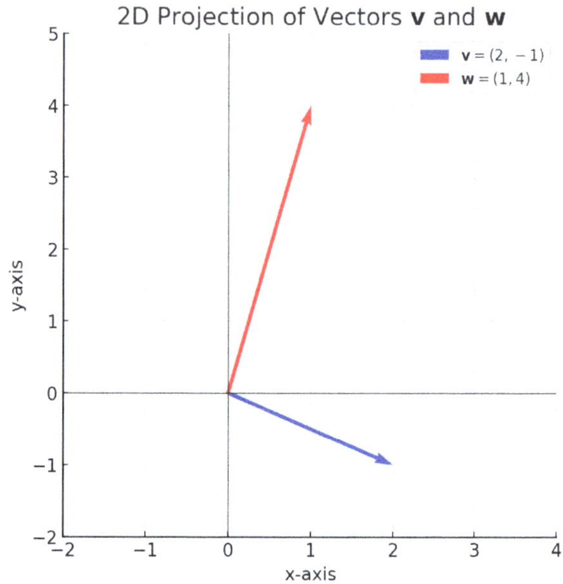

Fig. 2.1 Example vectors

Vector spaces arise in many guises in SciML. Parameter vectors typically live in $\mathbb{R}^n$. Matrices $\mathbb{R}^{m\times n}$ form a vector space under entrywise addition and scaling, and while we often flatten matrices into long vectors, remembering their shape is useful for interpreting Jacobians, kernels, and weights. Infinite-dimensional function spaces [1], such as $C([a, b]; \mathbb{R})$ of continuous functions on an interval, are also vector spaces: here addition is pointwise, $(f + g)(x) = f(x) + g(x)$, and scaling is $(\alpha f)(x) = \alpha f(x)$. The zero vector in this setting is simply the function that is identically zero. Many scientific models start in such function spaces before we discretize them.

Once we know how to add and scale vectors, we can ask: how do we measure size? For this we need a *norm*. A norm assigns a nonnegative length to every vector and ensures that scaling or combining vectors behaves in line with our geometric intuition.

Definition: Norm

A *norm* on a vector space V is a map $\|\cdot\| : V \to \mathbb{R}_{\geq 0}$ such that for all $u, v \in V$ and $\alpha \in \mathbb{F}$:

$$\|u\| = 0 \iff u = 0, \qquad \|\alpha u\| = |\alpha|\,\|u\|, \qquad \|u + v\| \leq \|u\| + \|v\|.$$

For our example $\mathbf{v} = (2, -1, 3)$, we can compute several common norms:

$$\|\mathbf{v}\|_2 = \sqrt{2^2 + (-1)^2 + 3^2} = \sqrt{14}, \qquad \|\mathbf{v}\|_1 = |2| + |-1| + |3| = 6,$$
$$\|\mathbf{v}\|_\infty = \max\{2, 1, 3\} = 3.$$

The 2-norm gives the familiar Euclidean length, the 1-norm sums magnitudes (useful in sparse modeling), and the ∞-norm measures the largest entry (important for uniform error bounds and step-size control).

Norms also extend naturally to matrices. With the Euclidean norm, the induced matrix norm $\|A\|_2$ equals the largest singular value of A, telling us how much A can stretch a vector. The Frobenius norm $\|A\|_F = \left(\sum_{i,j} A_{ij}^2\right)^{1/2}$ treats a matrix like a big vector of entries, and the ∞-norm $\|A\|_\infty = \max_i \sum_j |A_{ij}|$ measures the largest row sum. Weighted norms, such as $\|u\|_W = \sqrt{u^\top W u}$ for a symmetric positive definite matrix W, allow us to emphasize certain directions, encoding anisotropy or measurement fidelity.

While norms give us a notion of size, they do not yet tell us how to measure orientation. To compare vectors in terms of direction, we need an *inner product.*

Definition: Inner Product

An *inner product* on V is a map $\langle \cdot, \cdot \rangle : V \times V \to \mathbb{F}$ such that for all $u, v, w \in V$ and $\alpha \in \mathbb{F}$:

$$\langle u + v, w \rangle = \langle u, w \rangle + \langle v, w \rangle, \qquad \langle \alpha u, v \rangle = \alpha \langle u, v \rangle,$$
$$\langle v, u \rangle = \overline{\langle u, v \rangle}, \qquad \langle u, u \rangle > 0 \text{ if } u \neq 0.$$

In $\mathbb{R}^n$, the standard inner product is the dot product $u^\top v$. For our example,

$$\langle \mathbf{v}, \mathbf{w} \rangle = 2 \cdot 1 + (-1) \cdot 4 + 3 \cdot (-2) = -8.$$

The negative result tells us that $\mathbf{v}$ and $\mathbf{w}$ form an obtuse angle. More generally, the angle θ between two nonzero vectors satisfies

$$\cos\theta = \frac{\langle u, v \rangle}{\|u\| \, \|v\|}.$$

Orthogonality corresponds to $\cos\theta = 0$. The Cauchy–Schwarz inequality, $|\langle u, v \rangle| \leq \|u\| \, \|v\|$, guarantees that this definition of angle makes sense.

In function spaces, the L^2 inner product

$$\langle f, g \rangle = \int_a^b f(x)g(x)\,dx$$

induces the energy norm $\|f\|_{L^2} = \left(\int_a^b |f(x)|^2 dx\right)^{1/2}$, central to error analysis in differential equations and physics-informed learning. Orthogonality in this setting means two functions are uncorrelated over the domain. These concepts are the foundation for conditioning, stability, and approximation error, and they will support everything from projections and Krylov methods [2] to error analysis and algorithm design.

2.2.2 *Operators, Conditioning, and Sensitivity*

Now that we have a language for size and angle, we turn to *linear operators*. In finite dimensions, these are simply matrices acting on vectors. Operators are central to SciML: they encode the dynamics of physical systems, the transformations inside neural networks, and the Jacobians and Hessians that govern optimization. Understanding how operators amplify or distort vectors allows us to reason about stability, error growth, and conditioning.

Definition: Linear Operator

A **linear operator** $A : V \to W$ between vector spaces V and W satisfies

$$A(\alpha u + \beta v) = \alpha Au + \beta Av \quad \text{for all } u, v \in V,\ \alpha, \beta \in \mathbb{F}.$$

In finite dimensions, linear operators are represented by matrices.

Let us return to our running example. Consider the matrix

$$A = \begin{bmatrix} 1 & 2 & 0 \\ 0 & -1 & 1 \\ 2 & 0 & 3 \end{bmatrix}, \qquad \mathbf{v} = (2, -1, 3).$$

Applying A to $\mathbf{v}$ gives

$$A\mathbf{v} = \begin{bmatrix} 1 & 2 & 0 \\ 0 & -1 & 1 \\ 2 & 0 & 3 \end{bmatrix} \begin{bmatrix} 2 \\ -1 \\ 3 \end{bmatrix} = \begin{bmatrix} 0 \\ 4 \\ 13 \end{bmatrix}.$$

Here A acts as a transformation: it rotates, scales, and mixes the components of $\mathbf{v}$.

The natural question is: how much can A amplify a vector? This leads to the notion of an *operator norm*.

Definition: Operator Norm

Given a vector norm $\|\cdot\|$ on V, the induced operator norm of $A : V \to V$ is

$$\|A\| := \sup_{x\neq 0} \frac{\|Ax\|}{\|x\|}.$$

For the Euclidean norm, $\|A\|_2$ is the largest singular value of A. This number tells us the maximum stretch factor of A: for any x, $\|Ax\|_2 \leq \|A\|_2\|x\|_2$. In our example, computing the singular values of A shows that the largest one is about 3.92. This means A can at most amplify the length of a vector by a factor of 3.92.

Why does this matter? Because amplification governs *sensitivity* to perturbations. Suppose we want to solve the linear system $Ax = b$. If b is perturbed to $b + \delta b$, the solution changes by $x + \delta x$, and the relative error satisfies

$$\frac{\|\delta x\|}{\|x\|} \lesssim \kappa(A) \frac{\|\delta b\|}{\|b\|},$$

where $\kappa(A)$ is the *condition number*.

Definition: Condition Number

For a nonsingular matrix A, the condition number with respect to the Euclidean norm is

$$\kappa(A) = \|A\|_2 \, \|A^{-1}\|_2.$$

The condition number measures how sensitive the solution of $Ax = b$ is to small changes in b. If $\kappa(A)$ is close to 1, the system is well-conditioned: small perturbations in b cause small changes in x. If $\kappa(A)$ is very large, the system is ill-conditioned: tiny perturbations can lead to large errors in the solution.

For our matrix A, we compute $\kappa(A) \approx 9.7$. This means solving $Ax = b$ is moderately stable: a 1% error in b can translate into nearly a 10% error in x. In large-scale scientific problems, condition numbers can be orders of magnitude larger, which is why preconditioning and stability analysis become essential.

The same ideas extend beyond linear systems. In optimization, the conditioning of the Hessian matrix controls convergence rates of gradient methods. In SciML models, conditioning reflects how errors propagate through time-stepping, PDE discretizations, and deep architectures.

By combining the language of vector spaces, norms, and operators, we now have a way to quantify not only the size of vectors, but also the distortion caused by transformations. This framework will serve as the basis for adjoints, differentiable programming, and ultimately the stability of scientific machine learning algorithms.

2.2.3 Adjoints and Duality

Once we understand operators and conditioning, the next step is to consider their *adjoints*. Adjoints capture how operators interact with inner products, and they are central to sensitivity analysis [3, 4], gradient computation [5], and variational principles [6] in SciML. Almost every gradient-based method we encounter will (with exceptions of-course [7]), under the hood, use adjoints or their generalizations.

Definition: Adjoint Operator

Let $A : V \to V$ be a linear operator on an inner product space $(V, \langle \cdot, \cdot \rangle)$. The **adjoint** A^* is defined by the property

$$\langle Au, v \rangle = \langle u, A^* v \rangle \quad \text{for all } u, v \in V.$$

In $\mathbb{R}^n$ with the standard inner product, the adjoint is simply the transpose: $A^* = A^\top$. In complex spaces, A^* is the conjugate transpose $A^\dagger$.

Let us revisit our running example:

$$A = \begin{bmatrix} 1 & 2 & 0 \\ 0 & -1 & 1 \\ 2 & 0 & 3 \end{bmatrix}, \qquad \mathbf{v} = (2, -1, 3), \quad \mathbf{w} = (1, 4, -2).$$

The adjoint in this case is $A^* = A^\top$:

$$A^\top = \begin{bmatrix} 1 & 0 & 2 \\ 2 & -1 & 0 \\ 0 & 1 & 3 \end{bmatrix}.$$

Indeed, for these specific vectors we can check:

$$\langle A\mathbf{v}, \mathbf{w}\rangle = (0, 4, 13) \cdot (1, 4, -2) = -18,$$
$$\langle \mathbf{v}, A^\top \mathbf{w}\rangle = (2, -1, 3) \cdot (-7, 2, 10) = -18.$$

The two match exactly, confirming the defining property.

Adjoints are not only abstract concepts: they are computational workhorses. When we compute gradients through backpropagation, we are applying the adjoint of the Jacobian to the gradient signal. In time-dependent problems, adjoint equations give us sensitivities of outputs with respect to parameters at a fraction of the cost of forward sensitivity analysis. This efficiency is what makes adjoints indispensable in SciML.

Definition: Dual Space

For a vector space V, the **dual space** V^* is the set of all linear functionals $z : V \to \mathbb{F}$, i.e. maps that satisfy $z(\alpha u + \beta v) = \alpha z(u) + \beta z(v)$.

In finite dimensions, every linear functional can be represented as an inner product with some vector. That is, for any $z \in V^*$, there exists $w \in V$ such that $z(u) = \langle w, u\rangle$. This correspondence is called the Riesz representation theorem.

Returning to our example, if we define a functional $z(u) = u_1 + 4u_2 - 2u_3$, then $z(u) = \langle(1, 4, -2), u\rangle$, meaning the vector $(1, 4, -2)$ represents this functional. This is precisely our earlier vector $\mathbf{w}$, showing how dual vectors naturally appear in computations.

Adjoints and duality are therefore two sides of the same coin: adjoints describe how operators move across inner products, and duality tells us that linear functionals are themselves inner products with some vector. Together, they form the foundation of gradient-based optimization, backpropagation, and sensitivity analysis that drive modern scientific machine learning.

2.2.4 *Matrix Factorizations*

Matrix factorizations are one of the most powerful tools in numerical linear algebra. They allow us to decompose a matrix into simpler building blocks that reveal its structure, make computations more efficient, and improve stability. In SciML, factorizations appear everywhere: in solving linear systems, computing least-squares fits, reducing models, and analyzing stability of dynamical operators.

Definition: Matrix Factorization

A **matrix factorization** expresses a matrix A as a product of matrices $A = F_1 F_2 \cdots F_k$, where the factors F_i have special structure (triangular, orthogonal, diagonal, etc.) that makes them easier to work with.

To make this concrete, let us work with a small running example:

$$A = \begin{bmatrix} 2 & 1 & 1 \\ 4 & -6 & 0 \\ -2 & 7 & 2 \end{bmatrix}.$$

LU Factorization: The LU factorization writes A as $A = LU$, where L is lower triangular with ones on the diagonal and U is upper triangular [8]. For our example, performing Gaussian elimination gives

$$L = \begin{bmatrix} 1 & 0 & 0 \\ 2 & 1 & 0 \\ -1 & -1 & 1 \end{bmatrix}, \qquad U = \begin{bmatrix} 2 & 1 & 1 \\ 0 & -8 & -2 \\ 0 & 0 & -1 \end{bmatrix}.$$

This decomposition is valuable because solving $Ax = b$ reduces to two simpler triangular solves: first $Ly = b$, then $Ux = y$. Both can be done efficiently in $O(n^2)$ time.

QR Factorization: The QR factorization writes $A = QR$, where Q has orthonormal columns ($Q^\top Q = I$) and R is upper triangular [9]. For our example, applying Gram-Schmidt orthogonalization [10] gives

$$Q \approx \begin{bmatrix} 0.408 & 0.408 & 0.816 \\ 0.816 & -0.408 & 0.408 \\ -0.408 & 0.816 & -0.408 \end{bmatrix}, \qquad R \approx \begin{bmatrix} 4.898 & -4.082 & 0.000 \\ 0.000 & 8.165 & 1.633 \\ 0.000 & 0.000 & 1.633 \end{bmatrix}.$$

QR is the workhorse of least-squares problems: to solve $\min_x \|Ax - b\|_2$, we write $Ax = b$ as $QRx = b$, then use orthogonality to simplify the system. This avoids forming $A^\top A$, which can be ill-conditioned.

Eigen and Singular Value Decomposition (SVD): Two further factorizations reveal the deep geometry of a matrix. The eigen-decomposition (for diagonalizable A) writes $A = V \Lambda V^{-1}$, with eigenvalues Λ capturing growth rates and eigenvectors V capturing invariant directions. This is central in analyzing stability of dynamical systems.

The singular value decomposition applies to any matrix, writing

$$A = U \Sigma V^\top,$$

where U and V are orthogonal and Σ is diagonal with nonnegative entries (the singular values). The SVD tells us how A acts as a rotation, scaling, and rotation again. In SciML, SVD underpins reduced-order modeling, principal component analysis, and low-rank operator learning.

Matrix factorizations thus serve two roles: they give us efficient algorithms (triangular solves, least-squares, low-rank approximations), and they give us insight into the structure of operators (stability, rank, conditioning). For small systems, we can compute them exactly; for large-scale SciML models, we often exploit sparsity or approximate them iteratively. Either way, factorization is a fundamental tool that links linear algebra to scientific learning and computation.

2.2.5 *Sparsity and Graph-Structured Matrices*

Real-world scientific problems rarely produce dense matrices. Discretizations of partial differential equations (PDEs), finite element methods, and graph-based models all give rise to matrices where most entries are zero. This *sparsity* is not just a storage detail ; it encodes locality, geometry, and network structure. Exploiting sparsity is what makes large-scale SciML computations feasible.

Definition: Sparsity Pattern

A **sparsity pattern** of a matrix $A \in \mathbb{R}^{m \times n}$ is the set of index pairs (i, j) such that $A_{ij} \neq 0$. The *sparsity structure* describes the positions of nonzeros, independent of their values.

Let us take a small example. Suppose we discretize the one-dimensional Laplacian operator $u''(x)$ on three interior points with finite differences. The resulting matrix is

$$L = \begin{bmatrix} -2 & 1 & 0 \\ 1 & -2 & 1 \\ 0 & 1 & -2 \end{bmatrix}.$$

This matrix is tridiagonal: nonzeros appear only on the main diagonal and the first off-diagonals. The sparsity pattern directly reflects the local coupling of each grid point to its neighbors. If we extended this discretization to n points, L would remain sparse, with only $3n - 2$ nonzero entries instead of n^2.

Sparsity is naturally visualized as a graph. Each row/column corresponds to a node, and a nonzero entry A_{ij} corresponds to an edge between nodes i and j.

Definition: Graph of a Matrix

Given a square matrix $A \in \mathbb{R}^{n \times n}$, the **graph of** A is the graph $G = (V, E)$ with vertex set $V = \{1, \ldots, n\}$ and edge set $E = \{(i, j) : A_{ij} \neq 0 \text{ or } A_{ji} \neq 0\}$.

For our tridiagonal Laplacian L, the graph is simply a path graph with three nodes connected in a line. In higher dimensions, discretized Laplacians produce grid graphs, where each node represents a spatial point and edges represent local couplings. In SciML, such graph-structured matrices appear in graph neural networks, message-passing schemes, and operator learning.

Sparsity and graph structure have profound computational consequences: *Storage*: instead of storing all n^2 entries, we only store the nonzeros, often in compressed formats (CSR/CSC). *Speed*: sparse matrix-vector products (SpMV) scale with the number of nonzeros, enabling large-scale iterative solvers. *Structure*: the graph view informs reordering strategies (nested dissection, minimum degree) that reduce fill-in during factorization.

As an example, if we consider the adjacency matrix of a star graph with five nodes,

$$A = \begin{bmatrix} 0 & 1 & 1 & 1 & 1 \\ 1 & 0 & 0 & 0 & 0 \\ 1 & 0 & 0 & 0 & 0 \\ 1 & 0 & 0 & 0 & 0 \\ 1 & 0 & 0 & 0 & 0 \end{bmatrix},$$

then the sparsity pattern highlights the central node connected to all others. This structure suggests hub-and-spoke interactions, very different from the local coupling of the Laplacian. Both patterns are sparse, but their graphs encode very different geometries.

In SciML, sparsity and graph-structured matrices bridge linear algebra and modeling. They allow us to encode locality in PDEs, connectivity in networks, and modularity in large systems. Exploiting sparsity is not optional : it is the only way to scale algorithms to the massive problems we face in scientific computing and machine learning.

2.2.6 *Iterative Solvers and Preconditioning*

Direct methods such as LU or QR are invaluable for small to medium systems, but in large-scale SciML problems the matrices are often too large or too sparse for factorization to be practical. Instead, we rely on *iterative solvers*. These methods start with an initial guess and progressively refine it, requiring only matrix-vector products rather than explicit factorizations. Their performance, however, depends strongly on conditioning ; which brings us to the crucial role of preconditioning.

Definition: Iterative Solver

An **iterative solver** for a linear system $Ax = b$ generates a sequence $\{x_k\}$ such that $x_k \to x^\star$, the exact solution, by repeatedly applying updates of the form

$$x_{k+1} = x_k + M_k^{-1} r_k,$$

where $r_k = b - Ax_k$ is the residual and M_k is a chosen update operator.

Example (Jacobi Iteration): Consider the system $Ax = b$ with

$$A = \begin{bmatrix} 4 & 1 & 0 \\ 1 & 3 & -1 \\ 0 & -1 & 2 \end{bmatrix}, \quad b = \begin{bmatrix} 1 \\ 2 \\ 3 \end{bmatrix}.$$

Splitting $A = D + (L + U)$ into its diagonal part D and off-diagonal parts $L + U$, the Jacobi iteration reads

$$x_{k+1} = D^{-1}(b - (L + U)x_k).$$

Starting from $x_0 = (0, 0, 0)$, each step updates the components using only the previous iterate. This method converges provided the matrix is diagonally dominant. Although simple, it illustrates the principle of iterative refinement.

Krylov Subspace Methods [11]: Modern large-scale solvers use Krylov subspaces: $\mathcal{K}_k(A, r_0) = \text{span}\{r_0, Ar_0, \ldots, A^{k-1}r_0\}$. Conjugate Gradient (CG) solves symmetric positive definite systems by minimizing the error over successive Krylov subspaces. Generalized minimal residual method (GMRES) [12] handles nonsymmetric systems by minimizing residuals in these spaces. These methods converge rapidly if the spectrum of A is clustered ; but can stall when eigenvalues are spread out.

Preconditioning: To accelerate convergence, we use a *preconditioner* P that transforms the system into one with better conditioning. Ideally, P approximates A but is cheap to invert. We then solve

$$P^{-1}Ax = P^{-1}b,$$

where the new system has condition number $\kappa(P^{-1}A)$ much smaller than $\kappa(A)$.

Definition: Preconditioner

A **preconditioner** P for A is a matrix such that $P^{-1}A$ (or AP^{-1}, or a symmetric variant) is better conditioned than A, while applying P^{-1} is computationally efficient.

Returning to our example system, if we take $P = D$, the diagonal of A, then the preconditioned Jacobi iteration converges much faster: it rescales each equation to balance magnitudes. In practice, more sophisticated preconditioners ; incomplete LU (ILU), multigrid, domain decomposition ; are indispensable for PDE discretizations and large-scale SciML models.

Iterative solvers and preconditioning together make high-dimensional SciML problems tractable. Without them, training neural PDE solvers, simulating discretized dynamical systems, or solving inverse problems would be computationally prohibitive. By exploiting only matrix-vector products and structure-aware preconditioners, we can scale to millions of unknowns while keeping algorithms stable and efficient.

2.3 Calculus and Automatic Differentiation

2.3.1 *Matrix/Vector Calculus Essentials*

With vector spaces and operators in hand, we now turn to calculus. Scientific Machine Learning relies heavily on derivatives: gradients of loss functions, Jacobians of dynamical systems, and Hessians of optimization problems. To work consistently, we need a compact notation for differentiation in vector and matrix settings. This is the domain of *matrix calculus*.

Definition: Gradient

Let $f : \mathbb{R}^n \to \mathbb{R}$ be a scalar function. The **gradient** of f at $\mathbf{x}$ is the vector of partial derivatives

$$\nabla f(\mathbf{x}) = \begin{bmatrix} \frac{\partial f}{\partial x_1}(\mathbf{x}) \\ \vdots \\ \frac{\partial f}{\partial x_n}(\mathbf{x}) \end{bmatrix}.$$

As a running example, let

$$f(\mathbf{v}) = v_1^2 + v_2 v_3, \qquad \mathbf{v} = (v_1, v_2, v_3).$$

Then

$$\nabla f(\mathbf{v}) = (2v_1,\ v_3,\ v_2)\,.$$

At $\mathbf{v} = (2, -1, 3)$, the gradient evaluates to

$$\nabla f(\mathbf{v}) = (4, 3, -1).$$

The gradient points in the direction of steepest ascent of f. In optimization, moving opposite to the gradient is the basis of gradient descent.

Definition: Jacobian

Let $F : \mathbb{R}^n \to \mathbb{R}^m$ be a vector-valued function $F(\mathbf{x}) = (F_1(\mathbf{x}), \ldots, F_m(\mathbf{x}))^\top$. The **Jacobian** of F at $\mathbf{x}$ is the $m \times n$ matrix

$$J_F(\mathbf{x}) = \begin{bmatrix} \frac{\partial F_1}{\partial x_1} & \cdots & \frac{\partial F_1}{\partial x_n} \\ \vdots & \ddots & \vdots \\ \frac{\partial F_m}{\partial x_1} & \cdots & \frac{\partial F_m}{\partial x_n} \end{bmatrix}.$$

For example, define

$$F(\mathbf{v}) = \begin{bmatrix} v_1 v_2 \\ v_2^2 + v_3 \\ e^{v_1} \end{bmatrix}.$$

The Jacobian is

$$J_F(\mathbf{v}) = \begin{bmatrix} v_2 & v_1 & 0 \\ 0 & 2v_2 & 1 \\ e^{v_1} & 0 & 0 \end{bmatrix}.$$

At $\mathbf{v} = (2, -1, 3)$, we find

$$J_F(\mathbf{v}) = \begin{bmatrix} -1 & 2 & 0 \\ 0 & -2 & 1 \\ e^2 & 0 & 0 \end{bmatrix}.$$

The Jacobian describes the best linear approximation of F near a point. It generalizes the derivative to multivariate functions, and in SciML it appears as the sensitivity matrix of dynamical systems and the backpropagated Jacobian in neural networks.

Definition: Hessian

Let $f : \mathbb{R}^n \to \mathbb{R}$ be a scalar function. The **Hessian** of f at $\mathbf{x}$ is the $n \times n$ symmetric matrix

$$H_f(\mathbf{x}) = \begin{bmatrix} \frac{\partial^2 f}{\partial x_1^2} & \cdots & \frac{\partial^2 f}{\partial x_1 \partial x_n} \\ \vdots & \ddots & \vdots \\ \frac{\partial^2 f}{\partial x_n \partial x_1} & \cdots & \frac{\partial^2 f}{\partial x_n^2} \end{bmatrix}.$$

Returning to our scalar example $f(\mathbf{v}) = v_1^2 + v_2 v_3$, the Hessian is

$$H_f(\mathbf{v}) = \begin{bmatrix} 2 & 0 & 0 \\ 0 & 0 & 1 \\ 0 & 1 & 0 \end{bmatrix}.$$

This matrix is constant: the curvature of f does not depend on $\mathbf{v}$. In optimization, the Hessian captures second-order information. Positive definiteness of the Hessian indicates a local minimum, while indefiniteness indicates a saddle point.

Together, gradients, Jacobians, and Hessians form the core of matrix/vector calculus. They allow us to measure slopes, sensitivities, and curvature in high dimensions. In SciML, these objects drive optimization algorithms, stability analysis, and adjoint methods. Later, when we discuss automatic differentiation, we will see how modern software computes them efficiently through computational graphs, without symbolic manipulation or finite differences.

2.3.2 *Automatic Differentiation*

Calculus gives us gradients, Jacobians, and Hessians, but in practice we rarely compute them by hand. For small examples this is feasible, but in SciML models with millions of parameters and complex compositions, manual differentiation is impossible. Finite differences are also inadequate: they suffer from truncation and round-off errors, and their cost grows linearly with the number of parameters. The solution is *automatic differentiation (AD)*; the backbone of modern scientific machine learning.

Definition: Automatic Differentiation

Automatic differentiation (AD) is a set of techniques for evaluating derivatives of functions expressed as programs by systematically applying the chain rule to the operations in the computational graph.

The key idea is that every function we implement in code is a composition of simple operations: addition, multiplication, trigonometric functions, exponentials, etc. AD records these operations in a *computational graph*, then applies the chain rule mechanically, producing exact derivatives up to machine precision.

Running Example: Let us revisit our scalar function

$$f(\mathbf{v}) = v_1^2 + v_2 v_3, \qquad \mathbf{v} = (v_1, v_2, v_3).$$

We know the gradient is $\nabla f(\mathbf{v}) = (2v_1, v_3, v_2)$. At $\mathbf{v} = (2, -1, 3)$, the result is $(4, 3, -1)$. Automatic differentiation produces this result automatically by tracing the computational graph:

$$v_1 \xrightarrow{x^2} v_1^2, \quad (v_2, v_3) \xrightarrow{\times} v_2 v_3, \quad f = v_1^2 + v_2 v_3.$$

AD then propagates derivatives along the graph. Unlike symbolic differentiation, it does not generate large expressions; unlike finite differences, it is exact.

Modes of AD: There are two main ways to traverse the computational graph:

- *Forward mode AD*: derivatives are propagated alongside values from inputs to outputs. For each input, we track how perturbations flow through the program. This is efficient when the number of inputs n is small compared to the number of outputs m.
- *Reverse mode AD (backpropagation)*: derivatives are propagated backward from outputs to inputs. This is efficient when m is small (often just one scalar loss) and n is large. This is the regime of neural networks, where millions of weights influence a single loss value.

Comparison of AD Modes

For a function $F : \mathbb{R}^n \to \mathbb{R}^m$:

- Forward mode computes $J_F \cdot v$ (Jacobian-vector products).
- Reverse mode computes $u^\top J_F$ (vector-Jacobian products).

Example of Forward versus Reverse: Take $f(v_1, v_2) = \sin(v_1 v_2)$. In forward mode, we propagate derivatives with respect to v_1: $v_1 v_2 \mapsto v_2$, then $\sin(\cdot) \mapsto \cos(v_1 v_2) \cdot v_2$. In reverse mode, we start from $\frac{\partial f}{\partial f} = 1$, then propagate backward: $\frac{\partial f}{\partial (v_1 v_2)} = \cos(v_1 v_2)$, and distribute to inputs: $\frac{\partial f}{\partial v_1} = v_2 \cos(v_1 v_2)$, $\frac{\partial f}{\partial v_2} = v_1 \cos(v_1 v_2)$. Both arrive at the same result, but the efficiency differs depending on input/output dimensions.

Automatic differentiation is indispensable in SciML. It enables us to train neural operators, compute sensitivities in PDE solvers, and optimize controls with exact gradients. Libraries such as JAX [13, 14], PyTorch, TensorFlow, and Zygote [15] implement AD at scale, handling millions of operations per second. AD is therefore not just a convenience ; it is the infrastructure that makes scientific machine learning practical.

2.3.3 Implicit Differentiation and Fixed-Point Maps

Not all functions we deal with in SciML are given explicitly as formulas. Often, the quantity of interest u is defined only implicitly: as the solution of an equation $F(u, \theta) = 0$ or as the fixed point of an iteration $u = T(u, \theta)$. In these cases, we cannot simply apply the chain rule directly. Instead, we use *implicit differentiation*. This technique allows us to compute sensitivities of implicitly defined variables with respect to parameters, without re-solving the system from scratch for every parameter perturbation.

Definition: Implicit Differentiation

Suppose $u(\theta)$ satisfies $F(u, \theta) = 0$ with $F : \mathbb{R}^n \times \mathbb{R}^p \to \mathbb{R}^n$. Differentiating with respect to θ gives

$$\frac{\partial F}{\partial u}\frac{du}{d\theta} + \frac{\partial F}{\partial \theta} = 0,$$

so that

$$\frac{du}{d\theta} = -\left(\frac{\partial F}{\partial u}\right)^{-1}\frac{\partial F}{\partial \theta}.$$

Running Example: Consider the scalar equation

$$F(u, \theta) = u^2 - \theta = 0,$$

so that $u(\theta) = \sqrt{\theta}$. We can compute the derivative $\frac{du}{d\theta}$ without solving explicitly. Differentiating:

$$\frac{\partial F}{\partial u} = 2u, \qquad \frac{\partial F}{\partial \theta} = -1,$$

so

$$\frac{du}{d\theta} = -\frac{-1}{2u} = \frac{1}{2u}.$$

If $\theta = 4$, then $u = 2$ and $\frac{du}{d\theta} = \frac{1}{4}$. This matches the direct derivative of $\sqrt{\theta}$.

Fixed-Point Maps: Another common case is when u is defined as the fixed point of a map:

$$u = T(u, \theta).$$

Differentiating both sides with respect to θ gives

$$\frac{du}{d\theta} = \frac{\partial T}{\partial u}\frac{du}{d\theta} + \frac{\partial T}{\partial \theta}.$$

Rearranging:

$$\left(I - \frac{\partial T}{\partial u}\right)\frac{du}{d\theta} = \frac{\partial T}{\partial \theta},$$

so that

$$\frac{du}{d\theta} = \left(I - \frac{\partial T}{\partial u}\right)^{-1}\frac{\partial T}{\partial \theta}.$$

Example (Fixed-Point Iteration): Suppose u is defined by

$$u = \tanh(\theta u).$$

Here $T(u, \theta) = \tanh(\theta u)$. We compute

$$\frac{\partial T}{\partial u} = \theta \operatorname{sech}^2(\theta u), \qquad \frac{\partial T}{\partial \theta} = u \operatorname{sech}^2(\theta u).$$

Therefore,

$$\frac{du}{d\theta} = \frac{u \operatorname{sech}^2(\theta u)}{1 - \theta \operatorname{sech}^2(\theta u)}.$$

This gives the sensitivity of the fixed point $u(\theta)$ with respect to θ, without explicitly differentiating through the entire iterative solution process.

Implicit differentiation and fixed-point analysis are widely used in SciML. They allow us to backpropagate through optimization problems, equilibria, and PDE solvers without unrolling all iterations. This is the foundation of modern techniques such as *implicit layers*, *equilibrium models*, and *differentiable physics solvers*. By shifting the computation to Jacobian inverses or linear solves, we trade repeated differentiation for a single sensitivity solve ; making large-scale scientific learning problems tractable.

2.3.4 *Adjoint-Backprop Equivalence*

Adjoint methods in scientific computing and backpropagation in machine learning are not separate inventions; they are two faces of the same idea. Both compute gradients of scalar functionals with respect to high-dimensional inputs by propagating sensitivities backward through a computational graph. Recognizing this equivalence unifies decades of work in numerical analysis and modern deep learning.

Key Idea

Adjoint methods and backpropagation both implement reverse-mode automatic differentiation. They propagate derivatives of a scalar output with respect to many inputs by traversing the computational graph in reverse, applying the chain rule.

Scientific Computing View: Suppose u is the solution of a system of equations $F(u, \theta) = 0$, and we are interested in a scalar objective $J(u, \theta)$. The adjoint method introduces a variable λ (the adjoint state) satisfying

$$\left(\frac{\partial F}{\partial u}\right)^{\top} \lambda = \frac{\partial J}{\partial u},$$

so that the total gradient with respect to θ is

$$\frac{dJ}{d\theta} = \frac{\partial J}{\partial \theta} - \lambda^{\top} \frac{\partial F}{\partial \theta}.$$

Machine Learning View: In a neural network, we have a scalar loss $L(x)$ that depends on input x through layers of transformations. Backpropagation computes the gradient $\nabla_x L$ by traversing the network in reverse: each layer passes an "error signal" (the gradient of the loss with respect to its output) back to the previous layer, applying the chain rule. This error signal is exactly the adjoint variable.

Running Example: Consider the scalar function

$$f(v_1, v_2, v_3) = v_1^2 + v_2 v_3, \qquad \mathbf{v} = (2, -1, 3).$$

We already know $\nabla f(\mathbf{v}) = (4, 3, -1)$. Let us recompute this using the adjoint-backprop perspective.

- Forward evaluation: compute intermediate nodes $z_1 = v_1^2$, $z_2 = v_2 v_3$, and $f = z_1 + z_2$.

- Backward sweep:

$$\frac{\partial f}{\partial z_1} = 1, \quad \frac{\partial f}{\partial z_2} = 1.$$

Propagate to inputs:

$$\frac{\partial f}{\partial v_1} = \frac{\partial f}{\partial z_1} \cdot \frac{\partial z_1}{\partial v_1} = 1 \cdot 2v_1 = 4,$$

$$\frac{\partial f}{\partial v_2} = \frac{\partial f}{\partial z_2} \cdot \frac{\partial z_2}{\partial v_2} = 1 \cdot v_3 = 3,$$

$$\frac{\partial f}{\partial v_3} = \frac{\partial f}{\partial z_2} \cdot \frac{\partial z_2}{\partial v_3} = 1 \cdot v_2 = -1.$$

This reproduces the gradient $(4, 3, -1)$ exactly.

Here the "adjoint variables" $\frac{\partial f}{\partial z_i}$ are exactly the backpropagated sensitivities. In scientific computing, we would call these the adjoint states. In machine learning, they are called gradients of the loss signal. The mechanism is the same.

The equivalence has major implications. In SciML, when we say we are using adjoints to compute sensitivities of PDEs or dynamical systems, we are performing backpropagation through the solver. Conversely, when training neural networks, backpropagation is just the adjoint method applied to a discrete computational graph. Bridging these communities clarifies that both fields rely on the same fundamental idea: reverse-mode automatic differentiation.

2.4 Function Spaces and Variational Methods

2.4.1 Lebesgue and Sobolev Spaces

So far, we have worked mostly in finite-dimensional vector spaces such as $\mathbb{R}^n$. But many scientific problems ; especially PDEs and variational formulations ; live naturally in infinite-dimensional function spaces. To reason about convergence, stability, and error, we need norms and inner products that extend beyond discrete vectors. Two central families of function spaces in SciML are *Lebesgue spaces* [16] and *Sobolev spaces* [17].

Definition: Lebesgue Space

For a measurable domain $\Omega \subset \mathbb{R}^d$ and $1 \leq p < \infty$, the **Lebesgue space** $L^p(\Omega)$ is the set of measurable functions $f : \Omega \to \mathbb{R}$ such that

$$\|f\|_{L^p(\Omega)} := \left(\int_\Omega |f(x)|^p \, dx \right)^{1/p} < \infty.$$

For $p = \infty$, we define

$$\|f\|_{L^\infty(\Omega)} := \text{ess sup}_{x \in \Omega} |f(x)|.$$

Lebesgue spaces generalize the familiar ℓ^p norms on vectors to functions. For example, $L^2(\Omega)$ is the space of square-integrable functions, equipped with the inner product

$$\langle f, g \rangle_{L^2} = \int_\Omega f(x) g(x) \, dx.$$

This makes L^2 a Hilbert space, the natural setting for variational formulations of PDEs and physics-informed learning.

Example: On $\Omega = [0, 1]$, consider $f(x) = x$ and $g(x) = 1 - x$. We have

$$\|f\|_{L^2(0,1)} = \left(\int_0^1 x^2 \, dx \right)^{1/2} = \sqrt{\tfrac{1}{3}}, \qquad \|g\|_{L^2(0,1)} = \left(\int_0^1 (1-x)^2 \, dx \right)^{1/2} = \sqrt{\tfrac{1}{3}}.$$

Their inner product is-

$$\langle f, g \rangle_{L^2} = \int_0^1 x(1-x) \, dx = \tfrac{1}{6}.$$

This is the infinite-dimensional analogue of dot products in $\mathbb{R}^n$.

Definition: Sobolev Space

For a domain $\Omega \subset \mathbb{R}^d$, the **Sobolev space** $H^k(\Omega)$ consists of functions $f \in L^2(\Omega)$ whose weak derivatives up to order k also belong to $L^2(\Omega)$:

$$H^k(\Omega) = \left\{ f \in L^2(\Omega) : \partial^\alpha f \in L^2(\Omega) \text{ for all multi-indices } |\alpha| \le k \right\}.$$

Sobolev spaces measure not only the size of a function, but also the size of its derivatives. This is crucial in PDE analysis: stability and well-posedness often require control of derivatives as well as values.

Example: On $\Omega = [0, 1]$, consider $f(x) = x(1 - x)$. We have

$$\|f\|_{L^2(0,1)} = \left(\int_0^1 (x - x^2)^2 dx \right)^{1/2} = \sqrt{\tfrac{1}{30}},$$

while its derivative $f'(x) = 1 - 2x$ satisfies

$$\|f'\|_{L^2(0,1)} = \left(\int_0^1 (1 - 2x)^2 dx \right)^{1/2} = \sqrt{\tfrac{1}{3}}.$$

Thus $f \in H^1(0, 1)$, with the Sobolev norm

$$\|f\|_{H^1} = \left(\|f\|_{L^2}^2 + \|f'\|_{L^2}^2 \right)^{1/2}.$$

Lebesgue and Sobolev spaces are the natural arenas for variational methods. In SciML, they appear when we discretize PDEs, train physics-informed neural networks, or analyze approximation errors. L^2 encodes energy norms, while H^1 encodes smoothness. Together, they provide the functional analytic foundation that allows us to rigorously move between continuous problems and discrete algorithms.

2.4.2 *Weak Forms and Energy Principles*

One of the main reasons we introduce Sobolev spaces is that they provide a natural setting for the *weak formulation* of partial differential equations (PDEs). Instead of requiring classical differentiability, weak forms only require integrability of functions and their derivatives, allowing us to pose PDEs in L^2 or H^1 spaces. This relaxes smoothness requirements and makes problems well-posed for both analysis and computation.

Definition: Weak Formulation

Let $Lu = f$ be a PDE with differential operator L on domain Ω and boundary conditions on $\partial\Omega$. A **weak formulation** is obtained by multiplying the equation by a *test function* v, integrating over Ω, and applying integration by parts to shift derivatives from u to v.

Example (Poisson's Equation): Consider the one-dimensional Poisson problem

$$-u''(x) = f(x), \qquad x \in (0, 1), \quad u(0) = u(1) = 0.$$

Multiplying by a test function $v(x)$ with $v(0) = v(1) = 0$, and integrating:

$$\int_0^1 -u''(x)\,v(x)\,dx = \int_0^1 f(x)\,v(x)\,dx.$$

Integrating by parts removes one derivative from u:

$$\int_0^1 u'(x)\,v'(x)\,dx = \int_0^1 f(x)\,v(x)\,dx.$$

This is the weak form: find $u \in H_0^1(0,1)$ such that the above holds for all $v \in H_0^1(0,1)$. We no longer require u'' to exist in the classical sense, only that u' is square integrable.

Energy Principle: Many PDEs can also be derived from an energy minimization principle. For Poisson's equation, define the energy functional

$$E(u) = \frac{1}{2}\int_0^1 |u'(x)|^2\,dx - \int_0^1 f(x)u(x)\,dx.$$

The weak solution u minimizes $E(u)$ over $H_0^1(0,1)$. The Euler-Lagrange equations of this variational problem coincide with the weak form derived above.
SciML Perspective: Weak formulations and energy principles underpin many modern SciML techniques:

- Finite element methods approximate the weak form by restricting u and v to finite-dimensional subspaces.
- Physics-Informed Neural Networks (PINNs) can be viewed as enforcing weak forms or minimizing energy functionals.
- Variational autoencoders and neural operators often encode structure by working in weak or energy-based settings.

Weak forms are thus not only a relaxation of PDEs to Sobolev spaces ; they also provide a unifying perspective. By shifting derivatives, they expose the natural bilinear forms and energies underlying physical systems. This makes them the foundation for both classical numerical methods and modern SciML approaches that combine learning with physics.

2.4.3 *Well-Posedness via Lax–Milgram and Friends*

Formulating a PDE in weak form is only the first step. To ensure that our problem is meaningful, we must guarantee that a solution exists, that it is unique, and that it depends continuously on the data. This is the notion of *well-posedness*, introduced by Hadamard [18, 19]. Functional analysis provides the tools to establish well-posedness, and at the heart of this theory lies the Lax–Milgram theorem [20].

Theorem: Lax–Milgram

Let V be a Hilbert space with inner product $\langle \cdot, \cdot \rangle$ and norm $\| \cdot \|$. Suppose $a(\cdot, \cdot) : V \times V \to \mathbb{R}$ is a bilinear form satisfying:

1. **Continuity**: There exists $C > 0$ such that

$$|a(u, v)| \leq C\|u\|\|v\| \quad \forall u, v \in V.$$

2. **Coercivity**: There exists $\alpha > 0$ such that

$$a(u, u) \geq \alpha\|u\|^2 \quad \forall u \in V.$$

Then for any bounded linear functional $f \in V^*$, there exists a unique $u \in V$ such that

$$a(u, v) = f(v) \quad \forall v \in V.$$

Moreover, $\|u\| \leq \frac{1}{\alpha}\|f\|_{V^*}$.

Example (Poisson's Equation): In the weak form of the 1D Poisson problem,

$$\int_0^1 u'(x)v'(x)\,dx = \int_0^1 f(x)\,v(x)\,dx,$$

we define

$$a(u, v) = \int_0^1 u'(x)v'(x)\,dx, \qquad f(v) = \int_0^1 f(x)v(x)\,dx.$$

On $V = H_0^1(0, 1)$, $a(u, v)$ is continuous and coercive: by Poincaré's inequality,

$$a(u, u) = \int_0^1 |u'(x)|^2 dx \geq \alpha\|u\|_{H^1}^2.$$

Thus, the Lax–Milgram theorem guarantees a unique weak solution $u \in H_0^1(0, 1)$.

Friends of Lax–Milgram: While Lax–Milgram applies to coercive bilinear forms, many problems require other tools: The *Lax–Milgram lemma with Gåding's inequality* [21] extends the result to bilinear forms that are only coercive up to a compact perturbation. The *Lax–Milgram–Babuška theorem* (or Babuška-Nečas-Brezzi condition) handles saddle-point problems, such as incompressible flow (Stokes equations) [22]. The *Lax equivalence theorem* [23] in numerical analysis states that for finite-difference schemes, consistency plus stability implies convergence.

Each of these results provides a framework to guarantee well-posedness, ensuring that both continuous and discrete problems behave predictably.

SciML Perspective: For scientific machine learning, well-posedness is more than theory: it ensures that the learning problem is stable. In PINNs, coercivity guarantees that the loss functional has a well-defined minimizer. In operator learning, continuity ensures that small perturbations in training data lead to controlled errors in predictions. In hybrid physics-ML solvers, Babuška-Nečas conditions arise naturally in constrained optimization and variational formulations.

Thus, Lax–Milgram and its extensions form the backbone of well-posedness in SciML. They assure us that when we approximate, discretize, or learn, we are building on solid mathematical ground: the solutions we seek exist, are unique, and depend continuously on the data.

2.4.4 Galerkin Methods

Weak formulations allow us to pose PDEs in Hilbert spaces such as H^1, but in practice we cannot work in infinite dimensions. The Galerkin method provides a systematic way to approximate these problems in finite-dimensional subspaces [24]. It is the foundation of finite element methods (FEM), spectral methods, and many modern operator-learning techniques in SciML [25].

Definition: Galerkin Method

Given a weak problem: find $u \in V$ such that

$$a(u, v) = f(v) \quad \forall v \in V,$$

the **Galerkin method** seeks an approximate solution u_h in a finite-dimensional subspace $V_h \subset V$ such that

$$a(u_h, v_h) = f(v_h) \quad \forall v_h \in V_h.$$

The key principle is that we restrict both the trial functions (solutions) and the test functions to the same finite-dimensional subspace. This leads to a system of linear equations, where the matrix entries are given by inner products of basis functions.

Example (Poisson's Equation): Recall the weak form of the 1D Poisson problem:

$$\int_0^1 u'(x)v'(x)\,dx = \int_0^1 f(x)v(x)\,dx, \quad u, v \in H_0^1(0, 1).$$

Choose the finite-dimensional subspace $V_h = \text{span}\{\phi_1, \phi_2, \phi_3\}$, where ϕ_i are the standard piecewise linear "hat" functions on a uniform mesh with three interior nodes. The Galerkin method seeks $u_h(x) = \sum_{i=1}^{3} U_i \phi_i(x)$. Substituting into the weak form gives

$$\sum_{i=1}^{3} U_i \int_0^1 \phi_i'(x)\phi_j'(x)\,dx = \int_0^1 f(x)\phi_j(x)\,dx, \quad j = 1, 2, 3.$$

This yields the linear system

$$KU = F,$$

where $K_{ij} = \int_0^1 \phi_i'(x)\phi_j'(x)\,dx$ is the stiffness matrix and $F_j = \int_0^1 f(x)\phi_j(x)\,dx$ is the load vector.

Galerkin Orthogonality: A remarkable property of Galerkin methods is that the error $e = u - u_h$ is orthogonal to the test space:

$$a(e, v_h) = 0 \quad \forall v_h \in V_h.$$

This ensures quasi-optimality: the Galerkin solution is the best approximation in V_h up to a constant, and its error can be bounded in terms of the best possible error in the chosen subspace.

SciML Perspective: Galerkin methods have a deep connection to scientific machine learning: *Finite element methods* are Galerkin methods with piecewise polynomial basis functions. *Spectral methods* use global basis functions, such as Fourier or Chebyshev polynomials. *Physics-Informed Neural Networks (PINNs)* can be interpreted as Petrov–Galerkin methods, where test functions are encoded by collocation or residual minimization. *Neural operators* may be viewed as learning Galerkin projections directly from data, mapping input functions to solution functions via learned trial/test spaces.

Thus, the Galerkin method forms the bridge between weak formulations in function spaces and practical discretizations used in both classical numerical methods and SciML. By choosing different subspaces V_h, we recover a wide spectrum of methods: finite elements, spectral methods, collocation methods, and data-driven operator learning.

2.4.5 Additional Variational Tools and Error Estimates

Before leaving the world of function spaces and variational methods, it is useful to highlight a few additional results and generalizations that frequently appear in both numerical analysis and SciML. These extend the Galerkin framework and provide rigorous error guarantees.

Petrov–Galerkin Methods: In the standard Galerkin method, trial and test spaces are the same. A more general approach is the *Petrov–Galerkin method* [26], where the solution is sought in a trial space V_h and tested against a (possibly different) space W_h. The weak problem then reads

$$a(u_h, v_h) = f(v_h) \quad \forall v_h \in W_h.$$

This flexibility is useful when enforcing stability in convection-dominated problems, or when designing residual-based methods such as discontinuous Petrov–Galerkin schemes. In SciML, Petrov–Galerkin ideas are mirrored in physics-informed neural networks (PINNs), where test functions are encoded by collocation or residual minimization [26].

Céa's Lemma: One of the fundamental results in Galerkin analysis is Céa's lemma [27], which guarantees quasi-optimality of the approximation. If the bilinear form $a(\cdot, \cdot)$ is continuous and coercive on V, and $u \in V$ is the exact solution while $u_h \in V_h \subset V$ is the Galerkin solution, then

$$\|u - u_h\|_V \leq \frac{C}{\alpha} \inf_{w_h \in V_h} \|u - w_h\|_V,$$

where C is the continuity constant and α is the coercivity constant. This means the Galerkin solution is as close to the exact solution as the best possible approximation in V_h, up to a constant factor.

Ritz and Rayleigh-Ritz Methods: The Galerkin method can also be understood as an energy minimization principle. If the weak form arises from minimizing an energy functional $E(u)$, then restricting to a subspace V_h gives the *Ritz method*: find $u_h \in V_h$ minimizing $E(u)$. In eigenvalue problems, this leads to the *Rayleigh-Ritz method*, which approximates eigenvalues by minimizing the Rayleigh quotient $\frac{\langle Au,u \rangle}{\langle u,u \rangle}$ over V_h. In SciML, these variational perspectives connect directly to energy-based learning, where neural networks minimize physics-inspired energies or functionals.

These additional tools round out the variational framework. Petrov–Galerkin methods extend flexibility, Céa's lemma ensures error control, and Ritz principles highlight the energy viewpoint. Together, they provide a rigorous and versatile foundation before we turn to differential equations and discretization.

2.5 Differential Equations and Discretization

Differential equations are at the heart of scientific models. Ordinary differential equations (ODEs) describe time evolution in finite-dimensional systems, while partial differential equations (PDEs) govern spatially extended systems. Before we discretize, approximate, or learn solutions, we must understand whether the continuous problems themselves are well-posed and stable.

2.5.1 ODE Well-Posedness and Stability

We begin with initial value problems (IVPs) for ODEs:

$$\frac{du}{dt} = f(t, u), \qquad u(0) = u_0,$$

where $f : [0, T] \times \mathbb{R}^n \to \mathbb{R}^n$ is a vector field. The key questions are: does a solution exist, is it unique, and does it depend continuously on the data? These are the requirements of *well-posedness*.

Theorem: Picard–Lindelöf

If $f(t, u)$ is continuous in t and Lipschitz continuous in u, then for each initial condition u_0 there exists a unique solution $u(t)$ defined on some interval $[0, T]$. Moreover, the solution depends continuously on u_0.

Running Example: Consider the scalar ODE

$$\frac{du}{dt} = -\theta u, \qquad u(0) = 1,$$

with parameter $\theta > 0$. The exact solution is $u(t) = e^{-\theta t}$. The vector field $f(u) = -\theta u$ is globally Lipschitz with constant θ, so Picard–Lindelöf guarantees existence and uniqueness [28].

Stability: Well-posedness guarantees that solutions exist and are unique, but for long-time behavior we also care about *stability*: do small perturbations in initial conditions or inputs remain small in the solution?

Definition: Stability of ODE Solutions

A solution $u(t)$ of $\dot{u} = f(t, u)$ is **stable** if, for every $\varepsilon > 0$, there exists $\delta > 0$ such that if $\|u_0 - \tilde{u}_0\| < \delta$ then $\|u(t) - \tilde{u}(t)\| < \varepsilon$ for all $t \geq 0$.

For our exponential decay example, perturbing the initial condition $u(0) = 1$ to $u(0) = 1 + \eta$ yields the solution $(1 + \eta)e^{-\theta t}$. The perturbation decays like $|\eta|e^{-\theta t}$, showing stability. In fact, the equilibrium solution $u = 0$ is asymptotically stable: all trajectories converge to it.

Lyapunov Stability: A powerful tool for nonlinear systems is Lyapunov's method [29]. If there exists a function $V(u)$ such that

$$V(u) \geq 0, \quad V(0) = 0, \quad \frac{d}{dt}V(u(t)) \leq 0,$$

then the equilibrium at $u = 0$ is stable. If $\frac{d}{dt}V(u) < 0$ except at $u = 0$, the equilibrium is asymptotically stable.

For our example $\dot{u} = -\theta u$, choosing $V(u) = \frac{1}{2}u^2$ gives

$$\frac{dV}{dt} = u\frac{du}{dt} = -\theta u^2 \leq 0,$$

which confirms stability.

SciML Perspective: Well-posedness and stability of ODEs are crucial for SciML: Neural ODEs rely on uniqueness to ensure trajectories are meaningful. PINNs for ODEs implicitly assume the solution is stable with respect to perturbations. In data assimilation, continuous dependence on initial conditions guarantees that learning from noisy data is feasible.

Thus, the Picard–Lindelöf theorem provides the theoretical backbone of ODE well-posedness, and stability analysis ensures that solutions behave predictably. With these guarantees in place, we can now turn to discretization methods that approximate ODEs numerically while preserving their qualitative properties.

2.5.2 *Time-Stepping: Order, Consistency, Stability*

To solve ODEs numerically, we approximate continuous dynamics by discrete updates. Time-stepping methods replace the derivative $\dot{u} = f(t, u)$ with difference equations on a grid $t_0, t_1, \ldots, t_N$. Three properties govern their reliability: *consistency*, *order*, and *stability*.

Definition: One-Step Method

A **one-step method** advances the solution of $\dot{u} = f(t, u)$ by

$$u_{n+1} = \Phi_h(t_n, u_n),$$

where Φ_h approximates the exact flow $u(t_{n+1}) = \varphi_h(t_n, u_n)$ and $h = t_{n+1} - t_n$ is the step size.

Example 1 (Forward Euler): For $\dot{u} = -\theta u$, $u(0) = 1$, the exact solution is

$$u(t) = e^{-\theta t}.$$

Forward Euler updates as

$$u_{n+1} = u_n + hf(t_n, u_n) = u_n - h\theta u_n = (1 - h\theta)u_n.$$

After n steps:

$$u_n = (1 - h\theta)^n.$$

If $\theta = 1$ and $h = 0.1$, the numerical solution at $t = 1$ is $(0.9)^{10} \approx 0.35$, while the exact solution is $e^{-1} \approx 0.3679$. The error is small and decreases as h shrinks.

Consistency: Consistency ensures that as $h \to 0$, the discrete step approaches the exact solution's Taylor expansion.

Definition: Consistency

The **local truncation error** τ_{n+1} of a scheme is

$$\tau_{n+1} = \frac{u(t_{n+1}) - \Phi_h(t_n, u(t_n))}{h}.$$

A method is consistent if $\tau_{n+1} \to 0$ as $h \to 0$.

For forward Euler, Taylor expansion of $u(t+h) = u(t) + hf(t,u) + O(h^2)$ shows that the truncation error is $O(h)$. Thus, Euler is consistent.

Example 2 (Order of Accuracy): Consider again $\dot{u} = -u$, $u(0) = 1$. The exact solution at $t = 1$ is $e^{-1} \approx 0.3679$. - With step $h = 0.5$, forward Euler gives $u_2 = (1 - 0.5)^2 = 0.25$ (error ≈ 0.12). - With step $h = 0.25$, forward Euler gives $u_4 = (0.75)^4 = 0.316$ (error ≈ 0.052). - With step $h = 0.125$, forward Euler gives $u_8 = (0.875)^8 = 0.343$ (error ≈ 0.025).

The error shrinks roughly proportionally to h. This confirms forward Euler is first-order accurate ($p = 1$). By contrast, classical Runge-Kutta (RK4) would achieve $O(h^4)$, giving much smaller errors for the same h.

Definition: Order

A time-stepping method has **order** p if the global error satisfies

$$\max_{0 \le n \le N} \|u(t_n) - u_n\| = O(h^p), \quad h \to 0.$$

Stability: Even a consistent method of high order can fail if it is unstable. We test stability using the linear equation $\dot{u} = \lambda u$, with $\Re(\lambda) < 0$. A one-step method applied to this problem gives

$$u_{n+1} = R(h\lambda)u_n,$$

where $R(z)$ is the *stability function* of the method.

Definition: Absolute Stability

A method is **absolutely stable** for step size h if $|R(h\lambda)| < 1$ whenever $\Re(\lambda) < 0$.

- For forward Euler, $R(z) = 1 + z$. Stability requires $|1 + h\lambda| < 1$. For real $\lambda < 0$, this means $h < 2/|\lambda|$. If h is too large, the method produces growth instead of decay.
- For backward Euler, $R(z) = \frac{1}{1-z}$. If $\Re(z) < 0$, then $|R(z)| < 1$ for all $h > 0$. This makes backward Euler *A-stable*: unconditionally stable for stiff problems.

Example 3 (Stability): Let $\dot{u} = -10u$, $u(0) = 1$. The exact solution is $u(t) = e^{-10t}$, which decays rapidly. Forward Euler with $h = 0.1$ gives update factor $1 - h \cdot 10 = 0$. The method "kills" the solution in one step ; stable but inaccurate. With $h = 0.25$, the update factor is $1 - 2.5 = -1.5$. The solution oscillates and grows in magnitude; completely unstable. Backward Euler with $h = 0.25$ gives update factor $\frac{1}{1+2.5} = 0.2857$, which decays correctly.

This illustrates why implicit methods are indispensable for stiff systems.

SciML Perspective: In SciML contexts: Neural ODEs require stable integrators to avoid exploding gradients. Learned solvers often mimic stability regions of classical methods. PINNs and operator learners rely on residual norms that encode consistency and stability implicitly.

Thus, *Consistency* ensures correctness in the limit $h \to 0$; *Order* quantifies convergence speed; *Stability* guarantees that numerical solutions reflect the qualitative behavior of the true ODE. Together, these properties form the bedrock of reliable time-stepping.

2.5.3 PDE Taxonomy and Boundary/Initial Data

Partial differential equations (PDEs) describe systems where the unknown function depends on multiple independent variables, such as space and time. Unlike ODEs, which evolve in a single variable, PDEs encode both temporal dynamics and spatial interactions. To analyze or discretize a PDE, we first classify it and then specify boundary/initial conditions.

Canonical Second-Order Linear PDE
A second-order linear PDE in two variables takes the form

$$a(x, y)u_{xx} + 2b(x, y)u_{xy} + c(x, y)u_{yy} + \text{(lower-order terms)} = f(x, y).$$

The discriminant $\Delta = b^2 - ac$ determines its type:
- **Elliptic**: $\Delta < 0$ (steady-state or equilibrium problems).
- **Parabolic**: $\Delta = 0$ (diffusion-like problems).
- **Hyperbolic**: $\Delta > 0$ (wave or signal propagation).

This classification is not just algebraic ; it reflects physical behavior and the kind of numerical methods we need.

Elliptic PDEs: Elliptic equations model equilibrium states, where the solution is determined entirely by boundary conditions. They are associated with smoothness and stability.

$$\text{Example :} \quad -\Delta u = f \quad \text{on } \Omega \subset \mathbb{R}^d.$$

- Laplace's equation ($f = 0$) describes steady-state heat or electrostatic potential.
- Poisson's equation ($f \neq 0$) describes sources in a field.

Solutions of elliptic problems are "global": perturbations at one boundary point affect the solution everywhere.

Parabolic PDEs: Parabolic equations describe diffusion, smoothing, and evolution toward equilibrium. They require both initial data and boundary conditions.

$$\text{Example :} \quad u_t = \alpha \Delta u \quad \text{(Heat equation).}$$

Models heat diffusion, population spread, or financial option pricing. Solutions gradually lose memory of initial data, converging to steady state.

Hyperbolic PDEs: Hyperbolic equations describe wave propagation, transport, and finite-speed signal travel. They require initial data (often both u and u_t) and boundary conditions.

$$\text{Example :} \quad u_{tt} = c^2 u_{xx} \quad \text{(Wave equation).}$$

Models vibrations of a string, sound waves, seismic waves. Solutions preserve sharp features and propagate discontinuities along characteristic lines.

Key Distinction

- *Elliptic*: Boundary value problems only. Smooth solutions, no dynamics.
- *Parabolic*: Boundary + initial data. Diffusive smoothing in time.
- *Hyperbolic*: Boundary + initial data. Finite-speed propagation along characteristics.

Boundary and Initial Data: To fully specify a PDE, we must prescribe conditions:

- **Dirichlet**: prescribe u on the boundary, e.g. $u(0, t) = 0$.
- **Neumann**: prescribe normal derivative $\partial u/\partial n$, e.g. $u_x(1, t) = 0$.
- **Robin (mixed)**: prescribe $au + bu_n = g$ on the boundary.
- **Initial conditions**: for time-dependent PDEs, specify $u(x, 0) = u_0(x)$ (and possibly $u_t(x, 0)$).

Example (Heat Equation):

$$u_t = u_{xx}, \quad x \in (0, 1), \ t > 0,$$

with boundary conditions $u(0, t) = u(1, t) = 0$ (Dirichlet) and initial condition $u(x, 0) = \sin(\pi x)$. The solution is

$$u(x, t) = e^{-\pi^2 t} \sin(\pi x),$$

which decays to zero as $t \to \infty$ while satisfying both boundary and initial conditions.

SciML Perspective: Understanding PDE taxonomy is essential in SciML: PINNs treat boundary and initial conditions as constraints in the loss function. Neural operators learn mappings (BC/IC) $\mapsto$ solution, with different architectures suited for elliptic, parabolic, or hyperbolic problems. Stability of learned solvers depends on respecting the qualitative behavior of each PDE type: smoothing for parabolic, conservation for hyperbolic, and global equilibrium for elliptic.

Thus, PDE classification and data specification determine both the mathematical analysis and the learning strategies we apply. Elliptic, parabolic, and hyperbolic equations form the three archetypes, each with distinct requirements and behaviors that SciML methods must respect.

2.5.4 *Spatial Discretizations*

Once a PDE is classified and equipped with boundary/initial data, the next step toward computation is to approximate it on a finite-dimensional space. This process is called *spatial discretization.* The continuous spatial domain Ω is replaced by a mesh or basis, and the PDE operator is replaced by a finite-dimensional matrix or operator.

Goal of Spatial Discretization

Given a PDE in weak or strong form, spatial discretization reduces it to a system of ODEs:

$$M \frac{dU}{dt} = AU + F,$$

where U are the unknown coefficients, M is the mass matrix, A is the stiffness matrix, and F encodes forcing and boundary conditions.

Finite Difference Method (FDM): The simplest approach replaces derivatives by local difference quotients on a grid.
Example: For the 1D Laplacian u_{xx} on $(0, 1)$ with n interior grid points, using spacing $h = 1/(n+1)$ gives

$$\frac{d^2u}{dx^2}(x_i) \approx \frac{u_{i-1} - 2u_i + u_{i+1}}{h^2}.$$

This leads to the tridiagonal stiffness matrix

$$L = \frac{1}{h^2}\begin{bmatrix} -2 & 1 & 0 & \cdots & 0 \\ 1 & -2 & 1 & \ddots & \vdots \\ 0 & 1 & -2 & \ddots & 0 \\ \vdots & \ddots & \ddots & \ddots & 1 \\ 0 & \cdots & 0 & 1 & -2 \end{bmatrix}.$$

Finite differences are easy to implement and efficient, but limited in handling complex geometries.

Finite Element Method (FEM): FEM arises naturally from weak formulations. The domain is divided into elements (intervals, triangles, tetrahedra), and trial/test functions are chosen locally (e.g. piecewise polynomials).
Example: For the same 1D Laplacian, using piecewise linear "hat" basis functions $\{\phi_i\}$ leads to the stiffness matrix

$$K_{ij} = \int_0^1 \phi_i'(x)\phi_j'(x)\,dx.$$

The resulting K is also tridiagonal but derived variationally. FEM generalizes easily to irregular domains and higher dimensions.

Spectral Methods: Spectral methods expand the solution in global basis functions (e.g. Fourier modes, Chebyshev polynomials). Differentiation is exact in this basis, so accuracy can be exponential when solutions are smooth.
Example: On $(0, 2\pi)$, approximate $u(x, t)$ by

$$u(x,t) \approx \sum_{k=-N}^{N} \hat{u}_k(t)e^{ikx}.$$

The Laplacian acts diagonally in this basis: $-\partial_{xx} e^{ikx} = k^2 e^{ikx}$. Thus spatial discretization reduces to multiplying coefficients by $-k^2$.

Spectral methods excel at periodic or smooth problems, but can struggle with discontinuities (Gibbs phenomenon).

SciML Perspective: Spatial discretization is not only a numerical step but also a modeling choice: *PINNs* bypass explicit discretization but still approximate derivatives by neural networks. *Neural operators* can be seen as learning mappings between discretized function spaces. *Graph neural networks* naturally discretize PDEs on irregular domains, with adjacency matrices replacing classical stiffness matrices.

Thus, spatial discretizations transform infinite-dimensional PDEs into finite-dimensional ODE systems. Finite differences, finite elements, and spectral methods represent three classical pillars ; each with distinct strengths ; and modern SciML frameworks often reinterpret or hybridize them to achieve scalability and accuracy.

2.5.5 CFL Conditions and Error/Convergence

When discretizing PDEs in both time and space, stability depends on the interaction of time-step size h and spatial grid size Δx. The fundamental principle is the *Courant–Friedrichs–Lewy (CFL) condition*, which constrains the time step for explicit schemes to remain stable.

Theorem: CFL Condition

For a PDE with characteristic wave speed c, an explicit time-stepping scheme with spatial step Δx is stable only if

$$c \frac{\Delta t}{\Delta x} \leq C,$$

where C is a constant depending on the scheme (the *CFL number*).

Example (1D Wave Equation) Consider

$$u_t + c u_x = 0, \quad u(x, 0) = u_0(x).$$

Discretize with forward Euler in time and upwind differences in space:

$$u_j^{n+1} = u_j^n - \frac{c\Delta t}{\Delta x}(u_j^n - u_{j-1}^n).$$

Von Neumann analysis shows stability if

$$\frac{c\Delta t}{\Delta x} \leq 1.$$

This is the CFL condition: information cannot travel across more than one grid cell in a single time step.

Example (Heat Equation): For $u_t = \alpha u_{xx}$ discretized with forward Euler in time and second-order differences in space,

$$u_j^{n+1} = u_j^n + \frac{\alpha \Delta t}{(\Delta x)^2}\left(u_{j-1}^n - 2u_j^n + u_{j+1}^n\right).$$

Stability requires

$$\frac{\alpha \Delta t}{(\Delta x)^2} \leq \tfrac{1}{2}.$$

If the step size is too large, the solution oscillates and blows up.

Error and Convergence: Stability is only one part of the story ; we also care about accuracy.

Definition: Consistency, Stability, Convergence

For a PDE discretization scheme:
- **Consistency**: the local truncation error $\tau \to 0$ as $\Delta t, \Delta x \to 0$.
- **Stability**: numerical solutions remain bounded as the mesh is refined.
- **Convergence**: the discrete solution u_j^n approaches the exact solution $u(x_j, t_n)$.

The **Lax Equivalence Theorem** states: for a consistent linear scheme, *stability is equivalent to convergence* [30].

Example (Convergence): For the heat equation with initial data $u(x, 0) = \sin(\pi x)$, the exact solution is $u(x, t) = e^{-\pi^2 t} \sin(\pi x)$. Discretizing with $\Delta x = 0.1$ and $\Delta t = 0.002$, we compute the L^2 error at $t = 0.1$. Refining the mesh to $\Delta x = 0.05$ and halving Δt reduces the error by approximately a factor of four, confirming second-order spatial convergence.

SciML Perspective: CFL conditions and convergence analysis remain relevant even when neural networks are used: PINNs must implicitly satisfy stability-like constraints when training with explicit time marching. Learned PDE solvers often struggle if trained outside the CFL regime. Hybrid schemes combine neural operators with classical discretizations, where error and convergence rates are inherited from the underlying numerical method.

Thus, the CFL condition embodies the fundamental link between physics, grid resolution, and time-step size. Together with consistency and stability, it ensures that discretizations converge to the correct PDE solution, whether solved by traditional schemes or SciML-based models.

2.6 Optimization Basics for SciML

2.6.1 Smoothness, Convexity, and Geometry

Optimization algorithms depend on a few geometric properties of the objective: *convexity* (curvature always bending upward), *smoothness* (gradients not changing too abruptly), and *strong convexity* (a uniform upward bend). Together with the choice of norm, these properties determine step sizes, convergence rates, and the need for preconditioning.

Definition: Convexity

A function $f : \mathbb{R}^n \to \mathbb{R}$ is **convex** if

$$f(\theta x + (1-\theta)y) \le \theta f(x) + (1-\theta) f(y), \qquad \forall x, y,\ \theta \in [0, 1].$$

If f is differentiable, this is equivalent to

$$f(y) \ \ge\ f(x) + \nabla f(x)^\top (y - x),$$

meaning the tangent plane at any point lies below the graph of f.

Running Example: Consider the quadratic function

$$f(x) = \tfrac{1}{2} x^\top Q x - b^\top x, \qquad Q \succ 0.$$

Its level sets are ellipses aligned with eigenvectors of Q. If $Q = I$, the ellipses are circles (well-conditioned). If Q has eigenvalues with large spread, the ellipses are elongated (ill-conditioned). Convexity here follows from $Q \succeq 0$.

Definition: L-Smoothness and μ-Strong Convexity

Fix a norm $\|\cdot\|$ with dual norm $\|\cdot\|_*$. A differentiable f is:

- ***L*-smooth** if

$$\|\nabla f(x) - \nabla f(y)\|_* \le L\, \|x - y\| \ \ \forall x, y,$$

equivalently

$$f(y) \le f(x) + \nabla f(x)^\top (y - x) + \tfrac{L}{2} \|y - x\|^2.$$

- **μ-strongly convex** if

$$f(y) \geq f(x) + \nabla f(x)^\top (y - x) + \tfrac{\mu}{2}\|y - x\|^2.$$

If both hold, the ratio $\kappa = L/\mu$ is the *condition number.*

Smoothness bounds how steeply the function can bend upward; it prevents "sharp turns" in the gradient. Strong convexity guarantees uniform curvature: the function is bowl-shaped everywhere. The condition number κ measures anisotropy: $\kappa \approx 1$ means nearly spherical level sets (easy optimization), while $\kappa \gg 1$ means long, narrow valleys (slow gradient descent).

For our quadratic $f(x)$, smoothness and convexity constants are $L = \lambda_{\max}(Q)$ and $\mu = \lambda_{\min}(Q)$, so κ is exactly the eigenvalue condition number of Q.

Step Sizes and Convergence: From L-smoothness, a gradient descent step with $\eta \leq 1/L$ guarantees

$$f(x - \eta\nabla f(x)) \leq f(x) - \tfrac{1}{2L}\|\nabla f(x)\|_*^2,$$

so the method always descends. If f is also μ-strongly convex, gradient descent converges linearly:

$$\|x_{k+1} - x^\star\| \leq \rho\|x_k - x^\star\|, \qquad \rho = \tfrac{\kappa-1}{\kappa+1}.$$

SciML Perspective: In inverse problems, κ reflects the conditioning of the Jacobian; poorly scaled features slow optimization. In PINNs, gradients often vary in magnitude across residual terms; rescaling them effectively changes the norm and improves κ. Natural gradient methods replace the Euclidean norm by a metric from the Fisher information, reshaping the landscape into rounder level sets.

Beyond Convexity: Not all SciML problems are convex, but weaker properties still guide algorithms. The Polyak–Łojasiewicz (PL) inequality,

$$\tfrac{1}{2}\|\nabla f(x)\|^2 \geq \mu\,(f(x) - f^\star),$$

implies linear convergence of gradient descent without convexity. This holds in many over-parameterized least-squares problems common in SciML.

In summary, smoothness, convexity, and strong convexity give us quadratic models of the objective. The condition number L/μ is the key difficulty knob. Good algorithms either adapt step sizes to these constants or redesign geometry (preconditioning, natural gradients) to make the optimization landscape more favorable.

2.6.2 *First-Order Methods*

First-order methods rely only on function values and gradients. They scale to high dimensions, integrate naturally with automatic differentiation, and are the default choice in SciML. Their performance depends on smoothness, convexity, and conditioning of the objective, as well as the geometry chosen through scaling or preconditioning.

Gradient Descent (Steepest Descent)

The basic gradient descent update is

$$x_{k+1} = x_k - \eta \nabla f(x_k),$$

where $\eta > 0$ is the step size.

Example (Quadratic Minimization) Consider

$$f(x) = \tfrac{1}{2} x^\top Q x, \qquad Q = \begin{bmatrix} 10 & 0 \\ 0 & 1 \end{bmatrix}.$$

The minimizer is $x^\star = 0$, with condition number $\kappa = \frac{10}{1} = 10$. If we start from $x_0 = (1, 1)$ and use step size $\eta = 0.1$, then

$$x_1 = (1, 1) - 0.1(10, 1) = (0, 0.9).$$

The first coordinate shrinks rapidly, while the second decays slowly. Level sets are elongated ellipses, and gradient descent zig-zags across them. This illustrates the impact of conditioning: ill-conditioned problems slow convergence.

Choosing Step Size: For an L-smooth function, any $\eta \in (0, 2/L)$ guarantees descent. If f is also μ-strongly convex, the optimal fixed step is $\eta = 2/(L + \mu)$, yielding linear convergence with factor $\frac{\kappa - 1}{\kappa + 1}$.

Line Search: When L is unknown, backtracking line search adjusts η dynamically. Start from a trial η (often scaled up from the previous step), and shrink by factor $\beta \in (0, 1)$ until the Armijo condition holds:

$$f(x_k - \eta \nabla f(x_k)) \le f(x_k) - c\,\eta \|\nabla f(x_k)\|^2, \qquad c \in (0, 1).$$

Momentum: To accelerate convergence, we can add memory of past gradients.

Momentum and Nesterov Acceleration

- Heavy-ball momentum:

$$v_{k+1} = \beta v_k - \eta \nabla f(x_k), \qquad x_{k+1} = x_k + v_{k+1}.$$

- Nesterov's method:

$$y_k = x_k + \beta(x_k - x_{k-1}), \qquad x_{k+1} = y_k - \eta \nabla f(y_k).$$

Example: Using the quadratic $f(x) = \frac{1}{2}x^\top Qx$ with $Q = \text{diag}(10, 1)$, plain gradient descent zig-zags. Adding momentum with $\beta = 0.9$ smooths oscillations and accelerates convergence, as iterates follow the valley more directly. Nesterov's look-ahead stabilizes the update and yields the optimal $O(1/k^2)$ rate in convex settings.

Adaptive Methods: Modern SciML often uses adaptive diagonal scaling, such as AdaGrad, RMSProp, or Adam.

Adam Update Rule

Adam maintains exponential averages of gradients (m_k) and squared gradients (v_k):

$$m_k = \beta_1 m_{k-1} + (1 - \beta_1)\nabla f(x_k), \qquad v_k = \beta_2 v_{k-1} + (1 - \beta_2)\nabla f(x_k)^2.$$

The update is

$$x_{k+1} = x_k - \eta \frac{\widehat{m}_k}{\sqrt{\widehat{v}_k} + \epsilon},$$

with bias-corrected moments $\widehat{m}_k$ and $\widehat{v}_k$.

Example (Scale-Invariance): Suppose $f(\theta) = \frac{1}{2}(a\theta - 1)^2$ with $a = 1000$. Gradient descent requires very small η to avoid divergence, since gradients are scaled by a. Adam rescales automatically: large gradients are damped by $\sqrt{\widehat{v}_k}$, making training stable across scales.

Proximal and Projected Gradient: Many SciML problems combine smooth terms (data misfit) with non-smooth ones (sparsity, constraints). In such cases, proximal steps extend gradient descent:

Proximal Gradient Step

Given $f(x) = g(x) + r(x)$ with g smooth and r convex,

$$x_{k+1} = \mathrm{prox}_{\eta r}(x_k - \eta \nabla g(x_k)),$$

where $\mathrm{prox}_{\eta r}(z) = \arg\min_x \left(r(x) + \frac{1}{2\eta}\|x - z\|^2\right)$.

Example: For $r(x) = \lambda \|x\|_1$, the proximal operator is soft-thresholding:

$$\mathrm{prox}_{\eta r}(z) = \mathrm{sign}(z) \max\{|z| - \eta\lambda, 0\}.$$

This promotes sparsity, widely used in compressed sensing and SciML models with feature selection.

SciML Perspective: Neural ODE training often starts with Adam for stability, then switches to L-BFGS for precision. PINNs benefit from adaptive scaling to handle unbalanced residuals. In inverse problems, proximal steps enforce physical constraints or sparsity.

In summary, first-order methods are versatile and scalable. Their behavior is governed by smoothness, convexity, and conditioning, but can be dramatically improved by momentum, adaptive scaling, and proximal extensions. They remain the workhorse of optimization in SciML.

2.6.3 Second-Order and Quasi-Newton Methods

First-order methods rely only on gradients. Second-order methods exploit curvature information (the Hessian or its approximations) to adapt step directions and sizes. This often yields far faster convergence, especially for ill-conditioned problems. In SciML, where gradients come from automatic differentiation, Hessian-vector products (HVPs) are often available at nearly the cost of a gradient, making second-order ideas attractive.

Newton's Method

Given current iterate x_k, Newton's step solves the quadratic model

$$m_k(p) = f(x_k) + \nabla f(x_k)^\top p + \tfrac{1}{2} p^\top H_k p, \qquad H_k = \nabla^2 f(x_k).$$

The Newton direction is

$$p_k^{\mathrm{N}} = -H_k^{-1} \nabla f(x_k).$$

The update is $x_{k+1} = x_k + p_k^{\mathrm{N}}$.

Example (Quadratic Function): For $f(x) = \frac{1}{2} x^\top Q x - b^\top x$, the gradient is $\nabla f(x) = Qx - b$, and the Hessian is $H = Q$. The Newton step solves $Qp = -(Qx - b)$, i.e.

$$p = x^\star - x,$$

so Newton jumps directly to the minimizer in *one step*. This explains the quadratic convergence of Newton's method near a minimizer: once curvature is well estimated, error shrinks superlinearly.

Limitations: Pure Newton can fail if H is indefinite (nonconvex regions) or badly conditioned. In practice, we combine Newton steps with globalization strategies: *Line search*: only accept steps that decrease f. *Trust region*: restrict step size to a ball $\|p\| \leq \Delta$.

Gauss–Newton for Least Squares: Many SciML problems are nonlinear least squares:

$$f(x) = \tfrac{1}{2} \|F(x)\|_2^2, \qquad F : \mathbb{R}^n \to \mathbb{R}^m.$$

Here

$$\nabla f(x) = J(x)^\top F(x), \qquad \nabla^2 f(x) = J(x)^\top J(x) + \sum_i F_i(x) \nabla^2 F_i(x).$$

Near a residual-minimizing solution ($F(x^\star) \approx 0$), the second term is small, so the *Gauss–Newton step*

$$(J^\top J) p = -J^\top F$$

is a good approximation.

Example (Nonlinear Least Squares) Suppose $F(x) = \begin{bmatrix} x_1^2 - 1 \\ x_1 + x_2 - 2 \end{bmatrix}$. At $x = (1, 1)$, $F = (0, 0)$ and $J = \begin{bmatrix} 2 & 0 \\ 1 & 1 \end{bmatrix}$. Then $J^\top J = \begin{bmatrix} 5 & 1 \\ 1 & 1 \end{bmatrix}$, so the Gauss–Newton step solves $(J^\top J)p = -J^\top F = 0$, i.e. $p = 0$; we are already at a minimizer. At a nearby point, Gauss–Newton would provide a correction direction without needing the full Hessian.

Quasi-Newton Methods: Forming Hessians explicitly is costly. Quasi-Newton methods build low-rank approximations from gradients and iterates. The most famous is BFGS.

Quasi-Newton Update: BFGS

Given $s_k = x_{k+1} - x_k$ and $y_k = \nabla f(x_{k+1}) - \nabla f(x_k)$, BFGS updates an inverse Hessian approximation B_k by

$$B_{k+1} = \left(I - \rho_k s_k y_k^\top\right) B_k \left(I - \rho_k y_k s_k^\top\right) + \rho_k s_k s_k^\top, \quad \rho_k = \frac{1}{y_k^\top s_k}.$$

BFGS preserves positive definiteness (under curvature conditions) and converges superlinearly for smooth convex functions. Its limited-memory variant (L-BFGS) stores only the last m updates, making it practical for high dimensions.

Example (Logistic Regression): For $f(\theta) = \sum_i \log(1 + \exp(-y_i a_i^\top \theta)) + \frac{\lambda}{2}\|\theta\|^2$, the Hessian is $A^\top W A + \lambda I$, with $0 \preceq W \preceq \frac{1}{4} I$. Gradient descent struggles when A is ill-conditioned, but L-BFGS quickly adapts to curvature, often converging in tens of steps.

SciML Perspective: In PINNs, Gauss–Newton steps align gradients across collocation points, dramatically accelerating convergence compared to Adam. In inverse problems, Newton and Gauss–Newton connect naturally to adjoint methods, since J and $J^\top J$ appear in sensitivity equations. In large neural models, L-BFGS is often used after Adam to refine solutions once noise is low.

In summary, second-order and quasi-Newton methods trade heavier linear algebra for faster, curvature-aware steps. Newton is exact for quadratics, Gauss–Newton exploits least-squares structure, and quasi-Newton (BFGS/L-BFGS) sketch curvature on the fly. All of them embody the principle that geometry matters ; and SciML often provides the structure (Jacobians, adjoints, Hessian-vector products) to make these methods practical.

2.6.4 *Constraints and Duality*

Many SciML optimization problems come with constraints: physical conservation laws, box bounds, monotonicity requirements, sparsity, or PDEs that the state must satisfy. Constraints can be enforced explicitly by projection, implicitly through penalties, or systematically via Lagrangian duality.

General Constrained Problem

We consider

$$\min_{x \in \mathbb{R}^n} f(x) \quad \text{s.t.} \quad c(x) = 0, \;\; g(x) \leq 0,$$

where $c : \mathbb{R}^n \to \mathbb{R}^m$ are equality constraints and $g : \mathbb{R}^n \to \mathbb{R}^r$ are inequalities (interpreted componentwise).

Example (Constrained Quadratic): Minimize $f(x) = \frac{1}{2}\|Ax - b\|^2$ subject to $Cx = d$. The solution must both fit the data and satisfy the linear constraint.

Lagrangian and KKT Conditions: Constraints can be incorporated via the Lagrangian

$$\mathcal{L}(x, \lambda, \nu) = f(x) + \lambda^\top c(x) + \nu^\top g(x), \qquad \nu \geq 0,$$

where λ and ν are multipliers.

KKT Conditions
A point $(x^\star, \lambda^\star, \nu^\star)$ satisfies the Karush-Kuhn-Tucker (KKT) conditions if:

$$\begin{aligned} \nabla f(x^\star) + \nabla c(x^\star)^\top \lambda^\star + \nabla g(x^\star)^\top \nu^\star &= 0, \\ c(x^\star) &= 0, \\ g(x^\star) \leq 0, \;\; \nu^\star \geq 0, \;\; \nu_i^\star g_i(x^\star) &= 0 \; \forall i. \end{aligned}$$

The first line is *stationarity*. The second ensures equality constraints hold. The third enforces inequality feasibility and *complementary slackness*.

Example (Quadratic with Equality Constraint): Take $f(x) = \frac{1}{2}\|x\|^2$ subject to $1^\top x = 1$. The Lagrangian is $\mathcal{L}(x, \lambda) = \frac{1}{2}\|x\|^2 + \lambda(1^\top x - 1)$. Stationarity gives $x + \lambda 1 = 0$, i.e. $x = -\lambda 1$. Constraint $1^\top x = 1$ gives $-\lambda n = 1 \implies \lambda = -1/n$. Thus $x^\star = (1/n, 1/n, \ldots, 1/n)$: the uniform point.

Duality: Instead of solving the primal problem directly, we can form the dual:

$$q(\lambda, \nu) = \inf_x \mathcal{L}(x, \lambda, \nu),$$

and maximize q over $\lambda, \nu \geq 0$.

Weak and Strong Duality Weak duality: For any feasible dual (λ, ν), $q(\lambda, \nu) \leq f^\star$. **Strong duality**: If f and g_i are convex, c is affine, and a strictly feasible point exists (Slater's condition), then $\max q = f^\star$.

Example (LASSO Duality): Primal: $\min_x \frac{1}{2}\|Ax - b\|^2 + \lambda\|x\|_1$. This can be written as a constrained least-squares problem:

$$\min_x \frac{1}{2}\|Ax - b\|^2 \quad \text{s.t.} \quad \|x\|_1 \leq \tau,$$

with τ linked to λ. The dual involves maximizing over multipliers for the ℓ_1-ball constraint, leading to soft-thresholding operators in the primal. This illustrates how sparsity-inducing penalties and constraints are two sides of the same coin.

Enforcement Strategies: *Projection*: after each gradient step, project x back onto feasible set $\mathcal{C}$. *Penalty/Barrier*: add $\rho\|c(x)\|^2$ or $-\mu\sum\log(-g_i(x))$ to the objective. *Augmented Lagrangian / ADMM*: alternate between minimizing over x and updating multipliers, especially effective for decomposable problems.

SciML Perspective: In PINNs, boundary conditions are often treated as soft penalties; augmented Lagrangian approaches can enforce them more robustly. In inverse problems, multipliers quantify sensitivities: how much the solution changes if a constraint is perturbed. ADMM variants are used in distributed SciML to split data misfit, physics constraints, and regularization across modules.

In summary, constraints shape feasible solutions, and duality provides both theoretical guarantees and practical algorithms. Whether through projections, penalties, or Lagrange multipliers, SciML problems with physical laws or structural priors benefit from constraint-aware optimization.

2.6.5 Stochastic Optimization and Variance Reduction

Many SciML objectives are expectations or empirical averages. Evaluating the full gradient may require sweeping through all data points, all collocation nodes, or an entire spatio-temporal mesh ; too costly per step. *Stochastic optimization* replaces the exact gradient with a cheap, unbiased estimate. Variance reduction techniques then refine these estimates to recover faster convergence.

Stochastic Gradient Descent (SGD)

For $f(x) = \frac{1}{m}\sum_{i=1}^m f_i(x)$, SGD samples an index i_k and updates

$$x_{k+1} = x_k - \eta_k \nabla f_{i_k}(x_k).$$

This ensures $\mathbb{E}[\nabla f_{i_k}(x_k)] = \nabla f(x_k)$, but each step is noisy.

Example (Linear Regression): Consider $f(\theta) = \frac{1}{2m}\sum_{i=1}^m (a_i^\top\theta - y_i)^2$. The full gradient is $\nabla f(\theta) = \frac{1}{m}A^\top(A\theta - y)$. SGD samples a single row (a_i, y_i):

$$g(\theta) = (a_i^\top \theta - y_i)a_i,$$

with expectation equal to the full gradient. Each update is cheap, but noisy; convergence is slower than full gradient descent, though per-iteration cost is much lower.

Variance and Mini-Batching: The variance of SGD decreases if we average multiple samples per step. With batch size B,

$$g_B(\theta) = \tfrac{1}{B} \sum_{i \in \mathcal{B}} \nabla f_i(\theta),$$

the variance scales like σ^2/B. This explains why larger batches stabilize training, but with diminishing returns.

Variance Reduction Methods: Modern algorithms reduce gradient noise by reusing information.

Variance Reduction Examples

- **SVRG (Stochastic Variance Reduced Gradient)**: Maintain a reference point $\tilde{x}$ and full gradient $\tilde{\mu} = \nabla f(\tilde{x})$. Inside an epoch, use

$$g_k = \nabla f_{i_k}(x_k) - \nabla f_{i_k}(\tilde{x}) + \tilde{\mu}.$$

- **SAGA**: Keep a memory of past gradients $\{\phi_i\}$. At each step update

$$g_k = \nabla f_{i_k}(x_k) - \phi_{i_k} + \tfrac{1}{m} \sum_{j=1}^{m} \phi_j,$$

  and replace ϕ_{i_k} with $\nabla f_{i_k}(x_k)$.
- **SARAH/SPIDER**: Use recursive gradient estimates with periodic full refreshes, reducing cost while keeping variance low.

Example (Logistic Regression with SVRG): Suppose $f(\theta) = \frac{1}{m} \sum_{i=1}^{m} \log(1 + \exp(-y_i a_i^\top \theta))$. SGD gradients are noisy. With SVRG, at a snapshot $\tilde{\theta}$, we compute $\tilde{\mu} = \nabla f(\tilde{\theta})$. Then each update subtracts the stale gradient at $\tilde{\theta}$ and adds $\tilde{\mu}$. This keeps the expectation correct but reduces variance dramatically, often yielding linear convergence in convex problems.

Step Sizes and Noise Floors: In nonconvex settings, SGD with diminishing step sizes $\eta_k \sim 1/\sqrt{k}$ achieves $\mathbb{E}[\|\nabla f(x_k)\|^2] \to 0$. With constant step size, SGD stabilizes in a neighborhood of the minimizer, whose radius depends on the gradient variance σ^2. Variance reduction shrinks this neighborhood, allowing convergence closer to $x^\star$.

SciML Perspective: In PINNs, stochastic sampling of collocation points is just SGD in disguise. Variance reduction can prioritize "hard" regions (e.g. near boundaries or sharp gradients) by importance sampling, stabilizing training. Neural operators often rely on mini-batch SGD, and variance reduction helps when data are highly imbalanced across domains. Hybrid solvers combine stochastic gradients with second-order corrections (e.g. L-BFGS after SGD).

In summary, stochastic optimization reduces per-step cost by trading accuracy for noise. Variance reduction techniques (SVRG, SAGA, SARAH) balance this trade, yielding convergence guarantees close to full gradient descent while keeping stochastic efficiency. For SciML, they are indispensable when training involves vast datasets or discretized PDE residuals.

2.6.6 Nonconvex Analysis Touchpoints

Most SciML problems are nonconvex: training neural operators, fitting PINNs, or solving inverse problems with nonlinear PDEs. Nonconvexity means the landscape may contain many local minima, saddle points, or flat regions. Analysis therefore focuses not on finding global minima, but on understanding stationarity, curvature, and mild regularity properties that still guarantee convergence.

Stationary Points

For $f \in C^2$, a point $x^\star$ is:

- **First-order stationary** if $\nabla f(x^\star) = 0$.
- **Second-order stationary** if $\nabla f(x^\star) = 0$ and $\nabla^2 f(x^\star) \succeq 0$.
- A **strict saddle** if $\nabla f(x^\star) = 0$ and $\lambda_{\min}(\nabla^2 f(x^\star)) < 0$.

Example (Saddle in a Quadratic): Consider $f(x, y) = x^2 - y^2$. At $(0, 0)$, $\nabla f = (0, 0)$ but $\nabla^2 f = \mathrm{diag}(2, -2)$, so $(0, 0)$ is a saddle. Gradient descent can stall here if steps are tiny, but noise or curvature exploitation pushes iterates away along the negative direction.

Escaping Saddles: Modern analysis shows that gradient descent with random initialization avoids strict saddles with probability one. Stochasticity (as in SGD) or explicit negative-curvature steps helps escape quickly. This explains why deep-learning optimizers typically do not get stuck at saddles.

Polyak–Łojasiewicz (PL) Inequality: Even without convexity, a function can satisfy PL:

PL Inequality

A differentiable f satisfies the PL condition with $\mu > 0$ if

$$\frac{1}{2}\|\nabla f(x)\|^2 \geq \mu\,(f(x) - f^\star), \qquad \forall x.$$

PL implies linear convergence of gradient descent to the global minimum value $f^\star$, even if f has nonconvex level sets.

Example (Over-parameterized Least Squares): Let $f(\theta) = \frac{1}{2}\|A\theta - b\|^2$. If A has full column rank, f is strongly convex. If A is rank-deficient (as in over-parameterized models), f is nonconvex with many minimizers, but the PL inequality still holds:

$$\tfrac{1}{2}\|\nabla f(\theta)\|^2 = \tfrac{1}{2}\|A^\top(A\theta - b)\|^2 \geq \sigma_{\min}^2\,(f(\theta) - f^\star),$$

where $\sigma_{\min}$ is the smallest nonzero singular value of A. Thus gradient descent still converges linearly to the minimum value.

Quadratic Growth and Error Bounds: Another useful property is quadratic growth (QG):

$$f(x) - f^\star \;\geq\; \tfrac{\mu}{2}\,\|x - x^\star\|^2,$$

for all x near minimizers $x^\star$. QG implies PL for smooth functions, and both guarantee linear convergence.

Kurdyka-Łojasiewicz (KL) Property: Many nonconvex objectives in SciML are *semi-algebraic* (combinations of polynomials, exponentials, ReLUs). Such functions satisfy the KL property: near any critical point, a desingularizing function φ exists such that

$$\varphi'(f(x) - f(x^\star)) \cdot \|\nabla f(x)\| \;\geq\; 1.$$

This ensures that descent methods generate convergent sequences.

SciML Perspective: In PINNs, loss landscapes are highly nonconvex, but PL-type behavior often holds near interpolating solutions. In inverse problems, flat directions arise from underdetermined parameters; regularization improves QG constants. Neural operators and wide neural nets often exhibit benign landscapes where all local minima are global or PL holds globally. Algorithms that detect negative curvature (trust-region, cubic regularization) are valuable for escaping saddle plateaus in SciML training.

In summary, nonconvex analysis in SciML focuses not on ruling out bad critical points entirely, but on ensuring that algorithms converge to meaningful solutions. Stationarity, PL/QG inequalities, and the KL property provide the mathematical tools. Stochasticity and preconditioning make them practical: they explain why simple first-order methods succeed even in highly nonconvex landscapes.

2.7 Probability, Statistics, and Information

2.7.1 Random Variables, Moments, Concentration

A random variable formalizes uncertainty about a quantity; in SciML this might be sensor noise, randomized quadrature points, mini-batches, or draws from a physical ensemble. We use expectations to summarize typical behavior and *concentration* inequalities to control fluctuations around those summaries with explicit, high-probability bounds. These bounds are the backbone of error bars for Monte Carlo estimates, sample-size rules for training/validation splits, and stability guarantees for stochastic gradients.

Definition: Random variable, expectation, variance

A real-valued random variable X is a measurable map from a probability space to $\mathbb{R}$. The *expectation* is $\mathbb{E}[X]$, the *variance* is $\mathrm{Var}(X) = \mathbb{E}\big[(X - \mathbb{E}X)^2\big]$, and the *covariance* of X, Y is $\mathrm{Cov}(X, Y) = \mathbb{E}\big[(X - \mathbb{E}X)(Y - \mathbb{E}Y)\big]$. The moment generating function (mgf) is $M_X(\lambda) = \mathbb{E}[e^{\lambda X}]$ when finite.

Expectations are linear ($\mathbb{E}[aX + bY] = a\mathbb{E}X + b\mathbb{E}Y$), and variances add for independent, centered variables. Averages reduce noise: for i.i.d. $X_1, \ldots, X_n$ with mean μ and variance σ^2, the sample mean $\bar{X}_n = \frac{1}{n}\sum_i X_i$ satisfies $\mathbb{E}[\bar{X}_n] = \mu$ and $\mathrm{Var}(\bar{X}_n) = \sigma^2/n$. The law of large numbers says $\bar{X}_n \to \mu$; the central limit theorem says $\sqrt{n}(\bar{X}_n - \mu) \Rightarrow \mathcal{N}(0, \sigma^2)$. Concentration inequalities go further: they give *nonasymptotic* tail bounds that hold at finite n with explicit constants.

Definition: Sub-Gaussian (tails no heavier than Gaussian)

A centered X is *sub-Gaussian* with parameter σ^2 if

$$\mathbb{E}\big[e^{\lambda X}\big] \leq \exp\Big(\tfrac{\lambda^2\sigma^2}{2}\Big) \quad \forall\lambda \in \mathbb{R}.$$

Equivalently, $\mathbb{P}(|X| \geq t) \leq 2\exp\big(-t^2/(2\sigma^2)\big)$ for all $t \geq 0$.

Bounded and Gaussian variables are sub-Gaussian: if $X \in [a, b]$ almost surely and $\mathbb{E}X = 0$ then X is sub-Gaussian with $\sigma^2 = (b - a)^2/4$; if $X \sim \mathcal{N}(0, \sigma^2)$ then the parameter is exactly σ^2. Sub-Gaussianity is preserved by sums: if X_i are independent, centered, sub-Gaussian with parameters σ_i^2, then $\sum_i X_i$ is sub-Gaussian with parameter $\sum_i \sigma_i^2$.

Hoeffding's inequality (bounded summands)

Let $X_1, \ldots, X_n$ be independent with $X_i \in [a_i, b_i]$ and means μ_i. Then for all $t > 0$,

$$\mathbb{P}\left(\frac{1}{n}\sum_{i=1}^{n} X_i - \frac{1}{n}\sum_{i=1}^{n} \mu_i \geq t\right) \leq \exp\left(-\frac{2n^2t^2}{\sum_{i=1}^{n}(b_i - a_i)^2}\right),$$

and the same bound holds for the lower tail with $t \mapsto -t$.

This gives sample sizes at a glance: if $X_i \in [0, 1]$ and $\mu = \mathbb{E}X_i$, then $\mathbb{P}(|\bar{X}_n - \mu| \geq \varepsilon) \leq 2\exp(-2n\varepsilon^2)$, so $n \geq \frac{1}{2\varepsilon^2}\log\frac{2}{\delta}$ suffices for accuracy ε with confidence $1-\delta$. In SciML this is how many Monte Carlo points you need so that a randomized quadrature or a stochastic loss term concentrates to within a target tolerance.

Chernoff bounds (multiplicative tails for sums of nonnegative r.v.'s)

Let $S = \sum_{i=1}^{n} Y_i$ be a sum of independent Bernoulli(p_i) or, more generally, independent nonnegative variables dominated in mgf by Bernoulli/Poisson with mean $\mu = \mathbb{E}S$. Then for $\varepsilon \in (0, 1)$,

$$\mathbb{P}\big(S \geq (1+\varepsilon)\mu\big) \leq \exp\Big(-\mu\,\psi_+(\varepsilon)\Big),$$
$$\mathbb{P}\big(S \leq (1-\varepsilon)\mu\big) \leq \exp\Big(-\mu\,\psi_-(\varepsilon)\Big),$$

with $\psi_+(\varepsilon) = (1+\varepsilon)\log(1+\varepsilon) - \varepsilon$ and $\psi_-(\varepsilon) = -\log(1-\varepsilon) - \varepsilon$.

Chernoff's multiplicative form is sharper than Hoeffding when the mean μ is small, which is common for rare-event counts (e.g., misclassification rate over a validation set).

Sub-exponential and Bernstein inequality

A centered X is *sub-exponential* with parameters (ν^2, b) if

$$\mathbb{E}[e^{\lambda X}] \leq \exp\left(\tfrac{\lambda^2\nu^2}{2}\right) \quad \text{for } |\lambda| \leq 1/b.$$

If $X_1, \ldots, X_n$ are independent centered sub-exponential with common (ν^2, b), then for all $t > 0$,

$$\mathbb{P}\left(\frac{1}{n}\sum_{i=1}^{n} X_i \geq t\right) \leq \exp\left(-n \min\left\{\frac{t^2}{2\nu^2}, \frac{t}{2b}\right\}\right),$$

and similarly for the lower tail.

Bernstein interpolates smoothly between sub-Gaussian behavior for small deviations ($t \ll \nu^2/b$) and exponential tails for large deviations. It is the right tool for losses with bounded variance but occasional large values (e.g., heavy-tailed gradients truncated by clipping).

The vector case reduces to scalars by projecting onto directions. A random vector $Z \in \mathbb{R}^d$ is sub-Gaussian if $v^\top Z$ is sub-Gaussian for every unit v. Then, for i.i.d. centered sub-Gaussian vectors with parameter σ^2, the sample mean obeys

$$\mathbb{P}(\|\bar{Z}_n\|_2 \geq t) \;\leq\; 2\exp\left(-\frac{nt^2}{2\sigma^2}\right) \quad \text{(any fixed direction)},$$

and a union bound tightens coordinate-wise control:

$$\mathbb{P}(\|\bar{Z}_n\|_\infty \geq t) \;\leq\; 2d\;\exp\left(-\frac{nt^2}{2\sigma^2}\right).$$

This $\log d$ penalty explains why estimating all d coordinates uniformly well costs an extra $\log d$ factor in samples; you see this when choosing batch sizes to stabilize all components of a gradient or all outputs of a model simultaneously.

Two proof ideas reappear across inequalities. The first is *Hoeffding's lemma*: a centered $X \in [a, b]$ satisfies $\mathbb{E}[e^{\lambda X}] \leq \exp\big(\lambda^2(b-a)^2/8\big)$ for all λ, which makes bounded X sub-Gaussian. The second is the mgf trick plus Markov's inequality: $\mathbb{P}(S \geq t) = \mathbb{P}(e^{\lambda S} \geq e^{\lambda t}) \leq e^{-\lambda t}\mathbb{E}[e^{\lambda S}]$, and independence factorizes the mgf of a sum; optimizing over λ yields sharp tails. These two lines;bounding mgfs and optimizing λ;power most of the bounds above.

Three short examples tie these tools to practice. In randomized quadrature for $\int_\Omega \phi(x)\,dx$, if $\phi(X)$ is centered and bounded in $[a, b]$, then with n i.i.d. points the Monte Carlo estimator has error $\mathcal{O}\big(\sqrt{\log(1/\delta)/n}\big)$ with probability $1-\delta$;a dimension-free rate that improves to $\mathcal{O}(n^{-1})$ for smooth ϕ under low-discrepancy sequences but still benefits from Hoeffding-style tails. In mini-batch SGD, if per-sample gradients are sub-Gaussian with parameter σ_g^2 in every direction, the batch mean has parameter σ_g^2/B; choosing B so that $\sqrt{\sigma_g^2/B} \lesssim \varepsilon$ guarantees gradient noise at most ε with high probability, aligning the step-size and batch-size rules. In generalization testing, a validation error computed on m i.i.d. held-out samples concentrates around the true risk with deviations $\mathcal{O}(\sqrt{\log(1/\delta)/m})$ by Hoeffding; a union bound across K hyperparameter settings adds a $\log K$ term, which is why nested validation or corrections (Bonferroni, Benjamini-Hochberg) are necessary when scanning many models.

Two practical cautions help avoid misuse. Sub-Gaussian assumptions fail for truly heavy-tailed data (e.g., power-law noise); then robust mean estimators (median-of-means, Catoni) recover exponential-type concentration without boundedness. And "independence" should match how you sample: overlapping mini-batches or correlated trajectories reduce effective sample size. When in doubt, estimate variance (or a σ proxy) online and use empirical Bernstein bounds that plug in the observed variance in place of a worst-case one; they adapt to benign regimes while remaining valid.

Finally, a compact rule-of-thumb connects accuracy, confidence, and samples for bounded/sub-Gaussian averages:

$$n \gtrsim \frac{\sigma^2}{\varepsilon^2} \log\frac{2}{\delta} \quad \Rightarrow \quad \mathbb{P}\big(|\bar{X}_n - \mu| \le \varepsilon\big) \ge 1 - \delta.$$

Insert $\sigma^2 = (b-a)^2/4$ for bounded $X \in [a, b]$, or estimate σ^2 from pilot runs. This single formula drives many design choices in SciML: how many collocation points to sample per epoch, how many Monte Carlo paths for an expectation in the loss, and how large a validation set to make performance comparisons meaningful.

2.7.2 Bayesian Modeling and Gaussian Processes

Bayesian modeling treats unknowns as random variables and updates beliefs with data. Suppose a dataset $D = \{(x_i, y_i)\}_{i=1}^n$ and a parametric likelihood $p(y \mid x, \theta)$. A prior $p(\theta)$ encodes assumptions before observing D. After seeing data, information flows through Bayes' rule to produce a posterior over parameters and a predictive distribution for future observations; the result is both a point estimate and calibrated uncertainty.

Bayes' rule and posterior predictive

$$p(\theta \mid D) = \frac{p(D \mid \theta)\, p(\theta)}{p(D)}, \quad p(y_* \mid x_*, D) = \int p(y_* \mid x_*, \theta)\, p(\theta \mid D)\, d\theta,$$

where $p(D) = \int p(D \mid \theta) p(\theta)\, d\theta$ is the (marginal) evidence.

Point estimates like MLE/MAP optimize $p(D \mid \theta)$ or $p(\theta \mid D)$; full Bayes integrates over θ in the posterior predictive and yields credible intervals. Priors act as regularizers: a Gaussian prior shrinks coefficients like ridge; a Laplace prior promotes sparsity like LASSO. Hyperparameters of the prior/likelihood (e.g., variances) can be set by maximizing the evidence ("type-II ML") or by placing hyperpriors and integrating them out.

A canonical example is Bayesian linear regression. With $y = X\theta + \varepsilon$, $\varepsilon \sim \mathcal{N}(0, \sigma^2 I)$ and $\theta \sim \mathcal{N}(m_0, \Sigma_0)$, the posterior is Gaussian with

$$\Sigma_n = \left(\sigma^{-2}X^\top X + \Sigma_0^{-1}\right)^{-1}, \qquad m_n = \Sigma_n\left(\sigma^{-2}X^\top y + \Sigma_0^{-1}m_0\right),$$

and the predictive at x_* is $y_* \mid D \sim \mathcal{N}\left(x_*^\top m_n,\ x_*^\top \Sigma_n x_* + \sigma^2\right)$. With $m_0 = 0$ and $\Sigma_0 = \tau^2 I$, the posterior mean equals the ridge solution, and the evidence

$$\log p(D) = -\tfrac{1}{2}\left[y^\top(\sigma^2 I + X\Sigma_0 X^\top)^{-1}y + \operatorname{logdet}(\sigma^2 I + X\Sigma_0 X^\top) + n\log 2\pi\right]$$

guides σ, τ automatically. Hierarchical models let τ vary by group or feature (automatic relevance determination), improving adaptation in high dimensions.

Gaussian processes (GPs) place a prior *directly on functions*. Instead of choosing a finite θ, we specify that any finite collection of function values is jointly Gaussian with a mean and a kernel that encodes correlation and smoothness. This makes GPs ideal surrogates for expensive simulators and natural regularizers for scattered data.

Definition: Gaussian process prior

A stochastic process $f : \mathcal{X} \to \mathbb{R}$ is a Gaussian process, $f \sim \mathcal{GP}(m, k)$, if for any inputs $X = (x_1, \ldots, x_n)$,

$$\left(f(x_1), \ldots, f(x_n)\right)^\top \sim \mathcal{N}\left(m(X),\ K(X, X)\right),$$

where $m(x) = \mathbb{E}[f(x)]$ and $k(x, x') = \operatorname{Cov}\left(f(x), f(x')\right)$.

In GP regression, observations follow $y_i = f(x_i) + \varepsilon_i$ with $\varepsilon_i \sim \mathcal{N}(0, \sigma^2)$. Writing $K = K(X, X)$, $K_* = K(X, X_*)$, and $K_{**} = K(X_*, X_*)$, the posterior predictive at test locations X_* is Gaussian with

$$\mu_* = m(X_*) + K_*^\top\left(K + \sigma^2 I\right)^{-1}\left(y - m(X)\right),\ \Sigma_* = K_{**} - K_*^\top\left(K + \sigma^2 I\right)^{-1}K_*.$$

The mean is an interpolant/smoother; the variance quantifies epistemic uncertainty and collapses at observed inputs as $\sigma^2 \downarrow 0$. Choosing m and k is modeling: squared-exponential (RBF) $k(x, x') = \sigma_f^2 \exp\left(-\tfrac{1}{2}\|x - x'\|^2/\ell^2\right)$ gives very smooth functions; Matérn kernels add a roughness parameter ν to control differentiability;

periodic, linear, and polynomial kernels encode known structure; sums and products compose behavior; ARD variants assign one length-scale per input coordinate to perform feature scaling automatically. Derivatives of f are again jointly Gaussian and handled by differentiating the kernel, enabling direct learning from gradient/flux data.

Hyperparameters θ_k (length-scales, variances, noise) are fitted by maximizing the log marginal likelihood

$$\mathcal{L}(\theta_k) = -\tfrac{1}{2}(y-m)^\top (K_\theta + \sigma^2 I)^{-1}(y-m) - \tfrac{1}{2}\operatorname{logdet}(K_\theta + \sigma^2 I) - \tfrac{n}{2}\log 2\pi,$$

whose gradients follow from matrix calculus and AD. This objective automatically trades data-fit against model complexity via the log det term, and it typically yields calibrated predictive uncertainties when the kernel class is appropriate.

The GP posterior mean coincides with kernel ridge regression: $\mu_* = K_*^\top (K + \lambda I)^{-1} y$ with $\lambda = \sigma^2$, linking Bayesian and regularization viewpoints. The posterior variance is the principled error bar missing from pure interpolation; it guides active learning and experimental design. In Bayesian optimization, a GP surrogate powers acquisition rules such as expected improvement or UCB that balance exploration (large variance) and exploitation (low mean), accelerating the search for optimal parameters or designs.

Computational cost is the main limitation. Exact GP training scales as $\mathcal{O}(n^3)$ and memory as $\mathcal{O}(n^2)$ due to the Cholesky of $K + \sigma^2 I$. Several remedies extend GPs to large n: inducing-point methods (e.g., variational sparse GPs) replace K by a low-rank Nyström approximation with $m \ll n$ pseudo-points and optimize a tight evidence lower bound; structured kernels (Toeplitz/Kronecker) exploit grids for $\mathcal{O}(n \log n)$ algebra; iterative solvers (CG) apply $(K + \sigma^2 I)^{-1}$ via fast kernel-vector products with preconditioning; and stochastic variational GPs combine mini-batches with inducing points for streaming data. Multi-output and multi-fidelity models use coregionalization to couple related tasks; state-space representations of Matérn and certain periodic kernels turn GP inference into Kalman filtering for linear-time scaling in 1D time series.

Physics enters the kernel. If a linear differential operator $\mathcal{L}$ governs the system, kernels closed under $\mathcal{L}$ encode the physics: choosing k as the Green's kernel of $\mathcal{L}$ makes $\mathcal{L}f$ Gaussian with known covariance, so conditioning on data that involve $\mathcal{L}f$ imposes the PDE softly but analytically. Derivative observations (e.g., $\partial_n f$ on boundaries) are handled by replacing $k(x, x')$ with derivatives $\partial k/\partial n$ in the appropriate arguments. This "operator-aware" GP is a probabilistic analogue of weak-form solvers and is often competitive as a surrogate or for uncertainty quantification around a baseline simulator.

From a workflow standpoint, several checks keep Bayesian models honest. Prior predictive draws $y \sim p(y \mid \theta), \theta \sim p(\theta)$ reveal whether the prior encodes realistic scales. Posterior predictive checks compare simulated residuals to observed ones to catch mis-specification (e.g., heteroscedastic noise calls for input-dependent $\sigma(x)$ or warped GPs). Credible intervals reflect epistemic (data scarcity) rather than aleatoric uncertainty; if calibration is poor, revisit the kernel class or add hierarchical structure.

AD computes $\nabla_\theta \mathcal{L}$ and predictive means/variances, and the linear-algebra kernels (Cholesky or conjugate gradients) should be exposed with custom JVP/VJP rules so gradients flow through hyperparameter fits stably.

In short, Bayesian modeling provides a language for uncertainty and principled regularization; GPs are its function-space workhorse, turning smoothness and structure into kernels whose posteriors interpolate, quantify uncertainty, and guide decisions. When data are modest and fidelity is key, exact GPs excel; when data are large, sparse/structured approximations and variational training keep the same probabilistic semantics while scaling to modern SciML workloads.

2.7.3 Information Measures

Information measures quantify uncertainty, dissimilarity between distributions, and statistical dependence. They appear throughout SciML: cross-entropy as a training loss, Kullback–Leibler (KL) penalties in variational inference and policy optimization, mutual information as a regularizer, and log-determinants (information gain) in experimental design. Throughout, log denotes the natural logarithm (units of *nats*); base-2 logs yield *bits*.

Entropy and cross-entropy

For a discrete $X \sim p$, the (Shannon) entropy is

$$H(X) = -\sum_x p(x) \log p(x).$$

For continuous X with density p, the *differential* entropy is

$$h(X) = -\int p(x) \log p(x)\, dx,$$

which can be negative and is *not* invariant to invertible transforms. The cross-entropy between p and q is

$$H(p, q) = -\sum_x p(x) \log q(x) \quad \text{(discrete)},$$

$$h(p, q) = -\int p(x) \log q(x)\, dx \quad \text{(continuous)}.$$

Entropy measures intrinsic uncertainty; cross-entropy measures coding cost when q is used to encode data drawn from p. Minimizing empirical cross-entropy is exactly maximum-likelihood estimation: replacing p by the empirical distribution turns $H(p, q)$ into the average negative log-likelihood. In classification with one-hot labels y and predicted probabilities $\hat{p}_\theta$, the loss $-\sum_c y_c \log \hat{p}_\theta(c \mid x)$ is cross-entropy; for Bernoulli labels it reduces to logistic loss.

Kullback–Leibler divergence

For distributions p and q on the same space with p absolutely continuous w.r.t. q,

$$\mathrm{KL}(p\|q) = \sum_x p(x) \log \frac{p(x)}{q(x)} \quad \text{or} \quad \mathrm{KL}(p\|q) = \int p(x) \log \frac{p(x)}{q(x)}\, dx.$$

KL is nonnegative and equals zero iff $p = q$ almost everywhere (Gibbs' inequality). It is *not* symmetric and does not satisfy the triangle inequality. Unlike differential entropy, KL is invariant to smooth bijections of variables: if $y = T(x)$ is invertible, then $\mathrm{KL}(p_X\|q_X) = \mathrm{KL}(p_Y\|q_Y)$. Two identities are used constantly:

$$H(p, q) = H(p) + \mathrm{KL}(p\|q), \qquad \mathrm{KL}(p\|q) = \mathbb{E}_p[-\log q] - \mathbb{E}_p[-\log p].$$

The first shows cross-entropy decomposes into intrinsic uncertainty plus mismatch; the second reads "expected NLL under q minus the optimal NLL."

Mutual information (MI)

For random variables X and Y with joint $p(x, y)$ and marginals $p(x)$, $p(y)$,

$$\begin{aligned} I(X; Y) &= \mathrm{KL}\big(p(x, y) \,\|\, p(x)p(y)\big) = H(X) - H(X \mid Y) \\ &= H(Y) - H(Y \mid X). \end{aligned}$$

Mutual information is nonnegative and symmetric; it equals zero iff X and Y are independent. Chain rules mirror conditioning in probability:

$$H(X, Y) = H(X) + H(Y \mid X), \qquad I(X; Y, Z) = I(X; Y) + I(X; Z \mid Y).$$

The *data processing inequality* says that processing cannot create information: if $X \to Y \to Z$ is a Markov chain, then $I(X; Z) \le I(X; Y)$. In practice this justifies bottlenecking: any deterministic or noisy map can only lose dependence with the source.

Several distances and inequalities organize these measures. Pinsker's inequality links KL to total variation (TV):

$$\|p-q\|_{\mathrm{TV}} \;=\; \tfrac{1}{2}\int |p-q| \;\le\; \sqrt{\tfrac{1}{2}\,\mathrm{KL}(p\|q)}.$$

The Jensen-Shannon divergence $\mathrm{JS}(p,q) := \frac{1}{2}\mathrm{KL}(p\|m) + \frac{1}{2}\mathrm{KL}(q\|m)$ with $m = \frac{1}{2}(p+q)$ is symmetric, bounded, and induces a metric via $\sqrt{\mathrm{JS}}$; it underpins GAN training. More generally, f-divergences

$$D_f(p\|q) \;=\; \int q(x)\, f\Big(\frac{p(x)}{q(x)}\Big)\, dx, \quad f \text{ convex},\ f(1)=0,$$

include KL, χ^2, Hellinger, and TV; different choices emphasize different tail or mode discrepancies and lead to different robust objectives.

A variational view is essential in modern algorithms. The *Donsker-Varadhan* representation,

$$\mathrm{KL}(p\|q) \;=\; \sup_T \Big\{ \mathbb{E}_p[T(X)] - \log \mathbb{E}_q[e^{T(X)}] \Big\},$$

expresses KL as a supremum over functions T, enabling neural estimation of MI and divergences (e.g., MINE). For latent-variable models with joint $p_\theta(x,z)$ and variational posterior $q_\phi(z \mid x)$, the evidence lower bound (ELBO)

$$\begin{aligned}\log p_\theta(x) \;&\ge\; \mathbb{E}_{q_\phi(z|x)}\big[\log p_\theta(x,z) - \log q_\phi(z \mid x)\big]\\ &= \mathbb{E}_{q_\phi}\big[\log p_\theta(x \mid z)\big] - \mathrm{KL}\big(q_\phi(z \mid x)\,\|\,p_\theta(z)\big)\end{aligned}$$

trades data fit against a KL-to-prior regularizer; VAEs optimize this bound with reparameterized gradients. Mutual information admits similar lower bounds using variational conditionals $r(y \mid x)$:

$$I(X;Y) \;=\; \mathbb{E}_{p(x,y)}\big[\log \tfrac{p(y|x)}{p(y)}\big] \;\ge\; \mathbb{E}_{p(x,y)}\big[\log r(y \mid x)\big] - \mathbb{E}_{p(y)}\big[\log r(y)\big],$$

with estimators such as InfoNCE implementing tractable surrogates via negative sampling.

Concrete models fix ideas. For Gaussians $p = \mathcal{N}(\mu_p, \Sigma_p)$ and $q = \mathcal{N}(\mu_q, \Sigma_q)$ in $\mathbb{R}^d$,

$$\mathrm{KL}(p\|q) = \tfrac{1}{2}\Big[\mathrm{tr}(\Sigma_q^{-1}\Sigma_p) + (\mu_q - \mu_p)^\top \Sigma_q^{-1}(\mu_q - \mu_p) - d + \log\frac{\det \Sigma_q}{\det \Sigma_p}\Big],$$

and for a linear-Gaussian channel $Y = AX + \varepsilon$ with $X \sim \mathcal{N}(0, \Sigma)$ and $\varepsilon \sim \mathcal{N}(0, \sigma^2 I)$, the mutual information is

$$I(X; Y) = \tfrac{1}{2} \log \det\big(I + \sigma^{-2} A \Sigma A^\top\big),$$

which equals the log-volume shrinkage induced by the posterior and drives optimal sensor placement (maximize information gain) and active learning for GPs.

Several caveats prevent common mistakes. Differential entropy $h(X)$ is coordinate-dependent and can increase under one-to-one transforms, so it is not a reliable "uncertainty" measure for continuous variables; use *relative* quantities (KL, entropy differences) or determinant-based measures of covariance instead. Cross-entropy and KL depend on support: if $q(x) = 0$ where $p(x) > 0$, then $H(p, q) = \infty$; smoothing or temperature scaling avoids numerical infinities. Estimating MI from samples is hard in high dimensions; variational estimators must be carefully regularized to avoid bias from function class limitations or estimator saturation.

In SciML practice, these measures become coding rules for objectives. Cross-entropy implements maximum likelihood; KL penalties temper posteriors toward priors in VAEs and regularize policies in reinforcement learning (trust-region and PPO constrain $\mathrm{KL}(\pi_{\mathrm{new}} \| \pi_{\mathrm{old}})$). Mutual information encourages or discourages dependence;information bottlenecks penalize $I(Z; X)$ while preserving $I(Z; Y)$, and disentanglement losses push latent coordinates to carry independent factors. In experimental design and active learning, expected information gain selects new measurements; in Gaussian models it reduces to log-determinants that can be optimized with sparse linear algebra and submodular heuristics.

A compact checklist helps connect symbols to code: use cross-entropy for training probabilistic classifiers; monitor KL terms to balance fit and regularization; prefer MI *bounds* with explicit negatives/contrastive terms for representation learning; and when uncertainty matters, propagate not just means but distributions and compute information gains that align with your physics and sensors.

2.7.4 Reparameterization and Score-Function Gradients

We often need $\nabla_\theta \, \mathbb{E}_{X \sim p_\theta}[\, f(X, \theta)\,]$ where sampling is easy but analytic gradients are not. Two general estimators are used: the *pathwise (reparameterization) gradient*, which differentiates *through* a sample by expressing it as a differentiable transform of fixed noise, and the *score-function (likelihood-ratio/REINFORCE) gradient*, which differentiates *the log-density*. The pathwise estimator usually has lower variance but needs a differentiable sampler; the score-function estimator works for discrete or nondifferentiable cases but needs variance reduction.

Score-function (likelihood-ratio) gradient

If $X \sim p_\theta$ and f is integrable,

$$\nabla_\theta \mathbb{E}_{p_\theta}[f(X,\theta)] = \mathbb{E}_{p_\theta}\big[f(X,\theta)\,\nabla_\theta \log p_\theta(X) + \nabla_\theta f(X,\theta)\big].$$

With f independent of θ, this reduces to

$$\nabla_\theta \mathbb{E}_{p_\theta}[f(X)] = \mathbb{E}_{p_\theta}\big[f(X)\,\nabla_\theta \log p_\theta(X)\big].$$

A Monte Carlo estimator uses samples $X^{(i)} \sim p_\theta$:

$$\widehat{g}_{\mathrm{SF}} = \frac{1}{N}\sum_{i=1}^{N}\big(f(X^{(i)},\theta) - b\big)\,\nabla_\theta \log p_\theta\big(X^{(i)}\big) + \frac{1}{N}\sum_{i=1}^{N}\nabla_\theta f\big(X^{(i)},\theta\big).$$

Subtracting a *baseline* b that does not depend on X keeps the estimator unbiased and reduces variance; the variance-minimizing scalar baseline for a single parameter is $b^\star = \frac{\mathbb{E}[f\,s^2]}{\mathbb{E}[s^2]}$ with $s = \partial_\theta \log p_\theta(X)$ (component-wise analogues apply for vectors). In reinforcement learning, b is a learned value function; in general Monte Carlo, b can be a running average, a linear control variate, or the output of a small regressor predicting f from features of X (learned control variates). Additional variance cuts come from common random numbers, antithetic pairs, and Rao-Blackwellization (analytically integrating out parts of X).

Pathwise (reparameterization) gradient

If one can sample X by a differentiable map T_θ applied to a *parameter-free* noise $\varepsilon \sim p_0$,

$$X \overset{d}{=} T_\theta(\varepsilon), \qquad \varepsilon \sim p_0 \text{ (independent of } \theta),$$

then

$$\nabla_\theta \mathbb{E}_{p_\theta}[f(X,\theta)] = \mathbb{E}_{\varepsilon\sim p_0}\Big[\nabla_x f\big(T_\theta(\varepsilon),\theta\big)\,\underbrace{\partial_\theta T_\theta(\varepsilon)}_{\text{'pathwise' term}} + \nabla_\theta f\big(T_\theta(\varepsilon),\theta\big)\Big].$$

This estimator differentiates *through the sampler*; it is typically low-variance because it exploits smoothness of f. Location-scale families are immediate: if $X = \mu_\theta + \Sigma_\theta^{1/2}\varepsilon$ with $\varepsilon \sim \mathcal{N}(0, I)$, we get standard "Gaussian reparameterization." More generally, any sampler expressible as $X = F_\theta^{-1}(U)$ with $U \sim \text{Unif}(0, 1)$ is pathwise once we can differentiate F_θ^{-1}.

Some common distributions require workarounds because T_θ is not available in closed form or the support depends on θ. *Implicit reparameterization* differentiates through the *CDF* F_θ using the identity $U = F_\theta(X)$ with $U \sim \text{Unif}(0, 1)$ and the implicit function theorem:

$$\partial_\theta X = -\frac{\partial_\theta F_\theta(X)}{\partial_x F_\theta(X)} = -\frac{\partial_\theta F_\theta(X)}{p_\theta(X)}.$$

This yields pathwise gradients for Gamma, Beta, Dirichlet, and truncated distributions by computing $\partial_\theta F_\theta$ and p_θ and solving for X (via a quantile routine) while stopping gradients through the root finder. When the density is available but sampling is via rejection, *rejection reparameterization* augments with accept-reject noise to retain unbiasedness.

Discrete variables cannot be reparameterized exactly because their sample paths are not differentiable. Two popular approximations are used in practice. The *Gumbel-Softmax/Concrete* relaxation draws $G_i \sim \text{Gumbel}(0, 1)$ and sets

$$Y = \text{softmax}\left(\frac{\log \pi_\theta + G}{\tau}\right),$$

a differentiable proxy for a one-hot sample from $\text{Cat}(\pi_\theta)$; as temperature $\tau \downarrow 0$, Y becomes nearly one-hot (bias-variance trade-off). The *straight-through* estimator backpropagates through the soft sample but uses a hard arg max forward pass; it is biased but often effective. Unbiased discrete gradients revert to score-function estimators with strong control variates; *RELAX/REBAR* combine Gumbel relaxations with learned baselines to reduce variance while retaining unbiasedness.

Mixtures and support issues deserve care. If $p_\theta = \sum_k \alpha_k(\theta)\, p_k(\cdot \mid \theta)$, a naive pathwise gradient through a single sampled component ignores how choices of k depend on θ; correct gradients either (i) use score-function terms for the discrete mixture index plus pathwise terms inside components, or (ii) marginalize over k when feasible (Rao-Blackwellization). If the support depends on θ (e.g., truncation at a moving boundary), pathwise derivatives acquire boundary terms; implicit reparameterization via the CDF handles this correctly, whereas ignoring boundaries biases gradients.

In simulation-based settings (ODE/SDE solvers with random inputs, randomized quadrature), two pragmatic rules cut variance. First, *common random numbers*: reuse the *same* ε across neighboring parameter values (or across line-search candidates) so that differences cancel. Second, *control variates*: subtract a cheap surrogate $h_\theta(\varepsilon)$ with known expectation and add it back analytically, i.e.,

$$\widehat{g} = \frac{1}{N}\sum_{i=1}^{N}\Big(\nabla_\theta f(T_\theta(\varepsilon_i),\theta) - c\,\nabla_\theta h_\theta(\varepsilon_i)\Big) + c\,\nabla_\theta \mathbb{E}[h_\theta],$$

choosing c by regression to minimize sample variance; linearization of f around a reference θ is a common h_θ.

A compact "when to use what" guide is helpful. If X is continuous and you have (or can build) a differentiable sampler, use pathwise gradients; prefer implicit reparameterization when quantiles are available or when support moves. If X is discrete or f contains nondifferentiable pieces, start with score-function plus a baseline; add control variates and, if acceptable, a continuous relaxation for lower variance. If f is a least-squares or PDE residual, exploit structure: many uncertainties enter through linear maps or location-scale noise, making pathwise especially effective; gradients then propagate through solvers via the implicit/adjoint rules from ??.

Finally, two small formulas connect estimators to code. For a location-scale Gaussian $X = \mu + \sigma \odot \varepsilon$ with $\varepsilon \sim \mathcal{N}(0, I)$ and f not depending on θ explicitly,

$$\nabla_\mu \mathbb{E}[f(X)] = \mathbb{E}[\nabla_x f(X)], \qquad \nabla_\sigma \mathbb{E}[f(X)] = \mathbb{E}[\nabla_x f(X) \odot \varepsilon],$$

which is exactly what reverse-mode AD will compute if you sample via ε and backpropagate. For a categorical policy $\pi_\theta(a \mid s)$ and return R, the unbiased REINFORCE gradient is

$$\nabla_\theta\, \mathbb{E}[R] = \mathbb{E}\big[(R - b(s))\,\nabla_\theta \log \pi_\theta(a \mid s)\big],$$

with $b(s)$ any state-dependent baseline (a value net); the optimal $b(s)$ in the least-squares sense is the conditional expectation $\mathbb{E}[R \mid s]$, i.e., the true value function. These identities crystallize the two estimators and the role of baselines/control variates in making them practical.

2.8 Numerical Stability and Floating-Point Arithmetic

2.8.1 IEEE-754 and Machine Epsilon

Computers do not store real numbers exactly; they store nearby *floating-point* numbers from a finite set. The IEEE-754 standard specifies how these numbers are represented and how basic operations round. Most SciML code runs in binary64 ("double") or binary32 ("single"); knowing their limits prevents silent but consequential errors.

Definition: IEEE-754 floating-point (binary)

A normalized floating-point number has the form

$$x = (-1)^s \, m \, 2^e, \qquad m \in [1, 2), \; e \in \mathbb{Z},$$

encoded by a sign bit s, an exponent field (with a bias), and a fraction field that stores the fractional part of the significand $m = 1.\text{fraction}_2$. Special patterns represent $\pm\infty$ and NaN. *Subnormals* (a.k.a. denormals) use $m \in [0, 1)$ at the smallest exponent to provide *gradual underflow*.

In binary64 (double): 1 sign bit, 11 exponent bits (bias 1023), and 52 fraction bits. The *unit roundoff* is

$$u = \tfrac{1}{2} 2^{-52} = 2^{-53} \approx 1.11 \times 10^{-16},$$

and the *machine epsilon* (gap at 1) is

$$\varepsilon_{\text{mach}} = 2^{-52} \approx 2.22 \times 10^{-16}.$$

In binary32 (single): 1 sign, 8 exponent bits (bias 127), 23 fraction bits; $u = 2^{-24} \approx 5.96 \times 10^{-8}$ and $\varepsilon_{\text{mach}} = 2^{-23} \approx 1.19 \times 10^{-7}$. The smallest positive *normal* in double is $2^{-1022} \approx 2.23 \times 10^{-308}$, the largest is $\approx 1.80 \times 10^{308}$; the smallest positive *subnormal* is $2^{-1074} \approx 4.94 \times 10^{-324}$.

Rounding maps the exact result of an operation to a nearby representable number. The default mode is *round-to-nearest, ties-to-even* (RN), which minimizes average bias; other modes are toward $+\infty$, toward $-\infty$, and toward 0. With RN, the fundamental *error model* is

$$\text{fl}(a \circ b) \;=\; (a \circ b)\,(1 + \delta), \qquad |\delta| \le u,$$

for basic operations $\circ \in \{+, -, \times, \div\}$ and for many elementary functions implemented to near-correct rounding. A fused multiply-add (FMA) computes $\text{fl}(ab + c)$ with a *single* rounding, improving accuracy relative to separate multiply and add.

Definition: ulp and machine epsilon

The *unit in the last place* at x, denoted $\text{ulp}(x)$, is the distance between adjacent representable numbers surrounding x. For normalized double with exponent e, $\text{ulp}(x) = 2^{e-52}$; thus the absolute spacing grows with $|x|$ even though the *relative* spacing is $\approx \varepsilon_{\text{mach}}$. The machine epsilon is the smallest $\varepsilon > 0$ such that $\text{fl}(1 + \varepsilon) > 1$ under RN; in double this is 2^{-52}.

These facts have immediate, practical consequences. Relative error is the right scale: a computation that keeps *relative* perturbations $\lesssim \varepsilon_{\text{mach}}$ per elementary operation is as good as hardware allows. Absolute spacing depends on magnitude:

near $x = 2^{20} \approx 10^6$, the gap between adjacent doubles is $\mathrm{ulp}(x) = 2^{20-52} = 2^{-32} \approx 2.33 \times 10^{-10}$; any change smaller than this is invisible at that scale.

Micro-examples make the pitfalls concrete. Adding a tiny number can do nothing:

$$\mathrm{fl}\left(1 + 2^{-53}\right) = 1 \quad \text{(too small to cross the next representable >1).}$$

Associativity fails:

$$(10^{16} - 10^{16}) + 1 = 1, \qquad 10^{16} + (-10^{16} + 1) = 10^{16} + 0 = 10^{16}$$

in double, because $-10^{16} + 1$ rounds to -10^{16}. Catastrophic cancellation occurs when subtracting nearly equal numbers: if $x \approx y$, then $x - y$ loses most leading bits, leaving a result dominated by rounding noise; e.g., evaluating $\sqrt{x+1} - \sqrt{x}$ is inaccurate for large x unless rewritten as

$$\sqrt{x+1} - \sqrt{x} = \frac{1}{\sqrt{x+1} + \sqrt{x}},$$

which avoids the bad subtraction.

Special values signal exceptional events: overflow produces $\pm\infty$; invalid operations (e.g. $0/0$) produce NaN, which propagates through subsequent arithmetic. *Subnormals* extend the dynamic range below the smallest normal, allowing gradual underflow at the cost of reduced relative accuracy; some kernels flush subnormals to zero for speed, which can subtly change results near underflow.

Simple rules of thumb improve numerical behavior without changing algorithms. Work with *scaled, nondimensional* variables so typical magnitudes are $O(1)$; this maximizes effective precision. Restructure formulas to avoid cancellation (use algebraic identities as above). Prefer FMA when available to reduce rounding steps. Compare floats with tolerances,

$$|x - y| \le \tau_{\mathrm{abs}} + \tau_{\mathrm{rel}} \max\{|x|, |y|\},$$

instead of exact equality. Sum many terms with a stable scheme (pairwise or compensated) and accumulate in higher precision than the inputs when possible. Finally, be mindful that changing evaluation order changes rounding: parallel reductions, GPU kernels, and compiler vectorization legitimately reorder arithmetic and can yield bit-different (but still correctly rounded) answers; design tests using tolerances, not bitwise equality.

These IEEE-754 basics underpin the next subsections: we will separate *problem* sensitivity (conditioning) from *algorithm* sensitivity (stability), and then show practical techniques;compensated summation, mixed-precision accumulation;to control error growth in large SciML computations.

2.8.2 Conditioning Versus Algorithmic Stability

Numerical error comes from two distinct sources: the *problem* itself may be sensitive to tiny perturbations (poorly conditioned), and the *algorithm/implementation* may amplify rounding and truncation more than necessary (unstable). Separating these is the key to deciding whether you need a better *method* or a better *model/regularization*.

Conditioning (problem sensitivity)

Given a map $F : \mathbb{R}^m \to \mathbb{R}^n$ and a point $y = F(x)$, the (relative) *condition number* at x is

$$\kappa_F(x) = \lim_{\epsilon \to 0} \sup_{\|\delta x\| \le \epsilon \|x\|} \frac{\|F(x+\delta x) - F(x)\|/\|F(x)\|}{\|\delta x\|/\|x\|}.$$

For differentiable F this is $\kappa_F(x) = \frac{\|J_F(x)\| \, \|x\|}{\|F(x)\|}$, with J_F the Jacobian (and norms chosen consistently).

Forward/Backward error and backward stability

For a problem $F(x) = y$, let $\widehat{x}$ be the output of an algorithm on data y. *Forward error* measures the distance to the true answer: $\|\widehat{x} - x^\star\|/\|x^\star\|$, where $x^\star$ solves $F(x^\star) = y$. *Backward error* asks "for which nearby data is $\widehat{x}$ *exact*?": find the smallest Δy s.t. $F(\widehat{x}) = y + \Delta y$ and report $\|\Delta y\|/\|y\|$. An algorithm is *backward stable* if the backward error is $O(u)$, the unit roundoff.

These notions connect by a simple but powerful inequality:

$$\underbrace{\frac{\|\widehat{x} - x^\star\|}{\|x^\star\|}}_{\text{forward error}} \lesssim \underbrace{\kappa_F(x^\star)}_{\text{conditioning}} \times \underbrace{\frac{\|\Delta y\|}{\|y\|}}_{\text{backward error}} + \text{higher-order terms}.$$

Hence: *backward stable algorithm + well-conditioned problem* $\Rightarrow$ small forward error (near machine precision); if the problem is ill-conditioned ($\kappa \gg 1/u$), no algorithm using standard precision can deliver many correct digits without changing the problem (regularization, different model, higher precision).

Linear solves $Ax = b$. Here $F^{-1}(b) = x = A^{-1}b$ and the classical condition number is

$$\kappa(A) = \|A\| \, \|A^{-1}\| \quad \text{(in a compatible matrix/vector norm).}$$

It bounds error via

$$\frac{\|\widehat{x} - x\|}{\|x\|} \lesssim \kappa(A) \frac{\|r\|}{\|b\|}, \qquad r := b - A\widehat{x},$$

and (for backward-stable solvers) $\|r\|/\|b\| = O(u)$, so the attainable relative accuracy is about $\kappa(A)\,u$. Two immediate diagnostics follow: (i) *always* form the residual $r = b - A\widehat{x}$ to check quality;if it is not tiny, the solve or data are suspect; (ii) if $\kappa(A)\,u$ is not small, you must precondition, regularize, or increase precision.

Algorithmically, Gaussian elimination with partial pivoting (GEPP) and Householder QR are *backward stable*: they produce $\widehat{x}$ that solves $(A + \Delta A)\widehat{x} = b$ with $\|\Delta A\|/\|A\| \lesssim \gamma_n$, where $\gamma_n \approx cu\,n$ in practice (the constant depends on growth factors). By contrast, the normal-equations approach to least squares, $A^\top Ax = A^\top b$, squares the condition number:

$$\kappa(A^\top A) = \kappa(A)^2,$$

and is thus *algorithmically stable* (backward stable for the normal equations) but *problem-wise* much worse; prefer QR or SVD which are backward stable for the original least-squares problem.

Least squares and residual conditioning. For $\min_x \|Ax - b\|_2$, a backward-stable method (Householder QR) returns $\widehat{x}$ with $(A + \Delta A)^\top(A + \Delta A)\widehat{x} = (A + \Delta A)^\top b$, $\|\Delta A\|/\|A\| = O(u)$. The forward error satisfies

$$\frac{\|\widehat{x} - x\|}{\|x\|} \lesssim \kappa(A) \left(\frac{\|r\|}{\|A\|\,\|x\|} + O(u) \right),$$

where $r = b - Ax$ is the *true* residual. Large residuals amplify sensitivity (near rank-deficiency); SVD-based regularization (Tikhonov, truncated SVD) reduces effective κ at the cost of bias.

Polynomial roots and cancellation. Root finding is often *intrinsically* ill-conditioned: small coefficient perturbations can move roots a lot (Wilkinson's polynomial is the classic cautionary tale). Here, even a backward-stable algorithm may deliver few true digits of the roots; the right response is problem reformulation (work with factorizations, companion matrices with balancing), scaling, or higher precision;not "a different solver".

Componentwise vs normwise stability. Normwise backward stability controls $\|\Delta A\|/\|A\|$, but some problems require *componentwise* control (each entry perturbed by $O(u)$). Triangular solves and summations benefit from componentwise analysis; pairwise summation and Kahan compensation are *backward stable componentwise* for sums, explaining their superior accuracy over naive left-to-right addition.

Growth factors and pivoting. In GEPP the element growth factor $\rho = \max_{i,j,k} |a_{ij}^{(k)}| / \max_{i,j} |a_{ij}|$ appears in bounds: roughly $\|\Delta A\|/\|A\| \lesssim \rho\,\gamma_n$. Worst-case ρ can be exponential in n, but such matrices are rare; partial pivoting is reliable in practice, while complete pivoting controls ρ more tightly at extra cost. In least squares, Householder QR avoids growth issues altogether.

Nonlinear problems and Jacobian conditioning. For $F(x) = 0$ with solution $x^\star$, the local conditioning is governed by $\|J_F(x^\star)^{-1}\|$: perturbations δF cause $\|\delta x\| \lesssim \|J_F^{-1}\| \, \|\delta F\|$. Newton with line search/trust region is typically backward stable (it finds an exact zero of a nearby $\widetilde{F}$); if $\|J_F^{-1}\|$ is large, regularization (Levenberg-Marquardt, damping), scaling, or better parameterization is required to reduce effective conditioning.

What this means in practice

- *Diagnose conditioning first.* Estimate κ (power method on A and A^{-1}, or cheaply via equilibrated norms). If $\kappa \, u$ is not small, do not expect many correct digits;improve the model (regularize), rescale variables, or raise precision.
- *Choose backward-stable algorithms.* For linear/least-squares solves use Householder QR/SVD or GEPP (not normal equations). For summation use pairwise or compensated schemes. For orthogonal transforms prefer Householder/Givens over Gram-Schmidt (classical GS can be unstable; modified GS is better but still inferior to Householder).
- *Trust residuals.* A tiny relative residual with moderate κ implies a tiny forward error. Always check $r = b - A\widehat{x}$ (or the KKT residuals in constrained problems); they are cheap and decisive.
- *Precondition to change the problem.* Preconditioning replaces A by a better-conditioned $M^{-1}A$; this reduces *problem* sensitivity and improves both accuracy and convergence of iterative solvers. Scaling (row/column equilibration) is the simplest and often most effective preconditioner for dense problems.
- *Use mixed precision wisely.* A backward-stable outer iteration plus higher-precision residuals (iterative refinement) attains near-double accuracy with mostly single-precision work;provided $\kappa(A) \, u_{\text{low}} < 1$; otherwise refinement stagnates (see next subsection).

In short, conditioning is about the world; stability is about your algorithm. Backward stability buys you correctness up to $O(u)$ *for a nearby problem*; conditioning tells you how close that nearby problem must be to guarantee an accurate answer for the one you care about. Keep this product rule in mind:

$$\text{forward accuracy} \;\approx\; \text{conditioning} \times \text{backward error} \;\sim\; \kappa \times O(u),$$

and let it guide both modeling choices (scaling, regularization, preconditioning) and numerical choices (algorithms that are backward stable by design).

2.8.3 *Error Accumulation and Compensated Techniques*

Finite-precision arithmetic introduces a small relative error each time we add or multiply. Repeating these operations n times can amplify roundoff from $O(u)$ to

$O(nu)$ in the worst case. The goal of *compensated* techniques is to restructure basic computations so that most rounding happens on tiny "error terms," recovering near-full precision at modest cost. Three patterns reappear: reorder to avoid cancellation, accumulate in higher precision (or with error compensation), and fuse operations to reduce the number of roundings.

A canonical object is the sum $S = \sum_{i=1}^{n} x_i$. Naive left-to-right summation in precision with unit roundoff u has a forward error bounded by

$$|\,\mathrm{fl}(S) - S| \;\le\; \gamma_n \sum_{i=1}^{n} |x_i|, \qquad \gamma_k := \frac{k\,u}{1-k\,u} \;\approx\; k\,u \text{ if } ku \ll 1,$$

so the relative error can grow linearly with n (and catastrophically when cancellation makes $S \ll \sum |x_i|$). Pairwise (tree) summation lowers this to $O(u \log n)$ by adding numbers of similar magnitude first; it is also friendly to parallel reductions. Sorting by magnitude (ascending) helps for scalar code but costs $O(n \log n)$ and disrupts memory locality.

Kahan / Kahan-Babuška-Neumaier (compensated summation)

Maintain a running compensation c for lost low-order bits. For each term x, add the corrected value $y = x - c$ to the sum s, then update c to the new rounding error. Neumaier's variant improves robustness when $|x| > |s|$.

These one-pass algorithms reduce the error bound essentially to $O(u) \sum |x_i|$; independent of n up to constants; and are streaming (no reordering). They cost a few extra flops per term and a scalar compensation variable, which is negligible compared with memory traffic. In practice: use pairwise summation for parallel reductions and Kahan/Neumaier within each thread/block; the two combine well.

Dot products $d = \sum_i a_i b_i$ behave similarly. Naively,

$$|\,\mathrm{fl}(d) - d| \;\le\; \gamma_n \sum_i |a_i b_i|.$$

Using FMA computes each $a_i b_i + s$ with a single rounding, already improving accuracy; combining FMA with pairwise accumulation or Kahan compensation achieves near-$O(u)$ accuracy. When available, perform the *accumulation* in a wider type (e.g., FP16 inputs $\to$ FP32 accumulation, FP32 $\to$ FP64); this is the principle behind tensor-core "mixed-precision" GEMMs reaching close-to-double accuracy when $\kappa\, u_{\mathrm{acc}} \ll 1$ (see also iterative refinement below).

Error-free transformations (EFTs): TwoSum / TwoProd

For floating-point addition with round-to-nearest, one can compute the rounded sum and the exact rounding error:

$$[s, e] = \mathrm{TwoSum}(a, b): \quad s = \mathrm{fl}(a + b), \quad e = (a + b) - s.$$

Similarly with FMA,

$$[p, e] = \mathrm{TwoProd}(a, b): \quad p = \mathrm{fl}(ab), \quad e = \mathrm{fma}(a, b, -p) = ab - p.$$

EFTs are the building blocks of compensated algorithms (sums, dot products, Horner schemes).

Polynomials and recurrences are sensitive to cancellation if evaluated naively. Horner's method minimizes multiplies/adds but still suffers when coefficients and x produce near-cancellation. *Compensated Horner* augments each stage with EFTs that track the low-order parts of both the product and the addition, yielding near-double-precision results in single precision at about 2-3$\times$ the cost. The same idea applies to Bézier curves via compensated de Casteljau and to barycentric Lagrange interpolation.

Variance and covariance are archetypal "cancellation traps." The textbook one-pass formula $\mathrm{Var}(x) = \overline{x^2} - \bar{x}^2$ can lose all digits when the variance is small relative to the mean. Use numerically stable online updates such as Welford's algorithm: maintain running mean m_k and sum of squares S_k via

$$\delta = x_k - m_{k-1}, \quad m_k = m_{k-1} + \delta/k, \quad S_k = S_{k-1} + \delta(x_k - m_k),$$

and report $\mathrm{Var} = S_n/(n-1)$; adding Kahan compensation to the S_k update further improves robustness.

Iterative refinement turns low-precision factorizations into high-accuracy solves. Suppose $Ax = b$ with a factorization in low precision (unit roundoff u_ℓ) but residuals formed in high precision u_h. Starting from a low-precision solution $\widehat{x}$, iterate:

$$r = b - A\widehat{x} \ \text{(high precision)}, \qquad \text{solve } A\,\delta = r \ \text{(low precision)},$$
$$\widehat{x} \leftarrow \widehat{x} + \delta \ \text{(high precision accumulate)}.$$

If $\kappa(A)\, u_\ell < 1$ and $u_h \ll u_\ell$, refinement converges rapidly to near-high-precision accuracy while most flops run in low precision. This is a workhorse in mixed-precision linear algebra and PDE solvers.

Cancellation-aware reformulations are cheap wins. Replace $\sqrt{x+1} - \sqrt{x}$ by $1/(\sqrt{x+1} + \sqrt{x})$; compute $\log(1+z)$ with `log1p(z)` and $(e^z - 1)$ with `expm1(z)`; use `hypot(x,y)` for $\sqrt{x^2 + y^2}$; evaluate $1 - \cos z$ via `expm1`-based identities for small z. Libraries provide these *compensated intrinsics*; using them is better than rolling your own.

Mixed precision deserves special mention. Accumulating many small terms in a wider type (FP32 accumulator for FP16 inputs; FP64 for FP32) is often the highest "accuracy per byte" option, because it changes the effective u of the critical reduction while keeping input bandwidth low. When wider types are unavailable or too slow, Kahan/Neumaier and pairwise summation approximate the same benefit. For matrix products and convolutions, *blockwise* accumulation in higher precision;summing k products in a wider register before rounding back;reduces local error and maps well to GPU tensor cores.

Parallel reductions change the order of operations and therefore the error. Favor balanced trees (pairwise) over arbitrary chunking; perform local (thread/block) Kahan or high-precision accumulation, then a short pairwise merge at the top. If reproducibility across hardware runs matters (next subsection), fix the reduction tree and disable kernels that reorder sums nondeterministically; otherwise, accept tiny run-to-run differences and test with tolerances.

A compact checklist helps decide what to use:

- Sums/dot products: use FMA; prefer pairwise (tree) reduction; add Kahan/Neumaier in tight loops; accumulate in a wider type when possible.
- Statistics: use Welford for mean/variance; avoid $\overline{x^2} - \bar{x}^2$; compensate accumulators if dynamic range is large.
- Polynomials/interpolation: use compensated Horner/de Casteljau when cancellation is expected; consider Estrin's scheme with FMA on SIMD/GPU for speed.
- Linear solves: exploit mixed precision + iterative refinement when $\kappa\, u_\ell < 1$; otherwise precondition or raise precision.
- Transcendentals: call stable intrinsics (`log1p`, `expm1`, `hypot`, `atan2`); they embody the algebraic rewrites that avoid cancellation.

The theme is simple: most roundoff accrues in long accumulations and near-cancellations. If you either *change the order* (pairwise), *change the precision* (wider accumulator), or *track the lost bits* (compensation/EFTs), you recover nearly all the digits that hardware can represent;without changing the mathematical result.

2.8.4 Determinism and Reproducibility

Two runs of the "same" experiment rarely produce bitwise-identical results unless you ask for it. Sources of variation include pseudorandom draws (initialization, data augmentation, dropout), nondeterministic parallel kernels (race conditions in reductions or atomics), changes in evaluation order (threading, GPU streams), mixed-precision rounding differences, library autotuners that pick different algorithms, and even different hardware/driver/compiler versions. Decide *what level* of reproducibility you need:

- **Bitwise determinism**: every floating-point word matches across runs. Strongest, often costly.
- **Numerical reproducibility**: results agree within a tolerance (e.g., relative error $< 10^{-12}$). Usually enough for scientific conclusions.
- **Statistical reproducibility**: distributions of outcomes (metrics) agree across repeated runs. Appropriate when randomness is part of the method.

A simple mental model helps: the overall run is a function of *code, data, randomness, hardware, libraries, and execution order*. Control each to the extent required by your target level.

Control randomness: Use a single source of entropy and propagate it everywhere. Seed all relevant generators (framework RNG, NumPy, host language RNG, data-augmentation libraries) at the start of the process and on every worker. Record and checkpoint RNG *states* (not just seeds) alongside model weights; this lets you resume training *exactly*. Prefer counter-based or keyed PRNGs (functional style) when available; they make parallel seeding unambiguous because each stream is indexed by a key instead of relying on global mutable state.

Enforce deterministic algorithms: Many compute libraries offer switches to use deterministic implementations for reductions, convolutions, and other kernels; they may disable atomics or fix reduction trees at some performance cost. Turn off autotuning that can choose different algorithms across runs; pick a specific algorithm when the library exposes that choice. Be aware that some GPU kernels are inherently nondeterministic when they use floating-point atomics;replace with deterministic alternatives or accept numerical (not bitwise) reproducibility.

Fix execution order: Parallel reductions accumulate in different orders on different runs unless you constrain them. Prefer balanced, fixed reduction trees; avoid operations whose order depends on thread scheduling. In data pipelines, set a fixed shuffle seed, a fixed number of workers, and a deterministic worker initialization (each worker gets a unique, reproducible sub-seed). Disable asynchronous randomness in prefetch/augmentation stages if bitwise determinism is required.

Stabilize arithmetic modes: Mixed precision, fast-math compiler flags, and hardware-specific modes (e.g., tensor-core low-precision matmuls) can change rounding and thereby the exact path of training. If you need byte-for-byte reproducibility, standardize the precision across runs, disable aggressive fast-math options, fix loss-scaling policies, and ensure the same math mode on all devices. Otherwise, accept tiny numerical differences and test with tolerances.

Pin the environment: Record exact versions of compilers, drivers, BLAS/GPU libraries, and math libraries; containerize if possible. Log CPU/GPU model identifiers, thread counts, and affinity/NUMA settings. Small upgrades can change kernel selection or evaluation order, producing legitimate (though tiny) numerical drift.

Define and test with tolerances: Bitwise equality is fragile under floating-point; prefer checks such as

$$|x - y| \le \tau_{\text{abs}} + \tau_{\text{rel}} \max\{|x|, |y|\}.$$

Use the same tolerances in regression tests and across devices. For stochastic training, fix a budget of repeated runs and compare distributions of metrics (means with confidence intervals) rather than single numbers.

Checkpointing and resumes: A "reproducible checkpoint" includes: model/state dict, optimizer state, *all* RNG states (host, framework, device), the data-loader epoch/position, and configuration/hyperparameters. On resume, restore RNG states before recreating shuffles/augmentations; otherwise you will silently diverge.

Multi-device and distributed training: All-reduce operations may aggregate gradients in different orders across steps or ranks; if exact determinism is needed, use reduction algorithms with fixed order and stable datatypes, and avoid overlapping communication that can reorder partial sums. Keep per-rank seeds distinct but reproducible (derive from a global seed and the rank id), and make sure gradient accumulation, dropout, and data shuffles are keyed consistently across ranks.

A practical, workflow:

1. Choose the target: bitwise / numerical / statistical reproducibility.
2. Set a global seed; derive all other seeds from it; log them.
3. Enable deterministic kernels in your stack; disable autotuners that change algorithms across runs.
4. Fix precision/mode settings consistently; avoid fast-math if bitwise reproducibility is required.
5. Make the data pipeline deterministic (seeded shuffle, deterministic worker init, bounded prefetch).
6. Pin software/hardware versions; record them in run logs and checkpoints.
7. Write tests with tolerances; for stochastic pipelines, validate via repeated-run statistics.

2.9 Differentiable Programming and Adjoint Methods

2.9.1 Computational Graphs and Checkpointing

A differentiable program builds a *computational graph*: nodes are primitive operations (adds, multiplies, convolutions, solver steps), edges carry values ("activations"), and the forward pass evaluates the graph from inputs (parameters and data) to outputs (losses, constraints). Reverse-mode AD (backpropagation) traverses the graph backward and accumulates *adjoints*;sensitivities of the final objective with respect to every intermediate. If $y = f_L(\cdots f_2(f_1(x)))$ and $\bar{y} = \partial \ell / \partial y$, the chain rule composes local Jacobians *right-to-left*:

$$\bar{z}_{L-1} = \bar{y}\, J_L, \qquad \bar{z}_{L-2} = \bar{z}_{L-1}\, J_{L-1}, \quad \ldots, \qquad \bar{x} = \bar{z}_0\, J_1,$$

where $J_\ell = \partial f_\ell / \partial z_{\ell-1}$ evaluated at the *forward* values. Crucially, reverse-mode needs those forward intermediates to evaluate each J_ℓ; storing them all is memory-expensive, recomputing them is time-expensive. Managing this time-memory trade-off is the role of *checkpointing*.

The raw memory demand scales with the number of nodes whose outputs are needed by some backward rule. In a feedforward network with L layers, saving every activation costs $O(L)$ tensors; in time-unrolled dynamics (RNNs, integrators with N steps) it is $O(N)$. On modern accelerators, memory caps long-horizon training and fine time discretizations. The idea of checkpointing is simple: save only a subset of intermediates ("checkpoints") during the forward pass; during the backward pass, *recompute* missing intermediates from the last available checkpoint to the point where a local gradient is needed, then discard those temporaries. The schedule;that is, which values to save and when to recompute;determines the recompute overhead for a given memory budget.

A periodic schedule saves every k-th activation and recomputes at most $k - 1$ steps between checkpoints. This is easy to implement and already turns $O(N)$ memory into $O(N/k)$ at an overhead of $\approx 2 - 1/k$ forward passes in total flops. For deep nets, many frameworks expose this as `checkpoint` wrappers: mark a block so that its inputs are saved, its interior is recomputed on the backward pass, and its outputs are *not* saved. For time-steppers, a "save-every-k-steps" policy serves as a robust baseline.

There is, however, a near-optimal schedule with significantly less recomputation for the same memory. When the forward is a linear chain of N stages (layers or steps) and you can store m checkpoints at once, the *binomial* or *Revolve*-type schedules achieve logarithmic recomputation depth: roughly $O(\log_{m+1} N)$ nested segments, yielding a total recompute factor close to the information-theoretic optimum. Intuitively, you place checkpoints so that the backward sweep never needs to re-evaluate long prefixes; instead, it recursively divides the interval into subintervals anchored at saved states. In practice, libraries provide "online" planners that produce the next save/recompute decision on the fly given (N, m) without precomputing the full plan; they interleave forward segments and short *replays* during backprop so memory usage stays within the m-checkpoint cap at all times.

Checkpointing interacts with randomness and nondeterminism. If a forward block draws random numbers (dropout, data augmentation) or uses nondeterministic kernels, the recomputed forward during backprop must *replay* the same values; otherwise gradients are biased. The remedy is to save and restore RNG states at checkpoints (or use stateless/functional PRNGs keyed by a seed and a counter), and to prefer deterministic kernels inside checkpointed regions when exact replay is required.

For long-horizon training (RNNs, control, differentiable simulators) two additional levers complement checkpointing. Truncated backpropagation through time limits the backward horizon to the last H steps, turning memory from $O(N)$ to $O(H)$ and stabilizing gradients at the cost of bias;often a good trade when feedback is local in time. Reversible layers (e.g., orthogonal/volume-preserving blocks, certain ResNet variants) reconstruct inputs from outputs during backprop and store only a few boundary tensors; this drastically reduces memory but requires numerically stable inverses and careful handling of non-invertible operations (normalization, dropout, padding), which are then moved outside the reversible core or augmented with small side buffers.

Mixed precision changes the picture subtly. Saving activations in reduced precision (e.g., FP16) halves memory but may degrade backward accuracy if local Jacobians are ill-conditioned; a common compromise is to keep *master* parameters in higher precision and cast activations selectively (keep those feeding stiff operators in FP32). When recomputing, use the same precision as in the original forward to ensure identical values under round-to-nearest; otherwise, tiny rounding differences alter adjoints. If exact replay is unnecessary (you accept numerical, not bitwise, reproducibility), recomputation in a "better" precision can even *improve* gradient accuracy.

There is also a graph-structural dimension. Not every program is a simple chain: branches, skips, and shared subgraphs appear in U-Nets, transformers, and multiphysics pipelines. Reverse-mode visits nodes in a topologically reversed order; storing the minimal cut of tensors that separate "past" from "future" along this order bounds peak memory. Practical heuristics identify cheap-to-recompute subgraphs (pure arithmetic) and expensive ones (I/O, large FFTs/solves) and only checkpoint the former. For PDE-constrained pipclincs, store coarse-grid or low-rank summaries as checkpoints and reconstruct fine states on the fly through fast prolongation/interpolation; this gives "physics-aware" checkpointing with small recompute overhead.

Finally, two quick patterns are worth writing down because they solve 80% of use cases:

- *Block checkpointing for deep nets.* Partition the network into B contiguous blocks. Save the inputs to each block; during backward, recompute each block's forward once and free its temporaries immediately after use. Memory $\approx$ cost of B block inputs; time ≈ 1 extra forward.
- *Segmented checkpointing for time-steppers.* Choose a segment length k and save states every k steps. During backward over a segment, replay those k steps from its checkpoint. Memory $\approx O(\text{state size})$; overhead $\approx 1 + 1/k$ forward equivalents. Upgrade to a binomial/Revolve schedule when N is very large or m is very small.

The upshot is simple: reverse-mode wants all forward intermediates; memory forbids it. Checkpointing buys memory with extra flops, and well-chosen schedules keep that tax close to the theoretical minimum. As we move to adjoint methods for ODE/PDE solvers, the same ideas reappear with "time steps" as the chain: we can save a sparse set of states and re-integrate short windows during the adjoint sweep, or adopt reversible/adjoint integrators that reduce storage while controlling numerical error.

2.9.2 Discrete Versus Continuous Adjoint

Consider the parameterized ODE

$$\dot{u}(t) = f\big(u(t), \theta, t\big), \qquad u(0) = u_0(\theta),$$

and an objective

$$\mathcal{J}(\theta) = \Phi\big(u(T), \theta\big) \; + \; \int_0^T L\big(u(t), \theta, t\big)\, dt.$$

There are two principled routes to $\nabla_\theta \mathcal{J}$. The *continuous adjoint* differentiates the *differential equation* first and then discretizes the resulting adjoint ODE/PDE. The *discrete adjoint* differentiates the *discrete time-stepping scheme* used in the solver, i.e., it is exactly reverse-mode AD through the numerical integrator. Both are "correct," but they answer slightly different questions: the continuous adjoint targets the gradient of the *continuous* objective; the discrete adjoint returns the gradient of the *discrete* objective actually minimized by code. When the discretization is refined and adjoint-consistent, both agree to the order of the scheme; at finite resolution, they can differ noticeably.

Continuous adjoint for ODEs

Define the adjoint $\lambda(t)$ by the backward-in-time ODE

$$-\dot{\lambda}(t) = f_u(u, \theta, t)^\top \lambda(t) + L_u(u, \theta, t), \qquad \lambda(T) = \Phi_u\big(u(T), \theta\big),$$

where subscripts denote partial derivatives (e.g., $f_u = \partial f / \partial u$). Then

$$\nabla_\theta \mathcal{J} = \Phi_\theta\big(u(T), \theta\big) + \int_0^T \Big(L_\theta(u, \theta, t) \; + \; f_\theta(u, \theta, t)^\top \lambda(t)\Big)\, dt$$
$$+ \, \lambda(0)^\top u_{0,\theta}.$$

Implementation is forward solve for $u(t)$ (with storage or checkpointing), backward solve for $\lambda(t)$ using $u(t)$ along the path, and quadrature of the gradient terms.

The same construction extends to PDEs with inner products over space. If $F(u, \theta) = 0$ represents the PDE in weak form and $\mathcal{J} = \Phi(u) + \int_0^T L(u, \theta)\, dt$, the adjoint p solves the formal L^2-adjoint (typically backward in time), and $\nabla_\theta \mathcal{J}$ is an integral of $F_\theta^\top p + L_\theta$ plus boundary/initial terms determined by transposition. Boundary conditions for the adjoint are the *dual* of the forward's: Dirichlet vs Neumann swap roles in elliptic parts, and hyperbolic parts impose *inflow* on characteristics traveling backward.

Discrete adjoint (differentiate-the-discretization)

Let a one-step method advance $u_{n+1} = \Psi_{\Delta t}(u_n, \theta, t_n)$ and define a discrete objective

$$\mathcal{J}_h(\theta) = \Phi(u_N, \theta) + \sum_{n=0}^{N-1} \Delta t\, L(u_n, \theta, t_n).$$

The discrete adjoint λ_n is defined backward by

$$\lambda_n = \left(\partial_u \Psi_{\Delta t}(u_n, \theta, t_n)\right)^\top \lambda_{n+1} + \Delta t\, L_u(u_n, \theta, t_n), \qquad \lambda_N = \Phi_u(u_N, \theta),$$

and the gradient is

$$\nabla_\theta \mathcal{J}_h = \Phi_\theta(u_N, \theta) + \sum_{n=0}^{N-1} \left[\Delta t\, L_\theta(u_n, \theta, t_n) + \left(\partial_\theta \Psi_{\Delta t}(u_n, \theta, t_n)\right)^\top \lambda_{n+1}\right]$$
$$+ \lambda_0^\top u_{0,\theta}.$$

This is exactly reverse-mode AD through the time-stepping code.

A tiny example makes the algebra concrete. For explicit Euler $u_{n+1} = u_n + \Delta t\, f(u_n, \theta, t_n)$,

$$\lambda_n = \left(I + \Delta t\, f_u(u_n, \theta, t_n)\right)^\top \lambda_{n+1} + \Delta t\, L_u(u_n, \theta, t_n),$$

and the parameter term is $\partial_\theta \Psi_{\Delta t} = \Delta t\, f_\theta(u_n, \theta, t_n)$. For implicit Euler $u_{n+1} = u_n + \Delta t\, f(u_{n+1}, \theta, t_{n+1})$, the discrete adjoint uses the Jacobian at the *new* state:

$$\lambda_n = \left(I - \Delta t\, f_u(u_{n+1}, \theta, t_{n+1})\right)^{-\top} \left(\lambda_{n+1} + \Delta t\, L_u(u_n, \theta, t_n)\right),$$

which mirrors the linear solves inside the forward step. These formulas generalize: the discrete adjoint of a Runge-Kutta (RK) method is another RK method that traverses stages in reverse with Butcher coefficients transformed by

$$a_{ij}^* = b_j - a_{ji}, \qquad b_i^* = b_i, \qquad c_i^* = 1 - c_i,$$

so that backprop exactly matches the forward scheme's algebra.

The choice between continuous and discrete adjoints is pragmatic. If you are calibrating a *continuous* model and want mesh-independent sensitivities (e.g., w.r.t. physical parameters), the continuous adjoint reflects the PDE/ODE's true calculus, and discretization errors vanish under refinement;provided the adjoint discretization is *adjoint-consistent* (dual-consistent fluxes and boundary terms in FV/DG; consistent test spaces in FE). If you are training through a *specific solver* at a fixed resolution (as in neural ODE layers or differentiable simulators embedded in larger nets), the discrete adjoint is the right gradient: it is identical to reverse-mode through the code and captures solver-specific artifacts (stability limits, limiters, projections). Mixing the two can cause bias at finite h, Δt; it is common to start with discrete adjoints for correctness at the working grid and later validate with a continuous-adjoint implementation to assess discretization sensitivity.

Stiffness and stability carry over to adjoints. Backward integration often makes the adjoint *more* stiff than the forward, because $f_u^\top$ can have larger negative real parts and non-normality can amplify transients. Implicit, L-stable schemes that were chosen for the forward remain a good choice for the adjoint; when using discrete adjoints, this happens automatically since you are differentiating the same scheme. When using continuous adjoints with on-the-fly recomputation of forward states, it is safer to *checkpoint* sparsely spaced states and re-integrate short windows than to rely on dense output over very long intervals, especially in chaotic or highly dissipative systems.

Memory-compute trade-offs are the same as in reverse-mode AD. The discrete adjoint nominally needs all intermediate states; checkpointing (2.9.1) stores a subset and replays short segments. The continuous adjoint needs access to $u(t)$ along the path; you either store it densely, interpolate from saved checkpoints, or re-integrate the forward in windows during the backward solve. Using identical random seeds and deterministic kernels during replays avoids bias.

The PDE case follows the same pattern with inner products over space. For a parabolic model $u_t - \nabla\cdot(\kappa\nabla u) + cu = s(\theta)$ and tracking loss $\frac{1}{2}\int_0^T\int_\Omega(u - u_d)^2\,dx\,dt$, the continuous adjoint solves

$$-p_t - \nabla\cdot(\kappa\nabla p) + c\,p = u - u_d, \qquad p(T,\cdot) = 0,$$

with boundary conditions dual to those of u; the gradient w.r.t. a source parameter is $\nabla_\theta \mathcal{J} = \int_0^T\int_\Omega(\partial_\theta s)\,p\,dx\,dt$. A conforming FE or DG discretization that is dual-consistent yields a discrete adjoint whose algebra coincides with reverse-mode through the assembled timestepper; flux choices and boundary terms must be mirrored to avoid "adjoint mismatch."

Two practical rules summarize the engineering: use the discrete adjoint when differentiating *through* a particular solver configuration (this is backprop through time and guarantees exact gradients of what you actually compute); use the continuous adjoint when you need *model-level* sensitivities that are robust to solver changes, ensuring adjoint-consistent discretization and appropriate backward-time integrators. In both cases, pair the adjoint with checkpointing, choose schemes with stability properties that mirror the forward, and expose the linear/implicit solves as primitives with custom VJP/JVPs so that outer optimizers see clean, accurate gradients.

2.9.3 Differentiating Through Solvers

Many SciML pipelines invoke *solvers* as black boxes: linear system solves, Newton steps, fixed-point iterations, optimization subproblems, projection operators, or entire time-steppers. Differentiating naively by unrolling all inner iterations can be prohibitively expensive or numerically fragile. The remedy is to replace the unrolled computation by its *implicit sensitivity*: differentiate the optimality/fixed-point condition that the solver enforces at convergence, and solve a *single* linear system for each Jacobian-vector or vector-Jacobian product needed by the outer reverse/forward pass.

Suppose a solver returns $x^\star(\theta)$ that satisfies the residual equation

$$F(x^\star, \theta) = 0, \qquad F : \mathbb{R}^n \times \mathbb{R}^p \to \mathbb{R}^n,$$

and an outer loss depends on $x^\star$ via $\mathcal{J}(\theta) = \ell(x^\star(\theta), \theta)$. Under standard regularity (F is differentiable and $F_x(x^\star, \theta)$ is nonsingular), the implicit function theorem gives the *forward* sensitivity

$$F_x \frac{dx^\star}{d\theta} + F_\theta = 0 \implies \frac{dx^\star}{d\theta} = -F_x^{-1} F_\theta,$$

and the *reverse* (adjoint) formula for a given seed $\bar{x} = \partial\ell/\partial x^\star$:

$$F_x^\top \lambda = \bar{x} \quad \text{and} \quad \frac{d\mathcal{J}}{d\theta} = \ell_\theta - \lambda^\top F_\theta.$$

Thus, one *linear solve* with $F_x^\top$ (for λ) replaces backpropagating through all inner iterations.

Linear solves. For $A(\theta)\, x = b(\theta)$ with solution $x^\star = A^{-1}b$, set $F(x, \theta) = A(\theta)x - b(\theta)$. The reverse formula yields

$$A^\top \lambda = \bar{x}, \qquad \frac{d\mathcal{J}}{d\theta} = \ell_\theta - \lambda^\top \Big(A_\theta x^\star - b_\theta\Big).$$

You never differentiate the *algorithm* (LU/CG); you differentiate the *equation* it solves. In practice one exposes a primitive `solve(A,b)` with custom VJP/JVP that: (i) caches a factorization or an effective preconditioner from the forward; (ii) on the backward call solves $A^\top \lambda = \bar{x}$; (iii) evaluates $\lambda^\top(A_\theta x^\star - b_\theta)$ using only Jacobian-vector products to avoid forming dense A_θ.

Newton and fixed-point iterations. If $x^\star$ is a fixed point of $G(x,\theta)$, i.e., $x^\star = G(x^\star,\theta)$, define $F(x,\theta) = x - G(x,\theta)$. Then

$$\left(I - G_x\right)^\top \lambda = \bar{x}, \qquad \frac{d\mathcal{J}}{d\theta} = \ell_\theta + \lambda^\top G_\theta.$$

For a Newton step $x^{k+1} = x^k - \left(\nabla^2\phi(x^k,\theta)\right)^{-1}\nabla\phi(x^k,\theta)$ run to convergence, the optimality is the stationarity condition $F(x,\theta) = \nabla_x\phi(x,\theta)$ with $F_x = \nabla^2\phi$. The reverse solve uses the Hessian transpose (equal to itself if symmetric). Truncated inner iterations are treatable as *inexact solves*: either unroll a few steps (cheap) or treat the stopping tolerance $\|F\| \le \varepsilon$ as part of the model;then the gradient is exact for the *inexact* fixed point, which is usually what the outer objective actually sees.

Constrained inner problems (KKT systems). If the inner problem is a small optimization solved to stationarity,

$$\min_x \ \phi(x,\theta) \quad \text{s.t.} \quad c(x,\theta) = 0,$$

its KKT conditions form $F(z,\theta) = 0$ with $z = (x,\lambda)$:

$$\nabla_x\phi(x,\theta) + \nabla_x c(x,\theta)^\top\lambda = 0, \qquad c(x,\theta) = 0.$$

The sensitivity solve uses the KKT matrix

$$F_z = \begin{bmatrix} \nabla^2_{xx}\phi + \sum_i \lambda_i \nabla^2_{xx} c_i & \nabla_x c^\top \\ \nabla_x c & 0 \end{bmatrix}.$$

The reverse step solves $F_z^\top \eta = \bar{z}$ and accumulates $d\mathcal{J}/d\theta = \ell_\theta - \eta^\top F_\theta$. In many SciML settings the “second-derivative” block is approximated by Gauss–Newton, yielding sparse SPD structure that Krylov methods precondition well.

Time-steppers and black-box ODE solvers. Treating an ODE solver as a black box that maps $(u_0,\theta) \mapsto u_T$ fits the template with F equal to the discrete optimality of the stepper (discrete adjoint; 2.9.2). If you do not want to differentiate per-step algebra, you may use the continuous-adjoint interface instead: expose an operator that, given $\bar{u}_T$, performs one backward adjoint integration and returns vector-Jacobian products with respect to (u_0,θ) via the continuous formulas. Both interfaces reduce AD through long trajectories to a *handful* of linear/adjoint solves plus checkpointing.

Projectors and proximal maps. Projection onto a convex set $\Pi_{\mathcal{C}}(y)$ solves $\min_x \frac{1}{2}\|x - y\|^2$ s.t. $x \in \mathcal{C}$. Differentiation uses the optimality condition $x^\star - y + s = 0$ with $s \in N_{\mathcal{C}}(x^\star)$ (normal cone). The Jacobian of $\Pi_{\mathcal{C}}$ is the orthogonal projector onto the tangent cone $T_{\mathcal{C}}(x^\star)$; in reverse mode,

$$\bar{y} \ \mapsto \ P_{T_{\mathcal{C}}(x^\star)}\,\bar{y},$$

i.e., drop components that would push $x^\star$ out of the feasible manifold. Proximal maps $\mathrm{prox}_{\tau r}(y) = \arg\min_x \big(r(x) + \frac{1}{2\tau}\|x - y\|^2\big)$ are handled similarly via their optimality conditions; for separable r (e.g., ℓ_1) the Jacobian is diagonal with $0/1$ entries depending on which coordinates are active.

Linear algebra as differentiable primitives. Treat factorizations/solves/least-squares as atomic ops with custom JVP/VJP:

- *Cholesky*: if $A = LL^\top$, then for a JVP with perturbation $\dot{A}$, solve $LX = \dot{A}\,L^{-\top}$ and symmetrize: $\dot{L} = \mathrm{lower}\big(X/2\big)$; reverse uses a triangular Sylvester solve. Prefer exposing solves with L over differentiating through the factorization kernels step-by-step.
- *QR/SVD*: use established backward rules that solve structured triangular/Sylvester equations; avoid forming dense Kronecker products. Respect sign/phase conventions to keep derivatives continuous near repeated singular values.
- *Least squares*: for $\min_x \|Ax - b\|_2$, the VJP with seed $\bar{x}$ solves normal-like systems but should be implemented through QR/LSMR rather than $A^\top A$ for stability, mirroring 2.8.2.

Accuracy, stability, and stopping tolerances. Implicit differentiation assumes the inner problem is solved "well enough" that linearization at $x^\star$ is meaningful. If the inner solver stops early, the gradient corresponds to the *inexact* fixed point;often desirable since the outer objective also sees that inexactness. When you need exact sensitivities of the exact solution, drive the inner residual below the outer accuracy target and use consistent linear algebra (same preconditioners and tolerances) in the sensitivity solves. For stiff or ill-conditioned F_x, preconditioning is as crucial for $F_x^\top \lambda = \bar{x}$ as it is for the forward solve; reusing factorizations or Krylov subspaces from the forward often pays off.

Putting it into a VJP/JVP API. Given a primitive `solve(F, x0, θ)` returning $x^\star$, define:

$$\text{JVP}: \ (\dot{\theta}, \dot{x}_0) \ \mapsto \ \dot{x}^\star \text{ by solving } F_x\,\dot{x}^\star = -F_\theta \dot{\theta} - F_{x_0}\dot{x}_0,$$

$$\text{VJP}: \ \bar{x} \ \mapsto \ \big(\bar{\theta}, \bar{x}_0\big) \text{ by first solving } F_x^\top \lambda = \bar{x}, \text{ then } \bar{\theta} = -F_\theta^\top \lambda, \ \bar{x}_0 = -F_{x_0}^\top \lambda,$$

with all Jacobian-vector products computed by AD inside `F` (no dense Jacobians). This isolates differentiation logic, preserves numerical stability, and avoids unrolling.

The guiding principle is consistent across solvers: *differentiate the optimality condition, not the iterations*. Doing so turns a potentially long and unstable reverse pass into a small number of well-conditioned linear/adjoint solves that reuse your best numerical tools;factorizations, preconditioners, and checkpointing;and that deliver gradients faithful to the mathematics the solver enforces.

2.10 High-Performance Computing (HPC) Essentials

Performance lives at the intersection of hardware limits and algorithmic structure. A simple, predictive model starts with the roofline: every kernel is bounded by compute throughput and by memory bandwidth. If a kernel performs FLOPs floating-point operations while moving Bytes of data to/from main memory, its *arithmetic intensity* is $\mathrm{AI} = \mathrm{FLOPs}/\mathrm{Bytes}$, and its attainable performance obeys

$$P \leq \min\left\{P_{\max},\ \mathrm{AI}\cdot B_{\max}\right\},$$

where $P_{\max}$ is peak FLOP/s of the processor and $B_{\max}$ is sustained memory bandwidth. Low-AI kernels are *memory-bound*: optimizing them is mostly about reducing traffic and improving reuse; high-AI kernels are *compute-bound*: optimizing them is mostly about feeding vector units or tensor cores efficiently. Two mental pictures make this concrete. Dense matrix multiplication on $n \times n$ matrices performs $\approx 2n^3$ flops and, when suitably blocked, moves $\Theta(n^2)$ data, so $\mathrm{AI} = \Theta(n)$ and it is compute-bound for modest n. By contrast, sparse matrix-vector multiply (SpMV) performs $\approx 2\,\mathrm{nnz}$ flops while moving at least $16\,\mathrm{nnz}$ bytes for double-valued CSR (values+indices) plus vector traffic, so $\mathrm{AI} \approx 0.1$-0.2 and is firmly memory-bound; squeezing extra GFLOP/s out of it hinges on layout and bandwidth, not on instruction pipelines.

Dense linear algebra is built from three layers with escalating AI. Level-1 BLAS (vector ops) touch $O(n)$ data for $O(n)$ flops and are bandwidth-limited; use them sparingly and fuse where possible. Level-2 BLAS (matrix-vector) are similarly memory-bound unless the matrix fits in cache. Level-3 BLAS (matrix-matrix) are compute-friendly because blocking creates reuse: each loaded tile participates in many multiply-accumulate operations. The lesson is to cast work into Level-3 patterns whenever possible;block coordinate updates, fuse successive linear layers, or reorganize physics kernels to operate on tiles. On CPUs, performance follows vectorization and cache locality: lay out data to match the library's expectation (row/column-major consistency, leading dimensions correct), align arrays (32–64 byte boundaries), and choose blocking that fits L1/L2 caches. Use vendor-tuned libraries (MKL, BLIS/OpenBLAS, cuBLAS, cuSOLVER, MAGMA, CUTLASS) before writing bespoke kernels; the fastest code is a *call*.

Sparse kernels demand a different playbook because the memory system dominates. The choice of storage;CSR/CSC for generality, ELL/SELL-P for uniform row lengths, BCSR for block structure, HYB for mixed patterns;determines coalescing and vectorization opportunities. Reordering (RCM, METIS) can reduce bandwidth of the sparsity pattern and improve cache reuse. Blocking small dense substructures (turning 3×3 physics blocks into BCSR) increases AI and maps better to SIMD/SIMT. For SpMV on GPUs, arrange rows so that a warp processes consecutive nonzeros (ELL/SELL) to avoid divergence and to coalesce memory accesses; for SpMM (sparse-dense), prefer column-major dense tiles so each nonzero reuses an entire column tile in registers/shared memory. When possible, fuse sequences like

$y \leftarrow \alpha Ax + \beta y$ or accumulate multiple right-hand sides to amortize index traffic; the bytes saved count more than flops added.

GPUs trade latency for throughput: massive concurrency hides memory and pipeline latencies. Threads are scheduled in warps (typically 32 lanes) that execute in lockstep; divergent branches serialize, so keep threads in a warp on the same path. Global memory is fast only when accesses are coalesced: adjacent threads should read adjacent addresses; strided or scattered reads cost bandwidth. Shared memory is a user-managed L1 with banked access; stage tiles there to create reuse and avoid bank conflicts by padding. Registers are the fastest tier; keep frequently reused scalars there and watch register pressure, which reduces occupancy. A robust kernel structure double-buffers: load the next tile asynchronously (e.g., with `cp.async`), compute on the current tile, then swap; this overlaps memory with math. Tensor cores (matrix-multiply-accumulate units) deliver order-of-magnitude speedups when operands match supported shapes and types; map GEMM-like work to small matrix fragments (e.g., 16×16) and accumulate in a higher precision when available to protect accuracy. Mixed precision is most effective when accumulation happens in FP32 even if inputs are FP16/BF16; combine with loss scaling (for training) and iterative refinement (for linear solves) to recover near-FP64 accuracy in many PDE/optimization kernels.

Parallel speedup depends on how computation and data partition. Strong scaling shrinks time-to-solution for a fixed problem; weak scaling keeps per-rank work fixed and grows global problem size. Communication limits both; minimizing bytes moved is paramount. Data-parallel training replicas the model on P workers and all-reduces gradients each step; the ring all-reduce moves $2(P-1)/P$ times the gradient size per worker with bandwidth-optimal links, while tree-based variants lower latency for many small tensors. Overlap communication with computation by launching asynchronous all-reduces as soon as a gradient bucket is ready, fuse tiny tensors into larger buckets, and use gradient accumulation to reduce synchronization frequency when memory allows. Model parallel shreds the model across devices: tensor parallel splits individual layers (synchronous and bandwidth-heavy), pipeline parallel splits layers into stages (adds a "bubble" that microbatching must fill), and optimizer/state sharding (e.g., ZeRO) distributes states to fit larger models. For PDE/domain decomposition, align subdomains with hardware topology, use halo exchanges that aggregate faces into few messages, and overlap boundary communication with interior work; algebraic multigrid and Krylov methods benefit from communication-avoiding smoothers and s-step variants that trade extra flops for fewer global reductions.

Across nodes, MPI is the lingua franca for collectives and point-to-point; within GPU clusters, NCCL provides topology-aware all-reduces and broadcasts. Bind processes to NUMA domains, pin threads, and place data close to the device that uses them; *where* memory resides matters as much as *how fast* the kernel is. Overlap PCIe/NVLink transfers with compute using streams; prefer page-locked (pinned) host buffers for high-throughput DMA; avoid implicit page migration in unified memory for performance-critical loops unless profiling proves it beneficial.

Profiling turns guesses into facts. Start with a timeline to locate the *time* sink, then apply micro-metrics to uncover the *cause*. On CPUs, measure cache misses, branch mispredicts, vectorization ratios, and achieved bandwidth (STREAM as a baseline). On GPUs, inspect achieved occupancy, memory throughput vs peak, tensor core utilization, warp stall reasons, and shared-memory efficiency. Mark phases with NVTX (or equivalent) so traces are readable; profile representative, steady-state windows, not warmup. Tie findings back to the roofline: is the kernel below the bandwidth roof? then fix coalescing, tiling, and bytes moved; is it below the compute roof with high AI? then fix instruction mix, register pressure, and launch configuration. Iterate after each change; performance is empirical.

I/O and input pipelines are often the real bottleneck in end-to-end SciML systems. Reading, decoding, and transferring data can starve accelerators. Use formats that favor sequential access (e.g., TFRecord/RecordIO/LMDB) over millions of small files; shard datasets across workers to avoid hot files; prefetch and buffer on the host; use multiple loader workers with pinned memory; overlap host decode/augment with device compute; consider GPU-side augmentation (e.g., DALI/NPP) when CPU becomes saturated. Cache expensive intermediate representations (tokenized/decoded arrays) and leverage OS page cache or in-memory caches for small-to-medium datasets. For distributed jobs, prefer local NVMe over shared network storage for hot paths, or stage shards locally at job start. Measure the pipeline like any kernel: record per-batch latency and variance, track queue depths, and ensure the device never idles waiting for input.

A few compact rules of thumb close the loop. First, recast work to increase arithmetic intensity and reuse (tile, fuse, batch); Level-3 beats Level-1 every time. Second, match data layout to hardware: contiguous, aligned, and coalesced reads write the performance story. Third, move less: compress, quantize, block, reorder, and aggregate to save bytes; communication avoided is speed gained. Fourth, overlap whenever possible: compute with one hand, move data with the other. Fifth, trust tuned libraries and only fall back to custom kernels when profiling proves a gap;and then prototype with building blocks like CUTLASS or vendor sparse libraries to inherit portability. Finally, measure, change one thing, and measure again; let the roofline and traces guide which knob to turn next.

2.11 Conclusion

This chapter has established the mathematical foundations upon which Scientific Machine Learning rests. With these foundations in place, the next chapter turns to the software frameworks and computational paradigms that enable practical implementation.

References

1. Van Mill, J.: The Infinite-Dimensional Topology of Function Spaces, vol. 64, Gulf Professional Publishing (2001)
2. Liesen, J., Strakos, Z.: Krylov subspace methods: principles and analysis. Num. Math. Sci. (2013)
3. Hall, M.C., Cacuci, D.G.: Physical interpretation of the adjoint functions for sensitivity analysis of atmospheric models. J. Atmos. Sci. **40**(10), 2537–2546 (1983)
4. Xiang, R., van Schie, S.P., Scotzniovsky, L., Yan, J., Kamensky, D., Hwang, J.T.: Automating adjoint sensitivity analysis for multidisciplinary models involving partial differential equations. Struct. Multidiscip. Optim. **67**(8), 146 (2024)
5. Rackauckas, C., Ma, Y., Martensen, J., Warner, C., Zubov, K., Supekar, R., Skinner, D., Ramadhan, A., Edelman, A.: Universal differential equations for scientific machine learning. arXiv preprint arXiv:2001.04385 (2020)
6. Offen, C.: Machine learning of continuous and discrete variational odes with convergence guarantee and uncertainty quantification. Math. Comput. (2025)
7. Boullé, N., Halikias, D., Otto, S.E., Townsend, A.: Operator learning without the adjoint. J. Mach. Learn. Res. **25**(364), 1–54 (2024)
8. Cryer, C.W.: The lu-factorization of totally positive matrices. Linear Algebra Appl. **7**(1), 83–92 (1973)
9. Chan, T.F.: Rank revealing qr factorizations. Linear Algebra Appl. **88**, 67–82 (1987)
10. Björck, Å.: Numerics of gram-schmidt orthogonalization. Linear Algebra Appl. **197**, 297–316 (1994)
11. Kelley, C.T.: Iterative Methods for Linear and Nonlinear Equations. SIAM (1995)
12. Saad, Y., Schultz, M.H.: Gmres: A generalized minimal residual algorithm for solving nonsymmetric linear systems. SIAM J. Sci. Stat. Comput. **7**(3), 856–869 (1986)
13. Schoenholz, S., Cubuk, E.D.: Jax md: a framework for differentiable physics. Adv. Neural. Inf. Process. Syst. **33**, 11428–11441 (2020)
14. Lin, M.: Automatic functional differentiation in JAX. In: The Twelfth International Conference on Learning Representations
15. Guo, C., Poletti, D.: Scheme for automatic differentiation of complex loss functions with applications in quantum physics. Phys. Rev. E **103**(1), 013309 (2021)
16. Castillo, R.E., Rafeiro, H.: An Introductory Course in Lebesgue Spaces. Springer (2016)
17. Adams, R.A., Fournier, J.J.: Sobolev Spaces, vol. 140. Elsevier (2003)
18. Dontchev, A.L., Zolezzi, T.: Well-Posed Optimization Problems. Springer (2006)
19. Lucchetti, R., Patrone, F.: Hadamard and tyhonov well-posedness of a certain class of convex functions. J. Math. Anal. Appl. **88**(1), 204–215 (1982)
20. Schneider, C.: Theory and background material for PDES. Beyond Sobolev and Besov: Regularity of Solutions of PDEs and Their Traces in Function Spaces, pp. 115–143 (2021)
21. Agranovich, M.S.: Remarks on strongly elliptic systems in lipschitz domains. Russ. J. Math. Phys. **19**(4), 405–416 (2012)
22. Tello Fachin, C.: Nitsche-based enforcement of Dirichlet boundary conditions for the Hodge Laplacian, Technische Universität Wien. Ph.D. thesis (2025)
23. Tekriwal, M., Duraisamy, K., Jeannin, J.B.: A formal proof of the lax equivalence theorem for finite difference schemes. In: NASA Formal Methods Symposium, pp. 322–339. Springer (2021)
24. Thomée, V.: Galerkin Finite Element Methods for Parabolic Problems, vol. 25. Springer Science & Business Media (2007)
25. Li, J., Yue, J., Zhang, W., Duan, W.: The deep learning galerkin method for the general stokes equations. J. Sci. Comput. **93**(1), 5 (2022)
26. Roop, J.P.: A randomized neural network based petrov-galerkin method for approximating the solution of fractional order boundary value problems. Results Appl. Math. **23**, 100493 (2024)
27. Chen, W., Křížek, M.: What is the smallest possible constant in céa's lemma? Appl. Math. **51**(2), 129–144 (2006)

28. Ramos, J.I.: On the picard-lindelof method for nonlinear second-order differential equations. Appl. Math. Comput. **203**(1), 238–242 (2008)
29. Haddad, W.M., Chellaboina, V.: Nonlinear dynamical systems and control: a Lyapunov-based approach. Princeton University Press (2008)
30. Lang, A.: A lax equivalence theorem for stochastic differential equations. J. Comput. Appl. Math. **234**(12), 3387–3396 (2010)

Chapter 3
Scientific Computing Fundamentals

Talk is cheap. Show me the code.

Linus Torvalds

Abstract Translating mathematical formulations into working implementations requires mastery of modern scientific computing tools and paradigms. This chapter surveys the software ecosystem that enables Scientific Machine Learning, beginning with programming paradigms (imperative, functional, declarative) and their implications for code structure, performance, and differentiability. The core of the chapter provides detailed coverage of the major frameworks: NumPy/SciPy for foundational numerics, JAX for composable transformations and hardware acceleration, PyTorch for dynamic computation graphs and research flexibility, TensorFlow for production deployment, and Julia's SciML ecosystem for differentiable programming with native numerical solvers. Practical considerations of parallelism (data, model, pipeline), device placement, and memory management address the computational demands of large-scale SciML. We conclude with patterns for integrating numerical solvers (ODE/DAE integrators, sparse linear algebra) within differentiable pipelines.

3.1 Introduction

In 2012, a small group of researchers at MIT faced a frustrating dilemma. They were developing algorithms for technical computing (differential equations, optimization, machine learning) and they were stuck between two unsatisfying choices. They could write in Python, which was pleasant to use, easy to prototype in, and had excellent libraries. But Python was slow. Anything computationally intensive had to be written in C or Fortran, wrapped, compiled, and carefully integrated. The result was a "two-language problem": prototype in Python, then rewrite in C for performance. The rewriting took months and introduced bugs.

B. P. Bhuyan et al., *Scientific Machine Learning*,
https://doi.org/10.1007/978-981-95-7856-6_3

Alternatively, they could write everything in C++ from the start. Fast, yes. But the development cycle was painful: compile, run, crash, debug, repeat. Interactive exploration was impossible. Graduate students spent more time fighting the compiler than doing science.

"There has to be a better way," said Jeff Bezanson, one of the frustrated researchers. What they created was Julia—a language designed from the ground up to solve the two-language problem through just-in-time compilation, multiple dispatch, and a type system designed for numerical computing.

A decade later, the Scientific Machine Learning ecosystem in Julia has become one of the most sophisticated platforms for physics-informed learning, with capabilities for differentiable programming that other ecosystems are still working to match.

But Julia is not the only answer. The same period saw PyTorch emerge as the dominant research framework for deep learning, prized for its dynamic computation graphs and pythonic feel. JAX arrived from Google, bringing composable function transformations that make automatic differentiation, vectorization, and compilation feel like natural extensions of NumPy. TensorFlow evolved from a static graph framework to a more flexible ecosystem with production-grade deployment tools.

The modern SciML practitioner faces an embarrassment of riches; and a genuine challenge of choice.

3.1.1 Why Framework Choice Matters

The choice of computational framework is not merely a matter of personal preference. Different frameworks embody different philosophies about how computation should be structured, and these philosophies have profound implications for Scientific Machine Learning:

- **Differentiability**: Can you compute gradients through arbitrary code, including numerical solvers, control flow, and custom operations?
- **Performance**: Can your code run efficiently on GPUs, TPUs, and distributed clusters without extensive rewrites?
- **Composability**: Can you combine transformations (differentiation, vectorization, compilation) in arbitrary ways?
- **Interoperability**: Can you integrate with existing scientific computing libraries, legacy code, and domain-specific tools?
- **Ecosystem**: Are the solvers, architectures, and utilities you need available and well-tested?

This chapter provides the information you need to make informed choices and to use your chosen tools effectively.

We begin with programming paradigms (imperative, functional, declarative) and their implications for scientific code. We then examine automatic differentiation in detail, as the enabling technology for physics-informed learning. The core of the chapter surveys the major frameworks, with enough depth to write real code and enough breadth to understand trade-offs. We conclude with practical considerations of parallelism, performance, and the integration of numerical solvers within differentiable pipelines.

3.2 Programming Paradigms for Scientific Computing

Modern scientific computing frameworks differ not only in performance but also in the *paradigms of programming* they encourage. The three most common styles are **imperative**, **declarative**, and **functional**. Understanding their differences is essential for writing code that is efficient, reproducible, and compatible with automatic differentiation.

3.2.1 Imperative, Declarative, and Functional Styles

Definition: Imperative, Declarative, and Functional

- **Imperative programming**: Code specifies *how* to perform computations step by step. This is natural in Python/NumPy and PyTorch eager mode.
- **Declarative programming**: Code specifies *what* computation should be performed, leaving the framework to optimize the execution. TensorFlow graphs and SQL are typical examples.
- **Functional programming**: Programs are expressed as compositions of pure functions without side effects. JAX and Julia encourage this style for composability and reproducibility.

To understand the difference, consider a simple task: squaring a vector and summing its elements. In an imperative style, the programmer gives explicit instructions for each step. Using NumPy, we create an array, loop over each element, compute its square, and then add them together. This style is straightforward and easy to follow, but it closely resembles "manual" bookkeeping and does not exploit deeper optimizations.

```
import numpy as np

x = np.array([1.0, 2.0, 3.0])
squares = []
for xi in x:          # explicit steps
```

```
    squares.append(xi ** 2)
result = sum(squares)
print(result)  # 14.0
```

Listing 3.1 Imperative style with NumPy

In a declarative style, the programmer instead describes the desired computation, while the framework decides how to execute it. TensorFlow achieves this through `@tf.function`, which builds a static computation graph. Here we only state that we want to compute the sum of squares, and TensorFlow determines how to schedule and optimize the operations, potentially across different devices.

```
import tensorflow as tf

@tf.function    # builds a static computation graph
def compute(x):
    return tf.reduce_sum(x ** 2)

x = tf.constant([1.0, 2.0, 3.0])
result = compute(x)    # framework decides execution order
print(result.numpy())  # 14.0
```

Listing 3.2 Declarative style with TensorFlow (graph)

In a functional style, the program is written as a pure function without side effects. With JAX, we can express the computation as a single mathematical function, and the framework automatically supports differentiation, just-in-time (JIT) compilation, and parallel execution. This style is concise, composable, and well-suited for scientific machine learning, where gradients and performance are essential.

```
import jax
import jax.numpy as jnp

def compute(x):
    return jnp.sum(x ** 2)    # pure function, no side effects

x = jnp.array([1.0, 2.0, 3.0])
result = compute(x)           # works as a mathematical function
grad_fn = jax.grad(compute)   # automatic differentiation
print(result)                 # 14.0
print(grad_fn(x))             # [2.0, 4.0, 6.0]
```

Listing 3.3 Functional style with JAX

These three approaches all produce the same numerical result, but they represent fundamentally different ways of thinking about computation. Imperative style emphasizes explicit control, declarative style emphasizes abstraction and optimization, and functional style emphasizes purity and composability. Modern frameworks often combine aspects of all three, but functional programming has become especially prominent in scientific machine learning due to its compatibility with differentiation and parallelization.

3.2.2 Vectorization Versus Explicit Loops

A central principle in scientific computing is to avoid explicit Python loops when working with large arrays. This idea, called **vectorization**, means expressing computations in terms of whole-array operations rather than element-by-element instructions. Vectorized code not only reads more clearly as "mathematics in code," but also leverages low-level optimizations and parallel execution on modern CPUs and GPUs.

To illustrate, consider the same task as before: squaring a vector and summing the results. If we write this imperatively with a Python loop, each iteration is handled by the interpreter, which adds overhead. This approach is clear but inefficient for large arrays.

```
import numpy as np

x = np.arange(1, 1_000_001, dtype=float)  # 1 to 1,000,000
result = 0.0
for xi in x:
    result += xi ** 2
print(result)
```

Listing 3.4 Explicit Python loop

A vectorized implementation using NumPy expresses the same computation as a single array operation. Internally, NumPy delegates the work to highly optimized C and Fortran libraries (such as BLAS and LAPACK). This means that the computation is executed in compiled code, often using vectorized CPU instructions, and runs orders of magnitude faster.

```
import numpy as np

x = np.arange(1, 1_000_001, dtype=float)
result = np.sum(x ** 2)
print(result)
```

Listing 3.5 Vectorized NumPy implementation

With JAX, the same vectorized expression can be combined with just-in-time (JIT) compilation for further speedups. The JAX compiler (XLA) fuses operations into efficient kernels, automatically generating machine code for CPUs, GPUs, or TPUs. This combines the clarity of vectorized code with the efficiency of low-level compilation.

```
import jax
import jax.numpy as jnp

@jax.jit
def compute(x):
    return jnp.sum(x ** 2)
```

```

x = jnp.arange(1, 1_000_001, dtype=float)
result = compute(x)
print(result)
```

Listing 3.6 Vectorized JAX with JIT compilation

These three approaches produce the same mathematical answer, but their performance is dramatically different. The explicit loop runs slowly because each operation is executed in Python. The NumPy vectorized version is far faster because it offloads the work to optimized libraries. The JAX JIT-compiled version often achieves further acceleration, especially on GPUs, by generating specialized kernels. In scientific machine learning, vectorization and JIT compilation are indispensable for scaling computations to realistic problem sizes.

3.2.3 Compilation Versus Interpretation

Another important distinction in scientific computing frameworks is whether code is executed **interpreted** or **compiled**. In interpreted execution, each instruction is translated on the fly by an interpreter (like the Python runtime). This makes experimentation easy but can introduce significant overhead. In compiled execution, code is translated once into machine instructions, allowing the computer to run it directly with much greater efficiency.

To illustrate, consider again the task of summing the squares of a large vector. In plain NumPy, the code is interpreted by Python. While NumPy internally calls optimized C routines for array operations, the Python layer still incurs overhead, especially when many small operations are chained.

```
import numpy as np

x = np.arange(1, 1_000_001, dtype=float)
result = np.sum(x ** 2)   # executed through Python interpreter
print(result)
```

Listing 3.7 Interpreted execution with NumPy

With JAX, the same code can be compiled ahead of time using the `@jit` decorator. The computation is traced once, converted into an intermediate representation (XLA HLO), and then compiled into optimized machine code. Subsequent calls reuse the compiled kernel, often achieving speedups of 10–100x compared to repeated interpretation.

```
import jax
import jax.numpy as jnp

@jax.jit
def compute(x):
```

```
    return jnp.sum(x ** 2)

x = jnp.arange(1, 1_000_001, dtype=float)
result = compute(x)   # compiled to efficient machine code
print(result)
```

Listing 3.8 Compiled execution with JAX JIT

Julia takes compilation even further by making it the default. All Julia functions are compiled just-in-time (JIT) using the LLVM compiler infrastructure. This means that ordinary Julia code, written in a mathematical style, is translated into optimized machine instructions without requiring explicit decorators.

```
# Julia code (not Python)
x = collect(1.0:1_000_000.0)
result = sum(x .^ 2)    # compiled to native code automatically
println(result)
```

Listing 3.9 Compiled execution in Julia (using JIT)

Although NumPy, JAX, and Julia can all produce the same numerical results, the execution models are different. NumPy relies on interpretation with some internal optimizations, JAX allows optional compilation for high-performance workloads, and Julia provides compiled execution by default. For scientific machine learning, compiled execution is particularly important because it enables both high performance on large problems and efficient differentiation through complex solver pipelines.

3.3 Automatic Differentiation Essentials

Automatic differentiation (AD) is the backbone of scientific machine learning, enabling gradients of complex models, solvers, and simulations to be computed reliably and efficiently. Unlike numerical differentiation (which approximates derivatives using finite differences) or symbolic differentiation (which manipulates algebraic expressions), AD systematically applies the chain rule at the level of elementary operations.

3.3.1 Modes of Automatic Differentiation

Definition: Modes of Automatic Differentiation

- **Forward-mode AD**: Propagates derivatives from inputs to outputs alongside the computation. Efficient when the number of inputs is small.

- **Reverse-mode AD**: Propagates derivatives backwards from outputs to inputs. Efficient when the number of outputs is small (e.g., scalar loss functions).
- **Higher-order AD**: Applies AD recursively to obtain second or higher derivatives (Hessian, Jacobian-vector products, etc.).

To understand, let us consider a simple scalar function:

$$f(x) = x^3 + 2x.$$

In forward-mode AD, we propagate derivatives together with the function evaluation. This is natural for functions of many outputs but few inputs, since each input requires a separate derivative trace.

```
import jax
import jax.numpy as jnp

def f(x):
    return x**3 + 2*x

# forward-mode using jax.jvp (Jacobian-vector product)
x = 3.0
y, dy = jax.jvp(f, (x,), (1.0,))
print("f(x) =", y)    # 33.0
print("f'(x) =", dy) # 29.0
```

Listing 3.10 Forward-mode AD in JAX

In reverse-mode AD, the derivative is obtained by backpropagating sensitivities from the output to the inputs. This is highly efficient when there is a single output (such as a loss function) and many inputs, which is why reverse-mode dominates in deep learning.

```
grad_f = jax.grad(f)    # reverse-mode by default
print("f'(3.0) =", grad_f(3.0))  # 29.0
```

Listing 3.11 Reverse-mode AD in JAX

Finally, higher-order derivatives are obtained by applying AD recursively. For example, taking the gradient of a gradient produces the second derivative.

```
grad2_f = jax.grad(jax.grad(f))    # second derivative
grad3_f = jax.grad(grad2_f)        # third derivative

print("f''(3.0) =", grad2_f(3.0)) # 18.0
print("f'''(3.0) =", grad3_f(3.0))# 6.0
```

Listing 3.12 Higher-order derivatives with JAX

All three modes compute derivatives exactly to machine precision by systematically applying the chain rule. Forward-mode is best suited when the number of inputs is small, reverse-mode is best when the number of outputs is small, and higher-order AD is crucial in scientific computing for tasks like optimization with Hessians or uncertainty quantification.

3.3.2 Operator Overloading Versus Source Transformation

Automatic differentiation can be implemented in two main ways: **operator overloading** and **source transformation**. Both compute exact derivatives by applying the chain rule, but they differ in how they trace computations.

Definition: Operator Overloading versus Source Transformation

- **Operator overloading**: Each arithmetic operation is replaced ("overloaded") with a special version that records its contribution to the computation graph. This graph is then traversed forward (for forward-mode) or backward (for reverse-mode). PyTorch and early versions of TensorFlow use this strategy.
- **Source transformation**: The program itself is analyzed and rewritten into a new program that computes both the original result and its derivatives. This allows for more global optimizations. JAX (via XLA) and Julia (via Zygote and Enzyme) primarily use this approach.

To see the difference in practice, let us consider the simple function :

$$f(x, y) = x \cdot y + \sin(x).$$

With operator overloading, the framework builds a dynamic computation graph as operations are executed. Each addition, multiplication, or sine call appends a node to the graph. When we request a gradient, the graph is traversed in reverse to compute sensitivities.

```python
import torch

x = torch.tensor(2.0, requires_grad=True)
y = torch.tensor(3.0, requires_grad=True)

f = x * y + torch.sin(x)    # graph is built dynamically
f.backward()                # backprop through graph

print(x.grad)  # df/dx = y + cos(x) = 3 + cos(2.0)
print(y.grad)  # df/dy = x = 2.0
```

Listing 3.13 Operator overloading with PyTorch

With source transformation, the program is not traced step by step but instead transformed into a new function that directly encodes the derivative calculation. This avoids some of the overhead of building dynamic graphs and allows for compilation into highly optimized kernels.

```
import jax
import jax.numpy as jnp

def f(x, y):
    return x * y + jnp.sin(x)

grad_f = jax.grad(f, argnums=(0, 1))  # transforms source into
    derivative code
print(grad_f(2.0, 3.0))  # (y + cos(x), x) = (3 + cos(2), 2.0)
```

Listing 3.14 Source transformation with JAX

Both approaches give the same mathematical result, but their trade-offs differ. Operator overloading is intuitive and flexible, since the graph is constructed as the program runs. However, it may be less efficient, as the graph cannot always be optimized globally. Source transformation enables ahead-of-time optimizations, kernel fusion, and better performance on accelerators, but requires more complex compiler infrastructure.

In modern scientific machine learning, both techniques coexist: PyTorch uses operator overloading (with partial compiler support in `torch.compile`), while JAX and Julia primarily use source transformation to enable large-scale optimization and differentiable simulation.

3.3.3 Differentiation Methods

Automatic differentiation (AD) should be distinguished from two classical approaches: **numerical differentiation** and **symbolic differentiation**. Although all three methods aim to compute derivatives, their accuracy, efficiency, and practicality are very different.

Definition: Differentiation Methods

- **Numerical differentiation**: Approximates derivatives by finite differences, e.g.

$$f'(x) \approx \frac{f(x+h) - f(x)}{h},$$

for small h. Easy to implement but suffers from truncation error and numerical instability.

- **Symbolic differentiation**: Manipulates algebraic expressions symbolically to produce exact derivative formulas. Useful for small problems but may lead to expression explosion.
- **Automatic differentiation**: Applies the chain rule at the level of elementary operations, combining the accuracy of symbolic methods with the efficiency of numerical evaluation.

Let us consider the function:

$$f(x) = x^3 + 2x.$$

With numerical differentiation, we approximate the derivative using a finite difference. This requires evaluating f multiple times and is sensitive to the choice of step size.

```
import numpy as np

def f(x):
    return x**3 + 2*x

x = 3.0
h = 1e-5
numerical_grad = (f(x + h) - f(x - h)) / (2*h)
print("Approximate f'(3) =", numerical_grad)  # ~29.0
```

Listing 3.15 Numerical differentiation with finite differences

With symbolic differentiation, we ask a computer algebra system (CAS) to manipulate the formula for us. This gives exact expressions but is limited to functions that can be represented symbolically.

```
import sympy as sp

x = sp.symbols('x')
f = x**3 + 2*x
df = sp.diff(f, x)
print("Symbolic derivative:", df)  # 3*x**2 + 2
print("f'(3) =", df.subs(x, 3))    # 29
```

Listing 3.16 Symbolic differentiation with SymPy

With automatic differentiation, the derivative is computed alongside the function evaluation. This combines the efficiency of numerical evaluation with the accuracy of symbolic rules, and it scales to high-dimensional problems.

```
import jax
import jax.numpy as jnp

def f(x):
```

```
    return x**3 + 2*x

grad_f = jax.grad(f)
print("f'(3) =", grad_f(3.0))    # 29.0
```

Listing 3.17 Automatic differentiation with JAX

All three approaches produce the correct result for this simple function, but their behavior differs in practice. Numerical differentiation is approximate and can be unstable for stiff or noisy problems. Symbolic differentiation is exact but can produce unwieldy expressions and is not practical for complex simulations. Automatic differentiation strikes a balance: it is exact up to machine precision, efficient, and works seamlessly with modern scientific computing frameworks, making it the preferred choice in scientific machine learning.

3.4 Core Frameworks Overview

Modern scientific machine learning builds on a family of frameworks that combine numerical computing, automatic differentiation, and high-performance execution. We begin with NumPy and SciPy, the foundations of Python-based numerical computing, before moving to modern differentiable frameworks such as JAX, PyTorch, TensorFlow, and Julia's SciML ecosystem.

3.4.1 NumPy and SciPy

NumPy and SciPy are the classical backbone of scientific Python. NumPy offers fast array operations and broadcasting; SciPy provides a rich catalog of numerical routines (optimization, integration, ODEs, sparse linear algebra). This stack is mature and reliable, but it does not support automatic differentiation (AD) natively. As soon as we need sensitivities (gradients) of simulation outputs w.r.t. parameters or initial conditions, we must fall back to *finite differences*; which are approximate, step-size sensitive, and require multiple solver calls.

Strengths versus Limitations (NumPy/SciPy)

Strengths: battle-tested, broad coverage of algorithms, excellent NumPy interoperability.
Limitations: no built-in AD; CPU-centric; differentiating through solvers requires finite differences (extra solver runs, step-size trade-offs).

To explain, we use the scalar linear ODE

$$\frac{dy}{dt} = -\lambda\, y, \qquad y(0) = y_0,$$

with analytic solution $y(t) = y_0\, e^{-\lambda t}$. This lets us verify accuracy and visualize limitations directly.

(1) Solving the ODE with SciPy and comparing to the analytic solution:

The following script integrates on $[0, 5]$ and overlays the numerical and analytic curves.

```
import numpy as np
from scipy.integrate import solve_ivp
import matplotlib.pyplot as plt

lam = 0.7
y0  = 1.0

def f(t, y):
    return -lam * y

sol = solve_ivp(f, (0.0, 5.0), [y0], dense_output=True, rtol=1e-9,
    atol=1e-12)

t = np.linspace(0.0, 5.0, 200)
y_num = sol.sol(t)[0]
y_an  = y0 * np.exp(-lam * t)

plt.figure()
plt.plot(t, y_num, label="SciPy (solve_ivp)")
plt.plot(t, y_an, "--", label=r"Analytic $y_0 e^{-\lambda t}$")
plt.xlabel("t"); plt.ylabel("y(t)")
plt.legend(); plt.title("Exponential decay: numerical vs analytic")
plt.tight_layout(); plt.show()
```

Listing 3.18 SciPy solve and comparison to the analytic solution

The numerical solution tracks the analytic curve closely (with tight tolerances). Up to this point, NumPy+SciPy deliver exactly what we need: *fast, accurate forward solves* as shown in Fig. 3.1.

(2) Sensitivity w.r.t. the initial condition y_0 via finite differences.

Now suppose our downstream task needs $S_{y_0} = \partial y(T)/\partial y_0$ at $T = 5$. Analytically, $S_{y_0} = e^{-\lambda T}$. Without AD, SciPy users commonly approximate this with central differences:

$$S_{y_0} \approx \frac{y(T; y_0 + h) - y(T; y_0 - h)}{2h}.$$

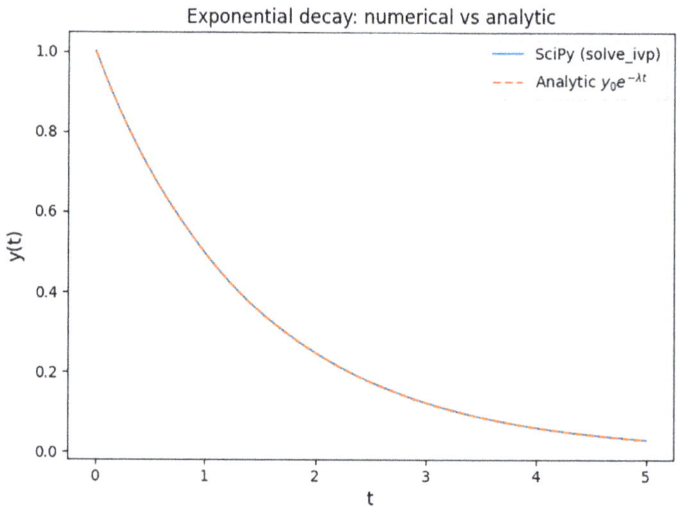

Fig. 3.1 Plot of SciPy solve and comparison to the analytic solution

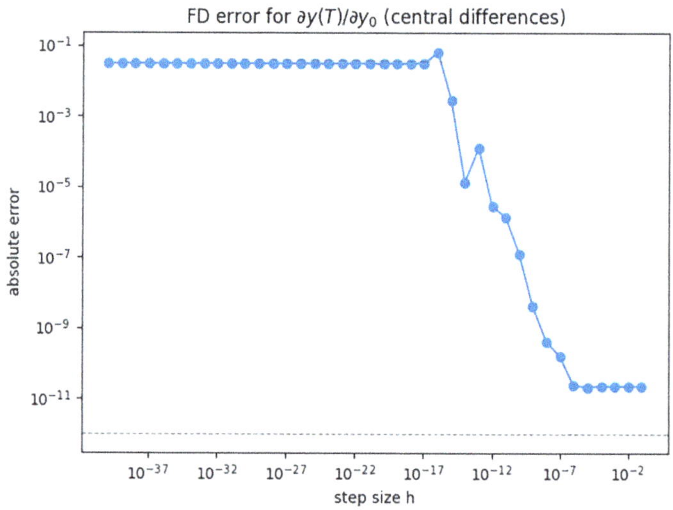

Fig. 3.2 Plot of finite-difference sensitivity

This requires two additional solves per y_0 and the accuracy depends precariously on h as shown in Fig. 3.2.

```
import numpy as np
from scipy.integrate import solve_ivp
import matplotlib.pyplot as plt

lam = 0.7
```

```
y0  = 1.0
T   = 5.0

def solve_yT(y0):
    def f(t, y): return -lam * y
    sol = solve_ivp(f, (0.0, T), [y0], dense_output=True, rtol=1e-9,
     atol=1e-12)
    return sol.sol(T)[0]

true = np.exp(-lam * T)  # exact dy/dy0 at T

hs = np.logspace(-40, -1, 40)
errs = []
for h in hs:
    approx = (solve_yT(y0 + h) - solve_yT(y0 - h)) / (2*h)
    errs.append(abs(approx - true))

plt.figure()
plt.loglog(hs, errs, marker="o")
plt.axhline(1e-12, color="gray", ls="--", lw=0.8)
plt.xlabel("step size h"); plt.ylabel("absolute error")
plt.title(r"FD error for $\partial y(T)/\partial y_0$ (central
     differences)")
plt.tight_layout(); plt.show()
```

Listing 3.19 Finite-difference sensitivity $\partial y(T)/\partial y_0$ and step-size sweep

The error curve shown in Fig. 3.3 typically has a "U"-shape: for large h, truncation error dominates (bad approximation); for tiny h, round-off and solver tolerances dominate (catastrophic cancellation). There is no single "safe" h across problems;

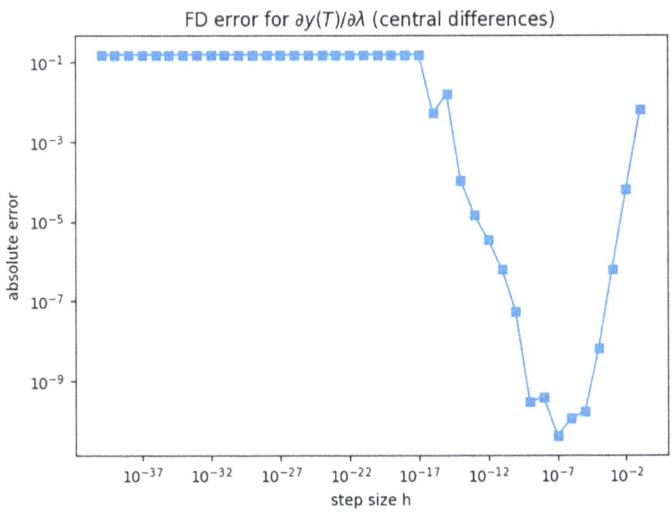

Fig. 3.3 Sensitivity w.r.t. the parameter λ via finite differences

tuning is manual and problem-specific. Moreover, every gradient component costs more solver calls.

(3) Sensitivity w.r.t. the parameter λ via finite differences.
We can repeat the experiment for $S_\lambda = \partial y(T)/\partial\lambda$. Analytically, $S_\lambda = -T\, y_0\, e^{-\lambda T}$. Again, finite differences require additional solves and delicate step-size tuning.

```python
import numpy as np
from scipy.integrate import solve_ivp
import matplotlib.pyplot as plt

y0  = 1.0
lam = 0.7
T   = 5.0

def solve_yT(lam):
    def f(t, y): return -lam * y
    sol = solve_ivp(f, (0.0, T), [y0], dense_output=True, rtol=1e-9,
     atol=1e-12)
    return sol.sol(T)[0]

true = -T * y0 * np.exp(-lam * T)  # exact dy/dlambda at T

hs = np.logspace(-40, -1, 40)
errs = []
for h in hs:
    approx = (solve_yT(lam + h) - solve_yT(lam - h)) / (2*h)
    errs.append(abs(approx - true))

plt.figure()
plt.loglog(hs, errs, marker="s")
plt.xlabel("step size h"); plt.ylabel("absolute error")
plt.title(r"FD error for $\partial y(T)/\partial \lambda$ (central
     differences)")
plt.tight_layout(); plt.show()
```

Listing 3.20 Finite-difference sensitivity $\partial y(T)/\partial\lambda$ and step-size sweep

The same trade-off appears. In higher dimensions (many parameters), finite differences scale poorly: each component requires at least one (often two) extra solves. This becomes a bottleneck for calibration, optimal control, or learning-in-the-loop.

Points to notice in the plots

- The forward solutions are accurate and cheap to compute with SciPy.
- Sensitivities via finite differences are *approximate* and *expensive*: extra solver calls per parameter and step-size tuning are required.
- Error versus step size exhibits the classic "U" shape (truncation error on the right; round-off/solver tolerance on the left).

Modern differentiable frameworks (such as JAX) compute the same sensitivities with a single reverse-mode pass, no step-size tuning, and exactness up to machine precision. For contrast, the JAX snippet below returns both $\partial y(T)/\partial y_0$ and $\partial y(T)/\partial \lambda$ from *one* definition of the ODE right-hand side.

```
import jax, jax.numpy as jnp
import diffrax as dfx  # pip install diffrax

def rhs(t, y, args):
    lam = args
    return -lam * y

def yT(y0, lam, T=5.0):
    term = dfx.ODETerm(lambda t, y, args: rhs(t, y, args))
    sol = dfx.diffeqsolve(term,
                          solver=dfx.Tsit5(),
                          t0=0.0, t1=T, dt0=1e-2,
                          y0=y0, args=lam, saveat=dfx.SaveAt(t1=True),
                          max_steps=4096)
    return sol.ys[0]  # Extract the scalar value

# Reverse-mode AD: gradients wrt y0 and lam
grad_y0  = jax.grad(lambda y0, lam: yT(y0, lam), argnums=0) # Specify argnums for clarity
grad_lam = jax.grad(lambda lam, y0: yT(y0, lam), argnums=0) # Specify argnums for clarity

print(grad_y0(1.0, 0.7))   # ~ e^{-lam T}
print(grad_lam(0.7, 1.0))  # ~ -T * y0 * e^{-lam T}
```

Listing 3.21 (Preview) Differentiable ODE end-point in JAX

We will build this up fully in the JAX subsection, but the message is simple: *NumPy/SciPy excel at forward solves; AD-capable frameworks excel when we need gradients through those solves.*

Key Takeaways

- Use SciPy for robust, fast forward integrations, prototyping, and comparisons against analytic baselines.
- Expect pain when you need large numbers of sensitivities: finite differences add solver calls and tuning knobs.
- Prefer AD-capable frameworks (JAX, PyTorch, Julia SciML) when gradients are first-class citizens in your workflow.

This limitation is precisely what motivated the development of modern differentiable frameworks like JAX, PyTorch, and Julia's SciML ecosystem.

3.4.2 JAX

JAX brings a NumPy-like API together with compilation and automatic differentiation. Four primitives cover most day-to-day SciML needs: `jit` (compile), `grad` (reverse-mode AD), `vmap` (vectorize without Python loops), and `pmap` (single-program, multiple-data across devices).

3.4.2.1 Core Primitives

Core Primitives

- **`jit`**: compile a pure function once and run it fast; enables kernel fusion and device offload.
- **`grad`**: take gradients of scalar outputs (reverse-mode); use `jvp`/`vjp` for JVP/VJP.
- **`vmap`**: batch a function over an axis without writing loops; preserves exact math semantics.
- **`pmap`**: run a function in parallel across multiple devices (GPU/TPU) with collectives.

We start with a small mathematical function, then progressively apply these primitives. The examples are self-contained and work on CPU; when a GPU/TPU is present, JAX offloads automatically.

```
import jax, jax.numpy as jnp

def f(x):
    # a tiny algebraic workload (pure function, no side effects)
    return jnp.sin(x) + 0.5 * x**2

f_jit = jax.jit(f)

x = jnp.linspace(-3.0, 3.0, 1_000)
y0 = f(x)          # interpreted trace + vectorized kernels
y1 = f_jit(x)      # first call: compiles; subsequent calls: fast
y2 = f_jit(x)      # reuses compiled program
print(y2.shape) # (1000,)
```

Listing 3.22 JIT compilation: first-call compile, then fast re-use

The next step is differentiation. Gradients in JAX are defined for scalar-output functions. If the function returns an array, we typically reduce (e.g., sum or mean) before calling `grad`. The result is another function that computes the gradient at any input.

```python
import jax, jax.numpy as jnp

def g(x):
    # scalar output (reduce), enabling reverse-mode
    y = jnp.sin(x) + 0.5 * x**2
    return jnp.sum(y)

grad_g = jax.grad(g)      # $\partial g/\partial x$ (same shape as x)
x = jnp.array([0.0, 1.0, 2.0])
print(grad_g(x))          # cos(x) + x
```

Listing 3.23 Reverse-mode gradients with `grad`

Vectorization eliminates Python loops and expresses "apply the same function to many items" as a single computation. With vmap, the batching dimension is explicit and composable with `grad`.

```python
import jax, jax.numpy as jnp

def loss_per_example(w, x, y):
    # simple linear model, squared error per example
    pred = jnp.dot(w, x)      # w: (d,), x: (d,)
    return 0.5 * (pred - y)**2

# Make a batched version over the sample axis (axis 0)
batched_loss = jax.vmap(loss_per_example, in_axes=(None, 0, 0))

# Per-example gradients wrt parameters w: vmap(grad(.))
grad_per_ex = jax.vmap(jax.grad(loss_per_example), in_axes=(None, 0,
    0))

key = jax.random.key(0)
d, n = 5, 1024
w = jax.random.normal(key, (d,))
X = jax.random.normal(key, (n, d))
Y = jax.random.normal(key, (n,))

L = batched_loss(w, X, Y)         # (n,)
G = grad_per_ex(w, X, Y)          # (n, d)
print(L.shape, G.shape)           # (1024,), (1024, 5)
```

Listing 3.24 Batching without loops: vmap and per-example gradients

We can combine `jit`, `grad`, and vmap freely. In practice, we JIT-compile whole training steps or solver calls, and rely on vmap to express batched simulation or per-sample loss computations.

```python
import jax, jax.numpy as jnp

def loss_per_example(w, x, y):
    return 0.5 * (jnp.dot(w, x) - y)**2

```

```python
@jax.jit
def step(w, X, Y, lr=1e-2):
    # mean gradient over the batch via vmap + mean reduce
    grad_w = jax.vmap(jax.grad(loss_per_example), in_axes=(None, 0,
    0))(w, X, Y).mean(axis=0)
    return w - lr * grad_w

# one SGD step (JIT-compiles on first call, then runs fast)
key = jax.random.key(0)
w = jax.random.normal(key, (5,))
X = jax.random.normal(key, (256, 5))
Y = jax.random.normal(key, (256,))
w_next = step(w, X, Y)
```

Listing 3.25 Composing `jit` + `vmap` + `grad`

For multi-device parallelism, `pmap` distributes a function across devices; each device processes a shard of the batch. Collectives such as `lax.psum` perform fast cross-replica reductions (e.g., gradient all-reduce). The snippet below runs even on a single device; on multi-GPU/TPU it executes in parallel.

```python
import jax, jax.numpy as jnp
from jax import lax

devices = jax.local_devices()
n_dev = len(devices)
print("devices:", n_dev)

def add_and_sum_shards(x):
    # each device adds 1 to its local shard, then sums across devices
    x = x + 1.0
    return lax.psum(x, axis_name="i")  # cross-replica sum

p_add_sum = jax.pmap(add_and_sum_shards, axis_name="i")

# replicate one value per device (shape: (n_dev,))
x0 = jnp.arange(n_dev, dtype=jnp.float32)
out = p_add_sum(x0)
print(out)  # each device sees the same global sum
```

Listing 3.26 Multi-device parallelism with pmap and collectives

Best practices with JAX primitives

- **Purity matters**: `jit`-compiled functions should be pure (no mutation, I/O, or Python-side effects).
- **Shapes are part of the program**: changing array shapes can trigger recompilation; prefer static shapes in hot paths.

- **Randomness is explicit**: use `PRNGKey` and splitting; avoid `np.random` inside JAX code.
- **Scalar outputs for grad**: differentiate scalar-output functions; reduce (e.g., mean/sum) when needed.
- **Batching first, loops last**: try `vmap` before writing Python loops; it composes with `grad` and `jit`.
- **Multi-device only when needed**: `pmap` introduces synchronization; start with single-device `jit/vmap`.

3.4.2.2 JAX Ecosystem

The JAX ecosystem is deliberately modular. Four libraries cover most SciML workflows:

JAX ecosystem

- **Equinox** - light neural network library that treats models as PyTrees of arrays; works naturally with `jit/grad/vmap`.
- **Optax** - functional optimizers and gradient transforms (Adam, SGD, clipping, weight decay).
- **Diffrax** - differentiable ODE/SDE/DDE solvers designed for JAX; supports adjoints.
- **Flax** - higher-level NN library (linen API) with modules, state, and common layers (CNNs, Transformers).

We begin with a tiny Equinox model and train it with Optax. Then we solve and differentiate an ODE with Diffrax. Finally, we show a minimal Flax example to indicate when you might prefer a higher-level API.

Equinox models are just dataclasses of parameters; we can `jit` their application and differentiate w.r.t. their PyTree fields. Optax provides the optimizer as a pure function that updates parameters.

```
import jax, jax.numpy as jnp, equinox as eqx, optax as ox

# Tiny MLP: 1 hidden layer
class MLP(eqx.Module):
    l1: eqx.nn.Linear
    l2: eqx.nn.Linear
    def __init__(self, key, in_dim=2, width=32, out_dim=1):
        k1, k2 = jax.random.split(key)
        self.l1 = eqx.nn.Linear(in_dim, width, key=k1)
        self.l2 = eqx.nn.Linear(width, out_dim, key=k2)
```

```
    def __call__(self, x):
        return self.l2(jax.nn.tanh(self.l1(x)))

key = jax.random.key(0)
model = MLP(key)

# Synthetic batch
X = jax.random.normal(key, (256, 2))
true_w = jnp.array([1.0, -2.0])
Y = (X @ true_w)[:, None] + 0.1 * jax.random.normal(key, (256, 1))

def loss_fn(m, x, y):
    pred = jax.vmap(m)(x)                  # batched forward
    return jnp.mean((pred - y)**2)

# Filter params vs. non-trainable/static fields
params, static = eqx.partition(model, eqx.is_array)

opt = ox.adam(1e-2)
opt_state = opt.init(params)

@jax.jit
def step(params, static, opt_state, x, y):
    m = eqx.combine(params, static)
    loss, grads = eqx.filter_value_and_grad(loss_fn)(m, x, y)
    updates, opt_state = opt.update(grads, opt_state, params)
    params = ox.apply_updates(params, updates)
    return params, opt_state, loss

for _ in range(200):
    params, opt_state, L = step(params, static, opt_state, X, Y)

print("Final loss:", float(L))
model = eqx.combine(params, static)       # reconstitute full model
```

Listing 3.27 Equinox model + Optax optimizer (one SGD step)

This illustrates three useful patterns: (i) models are pure callables, (ii) parameters are partitioned from static fields for clean differentiation, and (iii) the whole training step is `jit`-compiled.

Diffrax exposes ODE terms and solvers that are *natively differentiable*. We can compute sensitivities by reverse-mode AD without finite differences.

```
import jax, jax.numpy as jnp, diffrax as dfx

def rhs(t, y, lam):          # dy/dt = -lam*y
    return -lam * y

def yT(y0, lam, T=5.0):
    term = dfx.ODETerm(lambda t, y, args: rhs(t, y, args))
    sol  = dfx.diffeqsolve(term, dfx.Tsit5(),
```

```
                              t0=0.0, t1=T, dt0=1e-2,
                              y0=y0, args=lam, saveat=dfx.SaveAt(t1=True),
                              max_steps=4096)
    return sol.ys[0]              # y(T) - extract scalar value

# Gradients wrt y0 and lam (reverse-mode)
dy_dy0  = jax.grad(lambda y0: yT(y0, 0.7))(1.0)
dy_dlam = jax.grad(lambda lam: yT(1.0, lam))(0.7)

print("dy(T)/ dy0  =", float(dy_dy0),  "  expected exp(-lam T)")
print("dy(T)/ dlam =", float(dy_dlam), "  expected -T y0 exp(-lam T)")
```

Listing 3.28 Diffrax ODE solve and reverse-mode sensitivity

Because the solver is differentiable, we did not rerun it for multiple h values or tune finite-difference steps. This scales to high-dimensional parameters and can be combined with Optax for system identification.

```
key = jax.random.key(0)
t_obs = jnp.linspace(0., 5., 51)
y_true = jnp.exp(-0.7 * t_obs)
y_obs  = y_true + 0.01 * jax.random.normal(key, t_obs.shape)

def simulate(lam):
    term = dfx.ODETerm(lambda t, y, a: -a * y)
    sol  = dfx.diffeqsolve(term, dfx.Tsit5(),
                           t0=0., t1=5., dt0=1e-2,
                           y0=1.0, args=lam,
                           saveat=dfx.SaveAt(ts=t_obs))
    return sol.ys.squeeze()

def nll(lam):
    pred = simulate(lam)
    return jnp.mean((pred - y_obs)**2)

lam = jnp.array(0.3)              # bad initial guess
opt = ox.adam(1e-1)
opt_state = opt.init(lam)

@jax.jit
def step(lam, opt_state):
    val, g = jax.value_and_grad(nll)(lam)
    upd, opt_state = opt.update(g, opt_state, lam)
    lam = ox.apply_updates(lam, upd)
    return lam, opt_state, val

for _ in range(400):
    lam, opt_state, val = step(lam, opt_state)

print("Fitted lambda:", float(lam))
```

Listing 3.29 Parameter fit: learn λ from noisy observations

Flax offers a richer module system (linen API) useful for larger models (CNNs, Transformers) and stateful layers (BatchNorm). It integrates cleanly with Optax and JAX transforms.

```
import jax, jax.numpy as jnp
from flax import linen as nn

class MLP(nn.Module):
    width: int = 32
    @nn.compact
    def __call__(self, x):
        x = nn.tanh(nn.Dense(self.width)(x))
        x = nn.Dense(1)(x)
        return x

key = jax.random.key(0)
model = MLP(width=32)
x = jnp.ones((4, 2))                  # batch of 4, dim=2
params = model.init(key, x)           # random params (PyTree)
y = model.apply(params, x)            # forward pass
print(y.shape)                        # (4, 1)
```

Listing 3.30 Flax linen: minimal MLP forward + init

When to choose what ?

- Prefer **Equinox** when you want minimal ceremony, PyTree-native models, and direct control (good for research, SciML kernels).
- Prefer **Flax** when you need a batteries-included NN stack (initializers, common layers, linen patterns) or plan to scale to large models.
- Use **Optax** for all optimizers and gradient transforms (clipping, weight decay, schedules) in a functional style.
- Use **Diffrax** when your loop includes ODE/SDE solves and you need gradients through them (system ID, optimal control, neural ODEs).

Take-aways

- Keep *pure* functions inside `jit`; avoid Python side effects and mutation of arrays.
- With Equinox, split parameters from static fields via `partition/combine`; use `filter_value_and_grad`.
- With Diffrax, solver choice (`Tsit5`, `Dopri5`, `BDF`) and tolerances affect both accuracy and differentiability; keep shapes static for compilation.

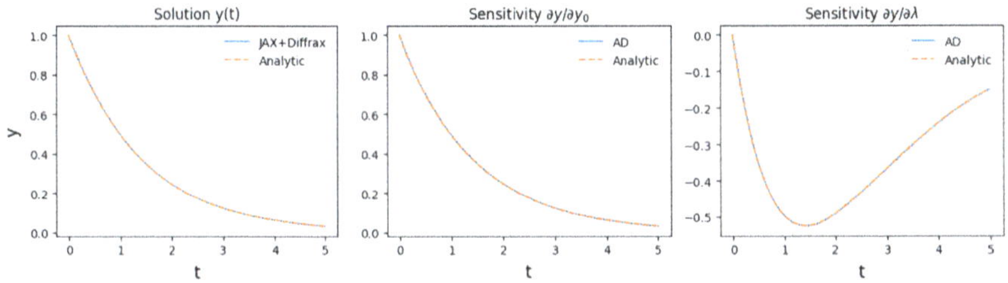

Fig. 3.4 Plots: **a** solution $y(t)$ versus analytic curve; **b** $\partial y/\partial y_0$ over time versus analytic; **c** $\partial y/\partial\lambda$ over time versus analytic

- Randomness is explicit (PRNG keys). Split keys per call; do not mix `numpy.random` with JAX code.

As an example, we consider the exponential decay model:

$$\frac{\mathrm{d}y}{\mathrm{d}t} = -\lambda\, y, \qquad y(0) = y_0,$$

whose analytic solution is $y(t) = y_0 e^{-\lambda t}$ with sensitivities

$$\frac{\partial y(t)}{\partial y_0} = e^{-\lambda t}, \qquad \frac{\partial y(t)}{\partial \lambda} = -t\, y_0\, e^{-\lambda t}.$$

The goal is to show that JAX + Diffrax can (i) solve the ODE, (ii) obtain *exact (to machine precision)* sensitivities by automatic differentiation (AD) *without* finite differences, and (iii) fit λ from noisy data using Optax.

The script below generates three plots as shown in Fig. 3.4: (a) solution $y(t)$ versus analytic curve; (b) $\partial y/\partial y_0$ over time vsersus analytic; (c) $\partial y/\partial\lambda$ over time versus analytic; and prints max errors. A small second part fits λ from synthetic data.

```
# pip install jax jaxlib diffrax optax matplotlib
import jax, jax.numpy as jnp
import diffrax as dfx
import optax as ox
import matplotlib.pyplot as plt

jax.config.update("jax_enable_x64", True)  # Numerical agreement

# --- Model and simulator
      ------------------------------------------------------
def rhs(t, y, lam):
    return -lam * y

def simulate(y0, lam, ts):
    term = dfx.ODETerm(lambda t, y, args: rhs(t, y, args))
```

```python
    sol  = dfx.diffeqsolve(term, solver=dfx.Tsit5(),
                           t0=ts[0], t1=ts[-1], dt0=1e-3,
                           y0=y0, args=lam, saveat=dfx.SaveAt(ts=ts),
                           max_steps=8192) #  max_steps
    return sol.ys.squeeze()  # shape: (len(ts),)

# --- Set parameters and grid --------------------------------------------------
y0_true = 1.0
lam_true = 0.7
ts = jnp.linspace(0.0, 5.0, 201)

# --- Forward solve and analytic reference ------------------------------------
ys_num = simulate(y0_true, lam_true, ts)
ys_an  = y0_true * jnp.exp(-lam_true * ts)

# --- Sensitivities via AD (full time-series Jacobians) ----------------------
# dy/dy0 over time:
dy_dy0_ts = jax.jacrev(lambda y0: simulate(y0, lam_true, ts))(y0_true)
# dy/dlambda over time:
dy_dlam_ts = jax.jacrev(lambda lam: simulate(y0_true, lam, ts))(lam_true)

# Analytic sensitivities:
dy_dy0_an  = jnp.exp(-lam_true * ts)
dy_dlam_an = -ts * y0_true * jnp.exp(-lam_true * ts)

# --- Error metrics ------------------------------------------------------------
sol_err  = jnp.max(jnp.abs(ys_num - ys_an))
s1_err   = jnp.max(jnp.abs(dy_dy0_ts - dy_dy0_an))
s2_err   = jnp.max(jnp.abs(dy_dlam_ts - dy_dlam_an))
print("max|y_num - y_an|           =", float(sol_err))
print("max|dy/dy0 (AD - analytic)| =", float(s1_err))
print("max|dy/dlamda  (AD - analytic)| =", float(s2_err))

# --- Plots: solution and sensitivities ---------------------------------------
fig, axs = plt.subplots(1, 3, figsize=(12, 3.4))

# (A) Solution
axs[0].plot(ts, ys_num, label="JAX+Diffrax")
axs[0].plot(ts, ys_an, "--", label="Analytic")
axs[0].set_title("Solution y(t)")
axs[0].set_xlabel("t"); axs[0].set_ylabel("y")
axs[0].legend()

# (B) dy/dy0
axs[1].plot(ts, dy_dy0_ts, label="AD")
axs[1].plot(ts, dy_dy0_an, "--", label="Analytic")
axs[1].set_title(r"Sensitivity $\partial y/\partial y_0$")
```

```
axs[1].set_xlabel("t")
axs[1].legend()

# (C) dy/dlambda
axs[2].plot(ts, dy_dlam_ts, label="AD")
axs[2].plot(ts, dy_dlam_an, "--", label="Analytic")
axs[2].set_title(r"Sensitivity $\partial y/\partial \lambda$")
axs[2].set_xlabel("t")
axs[2].legend()

plt.tight_layout(); plt.show()

# --- (Optional) Parameter fit for lambda from noisy data
    --------------------------

key = jax.random.key(0)
noise = 0.01 * jax.random.normal(key, ts.shape)
y_obs = ys_an + noise

def nll(lam):
    pred = simulate(y0_true, lam, ts)
    return jnp.mean((pred - y_obs)**2)

lam = jnp.array(0.3)            # poor initial guess
opt = ox.adam(3e-1)
opt_state = opt.init(lam)

@jax.jit
def step(lam, opt_state):
    val, g = jax.value_and_grad(nll)(lam)
    upd, opt_state = opt.update(g, opt_state, lam)
    lam = ox.apply_updates(lam, upd)
    return lam, opt_state, val

loss_hist = []
for _ in range(300):
    lam, opt_state, val = step(lam, opt_state)
    loss_hist.append(val)

print("True \lambda:", float(lam_true), "  Fitted \lambda:",
     float(lam))

plt.figure(figsize=(4.2, 3.2))
plt.plot(loss_hist); plt.xlabel("iteration"); plt.ylabel("loss")
plt.title("Fitting \lambda with AD (Optax)")
plt.tight_layout(); plt.show()
```

Listing 3.31 JAX + Diffrax: differentiable ODE solve, sensitivities, and parameter fit

The solution plot overlays the numerical integrator with the analytic curve; they match to solver tolerance. The two sensitivity plots show that AD reproduces the analytic $\partial y/\partial y_0$ and $\partial y/\partial \lambda$ across the whole time grid. The printed max errors are typically on the order of solver tolerances ($\sim 10^{-9}$–10^{-12} with `jax_enable_x64`). Finally, the parameter fit (in Fig. 3.5) demonstrates learning λ from noisy observations with a simple Adam loop, no finite differences.

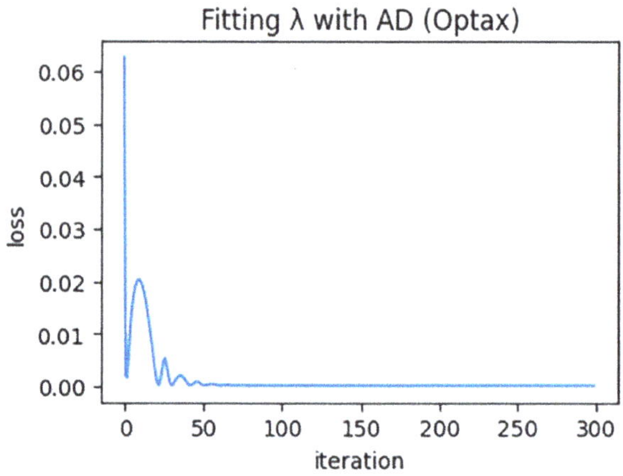

Fig. 3.5 Fitting λ with AD (Optax)

We observe that (i) AD gives time-series sensitivities directly (*one* reverse pass), avoiding extra solver calls and fragile step sizes. (ii) Sensitivities are as accurate as the forward solve (up to machine precision and solver tolerances). (iii) Differentiable solves integrate seamlessly with optimizers (Optax) for system identification and control.

3.4.3 PyTorch

PyTorch is a widely adopted ML framework with eager execution, dynamic computation graphs, and built-in reverse-mode automatic differentiation (Autograd). Since PyTorch 2.x, `torch.compile` can JIT-compile Python modules for speedups, while CUDA support enables GPU acceleration with simple device placement.

3.4.3.1 Autograd, `torch.compile`, and CUDA Basics

PyTorch basics

- **Autograd** builds a dynamic graph during the forward pass and computes gradients by calling `.backward()` on a scalar loss.
- **`torch.compile`** (PyTorch 2.x) captures and optimizes models to speed up training/inference (subject to shape specialization and graph-capture caveats).
- **CUDA** acceleration via `.to(device)` or `.cuda()`; gradients and tensors should live on the same device.

Below is a minimal example that: (i) defines an MLP, (ii) runs a forward pass, (iii) computes gradients with Autograd, and (iv) compiles the training step with `torch.compile`. If a GPU is available, it will be used automatically.

```
# pip install torch matplotlib
import torch, torch.nn as nn, torch.optim as optim
import matplotlib.pyplot as plt

device = torch.device('cuda' if torch.cuda.is_available() else 'cpu')
torch.manual_seed(0)

# Tiny MLP: R^2 -> R
class MLP(nn.Module):
    def __init__(self, in_dim=2, width=32, out_dim=1):
        super().__init__()
        self.net = nn.Sequential(
            nn.Linear(in_dim, width), nn.Tanh(),
            nn.Linear(width, out_dim)
        )
    def forward(self, x):
        return self.net(x)

model = MLP().to(device)
opt   = optim.Adam(model.parameters(), lr=1e-2)

# Synthetic batch
X = torch.randn(256, 2, device=device)
true_w = torch.tensor([1.0, -2.0], device=device)
Y = (X @ true_w).unsqueeze(-1) + 0.1*torch.randn(256,1,device=device)

def step(model, X, Y):
    model.train()
    opt.zero_grad(set_to_none=True)
    pred = model(X)
    loss = torch.mean((pred - Y)**2)
    loss.backward()           # autograd computes grads
    opt.step()
    return loss

compiled_step = torch.compile(step)      # PyTorch 2.x

loss_hist = []
for _ in range(200):
    loss = compiled_step(model, X, Y)
    loss_hist.append(loss.item())

plt.plot(loss_hist); plt.xlabel("iter"); plt.ylabel("loss");
     plt.title("Compiled training step")
plt.tight_layout(); plt.show()
```

Listing 3.32 Autograd, `torch.compile`, and CUDA device placement

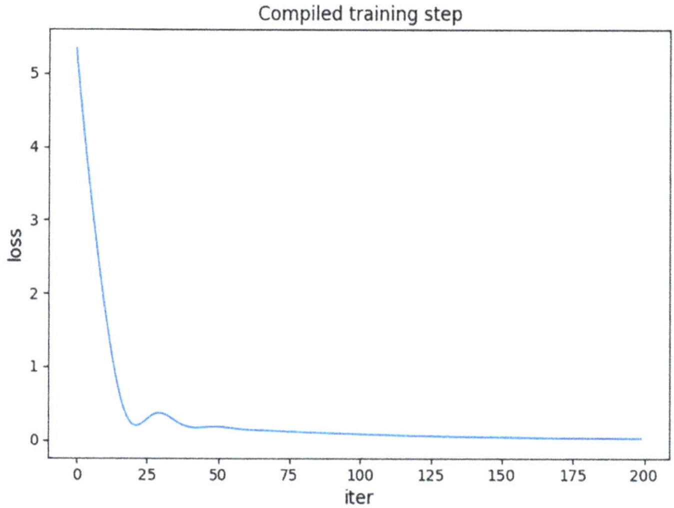

Fig. 3.6 Loss versus iteration curve for a simple MLP in PyTorch

The loss versus iteration curve is shown in Fig. 3.6. A few notes can be made here: (i) Gradients *accumulate* by default; always call `opt.zero_grad()` (or `set_to_none=True`). (ii) Use `model.train()` versus `model.eval()` to toggle layers like dropout/batchnorm. (iii) `torch.compile` may recompile if input shapes change; keep shapes stable in hot paths. (iv) Keep tensors and model on the same device; `.to(device)` is your friend. (v) For reproducibility: set `torch.manual_seed`, consider `torch.use_deterministic_algorithms (True)`.

3.4.3.2 PyTorch Ecosystem

PyTorch's ecosystem fills SciML needs with differentiable solvers, per-sample transforms, and high-level training scaffolds.

PyTorch ecosystem

- **`torchdiffeq`** : Neural ODEs and differentiable ODE solvers (adjoint method support).
- **`functorch`** : `vmap`, `jacrev`, `jvp`/`vjp`-like transforms for per-sample grads and vectorization.
- **PyTorch Lightning** : Structured training loops, logging, callbacks, multi-GPU strategies.

A quick functorch example: per-sample gradients of a simple loss without Python loops.

```
# pip install functorch  # (in PyTorch 2.x it's in torch.func)
import torch
from torch.func import vmap, grad

def loss_per_example(w, x, y):
    pred = (x @ w)            # x: (d,), w: (d,)
    return 0.5 * (pred - y)**2

w = torch.randn(5, requires_grad=True)
X = torch.randn(1024, 5)
Y = torch.randn(1024)

# grad wrt w for one sample:
g_single = grad(loss_per_example)(w, X[0], Y[0])   # (5,)

# batched (per-sample) gradients: shape -> (N, 5)
g_batch = vmap(lambda x,y: grad(loss_per_example)(w, x, y))(X, Y)
print(g_single.shape, g_batch.shape)
```

Listing 3.33 Per-sample gradients with functorch.vmap

We now fit a Neural ODE to a synthetic 2D spiral system. We generate data from a linear stable spiral $\dot{z} = Az, \quad A = \begin{bmatrix} -0.1 & -1.0 \\ 1.0 & -0.1 \end{bmatrix}$, then train an MLP-based vector field $f_\theta(t, z)$ so that the integrated trajectories match observations.

```
# pip install torchdiffeq matplotlib
import torch, torch.nn as nn, torch.optim as optim
import matplotlib.pyplot as plt
from torchdiffeq import odeint  # or odeint_adjoint for memory
    trade-offs

device = torch.device('cuda' if torch.cuda.is_available() else 'cpu')
torch.manual_seed(0)

# --- Generate synthetic data from a stable spiral
    ------------------------------
A = torch.tensor([[-0.1, -1.0],
                  [ 1.0, -0.1]], dtype=torch.float64, device=device)

def linear_rhs(t, z):
    return (A @ z.unsqueeze(-1)).squeeze(-1)

t = torch.linspace(0., 6.0, 121, dtype=torch.float64, device=device)
    # 121 time points
z0 = torch.tensor([1.0, 0.0], dtype=torch.float64, device=device)
with torch.no_grad():
    z_true = odeint(linear_rhs, z0, t, rtol=1e-9, atol=1e-12)  # (T, 2)
    z_true = z_true.to(torch.float32)  # training in float32
t = t.to(torch.float32); z0 = z0.to(torch.float32)
```

```python

# --- Neural ODE vector field f_theta(t,z) -----------------------------------
class ODEFunc(nn.Module):
    def __init__(self, width=64):
        super().__init__()
        self.net = nn.Sequential(
            nn.Linear(2+1, width), nn.Tanh(),
            nn.Linear(width, width), nn.Tanh(),
            nn.Linear(width, 2)
        )
    def forward(self, t, z):
        tz = torch.cat([z, t.expand_as(z[..., :1])], dim=-1)
        return self.net(tz)

f_theta = ODEFunc().to(device)
opt = optim.Adam(f_theta.parameters(), lr=3e-3)

# --- Loss: match full trajectory at sampled times ---------------------------
def simulate(f, z0, t):
    z_hat = odeint(f, z0, t, rtol=1e-5, atol=1e-7)  # (T, 2)
    return z_hat

def loss_fn(f, z0, t, z_true):
    z_hat = simulate(f, z0, t)
    return torch.mean((z_hat - z_true)**2)

loss_hist = []
for it in range(1500):
    opt.zero_grad(set_to_none=True)
    L = loss_fn(f_theta, z0, t, z_true)
    L.backward()
    opt.step()
    loss_hist.append(L.item())
    if (it+1) % 300 == 0:
        print(f"iter {it+1}: loss={L.item():.3e}")

# --- Plot: learned vs true trajectory ---------------------------------------
with torch.no_grad():
    z_learned = simulate(f_theta, z0, t).cpu().numpy()
z_true_np = z_true.cpu().numpy()

plt.figure(figsize=(10,4))
plt.subplot(1,2,1)
plt.plot(z_true_np[:,0], z_true_np[:,1], 'k--', label='true')
plt.plot(z_learned[:,0], z_learned[:,1], 'r', label='neural ODE')
plt.axis('equal'); plt.xlabel('z1'); plt.ylabel('z2'); plt.legend()
plt.title('Trajectory in phase space')

plt.subplot(1,2,2)
plt.plot(loss_hist)
```

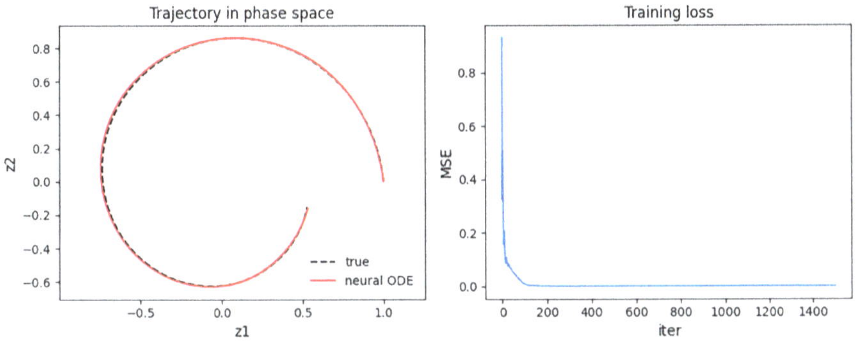

Fig. 3.7 Plot: **a** trajectory in phase space, **b** training loss

```
plt.xlabel('iter'); plt.ylabel('MSE'); plt.title('Training loss')
plt.tight_layout(); plt.show()
```

Listing 3.34 Neural ODE with `torchdiffeq`: fit a 2D spiral system

The plots are shown in Fig. 3.7. This experiment highlights PyTorch's strengths: (i) Autograd differentiates through the ODE solver seamlessly; (ii) the learned vector field reproduces the spiral trajectory; (iii) the training loop is straightforward to write and debug in eager mode. In larger projects, Lightning can structure the loop, logging, and checkpointing, while `torch.compile` can accelerate the *step*.

Best practices

- Prefer `odeint_adjoint` for long sequences to reduce memory footprint (trade-off: extra function evals).
- Keep tolerances reasonable (`rtol`, `atol`) to balance accuracy and speed; too-tight tolerances slow training.
- Stabilize training with gradient clipping (`torch.nn.utils.clip_grad_norm_`) when vector fields get stiff.
- If you use `torch.compile`, keep the input shapes stable across steps to avoid recompilation.

3.4.4 TensorFlow

TensorFlow (TF) executes computations as tensors flowing through ops. In eager mode it behaves imperatively, while `@tf.function` stages Python into a *graph* that TF can optimize and compile (optionally with XLA). TensorFlow Probability (TFP) adds distributions, bijectors, probabilistic inference (VI/MCMC), and a differentiable ODE module.

3.4.4.1 Functionalities

Common functionalities

- **Eager**: ops run immediately (easy debug).
- **`@tf.function`**: traces Python into a static TF graph; enables graph-level fusion and device portability.
- **XLA**: ahead-of-time kernel compilation; in TF 2.x use `@tf.function (jit_compile=True)` to request XLA.
- **TFP**: probability distributions, bijectors, optimizers for VI/MCMC, and differentiable ODE solvers under `tfp.math.ode`.

We show the same small function run eagerly versus compiled, plus gradients with `tf.GradientTape`. If a GPU is present, tensors are placed on it automatically when the device scope is set.

```
# pip install tensorflow
import tensorflow as tf

tf.random.set_seed(0)
device = "/GPU:0" if tf.config.list_physical_devices("GPU") else
    "/CPU:0"

def f(x):
    return tf.reduce_sum(tf.math.sin(x) + 0.5 * x**2)

@tf.function  # graph capture (no XLA)
def f_graph(x):
    return f(x)

@tf.function(jit_compile=True)  # request XLA compilation
def f_xla(x):
    return f(x)

x = tf.linspace(-3.0, 3.0, 1000)

with tf.device(device):
    # Eager value + gradient
    with tf.GradientTape() as tape:
        tape.watch(x)
        y = f(x)
    dy_dx = tape.gradient(y, x)

    # Compiled calls (first call traces/compiles; subsequent calls
     reuse)
    y_g  = f_graph(x)
    y_x  = f_xla(x)

print(y.numpy(), y_g.numpy(), y_x.numpy())  # same scalar result
```

```
print(dy_dx.shape)                          # gradient over the vector x
```

Listing 3.35 Eager vs compiled with `@tf.function` and XLA; gradients with `GradientTape`

Some notes can be made here: (a) Functions decorated with `@tf.function` should be *side-effect free* and shape-stable (changing shapes triggers retracing). (b) `jit_compile=True` requests XLA; not all graphs are XLA-compatible:fall back to plain `@tf.function` if needed. (c) TF gradients require scalar outputs; reduce losses before calling `tape.gradient`. (d) TFP integrates with TF graphs; distributions/bijectors are differentiable and run on the same device.

As an example, we infer a decay rate $\lambda > 0$ from noisy observations of the ODE

$$\frac{\mathrm{d}y}{\mathrm{d}t} = -\lambda y, \quad y(0) = y_0, \qquad y(t) = y_0 e^{-\lambda t}.$$

We: (i) solve the ODE with `tfp.math.ode`, (ii) place a simple probabilistic model over λ, (iii) run Hamiltonian Monte Carlo (HMC) with TFP to get a posterior, and (iv) plot posterior and posterior predictive trajectories.

```
# pip install tensorflow tensorflow_probability matplotlib
import tensorflow as tf
import tensorflow_probability as tfp
import matplotlib.pyplot as plt
tfd = tfp.distributions
tfb = tfp.bijectors
ode = tfp.math.ode

tf.random.set_seed(0)
dtype = tf.float64

# --- True process and noisy observations ------------------------------------
y0_true   = tf.constant(1.0, dtype)
lam_true  = tf.constant(0.7, dtype)
t_grid    = tf.linspace(tf.constant(0.0, dtype), tf.constant(5.0, dtype), 101)

def analytic_y(t, y0, lam):
    return y0 * tf.exp(-lam * t)

y_clean = analytic_y(t_grid, y0_true, lam_true)
noise_sd = tf.constant(0.02, dtype)
y_obs = y_clean + noise_sd * tf.random.normal(tf.shape(y_clean), dtype=dtype)

# --- Differentiable ODE solve with TFP ---------------------------------------
solver = ode.DormandPrince(rtol=1e-9, atol=1e-12)

@tf.function(jit_compile=True)
def simulate_y(y0, lam, ts):
```

```
    def rhs(t, y):
        return -lam * y
    sol = solver.solve(rhs,
                       initial_time=ts[0],
                       initial_state=y0,
                       solution_times=ts)
    # sol.states is a Tensor with shape (len(ts),) for this scalar ODE
    return tf.convert_to_tensor(sol.states, dtype)

# Quick check: solver vs analytic
y_num = simulate_y(y0_true, lam_true, t_grid)

# --- Probabilistic model: prior on eta (unconstrained), lam = softplus(eta) ---
# We place a N(0, 1) prior on eta; positivity enforced via lam = softplus(eta).
def target_log_prob_fn(eta):
    lam = tf.nn.softplus(eta)
    y_pred = simulate_y(y0_true, lam, t_grid)
    ll = tfd.Normal(loc=y_pred, scale=noise_sd).log_prob(y_obs)
    prior = tfd.Normal(loc=tf.constant(0.0, dtype), scale=tf.constant(1.0, dtype)).log_prob(eta)
    return tf.reduce_sum(ll) + prior

# --- HMC with step-size adaptation ------------------------------------------
num_burnin = 1000
num_samples = 1000
init_eta = tf.constant(0.0, dtype)  # softplus(0)=~0.693 close to true 0.7

hmc = tfp.mcmc.HamiltonianMonteCarlo(
    target_log_prob_fn=target_log_prob_fn,
    step_size=0.05,
    num_leapfrog_steps=8)

kernel = tfp.mcmc.SimpleStepSizeAdaptation(
    inner_kernel=hmc, num_adaptation_steps=int(0.8 * num_burnin))

@tf.function(jit_compile=True)
def run_chain():
    return tfp.mcmc.sample_chain(
        num_results=num_samples,
        num_burnin_steps=num_burnin,
        current_state=init_eta,
        kernel=kernel,
        trace_fn=lambda cs, kr: kr.inner_results.is_accepted)

eta_samples, is_acc = run_chain()
lam_samples = tf.nn.softplus(eta_samples)
acc_rate = tf.reduce_mean(tf.cast(is_acc, dtype))
print("Acceptance rate:", float(acc_rate))
```

```
# Posterior summary
lam_mean = tf.reduce_mean(lam_samples)
lam_ci = tfp.stats.percentile(lam_samples, [2.5, 97.5],
    interpolation='linear')
print("True lambda:", float(lam_true))
print("Posterior mean lambda:", float(lam_mean))
print("95% CI:", lam_ci.numpy())

# --- Posterior predictive trajectories
    ------------------------------------------

ns = 50
idx = tf.random.shuffle(tf.range(num_samples))[:ns]
lam_draws = tf.gather(lam_samples, idx)

@tf.function(jit_compile=True)
def simulate_many(lams):
    return tf.stack([simulate_y(y0_true, lam, t_grid) for lam in
    lams], axis=0)

y_draws = simulate_many(lam_draws).numpy()   # (ns, T)
y_obs_np = y_obs.numpy()
t_np = t_grid.numpy()

# --- Plots: posterior and predictive
    ------------------------------------------

plt.figure(figsize=(11,4))
plt.subplot(1,2,1)
plt.hist(lam_samples.numpy(), bins=40, density=True)
plt.axvline(float(lam_true), color='k', linestyle='--', label='true
    lambda')
plt.axvline(float(lam_mean), color='r', linestyle='-',
    label='posterior mean')
plt.title("Posterior over lambda (HMC)")
plt.legend()

plt.subplot(1,2,2)
for i in range(min(ns, 50)):
    plt.plot(t_np, y_draws[i], alpha=0.25)
plt.plot(t_np, analytic_y(t_np, y0_true, lam_true), 'k--',
    label='true')
plt.scatter(t_np, y_obs_np, s=10, color='tab:blue', label='obs')
plt.title("Posterior predictive trajectories")
plt.xlabel("t"); plt.ylabel("y(t)")
plt.legend()
plt.tight_layout(); plt.show()
```

Listing 3.36 TF + TFP: probabilistic ODE with HMC posterior over λ

The left panel of Fig. 3.8 shows the Hamiltonian Monte Carlo (HMC) posterior for the decay rate λ: the histogram shows the true value (black dashed line at $\lambda \approx 0.7$), with the posterior mean (red line). The right panel overlays posterior predictive trajectories of $y(t) = y_0 e^{-\lambda t}$ drawn from the posterior on top of the noisy observations (blue points) and the true curve (black dashed). The simulated trajectories

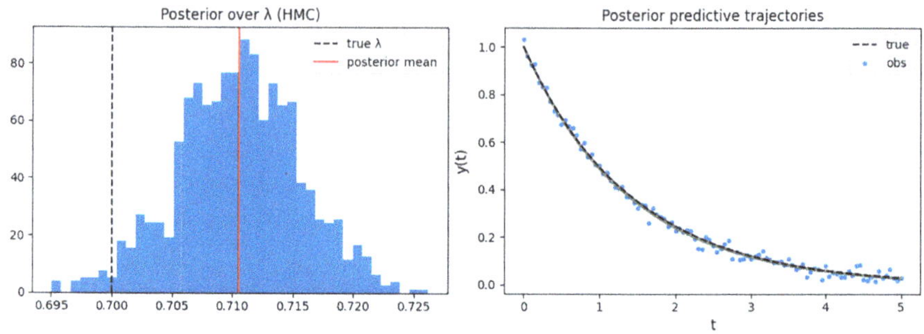

Fig. 3.8 Output of the tensorflow code 3.36

closely track both the observations and the ground truth across time, demonstrating that the inferred posterior yields accurate forward predictions with well-calibrated uncertainty.

Because the ODE solve runs *inside* `@tf.function` with XLA, HMC evaluations are compiled and vectorized where possible. Gradients of the log-probability flow through the solver via reverse-mode AD, so we avoid finite differences entirely.

Key takeaways

- Prefer `@tf.function(jit_compile=True)` for heavy inference loops; keep shapes fixed to limit retracing.
- Constrain parameters with bijectors (`Softplus`, `Exp`); putting priors on unconstrained variables simplifies MCMC (*no Jacobian terms* needed for the prior).
- Tolerances in `tfp.math.ode` (`rtol`, `atol`) control both forward accuracy and gradient quality; too-tight tolerances slow inference.
- For larger models, use `TransformedTransitionKernel` to sample constrained parameters directly or switch to VI (`tfp.vi.fit_surrogate_posterior`) for speed.

3.4.5 Julia and the SciML Ecosystem

Julia is designed for technical computing with performance close to C and the expressiveness of a high-level language. Two features are especially relevant for SciML: *multiple dispatch* (functions select method by argument types, enabling generic yet fast code) and *JIT compilation* (LLVM-based, specializing on concrete types at runtime). On top of the language, the `SciML` organization provides a unified suite for differential equations, sensitivity analysis, optimization, uncertainty quantification, and more.

3.4.5.1 Multiple Dispatch, JIT, and DifferentialEquations.jl

DifferentialEquations.jl offers a comprehensive solver stack: explicit/implicit Runge–Kutta, Rosenbrock/W-methods, BDF, symplectic, SDE/DDE/DAE solvers, event handling, callbacks, adaptive timestepping, ensemble simulations, and GPU offload [1]. Multiple dispatch lets a single user function work across problem types (ODE/SDE/DAE), scalar or array states, CPUs or GPUs, and single- or double-precision:specialized methods are chosen automatically.

```
# using Pkg; Pkg.add(["DifferentialEquations", "Plots"])
using DifferentialEquations, Plots

# dy/dt = -lambda y
function rhs!(du,u,p,t)
    lambda = p[1]
    du[1] = -lambda*u[1]
end

u0 = [1.0]; p = [0.7]; tspan = (0.0, 5.0)
prob = ODEProblem(rhs!, u0, tspan, p)

sol = solve(prob, Tsit5(); reltol=1e-9, abstol=1e-12,
     saveat=0:0.05:5.0)

plot(sol.t, sol[1,:], label="Tsit5", xlabel="t", ylabel="y(t)")
plot!(sol.t, u0[1] .* exp.(-p[1] .* sol.t), ls=:dash, label="analytic")
```

Listing 3.37 Julia: basic ODE solve with DifferentialEquations.jl. (Julia code)

Why this matters ?

- **Multiple dispatch** unifies APIs: the same rhs! works with different solvers, precisions, and devices.
- **JIT specialization** removes abstraction overhead: code is compiled to efficient machine code for the concrete types at hand.
- **Solver breadth** means you can match methods to physics (stiffness, conservation, symplecticity) without changing model code.

Julia supports several AD backends that integrate with SciMLSensitivity.jl:

- **ForwardDiff.jl** (dual-number forward-mode) : efficient when inputs are low-dimensional.
- **Zygote.jl** (source-to-source reverse-mode) : the default for many ML workloads.
- **Enzyme.jl** (LLVM-level AD) : low-level, performant, and increasingly used for differentiating compiled kernels.

`SciMLSensitivity.jl` provides adjoint methods for ODE/SDE/DAE solves and can select the best combination of adjoint (`InterpolatingAdjoint`, `QuadratureAdjoint`, `BacksolveAdjoint`) and AD backend (`ZygoteVJP`, `EnzymeVJP`, `ReverseDiffVJP`, `ForwardDiffSensitivity`). This enables stable, memory-efficient gradients through long simulations and stiff systems.

Choosing an adjoint in practice

- **InterpolatingAdjoint** : fast and memory-light for many ODEs; uses solution interpolation for backward pass.
- **QuadratureAdjoint** : robust and accurate; treats gradient as an integral (good with events/nonsmooth terms).
- **BacksolveAdjoint** : classic backsolve; can be accurate but memory-heavy for long horizons.

Note. Although we do not cover Julia code in this book, the adjoint choices above are *framework-agnostic*. The same trade-offs appear in JAX (e.g., differentiable ODE libraries), PyTorch (e.g., continuous adjoint vs. direct backprop through the solver), and TensorFlow/TFP. Think of the options as points on a spectrum trading **memory**, **time**, and **robustness to non-smooth dynamics** (events, discontinuous controls, switches).

If our loss is scalar and we have many parameters (the common case in learning and calibration), reverse-mode adjoints are preferred regardless of framework. For *short* horizons where memory is not a constraint, directly backpropagating through the solver (storing states) is simplest and usually fastest, but it costs $O(N_{\text{steps}})$ memory and can struggle with adaptive step sizes or event handling. For *long* horizons or large state dimensions, prefer an adjoint that limits memory: **InterpolatingAdjoint** stores a compact representation of the forward trajectory (checkpoints or interpolation coefficients) and recomputes as needed, providing an excellent memory–time compromise. **BacksolveAdjoint** integrates an adjoint ODE backward in time; it is memory-light but can be sensitive to solver reversibility and step reconstruction when the forward pass used adaptivity or implicit methods.

Julia is a strong choice when you need (i) extremely broad solver coverage with fine control over numerics, (ii) a single-language stack where models, solvers, AD, and optimization live together with minimal glue, or (iii) performance-critical kernels that benefit from compiler specialization (e.g., small-matrix dynamics, structured Jacobians, custom array types). It is also attractive for large parameter sweeps and uncertainty quantification thanks to ergonomic ensemble APIs and first-class parallelism. Typical bottlenecks addressed well by Julia include stiff multi-physics models that require implicit methods, event-rich systems where adjoints must remain robust, and workflows that alternate between forward simulation and gradient-based calibration or control.

Most SciML projects coexist with Python tooling (data pipelines, visualization, experiment tracking). Julia–Python bridges (e.g., `PyCall.jl`, `diffeqpy`, and `PyJulia`) allow you to call Julia solvers from Python or vice versa, so you can place Julia's solvers at the heart of a Python ML pipeline or embed Python models inside Julia optimization loops. In day-to-day practice, the biggest differences are cultural rather than conceptual: Julia emphasizes types and dispatch to express structure (which pays off in performance), while Python emphasizes libraries and composition through array APIs. Both converge on the same differentiable-programming picture: pure functions, well-posed numerics, and explicit control over AD and solver choices.

As with any compiled language, "time-to-first-plot" (initial JIT latency) can be noticeable in interactive sessions; caching strategies and precompiled system images mitigate this in production. Performance hinges on *type stability*: functions should specialize on concrete types to avoid dynamic dispatch in hot loops. Package maturity is excellent in solver stacks but varies in specialized ML subdomains; when a required component is Python-only, the interop paths above are typically sufficient.

Julia's SciML ecosystem demonstrates what a solver-centric, differentiable programming stack looks like when the language semantics are aligned with numerical computing. Even if you implement your project in JAX, PyTorch, or TensorFlow, the *ideas* that Julia makes explicit such as; problem/algorithm separation, method-of-lines composability, rigorous adjoint selection, and structure-preserving numerics; are directly portable and should guide how you structure differentiable simulations in any framework. Table 3.1 shows the comprehensive comparision of all the frameworks covered in this book.

Reading guide. NumPy/SciPy remain the dependable baseline for forward solves and numerical routines but lack built-in AD. JAX emphasizes functional purity, `jit`, and vectorization, making differentiable simulations concise and fast. PyTorch offers dynamic graphs and an extensive ML ecosystem, with differentiable ODEs via `torchdiffeq` and compilation via `torch.compile`. TensorFlow combines eager/-graph modes with `@tf.function` and integrates tightly with TensorFlow Probability (TFP) for Bayesian SciML. Julia's SciML stack is solver-centric with first-class adjoints and method breadth, benefiting from multiple dispatch and JIT specialization.

3.5 Parallelism and Performance in Practice

This section is a *cross-framework* field guide to scaling SciML workloads already given in the JAX, PyTorch, TF/TFP, and Julia subsections. It focuses on **device placement**, **memory**, and **parallel patterns** (data/model/pipeline), with concise tables and checklists one can apply regardless of framework.

Table 3.1 Feature matrix of core scientific computing frameworks

Framework	Execution model	AD (modes and implementation)	Differentiable solvers/SciML features	Hardware and parallelism	Notable ecosystem (selected)	**Typical strengths/limitations**
NumPy and SciPy	• Python eager (interpreted) • Vectorized ufuncs in C/Fortran	• No built-in AD • Finite differences or external tools	• ODE/PDE: `solve_ivp`, optimizers, sparse LA • *Not* differentiable end-to-end	• CPU-focused • GPU via CuPy (separate lib)	• SciPy (optimize, integrate, sparse) • NumPy ecosystem baseline	• + Mature, stable, broad algorithms • – No AD; limited accelerator story
JAX	• Functional, pure functions • `jit` (XLA) compilation • Eager-like semantics with tracing	• Forward & reverse AD (`grad`, `jvp/vjp`) • Source transformation + XLA	• Differentiable ODE/SDE via **Diffrax** • Neural nets as PyTrees (Equinox/Flax) • End-to-end differentiable sims	• GPU/TPU support • `vmap` (batch), `pmap` (SPMD)	• Equinox (NNs), Optax (optim) • Diffrax (ODE/SDE), Flax (NNs)	• + Clean AD, JIT, batching; great for SciML • – Purity/shape constraints; compile time
PyTorch	• Eager, dynamic graphs • `torch.compile` (2.x) for JIT	• Reverse-mode AD (Autograd) • Operator overloading + graph capture	• Differentiable ODEs via **torchdiffeq** • Ecosystem for Neural ODEs/controlled diff.	• GPU/CPU; DistributedDataParallel • functorch: `vmap`, jacobians	• torchdiffeq, functorch • Lightning (training), torchvision/torchaudio	• + Ubiquitous, easy debugging, rich ML stack • – Graph capture/JIT less automatic than JAX; shape changes can recompile
TensorFlow	• Eager or graph via `@tf.function` • Optional XLA JIT	• Reverse-mode AD (`GradientTape`) • Graph-based execution	• Differentiable ODEs via **tfp.math.ode** • Probabilistic modeling with **TFP**	• GPU/TPU support • `tf.distribute` for multi-device	• TFP (distributions, VI/MCMC) • Keras (high-level NN)	• + Production pipelines, probabilistic tools • – Graph retracing pitfalls; XLA compatibility varies
Julia	• JIT by default (LLVM) • Multiple dispatch; type specialization	• ForwardDiff (forward) • Zygote (reverse) • Enzyme (LLVM-level AD)	• **DifferentialEquations.jl**: ODE/SDE/DAE/DDE • **SciMLSensitivity**: adjoints (interpolating, quadrature, backsolve) • Structure-preserving, stiff solvers	• CPU/GPU (CUDA.jl), multi-threading • Distributed + ensembles	• SciML suite, Optimization.jl • ModelingToolkit, OrdinaryDiffEq	• + Solver breadth, performance, adjoint control • – JIT latency; ecosystem mix outside numerics

AD Automatic differentiation, *JIT* Just-In-Time, *SPMD* Single-Program Multiple-Data, *GPU* Graphics Processing Unit, *TPU* Tensor Processing Unit

3.5.1 Device Placement and Memory Management

Rule of thumb

Move data once, compute many times. Keep tensors/arrays and parameters on the same device; avoid host↔device ping-pong; prefer static shapes in hot paths so compilers can fuse and schedule aggressively.

The common pitfalls and fixes are shown in the table below:

Symptom	Likely cause	Fix
High CPU usage, low GPU util	Host→device copies per step; Python loops creating many small kernels	Preload/batch on device; vectorize (`vmap`/tensor ops); fuse with `jit`/`torch.compile`/`@tf.function`; increase batch size
Random slowdowns after first call	JIT warm-up; shape changes trigger recompilation	Keep shapes/dtypes static; stage warm-up iterations; cache compiled step
Unexpected OOM	Mixed devices for inputs/params; fragmented memory; storing full trajectories	Ensure single device; use gradient checkpointing/adjoints; clear caches between runs; lower `saveat`
DataLoader bottleneck	Synchronous IO; small workers; no pinned memory (CUDA)	Enable async prefetch; increase workers; use pinned memory; cache/TFRecords/WebDataset where applicable
NaNs at scale only	Different precision paths; reduction non-associativity; too-tight tolerances	Use consistent dtypes; enable deterministic kernels when needed; relax `rtol`/`atol` slightly; add gradient clipping

3.5.2 Data, Model, and Pipeline Parallelism

Three patterns at a glance

- **Data parallel** (SPMD): replicate parameters; shard batch; all-reduce gradients. Best default for most training loops.
- **Model parallel**: shard parameters/activations across devices. Use when the model does not fit on one device.
- **Pipeline parallel**: split layers into stages; micro-batch and stream through the pipeline. Use when depth is large and stage balance is possible.

A small informal selection guide is shown below:

Constraint	Prefer	Notes
Fits on one device	Data parallel	Simplest, scales well until global batch harms convergence
Model too large	Model parallel (tensor/parameter sharding)	Requires careful partitioning; watch communication patterns
Very deep network/long solver unroll	Pipeline parallel (+ data parallel)	Balance stage times; use micro-batching to keep devices busy
Multi-node cluster	Data parallel first	Use NCCL/XLA collectives/MPI backends; measure bandwidth

In short, for JAX:

- **`vmap`** for single-device batching: write a scalar function and lift to batch without Python loops; composes with `grad` and `jit`.
- **`pmap`** for multi-device SPMD: shard inputs across devices and synchronize with collectives (e.g., `lax.psum` for all-reduce).
- **`xmap/pjit`** for advanced sharding: map named axes to device meshes; useful for large models and operator/tensor sharding.

NOTE: Keep functions pure; set static shapes; minimize cross-replica communication inside the step; prefer one all-reduce per gradient bucket; seed via explicit key-splitting per device.

In short for PyTorch:

- **DDP** (DistributedDataParallel) is the standard: one process per GPU; gradients all-reduced via NCCL; works with `torch.compile`.
- **functorch/torch.func** offers `vmap`, per-sample grads, Jacobians without Python loops.
- **FSDP/ZeRO** for large models: shard optimizer states/parameters/gradients to fit memory.

NOTE: Use `DistributedSampler`; keep input shapes stable; tune gradient bucketing; avoid small-kernel explosions; set `find_unused_parameters=False` unless needed; ensure all params take part in backward.

In short for Julia:

- **Threads.@threads** for shared-memory loops; **Distributed** + `pmap` for process-level parallelism.
- **CUDA.jl** for GPU kernels; **MPI.jl** for multi-node collectives.
- **EnsembleProblem** in SciML: embarrassingly parallel simulation sweeps across threads/GPU workers with minimal boilerplate.

NOTE: Keep computations type-stable; avoid allocations in hot loops; pin task affinities when competing with BLAS threads; precompile kernels where possible.

A cross-framework cheat sheet now can be give as:

Framework	Single-device batching	Multi-device SPMD	Multi-node	Caveats
JAX	vmap[1]	pmap[2]/pjit[3]	XLA SPMD over hosts[4]	Static shapes; explicit PRNG; minimize collectives
PyTorch	Tensor ops/vmap	DDP/FSDP[5]	NCCL + torch.distributed[6]	Samplers, bucket sizes, determinism flags
TensorFlow	Vectorized ops[7]	tf.distribute. MirroredStrategy[8]	MultiWorkerMirrored Strategy[9]	Retracing on shape change; XLA compat varies
Julia	Broadcast/loops	CUDA.jl multi-GPU[10]; Threads.@threads[11]	MPI.jl/Distributed[12]	Type stability; kernel launch granularity

Performance checklist

1. **Vectorize** inner loops; use whole-array ops; hoist scalars.
2. **Compile** the step (`jit/torch.compile/@tf.function`); stabilize shapes.
3. **Batch** work to avoid tiny kernels; fuse pointwise ops.
4. **Profile** once on CPU and once on GPU to find the real bottleneck.
5. **Pin data path**: async prefetch; pinned memory (CUDA); pre-shard.
6. **Reduce traffic**: one all-reduce per bucket; avoid host↔device churn.
7. **Tune tolerances** (`rtol/atol`) for solver workloads; check gradient stability.
8. **Control precision**: prefer FP32/TF32 unless accuracy needs FP64; keep dtypes consistent.
9. **Seed deterministically** and record library versions, devices, and kernels.
10. **Warm up** JIT/graphs; then measure steady-state throughput and memory.

[1] https://docs.jax.dev/en/latest/_autosummary/jax.vmap.html.
[2] https://docs.jax.dev/en/latest/_autosummary/jax.pmap.html.
[3] https://docs.jax.dev/en/latest/jax.experimental.pjit.html.
[4] https://docs.jax.dev/en/latest/sharded-computation.html.
[5] https://docs.pytorch.org/tutorials/intermediate/FSDP_tutorial.html.
[6] https://docs.pytorch.org/docs/stable/distributed.html.
[7] https://www.tensorflow.org/guide/create_op.
[8] https://www.tensorflow.org/api_docs/python/tf/distribute/MirroredStrategy.
[9] https://www.tensorflow.org/api_docs/python/tf/distribute/MultiWorkerMirroredStrategy.
[10] https://cuda.juliagpu.org/stable/usage/multigpu/.
[11] https://docs.julialang.org/en/v1/manual/multi-threading/.
[12] https://docs.julialang.org/en/v1/manual/distributed-computing/.

3.6 Numerical Solvers Within Frameworks

This section is a *cross-framework* guide for picking and operating numerical solvers in SciML. We organize it into: (i) method selection for ODE/DAE tasks; (ii) linear-algebra backends and sparse operations; and (iii) patterns and caveats for *differentiable* solvers.

3.6.1 ODE/DAE Solvers

Method selection-what to use and when

Regime/requirement	Recommended method class	Notes/typical libraries
Nonstiff, smooth, moderate accuracy	Explicit RK (e.g., 4/5, 5th-order) with adaptive steps	Fast, low overhead. *Diffrax*: Tsit5/Dopri; *torchdiffeq*: `dopri5`; *TFP*: Dormand–Prince; *Julia*: Tsit5
Mild stiffness or mixed scales	Linearly implicit RK/Rosenbrock–W	Jacobian or Jacobian–vector products helpful; stable with moderate cost
Strong stiffness (reaction networks, diffusion-dominated)	BDF (backward differentiation), Radau, TR-BDF2	Implicit solves each step; pay attention to linear solver & preconditioning
Hamiltonian/long-time energy behavior	Symplectic (Verlet, Stormer–Verlet, symplectic RK)	Controls energy drift; often fixed step. Combine with structure-preserving losses
Events/switches/impacts	Any adaptive method + robust event handling	Segment the timeline at events; for gradients, prefer quadrature-style adjoints
SDEs (noise-driven)	Strong/weak SDE schemes (Euler—Maruyama, Milstein)	Decide: strong versus weak accuracy; fix RNG/seeding across runs for comparability
DAEs/constraints	Index-aware DAE integrators, projection methods	Provide consistent ICs; mass matrices; constraint stabilization may be needed

We can choose the tolerances & error control (applies across frameworks) as:

- Start with `rtol` $\in [10^{-5}, 10^{-7}]$, `atol` $\approx$ `rtol` $\cdot \text{scale}(y)$. Tighten until both solution *and* gradients stabilize.
- Monitor step counts and rejected steps; exploding rejections often indicate stiffness or poor Jacobian information.
- Prefer supplying Jacobian (or JVPs/VJPs) for stiff/implicit methods; it can dramatically reduce Newton/linear iterations.

For a quick map one can refer the Table below:

Framework	Family	Highlights
JAX	Diffrax	Explicit/implicit ODE/SDE, adjoints, JVP/VJP-friendly; integrates with `jit/vmap`
PyTorch	torchdiffeq	Neural ODEs; continuous adjoint; integrates with Autograd, works with `DDP`
TensorFlow	TFP `tfp.math.ode`	Graph- and XLA-friendly ODEs; integrates with TFP distributions/inference
Julia	DifferentialEquations.jl	Very broad solver coverage (ODE/SDE/-DAE/DDE), events, stiffness, adjoints via SciMLSensitivity
NumPy/SciPy	`integrate.solve_ivp`	Reliable forward solves; no built-in adjoints (use finite differences if needed)

3.6.2 *Linear Algebra Backends and Sparse Ops*

Modern solver performance is often dominated by linear algebra, especially for implicit methods and large systems. This subsection lists *what exists* in each ecosystem and *how to use it effectively* without getting lost in API details.

Dense linear algebra: do's and don'ts

- Prefer **factorizations** (Cholesky, QR, LU) over raw `inv`: solve $Ax = b$ via factors, not A^{-1}.
- Exploit **structure**: SPD $\Rightarrow$ Cholesky; tall/skinny $\Rightarrow$ QR; ill-conditioned $\Rightarrow$ SVD/regularization.
- Batch when possible: use batched solve/cholesky to amortize launch overhead and enable fusion.
- Mind precision: FP64 may be *much* slower on some GPUs; verify if FP32/TF32 suffices for your error budget.

The common dense ops across frameworks are show below:

Framework	Dense LA entry points	Notes
NumPy/SciPy	`numpy.linalg.*`, `scipy.linalg.*`	CPU BLAS/LAPACK (MKL/OpenBLAS); robust SVD/QR; SciPy has `solve`, `cho_factor/solve`
JAX	`jax.numpy.linalg.*`, `jax.scipy.linalg.*`	XLA-backed; `jit`-friendly; static shapes help; watch recompilation with changing matrix sizes
PyTorch	`torch.linalg.*`	GPU/CPU; batched solvers; `lstsq`, `svd`; gradients defined for many ops (check differentiability domains)
TensorFlow	`tf.linalg.*`	Graph/XLA friendly; supports batched SVD/QR/solve; use `@tf.function` for performance
Julia	`LinearAlgebra`, `SuiteSparse`	Multiple dispatch selects kernels; BLAS/LAPACK; factorization types (`Cholesky`, `QR`, `LU`)

For sparse matrices & iterative solvers:

- **Formats**: CSR/CSC for SpMV; COO for construction; BSR for block structure. Choose to match your sparsity pattern.
- **Iterative Krylov**: CG (SPD), MINRES (symmetric), GMRES (general) [2]. Stop by *relative* residual tolerance; cap iterations.
- **Preconditioning**: Jacobi/ILU/AMG or problem-aware preconditioners (physics-based). Preconditioner quality often dominates time-to-solution.
- **Batched sparse**: limited and backend-dependent; when unavailable, prefer block-structured dense batches or batched small solves.

The common sparse/iterative ops across frameworks are show below:

Framework	Sparse/iterative entry points	Notes
SciPy	`scipy.sparse`, `scipy.sparse.linalg`	CSR/CSC/COO; `cg`, `gmres`, `spsolve`; baseline reference on CPU
JAX	`jax.experimental.sparse`, `jax.scipy.sparse.linalg`	Sparse support is evolving; CG/GMRES wrappers exist; confirm performance on your backend
PyTorch	`torch.sparse`, `torch.sparse.mm`	Sparse matmul and ops; iterative solvers may require custom loops or libraries; check version capabilities
TensorFlow	`tf.sparse`	Sparse tensors and ops; iterative solvers often implemented at user level; consider XLA compatibility
Julia	`SparseArrays`, `IterativeSolvers.jl`, `AlgebraicMultigrid.jl`	Mature sparse stack; good ILU/AMG ecosystem; dispatch integrates with solvers

3.6.3 Differentiable Solvers: Patterns and Caveats

End-to-end gradients through numerical solvers enable calibration, control, and learning-in-the-loop. Two families of techniques are common: (i) **backpropaga-**

tion through the unrolled computation, and (ii) **implicit/adjoint differentiation** of the continuous/steady-state system.

Design patterns for differentiable solvers

1. **Direct backprop (unroll)**: record all intermediate states; differentiate through each step. *Pros*: simple, accurate. *Cons*: $O(N_{\text{steps}})$ memory; brittle with adaptive steps/events.
2. **Continuous adjoint (ODEs)**: solve an adjoint ODE backward; compute $\frac{d\mathcal{L}}{d\theta} = -\int \lambda^\top \frac{\partial f}{\partial \theta}\, dt$, where $\dot{\lambda} = -(\partial f/\partial y)^\top \lambda$. *Pros*: memory-light. *Cons*: sensitive to solver reversibility and events; pair with interpolation or quadrature.
3. **Implicit differentiation (steady state, fixed points)**: for $F(u, \theta) = 0$, the gradient follows from

$$\frac{d\mathcal{L}}{d\theta} = -\lambda^\top \frac{\partial F}{\partial \theta}, \quad \text{where } (\partial F/\partial u)^\top \lambda = \frac{\partial \mathcal{L}}{\partial u}.$$

   Solve the *linear* adjoint system once (often via Krylov) rather than re-unrolling iterations.
4. **Optimization-in-the-loop (argmin layers)**: if $u(\theta) = \arg\min_u \phi(u, \theta)$, apply the implicit-function form with $F = \nabla_u \phi$; re-use factorizations or preconditioners from the forward solve.

For caveats and stability checks; it is suggested to do these first:

- **Tolerance consistency**: gradients reflect solver accuracy; sweep `rtol/atol` to ensure $\nabla\mathcal{L}$ is stable.
- **Events & non-smoothness**: split trajectories at events; prefer quadrature adjoints or piecewise differentiation; add smoothing if physically acceptable.
- **Linear solves inside gradients**: iterative tolerances propagate to gradient error; use *relative* stopping and report residuals in logs.
- **Checkpointing**: for long horizons, combine checkpointing (forward) with adjoints (backward) to balance memory versus recomputation.
- **Custom VJP/JVP**: expose mathematically stable derivatives for brittle kernels (e.g., root finders, projections) via framework hooks (custom VJP/JVP/primitive rules).

Minimal diagnostic

- **Directional check**: compare AD $g^\top v$ to finite difference $[\mathcal{L}(\theta + \epsilon v) - \mathcal{L}(\theta - \epsilon v)]/(2\epsilon)$ for random unit v.

- **Convergence table**: record (steps, rejected steps, Newton/linear its) and gradient norms versus tolerances.
- **Conservation monitors**: if physics has invariants, track them in forward *and* backward passes to catch silent drift.

In Summary: Pick the *method class* to match physics (nonstiff/stiff, events, invariants), ensure the *linear algebra path* is healthy (factorizations, preconditioning, batching), and select a *gradient strategy* (direct, adjoint, implicit) that fits horizon length and smoothness. Validate with small tolerance/step sweeps before scaling up.

3.7 Conclusion

This chapter has surveyed the computational tools and paradigms that transform mathematical formulations into working SciML implementations. We have seen that programming paradigm choices have profound implications, that automatic differentiation is the enabling technology for physics-informed learning, and that framework selection involves genuine trade-offs between flexibility, performance, and ecosystem maturity. With mathematical foundations and computational tools established, the next chapter introduces the machine learning concepts that complete the SciML synthesis.

Key Takeaways:

1. **Differentiability Is the Foundation**: Physics-informed learning requires gradients through PDEs. Your framework choice determines what you can differentiate and how easily.
2. **JAX = Composable Transformations**: `grad`, `vmap`, `jit`, `pmap` compose arbitrarily. This philosophy enables elegant SciML code but requires functional thinking.
3. **PyTorch = Research Velocity**: Dynamic graphs and pythonic design accelerate experimentation. The ecosystem is vast but sometimes less rigorous for scientific computing.
4. **Julia = Native Differentiable Programming**: The SciML ecosystem offers the tightest integration between numerical solvers and machine learning, but requires learning a new language.
5. **Solver Integration Is Hard**: Differentiating through ODE/PDE solvers requires understanding discrete versus continuous adjoints, checkpointing, and solver-specific patterns.

References

1. Rackauckas, C., Nie, Q.: Differentialequations.jl—a performant and feature-rich ecosystem for solving differential equations in Julia. J. Open Res. Software **5**(1), 15 (2017)
2. Dimola, N., Coclite, A., Zunino, P.: Neural preconditioning via Krylov subspace geometry. Bollettino dell'Unione Matematica Italiana, pp. 1–23 (2025)

Chapter 4
Machine Learning Essentials for Scientists

All we have to decide is what to do with the time that is given us.

Gandalf, The Lord of the Rings

Abstract This chapter bridges the gap between scientific computing and machine learning, presenting ML concepts through the lens of scientific applications. We begin by reframing scientific questions as learning tasks: regression for function approximation, classification for phase identification, density estimation for sampling complex distributions, surrogate modeling for simulation acceleration, anomaly detection for experimental diagnostics, and inverse problems for parameter inference. The treatment of scientific data emphasizes the unique challenges of structured, spatiotemporal, and multi-scale datasets, along with appropriate normalization, splitting, and validation protocols. Representation learning covers linear methods and nonlinear embeddings, with attention to feature engineering that respects physical invariants, symmetries, and domain structure. A neural network primer develops feedforward architectures, activation functions, and training dynamics with emphasis on aspects most relevant to SciML (smoothness for PDE residuals, expressivity for function approximation). We conclude with causality and experimental design, distinguishing the correlational patterns that ML captures from the causal relationships that science seeks.

4.1 Introduction

Professor James Wright (an imaginary character) had spent thirty years studying turbulence. He knew the Navier-Stokes equations by heart. He could derive the Kolmogorov scaling laws on a napkin. His students joked that he dreamed in Reynolds numbers.

B. P. Bhuyan et al., *Scientific Machine Learning*,
https://doi.org/10.1007/978-981-95-7856-6_4

In 2018, a junior colleague showed him something that shook his foundations: a neural network that predicted turbulent flow fields faster than his carefully optimized spectral code, and with comparable accuracy.

"But it doesn't know any physics!" Wright protested. "It's just curve fitting!"

His colleague smiled. "Is it? Let me show you what happens when we train it."

Over the following weeks, Wright's skepticism slowly transformed into fascination. The neural network, he realized, was not ignoring physics; it was *learning* physics. From thousands of examples, it had discovered patterns that encoded the same conservation laws and scaling behaviors he had spent decades studying. It had learned the Kolmogorov spectrum not from equations but from data.

More striking was what the network could do that his equations couldn't. Turbulent flow in complex geometries (around buildings, through blood vessels, inside jet engines) defeated analytical methods and strained computational resources. The neural network, once trained, produced answers in milliseconds.

"I'm not abandoning physics," Wright wrote in his research notebook that December. "I'm gaining a new tool for doing physics. But I need to understand how this tool works."

This chapter is not only for Professor Wright and the many scientists like him: researchers who know their domains deeply but are new to machine learning; but also for young readers. It's also for machine learning practitioners moving into scientific applications who need to understand why scientific data and problems differ from ImageNet and chatbots.

4.1.1 Science Through the Lens of Learning

The first conceptual leap is recognizing that scientific questions can be reframed as learning tasks:

- **Regression**: Predicting continuous quantities—temperature fields, stress distributions, reaction rates—from input conditions
- **Classification**: Identifying discrete outcomes—phase transitions, stable versus unstable flows, healthy versus diseased tissue
- **Density estimation**: Learning probability distributions over physical states—molecular configurations, turbulent fluctuations, measurement uncertainties
- **Surrogate modeling**: Replacing expensive simulations with fast approximations that preserve accuracy where it matters
- **Anomaly detection**: Identifying unusual events in experimental data—instrument failures, unexpected physics, discovery opportunities
- **Inverse problems**: Inferring causes from effects—determining material properties from measurements, reconstructing sources from observations

The translation is not always straightforward. Scientific data has structure that generic ML methods may ignore or violate. Physical constraints impose requirements that standard loss functions don't capture. Extrapolation beyond training conditions—essential for science, dangerous for ML—requires special attention.

This chapter builds the bridge.

We begin with scientific questions as learning tasks, making explicit the translation from domain problems to ML formulations. We then address scientific data: its unique characteristics, appropriate preprocessing, and validation strategies that respect physical structure. Feature engineering for scientific applications follows, emphasizing representations that encode domain knowledge. A neural network primer covers the architectures and training dynamics most relevant to SciML. We conclude with causality—the distinction between the patterns ML captures and the mechanisms science seeks.

4.2 Scientific Questions as Machine Learning Tasks

4.2.1 Can Scientific Problems Be Solved Using Learning from Data?

Many scientific problems have traditionally been solved through first-principles modeling: partial differential equations (PDEs), conservation laws, or mechanistic simulations. However, advances in experimental instrumentation, large-scale simulations, and data collection have created an abundance of *data-rich scientific domains*. This raises a central question:

Guiding Question

Can scientific problems be solved using learning from data, rather than solely from first principles?

The answer is nuanced. We believe data-driven learning does not replace physical models but provides a complementary paradigm. In cases where governing equations are incomplete, too complex, or computationally expensive, machine learning can serve as:

- a **predictor** for mapping inputs to outputs directly,
- a **surrogate** for expensive simulations,
- a **detector** for rare events or anomalies,
- an **imputer** for reconstructing missing or noisy measurements,
- or an **explorer** of high-dimensional data structures.

Scientific ML therefore begins with recasting a domain-specific question into a *learning task*. For example:

- *Predict the yield of a chemical process from concentration and temperature* → regression.
- *Identify whether a material sample is crystalline or amorphous* → classification.
- *Approximate a turbulent flow solver* → surrogate modeling.

In the following subsections, we review the major categories of tasks such as regression, classification, density estimation, surrogate modeling, anomaly detection, and imputation; each explained with scientific examples, algorithms, and code.

4.2.2 Regression Problems

Regression is the fundamental task of predicting a *continuous* output variable from one or more input features. In scientific machine learning, regression problems arise naturally: forecasting temperature at a given depth, estimating pressure in a fluid system, or predicting the rate of a chemical reaction as a function of concentration and temperature. Unlike classification, which predicts discrete labels, regression focuses on approximating a real-valued functional relationship.

Definition: Regression Problem

Given a dataset $\{(x_i, y_i)\}_{i=1}^N$ where $x_i \in \mathbb{R}^d$ are input features and $y_i \in \mathbb{R}$ are continuous targets, regression seeks a function $f : \mathbb{R}^d \to \mathbb{R}$ that minimizes the expected prediction error

$$\mathcal{L}(f) = \mathbb{E}_{(x,y)}\big[\ell(f(x), y)\big],$$

where ℓ is a suitable loss function such as the mean squared error (MSE) or mean absolute error (MAE).

Two widely used regression losses are:

- **Mean Squared Error (MSE)**:

$$\text{MSE}(f) = \frac{1}{N}\sum_{i=1}^{N}\big(f(x_i) - y_i\big)^2.$$

It is sensitive to outliers, as large deviations are quadratically penalized.

- **Mean Absolute Error (MAE)**:

$$\text{MAE}(f) = \frac{1}{N}\sum_{i=1}^{N}|f(x_i) - y_i|.$$

It is more robust to outliers, but less smooth for optimization.

The simplest regression model assumes a linear relationship between inputs and outputs:

$$f(x) = w^\top x + b,$$

where $w \in \mathbb{R}^d$ are the coefficients and $b \in \mathbb{R}$ is the intercept. The optimal parameters are obtained by minimizing the error loss (shown in Algorithm 1).

Algorithm 1: Linear Regression via Ordinary Least Squares (OLS)

Input: Data matrix $X \in \mathbb{R}^{N\times d}$, target vector $y \in \mathbb{R}^N$
Output: Weight vector $w \in \mathbb{R}^d$, intercept $b \in \mathbb{R}$

$\tilde{X} \leftarrow [X\ \mathbf{1}]$ `// Augment X with ones for intercept`
$A \leftarrow \tilde{X}^\top \tilde{X}$ `// Gram matrix`
$b' \leftarrow \tilde{X}^\top y$ `// Feature-target correlations`
$\hat{w} \leftarrow A^{-1}b'$ `// Solve normal equations`
Split $\hat{w}$ into coefficients w and intercept b ;
return (w, b)

The OLS procedure works by first adding a column of ones to the data matrix so that the intercept term can be estimated alongside the coefficients. It then computes the correlations between features (A) and between features and the target (b'). Solving the linear system $A\hat{w} = b'$ yields the best-fit parameters in the least-squares sense. Finally, the solution vector is separated into slope coefficients and the intercept term. This method is exact and computationally efficient for small to moderate data sizes, though it may become unstable when features are highly correlated or when d is very large.

For large-scale problems or ill-conditioned systems, iterative methods such as *gradient descent* are preferred. Instead of solving normal equations directly, we start with an initial guess and update the parameters gradually in the direction that reduces the loss (shown in Algorithm 2).

Algorithm 2: Linear Regression via Gradient Descent

Input: Data $\{(x_i, y_i)\}_{i=1}^N$, learning rate η, iterations T
Output: Weight vector w, intercept b

Initialize $w \leftarrow 0, b \leftarrow 0$;
for $t = 1$ **to** T **do**
  Compute gradients:
  $\nabla_w \leftarrow -\frac{2}{N} \sum_{i=1}^N (y_i - (w^\top x_i + b))x_i$
  $\nabla_b \leftarrow -\frac{2}{N} \sum_{i=1}^N (y_i - (w^\top x_i + b))$
  Update parameters:
  $w \leftarrow w - \eta \nabla_w$
  $b \leftarrow b - \eta \nabla_b$
return (w, b)

Gradient descent begins with random or zero-initialized parameters and improves them incrementally. At each iteration, the algorithm computes the gradient of the mean squared error with respect to the weights and intercept, then shifts the parameters in the opposite direction of the gradient scaled by a learning rate η. Unlike OLS, which gives an exact solution in one step, gradient descent converges gradually, making it suitable for high-dimensional data, streaming scenarios, or cases where the closed-form inversion is computationally infeasible.

```python
import numpy as np
from sklearn.linear_model import LinearRegression

# Synthetic data: temperature as a linear function of depth + noise
np.random.seed(0)
X = np.linspace(0, 10, 50).reshape(-1, 1)    # depth (cm)
y = 20 + 2.5 * X.flatten() + np.random.randn(50)  # temperature (C)

# Fit linear regression model
model = LinearRegression()
model.fit(X, y)

print("Coefficient (slope):", model.coef_[0])
print("Intercept:", model.intercept_)

# Predict at new depth
x_new = np.array([[12]])
y_pred = model.predict(x_new)
print(f"Predicted temperature at depth 12 cm: {y_pred[0]:.2f} C")
```

Listing 4.1 Linear regression with scikit-learn on synthetic data

Not all scientific phenomena exhibit linear relationships. Many problems are inherently nonlinear; for instance, chemical reaction rates often follow exponential laws, and population growth may follow logistic dynamics. In such cases, one must either apply nonlinear regression models (e.g., polynomial regression, kernel regression, neural networks) or transform the data into a representation where linear models become effective.

4.2.3 *Classification Problems*

Classification is the task of predicting a *categorical label* given input features. Unlike regression, where the output is continuous, classification seeks to assign each input to one of a finite set of classes. Scientific applications of classification include identifying whether a material sample is crystalline or amorphous, predicting disease states from biomedical signals, or classifying phases of matter in condensed-matter physics.

Definition: Classification Problem

Given a dataset $\{(x_i, y_i)\}_{i=1}^N$ where $x_i \in \mathbb{R}^d$ are feature vectors and $y_i \in \{1, 2, \ldots, K\}$ are categorical labels, classification seeks a function $f : \mathbb{R}^d \to \{1, 2, \ldots, K\}$ that minimizes the misclassification risk

$$\mathcal{L}(f) = \mathbb{E}_{(x,y)}\big[\mathbf{1}\{f(x) \neq y\}\big],$$

where $\mathbf{1}\{\cdot\}$ is the indicator function. In practice, surrogate losses such as the cross-entropy are used for optimization.

Two common losses for classification are:

- **Zero-One Loss**:
$$\ell(f(x), y) = \mathbf{1}\{f(x) \neq y\},$$
which directly counts misclassifications but is not differentiable.
- **Cross-Entropy Loss**:
$$\ell(f(x), y) = -\sum_{k=1}^{K} \mathbf{1}\{y = k\} \log p_k(x),$$
where $p_k(x)$ is the predicted probability for class k. This smooth surrogate is widely used in practice.

Consider a dataset of mineral samples characterized by their spectral features. The task is to classify each sample as belonging to a known mineral type (e.g., quartz, feldspar, mica). A linear classifier can separate classes if the spectral signatures are sufficiently distinct.

Algorithm 3: Binary Logistic Regression Classifier

Input: Data $\{(x_i, y_i)\}_{i=1}^N$, with $x_i \in \mathbb{R}^d$, labels $y_i \in \{0, 1\}$, learning rate η, iterations T
Output: Weight vector $w \in \mathbb{R}^d$, intercept $b \in \mathbb{R}$

Initialize $w \leftarrow 0,\ b \leftarrow 0$;
for $t = 1$ **to** T **do**
  Compute predictions: $\hat{y}_i = \sigma(w^\top x_i + b)$ for all i // $\sigma(z) = \frac{1}{1+e^{-z}}$
  Compute gradients:
  $\nabla_w \leftarrow -\frac{1}{N}\sum_{i=1}^N (y_i - \hat{y}_i)x_i$
  $\nabla_b \leftarrow -\frac{1}{N}\sum_{i=1}^N (y_i - \hat{y}_i)$
  Update parameters:
  $w \leftarrow w - \eta\nabla_w$
  $b \leftarrow b - \eta\nabla_b$
return (w, b)

The logistic regression Algorithm 3 models the probability of the positive class as a sigmoid function of a linear combination of features. Training proceeds by minimizing the cross-entropy loss with respect to the model parameters. At each iteration, predicted probabilities are computed, the gradients are evaluated, and parameters are updated accordingly. Once trained, the classifier assigns a new sample x to class 1 if $\sigma(w^\top x + b) \geq 0.5$ and to class 0 otherwise.

```python
import numpy as np
from sklearn.linear_model import LogisticRegression

# Synthetic data: two classes based on a spectral feature
np.random.seed(1)
X_class0 = np.random.normal(loc=0.0, scale=1.0, size=(25, 2))
X_class1 = np.random.normal(loc=2.5, scale=1.0, size=(25, 2))
X = np.vstack([X_class0, X_class1])
y = np.array([0]*25 + [1]*25)

# Fit logistic regression
model = LogisticRegression()
model.fit(X, y)

print("Coefficients:", model.coef_)
print("Intercept:", model.intercept_)

# Predict on a new point
x_new = np.array([[1.0, 1.5]])
y_pred = model.predict(x_new)
print(f"Predicted class for {x_new}: {y_pred[0]}")
```

Listing 4.2 Binary classification with logistic regression on synthetic data

While logistic regression handles binary outcomes, scientific datasets often involve multiple categories. Multiclass classification can be addressed using extensions such as *softmax regression*, one-vs-rest strategies, or more flexible models like decision trees and neural networks. Furthermore, many classification boundaries are nonlinear; in such cases, kernel methods or neural architectures provide richer decision surfaces suitable for complex scientific data.

It is worth emphasizing that both regression and classification are fundamentally about *fitting equations to data*. In regression, the fitted function produces a continuous numerical prediction, while in classification, the fitted function produces a decision rule that partitions the input space into discrete regions. In both cases, machine learning seeks to approximate an underlying mapping from inputs to outputs using data, balancing accuracy, generalization, and (hopefully) interpretability.

4.2.4 Density Estimation and Generative Modeling

While regression and classification focus on predicting specific outputs, many scientific applications require modeling the entire *probability distribution* of data. This task is called *density estimation*. By learning the distribution $p(x)$ or the conditional distribution $p(y \mid x)$, one can not only make predictions but also quantify uncertainty, simulate new data, and identify rare or anomalous events.

Definition: Density Estimation

Given samples $\{x_i\}_{i=1}^N$ from an unknown distribution $p(x)$, density estimation seeks an approximation $\hat{p}(x)$ such that $\hat{p}(x) \approx p(x)$ for all relevant x. This may be parametric (e.g., Gaussian family) or non-parametric (e.g., kernel density estimation).

In molecular dynamics, researchers often wish to approximate the distribution of atomic configurations at thermal equilibrium. Similarly, in astrophysics, one may estimate the distribution of galaxy velocities or stellar masses. Density estimation allows scientists to both summarize observed data and generate synthetic samples that mimic the true system.

Its important at this point to differentiate between parametric and nonparametric methods:

- **Parametric**: Assume the data follow a distribution with fixed functional form (e.g., Gaussian with mean μ and variance σ^2). The task reduces to estimating parameters from data, usually by maximum likelihood.
- **Nonparametric**: Make minimal assumptions about the distribution. Kernel Density Estimation (KDE) is a classic method that smooths observed data points with kernel functions to approximate the density.

Algorithm 4: Kernel Density Estimation (KDE)

Input: Samples $\{x_i\}_{i=1}^N$, kernel function $K(\cdot)$, bandwidth $h > 0$
Output: Estimated density function $\hat{p}(x)$

for *each query point* x **do**
$\quad \hat{p}(x) \leftarrow \frac{1}{Nh} \sum_{i=1}^N K\left(\frac{x-x_i}{h}\right)$
return $\hat{p}(x)$

Kernel Density Estimation (Algorithm 4) works by centering a smooth kernel (often Gaussian) at each data point and averaging the contributions. The bandwidth h controls the trade-off between smoothness and fidelity: small h captures fine details but may overfit noise, while large h smooths too much and may miss structure.

```python
import numpy as np
from sklearn.neighbors import KernelDensity
import matplotlib.pyplot as plt

# Synthetic data: particle velocities from a Gaussian mixture
np.random.seed(2)
data = np.concatenate([np.random.normal(-2, 0.5, 100),
                       np.random.normal(2, 0.8, 150)])[:, np.newaxis]

# Fit KDE with Gaussian kernel
kde = KernelDensity(kernel="gaussian", bandwidth=0.5).fit(data)

# Evaluate on a grid
x_grid = np.linspace(-5, 5, 200)[:, np.newaxis]
log_density = kde.score_samples(x_grid)

plt.hist(data, bins=30, density=True, alpha=0.5, label="Histogram")
plt.plot(x_grid, np.exp(log_density), label="KDE estimate")
plt.legend(); plt.show()
```

Listing 4.3 Kernel density estimation on synthetic data

Beyond estimating densities, one can also construct *generative models* that sample new instances resembling the data. Classical approaches include Gaussian mixture models, while modern deep learning introduces powerful frameworks such as Variational Autoencoders (VAEs) [1], Generative Adversarial Networks (GANs) [2], and normalizing flows. These models are increasingly applied in scientific contexts such as generating plausible molecular structures or synthesizing climate scenarios (Fig. 4.1).

NOTE: Density estimation and generative modeling provide a bridge between purely predictive tasks and simulation. They are particularly valuable in scientific domains where obtaining real data is expensive or impossible, making synthetic data generation a practical tool for exploration and validation.

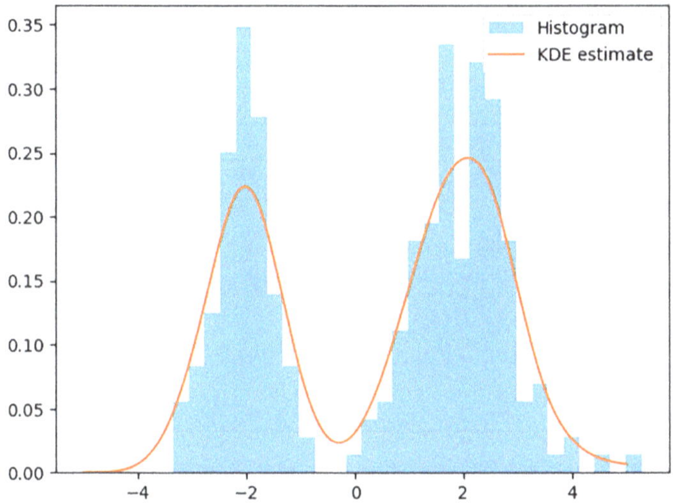

Fig. 4.1 Histogram and KDE estimate as an output of Listing 4.3

4.2.5 Surrogate Modeling and Emulation

In many areas of science and engineering, the governing models are known but extremely expensive to evaluate. Computational fluid dynamics (CFD), climate models, and molecular dynamics simulations can require hours or days for a single run. In such cases, one seeks to construct a *surrogate model* [3]: a fast statistical or machine learning approximation that emulates the input-output behavior of the underlying simulator at a fraction of the cost.

Definition: Surrogate Model

Given a computational model $F : \mathcal{X} \to \mathcal{Y}$ that is expensive to evaluate, a surrogate model $\hat{F} : \mathcal{X} \to \mathcal{Y}$ is a learned approximation trained on a limited set of evaluations $\{(x_i, F(x_i))\}_{i=1}^{N}$ such that $\hat{F}(x) \approx F(x)$ for all $x \in \mathcal{X}$.

Consider an aerodynamics simulation where $F(x)$ maps wing design parameters x to drag coefficients. Running CFD for each new wing design may take hours on a supercomputer. A Gaussian Process surrogate trained on a carefully chosen set of simulations can predict drag coefficients in milliseconds, enabling rapid exploration of design alternatives.

Surrogate modeling follows a simple workflow:

1. Select a set of training points (design of experiments, Latin hypercube, etc.).
2. Run the expensive simulator F on these points to generate training data.

3. Fit a regression model (Gaussian Process, polynomial regression, neural net) as the surrogate.
4. Use the surrogate for prediction, optimization, or uncertainty quantification.

Algorithm 5: Generic Surrogate Modeling Procedure

Input: Simulator F, input domain $\mathcal{X}$, sample size N, regression model class $\mathcal{M}$

Output: Surrogate model $\hat{F}$

Select training points $\{x_i\}_{i=1}^{N} \subset \mathcal{X}$ `// via DoE or sampling`
Evaluate simulator: $y_i \leftarrow F(x_i)$ for all i ;

Fit regression model $\hat{F} \in \mathcal{M}$ to dataset $\{(x_i, y_i)\}_{i=1}^{N}$;

return $\hat{F}$

The Algorithm 5 is deliberately abstract: the choice of sampling strategy and surrogate model class depends on the problem. Gaussian Processes are favored when N is small and uncertainty estimates are required; polynomial chaos expansions are useful for parametric uncertainty quantification; neural networks scale better to high-dimensional inputs.

```python
import numpy as np
from sklearn.gaussian_process import GaussianProcessRegressor
from sklearn.gaussian_process.kernels import RBF, WhiteKernel

# Toy expensive simulator: nonlinear function
def expensive_simulator(x):
    return np.sin(x) + 0.1 * x

# Training data (simulate expensive runs)
X_train = np.linspace(-5, 5, 12).reshape(-1, 1)
y_train = expensive_simulator(X_train)

# Fit GP surrogate
kernel = RBF(length_scale=1.0) + WhiteKernel(noise_level=1e-3)
gp = GaussianProcessRegressor(kernel=kernel)
gp.fit(X_train, y_train)

# Predict on finer grid
X_test = np.linspace(-5, 5, 200).reshape(-1, 1)
y_pred, y_std = gp.predict(X_test, return_std=True)
```

Listing 4.4 Gaussian Process surrogate modeling for an expensive simulator (toy example)

Surrogate models enable rapid prediction, sensitivity analysis, and optimization in settings where direct simulation is infeasible. They are widely used in aerospace, materials science, climate modeling, and systems biology. However, their accuracy depends critically on the coverage of training data, and extrapolation outside the training domain can be unreliable. Active learning or adaptive sampling strategies are often employed to refine surrogate accuracy in regions of interest.

4.2.6 Anomaly Detection and Outlier Identification

Anomaly detection refers to identifying patterns in data that do not conform to expected behavior. In science, such anomalies may correspond to rare but scientifically significant events (e.g., sudden stellar flares, novel chemical reactions), or to undesirable artifacts such as faulty sensor readings or experimental errors. Detecting outliers is therefore essential both for ensuring data quality and for discovering new phenomena.

Definition: Anomaly Detection

Given a dataset $\{x_i\}_{i=1}^N$ drawn from an unknown distribution $p(x)$, an anomaly detection method assigns to each x_i a score $s(x_i)$ indicating how unusual it is relative to the bulk of the data. A threshold τ is then used to classify points as *normal* ($s(x_i) \leq \tau$) or *anomalous* ($s(x_i) > \tau$).

In particle physics, collision experiments produce vast amounts of events, most of which follow known patterns. An anomaly detector can flag unusual collision signatures that may indicate new particles or physics beyond the Standard Model. In climate science, anomalies in temperature or pressure readings may signal instrument errors or emerging extreme events.

Algorithm 6 shows the Z-Score Based Outlier Detection. A simple statistical method assumes the data follow an approximately Gaussian distribution. Outliers are identified by deviations from the mean measured in units of standard deviation.

Algorithm 6: Z-Score Based Anomaly Detection

Input: Data $\{x_i\}_{i=1}^N$, threshold k (e.g., $k = 3$)
Output: Set of anomalies $\mathcal{A}$

Compute mean: $\mu \leftarrow \frac{1}{N}\sum_{i=1}^N x_i$;

Compute standard deviation: $\sigma \leftarrow \sqrt{\frac{1}{N}\sum_{i=1}^N (x_i - \mu)^2}$;

Initialize $\mathcal{A} \leftarrow \emptyset$;

for $i = 1$ **to** N **do**
  $z_i \leftarrow \frac{|x_i - \mu|}{\sigma}$;
  if $z_i > k$ **then**
    $\mathcal{A} \leftarrow \mathcal{A} \cup \{x_i\}$ `// Mark as anomaly`

return $\mathcal{A}$

This method is fast and interpretable, but it assumes Gaussianity and does not handle high-dimensional data well.

For complex or high-dimensional data, tree-based approaches such as Isolation Forest are effective (as shown in Algorithm 7). The key idea is that anomalies are easier to isolate than normal points: they tend to require fewer random splits in a decision tree. Figure 4.2 shows the plot on a synthetic data.

Algorithm 7: Isolation Forest (Conceptual)

Input: Data $\{x_i\}_{i=1}^N$, number of trees T, subsample size m
Output: Anomaly scores $\{s(x_i)\}$

for $t = 1$ **to** T **do**
  Draw random subsample of size m from the data ;
  Build a random binary tree by recursively splitting on random features and thresholds ;
  Record the path length $h(x)$ for each point x (depth at which it is isolated)
Average path lengths over all trees: $\bar{h}(x)$;
Assign anomaly score $s(x) = 2^{-\frac{\bar{h}(x)}{c(m)}}$ `//` $c(m)$ `normalizes expected depth`
return $\{s(x_i)\}$

Points with shorter average path lengths receive higher anomaly scores. This approach scales well, handles multivariate data, and requires fewer assumptions about distributional form.

```
import numpy as np
from sklearn.ensemble import IsolationForest
import matplotlib.pyplot as plt

# Synthetic data: mostly Gaussian, with a few anomalies
np.random.seed(3)
X_normal = 0.3 * np.random.randn(100, 2)
X_outliers = np.random.uniform(low=-4, high=4, size=(5, 2))
X = np.vstack([X_normal, X_outliers])

# Fit Isolation Forest
clf = IsolationForest(contamination=0.05, random_state=3)
y_pred = clf.fit_predict(X)  # -1 for anomaly, +1 for normal

# Plot results
plt.scatter(X[:,0], X[:,1], c=y_pred, cmap="coolwarm", edgecolor="k")
plt.title("Anomaly detection with Isolation Forest")
plt.show()
```

Listing 4.5 Anomaly detection using Isolation Forest on synthetic data

Anomaly detection is crucial for both scientific discovery and data reliability. While simple statistical methods like Z-scores are useful for univariate data, modern

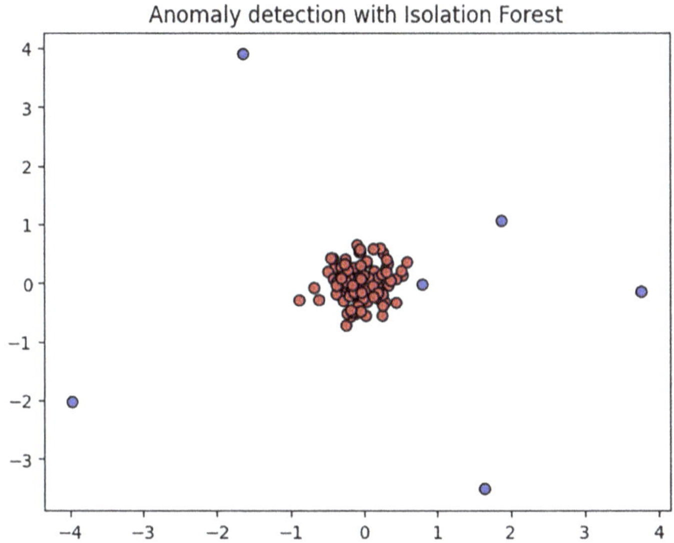

Fig. 4.2 Plot: Anomaly detection with Isolation Forest

machine learning approaches such as Isolation Forest, One-Class SVMs, or Autoencoders are better suited for high-dimensional or nonlinear datasets. The choice of method depends on the data characteristics and whether the goal is *data cleaning* or *scientific discovery of rare events*.

4.2.7 *Data Imputation and Reconstruction*

Scientific datasets are often incomplete: sensors may fail, experimental instruments may record missing values, or simulation outputs may be partially corrupted. *Data imputation* refers to filling in missing entries, while *reconstruction* focuses on recovering signals from noisy or incomplete measurements. These tasks are essential for ensuring that downstream analysis and modeling are reliable.

Definition: Data Imputation

Given a dataset with missing or corrupted values, the goal of imputation is to estimate those values in a way that is consistent with the observed data and preserves the statistical structure of the underlying process.

In climate modeling, weather stations may fail intermittently, leaving gaps in temperature or precipitation records. In biomedical imaging, MRI scans may be corrupted by motion artifacts. Data imputation and reconstruction provide complete and consistent datasets that enable accurate scientific inference.

The simplest methods replace missing values with either the mean (for numerical data) or the previous observation (for time series), as shown in Algorithm 8. While crude, these methods are fast and often used as baselines.

Algorithm 8: Mean Imputation for Numerical Data

Input: Dataset $\{x_{i,j}\}$ with missing values, column index j
Output: Completed dataset $\{\tilde{x}_{i,j}\}$

Compute column mean: $\mu_j \leftarrow \text{mean}\{x_{i,j} : x_{i,j} \text{ is observed}\}$;
for *each i* **do**
    if $x_{i,j}$ *is missing* **then**
        $\tilde{x}_{i,j} \leftarrow \mu_j$ `// Replace missing entry`
    else
        $\tilde{x}_{i,j} \leftarrow x_{i,j}$
return Completed dataset $\{\tilde{x}_{i,j}\}$

This method is simple but ignores correlations across features.

A more powerful approach models each feature with missing values as a function of the other features, filling in values iteratively, as shown in Algorithm 9.

Algorithm 9: Iterative Multivariate Imputation (MICE-style)

Input: Dataset with missing values
Output: Imputed dataset

Initialize missing values with simple imputation (e.g., column mean) ;
for *iteration* $t = 1$ *to* T **do**
    for *each feature j with missing values* **do**
        Fit regression model f_j predicting $x_{:,j}$ from other features ;
        Update missing entries in column j using predictions from f_j
return Completed dataset

This method, known as MICE (Multiple Imputation by Chained Equations), leverages correlations between variables, yielding more realistic imputations.

```python
import numpy as np
from sklearn.impute import SimpleImputer, IterativeImputer

# Dataset with missing entries
X = np.array([[1.0, 2.0, np.nan],
              [3.0, np.nan, 1.0],
              [np.nan, 4.0, 2.0],
              [5.0, 6.0, 3.0]])

# Mean imputation
mean_imputer = SimpleImputer(strategy="mean")
```

```
X_mean = mean_imputer.fit_transform(X)
print("Mean imputation:\n", X_mean)

# Iterative multivariate imputation
iter_imputer = IterativeImputer(random_state=0)
X_iter = iter_imputer.fit_transform(X)
print("Iterative imputation:\n", X_iter)
```

Listing 4.6 Data imputation with mean and iterative methods

Beyond filling in missing entries, reconstruction deals with recovering signals from incomplete or noisy measurements. Examples include image inpainting (reconstructing missing pixels), compressed sensing (recovering high-dimensional signals from few measurements), and denoising. These tasks often employ more advanced methods such as matrix completion, low-rank approximations, or neural-network-based autoencoders.

Imputation and reconstruction are indispensable tools in scientific workflows. While simple methods are easy to implement, they may distort correlations and bias downstream analysis. Iterative and model-based imputations yield more realistic reconstructions but require careful validation. Modern approaches combine statistical methods with machine learning, ensuring that missing data handling does not compromise scientific validity.

4.2.8 Inverse Problems

Many scientific questions are not about predicting outcomes from known inputs, but rather about inferring the *hidden causes* of observed effects. Such problems are called *inverse problems*. Formally, if a forward model maps parameters u to observations y via $y = F(u)$, then the inverse problem is to recover u from y. In practice, this inversion is often ill-posed: multiple parameter values may explain the same data, or small errors in y may lead to large uncertainties in u.

Definition: Inverse Problem

An inverse problem seeks to recover hidden parameters u from observed data y, given a forward mapping F such that $y \approx F(u)$. Unlike forward problems, inverse problems are often ill-posed and require regularization or additional constraints for stable solutions.

In geophysics, seismic measurements at the Earth's surface (y) are used to infer subsurface properties such as rock density and structure (u). In medical imaging, X-ray projections are inverted to reconstruct internal anatomical structures. In astronomy, telescope observations are inverted to estimate the underlying distribution of celestial objects. These are all examples of problems in the scientific domain.

A typical approach to solving inverse problems is to cast them as optimization tasks:

$$u^* = \arg\min_u \ \|F(u) - y\|^2 + \lambda R(u),$$

where the first term enforces data fidelity and the second term $R(u)$ is a regularization penalty that encodes prior knowledge (e.g., smoothness, sparsity).

Algorithm 10: Inverse Problem via Regularized Optimization

Input: Forward model F, observed data y, regularization R, parameter λ
Output: Estimated parameters u^*

Initialize guess u_0 ;
for $t = 1$ **to** T **do**
  Compute residual: $r_t \leftarrow F(u_t) - y$;
  Compute gradient: $g_t \leftarrow \nabla_u\left(\|r_t\|^2 + \lambda R(u_t)\right)$;
  Update: $u_{t+1} \leftarrow u_t - \eta g_t$ `// Gradient descent step`
return $u^* \leftarrow u_T$

The procedure in Algorithm 10; alternates between matching the observed data and enforcing regularization, balancing fidelity with stability.

Inverse problems are ubiquitous in science and engineering but are notoriously challenging due to ill-posedness. Classical solutions rely on regularization, while modern machine learning introduces data-driven priors and physics-informed neural networks (PINNs) for stabilization. A full treatment including Bayesian approaches, variational formulations, and physics-informed learning is given in Chap. 2.

4.3 Scientific Data and Preprocessing

4.3.1 Types of Scientific Data

Before applying machine learning techniques, it is important to recognize the different kinds of data encountered in scientific research. Unlike standard benchmark datasets, scientific data often arise from heterogeneous sources, with different structures, noise characteristics, and sampling protocols.

Scientific Data Types

Broadly, scientific data can be categorized as:

1. **Experimental data**: Measurements obtained from instruments, sensors, or laboratory experiments. Often subject to noise, missing values, or calibration errors.

2. **Simulation data**: Outputs generated from mechanistic models such as PDE solvers, climate models, or molecular dynamics simulations. These can be extremely accurate but computationally expensive.
3. **Multimodal data**: Data arising from multiple modalities (e.g., combining spectroscopy with microscopy, or simulation outputs with experimental measurements).

Scientific examples of each type can be:

- *Experimental*: Particle collision data from the Large Hadron Collider, gene expression measurements in biology, or seismic wave recordings.
- *Simulation*: Climate projections under different CO_2 scenarios, computational fluid dynamics models of airfoil designs, or ab initio molecular dynamics.
- *Multimodal*: Integrating MRI scans with genomic sequences in medicine, or combining satellite imagery with weather station data in environmental science.

Experimental data are typically limited in quantity and contaminated by noise, while simulation data may be abundant but biased by modeling assumptions. Multimodal integration introduces alignment issues (different spatial/temporal resolutions) and requires careful preprocessing before ML models can be applied.

Recognizing the origin and nature of data is crucial for designing ML workflows. For example, simulation datasets often lend themselves to surrogate modeling (see Sect. 4.2.5), while experimental datasets demand careful error modeling and preprocessing. Many modern scientific ML pipelines combine both sources: simulations provide large training sets, while experiments supply ground-truth validation.

4.3.2 Structured and Spatiotemporal Data

A distinguishing feature of scientific datasets is that observations are rarely independent and identically distributed (i.i.d.). Instead, they exhibit *structure*; correlations across space, time, or network topologies that must be respected for valid machine learning. Treating structured data as i.i.d. can lead to misleading conclusions and poor generalization.

Structured Scientific Data

Scientific data often have intrinsic structure:

- **Time series**: Sequential measurements (e.g., sensor data, molecular dynamics trajectories).
- **Spatial fields**: Data on regular or irregular grids (e.g., climate maps, medical images, fluid flow simulations).

- **Graphs and networks**: Data with explicit relational structure (e.g., molecular graphs, protein interaction networks, transportation systems).

We note some scientific examples of each type as:

- *Time series*: ECG recordings in cardiology, or temperature measurements from weather stations across years.
- *Spatial fields*: Ocean surface temperatures sampled on a global grid, or X-ray diffraction patterns in materials science.
- *Graphs*: Atom-level molecular structures, power grid connectivity, or ecological food webs.

The question is Why i.i.d. Fails? Standard ML algorithms assume samples are independent, but in scientific settings:

- Temporal data exhibit autocorrelation (yesterday's temperature is a predictor of today's).
- Spatial fields obey continuity and smoothness constraints.
- Graphs encode non-Euclidean structure where relationships matter more than coordinates.

Ignoring these dependencies can artificially inflate accuracy estimates or lead to invalid uncertainty quantification.

Recognizing spatiotemporal or network structure guides the choice of appropriate ML models. Sequential models (RNNs, Transformers) handle time series, convolutional architectures are suited for spatial grids, and graph neural networks (GNNs) leverage connectivity patterns. Even for classical models, data splitting protocols must respect structure—for example, ensuring that training and test sets are separated by time or space rather than randomly shuffled.

4.3.3 Normalization and Scaling

Scientific datasets often mix variables measured in very different physical units: temperature in degrees, pressure in Pascals, velocity in meters per second, and concentration in mol/L. Feeding such heterogeneous quantities directly into a machine learning model can be problematic, because algorithms that rely on distance measures or gradient updates are sensitive to the relative magnitude of features. A variable with values in the thousands can dominate another measured in fractions, even if the latter carries more predictive signal. For this reason, *normalization and scaling* are indispensable preprocessing steps.

Definitions: Normalization and Scaling

- **Standardization**: Transforming a feature x to have zero mean and unit variance:
$$x' = \frac{x - \mu}{\sigma}.$$
- **Min–Max Scaling**: Rescaling a feature x to lie in the interval $[0, 1]$:
$$x' = \frac{x - \min(x)}{\max(x) - \min(x)}.$$
- **Robust Scaling**: Using the median and interquartile range (IQR) instead of mean and variance, to reduce sensitivity to outliers:
$$x' = \frac{x - \text{median}(x)}{\text{IQR}(x)}.$$

Normalization typically refers to transforming variables so that they share a common range or statistical profile. Two of the most widely used strategies are *standardization* and *min–max scaling*. Standardization subtracts the mean and divides by the standard deviation, producing variables with zero mean and unit variance. This is particularly effective when features are approximately Gaussian and when algorithms such as linear regression or logistic regression are applied. Min–max scaling, by contrast, rescales variables to a fixed interval, often $[0, 1]$, preserving the shape of the distribution but constraining its range. This method is well-suited when inputs must lie in bounded domains, such as in neural networks with sigmoid or tanh activations.

A crucial consideration in scientific contexts is that scaling should respect the physical meaning of variables. Blindly normalizing may obscure interpretability. For instance, reporting regression coefficients on standardized scales makes them comparable but may detach them from physical units. Therefore, scientists often maintain two versions of the dataset: a scaled version for model training and an unscaled version for interpretability and validation.

Another challenge is that scaling must be performed in a way that avoids information leakage. For example, the mean and variance used in standardization should be computed only on the training set, not the full dataset, since incorporating information from the test set would bias evaluation. This distinction is particularly important in temporal or spatially correlated scientific data, where leakage may otherwise occur subtly. Proper preprocessing pipelines (e.g., using `scikit-learn` transformers) ensure that scaling is applied consistently across training, validation, and test sets.

Finally, scaling interacts with other preprocessing steps. Missing values must be imputed before scaling, as undefined entries disrupt statistics. Similarly, outliers—already discussed in Sect. 4.2.6—can heavily distort scaling factors, requiring robust approaches such as median–absolute–deviation scaling. When applied carefully, normalization and scaling transform raw scientific measurements into features that are numerically stable, interpretable, and well-suited for learning algorithms.

4.3.4 Splitting Protocols for Scientific Data

Model evaluation requires dividing available data into training, validation, and test sets. In standard machine learning, this is often done by random shuffling, assuming samples are independent and identically distributed (i.i.d.). However, in scientific contexts, this assumption frequently fails: measurements are temporally autocorrelated, spatially clustered, or grouped by experimental conditions. A naive random split can therefore leak information from the training set into the test set, producing overly optimistic performance estimates.

Splitting Protocols for Scientific Data

- **Random split**: Suitable only when samples are i.i.d.
- **Blocked (temporal) split**: Training on early times, testing on later times.
- **Spatial split**: Training on one region, testing on another.
- **Group split**: Ensuring all samples from the same experiment or subject remain together in either training or test sets.

In climate science, predicting future temperatures cannot be validated by randomly shuffling past and future data—doing so would allow the model to "peek" at the future. Similarly, in genomics, all measurements from the same patient must be kept in the same fold to prevent data leakage.

One robust strategy for structured datasets is blocked cross-validation (shown in Algorithm 11), which partitions data into contiguous blocks rather than random samples.

Algorithm 11: Blocked Cross-Validation (temporal or spatial)

Input: Ordered dataset $\{x_1, \ldots, x_N\}$, number of folds K
Output: Training/test partitions $\{(\mathcal{T}_k, \mathcal{S}_k)\}_{k=1}^{K}$

Divide dataset into K contiguous blocks of equal size ;

for $k = 1$ **to** K **do**
  Assign block k to test set: $\mathcal{S}_k$;
  Assign all other blocks to training set: $\mathcal{T}_k$;
return $\{(\mathcal{T}_k, \mathcal{S}_k)\}$

This ensures that the test data are separated in time or space from the training data, better reflecting the real-world task of forecasting or extrapolation.

The following code illustrates a temporal split, where early data are used for training and later data for testing:

```
import numpy as np
from sklearn.model_selection import TimeSeriesSplit

# Synthetic time series: 100 sequential measurements
X = np.arange(100).reshape(-1, 1)
y = np.sin(X / 10)

# Blocked cross-validation with 3 folds
tscv = TimeSeriesSplit(n_splits=3)

for train_index, test_index in tscv.split(X):
    print("Train indices:", train_index, "Test indices:", test_index)
```

Listing 4.7 Blocked temporal train/test split

Careful splitting protocols are essential in scientific machine learning. While blocked cross-validation is suitable for temporal or spatial data, other domains may require domain-specific strategies, such as leave-one-subject-out validation in medicine. The key principle is to respect the structure of the data, ensuring that test performance reflects the true generalization ability of the model.

4.4 Representations and Dimensionality Reduction

4.4.1 Linear Methods

Many scientific datasets are high-dimensional: spectra with thousands of wavelengths, gene expression arrays with tens of thousands of genes, or climate models with millions of grid points. High dimensionality not only increases computational cost but also complicates interpretation. Dimensionality reduction methods aim to find lower-dimensional representations that preserve the essential structure of the data.

Definition: Principal Component Analysis (PCA)

Principal Component Analysis (PCA) is a linear dimensionality reduction method that finds orthogonal directions (principal components) capturing the maximum variance in the data. Given a centered data matrix $X \in \mathbb{R}^{N \times d}$, PCA seeks vectors $v_1, v_2, \ldots, v_k$ such that:

$$v_j = \arg\max_{\|v\|=1} \mathrm{Var}(Xv), \quad j = 1, \ldots, k.$$

These directions correspond to the leading eigenvectors of the covariance matrix $C = \frac{1}{N} X^\top X$, or equivalently the left singular vectors in the singular value decomposition (SVD) of X.

In spectroscopy, thousands of intensity values are measured at different wavelengths. PCA can compress these measurements into a handful of principal components that capture major absorption patterns. In climate science, PCA applied to sea surface temperature fields yields *empirical orthogonal functions*, which reveal dominant spatial patterns such as El Niño.

PCA is efficiently computed using the singular value decomposition as shown in Algorithm 12.

Algorithm 12: Principal Component Analysis via SVD

Input: Data matrix $X \in \mathbb{R}^{N \times d}$, target dimension k
Output: Principal components $V_k \in \mathbb{R}^{d \times k}$, projected data $Z \in \mathbb{R}^{N \times k}$

Center data: $X \leftarrow X - \mathrm{mean}(X)$ // `Subtract column means`
;

Compute SVD: $X = U \Sigma V^\top$;

Select top k right singular vectors: V_k ;

Project data: $Z = X V_k$;

return (V_k, Z)

The right singular vectors V_k define the directions of maximum variance, while the projected data Z provide a reduced representation of the original dataset.

```python
import numpy as np
from sklearn.decomposition import PCA
import matplotlib.pyplot as plt

# Synthetic data: correlated variables
np.random.seed(42)
x = np.linspace(0, 10, 100)
y = 2 * x + np.random.normal(scale=2, size=100)
X = np.vstack([x, y]).T

# Apply PCA
pca = PCA(n_components=2)
X_pca = pca.fit_transform(X)

print("Explained variance ratio:", pca.explained_variance_ratio_)
print("Principal components:\n", pca.components_)

```

```
# Visualize projection onto first principal component
plt.scatter(X[:,0], X[:,1], alpha=0.5, label="Original data")
plt.quiver(pca.mean_[0], pca.mean_[1],
           pca.components_[0,0], pca.components_[0,1],
           scale=3, color="r", label="First PC")
plt.legend(); plt.show()
```

Listing 4.8 PCA on synthetic scientific data

PCA provides a compact, interpretable representation of data, revealing dominant modes of variation and reducing noise. However, it is limited to linear structures: if the data lie on a nonlinear manifold (e.g., curved surfaces or clusters), PCA may fail to capture the true geometry. In such cases, nonlinear methods such as kernel PCA or diffusion maps (see Sect. 4.4.2) are more appropriate.

4.4.2 Nonlinear Embeddings

While PCA captures directions of maximum variance in a linear subspace, many scientific datasets lie on inherently nonlinear manifolds. For example, protein conformations occupy curved regions of configuration space, and astrophysical spectra can cluster in nonlinear ways. In such cases, linear dimensionality reduction fails to reveal the true structure, and nonlinear embedding techniques are needed.

Definition: Nonlinear Embedding

A nonlinear embedding maps high-dimensional data points $\{x_i \in \mathbb{R}^d\}_{i=1}^N$ into a low-dimensional space $\{z_i \in \mathbb{R}^k\}_{i=1}^N$ while preserving *local or global structure* according to some similarity measure. Unlike PCA, embeddings are not restricted to linear projections.

In molecular dynamics, thousands of atomic coordinates describe protein folding. Diffusion maps can uncover a few latent variables representing collective motions such as folding angles. In astronomy, nonlinear embeddings separate galaxies into clusters with different morphologies, even when linear projections cannot distinguish them.

Some of the methods are:

- **Kernel PCA**: Extends PCA by implicitly mapping data into a high-dimensional feature space using a kernel function (e.g., Gaussian RBF), then performing PCA in that space.
- **Diffusion Maps**: Constructs a random walk on the data graph to capture connectivity and smooth variations, well-suited for dynamical systems.
- **t-SNE and UMAP**: Emphasize preserving local neighborhood structure, often used for visualization of high-dimensional biological or physical datasets.

Algorithm 13: Kernel Principal Component Analysis

Input: Data $\{x_i\}_{i=1}^N$, kernel function $k(\cdot, \cdot)$, dimension k
Output: Embedded coordinates $\{z_i \in \mathbb{R}^k\}$

Compute kernel matrix $K_{ij} = k(x_i, x_j)$;

Center kernel matrix: $\tilde{K} = K - \mathbf{1}K - K\mathbf{1} + \mathbf{1}K\mathbf{1}$;

Solve eigenproblem: $\tilde{K}v_j = \lambda_j v_j$;

Form embeddings: $z_i = (v_{1i}, v_{2i}, \ldots, v_{ki})$;

return $\{z_i\}_{i=1}^N$

The Algorithm 13 generalizes PCA by replacing inner products with kernel evaluations, allowing nonlinear structure to be represented in the embedding space.

```
import numpy as np
from sklearn.datasets import make_swiss_roll
from sklearn.decomposition import KernelPCA
import matplotlib.pyplot as plt

# Generate synthetic "Swiss roll" dataset
X, _ = make_swiss_roll(n_samples=1000, noise=0.05)

# Apply kernel PCA with RBF kernel
kpca = KernelPCA(n_components=2, kernel="rbf", gamma=0.01)
X_kpca = kpca.fit_transform(X)

# Plot embedding
plt.scatter(X_kpca[:,0], X_kpca[:,1], c=X[:,0], cmap=plt.cm.Spectral)
plt.title("Kernel PCA embedding of Swiss roll")
plt.show()
```

Listing 4.9 Kernel PCA on a nonlinear dataset (Swiss roll)

Nonlinear embeddings are powerful for uncovering low-dimensional manifolds, but they come with caveats. Unlike PCA, they often lack global interpretability: distances in the embedding space may not correspond to true distances in the original space. Furthermore, methods like t-SNE and UMAP are excellent for visualization but can distort global geometry, making them unsuitable for quantitative modeling. For scientific workflows, embeddings should be validated carefully to ensure that discovered structures correspond to physically meaningful patterns rather than artifacts of the algorithm.

4.4.3 Feature Selection and Sparsity

High-dimensional scientific datasets often contain many irrelevant or redundant features. For example, in genomics, tens of thousands of gene expression levels are measured, but only a small subset may drive a particular disease state. In spectroscopy, only certain wavelengths carry diagnostic information. Feature selection aims to identify the most informative variables, leading to simpler, more interpretable, and more robust models.

Definition: Feature Selection and Sparsity

- **Feature selection**: The process of identifying a subset of relevant variables that contribute most to predicting the target.
- **Sparsity**: A property of models that use only a small number of nonzero coefficients, effectively performing implicit feature selection.

In neuroscience, brain imaging datasets may include measurements from hundreds of regions. A sparse regression model can highlight which regions are most predictive of cognitive tasks. In chemistry, sparse models can identify which molecular descriptors determine solubility.

The Lasso (Least Absolute Shrinkage and Selection Operator) Algorithm (shown in Algorithm 14)introduces an ℓ_1 penalty to the regression loss, encouraging sparsity.

Algorithm 14: Lasso Regression for Sparse Feature Selection

Input: Data $\{(x_i, y_i)\}_{i=1}^{N}$, regularization parameter λ
Output: Weight vector w

Initialize $w \leftarrow 0$;

for $t = 1$ *to* T **do**

Compute gradient of squared loss: $\nabla \leftarrow -\frac{2}{N}\sum_i (y_i - w^\top x_i)x_i$;

Update weights with learning rate η: $w \leftarrow w - \eta\nabla$;

Apply soft-thresholding (L1 shrinkage):

$$w_j \leftarrow \text{sign}(w_j) \cdot \max\{|w_j| - \eta\lambda, 0\}, \quad \forall j$$

return w

The soft-thresholding operator drives small coefficients to exactly zero, yielding a sparse model that selects only the most relevant features.

```python
import numpy as np
from sklearn.linear_model import Lasso

# Synthetic dataset: 100 samples, 10 features
# Only 3 features truly influence the target
np.random.seed(0)
X = np.random.randn(100, 10)
```

```
true_coef = np.zeros(10)
true_coef[[1, 4, 7]] = [2.5, -1.5, 3.0]
y = X @ true_coef + np.random.randn(100) * 0.5

# Fit Lasso with L1 regularization
lasso = Lasso(alpha=0.1)
lasso.fit(X, y)

print("Estimated coefficients:", lasso.coef_)
print("Selected features:", np.nonzero(lasso.coef_)[0])
```

Listing 4.10 Lasso regression for feature selection

Sparsity-based methods provide both predictive power and interpretability, making them attractive for scientific discovery. However, feature selection must be applied cautiously: correlations among predictors can cause instability, where small changes in data yield different selected subsets. Methods such as stability selection and elastic-net regression (combining ℓ_1 and ℓ_2 penalties) mitigate these issues. Ultimately, sparsity offers a principled way to distill high-dimensional measurements into scientifically meaningful signals.

4.5 Feature Engineering for Scientific Structure

4.5.1 Basis Expansions

Many scientific relationships are nonlinear, yet simple linear models remain attractive for their interpretability and stability. A powerful way to bridge this gap is through *basis expansions*, where original features are transformed into a richer set of basis functions. Linear models applied in this expanded space can approximate complex nonlinear phenomena.

Definition: Basis Expansion

Given input $x \in \mathbb{R}^d$, a basis expansion applies a set of functions $\{\phi_j\}_{j=1}^M$ to generate new features

$$\Phi(x) = \big(\phi_1(x), \phi_2(x), \ldots, \phi_M(x)\big).$$

Common bases include polynomials, Fourier series, and wavelets, chosen according to the scientific domain.

In chemistry, reaction rates often depend polynomially on concentrations, making polynomial expansions natural. In signal processing, periodic phenomena such as molecular vibrations or climate oscillations are well captured by Fourier expansions. In fluid dynamics, wavelets are used to capture localized turbulence at multiple scales.

Algorithm: Polynomial Basis Expansion

Algorithm 15: Polynomial Basis Expansion (univariate case)

Input: Data $\{x_i\}_{i=1}^{N}$, polynomial degree p
Output: Expanded feature matrix $\Phi \in \mathbb{R}^{N\times(p+1)}$

for $i = 1$ *to* N **do**
  $\Phi[i, 0] \leftarrow 1$ `// bias term`
  for $j = 1$ *to* p **do**
    $\Phi[i, j] \leftarrow x_i^j$ `// monomial basis`
return Φ

This expansion transforms scalar inputs into polynomial features, enabling linear regression models to approximate nonlinear trends.

Python Example

```python
import numpy as np
from sklearn.preprocessing import PolynomialFeatures
import matplotlib.pyplot as plt

# Synthetic data: nonlinear relationship
np.random.seed(0)
X = np.linspace(0, 2*np.pi, 30).reshape(-1, 1)
y = np.sin(X).ravel() + 0.1*np.random.randn(30)

# Polynomial basis expansion
poly = PolynomialFeatures(degree=5)
X_poly = poly.fit_transform(X)

# Fit linear regression in expanded space
from sklearn.linear_model import LinearRegression
model = LinearRegression().fit(X_poly, y)

# Fourier expansion (manual)
def fourier_features(x, k=3):
    return np.column_stack([np.sin(j*x) for j in range(1, k+1)] +
                           [np.cos(j*x) for j in range(1, k+1)])

X_fourier = fourier_features(X, k=3)
model_fourier = LinearRegression().fit(X_fourier, y)
```

Listing 4.11 Polynomial and Fourier basis expansions

Discussion
Basis expansions are domain-driven: the choice of basis reflects prior knowledge of the system. Polynomial expansions approximate smooth trends, Fourier bases capture periodicity, and wavelets model localized phenomena at multiple scales. While expansions increase expressive power, they also inflate dimensionality, potentially causing overfitting. Regularization methods such as ridge regression or Lasso are therefore essential companions when working with rich bases.

4.5.2 *Physical Invariants and Non-Dimensionalization*

A central theme in the physical sciences is that many phenomena can be described more simply when expressed in terms of invariant or dimensionless quantities. Such representations reduce redundancy, reveal scaling laws, and often improve generalization across systems of different scales. In machine learning, embedding these principles into feature engineering can dramatically enhance model efficiency and interpretability.

Definition: Invariants and Non-Dimensionalization

- **Invariant**: A quantity that remains unchanged under certain transformations (e.g., energy conservation under time translation).
- **Dimensionless group**: A combination of physical variables with no units, often exposing fundamental similarity laws across systems.
- **Non-dimensionalization**: The process of rescaling variables using characteristic quantities (length, time, energy) to eliminate units.

Scientific Example
In fluid dynamics, the Reynolds number

$$\mathrm{Re} = \frac{\rho U L}{\mu}$$

is a dimensionless group combining density ρ, velocity U, length scale L, and viscosity μ. It governs the transition between laminar and turbulent flow, providing a unifying description across fluids of different sizes and properties. Models trained on dimensionless groups generalize better than those trained on raw dimensional inputs.

Algorithm Sketch: Buckingham Pi Theorem
The Buckingham Pi theorem formalizes the process of constructing dimensionless groups from dimensional variables.

Algorithm 16: Buckingham Pi Dimensional Analysis

Input: Set of n variables $\{q_1, \ldots, q_n\}$ with physical dimensions, fundamental dimension set $\{M, L, T, \ldots\}$
Output: Set of dimensionless groups $\{\pi_1, \pi_2, \ldots, \pi_{n-k}\}$

Construct dimensional matrix $D \in \mathbb{Z}^{k \times n}$, where each column encodes dimensions of q_i ;
Compute null space of D: find vectors α such that $D\alpha = 0$;
For each null-space vector α, form a dimensionless product:

$$\pi = \prod_{i=1}^{n} q_i^{\alpha_i}$$

Return all independent π groups

Here k is the number of fundamental dimensions (e.g., mass, length, time). The null space encodes all valid combinations of variables that yield dimensionless groups.

Discussion

Feature engineering based on invariants and dimensionless groups aligns machine learning with physical principles. Non-dimensionalization ensures that models are not biased by arbitrary unit choices, while invariants encourage models to respect symmetries such as rotation or translation. This approach reduces the effective hypothesis space, often requiring fewer samples to learn accurate relationships. However, automatic construction of invariants remains challenging, especially in complex multiphysics systems. Recent research explores integrating symbolic reasoning and machine learning to identify such structures automatically.

4.5.3 Symmetry-Aware Features

Symmetry principles are foundational in the natural sciences: laws of physics remain unchanged under translations, rotations, reflections, or permutations of identical particles. Embedding such symmetries into machine learning features is a powerful form of inductive bias. By ensuring that models respect the same invariances as the underlying scientific system, we reduce the risk of overfitting, improve generalization, and increase interpretability.

Definition: Symmetry-Aware Features

A symmetry-aware feature representation is one that remains invariant (or equivariant) under a prescribed group of transformations:

- **Invariant**: $f(Tx) = f(x)$ for transformation T (e.g., energy of a molecule is invariant to atom permutations).
- **Equivariant**: $f(Tx) = Tf(x)$, meaning the feature transforms in the same way as the input (e.g., velocity vectors rotate under spatial rotations).

Scientific Example

In molecular chemistry, the potential energy of a molecule should not depend on the absolute orientation of the molecule in space. Rotating the molecule should yield the same energy. Similarly, in astronomy, the dynamics of galaxies are invariant to translations in position. In material science, swapping identical atoms in a crystal does not change its properties. Symmetry-aware features encode these principles explicitly, ensuring models reflect the true physics.

Feature Constructions

Several strategies exist to encode symmetries:

- **Translation invariance**: Use relative positions (distances between particles) instead of absolute coordinates.
- **Rotation invariance**: Use scalar features like dot products or norms of vectors, rather than raw vector components.
- **Permutation invariance**: Represent a set by symmetric functions (sum, mean, max) over its elements.
- **Equivariance**: Preserve directional information using vector-valued features that transform consistently under symmetry operations.

Algorithm Sketch: Constructing Invariant Features for Molecules

Algorithm 17: Symmetry-Aware Feature Construction for Molecular Data

Input: Atomic positions $\{r_i \in \mathbb{R}^3\}$, atomic types $\{Z_i\}$

Output: Feature representation Φ invariant to translation, rotation, and permutation

Compute all pairwise distances: $d_{ij} = \|r_i - r_j\|$;

Apply radial basis expansion: $\phi_{ij} = [\exp(-(d_{ij} - \mu_k)^2/\sigma^2)]_{k=1}^{K}$;

Aggregate features per atom: $\Phi_i = \sum_j g(Z_j)\phi_{ij}$;

Form molecular representation: $\Phi = \sum_i h(Z_i)\Phi_i$;

return Φ

This construction uses distances (translation- and rotation-invariant) and symmetric aggregations (permutation-invariant), producing a representation suitable for predicting molecular energies.

Python Example

```
import numpy as np

# Atomic coordinates of a toy molecule (3 atoms)
R = np.array([[0.0, 0.0, 0.0],
              [1.0, 0.0, 0.0],
              [0.5, 0.8, 0.0]])

# Pairwise distances (invariant to translation/rotation)
def pairwise_distances(R):
    N = len(R)
    D = np.zeros((N, N))
    for i in range(N):
        for j in range(i+1, N):
            d = np.linalg.norm(R[i] - R[j])
            D[i, j] = D[j, i] = d
    return D

D = pairwise_distances(R)
print("Pairwise distance matrix:\n", D)

```

```
# Simple symmetric aggregation: sorted distances per atom
features = np.sort(D, axis=1)
print("Symmetry-aware features:\n", features)
```

Listing 4.12 Constructing translation- and rotation-invariant features

Discussion
Symmetry-aware features are not merely a convenience but a necessity for robust scientific modeling. They encode physical priors directly into the data representation, drastically reducing the search space of possible models. This principle underlies the success of specialized architectures such as convolutional neural networks (translation equivariance), graph neural networks (permutation invariance), and equivariant neural networks (SE(3)-equivariance for molecules and proteins). By aligning ML features with scientific symmetries, we ensure that models are both efficient and physically consistent.

4.5.4 Graph and Molecular Featurizations

Many scientific systems are most naturally represented not as fixed vectors, but as *graphs*, where nodes represent entities and edges represent interactions. Molecules, protein complexes, ecosystems, transportation networks, and power grids all fall into this category. Designing effective graph-based feature representations allows machine learning models to leverage connectivity and relational structure that would otherwise be lost in flat feature vectors.

Definition: Graph-Based Featurization

A graph $G = (V, E)$ consists of a set of nodes V and edges E. Graph featurization maps (V, E) into numerical descriptors, where:

- **Node features** encode intrinsic properties of entities (e.g., atomic number, mass, degree).
- **Edge features** encode pairwise relations (e.g., bond type, distance, interaction strength).
- **Graph-level features** summarize global structure (e.g., diameter, clustering coefficient).

Scientific Example
In molecular chemistry, atoms are represented as nodes and bonds as edges. Featurizations include node attributes (atomic number, hybridization state, partial charge) and edge attributes (bond order, aromaticity). In neuroscience, brain connectomes are graphs where nodes are regions of interest and edges represent structural or functional connectivity. In ecology, species form nodes while predation links form edges.

Algorithm Sketch: Molecular Graph Featurization

Algorithm 18: Molecular Graph Featurization

Input: Molecule with atoms $\{a_i\}$ and bonds $\{(a_i, a_j)\}$
Output: Node feature matrix X, adjacency matrix A, edge feature matrix E

for *each atom* a_i **do**
    Encode atomic properties (atomic number, valence, aromaticity) into X_i
for *each bond* (a_i, a_j) **do**
    Set $A_{ij} = 1$ if bond exists, else 0 ;
    Encode bond properties (bond type, length) into E_{ij}
Return (X, A, E)

This representation is directly usable by graph neural networks (GNNs) or can be aggregated into handcrafted descriptors.

Python Example

```
from rdkit import Chem
import networkx as nx

# Load molecule (benzene as example)
mol = Chem.MolFromSmiles("c1ccccc1")

# Convert to NetworkX graph
G = nx.Graph()
for atom in mol.GetAtoms():
    G.add_node(atom.GetIdx(), Z=atom.GetAtomicNum())
for bond in mol.GetBonds():
    i, j = bond.GetBeginAtomIdx(), bond.GetEndAtomIdx()
    G.add_edge(i, j, bond_type=bond.GetBondType())

# Inspect features
print("Nodes with atomic numbers:", G.nodes(data="Z"))
print("Edges with bond types:", G.edges(data="bond_type"))
```

Listing 4.13 Featurizing a molecule as a graph with RDKit and NetworkX

Discussion
Graph and molecular featurizations capture the relational nature of scientific data that vector-based approaches miss. They enable models to generalize across systems with varying size and topology. Beyond molecules, graph representations are used for materials discovery, epidemiological networks, power system stability, and ecological modeling. Choosing meaningful node and edge descriptors is a critical step: handcrafted features can embed domain expertise, while learned representations via GNNs provide flexibility. Both approaches highlight the importance of aligning ML features with the intrinsic structure of scientific systems.

4.6 Neural Networks: A Primer

Neural networks are among the most versatile and widely used machine learning models. They consist of interconnected layers of simple computational units (neurons), which together can approximate highly nonlinear functions. Their strength lies in their ability to automatically learn representations of data rather than relying solely on handcrafted features. For scientists, neural networks provide a flexible framework that can be adapted to many data types, from tabular measurements and time series to images, graphs, and text.

Scientific problems often involve complex, nonlinear relationships that defy simple statistical models. Neural networks offer a data-driven way to approximate such relationships with high accuracy, provided sufficient training data and computational resources are available. Beyond prediction, neural networks also serve as *surrogates* for expensive simulations, *classifiers* for categorizing experimental outcomes, and *generators* for proposing new molecules or materials.
Scientific Examples

- *Astronomy*: Neural networks classify galaxies from telescope images, distinguish stars from quasars, and help detect gravitational wave signals.
- *Biology*: Deep architectures predict protein structures (e.g., AlphaFold), identify cell types from single-cell sequencing data, and analyze microscopy images of tissues.
- *Materials Science*: Neural networks approximate potential energy surfaces, enabling molecular simulations that are orders of magnitude faster than quantum mechanical solvers.

Key Idea
At their core, neural networks learn a parametric mapping from inputs to outputs:

$$f(x; \theta) \approx y,$$

where θ denotes the collection of weights and biases. By adjusting θ to minimize a loss function on training data, the network adapts to capture the underlying relationship. With sufficient layers and neurons, neural networks can approximate any continuous function (the *universal approximation theorem*), making them especially powerful for scientific machine learning.

4.6.1 Feedforward Architecture

The most basic type of neural network is the *feedforward neural network*, also known as a *multilayer perceptron* (MLP). In this architecture, data flow strictly in one direction: from the input layer through one or more hidden layers to the output layer, with no feedback or recurrence. Each layer consists of simple computational units (neurons) that apply an affine transformation followed by a nonlinear activation function.

Definition: Feedforward Neural Network (MLP)

Given input $x \in \mathbb{R}^d$, a feedforward neural network with L layers is defined recursively as:

$$h^{(0)} = x, \quad h^{(\ell)} = \sigma\left(W^{(\ell)} h^{(\ell-1)} + b^{(\ell)}\right), \quad \ell = 1, \dots, L-1,$$

$$f(x) = W^{(L)} h^{(L-1)} + b^{(L)},$$

where $W^{(\ell)}$ are weight matrices, $b^{(\ell)}$ are bias vectors, $\sigma(\cdot)$ is a nonlinear activation function, and $f(x)$ is the final network output.

Each layer transforms its input through a linear mapping $Wh + b$, followed by a nonlinear activation. The weights W determine the strength of connections between neurons, and the biases b shift the activation thresholds. The collection of all weights and biases across layers constitutes the trainable parameters θ of the network.

The intermediate vectors $h^{(\ell)}$ are known as *hidden representations* or *latent features*. They are not specified by the user but are learned automatically to capture useful patterns in the data. This ability to discover hidden structure is one of the key reasons neural networks are so effective in scientific domains, where the underlying mechanisms may be complex or unknown.

The number of layers L defines the *depth* of the network. Even shallow networks (with a single hidden layer) are universal approximators, but in practice, deeper architectures are more parameter-efficient and capture hierarchical structures. For instance, in image analysis, early layers may detect edges, while later layers detect shapes or textures. In scientific ML, deeper layers can capture progressively higher-order interactions among variables.

From a scientific perspective, a feedforward neural network can be viewed as a functional approximation pipeline: each layer applies a basis expansion followed by a nonlinear filtering. Unlike handcrafted basis functions (Sect. 4.5.1), neural networks learn these transformations from data. This flexibility is powerful but comes at the cost of interpretability, since the learned features may not have direct physical meaning.

4.6.2 Activation Functions

Without nonlinearities, a neural network would collapse into a simple linear transformation, regardless of its depth. *Activation functions* introduce nonlinear mappings at each neuron, enabling the network to approximate arbitrarily complex functions.

The choice of activation affects both the expressive power and the trainability of the network.

Definition: Activation Function

An activation function is a nonlinear transformation $\sigma : \mathbb{R} \to \mathbb{R}$ applied elementwise to a neuron's pre-activation value $z = w^\top x + b$. The neuron's output is:

$$h = \sigma(z).$$

Its worth mentioning some of the basic activation functions popularly used. The classical sigmoid function:

$$\sigma(z) = \frac{1}{1 + e^{-z}}$$

squashes real values into $(0, 1)$, making it suitable for probabilities. It was historically popular but suffers from *vanishing gradients* for large positive or negative inputs, which hinders deep network training.

The Hyperbolic Tangent tanh function

$$\sigma(z) = \tanh(z) = \frac{e^z - e^{-z}}{e^z + e^{-z}}$$

maps values into $(-1, 1)$, centered at zero. It alleviates some of the issues of sigmoid activations but still suffers from saturation at large $|z|$.

The most widely used activation today is the Rectified Linear Unit (ReLU):

$$\sigma(z) = \max(0, z).$$

ReLU is computationally simple, mitigates vanishing gradients for positive values, and encourages sparse activations. However, neurons can become *inactive* ("dying ReLUs") if they only receive negative inputs.

Also worth mentioning that there are multiple variants of ReLU. Leaky ReLU for example, allows a small slope for negative values:

$$\sigma(z) = \begin{cases} z & z \geq 0, \\ \alpha z & z < 0, \end{cases} \quad \alpha \in (0, 1).$$

Other modern variants include *Parametric ReLU* (learnable α) and *GELU* (Gaussian Error Linear Unit), which smoothly weights inputs by their probability under a Gaussian distribution.

From a scientific perspective, activations can be interpreted as *response functions*: they dictate how a system reacts to input stimuli. A linear system responds proportionally, while nonlinear activations introduce thresholds, saturations, or graded responses that mimic biological neurons, chemical kinetics, or physical switches.

The choice of activation strongly influences the learning dynamics:

- Sigmoid/tanh are suitable for small networks or probabilistic outputs, but rarely used in deep models due to vanishing gradients.
- ReLU and its variants dominate modern practice, striking a balance between simplicity and performance.
- Smooth activations like GELU are increasingly favored in large architectures (e.g., Transformers).

In scientific ML, activations are often selected to match the physical characteristics of the system: for instance, enforcing positivity of concentrations or rates via ReLU, or modeling saturating responses via sigmoid or tanh.

4.6.3 Forward and Backpropagation

Training a neural network involves two complementary procedures: the *forward pass*, where inputs are propagated through the network to produce predictions, and the *backward pass* (backpropagation), where gradients of the loss with respect to all parameters are computed via the chain rule. Together, these form the backbone of neural network learning.

Forward and Backpropagation

- **Forward pass**: Compute outputs layer by layer:

$$h^{(\ell)} = \sigma(W^{(\ell)} h^{(\ell-1)} + b^{(\ell)}).$$

- **Backward pass**: Compute gradients of the loss $\mathcal{L}$ with respect to parameters using the chain rule:

$$\frac{\partial \mathcal{L}}{\partial W^{(\ell)}} = \delta^{(\ell)} \big(h^{(\ell-1)}\big)^{\top}, \quad \delta^{(\ell-1)} = \big(W^{(\ell)}\big)^{\top} \delta^{(\ell)} \odot \sigma'(z^{(\ell-1)}),$$

where $\delta^{(\ell)}$ are error signals and $\odot$ denotes elementwise multiplication.

The forward pass is straightforward: each layer applies an affine transformation followed by a nonlinearity. Inputs x are transformed into hidden representations $h^{(\ell)}$, until the network produces an output $\hat{y}$. For regression, $\hat{y}$ may be continuous; for classification, it may be probabilities via a softmax function.

Backpropagation computes how the loss function changes with respect to each parameter. It works by propagating *error signals* backward from the output layer to the input. At each layer, gradients are computed locally and combined via the chain rule. This reuses intermediate computations from the forward pass, making it highly efficient.

Algorithm 19 shows the forward and backpropagation in a Two-Layer MLP.

Algorithm 19: Forward and Backpropagation

Input: Data (x, y), parameters $W^{(1)}, b^{(1)}, W^{(2)}, b^{(2)}$, learning rate η

Output: Updated parameters

```
// Forward pass
```
$z^{(1)} = W^{(1)}x + b^{(1)}$;

$h^{(1)} = \sigma(z^{(1)})$;

$z^{(2)} = W^{(2)}h^{(1)} + b^{(2)}$;

$\hat{y} = z^{(2)}$ `// (for regression)`

;

```
// Compute loss
```
$L = \ell(\hat{y}, y)$;

```
// Backward pass
```
$\delta^{(2)} = \frac{\partial L}{\partial \hat{y}}$;

$\nabla W^{(2)} = \delta^{(2)}(h^{(1)})^\top, \quad \nabla b^{(2)} = \delta^{(2)}$;

$\delta^{(1)} = (W^{(2)})^\top \delta^{(2)} \odot \sigma'(z^{(1)})$;

$\nabla W^{(1)} = \delta^{(1)} x^\top, \quad \nabla b^{(1)} = \delta^{(1)}$;

```
// Parameter update
```
$W^{(\ell)} \leftarrow W^{(\ell)} - \eta \nabla W^{(\ell)}, b^{(\ell)} \leftarrow b^{(\ell)} - \eta \nabla b^{(\ell)}$, for $\ell = 1, 2$.

Naively computing derivatives for every parameter would be prohibitively expensive. Backpropagation avoids this by reusing partial derivatives through dynamic programming: each layer passes an “error signal” backward, eliminating redundant computation. This makes gradient computation scale linearly with the number of parameters, which is essential for modern deep learning.

In scientific machine learning, backpropagation is not only used for training neural networks but also for differentiating through complex simulations. When combined with automatic differentiation (AD), it allows gradients to propagate through PDE solvers, molecular dynamics engines, or optimization routines, enabling *end-to-end learning* of scientific models. This capability is the cornerstone of methods such as physics-informed neural networks (PINNs), discussed in Chap. ?.

4.6.4 Training Neural Networks

Neural networks are not predefined models; they must be trained from data. Training involves optimizing the network parameters (weights and biases) so that the predicted outputs $\hat{y}$ approximate the true targets y. This is typically formulated as minimizing a loss function using stochastic optimization algorithms.

Definition: Neural Network Training

Training a neural network seeks parameters θ that minimize the empirical loss:

$$\theta^* = \arg\min_{\theta} \frac{1}{N} \sum_{i=1}^{N} \ell(f(x_i; \theta), y_i),$$

where $\ell(\cdot, \cdot)$ is a loss function (e.g., mean squared error, cross-entropy) and $f(x; \theta)$ is the network prediction.

The loss function quantifies the discrepancy between predictions and observations.

- **Regression**: Mean Squared Error (MSE)

$$\ell(\hat{y}, y) = \frac{1}{N} \sum_{i} (\hat{y}_i - y_i)^2$$

  penalizes large deviations and provides smooth optimization.
- **Classification**: Cross-Entropy Loss

$$\ell(\hat{y}, y) = -\sum_{i=1}^{N} \sum_{k=1}^{K} \mathbf{1}\{y_i = k\} \log \hat{p}_{i,k}$$

  measures the divergence between predicted probabilities $\hat{p}_{i,k}$ and true labels.

Loss minimization is typically achieved via iterative gradient-based methods.

- **Stochastic Gradient Descent (SGD)**: Updates parameters using gradients computed on small minibatches, improving efficiency and generalization.
- **Momentum and Nesterov Accelerated Gradient**: Add memory of past updates to accelerate convergence.
- **Adam Optimizer**: Combines momentum with adaptive learning rates per parameter, widely used in practice due to robustness.

Algorithm: SGD Training Loop

Algorithm 20: Training with Stochastic Gradient Descent

Input: Training data $\{(x_i, y_i)\}$, learning rate η, batch size B, epochs E
Output: Learned parameters θ

```
Initialize parameters θ randomly ;
for e = 1 to E do
    Shuffle dataset ;
    Partition data into batches of size B ;
    for each batch (X_b, y_b) do
        Compute predictions ŷ_b = f(X_b; θ) ;
        Compute loss L = ℓ(ŷ_b, y_b) ;
        Compute gradient ∇_θ L via backpropagation ;
        Update parameters: θ ← θ − η∇_θ L ;
return θ
```

Training success depends on hyperparameters:

- **Learning rate**: Too large leads to divergence, too small leads to slow convergence.
- **Batch size**: Small batches improve generalization but increase gradient noise; large batches are more stable but may overfit.
- **Epochs and early stopping**: Training proceeds until performance on validation data stops improving, avoiding overfitting.

From a scientific viewpoint, neural network training resembles parameter estimation in nonlinear models. Just as fitting physical models often requires careful regularization or constraints, neural networks also require strategies (dropout, weight decay, early stopping) to prevent overfitting. Moreover, scientific ML often combines data-driven loss terms with physics-based constraints, leading to hybrid training objectives.

4.7 Causality and Experimental Design

4.7.1 Correlation Versus Causation

A central challenge in science is distinguishing between patterns that are merely correlated and those that reflect true causal mechanisms. Machine learning excels at detecting correlations, but without care, models may exploit spurious associations that do not generalize. Scientific reasoning requires an understanding of when statistical associations can be interpreted causally, and when experimental intervention is necessary.

Definition: Correlation versus Causation

- **Correlation**: A statistical relationship between two variables, e.g., $\mathrm{Cov}(X, Y) \neq 0$. Correlation does not imply that X causes Y.
- **Causation**: A relationship where intervening on X changes the distribution of Y, formally $P(Y \mid \mathrm{do}(X)) \neq P(Y)$.

Scientific Examples

- *Climate science*: Rising ice cream sales and heat-related hospitalizations are correlated, but ice cream does not cause hospital visits—temperature is a confounder.
- *Biomedicine*: Gene expression levels may correlate with disease, but interventions (e.g., gene knockouts) are needed to establish causal roles.
- *Astronomy*: A correlation between galaxy size and star formation rate may be confounded by redshift; only causal modeling reveals underlying physics.

Algorithm Sketch: From Correlation to Causation

Algorithm 21: Causal Discovery Framework (Conceptual)

Input: Observed data $\{(x_i, y_i)\}$, potential confounders Z
Output: Causal conclusion about $X \rightarrow Y$

1. Detect correlation: estimate $\mathrm{Cov}(X, Y)$;
2. Control for confounders: compute conditional independence $X \perp Y \mid Z$;
3. If independence holds, reject causal link $X \rightarrow Y$;
4. Otherwise, design intervention: apply $\mathrm{do}(X = x)$ and measure response in Y ;
5. Compare $P(Y|X)$ versus $P(Y|\mathrm{do}(X))$ to confirm causation ;

return causal conclusion

This procedure emphasizes that observational correlations are only a starting point. Establishing causality requires either statistical controls (adjusting for confounders) or interventional/experimental data.

Correlation-driven models may perform well within a dataset but fail under distributional shifts where spurious associations break down. Causal models, in contrast, capture mechanisms that remain stable across environments, making them more reliable for scientific inference. For machine learning in science, this means that models should not only fit data but also respect plausible causal structures, either by explicitly including known physical mechanisms or by validating predictions through designed experiments. The next subsections build on this idea by addressing domain shift, experimental design, and active learning strategies.

Traditional probability expresses associations as conditional distributions such as $P(Y \mid X = x)$, meaning "the probability of Y given that we observed $X = x$."

However, this does not imply that manipulating X will change Y. The *do-operator*, introduced by Judea Pearl, formalizes interventions:

$$P(Y \mid \text{do}(X = x)).$$

Here, $\text{do}(X = x)$ represents an intervention that sets X to x externally, disconnecting it from its usual causes. This distinguishes between observing X and actively manipulating it.

Lets take the example of Ice Cream and Hospital Visits. Consider $X =$ "ice cream sales," $Y =$ "heat-related hospital visits," and $Z =$ "temperature." Observationally, we may find

$$P(Y \mid X) \neq P(Y),$$

suggesting a correlation. But if we intervene by artificially increasing ice cream sales ($\text{do}(X = \text{high})$) while holding temperature fixed, hospital visits will not change. Thus

$$P(Y \mid \text{do}(X)) = P(Y),$$

showing that X is not a cause of Y; the correlation was due to confounding by Z.

Pearl's *do-calculus* provides rules for transforming expressions involving interventions ($P(Y \mid \text{do}(X))$) into ones involving only observable quantities ($P(Y \mid X)$, $P(X, Y, Z)$, etc.), under assumptions encoded in a causal graph. This enables scientists to estimate causal effects from data without always needing to perform new experiments.

Do-Calculus (Simplified View)

The do-calculus consists of rules that allow the replacement of interventional distributions with observational ones when certain conditional independencies hold in the causal graph. For example:

$$P(Y \mid \text{do}(X)) = \sum_z P(Y \mid X, Z = z) P(Z = z),$$

if Z blocks all backdoor paths from X to Y.

The do-operator is especially relevant in domains where interventions are expensive or impossible (astronomy, geology). In such cases, causal graphs and do-calculus provide a way to infer causal effects indirectly. In experimental sciences, the framework helps design targeted interventions that disambiguate correlation from causation, ensuring that machine learning models are grounded in real mechanisms rather than spurious patterns.

4.7.2 Confounding and Backdoor Paths

One of the greatest challenges in causal inference is dealing with *confounders*—variables that influence both the putative cause X and the effect Y. If not properly accounted for, confounders can create spurious correlations, leading machine learning models to identify patterns that are statistically strong but scientifically misleading.

Definition: Confounder and Backdoor Path

- A **confounder** is a variable Z that causally influences both X and Y, potentially creating a spurious association between them.
- A **backdoor path** is any path from X to Y in a causal graph that begins with an arrow into X. These paths represent non-causal correlations that must be blocked to estimate causal effects.

Suppose we want to estimate the effect of exercise (X) on heart disease risk (Y). Dietary habits (Z) may act as a confounder: individuals with healthier diets are both more likely to exercise and less likely to develop heart disease. Failing to adjust for Z would overestimate the causal effect of exercise.

Pearl's backdoor criterion provides a systematic way to control for confounders:

$$P(Y \mid \mathrm{do}(X)) = \sum_{z} P(Y \mid X, Z = z)\, P(Z = z),$$

if the set Z blocks all backdoor paths between X and Y. This formula expresses the causal effect as a weighted average of conditional associations, adjusted for confounders.

Algorithm Sketch: Controlling for Confounding

Algorithm 22: Estimating Causal Effect via Backdoor Adjustment

Input: Data on variables X, Y, Z
Output: Estimate of causal effect $P(Y \mid \mathrm{do}(X))$

1. Identify candidate confounders Z using domain knowledge ;
2. Check that Z blocks all backdoor paths from X to Y in the causal graph ;
3. Compute conditional distribution $P(Y \mid X, Z)$ from data ;
4. Compute $P(Z)$ from data ;
5. Estimate causal effect as:

$$P(Y \mid \mathrm{do}(X)) = \sum_{z} P(Y \mid X, Z = z)\, P(Z = z).$$

In scientific ML, confounding is pervasive: observational data are rarely collected under randomized conditions. The backdoor adjustment ensures that models estimate causal effects rather than spurious associations. However, identifying confounders requires domain expertise and assumptions about the causal graph. In high-dimensional settings, statistical methods such as propensity score matching, inverse probability weighting, and causal discovery algorithms attempt to automate confounder control, but careful scientific judgment remains essential.

4.7.3 Domain Shift and Invariance

A key difficulty in applying machine learning to scientific problems is that training and deployment conditions often differ. This phenomenon is known as *domain shift*. Models that exploit spurious correlations may perform well on training data but fail catastrophically in new environments. Causal reasoning and invariance principles provide strategies to overcome this challenge by focusing on mechanisms that remain stable across domains.

Definition: Domain Shift and Invariance

- **Domain shift**: A situation where the joint distribution $P(X, Y)$ during deployment differs from the one during training.
- **Invariant predictor**: A function f that maintains predictive validity across multiple environments by capturing causal relationships rather than spurious correlations.

Scientific Example
In biology, gene expression profiles collected in one laboratory may not directly generalize to another due to differences in measurement protocols. In climate science, models trained on historical data may fail when greenhouse gas concentrations exceed past ranges. In materials science, a model trained on simulations may not transfer directly to experimental data, as the simulation omits real-world noise.
Invariant Causal Prediction (ICP)
One approach to handle domain shift is to identify subsets of features whose predictive relationship to the target is invariant across environments. If Y is the effect of X, then

$$P(Y \mid \text{do}(X)) \quad \text{remains invariant across domains,}$$

even if $P(X)$ changes. Learning models that capture this causal relationship ensures robustness under distribution shifts.

Algorithm Sketch: Invariant Feature Selection

Algorithm 23: Invariant Causal Prediction (simplified)

Input: Datasets from multiple environments $\{\mathcal{D}_e\}$
Output: Invariant feature set S

Initialize $S \leftarrow \emptyset$;
for *each candidate subset of features* F **do**
  Fit predictor f_F on pooled data ;
  Test whether residuals are independent of environment e ;
  If independence holds, mark F as invariant ;
Return the largest invariant set S

This algorithm identifies features whose effect on the outcome is stable across environments, isolating causal predictors.

Discussion
Domain shift is the rule rather than the exception in science. Robust generalization requires identifying invariances—properties that remain consistent across environments. These may arise from physical laws, conservation principles, or biological mechanisms. By focusing on invariance rather than correlation, scientific ML models can extrapolate more reliably, bridging gaps between simulations and experiments, past and future climates, or one instrument and another.

4.8 Conclusion

This chapter has introduced machine learning through the lens of scientific applications, establishing the conceptual bridges that enable the SciML synthesis. We have seen that scientific questions map systematically to ML tasks, that scientific data has distinctive structure requiring adapted strategies, and that the distinction between correlation and causation remains fundamental.

Key Takeaways

1. **Translate, Don't Force**: Scientific problems map to ML tasks, but the translation requires thought. Regression, classification, and generation have specific meanings in scientific contexts.
2. **Respect Data Structure**: Scientific data has spatiotemporal correlations, symmetries, and physical constraints. Splitting protocols must respect these structures to avoid information leakage.

3. **Encode Domain Knowledge**: Features that respect physical invariants (dimensionless numbers, conserved quantities, symmetries) dramatically improve learning efficiency.
4. **Neural Networks Are Universal**: They can approximate any continuous function—but expressivity is not accuracy. Architecture, activation, and training all matter.
5. **Correlation $\neq$ Causation**: ML finds statistical patterns. Science seeks causal mechanisms. The gap is real and consequential for interpretation.

References

1. Doersch, C.: Tutorial on variational autoencoders (2016). arXiv:1606.05908 arXiv preprint
2. Saxena, D., Cao, J.: Generative adversarial networks (GANs) challenges, solutions, and future directions. ACM Computing Surveys (CSUR) **54**(3), 1–42 (2021)
3. Alizadeh, R., Allen, J.K., Mistree, F.: Managing computational complexity using surrogate models: a critical review. Res. Eng. Design **31**(3), 275–298 (2020)

Part II
Algorithms

Chapter 5
Physics-Informed Machine Learning and Hybrid Modeling

The Force is what gives a Jedi his power. It's an energy field created by all living things. It surrounds us and penetrates us; it binds the galaxy together.

Obi-Wan Kenobi, Star Wars: A New Hope (1977)

Abstract Physics-informed neural networks (PINNs) embed governing equations directly into the learning process, enabling solutions to differential equations without labeled data and with seamless incorporation of sparse observations. This chapter provides comprehensive coverage of the PINN methodology, beginning with motivation and the fundamental architecture that combines data loss, physics residual loss, and boundary/initial condition enforcement. We develop both forward problems (solving PDEs given parameters) and inverse problems (inferring parameters from observations) through the running example of heat conduction. Best practices address loss weighting, collocation point sampling, network architecture, activation functions, and training strategies that determine practical success. A detailed taxonomy organizes the expanding PINN literature by learning objective (residual, variational, energy-based), domain decomposition (spatial, temporal, conservative), constraint enforcement (soft, hard, augmented Lagrangian), uncertainty quantification (Bayesian, ensemble, dropout), and model integration (hybrid, neural closure, universal differential equations). We conclude with design guidelines and critical considerations for practitioners.

5.1 Introduction

Let us start this chapter with a fiction. Dr. Amanda Torres stared at the finite element mesh on her screen and sighed. Three months of work, and she still couldn't get a converged solution.

B. P. Bhuyan et al., *Scientific Machine Learning*,
https://doi.org/10.1007/978-981-95-7856-6_5

The problem seemed simple enough: model the thermal stress in a new battery design for electric vehicles. The geometry was complex with layered electrodes, serpentine cooling channels, irregular interfaces; but that was manageable. The physics was standard: heat conduction coupled to thermal expansion.

The killer was the data. Experimental measurements from fourteen sensor locations, noisy and sparse, that her simulation had to match. Every time she adjusted the material parameters to fit one sensor, she destroyed the fit at another. The inverse problem was ill-posed, the optimization landscape was pathological, and her deadline was in two weeks.

"Have you tried a PINN?" asked her postdoc, Dr. Chen Wei, who had joined from a machine learning group.

Torres had heard of physics-informed neural networks but dismissed them as academic toys. "Neural networks can't solve PDEs. They're for cat pictures."

"Actually," Wei said, "they solve this exact problem. Watch."

An hour later, Torres watched in disbelief as Wei's code converged to a solution that matched all fourteen sensors while satisfying the heat equation everywhere in the domain. The neural network had learned a temperature field consistent with both the physics and the data—something her finite element approach couldn't do without heroic manual intervention.

"How does it know the physics?" Torres asked.

"We tell it. The loss function includes the PDE residual. At every point in space, we penalize deviations from the heat equation. The network learns to satisfy the physics."

"And the data?"

"Another term in the loss. We penalize deviations from the measurements. The network finds the temperature field that best balances physical consistency with data fit."

Torres leaned back. "So instead of solving the equations, you're training a network to represent the solution, using the equations as constraints on what it can learn?"

Wei nodded. "Exactly. And because it's a neural network, it handles the complex geometry naturally. No mesh generation. No matrix assembly. Just function evaluation and automatic differentiation."

That evening, Torres downloaded the DeepXDE library and began reading the 2019 Raissi paper that had started it all. By morning, she had a working prototype for her battery problem.

The story of physics-informed neural networks begins with a simple observation: neural networks are function approximators, and the solutions to differential equations are functions. What if we could train a network to represent the solution function, using the governing equations themselves as the training signal?

Consider the heat equation:

$$\frac{\partial u}{\partial t} = \alpha \nabla^2 u + f(x, t) \tag{5.1}$$

Classically, we discretize: replace continuous derivatives with finite differences, the continuous domain with a mesh, and solve the resulting algebraic system.

The PINN approach is fundamentally different. We represent the temperature field $u(x,t)$ as a neural network $u_\theta(x,t)$, where θ denotes the network parameters. We then define a loss function:

$$\mathcal{L}(\theta) = \mathcal{L}_{\text{physics}} + \lambda_{\text{data}}\mathcal{L}_{\text{data}} + \lambda_{\text{bc}}\mathcal{L}_{\text{bc}} \tag{5.2}$$

where:

- $\mathcal{L}_{\text{physics}}$ measures the PDE residual at sampled collocation points
- $\mathcal{L}_{\text{data}}$ measures deviation from observed data
- $\mathcal{L}_{\text{bc}}$ measures boundary and initial condition violations

Training the network minimizes this loss, producing a continuous function that approximately satisfies the governing equations, matches the data, and respects the boundary conditions.

This simple idea has profound implications. No mesh generation. Seamless handling of complex geometry. Natural incorporation of scattered observations. Unified treatment of forward and inverse problems. And with appropriate architecture choices; physics encoded as inductive bias rather than external constraint.

This chapter develops the theory and practice of physics-informed neural networks. We begin with motivation and formulation, establishing the mathematical framework. We then work through forward and inverse problems using heat conduction as our running example, developing intuition through concrete implementations.

Best practices address the "dark art" of PINN training: loss weighting, collocation sampling, architecture selection, and optimization strategy. A comprehensive taxonomy organizes the expanding literature by how physics enters the learning process. We conclude with hybrid approaches that combine the flexibility of PINNs with the stability of established numerical methods.

5.2 Motivation and Definitions

Two mature paradigms dominate scientific modeling: (i) *first-principles* models (ODEs/PDEs, conservation laws, symmetries) that are interpretable but costly and imperfect, and (ii) *data-driven* models (neural networks, kernels) that learn patterns but can violate physics and extrapolate poorly. *Physics-Informed Machine Learning (PIML)* [1] and *hybrid (gray-box) modeling* fuse these views: we *learn* from data while *respecting* known structure. This fusion yields better data efficiency, physical plausibility, and credible extrapolation [2].

Problem Set-up

Let $\Omega \subset \mathbb{R}^d$ be a spatial domain, $[0, T]$ a time interval, and $u : \Omega \times [0, T] \rightarrow \mathbb{R}^m$ a state field. Governing physics is expressed by a (possibly nonlinear) operator

$$\mathcal{N}[u](x, t; \lambda) = 0, \qquad (x, t) \in \Omega \times (0, T],$$

with parameters $\lambda \in \Lambda$ (e.g., diffusivity), boundary conditions $\mathcal{B}[u] = 0$ on $\partial\Omega$, and initial condition $u(\cdot, 0) = u_0$.

We observe data $\mathcal{D} = \{(z_i, y_i)\}_{i=1}^n$, where $z_i = (x_i, t_i)$ and y_i are noisy measurements of u or task-specific outputs (e.g., $g(u)$). Let $\mathcal{F}$ denote an admissible hypothesis space (e.g., neural networks). Our goal is to learn $u_\theta \in \mathcal{F}$ (and possibly λ) that fits data *and* respects physics.

Let us consider a thin, straight metal rod (see Fig. 5.1) of length $L = 1$ m, which forms our spatial domain $\Omega \subset \mathbb{R}^d$. Since heat conduction is studied along the length of the rod only, the dimension is $d = 1$. The spatial coordinate $x \in \Omega$ varies from $x = 0$ at the left end to $x = 1$ at the right end. The process is observed over a time interval $[0, T]$ with $T = 10$ s, so the time variable t lies in $[0, 10]$. The state field $u : \Omega \times [0, T] \rightarrow \mathbb{R}^m$ represents the physical quantity of interest, which in this case is temperature in degrees Celsius, so $m = 1$. For instance, $u(0.4, 5)$ denotes the temperature at position $x = 0.4$ m after $t = 5$ s.

The governing physics is given by the one-dimensional heat equation, expressed in operator form as

$$\mathcal{N}[u] = \frac{\partial u}{\partial t} - \alpha \frac{\partial^2 u}{\partial x^2} = 0,$$

where α is the thermal diffusivity of the material. For copper, $\alpha \approx 1.11 \times 10^{-4}$ m^2/s. In this problem, the set of parameters $\lambda \in \Lambda$ consists of α. The boundary conditions $\mathcal{B}[u] = 0$ specify that the left end ($x = 0$) is held at 100°C and the right end ($x = 1$) at 25°C for all $t \in [0, T]$. The initial condition $u(x, 0) = u_0(x)$ is a uniform 25°C along the rod.

In real measurements, we observe u only at discrete points in space and time. Let $\mathcal{D} = \{(z_i, y_i)\}_{i=1}^n$ denote the dataset, where $z_i = (x_i, t_i)$ and y_i is the temperature measured at position x_i and time t_i. Suppose three temperature sensors are placed at $x = 0.25, 0.50$, and 0.75 m. Each sensor records data at times $t = 0, 2, 4, 6, 8, 10$ seconds. For example, $z_i = (0.50, 4)$ and $y_i = 47.2$ mean that at $x = 0.50$ m after 4 seconds, the recorded temperature was 47.2°C.

We choose $\mathcal{F}$ to be a family of neural networks $u_\theta(x, t)$ parameterized by θ, with inputs (x, t) and output temperature. The goal is to find parameters θ so that u_θ matches the measured data while also satisfying the heat equation at a set of collocation points (x, t) across the space–time domain. The boundary and initial conditions are incorporated either as hard constraints or as penalty terms in the loss function.

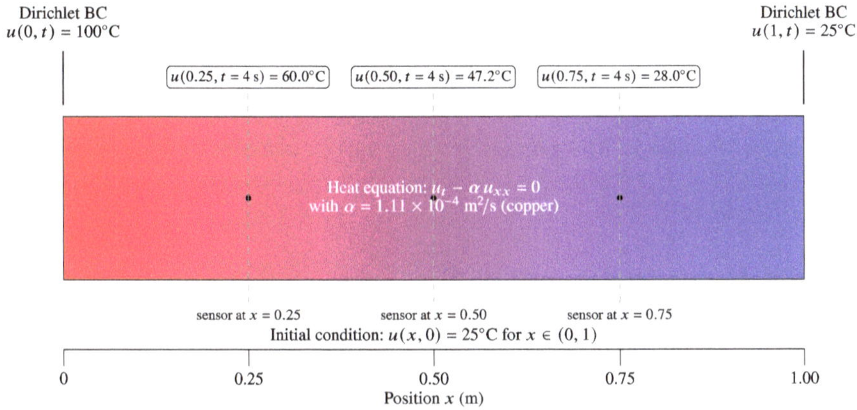

Fig. 5.1 One-dimensional heat conduction in a 1 m metal rod. The left boundary is held at 100°C and the right boundary at 25°C. Sensors at $x = 0.25,\ 0.50,\ 0.75$ m record temperatures at discrete times; three illustrative readings are shown here at $t = 4$ s. The physics is governed by $u_t - \alpha u_{xx} = 0$ with thermal diffusivity α

At $t = 0$, the rod is uniformly at 25°C. After $t = 2$ seconds, the leftmost sensor ($x = 0.25$ m) might read 60°C, reflecting rapid heating near the hot boundary, while the rightmost sensor ($x = 0.75$ m) might record only 28°C. A physics-informed neural network learns a smooth temperature field $u_\theta(x, t)$ that interpolates these sparse observations and evolves according to the heat equation, enabling accurate predictions at all positions and times, even where no sensors are placed.

We return to the one-dimensional heat conduction problem in a metal rod of length $L = 1$ m, but now we assume that the thermal diffusivity α of the material is *unknown*. This transforms our task from a *forward problem*: predicting $u(x, t)$ given α into an *inverse problem* in which both the temperature field $u(x, t)$ and the parameter α must be determined from limited and noisy measurements.

The governing equation remains

$$\frac{\partial u}{\partial t} - \alpha \frac{\partial^2 u}{\partial x^2} = 0,$$

subject to Dirichlet boundary conditions

$$u(0, t) = 100°\mathrm{C}, \quad u(1, t) = 25°\mathrm{C}, \quad t \in [0, T],$$

and the initial condition

$$u(x, 0) = 25°\mathrm{C}, \quad x \in (0, 1).$$

However, α is now an *unknown constant* to be estimated. The set of parameters becomes $\lambda = \{\alpha\}$ with $\Lambda \subset \mathbb{R}^+$.

The measurement setup is identical to the forward problem: three sensors are placed at $x = 0.25$, 0.50, and 0.75 m, recording temperatures at times $t = 0, 2, 4, 6, 8, 10$ seconds. The dataset $\mathcal{D} = \{(z_i, y_i)\}_{i=1}^n$ contains pairs of space–time coordinates $z_i = (x_i, t_i)$ and measured temperatures y_i. For example, at $t = 4$ s the readings might be

$$u(0.25, 4) = 60.0°\text{C}, \quad u(0.50, 4) = 47.2°\text{C}, \quad u(0.75, 4) = 28.0°\text{C}.$$

The learning objective is now twofold:

1. Determine the parameter α such that the PDE residual

$$\mathcal{R}_{\text{PDE}}(\theta, \alpha) = \mathbb{E}_{(x,t)}\left[\left(\frac{\partial u_\theta}{\partial t} - \alpha \frac{\partial^2 u_\theta}{\partial x^2}\right)^2\right]$$

is minimized across collocation points.
2. Learn a surrogate $u_\theta(x, t)$ from $\mathcal{F}$ (e.g., a neural network) that matches the observed data:

$$\mathcal{R}_{\text{data}}(\theta) = \frac{1}{n}\sum_{i=1}^{n} (u_\theta(z_i) - y_i)^2 .$$

The total loss combines both terms:

$$\min_{\theta,\alpha} \mathcal{R}_{\text{data}}(\theta) + \beta\, \mathcal{R}_{\text{PDE}}(\theta, \alpha),$$

where $\beta > 0$ balances the influence of physics and data.

In this inverse setting, the challenge lies in the *identifiability* of α: if the measurement locations are too sparse, or if noise is high, different values of α may fit the data nearly equally well. Physics-informed neural networks handle this by simultaneously fitting u_θ to the data and enforcing the PDE structure, which regularizes the estimation and often yields more accurate and physically consistent α values.

Now, in yet another example, we keep the one-dimensional rod of length $L = 1$ m and time window $[0, T]$ with $T = 10$ s. The temperature field $u : (0, 1) \times [0, 10] \to \mathbb{R}$ obeys Dirichlet boundaries $u(0, t) = 100°\text{C}$ and $u(1, t) = 25°\text{C}$ and the initial condition $u(x, 0) = 25°\text{C}$. In the forward and inverse formulations we assumed that the physics is perfectly captured by the heat equation $u_t - \alpha u_{xx} = 0$. In practice, unmodeled effects (small heat sources, contact resistance, mild convection, radiation losses) introduce a systematic discrepancy. We therefore posit a *hybrid model* that retains a mechanistic core but augments it with a learned correction,

$$u_t - \alpha u_{xx} = q_\theta(x, t, u), \qquad (x, t) \in (0, 1) \times (0, T],$$

where q_θ is a low-capacity neural field representing the *missing physics* (e.g., net localized gain/loss). The boundary and initial conditions are enforced as before. When $q_\theta \equiv 0$ the model reduces to the pure PDE; when α is unknown we may estimate it jointly with θ.

For computation, we introduce a coarse, stable time–space discretization that we treat as the mechanistic core. Let $\Delta x = 1/25$ (26 grid nodes including boundaries), $\Delta t = 0.2\,\mathrm{s}$, and denote by $L \in \mathbb{R}^{24\times 24}$ the standard second-difference Laplacian on interior nodes with the Dirichlet boundaries imposed. Using backward Euler, the discrete mechanistic update is

$$(\mathrm{I} - \Delta t\, \alpha L)\, u^{k+1} = u^k + \Delta t\, q_\theta^k, \qquad k = 0, 1, \ldots, 49,$$

where $u^k \in \mathbb{R}^{24}$ collects interior temperatures at $t_k = k\Delta t$ and $q_\theta^k \in \mathbb{R}^{24}$ is the learned correction evaluated on the grid. This yields the gray-box time-$k \to k{+}1$ map

$$u^{k+1} = f_{\mathrm{phys}}(u^k; \alpha, \Delta t, \Delta x) + r_\theta(u^k, t_k), \quad r_\theta := (\mathrm{I} - \Delta t\, \alpha L)^{-1} \Delta t\, q_\theta^k.$$

Thus the *physics* is preserved in the left-hand operator $(\mathrm{I} - \Delta t\, \alpha L)$ while the *learned part* injects only a small, structured residual through q_θ.

Measurements remain sparse: three sensors at $x = 0.25, 0.50, 0.75\,\mathrm{m}$ record temperatures at $t \in \{0, 2, 4, 6, 8, 10\}\,\mathrm{s}$. Let $S \in \mathbb{R}^{m\times 24}$ be the sampling matrix that picks the grid nodes closest to the sensor locations (with bilinear interpolation if sensors fall between nodes). With dataset $\mathcal{D} = \{(x_i, t_i, y_i)\}_{i=1}^n$, the hybrid learning problem estimates network parameters θ (and optionally α) by simulating the gray-box model forward from u^0 and minimizing a composite objective

$$\min_{\theta,\alpha} \underbrace{\frac{1}{n}\sum_{i=1}^{n} \left(Su^{k_i} - y_i\right)^2}_{\text{data misfit}} + \underbrace{\lambda_{\mathrm{reg}} \frac{1}{T}\sum_{k=0}^{49}\left(\|q_\theta^k\|_2^2 + \mu\, \|\nabla_x q_\theta^k\|_2^2\right)}_{\text{parsimony and smoothness of the learned physics}},$$

where k_i is the index with $t_{k_i} = t_i$. The first term enforces agreement with sensors; the second discourages the learned field from absorbing noise or violating Occam's razor, while $\|\nabla_x q_\theta^k\|_2$ promotes spatial smoothness consistent with diffusive phenomena. If energy conservation is relevant, one may add an integral constraint $\left(\int_0^1 q_\theta(x, t_k, u)\, dx\right)^2$ per step, which penalizes net artificial heat creation and helps identifiability when data are sparse.

This hybridization has two advantages in the rod setting. First, even with a coarse grid ($\Delta x = 4\,\mathrm{cm}$ and $\Delta t = 0.2\,\mathrm{s}$), the mechanistic core captures the dominant diffusion and enforces boundary conditions exactly; the network only learns the small bias that the coarse model cannot resolve (for instance, a mild localized heat source around $x \approx 0.6\,\mathrm{m}$ active between $t = 3$ and $t = 6\,\mathrm{s}$). Second, because q_θ enters additively and is regularized, the learned correction extrapolates more credibly than a black-box predictor: when sensors go silent (e.g., between $t = 6$ and $t = 10\,\mathrm{s}$), the simulation remains anchored by the PDE and boundary data rather than drifting.

In practice, we initialize u^0 from the known initial condition, roll the gray-box model forward to produce $\{u^k\}_{k=1}^{50}$, and backpropagate through the linear solves to update θ (and α if unknown). When the noise level is nontrivial, calibrating λ_{reg} and μ via a small validation split of the sensor time series prevents the network from over-explaining stochastic fluctuations. When α is estimated jointly, a weak Gaussian prior or a bounded box prior $\alpha \in [0.5,\ 2.0] \times 10^{-4}\,\mathrm{m}^2/\mathrm{s}$ stabilizes the fit and keeps the correction q_θ physically interpretable as a small residual rather than a surrogate for misspecified diffusivity.

Definition 5.1 (Physics-Informed Learning)

Physics-Informed Learning

Given data $\mathcal{D}$ and physics $\{\mathcal{N}, \mathcal{B}, u_0\}$, a *physics-informed learner* seeks

$$\min_{\theta,\lambda} \underbrace{\mathcal{R}_{\text{data}}(\theta)}_{\text{data fit}} + \alpha\, \underbrace{\mathcal{R}_{\text{PDE}}(\theta, \lambda)}_{\text{physics residual}} + \beta\, \underbrace{\mathcal{R}_{\text{BC/IC}}(\theta)}_{\text{boundary/initial}},$$

where $\mathcal{R}_{\text{data}} = \frac{1}{n}\sum_i \ell\big(u_\theta(z_i), y_i\big)$, $\mathcal{R}_{\text{PDE}} = \mathbb{E}_{(x,t)\sim\mu}\big[\|\mathcal{N}[u_\theta](x,t;\lambda)\|^2\big]$ (estimated at collocation points), and $\mathcal{R}_{\text{BC/IC}}$ penalizes violations of boundary/initial conditions. We call $\alpha, \beta > 0$ *balancing weights.*

Explanation: This is the most general formulation of physics-informed learning. It shows how we balance three goals: fitting the data, respecting the governing equations, and satisfying the boundary and initial conditions. The weights α and β control the trade-off between data fidelity and physics enforcement.

Definition 5.2 (Physics-Informed Neural Network (PINN))

Physics-Informed Neural Network (PINN)

A *Physics-Informed Neural Network (PINN)* represents u_θ by a neural network and computes $\partial_x u_\theta$, $\partial_t u_\theta$, etc., by automatic differentiation to evaluate $\mathcal{N}[u_\theta]$ and form $\mathcal{R}_{\text{PDE}}$.

Explanation: In practice, PINNs use the fact that modern autodiff frameworks (e.g., TensorFlow, PyTorch, JAX) can compute derivatives of the network output with respect to its inputs. This makes it possible to insert PDE residuals directly into the loss function without discretizing derivatives.

Definition 5.3 (Hybrid (Gray-Box) Model)

Hybrid (Gray-Box) Model

Let $f_{\text{phys}}(u; \lambda)$ denote a mechanistic model (e.g., time-stepping of a discretized PDE) and r_θ a learned correction. A *hybrid model* composes them as

$$u_{k+1} = f_{\text{phys}}(u_k; \lambda) \ + \ r_\theta(u_k, \phi_k),$$

where ϕ_k are exogenous inputs (e.g., forcings). When $r_\theta \equiv 0$ we recover a white-box model; when f_{phys} is absent we recover a black-box model.

Explanation: Hybrid modeling leverages strengths of both physics-based simulators and machine learning. The physics core enforces known dynamics, while the learned correction captures unmodeled effects. This reduces overfitting and improves extrapolation compared to pure black-box learning.

Definition 5.4 Forward Versus Inverse Problems

Forward Versus Inverse Problems

A *forward* PIML problem learns u_θ given λ. An *inverse* PIML problem estimates λ (and possibly u_θ) from data. Inverse problems raise identifiability and regularization issues.

Explanation: Forward problems are predictive: parameters are known and we simulate the system. Inverse problems are diagnostic: we use observed behavior to infer unknown parameters. The latter are often ill-posed and benefit greatly from physics-informed regularization.

Definition 5.5 (Hard Versus Soft Constraints)

Hard Versus Soft Constraints

Soft constraints penalize violations via loss terms (as above). *Hard constraints* enforce feasibility by construction, e.g., using a trial solution $\tilde{u}_\theta$ that satisfies $\mathcal{B}[\tilde{u}_\theta] = 0$:

$$\tilde{u}_\theta(x, t) = g(x, t) + h(x, t)\, v_\theta(x, t), \quad \text{with } \mathcal{B}[g] = 0, \ \mathcal{B}[h] = 0.$$

Explanation: Soft constraints give flexibility but allow small violations, while hard constraints guarantee feasibility for all predictions. Choosing between them depends on how critical exact satisfaction of physics is for the application.

Definition 5.6 (Operator Learning)

Operator Learning

Let $(\mathcal{X}, \|\cdot\|)$ and $(\mathcal{Y}, \|\cdot\|)$ be function spaces and $\mathcal{G} : \mathcal{X} \to \mathcal{Y}$ map inputs (e.g., coefficients/forcing) to solutions of a PDE. *Operator learning* fits a parametric $\mathcal{G}_\theta$ to approximate $\mathcal{G}$ from observed pairs $\{(a_j, \mathcal{G}(a_j))\}$, enabling cheap surrogates across parameter fields.

Explanation: Instead of learning a single solution for fixed parameters, operator learning learns a *mapping* from parameters or forcing functions to full solutions. This makes it ideal for rapid evaluation over many different scenarios.

5.3 Physics-Informed Neural Networks

5.3.1 Architecture and Training

A *Physics-Informed Neural Network* (PINN) [3] approximates the unknown solution $u : \Omega \times [0, T] \to \mathbb{R}^m$ by a neural network u_θ with trainable parameters θ. The inputs (x, t) are fed into the network, which outputs an approximation of the state $u_\theta(x, t)$. Two kinds of training points are used:

- *Collocation points* $(x_i^{\mathrm{col}}, t_i^{\mathrm{col}})$ sampled in the space-time domain to evaluate the PDE residual $\mathcal{N}[u_\theta]$.
- *Data points* $(x_j^{\mathrm{data}}, t_j^{\mathrm{data}}, y_j)$ where measurements are available.

Boundary and initial conditions are enforced either as additional loss terms (*soft constraints*) or by building them into the trial solution (*hard constraints*).

PINN Loss Function

Given data $\mathcal{D}$ and physics $\{\mathcal{N}, \mathcal{B}, u_0\}$, a standard PINN minimizes

$$L(\theta) = \underbrace{\frac{1}{n_{\mathrm{data}}} \sum_{j=1}^{n_{\mathrm{data}}} \ell\big(u_\theta(z_j^{\mathrm{data}}), y_j\big)}_{\mathcal{R}_{\mathrm{data}}} + \alpha \underbrace{\frac{1}{n_{\mathrm{col}}} \sum_{i=1}^{n_{\mathrm{col}}} \|\mathcal{N}[u_\theta](z_i^{\mathrm{col}})\|^2}_{\mathcal{R}_{\mathrm{PDE}}}$$

$$+\beta \underbrace{\mathcal{R}_{\mathrm{BC/IC}}(\theta)}_{\text{boundary/initial penalty}},$$

where $\alpha, \beta > 0$ are balancing weights.

Explanation: The first term matches the network predictions to observed data. The second term enforces the governing PDE at collocation points using automatic differentiation to compute $\partial_x u_\theta$, $\partial_t u_\theta$, and higher derivatives. The third term penalizes violations of boundary and initial conditions when they are not enforced exactly. By tuning α and β, we control how strongly the model adheres to physics versus fitting the data.

Figure 5.2 illustrates the PINN architecture: a fully connected feedforward neural network takes (x, t) as input, produces $u_\theta(x, t)$, and branches into separate loss components for data, PDE residual, and boundary/initial terms.

5.3.2 *Forward Problem: Physics-Informed Neural Network for the Heat Equation*

We begin with the one-dimensional heat conduction problem introduced earlier in Example 1 (see Fig. 5.1). The spatial domain is $\Omega = (0, 1)$ meters and the time interval is $[0, 10]$ seconds. The temperature field $u : \Omega \times [0, 10] \to \mathbb{R}$ satisfies the heat equation

$$\frac{\partial u}{\partial t} - \alpha \frac{\partial^2 u}{\partial x^2} = 0,$$

with $\alpha = 1.11 \times 10^{-4}\,\mathrm{m}^2/\mathrm{s}$ (thermal diffusivity for copper). We impose Dirichlet boundary conditions [4]:

$$u(0, t) = 100^\circ\mathrm{C}, \quad u(1, t) = 25^\circ\mathrm{C}, \quad t \in [0, 10],$$

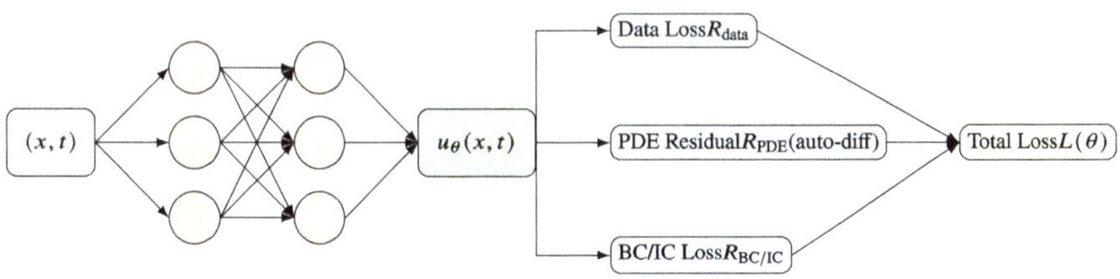

Fig. 5.2 Architecture of a Physics-Informed Neural Network (PINN). Inputs (x, t) are processed by a fully connected feedforward neural network to produce $u_\theta(x, t)$. Automatic differentiation is applied to compute PDE residuals at collocation points. Three loss components-data fit, PDE residual, and boundary/initial conditions-are combined into the total loss $L(\theta)$ used for training

and the initial condition:

$$u(x, 0) = 25°\text{C}, \quad x \in (0, 1).$$

Our goal is to learn a neural approximation $u_\theta(x, t)$ that:

1. Matches the known boundary and initial conditions.
2. Minimizes the PDE residual over the space–time domain.
3. Fits available sensor data, if provided.

This is a *forward* problem because α is assumed known.

PINN Loss Function for the Forward Heat Equation

We define the total loss

$$L(\theta) = \mathcal{R}_{\text{data}}(\theta) + \alpha_w \, \mathcal{R}_{\text{PDE}}(\theta) + \beta_w \, \mathcal{R}_{\text{BC/IC}}(\theta),$$

where:

$$\mathcal{R}_{\text{data}} = \frac{1}{n_{\text{data}}} \sum_{j=1}^{n_{\text{data}}} \left(u_\theta(z_j^{\text{data}}) - y_j\right)^2,$$

$$\mathcal{R}_{\text{PDE}} = \frac{1}{n_{\text{col}}} \sum_{i=1}^{n_{\text{col}}} \left(u_t(x_i^{\text{col}}, t_i^{\text{col}}) - \alpha \, u_{xx}(x_i^{\text{col}}, t_i^{\text{col}})\right)^2,$$

$$\mathcal{R}_{\text{BC/IC}} = \frac{1}{n_{\text{BC}}} \sum_{k=1}^{n_{\text{BC}}} \left(u_\theta(z_k^{\text{BC}}) - u_{\text{BC}}(z_k^{\text{BC}})\right)^2 + \frac{1}{n_{\text{IC}}} \sum_{\ell=1}^{n_{\text{IC}}} \left(u_\theta(z_\ell^{\text{IC}}) - u_{\text{IC}}(z_\ell^{\text{IC}})\right)^2.$$

Here, α_w and β_w are balancing weights between the physics residual and the boundary/initial condition terms.

The implementation below uses JAX for efficient automatic differentiation. It is presented in four steps:

1. Neural network architecture.
2. Derivatives and PDE residual.
3. Loss function components.
4. Training loop and visualization.

```
def init_mlp(key, layers):
    """Initialize MLP weights with Xavier initialization."""
    params = []
    keys = jax.random.split(key, num=len(layers)-1)
    for k, (m, n) in zip(keys, zip(layers[:-1], layers[1:])):
```

```
        W = jax.random.uniform(k, (m, n), minval=-1, maxval=1) *
    jnp.sqrt(6/(m+n))
        b = jnp.zeros((n,))
        params.append((W, b))
    return params

def mlp_apply(params, x):
    """Forward pass through the MLP."""
    for W, b in params[:-1]:
        x = jnp.tanh(x @ W + b)
    W, b = params[-1]
    return x @ W + b

def u_theta(params, x, t):
    """Neural network approximation u_theta(x,t)."""
    return mlp_apply(params, jnp.stack([x, t], axis=-1)).squeeze(-1)
```

Listing 5.1 Initialization and forward pass of a fully-connected neural network in JAX.

This defines a simple feed-forward neural network with $tanh$ activations. The network takes spatial-temporal inputs (x, t) and returns the predicted temperature $u_\theta(x, t)$.

```
u_x  = grad(lambda p, x, t: u_theta(p, x, t), argnums=1)
u_t  = grad(lambda p, x, t: u_theta(p, x, t), argnums=2)
u_xx = grad(lambda p, x, t: u_x(p, x, t), argnums=1)

def pde_residual(params, x, t):
    """Heat equation residual: u_t - alpha u_xx."""
    return u_t(params, x, t) - alpha_true * u_xx(params, x, t)
```

Listing 5.2 Derivatives and PDE residual computation using JAX automatic differentiation.

JAX's grad computes exact derivatives of the neural network output with respect to x and t. The PDE residual $u_t - \alpha u_{xx}$ is then evaluated at collocation points.

```
def data_loss(params, Z_data, y_data):
    pred = vmap(lambda z: u_theta(params, z[0], z[1]))(Z_data)
    return jnp.mean((pred - y_data)**2)

def pde_loss(params, Z_col):
    R = vmap(lambda z: pde_residual(params, z[0], z[1]))(Z_col)
    return jnp.mean(R**2)

def bcic_loss(params, Z_b, u_b, Z_i, u_i):
    ub_pred = vmap(lambda z: u_theta(params, z[0], z[1]))(Z_b)
    ui_pred = vmap(lambda z: u_theta(params, z[0], z[1]))(Z_i)
    return jnp.mean((ub_pred - u_b)**2) + jnp.mean((ui_pred - u_i)**2)
```

Listing 5.3 Loss components: data fit, PDE residual, and boundary/initial conditions.

Data loss matches network predictions to sensor measurements. PDE loss enforces the governing equation at collocation points. BC/IC loss enforces the boundary and initial conditions.

```
opt = optax.adam(1e-3)
opt_state = opt.init(params)

@jit
def total_loss(params):
    return gamma_w * data_loss(params, Z_data, y_data) \
         + alpha_w * pde_loss(params, Z_col) \
         + beta_w * bcic_loss(params, Z_b, u_b, Z_i, u_i)

@jit
def step(params, opt_state):
    loss_val, grads = jax.value_and_grad(total_loss)(params)
    updates, opt_state = opt.update(grads, opt_state, params)
    params = optax.apply_updates(params, updates)
    return params, opt_state, loss_val

for epoch in range(4000):
    params, opt_state, L = step(params, opt_state)
    if epoch % 200 == 0:
        print(f"Epoch {epoch}: loss = {L:.4e}")
```

Listing 5.4 Training loop using Optax Adam optimizer.

The network is trained with the Adam optimizer. The total loss combines the three components with balancing weights γ_w, α_w, β_w. JAX's jit and vmap ensure efficient vectorized and compiled execution. After training, we visualize:

1. Surface plot of $u_\theta(x, t)$.
2. Line cuts at fixed t compared to the reference steady state.
3. Loss curves for total loss and individual components.

Figures 5.3, 5.4 and 5.5 show the trained PINN solution for the one-dimensional heat conduction problem, along with convergence behavior.

The surface plot in Fig. 5.3 shows the evolution of temperature from the initial condition toward the steady state. The model enforces boundary conditions precisely: the hot left boundary and cool right boundary remain fixed at 100°C and 25°C respectively. The loss curve in Fig. 5.4 highlights two regimes: an initial rapid drop as the network learns the gross profile, and a slower decay as it fine-tunes to satisfy the PDE residual. Finally, Fig. 5.5 confirms that the PINN interpolates intermediate states faithfully, with curves smoothly approaching the steady-state profile over time.

These results demonstrate the core advantage of PINNs: even with sparse data, the learned solution respects both data and physics, producing a physically consistent field over the full spatio-temporal domain.

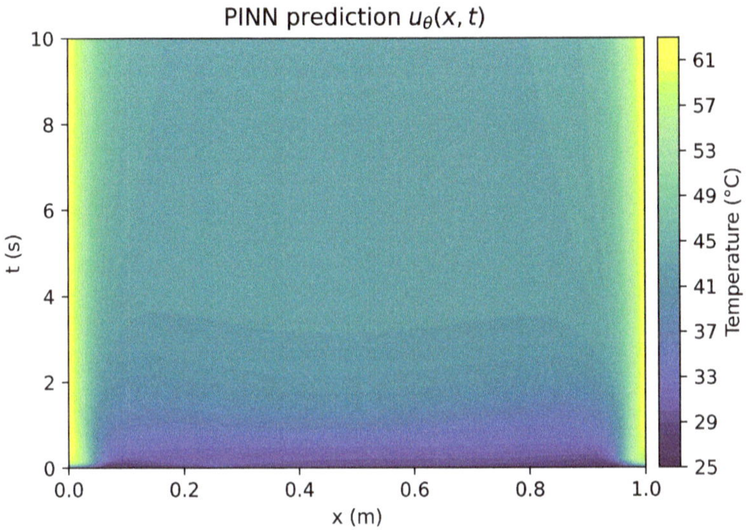

Fig. 5.3 Predicted temperature field $u_\theta(x, t)$ from the trained PINN. The model captures the expected diffusive smoothing of the temperature profile, with hot and cold boundaries maintained throughout the time horizon

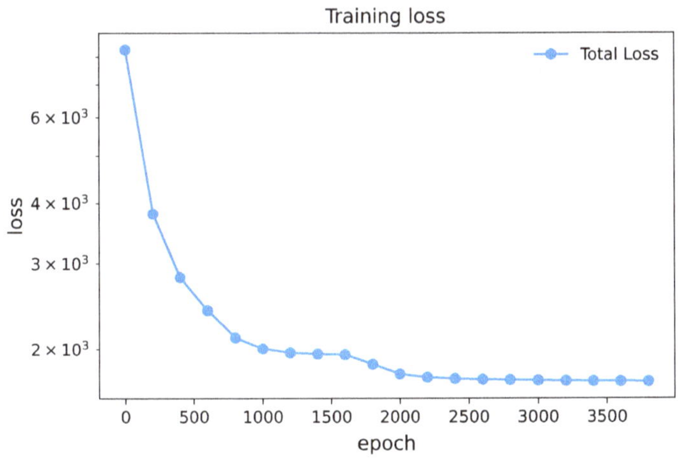

Fig. 5.4 Training loss versus epoch for the PINN. The loss decreases rapidly in the first 500 iterations, followed by a slower convergence phase as the network refines the PDE residual and boundary condition satisfaction

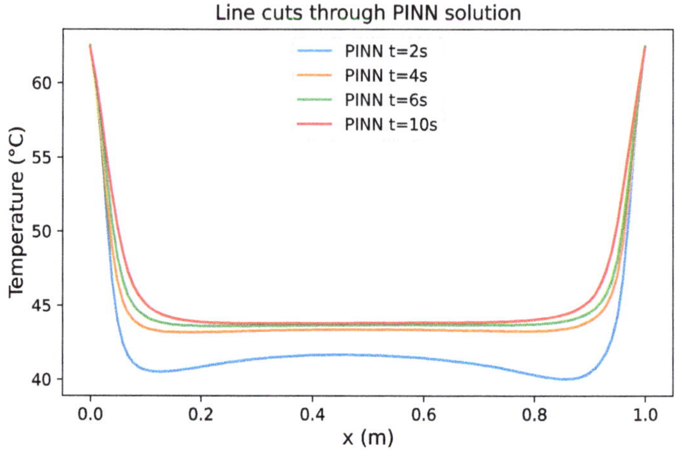

Fig. 5.5 Line cuts through the predicted temperature field at selected times. The profiles are smooth and consistent with diffusive heat transfer, showing monotonic relaxation toward the steady-state linear profile

5.3.3 *Inverse Problem: Learning Thermal Diffusivity*

We now consider the same heat conduction setup as in the forward problem, but with the thermal diffusivity α unknown. The task is to estimate α from sparse measurements of $u(x, t)$, while still enforcing the governing PDE and boundary/initial conditions.

The formulation remains:

$$u_t - \alpha\, u_{xx} = 0,$$

with the same domain, time horizon, and boundary/initial conditions as before. Unlike the forward case, α becomes a *trainable parameter* alongside the neural network weights θ.

Inverse PINN Loss Function

In the inverse problem, the total loss becomes

$$L(\theta, \alpha) = \mathcal{R}_{\text{data}}(\theta) + \alpha_w\, \mathcal{R}_{\text{PDE}}(\theta, \alpha) + \beta_w\, \mathcal{R}_{\text{BC/IC}}(\theta).$$

Here α is optimized jointly with θ; $\mathcal{R}_{\text{PDE}}$ explicitly depends on α.

```
# Derivatives
u_x  = grad(lambda p, x, t: u_theta(p, x, t), argnums=1)
u_t  = grad(lambda p, x, t: u_theta(p, x, t), argnums=2)
u_xx = grad(lambda p, x, t: u_x(p, x, t), argnums=1)

# PDE residual now uses alpha as input
def pde_residual(params, alpha, x, t):
    return u_t(params, x, t) - alpha * u_xx(params, x, t)
```

Listing 5.5 Defining PDE residual with trainable α.

The derivatives u_x, u_t, and u_{xx} are computed using JAX automatic differentiation. The PDE residual now includes the *unknown* parameter α, which will be optimized jointly with the neural network weights θ.

```
def data_loss(params, Z_data, y_data):
    pred = vmap(lambda z: u_theta(params, z[0], z[1]))(Z_data)
    return jnp.mean((pred - y_data)**2)

def pde_loss(params, alpha, Z_col):
    R = vmap(lambda z: pde_residual(params, alpha, z[0], z[1]))(Z_col)
    return jnp.mean(R**2)

def bcic_loss(params, Z_b, u_b, Z_i, u_i):
    ub_pred = vmap(lambda z: u_theta(params, z[0], z[1]))(Z_b)
    ui_pred = vmap(lambda z: u_theta(params, z[0], z[1]))(Z_i)
    return jnp.mean((ub_pred - u_b)**2) + jnp.mean((ui_pred - u_i)**2)
```

Listing 5.6 Loss functions for the inverse PINN.

Data loss matches the predicted values to the sensor measurements. *PDE loss* enforces the heat equation using the current estimate of α. *BC/IC loss* ensures that the learned solution satisfies the boundary and initial conditions.

```
# Initialize network parameters and alpha guess
layers = [2, 64, 64, 64, 1]
params = init_mlp(k4, layers)
alpha_est = jnp.array(1e-5)  # initial guess far from true

# Total loss
def total_loss(params, alpha):
    return gamma_w * data_loss(params, Z_data, y_data) \
         + alpha_w * pde_loss(params, alpha, Z_col) \
         + beta_w * bcic_loss(params, Z_b, u_b, Z_i, u_i)

# Optimizer setup for both params and alpha
opt = optax.adam(1e-3)
opt_state = opt.init((params, alpha_est))

@jit
def step(params, alpha, opt_state):
    loss_val, grads = jax.value_and_grad(total_loss,
     argnums=(0,1))(params, alpha)
```

```
    updates, opt_state = opt.update(grads, opt_state, (params, alpha))
    params, alpha = optax.apply_updates((params, alpha), updates)
    return params, alpha, opt_state, loss_val

# Training loop
history = []
for epoch in range(4000):
    params, alpha_est, opt_state, L = step(params, alpha_est,
     opt_state)
    if epoch % 200 == 0:
        history.append((epoch, float(L), float(alpha_est)))
        print(f"Epoch {epoch}: loss={L:.4e}, alpha={alpha_est:.6e}")
```

Listing 5.7 Joint training of neural network parameters and α.

Here we train both the neural network parameters θ and the diffusivity α. Gradients are computed with respect to both sets of parameters, and the optimizer updates them simultaneously. The initial guess for α is deliberately far from the true value to test the method's ability to converge. After training:

- The estimated α should converge to the true value 1.11×10^{-4}.
- The predicted $u_\theta(x, t)$ matches the forward problem solution.
- Loss curves and α-vs-epoch plots show stable convergence.

Figures 5.6, 5.7 and 5.8 collectively demonstrate that the inverse PINN can successfully reconstruct both the temperature field and the unknown diffusivity from sparse measurements. The surface plot in Fig. 5.6 confirms that the learned solution adheres to the expected diffusive dynamics. The training loss curve in Fig. 5.7 shows

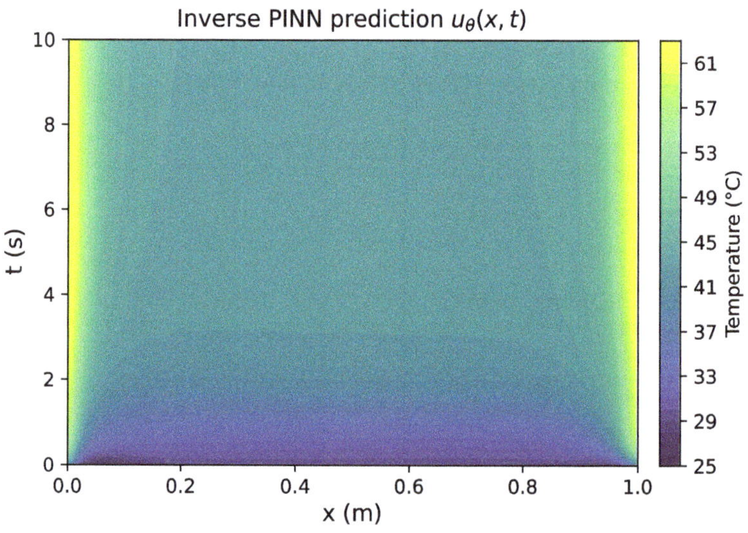

Fig. 5.6 Predicted temperature field $u_\theta(x, t)$ from the trained inverse PINN. The model accurately captures the diffusive heat transfer behavior over time while maintaining the boundary temperatures at 100°C (left) and 25°C (right). The smooth gradient from hot to cold regions demonstrates the PDE constraint is well enforced even when α is unknown

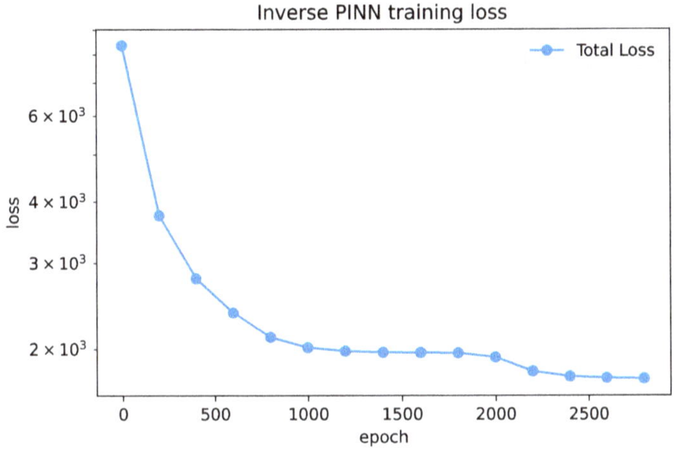

Fig. 5.7 Inverse PINN training loss as a function of epochs. The initial sharp decrease reflects rapid learning of the coarse solution structure. The subsequent gradual decline corresponds to fine-tuning of both the network weights and the diffusivity estimate. The stability of the curve indicates well-conditioned joint optimization

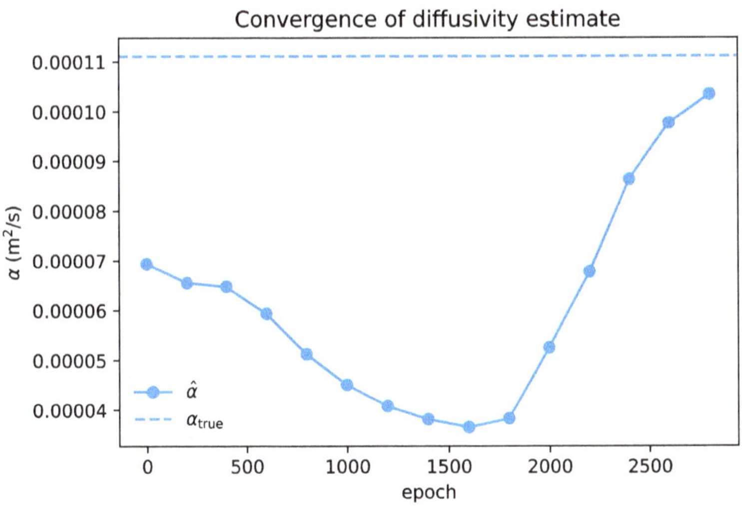

Fig. 5.8 Convergence of the estimated diffusivity $\hat{\alpha}$ toward the true value $\alpha_{\text{true}} = 1.11 \times 10^{-4}$ m^2/s. The transient undershoot during early training arises from the optimizer balancing PDE residual minimization with data fitting. Eventually, $\hat{\alpha}$ stabilizes near the true value, validating the inverse PINN's ability to recover unknown physical parameters

a two-phase convergence pattern: rapid error reduction followed by gradual refinement. Most importantly, Fig. 5.8 highlights the reliable recovery of α without prior knowledge, underscoring the potential of PINNs for parameter identification tasks in real-world scenarios.

5.3.4 *Best Practices*

While the PINN framework is conceptually simple, training can be delicate : especially for stiff PDEs, multiscale dynamics, or poorly scaled variables. The following practices have proven effective across forward and inverse settings.

5.3.4.1 Non-dimensionalization and Scaling

A common source of instability in training PINNs is the mismatch in the numerical scales of the spatial coordinates x, the temporal coordinate t, and the physical field u. If one variable varies over a much larger range than the others (e.g., $x \in [0, 1]$ but $t \in [0, 10^4]$), the neural network will tend to allocate disproportionate capacity to fitting the larger-scale variable, making it harder to learn fine variations in the smaller-scale one. This is particularly acute in PDEs where the coefficients themselves have small or large magnitudes (e.g., $\alpha \approx 10^{-4}$ for thermal diffusivity), causing the PDE residual $\mathcal{R}_{\text{PDE}}$ to have an unbalanced scale compared to the data or boundary condition terms.

A standard remedy is to *non-dimensionalize* the governing PDE before training. Let L be a characteristic length, T_c a characteristic time scale, and U_c a characteristic temperature (or field magnitude). Define dimensionless variables

$$x^\star = \frac{x}{L}, \qquad t^\star = \frac{t}{T_c}, \qquad u^\star = \frac{u}{U_c}.$$

For the heat equation

$$u_t - \alpha\, u_{xx} = 0,$$

choosing $T_c = L^2/\alpha_{\text{ref}}$ with a reference diffusivity α_{ref} reduces the PDE to

$$\frac{\partial u^\star}{\partial t^\star} - \alpha^\star \frac{\partial^2 u^\star}{\partial x^{\star 2}} = 0,$$

where $\alpha^\star = \alpha/\alpha_{\text{ref}}$ is an $\mathcal{O}(1)$ dimensionless parameter. This scaling makes all coordinates and coefficients dimensionless and of similar numerical magnitude, which improves the conditioning of the optimization problem.

Once the PDE is non-dimensionalized, all inputs to the neural network are normalized to a common range, typically $[-1, 1]$ or $[0, 1]$. Similarly, the outputs $u^\star$ are often centered (zero mean) and scaled to unit variance before training, especially when the network uses activations like `tanh` that are sensitive to input scale. After training, predictions in dimensionless variables are mapped back to the physical units via the inverse transform:

$$u(x,t) = U_c\, u^{\star}\left(\frac{x}{L}, \frac{t}{T_c}\right).$$

Benefit: Non-dimensionalization and scaling ensure that all terms in the loss function contribute comparably during training, prevent domination by one coordinate or variable, and allow the optimizer to make uniform progress across the entire spatio-temporal domain.

5.3.4.2 Residual Normalization (Loss Balancing)

In a PINN, the total loss

$$L(\theta,\lambda) = \underbrace{\mathcal{R}_{\text{data}}(\theta)}_{\text{data fit}} + \alpha_w \underbrace{\mathcal{R}_{\text{PDE}}(\theta,\lambda)}_{\text{physics residual}} + \beta_w \underbrace{\mathcal{R}_{\text{BC/IC}}(\theta)}_{\text{boundary/initial penalty}}$$

typically involves terms with different units and vastly different magnitudes. For instance, in the heat equation example, the data loss $\mathcal{R}_{\text{data}}$ might be of order 10^{-1} (degrees Celsius squared), whereas the PDE residual $\mathcal{R}_{\text{PDE}}$ can start at 10^{+3} or higher depending on α and the scaling of derivatives. If left unbalanced, the optimizer will prioritize reducing the largest term, often at the expense of others, leading to a solution that fits one objective well but violates other constraints (e.g., fits the data but ignores the PDE).

Residual normalization seeks to keep all terms in the loss function numerically comparable. A simple manual approach is to pre-compute the initial magnitudes of each term on a small batch of points and choose weights α_w and β_w to bring their scales to the same order of magnitude:

$$\alpha_w \approx \frac{\text{scale}(\mathcal{R}_{\text{data}})}{\text{scale}(\mathcal{R}_{\text{PDE}})}, \quad \beta_w \approx \frac{\text{scale}(\mathcal{R}_{\text{data}})}{\text{scale}(\mathcal{R}_{\text{BC/IC}})}.$$

A more principled approach is *adaptive loss balancing*, where α_w and β_w are treated as additional trainable parameters updated alongside θ and λ. One popular method is the *uncertainty weighting* scheme:

$$\begin{aligned} L(\theta,\lambda,\sigma_d,\sigma_p,\sigma_b) = {} & \frac{1}{2\sigma_d^2}\mathcal{R}_{\text{data}} + \frac{1}{2\sigma_p^2}\mathcal{R}_{\text{PDE}} + \frac{1}{2\sigma_b^2}\mathcal{R}_{\text{BC/IC}} \\ & + \log\sigma_d + \log\sigma_p + \log\sigma_b, \end{aligned}$$

where $\sigma_d, \sigma_p, \sigma_b > 0$ are learned noise scales representing the relative confidence in each component. This formulation naturally balances the gradients so that no single term dominates, and it has been shown to accelerate convergence in inverse problems.

In practice, residual normalization can be combined with non-dimensionalization: first bring variables and coefficients to comparable physical scales, then balance the loss components numerically. The result is a more stable optimization landscape where all physical constraints are enforced in harmony.

Benefit: By keeping the contributions of each loss term balanced, residual normalization prevents the optimizer from "forgetting" important constraints and improves both convergence speed and final accuracy.

5.3.4.3 Hard Constraints When Feasible

When boundary and initial conditions are simple and known exactly, it is often advantageous to enforce them *by construction* rather than as soft penalties in the loss. This is done by designing the trial solution u_θ to satisfy the conditions for all θ, a strategy sometimes referred to as the *method of constrained trial functions*.

For example, suppose the PDE is defined on $x \in [0, 1]$, $t \in [0, T]$, with Dirichlet boundary conditions

$$u(0, t) = g_0(t), \quad u(1, t) = g_1(t),$$

and initial condition

$$u(x, 0) = u_0(x).$$

We can write a modified network output:

$$\tilde{u}_\theta(x, t) = g_0(t) + x\,[g_1(t) - g_0(t)] + x(1 - x)\,N_\theta(x, t),$$

where N_θ is an unconstrained neural network. The first two terms interpolate the boundary values exactly, while the multiplicative factor $x(1 - x)$ ensures that the correction N_θ vanishes at both $x = 0$ and $x = 1$. To incorporate the initial condition exactly, one may further define:

$$\tilde{u}_\theta(x, t) = u_0(x) + t\,B(x)\,N_\theta(x, t),$$

where $B(x)$ is a spatial mask (e.g., $x(1 - x)$) that enforces both spatial and temporal boundary conditions. This guarantees $\tilde{u}_\theta(x, 0) = u_0(x)$ for all θ.

By enforcing constraints in this way, the BC/IC loss $\mathcal{R}_{\mathrm{BC/IC}}$ is identically zero, removing one source of stiffness in the optimization and reducing the number of terms that need balancing. Moreover, the network capacity is focused entirely on satisfying the PDE residual and fitting the data, rather than spending optimization effort on re-learning fixed physical facts.

Benefit: Hard constraints eliminate systematic violations of known conditions, reduce the size of the loss landscape, and often improve generalization, particularly in long-time extrapolation where small boundary errors could otherwise accumulate.

5.3.4.4 Curriculum Learning

The idea of curriculum learning in PINNs [5] is to begin training with an *easier* version of the problem and gradually increase the difficulty, much like how humans learn progressively from simple to complex tasks. For PDE-constrained learning, "difficulty" can be defined in several ways:

1. **Number of collocation points**: Start with a relatively small set of collocation points (x, t) where the PDE residual is evaluated. The network first learns a coarse approximation of the solution. Over successive training stages, the number of collocation points is increased, refining the approximation and forcing the network to respect the PDE more strictly across the entire domain.
2. **Physics penalty weight**: Begin with a small value of the PDE residual weight α_w in the total loss:

$$L(\theta) = \mathcal{R}_{\text{data}}(\theta) + \alpha_w \, \mathcal{R}_{\text{PDE}}(\theta) + \beta_w \, \mathcal{R}_{\text{BC/IC}}(\theta).$$

   Initially, a small α_w allows the optimizer to fit observed data without being overly constrained by physics. As training progresses, α_w is increased, shifting emphasis toward strict PDE satisfaction.
3. **Domain expansion**: In time-dependent problems, start training over a shorter time window (e.g., $t \in [0, T_0]$ with $T_0 < T$) where dynamics are simpler. Once the model has learned this subproblem, extend the domain to the full time horizon, reusing the learned parameters as initialization.
4. **Complexity of physics**: For hybrid models or PDEs with multiple terms, train first on a simplified PDE (e.g., omit certain nonlinear terms) and gradually reintroduce the full complexity.

Formally, curriculum learning in PINNs can be expressed as a sequence of optimization problems:

$$\{(\theta^{(k)}, \lambda^{(k)})\} = \arg\min_{\theta,\lambda} L^{(k)}(\theta, \lambda), \quad k = 1, 2, \ldots, K,$$

where $L^{(k)}$ represents the k-th stage with its own collocation set, weights, or domain. The solution from stage k is used to initialize stage $k+1$.
Benefit: This progressive strategy helps the optimizer avoid poor local minima and reduces training instability, especially in stiff or multi-scale PDEs where enforcing the full physics from the outset can lead to divergence or slow convergence.

5.3.4.5 Adaptive Collocation

In the standard PINN workflow, collocation points (x_i, t_i) used to evaluate the PDE residual $\mathcal{R}_{\text{PDE}}$ are drawn from a fixed distribution, often uniform over the domain $\Omega \times [0, T]$. However, PDE solutions are rarely uniform in complexity across the domain: sharp gradients, boundary layers, shocks, or localized sources require much finer resolution to enforce the physics accurately. *Adaptive collocation* dynamically reallocates points to regions where the network's current prediction violates the PDE most severely.

The key steps are:

1. **Initial sampling**: Begin with a baseline set of collocation points $\mathcal{C}_0$ sampled uniformly or via low-discrepancy sequences (e.g., Sobol, Halton).
2. **Error estimation**: After a given number of training epochs, compute the PDE residual
$$r(x, t) = \mathcal{N}[u_\theta](x, t; \lambda)$$
on a dense probe grid or a large candidate set of points.
3. **Residual-based resampling**: Identify points where $|r(x, t)|$ is large, indicating poor physics satisfaction. These points are then selected with higher probability for the next collocation set $\mathcal{C}_{k+1}$.
4. **Iterative refinement**: Alternate between training and residual-based resampling until convergence. Optionally, retain a fraction of the old points to maintain coverage of the entire domain.

Formally, adaptive collocation can be seen as minimizing the PDE residual norm

$$\mathcal{R}_{\text{PDE}}(\theta, \lambda) = \frac{1}{|\mathcal{C}_k|} \sum_{(x,t)\in\mathcal{C}_k} r(x, t)^2$$

under a sampling distribution $p_k(x, t)$ that changes with the iteration k to concentrate mass where $|r(x, t)|$ is largest.

Benefit: By focusing training effort on the "hard" regions of the solution space, adaptive collocation improves accuracy without increasing the total number of collocation points. This is particularly effective for problems with sharp transients, localized phenomena, or moving fronts, where a uniform sampling would waste resources on regions that are already well approximated.

5.3.4.6 Robust Optimizers

Optimization in PINNs is inherently more challenging than in purely data-driven regression because the loss surface combines multiple heterogeneous components : data misfit, PDE residual, and boundary/initial penalties : that may differ in scale and curvature. These components can cause ill-conditioning and slow convergence if a single optimizer or step size is used throughout.

A widely adopted strategy is to combine the strengths of *first-order adaptive methods* and *second-order refinements*:

- **Stage 1: Rapid exploration**: Use an adaptive first-order optimizer such as `Adam` or `AdamW` for the initial phase. These optimizers rescale gradients based on their moving averages, enabling rapid reduction of large-scale errors and quick traversal toward promising regions of parameter space.
- **Stage 2: Precise convergence**: Once the loss plateau is reached, switch to a second-order or quasi-Newton method such as `L-BFGS`. These methods leverage curvature information to refine the solution and are especially effective at reducing the PDE residual to machine precision.

In practice, this staged approach can be formalized as:

$$\theta^{(k+1)} = \begin{cases} \theta^{(k)} - \eta_{\text{Adam}} \cdot \hat{g}_{\text{Adam}}(\theta^{(k)}), & k \leq k_{\text{switch}}, \\ \text{LBFGS}(\theta^{(k)}), & k > k_{\text{switch}}, \end{cases}$$

where k_{switch} is the iteration or epoch at which the optimizer changes, chosen via validation loss monitoring or after a fixed burn-in period.
Benefit: First-order methods are computationally cheap and robust to noisy gradients from mini-batches or imperfect residual sampling. Second-order refinements exploit the smoothness of the PDE residual landscape in later stages, achieving tighter physics satisfaction and improved generalization.

5.3.4.7 Architectures and Activations

The choice of neural network architecture and activation function has a direct impact on the ability of a Physics-Informed Neural Network (PINN) to represent smooth, physically plausible solutions.

- **Depth and width**. For smooth PDE solutions such as those from the heat equation, moderately deep fully-connected networks (e.g., 3–8 hidden layers) with sufficient width (e.g., 50–200 neurons per layer) are often adequate. Excessive depth may lead to vanishing gradients, while too few layers or neurons may limit expressiveness, particularly for multi-scale or nonlinear PDEs.
- **Activation functions**. Classical `tanh` activations are widely used in PINNs because they are smooth, infinitely differentiable, and capable of representing low- and high-frequency components. Alternatives such as `sin` activations (*SIREN* networks) can capture highly oscillatory solutions (e.g., wave equations) more efficiently. Piecewise-linear functions like `ReLU` are less ideal for PINNs, as their non-smooth derivatives can hinder PDE residual convergence : especially when higher-order derivatives appear in $\mathcal{N}[u_\theta]$.

- **Spectral bias mitigation**. Standard feedforward networks exhibit *spectral bias*: they learn low-frequency modes before high-frequency modes. For PDEs with sharp gradients or multi-frequency structure, techniques such as *Fourier feature embeddings*, *positional encoding*, or *frequency annealing* can help balance the representation of all modes.
- **Physics-guided architectures**. When feasible, hard constraints (Section 5.3.4.3) can be embedded into the architecture to exactly satisfy boundary or conservation laws. Examples include trial functions $u_\theta(x, t) = g(x, t) + h(x, t)v_\theta(x, t)$ where g and h enforce boundary conditions by construction.

Benefit: Choosing smooth, differentiable activations ensures stable computation of higher-order derivatives via automatic differentiation, while physics-guided architectures can drastically reduce the size of the search space and improve convergence speed.

5.3.4.8 Batching and Compiled Autodiff

Efficient computation of PDE residuals is essential in PINN training, as it involves repeated evaluation of the neural network and its derivatives at many collocation points.

- **Mini-batching collocation points**. Evaluating the PDE residual at all collocation points in one pass can exceed memory limits, especially for large n_{col} or when computing higher-order derivatives. A practical solution is to split the collocation set into mini-batches and accumulate loss over them, similar to stochastic gradient descent in classical machine learning.
- **Vectorized mapping (`vmap`)**. Frameworks such as JAX offer vectorized transformations (`jax.vmap`) to apply the same function to many inputs in parallel without writing explicit loops. This is particularly beneficial for derivative computations, where repeated scalar differentiation would otherwise be prohibitively slow.
- **Compiled computation (`jit`)**. Just-in-Time (JIT) compilation (e.g., `jax.jit`) transforms Python-level code into highly optimized, fused kernels. This can yield significant speedups by removing Python overhead and exploiting backend optimizations (e.g., XLA for JAX, TorchScript for PyTorch).
- **Memory–computation trade-off**. Automatic differentiation can be memory-intensive because intermediate values are stored for the backward pass. When memory is limited, techniques such as *checkpointing* (recomputing certain intermediate results during backpropagation) can balance memory usage and runtime.

Benefit: Batching and compiled autodiff allow PINNs to scale to larger domains and finer collocation grids without exhausting GPU memory, while maintaining the flexibility of automatic differentiation for complex PDEs.

5.3.4.9 Diagnostics and Early Stopping

Training a PINN involves a delicate balance between fitting the data, enforcing the PDE residual, and satisfying boundary/initial conditions. Unlike purely data-driven models, overfitting in PINNs may manifest as *overfitting to noise in the data or* as *degenerate solutions that satisfy the data but violate the PDE*. Robust diagnostics and stopping criteria help detect and prevent such outcomes.

- **Separate loss monitoring**. Track $\mathcal{R}_{\text{data}}$, $\mathcal{R}_{\text{PDE}}$, and $\mathcal{R}_{\text{BC/IC}}$ individually. A decrease in $\mathcal{R}_{\text{data}}$ with a simultaneous increase in $\mathcal{R}_{\text{PDE}}$ often signals that the network is sacrificing physics consistency for data fit.
- **Validation sets for physics and data**. In addition to a standard validation set for data points, define a set of *validation collocation points* not used during training. Monitoring PDE residuals on this set helps detect physics overfitting or poor generalization to unseen regions.
- **Early stopping**. Apply early stopping when both the validation data loss and validation PDE residual stop improving for a predefined number of epochs. This prevents unnecessary training and reduces the risk of the network drifting toward unphysical minima.
- **Gradient-path diagnostics**. Inspect the magnitudes of gradients with respect to network parameters and trainable physical parameters (e.g., α in the inverse problem). Vanishing or exploding gradients can indicate improper scaling, poor initialization, or an overly stiff loss landscape.
- **Residual visualization**. Periodically visualize the PDE residual field over the domain. Localized spikes can reveal regions where the network struggles to fit the physics, guiding adaptive collocation or targeted reweighting.

Benefit: By systematically monitoring loss components and residuals, practitioners can terminate training at a physically meaningful optimum, avoiding both overfitting and underfitting while saving computational resources.

5.3.4.10 Inverse Problems: Identifiability and Constraints

In inverse PINN formulations, one seeks to recover unknown physical parameters $\lambda \in \Lambda$ (e.g., diffusivity α) alongside the state field $u_\theta(x, t)$. Although conceptually straightforward, these problems are often *ill-posed* in the sense of Hadamard: small perturbations in the data can cause large variations in the inferred parameters, and multiple parameter sets may fit the observations equally well. Such issues are referred to as *identifiability* problems.

- **Sufficiency of data coverage**. Parameters that influence the solution globally (e.g., wave speed, diffusivity) require measurements that are both *spatially* and *temporally* distributed. Sparse or clustered measurements may fail to constrain λ, leading to multiple equally plausible fits.

- **Sensitivity analysis**. Before training, examine the sensitivity $\partial u/\partial \lambda$ by solving the PDE for perturbed values of λ. Low sensitivity suggests that λ cannot be reliably inferred from the available measurements, regardless of network capacity.
- **Regularization and priors**. Add penalty terms (e.g., $\|\lambda\|^2$) or enforce bounds $\lambda \in [\lambda_{\min}, \lambda_{\max}]$ to restrict the search space. In Bayesian PINNs, place informative priors on λ to encode physical knowledge and stabilize inference.
- **Joint versus staged training**. When λ and θ are both unknown, it can help to *stage* the training: (i) fit u_θ with a fixed initial guess for λ, (ii) update λ while holding θ fixed, (iii) jointly fine-tune both. This avoids situations where poor early estimates of λ derail the network's learning of the state field.
- **Physics-based constraints**. If λ has known physical meaning, incorporate constraints directly into the model: for example, enforce $\alpha > 0$ by parameterizing $\alpha = \text{softplus}(\beta)$, or impose conservation laws to reject unphysical combinations of parameters and states.

Remark: Identifiability is not just a numerical issue but a structural one: if the PDE and data acquisition setup do not uniquely determine λ, no amount of optimization will recover the correct value. Well-designed experiments, informed priors, and constraint enforcement are often as important as the PINN architecture itself in successful inverse problem solving.

5.3.4.11 Domain Decomposition and Time Marching

When the spatio–temporal domain $\Omega \times [0, T]$ is large, training a single PINN over the entire domain can be inefficient or even unstable. Two complementary strategies: *domain decomposition* and *time marching* : help alleviate this by breaking the problem into smaller, more manageable subproblems.

- **Spatial domain decomposition**. Partition the spatial domain Ω into subdomains $\{\Omega_j\}_{j=1}^J$ and train a separate PINN u_{θ_j} on each. To ensure global consistency, impose *interface conditions* (continuity of u and, if required, of fluxes $\partial_n u$) as additional loss terms:

$$\mathcal{R}_{\text{interface}} = \sum_{j=1}^{J-1} \frac{1}{n_{\text{int}}} \sum_{(x,t)\in\Gamma_{j,j+1}} \left(u_{\theta_j}(x,t) - u_{\theta_{j+1}}(x,t)\right)^2 + \eta\left(\partial_n u_{\theta_j}(x,t) - \partial_n u_{\theta_{j+1}}(x,t)\right)^2.$$

  This approach reduces the number of collocation points per network and allows parallel training.
- **Time marching (sequential training)**. Instead of solving for $u(x, t)$ over $[0, T]$ in one network, divide the interval into slabs $[t_k, t_{k+1}]$ and train a PINN on each slab, using $u(x, t_k)$ from the previous step as the initial condition for the next. This can significantly improve stability for long-time integrations, especially for stiff or chaotic systems.

- **Overlapping versus non-overlapping domains**. Overlapping domains ($\Omega_j \cap \Omega_{j+1} \neq \emptyset$) improve stability by providing shared context for neighboring sub-PINNs, at the cost of more training data. Non-overlapping decompositions are cheaper but require stronger interface condition enforcement.
- **Hybrid approaches**. Domain decomposition and time marching can be combined : e.g., split a large 3D transient problem into spatial subdomains and then apply time marching in each. This is particularly effective on distributed hardware, where each subdomain–time-slab pair can be trained on a separate device.

Remark: These strategies are closely related to classical numerical methods such as the *Schwarz alternating method* for PDEs and *marching-on-in-time* schemes. In the PINN setting, they reduce the complexity of each subproblem, make training more robust, and naturally enable parallelism : often essential for scaling to industrial-scale PDE applications.

5.3.4.12 Transfer Learning and Warm Starts

In many scenarios, the PDE to be solved (or its solution manifold) changes only slightly between related problems:for example, when boundary conditions, parameters, or geometries vary within a family. In such cases, *transfer learning* and *warm starts* can dramatically reduce training time and improve convergence.

- **Transfer learning across problems**. Suppose we have already trained a PINN u_{θ^*} for a PDE with parameters λ^*. For a new problem with $\lambda' \approx \lambda^*$, we can initialize the new network with θ^* instead of random weights. This reuses learned low-level features (e.g., smooth spatial–temporal basis functions) and often requires only fine-tuning:

$$\theta^{(0)} \leftarrow \theta^*, \quad \lambda^{(0)} \leftarrow \lambda^* \text{ (if known or to be estimated).}$$

- **Warm starts in continuation methods**. For difficult PDEs or high-frequency regimes, start from a simpler, solvable instance (e.g., lower Reynolds number [6], coarser grid, smaller time horizon) and progressively increase complexity. At each stage, initialize from the converged network of the previous stage:

$$\theta^{(k+1)} \leftarrow \arg\min_{\theta} L(\theta;\ \text{stage } k+1) \quad \text{with} \quad \theta^{(k)} \text{ as initial guess.}$$

  This is a form of *curriculum learning* in parameter space.
- **Cross-domain initialization**. When solving PDEs on similar geometries (e.g., different rod lengths or material properties), initialize the early layers from a previously trained PINN on a reference domain, while reinitializing the last layer(s) to adapt to the new output scale or boundary conditions.
- **Regularization during fine-tuning**. When transferring from one problem to another, a small *elastic weight consolidation* term

$$\mathcal{R}_{\mathrm{EWC}} = \mu \sum_{\ell} F_{\ell}(\theta_{\ell} - \theta_{\ell}^{*})^{2}$$

can prevent catastrophic forgetting, where F_ℓ is the estimated Fisher information for parameter θ_ℓ.

Remark: In the context of domain decomposition or time marching (see previous subsection), transfer learning can initialize each subdomain or time-slab network from a neighboring one, effectively propagating learned features through space and time. This mirrors the idea of "recycling" Krylov subspaces in numerical linear algebra: knowledge from one solve accelerates the next.

5.3.4.13 Multi-fidelity and Hybrid Modeling

Pure PINNs can become computationally expensive or data-hungry when the PDE is complex, multi-scale, or when the parameter space is large. *Multi-fidelity* and *hybrid (gray-box)* modeling leverage additional, cheaper sources of information to accelerate and stabilize training.

- **Multi-fidelity training**. Let u_{HF} denote the high-fidelity (accurate but costly) solution and u_{LF} a low-fidelity approximation (e.g., coarse-grid simulation, simplified physics). Multi-fidelity PINNs integrate both sources:

$$L(\theta) = w_{\mathrm{HF}} \mathcal{R}_{\mathrm{data}}^{\mathrm{HF}} + w_{\mathrm{LF}} \mathcal{R}_{\mathrm{data}}^{\mathrm{LF}} + \alpha\, \mathcal{R}_{\mathrm{PDE}} + \beta\, \mathcal{R}_{\mathrm{BC/IC}}.$$

  Low-fidelity data guides the network toward the correct qualitative structure, while high-fidelity data refines the details.
- **Correction networks**. A simple approach is to train the PINN to model only the *residual* between low- and high-fidelity solutions:

$$r_\theta(x, t) \approx u_{\mathrm{HF}}(x, t) - u_{\mathrm{LF}}(x, t).$$

  The final prediction is $u_{\mathrm{LF}} + r_\theta$, focusing the learning capacity on discrepancies rather than reconstructing the full solution.
- **Hybrid (gray-box) PINNs**. Embed a mechanistic solver $f_{\mathrm{phys}}(u; \lambda)$ inside the PINN framework:

$$u_{k+1} = f_{\mathrm{phys}}(u_k; \lambda) \;+\; r_\theta(u_k, t_k),$$

  where r_θ is a small-capacity neural correction term. This preserves the main dynamics in f_{phys} while learning unmodeled effects (e.g., subgrid physics, missing forcing).
- **Co-training with model outputs**. When a fast reduced-order model (ROM) or empirical fit is available, use it to generate abundant pseudo-labels in regions without measurements. These pseudo-labels can be assigned a lower weight in the loss to prevent overfitting to low-accuracy predictions.

- **Regularization toward the mechanistic prior**. When the mechanistic model is trusted but imperfect, penalize deviations from it:

$$\mathcal{R}_{\text{prior}}(\theta) = \frac{1}{n_{\text{col}}} \sum_i \left(u_\theta(z_i) - u_{\text{phys}}(z_i)\right)^2.$$

  The penalty weight can be decreased over training to allow the network to correct biases without discarding useful prior knowledge.

Remark: Multi-fidelity and hybrid approaches align with engineering practice, where fast but approximate solvers are refined using sparse experimental data. In the PINN context, they improve data efficiency, reduce optimization stiffness, and preserve interpretability by keeping the physical model in the loop.

Summary of Best Practices for PINNs

Effective training of Physics-Informed Neural Networks requires a careful balance between *data fidelity*, *physics enforcement*, and *numerical stability*. The strategies discussed can be grouped into four complementary themes:

1. **Preconditioning and scaling**: Techniques such as non-dimensionalization, residual normalization, and variable scaling ensure that gradients are well-behaved and loss terms are balanced.
2. **Constraint enforcement and sampling**: Hard constraints, curriculum learning, and adaptive collocation focus the network's capacity where it matters most, improving accuracy and physical consistency.
3. **Optimization strategy**: Robust optimizers, batching with compiled autodiff, and diagnostic tools (including early stopping) help navigate challenging loss landscapes and prevent overfitting.
4. **Structural enhancements**: Domain decomposition, time marching, transfer learning, multi-fidelity data integration, and hybrid modeling extend the applicability of PINNs to larger, stiffer, and more complex systems.

By combining these principles (starting with good scaling, enforcing constraints where possible, refining sampling adaptively, and leveraging prior models) practitioners can improve convergence speed, stability, and generalization of PINNs across both forward and inverse problems.

5.4 A Taxonomy of Physics-Informed Neural Networks

Physics-Informed Neural Networks (PINNs) constitute a broad family of neural architectures in which the learning process is explicitly guided by a set of physical laws, typically expressed as partial differential equations (PDEs), together with boundary and initial conditions. While the general framework was introduced in [9], the landscape of PINN methodologies has rapidly evolved, giving rise to a variety of formulations, architectures, and training paradigms. We organize Physics-Informed Neural Networks (PINNs) into five complementary axes that reflect the *learning objective*, *domain decomposition*, *constraint enforcement*, *uncertainty treatment*, and *model integration*. Throughout, we retain the notation introduced earlier: $\Omega \subset \mathbb{R}^d$ (spatial domain), $[0, T]$ (time horizon), state field $u : \Omega \times [0, T] \to \mathbb{R}^m$, governing physics $\mathcal{N}[u](x, t; \boldsymbol{\lambda}) = 0$ with boundary/initial operators $\mathcal{B}[u] = 0$ and $u(\cdot, 0) = u_0$, data $\mathcal{D} = \{(z_i, y_i)\}_{i=1}^n$, and a parametric surrogate $u_\theta \in \mathcal{F}$. Loss terms follow Sect. 5.3:

$$\mathcal{R}_{\text{data}}, \quad \mathcal{R}_{\text{PDE}}, \quad \mathcal{R}_{\text{BC/IC}},$$

with positive weights $w_{\text{data}}, w_{\text{PDE}}, w_{\text{BC}}, w_{\text{IC}}$ (we use $w_\bullet$ to avoid overloading physical parameters such as α).

(A) By learning objective *What is being inferred?* (state only, state+parameters, or even the operator itself)

- **Forward PINNs (state estimation)**. The physical parameters $\boldsymbol{\lambda}$ are known. One minimizes

$$\min_{\theta} \; w_{\text{data}} \mathcal{R}_{\text{data}}(\theta) + w_{\text{PDE}} \mathcal{R}_{\text{PDE}}(\theta; \boldsymbol{\lambda}) + w_{\text{BC}} \mathcal{R}_{\text{BC/IC}}(\theta).$$

  This corresponds to our rod forward problem with fixed α.
- **Inverse PINNs (parameter identification)**. The parameters $\boldsymbol{\lambda}$ are unknown and are learned jointly:

$$\min_{\theta, \boldsymbol{\lambda}} \; w_{\text{data}} \mathcal{R}_{\text{data}}(\theta) + w_{\text{PDE}} \mathcal{R}_{\text{PDE}}(\theta; \boldsymbol{\lambda}) + w_{\text{BC}} \mathcal{R}_{\text{BC/IC}}(\theta) \quad \text{s.t. } \boldsymbol{\lambda} \in \Lambda.$$

  Optional priors or bounds (e.g., $\alpha \in [0.5, 2.0] \times 10^{-4}$) improve identifiability (see Sect. 5.3.4).
- **Discovery PINNs (structure/term discovery)**. Beyond coefficients, infer missing terms in $\mathcal{N}$. A common setting posits a library $\Phi(u)$ of candidate terms and sparse coefficients $c \in \mathbb{R}^M$:

$$\mathcal{N}[u_\theta](x, t; \; c) \equiv u_t - \sum_{j=1}^{M} c_j \, \Phi_j\big(u_\theta, \nabla u_\theta, \nabla^2 u_\theta, \dots\big) = 0,$$

  with a sparsity penalty $\lambda_{\text{sp}} \|c\|_1$:

$$\min_{\theta, c} \; w_{\text{data}} \mathcal{R}_{\text{data}}(\theta) + w_{\text{PDE}} \mathbb{E}_{(x,t)\sim\mu}\big[\mathcal{N}[u_\theta](x, t; \; c)^2\big]$$
$$+ \lambda_{\text{sp}} \|c\|_1 + w_{\text{BC}} \mathcal{R}_{\text{BC/IC}}(\theta).$$

(B) By domain decomposition *How is* $\Omega \times [0, T]$ *represented?* (single network versus tiled/partitioned networks)

- **Vanilla PINNs**. A single u_θ over $\Omega \times [0, T]$; objective as above.
- **XPINNs (extended PINNs)** [7]. Partition the domain into subdomains $\{\Omega_j\}_{j=1}^J$ (and optionally time slabs). Each subdomain has its own network u_{θ_j}. Add interface consistency on $\Gamma_{j\ell} = \partial\Omega_j \cap \partial\Omega_\ell$:

$$\mathcal{R}_{\text{int}} = \sum_{(j,\ell)} \mathbb{E}_{(x,t)\in\Gamma_{j\ell}}\Big[\big(u_{\theta_j} - u_{\theta_\ell}\big)^2 + \eta\big(\mathbf{n}_j \cdot \mathcal{F}(u_{\theta_j}) - \mathbf{n}_\ell \cdot \mathcal{F}(u_{\theta_\ell})\big)^2\Big],$$

where $\mathcal{F}$ denotes the physical flux (e.g., for diffusion $\mathcal{F} = -\alpha\nabla u$). The objective becomes

$$\min_{\{\theta_j\}} \sum_{j=1}^{J} \Big[w_{\text{data}}\mathcal{R}_{\text{data}}^{(j)} + w_{\text{PDE}}\mathcal{R}_{\text{PDE}}^{(j)} + w_{\text{BC}}\mathcal{R}_{\text{BC/IC}}^{(j)}\Big] + w_{\text{int}}\mathcal{R}_{\text{int}}.$$

- **cPINNs (conservative PINNs)** [8]. A special XPINN variant that *prioritizes flux conservation* across interfaces by emphasizing the flux term above (choose larger w_{int} and η), which is useful for conservation laws and transport.

(C) By constraint enforcement *How are* $\mathcal{B}[u] = 0$ *and* $u(\cdot, 0) = u_0$ *enforced?*

- **Soft constraints**. Include penalties in the loss:

$$\mathcal{R}_{\text{BC/IC}}(\theta) = \mathbb{E}_{(x,t)\in\partial\Omega\times(0,T]}\big[\mathcal{B}[u_\theta](x,t)^2\big] + \mathbb{E}_{x\in\Omega}\big[\big(u_\theta(x,0) - u_0(x)\big)^2\big].$$

- **Hard constraints**. Construct $\tilde{u}_\theta$ that *always* satisfies BC/IC. For Dirichlet boundaries $u|_{\partial\Omega} = g$ and initial $u(x, 0) = u_0(x)$, one template is

$$\tilde{u}_\theta(x,t) = g(x,t) + \psi(x,t)\, v_\theta(x,t), \qquad \psi|_{\partial\Omega} = 0,$$

and to encode the initial condition exactly,

$$\tilde{u}_\theta(x,t) = u_0(x) + t\,\phi(x,t)\, v_\theta(x,t), \qquad \phi|_{\partial\Omega} = 0,$$

so that $\tilde{u}_\theta(x, 0) = u_0(x)$ by construction. Then set $u_\theta \leftarrow \tilde{u}_\theta$ and drop the BC/IC penalty.

(D) By uncertainty treatment and variational form *How are epistemic/aleatoric uncertainties or weak forms handled?*

- **Bayesian PINNs**. Treat $(\theta, \boldsymbol{\lambda})$ as random, with priors $p(\theta)$, $p(\boldsymbol{\lambda})$ and Gaussian error models for each residual. Maximizing the posterior is equivalent to minimizing the negative log-posterior:

$$\min_{\theta,\boldsymbol{\lambda},\sigma_d,\sigma_p,\sigma_b} \frac{1}{2\sigma_d^2}\mathcal{R}_{\text{data}} + \frac{1}{2\sigma_p^2}\mathcal{R}_{\text{PDE}} + \frac{1}{2\sigma_b^2}\mathcal{R}_{\text{BC/IC}}$$
$$+ \log\sigma_d + \log\sigma_p + \log\sigma_b - \log p(\theta) - \log p(\boldsymbol{\lambda}).$$

Posteriors yield credible intervals on u and $\boldsymbol{\lambda}$.

- **Weak/variational-form PINNs**. Enforce physics in a weak sense: for test functions $v \in \mathcal{V}$,

$$\int_{\Omega} \mathcal{N}[u_\theta](x,t;\boldsymbol{\lambda})\, v(x)\, dx = 0 \quad \forall v \in \mathcal{V},$$

approximated via quadrature. After integration by parts, boundary flux terms appear explicitly and high-order derivatives on u_θ are reduced, improving stability for certain PDEs. Define a weak residual

$$\mathcal{R}_{\text{PDE}}^{\text{weak}}(\theta) = \sum_{q=1}^{Q} \Big(\int_{\Omega} \mathcal{N}[u_\theta]\, v_q\, dx \Big)^2$$

and substitute it for $\mathcal{R}_{\text{PDE}}$ in the total loss.

(E) By model integration (hybrid and multi-fidelity) *How does learning interact with mechanistic solvers or data of different fidelities?*

- **Hybrid (gray-box) PINNs**. Embed a trusted mechanistic core and learn corrections:

$$u_t - \mathcal{D}[u;\boldsymbol{\lambda}] = q_\theta(x,t,u),$$
$$\min_{\theta,\boldsymbol{\lambda}} w_{\text{data}}\mathcal{R}_{\text{data}} + w_{\text{PDE}}\mathbb{E}\big[(u_t - \mathcal{D}[u_\theta] - q_\theta)^2\big] + \lambda_{\text{reg}}\mathcal{R}_{\text{reg}}(q_\theta),$$

where $\mathcal{D}$ denotes the mechanistic operator (e.g., αu_{xx} for diffusion) and $\mathcal{R}_{\text{reg}}$ encodes parsimony/smoothness (see the rod hybrid model in the previous subsection).

- **Multi-fidelity PINNs**. Combine low- and high-fidelity observations:

$$\min_{\theta} w_{\text{HF}}\mathcal{R}_{\text{data}}^{\text{HF}}(\theta) + w_{\text{LF}}\mathcal{R}_{\text{data}}^{\text{LF}}(\theta) + w_{\text{PDE}}\mathcal{R}_{\text{PDE}}(\theta) + w_{\text{BC}}\mathcal{R}_{\text{BC/IC}}(\theta),$$

or learn the LF→HF discrepancy r_θ with $u_{\text{pred}} = u_{\text{LF}} + r_\theta$ to focus capacity on corrections.

Most variants are captured by selecting (i) which variables are optimized (only θ, or θ and $\boldsymbol{\lambda}$, or also c in discovery), (ii) how $\mathcal{R}_{\text{PDE}}$ is formed (strong versus weak form, subdomain versus global, hybrid residuals), and (iii) which additional priors/constraints are active (interface conservation, sparsity, Bayesian noise scales). Table 5.1 and Fig. 5.9 depicts the taxonomy.

Table 5.1 A consolidated taxonomy of PINNs along five axes. The "base loss" refers to the forward formulation with $\mathcal{R}_{\text{data}}$, $\mathcal{R}_{\text{PDE}}$, $\mathcal{R}_{\text{BC/IC}}$; each variant adds, replaces, or constrains components accordingly

Axis	Variant	Canonical objective (increment over base loss)	Notes/Typical use
Learning objective	Forward PINN	$\min_\theta \ w_{\text{data}}\mathcal{R}_{\text{data}} + w_{\text{PDE}}\mathcal{R}_{\text{PDE}} + w_{\text{BC}}\mathcal{R}_{\text{BC/IC}}$	State reconstruction with known $\boldsymbol{\lambda}$
	Inverse PINN	$+ \min_{\boldsymbol{\lambda}\in\Lambda}$ (jointly with θ)	Parameter ID (e.g., α in heat)
	Discovery PINN	$+ \lambda_{\text{sp}}\\|c\\|_1$, $\mathcal{N}[u_\theta] = u_t - \sum_j c_j \Phi_j(\cdot)$	Structure/term discovery, sparse libraries
Domain decomposition	Vanilla PINN	Single u_θ on $\Omega \times [0, T]$	Simple domains
	XPINN	$+ w_{\text{int}}\mathcal{R}_{\text{int}}$ (state & flux continuity on interfaces)	Large/multiscale domains; parallel training
	cPINN	Emphasis on flux match in $\mathcal{R}_{\text{int}}$	Conservation laws, transport
Constraint enforcement	Soft constraints	Include $\mathcal{R}_{\text{BC/IC}}$ as penalties	Flexible; small BC/IC violations allowed
	Hard constraints	Replace u_θ by $\tilde{u}_\theta$ satisfying BC/IC by design	Zero BC/IC residual; easier optimization
Uncertainty/variational	Bayesian PINN	$+ \sum_\bullet \left(\frac{1}{2\sigma_\bullet^2}\mathcal{R}_\bullet + \log\sigma_\bullet\right) - \log p(\theta) - \log p(\boldsymbol{\lambda})$	UQ on u, $\boldsymbol{\lambda}$; robust inverse problems
	Weak/variational PINN	Replace $\mathcal{R}_{\text{PDE}}$ by $\mathcal{R}_{\text{PDE}}^{\text{weak}}$ with quadrature/test functions	Lowers derivative order; better for rough solutions
Model integration	Hybrid (gray-box)	Residual form $u_t - \mathcal{D}[u_\theta] = q_\theta$, $+ \lambda_{\text{reg}}\mathcal{R}_{\text{reg}}(q_\theta)$	Missed physics as low-capacity residual
	Multi-fidelity	$+ w_{\text{LF}}\mathcal{R}_{\text{data}}^{\text{LF}}$ or discrepancy learning $u_{\text{LF}} + r_\theta$	Data-scarce regimes; leverage simulators

5.4.1 *Learning Objective in PINNs*

In the previous sections, we presented the two most common learning objectives for Physics-Informed Neural Networks (PINNs): *Forward PINNs* (state estimation with known physical parameters) and *Inverse PINNs* (joint estimation of the state and unknown physical parameters).

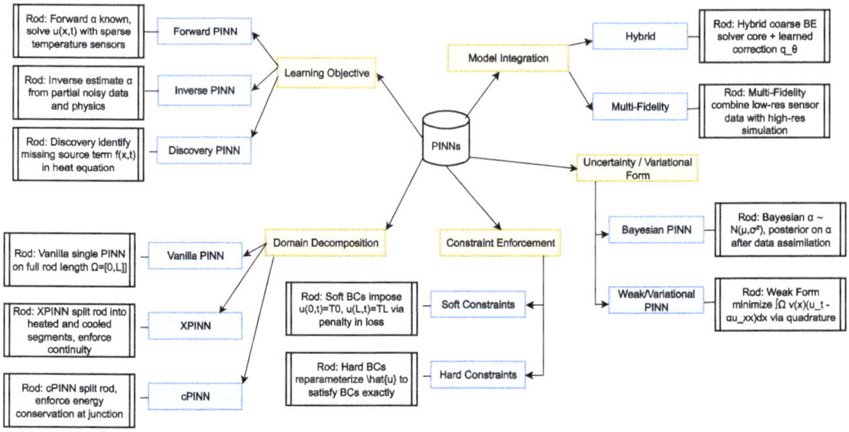

Fig. 5.9 Taxonomy of PINNs with our rod example

We now develop a third and more ambitious objective:

Discovery PINNs (structure/term identification). Here the governing operator is only partially known. Instead of assuming a fixed form for the physics $\mathcal{N}[u] = 0$, we posit that $\mathcal{N}$ is a sparse combination of candidate terms (a "library"), and we learn both the network parameters θ and the coefficients $c \in \mathbb{R}^M$ that select the active terms. Concretely, we write

$$u_t - \sum_{j=1}^{M} c_j \, \Phi_j\big(u_\theta, \nabla u_\theta, \nabla^2 u_\theta, \dots\big) = 0 \qquad \text{on } \Omega \times (0, T], \tag{5.3}$$

where $\{\Phi_j\}_{j=1}^M$ is a pre-defined dictionary (e.g., $u, u_x, u_{xx}, u\,u_x, u^2, \dots$). To promote parsimony and identifiability, we impose a sparsity prior on c (e.g., ℓ_1 penalty). Two common settings for the heat-rod example are:

- *Known diffusivity α, unknown source*: take $\{\Phi_j\} = \{\alpha u_{xx},\ 1,\ u,\ u_x,\ u_{xx},\ u^2,\ \dots\}$ and expect the optimizer to keep the known αu_{xx} term and select a small number of additional terms that act as sources/sinks.
- *Unknown structure including diffusion*: include u_{xx} itself in the library and let its coefficient c_{diff} play the role of an (estimated) diffusivity.

Discovery PINN Loss Function

The Discovery PINN minimizes the composite objective

$$\min_{\theta, c} \; w_{\text{data}} \mathcal{R}_{\text{data}}(\theta) \; + \; w_{\text{PDE}} \underbrace{\mathbb{E}_{(x,t)}\big[\mathcal{N}_c[u_\theta](x,t)^2\big]}_{\mathcal{R}_{\text{PDE}}(\theta, c)} \tag{5.4}$$
$$+ \, w_{\text{BC}} \mathcal{R}_{\text{BC/IC}}(\theta) \; + \; \lambda_{\text{sp}} \|c\|_1,$$

where $\mathcal{N}_c$ denotes the residual in (5.3) and $w_\bullet > 0$ are balancing weights. As in the inverse PINN, BC/IC may be imposed softly via $\mathcal{R}_{\text{BC/IC}}$ or *hard* by reparameterizing the trial solution.

Rod setup for discovery (recalled): We keep $\Omega = (0, 1)$, $T = 10\,\text{s}$, Dirichlet boundaries $u(0, t) = 100$, $u(1, t) = 25$, and initial condition $u(x, 0) = 25$. Sensor data $\mathcal{D} = \{(z_i, y_i)\}$ may be the same as earlier (e.g., three sensors at $x \in \{0.25, 0.5, 0.75\}$ sampled at $t \in \{0, 2, 4, 6, 8, 10\}$). Collocation points (x, t) are drawn in $\Omega \times (0, T]$ to estimate the residual expectation.

```
import jax, jax.numpy as jnp
from jax import grad, vmap, jit
import optax

--- Network, forward pass, and derivatives (as before) ---
def init_mlp(key, layers):
params, keys = [], jax.random.split(key, len(layers)-1)
for k, (m, n) in zip(keys, zip(layers[:-1], layers[1:])):
lim = jnp.sqrt(6/(m+n))
W = jax.random.uniform(k, (m, n), minval=-lim, maxval=lim)
b = jnp.zeros((n,))
params.append((W, b))
return params

def mlp_apply(params, x): # x: (..., d_in)
for W, b in params[:-1]:
x = jnp.tanh(x @ W + b)
W, b = params[-1]
return x @ W + b

def u_theta(params, x, t):
return mlp_apply(params, jnp.stack([x, t], axis=-1)).squeeze(-1)

u_x = grad(lambda p, x, t: u_theta(p, x, t), argnums=1)
u_t = grad(lambda p, x, t: u_theta(p, x, t), argnums=2)
u_xx = grad(lambda p, x, t: u_x(p, x, t), argnums=1)

--- Library of candidate terms ---
Minimal examples for diffusion/source discovery; expand as needed.
def phi_library(params, x, t, alpha_known=None):
u = u_theta(params, x, t)
```

```
ux = u_x(params, x, t)
uxx = u_xx(params, x, t)
# Include known alpha as a fixed term if alpha_known is provided
terms = []
if alpha_known is not None:
terms.append(alpha_known * uxx) # known diffusion contribution
# Candidate terms for discovery (examples)
terms.extend([
u, # linear source/sink
uxx, # diffusion with unknown prefactor (if alpha unknown)
ux, # convective-like term (1D surrogate)
uu, # quadratic reaction
uux, # nonlinear advection-like
jnp.ones_like(u), # constant source
])
# Stack as vector
return jnp.stack(terms, axis=0) # shape (M,)
```

Listing 5.8 Derivatives and a candidate-term library for Discovery PINN.

Explanation. This block builds a differentiable library $\Phi(u_\theta, \ldots)$ on-the-fly from the network prediction and its derivatives. You can tailor the dictionary to your physics (polynomials in u, derivatives, mixed products, trigs, etc.). When α is trusted, pass it as `alpha_known` so the diffusion term is *not* re-identified; otherwise omit it and let the optimizer learn the diffusion coefficient through the u_{xx} column.

```
--- Data loss (unchanged) ---
def data_loss(params, Z_data, y_data):
pred = vmap(lambda z: u_theta(params, z[0], z[1]))(Z_data)
return jnp.mean((pred - y_data)**2)

--- BC/IC penalty (soft constraints) ---
def bcic_loss(params, Z_b, u_b, Z_i, u_i):
ub_pred = vmap(lambda z: u_theta(params, z[0], z[1]))(Z_b) # boundary
    points
ui_pred = vmap(lambda z: u_theta(params, z[0], z[1]))(Z_i) # initial
    points
return jnp.mean((ub_pred - u_b)**2) + jnp.mean((ui_pred - u_i)**2)

--- PDE residual using the sparse library $\Phi$ and coefficients c
    ---
def pde_residual_disc(params, c, x, t, alpha_known=None):
ut = u_t(params, x, t)
Phi = phi_library(params, x, t, alpha_known=alpha_known) # (M,)
return ut - jnp.dot(c, Phi) # u_t - sum c_j $\Phi$_j

def pde_loss(params, c, Z_col, alpha_known=None):
R = vmap(lambda z: pde_residual_disc(params, c, z[0], z[1],
    alpha_known))(Z_col)
return jnp.mean(R**2)
```

Listing 5.9 Loss functions for the Discovery PINN.

Explanation. $\mathcal{R}_{\text{data}}$ and $\mathcal{R}_{\text{BC/IC}}$ follow our earlier definitions. The physics residual replaces the fixed operator by a learned sparse combination $\sum_j c_j \Phi_j$. Passing `alpha_known` includes a fixed diffusion contribution in the library; otherwise the optimizer decides whether and how much diffusion is needed via the coefficient on the u_{xx} term.

```
--- Proximal operator for L1 sparsity: soft-thresholding ---
def soft_threshold(v, lam):
return jnp.sign(v) * jnp.maximum(0.0, jnp.abs(v) - lam)

--- One proximal-gradient step for c (holding $\theta$ fixed) @jit ---
def prox_c_step(params, c, Z_col, alpha_known, lr_c, lam_l1):
# gradient of pde_loss wrt c
def loss_c(c_):
return pde_loss(params, c_, Z_col, alpha_known)
g = jax.grad(loss_c)(c)
# gradient step, then proximal soft-threshold
c_new = c - lr_c * g
c_new = soft_threshold(c_new, lam_l1 * lr_c)
return c_new
```

Listing 5.10 Sparsity via proximal (soft-threshold) update on c.

Explanation. Discovery is a *sparse regression* problem embedded inside a PINN. The ℓ_1 penalty is implemented efficiently by a proximal step: a gradient step on c followed by coordinate-wise soft-thresholding. This encourages exact zeros in c, selecting only the terms truly required by the data and physics.

```
--- Total loss with explicit L1 on c (for logging; c-update uses
     proximal step) ---
def total_loss(params, c, Z_col, Z_data, y_data, Z_b, u_b, Z_i, u_i,
w_data=1.0, w_pde=1.0, w_bc=1.0, lam_l1=1e-4, alpha_known=None):
Rd = data_loss(params, Z_data, y_data)
Rp = pde_loss(params, c, Z_col, alpha_known)
Rb = bcic_loss(params, Z_b, u_b, Z_i, u_i)
return w_dataRd + w_pdeRp + w_bcRb + lam_l1jnp.linalg.norm(c, 1)

--- $\theta$-step (Adam) with c fixed @jit ---
def theta_step(params, c, opt_state, opt, batch, weights, lam_l1,
     alpha_known):
Z_col, Z_data, y_data, Z_b, u_b, Z_i, u_i = batch
w_data, w_pde, w_bc = weights
loss_fun = lambda P: total_loss(P, c, Z_col, Z_data, y_data, Z_b,
     u_b, Z_i, u_i,
w_data, w_pde, w_bc, lam_l1, alpha_known)
L, grads = jax.value_and_grad(loss_fun)(params)
updates, opt_state = opt.update(grads, opt_state, params)
params = optax.apply_updates(params, updates)
return params, opt_state, L
```

```
--- Example minimal setup (replace with your loaders) ---
key = jax.random.PRNGKey(0)
layers = [2, 64, 64, 64, 1]
params = init_mlp(key, layers)

Sampling (as in earlier sections)
def sample_collocation(key, n, L=1.0, T=10.0):
x = jax.random.uniform(key, (n,), 0.0, L)
t = jax.random.uniform(key, (n,), 0.0, T)
return jnp.stack([x, t], axis=-1)

def sample_boundary_points(key_t, n_t, T=10.0):
t = jax.random.uniform(key_t, (n_t,), 0.0, T)
Z0 = jnp.stack([jnp.zeros_like(t), t], axis=-1)
ZL = jnp.stack([jnp.ones_like(t), t], axis=-1)
u0 = jnp.full((n_t,), 100.0)
uL = jnp.full((n_t,), 25.0)
return jnp.concatenate([Z0, ZL], axis=0), jnp.concatenate([u0, uL],
     axis=0)

def sample_initial_points(key_x, n_x, L=1.0):
x = jax.random.uniform(key_x, (n_x,), 0.0, L)
Z_i = jnp.stack([x, jnp.zeros_like(x)], axis=-1)
u_i = jnp.full((n_x,), 25.0)
return Z_i, u_i

k1,k2,k3,k4 = jax.random.split(key, 4)
Z_col = sample_collocation(k1, 4096)
Z_b, u_b = sample_boundary_points(k2, 256)
Z_i, u_i = sample_initial_points(k3, 256)

--- Example sensor grid (replace y_data with measurements) ---
xs = jnp.array([0.25, 0.5, 0.75])
ts = jnp.array([0., 2., 4., 6., 8., 10.])
Z_grid = jnp.stack(jnp.meshgrid(xs, ts, indexing="ij"),
     axis=-1).reshape(-1,2)
y_data = 25.0 + (100.0-25.0)*(1.0 - Z_grid[:,0]) *
     jnp.exp(-Z_grid[:,1]/10.0)

batch = (Z_col, Z_grid, y_data, Z_b, u_b, Z_i, u_i)

--- Initialize coefficients c (size M equals the library length
     without alpha_known) ---
alpha_known = 1.11e-4 # set to None if you want to learn diffusion too
M = phi_library(params, 0.5, 5.0, alpha_known).shape[0]
c = jnp.zeros((M,))

--- Optimizers and hyperparameters ---
opt = optax.adam(1e-3); opt_state = opt.init(params)
weights = (1.0, 1.0, 1.0)
lam_l1 = 1e-4
lr_c = 1e-2
```

```
--- Alternating minimization ---
for epoch in range(2000):
# (1) several theta-steps per epoch (keep c fixed)
for _ in range(5):
params, opt_state, L = theta_step(params, c, opt_state, opt, batch,
    weights, lam_l1, alpha_known)
# (2) one proximal c-step (keep theta fixed)
c = prox_c_step(params, c, Z_col, alpha_known, lr_c, lam_l1)
if epoch % 200 == 0:
print(f"Epoch {epoch:4d} | Loss={float(L):.4e} |
    nnz(c)={(jnp.count_nonzero(jnp.abs(c)>1e-6))}")
```

Listing 5.11 Joint training loop (alternating updates of θ and c).

Explanation. We alternate between (i) θ*-updates* (standard PINN training with c fixed) and (ii) c*-updates* (a proximal gradient step with soft-thresholding). This simple scheme works well in practice:

- The θ-step aligns u_θ to data/BC/IC and provides stable derivatives for the library.
- The c-step performs *sparse model selection* for the operator.

One may increase the number of θ-steps between c-updates to stabilize the alternation. Once training stabilizes, one can *prune* tiny coefficients (e.g., $|c_j| < 10^{-5}$), *freeze* the active set, and fine-tune θ with fixed c. However there are some practical considerations:

- **Library scaling**. Normalize each library column Φ_j to unit variance over a warmup set of collocation points; this prevents the optimizer from preferring terms solely because they are large in magnitude.
- **Sparsity schedule**. Start with a modest λ_{sp} (or `lam_l1`) and gradually increase to encourage pruning after the network has learned a decent state approximation.
- **Identifiability**. Ensure the candidate terms are not collinear (e.g., u and u^2 can correlate on narrow ranges). Diverse collocation and sensor coverage in both x and t improves term separation.
- **Known versus unknown diffusion**. If α is trusted, include αu_{xx} as a fixed (non-learned) contribution by passing `alpha_known`; otherwise omit it and let the coefficient on u_{xx} be learned as part of c.
- **Validation.** Track the PDE residual on a separate collocation set, the number of active terms $|\hat{c}|0$, and out-of-sample error on sensors not used in training.
- **Hard constraints (optional)**. If BC/IC are simple and known exactly, reparameterize $u\theta$ to satisfy them identically (see Section ??) and drop $\mathcal{R}_{BC/IC}$.

Outcome. The Discovery PINN yields a sparse vector $\hat{c}$ selecting a small set of physically meaningful terms in the library and a function u_θ that both fits data and satisfies the learned operator. In the rod scenario, this can reveal, e.g., a weak localized source (constant or proportional to u) in addition to diffusion, or confirm that a pure diffusion model suffices within sensor noise levels.

5.4.2 *Domain Decomposition in PINNs*

In some scenarios, solving a PDE over the entire spatio–temporal domain $\Omega \times (0, T]$ using a single neural network may be inefficient or inaccurate. This is particularly true when:

- the solution exhibits strong local variations or discontinuities,
- material properties are heterogeneous across the domain,
- the spatial or temporal extent is large, making optimization difficult.

Domain Decomposition addresses this by partitioning $\Omega \times (0, T]$ into smaller subdomains

$$\{\mathcal{D}_1, \mathcal{D}_2, \ldots, \mathcal{D}_K\},$$

with each subdomain assigned its own neural network $u_{\theta_k}^{(k)}$. Interface conditions between subdomains enforce physical continuity of the solution and, where required, its flux.

Vanilla PINN

Vanilla PINN uses a *single* network for the entire domain.

$$u_\theta : \Omega \times (0, T] \to \mathbb{R}$$

All residuals (data, PDE, and BC/IC) are computed globally. This approach is simple but can struggle with multi-scale or heterogeneous problems.

XPINN (Extended PINN)

XPINN splits $\Omega \times (0, T]$ into multiple non-overlapping subdomains. For each $\mathcal{D}_k$, a network u_{θ_k} is trained with:

$$\mathcal{R}^{(k)}_{\text{PDE}}, \quad \mathcal{R}^{(k)}_{\text{BC/IC}}, \quad \mathcal{R}^{(k)}_{\text{interface}}$$

The interface loss enforces:

$$u_{\theta_i}(x,t) = u_{\theta_j}(x,t), \quad \nabla u_{\theta_i}(x,t) = \nabla u_{\theta_j}(x,t), \quad \forall (x,t) \in \Gamma_{ij}.$$

cPINN (Conservative PINN)

cPINN builds on XPINN but strengthens the interface constraints to ensure *exact conservation laws* (e.g., mass, energy). Flux continuity becomes:

$$\mathbf{F}^{(i)}(u_{\theta_i}) \cdot \mathbf{n}_{ij} = \mathbf{F}^{(j)}(u_{\theta_j}) \cdot \mathbf{n}_{ji}, \quad \forall (x,t) \in \Gamma_{ij}$$

where $\mathbf{F}$ is the PDE flux vector and $\mathbf{n}$ is the interface normal.

General Loss Structure

Across all domain decomposition strategies, the total loss takes the form:

$$\mathcal{L} = \sum_{k=1}^{K} \left[w_{\text{data}}^{(k)} \mathcal{R}_{\text{data}}^{(k)} + w_{\text{PDE}}^{(k)} \mathcal{R}_{\text{PDE}}^{(k)} + w_{\text{BC/IC}}^{(k)} \mathcal{R}_{\text{BC/IC}}^{(k)} \right] + \sum_{\Gamma_{ij}} w_{\text{int}}^{(i,j)} \mathcal{R}_{\text{interface}}^{(i,j)}.$$

In our Rod Example, we revisit the 1D heat equation $u_t = \alpha(x) u_{xx}$ on $\Omega = (0, 1)$. Let $\alpha(x) = \alpha_1$ for $x \in (0, 0.5]$ and $\alpha_2 = 2\alpha_1$ for $x \in (0.5, 1)$. We define:

$$\mathcal{D}_1 = (0, 0.5], \quad \mathcal{D}_2 = (0.5, 1)$$

and train u_{θ_1} and u_{θ_2} separately, with interface continuity enforced at $x = 0.5$.

```
def interface_loss(params1, params2, x_int, t_int, alpha1, alpha2,
    beta):
    # Temperature match
    u1 = u_theta1(params1, x_int, t_int)
    u2 = u_theta2(params2, x_int, t_int)
    temp_mismatch = jnp.mean((u1 - u2)**2)
    # Flux match
    ux1 = u_x(params1, x_int, t_int)
    ux2 = u_x(params2, x_int, t_int)
    flux_mismatch = jnp.mean((alpha1*ux1 - alpha2*ux2)**2)
    return temp_mismatch + beta * flux_mismatch
```

Listing 5.12 Interface loss for two-domain rod problem.

Advantages and Challenges:

- **Advantages**: better handling of heterogeneous coefficients, parallelizable training, improved convergence for multi-scale problems.

- **Challenges**: selection of interface points, balancing of interface weights, potential instability for highly coupled subdomains.

5.4.3 *Constraint Enforcement in PINNs*

A central design choice in PINNs is *how* to enforce boundary/initial conditions (BC/IC) and other side constraints. We distinguish two complementary strategies:

Soft Constraints (penalty-based)

Soft constraints add penalty terms to the loss:

$$\mathcal{R}_{\text{BC/IC}}(\theta) = \mathbb{E}_{(x,t)\in\partial\Omega\times(0,T]}\big[\mathcal{B}[u_\theta](x,t)^2\big] + \mathbb{E}_{x\in\Omega}\big[(u_\theta(x,0) - u_0(x))^2\big].$$

They are simple and general (Dirichlet/Neumann/Robin, mixed/time-dependent), but permit small violations whose magnitude depends on weights.

Hard Constraints (by construction)

Hard constraints reparameterize the trial solution so BC/IC hold *exactly* for all θ:

$$\tilde{u}_\theta(x,t) = g(x,t) + \psi(x,t)\, v_\theta(x,t),\ \psi|_{\partial\Omega} = 0,\ \tilde{u}_\theta(x,0) = u_0(x) \text{ by design.}$$

They eliminate the BC/IC loss and improve optimization, at the cost of engineering suitable masks ψ (and analogous constructs for Neumann/Robin).

We now give compact JAX patterns for both strategies, using the rod setup from earlier: $\Omega = (0, 1), T = 10\,\text{s}$; Dirichlet $u(0, t) = 100, u(1, t) = 25$; initial $u(x, 0) = 25$.

```
# --- Hard Dirichlet BCs: enforce u(0,t)=T0, u(1,t)=TL for all t ---
def hard_bc_ansatz(params_core, x, t, T0=100.0, TL=25.0):
    """
    Template: u(x,t) = linear_interp(T0, TL) + B(x) * v_theta(x,t)
    with B(0)=B(1)=0 so the correction vanishes on the boundary.
    """
    # Linear interpolant
```

```
    g = (1.0 - x)*T0 + x*TL
    # Boundary mask: zero at x=0,1 (tanh-smooth option helps gradients)
    B = x * (1.0 - x)                         # simple polynomial mask
    core = mlp_apply(params_core, jnp.stack([x,t], axis=-1)).squeeze(-1)
    return g + B * core

# --- Add exact initial condition: u(x,0)=u0(x) ---
def hard_bcic_ansatz(params_core, x, t, u0_fun, T0=100.0, TL=25.0):
    """
    Enforce IC exactly by blending with time factor: at t=0 return u0(x);
    for t>0, transition to the boundary-satisfying form.
    """
    # base hard-BC form (will be used for t>0)
    g = (1.0 - x)*T0 + x*TL
    B = x * (1.0 - x)
    core = mlp_apply(params_core, jnp.stack([x,t], axis=-1)).squeeze(-1)

    # smooth time gate: s(t)=t (or tanh-based) ensures s(0)=0, s(t)>0 for t>0
    s = t
    # exact IC at t=0; for t>0 blend toward BC-safe correction
    return (1.0 - s) * u0_fun(x) + s * (g + B * core)

# If you adopt a hard form, REPLACE u_theta by hard_bcic_ansatz in all residuals:
def u_theta_hard(params_core, x, t):
    return hard_bcic_ansatz(params_core, x, t, u0_fun=lambda xx: 25.0, T0=100.0, TL=25.0)

# Then the BC/IC penalty is unnecessary:
def total_loss_hard(params_core, Z_col, alpha, Z_data, y_data, w):
    # PDE and data only (BC/IC are satisfied exactly by construction)
    def pde_residual_hard(p, x, t, a):
        # Use u_theta_hard and its derivatives
        uh = lambda xx, tt: u_theta_hard(p, xx, tt)
        ut = grad(lambda tt: uh(x, tt))(t)
        ux = grad(lambda xx: uh(xx, t))(x)
        uxx = grad(lambda xx: grad(lambda xxx: uh(xxx, t))(xx))(x)
        return ut - a * uxx

    Rp = jnp.mean(vmap(lambda z: pde_residual_hard(params_core, z[0], z[1], alpha))(Z_col)**2)
    Rd = jnp.mean((vmap(lambda z: u_theta_hard(params_core, z[0], z[1]))(Z_data) - y_data)**2)
    return w['data']*Rd + w['pde']*Rp
```

Listing 5.13 Hard Dirichlet BCs via spatial mask; exact IC via time factor.

For a Neumann condition $\partial_n u = h$ on $\partial\Omega$, one can encode the flux at the boundary by constructing a function with prescribed normal derivative. In 1D, enforce at $x = 0$

and $x = 1$ with boundary-local basis functions whose derivatives match h. A practical hybrid is: *hard Dirichlet, soft Neumann* (or vice versa), depending on which is more critical.

5.4.4 Uncertainty Quantification and Variational PINNs

Standard PINNs produce a single deterministic approximation $\hat{u}_\theta(x, t)$, which offers no direct measure of the model's confidence. In many scientific and engineering settings, *quantifying uncertainty* is as important as predicting the mean state. This is critical when:

- Data is sparse, noisy, or partially missing.
- The PDE model contains uncertain parameters or terms.
- We require robust decision-making based on model predictions.

We distinguish two main sources of uncertainty:

Epistemic Uncertainty (model uncertainty)

Represents lack of knowledge due to finite data or limited model capacity. It *can be reduced* by collecting more data or improving the model.

Aleatoric Uncertainty (data noise)

Represents inherent variability in the process (sensor noise, random forcing). It *cannot be eliminated* by more data; it must be modelled explicitly.

We now present three PINN-based strategies for uncertainty estimation:

1. **Monte Carlo (MC) Dropout**: Approximate Bayesian inference by enabling dropout during inference and sampling multiple predictions.
2. **Ensemble PINNs**: Train multiple PINNs with different initializations; variability across models approximates uncertainty.
3. **Bayesian / Variational PINNs (VPINNs)**: Place probability distributions over parameters (or outputs) and optimize via variational inference.

For Bayesian/variational PINNs, we treat parameters θ as random variables with posterior:

$$p(\theta \mid \mathcal{D}) \propto p(\mathcal{D} \mid \theta)\, p(\theta),$$

where $\mathcal{D}$ includes both data and PDE constraints. Variational inference approximates $p(\theta \mid \mathcal{D})$ with $q_\phi(\theta)$ by minimizing:

$$\mathrm{KL}\big(q_\phi(\theta) \parallel p(\theta \mid \mathcal{D})\big) \iff \mathcal{L}_{\mathrm{ELBO}} = \mathbb{E}_{q_\phi(\theta)}[\log p(\mathcal{D} \mid \theta)] - \mathrm{KL}(q_\phi(\theta) \parallel p(\theta)).$$

In PINNs, the PDE residual term is incorporated inside $\log p(\mathcal{D} \mid \theta)$.

Three strategies at a glance

1. **MC Dropout (Epistemic)**: keep dropout *on* at inference, sample many forward passes ⇒ predictive mean ± std.
2. **Ensemble PINNs (Epistemic)**: train M independent PINNs (different seeds/bootstraps); dispersion across models ⇒ uncertainty.
3. **Variational PINNs (Bayesian/Variational)**: place distributions over weights; optimize ELBO with a KL prior ⇒ posterior predictive.

We interpret the composite PINN loss as (negative) log-likelihood terms for data, PDE residual, and BC/IC, possibly with different noise scales. The uncertainty methods below differ in *how* they treat model parameters as random (dropout masks; multiple initializations; variational distributions over weights), but they reuse the same residuals you already defined.

```
# ============================================
# Uncertainty for PINNs: MC Dropout, Ensemble, Variational
# ============================================
import jax, jax.numpy as jnp
from jax import jit, grad, vmap
import optax

# ---------- Base MLP & derivatives ----------
def init_mlp(key, layers):
    params, keys = [], jax.random.split(key, len(layers)-1)
    for k,(m,n) in zip(keys, zip(layers[:-1], layers[1:])):
        lim = jnp.sqrt(6/(m+n))
        W = jax.random.uniform(k, (m,n), minval=-lim, maxval=lim)
        b = jnp.zeros((n,))
        params.append((W,b))
    return params

def mlp_apply(params, x):  # x: (..., d_in)
    for W,b in params[:-1]:
        x = jnp.tanh(x @ W + b)
    W,b = params[-1]
    return x @ W + b

def u_theta(params, x, t):
    return mlp_apply(params, jnp.stack([x,t], axis=-1)).squeeze(-1)

```

```
u_x  = grad(lambda p,x,t: u_theta(p,x,t), argnums=1)
u_t  = grad(lambda p,x,t: u_theta(p,x,t), argnums=2)
u_xx = grad(lambda p,x,t: u_x(p,x,t),     argnums=1)

# ---------- PINN residuals (as in earlier sections) ----------
def data_loss(params, Z_data, y_data):
    pred = vmap(lambda z: u_theta(params, z[0], z[1]))(Z_data)
    return jnp.mean((pred - y_data)**2)

def pde_loss(params, alpha, Z_col):
    R = vmap(lambda z: u_t(params, z[0], z[1]) - alpha*u_xx(params, z[0], z[1]))(Z_col)
    return jnp.mean(R**2)

def bcic_loss_dirichlet(params, Z_b, u_b, Z_i, u_i):
    ub = vmap(lambda z: u_theta(params, z[0], z[1]))(Z_b)
    ui = vmap(lambda z: u_theta(params, z[0], z[1]))(Z_i)
    return jnp.mean((ub - u_b)**2) + jnp.mean((ui - u_i)**2)

def total_loss_pinn(params, alpha, Z_col, Z_data, y_data, Z_b, u_b, Z_i, u_i, w):
    return (w['data']*data_loss(params, Z_data, y_data) +
            w['pde'] *pde_loss (params, alpha, Z_col)    +
            w['bc']  *bcic_loss_dirichlet(params, Z_b, u_b, Z_i, u_i))

# ==========================================
# (1) MC DROPOUT: keep dropout active at inference
# ==========================================
def mlp_with_dropout(params, x, t, rng, p_drop=0.1, force_on=True):
    """Dropout MLP: if force_on=True, apply dropout both in training and inference."""
    h = jnp.stack([x,t], axis=-1)
    for W,b in params[:-1]:
        h = jnp.tanh(h @ W + b)
        if force_on:
            keep = jax.random.bernoulli(rng, 1.0 - p_drop, h.shape)
            h = h * keep / (1.0 - p_drop)
    W_last, b_last = params[-1]
    return (h @ W_last + b_last).squeeze(-1)

def predict_mc_dropout(params, Z_eval, rng, n_samples=100, p_drop=0.1):
    """Return mean and std over n_samples stochastic forward passes."""
    preds = []
    for s in range(n_samples):
        rng_s = jax.random.fold_in(rng, s)
        pred_s = vmap(lambda z: mlp_with_dropout(params, z[0], z[1], rng_s, p_drop, True))(Z_eval)
        preds.append(pred_s)
    preds = jnp.stack(preds)  # (n_samples, n_points)
    return preds.mean(axis=0), preds.std(axis=0)
```

```
# ==========================================
# (2) ENSEMBLE: multiple independently trained PINNs
# ==========================================
@jit
def step_train(params, opt_state, opt, alpha, batch, w):
    Zc, Zd, yd, Zb, ub, Zi, ui = batch
    loss = lambda P: total_loss_pinn(P, alpha, Zc, Zd, yd, Zb, ub, Zi, ui, w)
    L, grads = jax.value_and_grad(loss)(params)
    updates, opt_state = opt.update(grads, opt_state, params)
    params = optax.apply_updates(params, updates)
    return params, opt_state, L

def train_single_model(seed, alpha, batch, w, layers=[2,64,64,64,1], lr=1e-3, epochs=2000):
    key = jax.random.PRNGKey(seed)
    params = init_mlp(key, layers)
    opt = optax.adam(lr); opt_state = opt.init(params)
    for ep in range(epochs):
        params, opt_state, L = step_train(params, opt_state, opt, alpha, batch, w)
    return params

def ensemble_predict(models, Z_eval):
    preds = [vmap(lambda z: u_theta(p, z[0], z[1]))(Z_eval) for p in models]
    preds = jnp.stack(preds)  # (n_models, n_points)
    return preds.mean(axis=0), preds.std(axis=0)

# ==========================================
# (3) VARIATIONAL PINN: variational distributions over weights
# ==========================================
# Represent each weight/bias with (mu, rho) so std = softplus(rho).
def init_variational_params(key, layers):
    params_mu, params_rho = [], []
    keys = jax.random.split(key, len(layers)-1)
    for k,(m,n) in zip(keys, zip(layers[:-1], layers[1:])):
        lim = jnp.sqrt(6/(m+n))
        W_mu = jax.random.uniform(k, (m,n), minval=-lim, maxval=lim)
        b_mu = jnp.zeros((n,))
        # initialize rho small negative => small std
        W_rho = -2.0 * jnp.ones((m,n))
        b_rho = -2.0 * jnp.ones((n,))
        params_mu.append((W_mu, b_mu)); params_rho.append((W_rho, b_rho))
    return params_mu, params_rho

def sample_params(mu_list, rho_list, rng):
    """Reparameterization: theta = mu + softplus(rho) * eps."""
    params = []
    for (W_mu, b_mu), (W_rho, b_rho) in zip(mu_list, rho_list):
        kW, kb = jax.random.split(rng, 2)
        W_eps = jax.random.normal(kW, W_mu.shape)
```

```
        b_eps = jax.random.normal(kb, b_mu.shape)
        W_std = jax.nn.softplus(W_rho); b_std = jax.nn.softplus(b_rho)
        W = W_mu + W_std * W_eps
        b = b_mu + b_std * b_eps
        params.append((W,b))
        rng = jax.random.fold_in(rng, 1)
    return params

def kl_normal(mu_list, rho_list, prior_std=1.0):
    """KL(q||p) for factorized Gaussians q=N(mu,softplus(rho)^2), p=N(0, prior_std^2)."""
    kl = 0.0
    for (W_mu, b_mu), (W_rho, b_rho) in zip(mu_list, rho_list):
        for mu, rho in [(W_mu, W_rho), (b_mu, b_rho)]:
            std_q = jax.nn.softplus(rho)
            kl += jnp.sum(jnp.log(prior_std/std_q) + (std_q**2 + mu**2)/(2*prior_std**2) - 0.5)
    return kl

def elbo_loss(mu_list, rho_list, alpha, batch, w, prior_std=1.0, n_mc=3, rng=jax.random.PRNGKey(0)):
    """Negative ELBO: average NLL (PINN residuals) over MC samples + KL."""
    Zc, Zd, yd, Zb, ub, Zi, ui = batch
    # Likelihood part (negative log-likelihood ~ weighted residuals)
    nll = 0.0
    for s in range(n_mc):
        rng_s = jax.random.fold_in(rng, s)
        params_s = sample_params(mu_list, rho_list, rng_s)
        nll += (w['data']*data_loss(params_s, Zd, yd) +
                w['pde'] *pde_loss (params_s, alpha, Zc) +
                w['bc']  *bcic_loss_dirichlet(params_s, Zb, ub, Zi, ui))
    nll = nll / n_mc
    # KL regularization to the prior
    kl = kl_normal(mu_list, rho_list, prior_std)
    return nll + kl

@jit
def step_variational(mu_list, rho_list, opt_state, opt, alpha, batch, w, prior_std, n_mc, rng):
    loss = lambda MU,RH: elbo_loss(MU, RH, alpha, batch, w, prior_std, n_mc, rng)
    L, grads = jax.value_and_grad(loss, argnums=(0,1))(mu_list, rho_list)
    updates, opt_state = opt.update(grads, opt_state, (mu_list, rho_list))
    mu_list, rho_list = optax.apply_updates((mu_list, rho_list), updates)
    return mu_list, rho_list, opt_state, L

def predict_variational(mu_list, rho_list, Z_eval, n_samples=100, rng=jax.random.PRNGKey(0)):
```

```
    preds = []
    for s in range(n_samples):
        rng_s = jax.random.fold_in(rng, s)
        params_s = sample_params(mu_list, rho_list, rng_s)
        preds.append(vmap(lambda z: u_theta(params_s, z[0],
    z[1]))(Z_eval))
    preds = jnp.stack(preds)  # (n_samples, n_points)
    return preds.mean(axis=0), preds.std(axis=0)
```

Listing 5.14 One unified JAX scaffold covering MC Dropout, Ensembles, and Variational PINNs.

How to use this unified scaffold:

- **Shared setup**. Prepare your batch $(Z_{\text{col}}, Z_{\text{data}}, y_{\text{data}}, Z_b, u_b, Z_i, u_i)$ and weights $w = \{\text{data, pde, bc}\}$ exactly as in the forward/inverse subsections.
- **MC Dropout**. Train a standard PINN (no code change to the loss). At inference, call `predict_mc_dropout(params, Z_eval, rng)` to obtain mean/std. The stochasticity comes from dropout masks (kept active).
- **Ensemble**. Train M independent models via `train_single_model(seed,...)` for different seeds; then aggregate with `ensemble_predict(models, Z_eval)`. Dispersion $\Rightarrow$ epistemic uncertainty.
- **Variational**. Initialize (μ, ρ) with `init_variational_params`, optimize the negative ELBO via `step_variational`, and predict with `predict_variational`. KL to a zero-mean Gaussian prior regularizes the weights.

What the standard deviations mean:

- **MC Dropout/Ensemble**: the reported $\sigma(x, t)$ approximates *epistemic* uncertainty (model uncertainty). It shrinks as you add informative data.
- **Variational**: the predictive spread includes epistemic effects via weight posteriors. If you also want *aleatoric* (data) noise, extend u_θ to output $(\mu_\theta, \log \sigma_\theta^2)$ and replace the MSE data loss by a Gaussian NLL; do likewise for PDE/BC residuals if you want per-component noise scales (see uncertainty weighting in Sect. 5.3.4).

Practical tips

- **Calibration**: Assess whether predicted intervals have correct coverage using held-out sensors or synthetic ground truth.
- **Cost versus fidelity**: MC Dropout is cheapest, Ensembles cost $M\times$, Variational is in between but needs KL tuning (prior_std) and sufficient MC samples in the ELBO.
- **Physics awareness**: Always compute uncertainty on *physics-respecting* solutions: the code above samples models and still enforces PDE/BC terms in the objective, so uncertainty reflects both data and physics constraints.

5.4.5 Model Integration

We focus here on the two integration modes used throughout this chapter–*Hybrid (gray-box)* and *Multi-Fidelity*—and provide succinct, non-redundant implementations that complement earlier material. We reuse the rod notation and losses defined before (data, PDE, BC/IC), and only introduce what is new for each mode.

Hybrid (Gray-Box) PINN

Embed a trusted mechanistic core and learn a small, structured correction:

$$u_t - \alpha\, u_{xx} = q_\theta(x, t, u_\theta) \quad \Rightarrow \quad \mathcal{R}^{\text{hyb}}_{\text{PDE}} = \mathbb{E}\big[(u_t - \alpha u_{xx} - q_\theta)^2\big].$$

Regularize q_θ (parsimony, smoothness, optional energy neutrality) so it captures only *missing physics*.

Multi-Fidelity PINN

Fuse a cheap low-fidelity (LF) model u_{LF} with a learned discrepancy r_θ:

$$u_{\text{HF}}(x, t) \approx u_{\text{LF}}(x, t) + r_\theta(x, t),$$
$$\mathcal{R}_{\text{data}} = w_{\text{HF}}\|u_\theta - y^{\text{HF}}\|^2 + w_{\text{LF}}\|u_{\text{LF}} + r_\theta - y^{\text{LF}}\|^2,$$

and enforce physics on the *combined* prediction.

For the setup we keep $\Omega = (0, 1)$, $T = 10\,\text{s}$; $u(0, t) = 100$, $u(1, t) = 25$, $u(x, 0) = 25$; diffusivity α known unless noted. We write u_θ for the state network, and introduce a small-capacity q_θ (hybrid) or r_θ (multi-fidelity).

```
import jax, jax.numpy as jnp
from jax import grad, vmap, jit

--- state network u_theta (as earlier) ---
def u_theta(params_u, x, t):
return mlp_apply(params_u, jnp.stack([x, t], axis=-1)).squeeze(-1)

--- small correction network q_theta (low capacity recommended) ---
def q_theta(params_q, x, t, u_val):
xtu = jnp.stack([x, t, u_val], axis=-1) # allow state-dependence
return mlp_apply(params_q, xtu).squeeze(-1)

#Autodiff on u_theta
```

```
u_x = grad(lambda p,x,t: u_theta(p,x,t), argnums=1)
u_t = grad(lambda p,x,t: u_theta(p,x,t), argnums=2)
u_xx = grad(lambda p,x,t: u_x(p,x,t), argnums=1)

#PDE residual: u_t - \alpha u_xx = q_theta
def pde_residual_hybrid(params_u, params_q, x, t, alpha):
uval = u_theta(params_u, x, t)
return u_t(params_u, x, t) - alpha*u_xx(params_u, x, t) -
     q_theta(params_q, x, t, uval)

def pde_loss_hybrid(params_u, params_q, Z_col, alpha):
R = vmap(lambda z: pde_residual_hybrid(params_u, params_q, z[0],
     z[1], alpha))(Z_col)
return jnp.mean(R**2)

#Data and BC/IC (reuse your earlier implementations)
def data_loss(params_u, Z_data, y_data):
pred = vmap(lambda z: u_theta(params_u, z[0], z[1]))(Z_data)
return jnp.mean((pred - y_data)**2)

def bcic_loss(params_u, Z_b, u_b, Z_i, u_i):
ub = vmap(lambda z: u_theta(params_u, z[0], z[1]))(Z_b)
ui = vmap(lambda z: u_theta(params_u, z[0], z[1]))(Z_i)
return jnp.mean((ub - u_b)**2) + jnp.mean((ui - u_i)**2)

--- Regularization on the learned physics q_theta ---
def q_parsimony(params_q, params_u, Z_col):
def q_at(z):
x,t = z
return q_theta(params_q, x, t, u_theta(params_u, x, t))
qvals = vmap(q_at)(Z_col)
return jnp.mean(qvals**2)

def q_smoothness_x(params_q, params_u, Z_col):
def dqdx_at(z):
x,t = z
def qx(xx): # hold t fixed
return q_theta(params_q, xx, t, u_theta(params_u, xx, t))
return grad(qx)(x)
dqdx = vmap(dqdx_at)(Z_col)
return jnp.mean(dqdx**2)

#Gaussian quadrature on [0,1] to approximate \int_0^1 q(x,t) dx
def gauss_legendre(n=32):
xs, ws = jnp.polynomial.legendre.leggauss(n)
x = 0.5*(xs+1.0); w = 0.5*ws
return x, w

def q_energy_neutrality(params_q, params_u, t_samples):
xq, wq = gauss_legendre(48)
def integral_at_t(t):
qx = vmap(lambda x: q_theta(params_q, x, t, u_theta(params_u, x,
     t)))(xq)
```

```
return jnp.sum(qx * wq)
vals = vmap(integral_at_t)(t_samples)
return jnp.mean(vals**2)

#Total loss (hybrid)
def total_loss_hybrid(params_u, params_q, batch, alpha, w, lam=(1e-3,
    1e-3, 0.0)):
Zc, Zd, yd, Zb, ub, Zi, ui, t_neu = batch
Rpde = pde_loss_hybrid(params_u, params_q, Zc, alpha)
Rdata = data_loss(params_u, Zd, yd)
Rbcic = bcic_loss(params_u, Zb, ub, Zi, ui)
Rpars = q_parsimony(params_q, params_u, Zc)
Rsmoo = q_smoothness_x(params_q, params_u, Zc)
Rener = q_energy_neutrality(params_q, params_u, t_neu) if t_neu is
    not None else 0.0
lam_p, lam_s, lam_e = lam
return (w['pde']Rpde + w['data']Rdata + w['bc']Rbcic
+ lam_pRpars + lam_sRsmoo + lam_eRener)
```

Listing 5.15 Hybrid (gray-box) PINN: learned correction q_θ with parsimony, smoothness, and optional energy neutrality.

Explanation (Hybrid): The mechanistic core $u_t - \alpha u_{xx}$ anchors the dynamics. The neural q_θ is restricted by (i) parsimony, (ii) spatial smoothness, and optionally (iii) net-zero energy injection, so that it explains only what the coarse/idealized physics misses (see earlier gray-box discussion for discrete variants).

```
#Assume a low-fidelity model u_LF(x,t) (e.g., coarse solver or
    empirical fit)
def u_LF(x, t, params_lf=None):
# placeholder; replace by your coarse simulation or surrogate
return 25.0 + (100.0-25.0)*(1.0 - x) * jnp.exp(-t/10.0)

#Discrepancy network r_theta; final prediction u_pred = u_LF + r_theta
def r_theta(params_r, x, t):
return mlp_apply(params_r, jnp.stack([x, t], axis=-1)).squeeze(-1)

def u_pred(params_r, x, t):
return u_LF(x,t) + r_theta(params_r, x, t)

#Compute data losses on both HF and (optionally) LF pairs
def data_loss_mf(params_r, Z_HF, y_HF, Z_LF=None, y_LF=None,
    w_HF=1.0, w_LF=0.2):
# High-fidelity mismatch on u_pred
pred_HF = vmap(lambda z: u_pred(params_r, z[0], z[1]))(Z_HF)
Rd_HF = jnp.mean((pred_HF - y_HF)**2)
# Low-fidelity self-consistency (optional)
if Z_LF is not None and y_LF is not None:
pred_LF = vmap(lambda z: u_LF(z[0], z[1]))(Z_LF)
Rd_LF = jnp.mean((pred_LF - y_LF)**2)
else:
Rd_LF = 0.0
return w_HFRd_HF + w_LFRd_LF
```

```

#Enforce physics on the combined prediction u_LF + r_theta
def pde_loss_mf(params_r, Z_col, alpha):
def ut_pred(x,t):
return grad(lambda tt: u_pred(params_r, x, tt))(t)
def uxx_pred(x,t):
return grad(lambda xx: grad(lambda xxx: u_pred(params_r, xxx,
    t))(xx))(x)
R = vmap(lambda z: ut_pred(z[0], z[1]) - alpha * uxx_pred(z[0],
    z[1]))(Z_col)
return jnp.mean(R**2)

#Boundary/initial on the combined field (soft penalties shown here)
def bcic_loss_mf(params_r, Z_b, u_b, Z_i, u_i):
ub = vmap(lambda z: u_pred(params_r, z[0], z[1]))(Z_b)
ui = vmap(lambda z: u_pred(params_r, z[0], jnp.array(0.0)))(Z_i)
return jnp.mean((ub - u_b)**2) + jnp.mean((ui - u_i)**2)

#Optional trust schedule for w_LF (downweight LF over epochs)
def trust_weight(epoch, warm=500):
return jnp.clip(1.0 - epoch/warm, 0.0, 1.0)

#Total loss (multi-fidelity)
def total_loss_mf(params_r, batch, alpha, w, epoch=None):
(Zc, Zhf, yhf, Zb, ub, Zi, ui, Zlf, ylf) = batch
w_LF = trust_weight(epoch) if epoch is not None else 0.2
Rd = data_loss_mf(params_r, Zhf, yhf, Zlf, ylf, w_HF=1.0, w_LF=w_LF)
Rp = pde_loss_mf(params_r, Zc, alpha)
Rb = bcic_loss_mf(params_r, Zb, ub, Zi, ui)
# small regularizer on r_theta to avoid overfitting HF noise
def r_vals(z): return r_theta(params_r, z[0], z[1])
Rr = jnp.mean(vmap(r_vals)(Zc)**2)
return w['data']*Rd + w['pde']*Rp + w['bc']Rb + 1e-4Rr
```

Listing 5.16 Multi-Fidelity PINN: discrepancy learning r_θ with physics on the combined prediction and HF/LF data fusion.

Explanation (Multi-Fidelity). The prediction is $LF + u_{\mathrm{LF}} + r_\theta$. Physics is enforced on the *combined* field, so the correction cannot violate the PDE. HF data carry full weight; LF data act as a gentle guide early in training (the *trust schedule* decays w_LF over epochs). A light penalty on r_θ discourages over-explaining HF noise. How they differ (at a glance):

- **Hybrid** targets missing *dynamics* as a source term q_θ in the PDE residual. Use when a mechanistic core is credible but incomplete; add structure priors (parsimony/smoothness/neutrality) on q_θ.
- **Multi-Fidelity** targets missing *resolution or bias* relative to a cheaper model via a discrepancy r_θ added to u_{LF}. Use when you have abundant LF outputs and sparse HF data.

Practical guidance

1. Start with conservative regularization on q_θ or r_θ; relax as residuals stabilize.
2. In MF, calibrate u_{LF} (offset/scale) before discrepancy learning to avoid wasting r_θ on trivial biases.
3. Always compute validation residuals on the full field you deploy (hybrid: u_θ with q_θ; MF: $u_{\mathrm{LF}} + r_\theta$).

5.5 Conclusion

This chapter has developed the theory and practice of physics-informed neural networks, the foundational methodology of Scientific Machine Learning. We have seen that PINNs embed physics as soft constraints through residual terms in the loss function, that practical success requires careful attention to the "dark art" of training, and that the PINN taxonomy reveals a rich design space of variants.

Key Takeaways

1. **Physics in the Loss**: PINNs enforce governing equations by penalizing PDE residuals at collocation points. No mesh required—the network *is* the solution.
2. **Forward + Inverse Unified**: The same architecture solves forward problems (simulate given parameters) and inverse problems (infer parameters from data) by adjusting what's known versus learned.
3. **Loss Weighting Matters**: Balancing physics, data, and boundary terms is critical. Adaptive weighting strategies (GradNorm, NTK-based) often outperform fixed weights.
4. **Architecture Encodes Bias**: Activation functions (tanh, sin), positional encodings, and multiplicative structures affect what physics the network can represent.
5. **Hybrid for Stability**: Universal Differential Equations embed neural networks within classical solvers, combining learning flexibility with numerical stability guarantees.

References

1. Karniadakis, G.E., Kevrekidis, I.G., Lu, L., Perdikaris, P., Wang, S., Yang, L.: Physics-informed machine learning. Nat. Rev. Phys. **3**(6), 422–440 (2021)
2. Wang, Y., Wang, H., Zhang, Y., Xie, M.: Gray-box dynamic model for wave glider driven by a hybrid of deep learning and physics-based models. Eng. Appl. Artif. Intell. **162**, 112730 (2025)
3. Cuomo, S., Di Cola, V.S., Giampaolo, F., Rozza, G., Raissi, M., Piccialli, F.: Scientific machine learning through physics-informed neural networks: Where we are and what's next. J. Sci. Comput. **92**(3), 88 (2022)
4. Berrone, S., Canuto, C., Pintore, M., Sukumar, N.: Enforcing Dirichlet boundary conditions in physics-informed neural networks and variational physics-informed neural networks. Heliyon **9**(8) (2023)
5. Krishnapriyan, A., Gholami, A., Zhe, S., Kirby, R., Mahoney, M.W.: Characterizing possible failure modes in physics-informed neural networks. Adv. Neural. Inf. Process. Syst. **34**, 26548–26560 (2021)
6. Ren, X., Hu, P., Su, H., Zhang, F., Yu, H.: Physics-informed neural networks for transonic flow around a cylinder with high Reynolds number. Phys. Fluids 36(3), (2024)
7. Jagtap, A.D., Karniadakis, G.E.: Extended physics-informed neural networks (xpinns): a generalized space-time domain decomposition based deep learning framework for nonlinear partial differential equations. Commun. Comput. Phys 28(5), (2020)
8. Wang, J., Peng, X., Chen, Z., Zhou, B., Zhou, Y., Zhou, N.: Surrogate modeling for neutron diffusion problems based on conservative physics-informed neural networks with boundary conditions enforcement. Ann. Nucl. Energy **176**, 109234 (2022)
9. Raissi, M., Perdikaris, P., Karniadakis, G.E.: Physics-informed neural networks: A deep learning framework for solving forward and inverse problems involving nonlinear partial differential equations. J. Comput. Phys. **378**, 686–707 (2019)

Chapter 6
Optimization in Scientific ML

Perfection is achieved not when there is nothing more to add, but when there is nothing left to take away.

– Antoine de Saint-Exupéry, Airman's Odyssey

Abstract Training Scientific Machine Learning models presents optimization challenges beyond standard deep learning: physics residuals create ill-conditioned loss landscapes; competing loss terms require careful balancing; constraints must be satisfied to physical tolerances; and expensive forward solves limit gradient computation. This chapter addresses these challenges comprehensively. We begin with gradient-based methods, covering classical approaches (SGD, momentum, Adam) and their behavior on SciML loss landscapes, followed by quasi-Newton methods (L-BFGS, trust-region) that often outperform first-order methods for physics-informed training. Constrained optimization develops penalty methods, augmented Lagrangian approaches, and variational formulations that handle equality and inequality constraints. Gradient-free methods (evolutionary algorithms, Bayesian optimization, surrogate-based approaches) address non-differentiable objectives and discrete decisions. Adjoint-based optimization enables efficient gradients through PDE solvers, connecting classical optimal control to modern differentiable programming. We conclude with hybrid and multi-fidelity strategies that combine models of varying cost and accuracy.

6.1 Introduction

In the classic film *The Princess Bride*, the hero Westley must navigate the Fire Swamp, a treacherous landscape of flame spurts, lightning sand, and Rodents of Unusual Size. "It's not that bad," he assures Buttercup after they survive. "I'm not saying I'd like to build a summer home here, but the trees are actually quite lovely."

Dr. Kenji Tanaka (a fictional character), training his first physics-informed neural network in 2026, would have appreciated the metaphor. The loss landscape he was

B. P. Bhuyan et al., *Scientific Machine Learning*,
https://doi.org/10.1007/978-981-95-7856-6_6

navigating had its own hazards: regions where the physics loss was low but the data loss was catastrophic; narrow ravines where both seemed acceptable but the boundary conditions were violated; and saddle points where gradient descent stalled indefinitely, unable to decide which way to go.

"Adam works fine on ImageNet," Tanaka complained to his advisor. "Why is it useless here?"

The answer, as his advisor explained, lay in the structure of the loss function itself. Standard deep learning losses are designed to be well-behaved: cross-entropy for classification, MSE for regression, carefully scaled and normalized. Physics-informed losses are not. The PDE residual involves second derivatives of the network; dramatically amplifying high-frequency components. The physics and data terms live at different scales. The optimization landscape is fundamentally different from anything Adam was designed for.

"Try L-BFGS"; his advisor suggested. "It uses curvature information. Often works better for these problems."

Tanaka was skeptical. L-BFGS was his grandfather's optimizer, a relic from the 1980s. But he tried it, and to his surprise, it worked. Not perfectly (nothing worked perfectly) but better than the adaptive gradient methods he had assumed were state of the art.

The lesson stuck: *SciML optimization is not standard deep learning optimization*. The loss landscapes are different, the failure modes are different, and the solutions are different. Understanding why is essential for practical success.

Several factors conspire to make Scientific Machine Learning optimization challenging:

- **Ill-conditioned losses**: PDE residuals involve high-order derivatives. Training on $\nabla^2 u$ amplifies high-frequency components, creating loss landscapes with very different curvatures in different directions.
- **Competing objectives**: Physics loss, data loss, and boundary condition loss may conflict. Reducing one often increases another, requiring careful balancing.
- **Hard constraints**: Physical solutions must satisfy conservation laws exactly, not approximately. Penalty methods may never achieve true feasibility.
- **Expensive evaluations**: Computing physics residuals requires forward and backward passes through complex operations. Each loss evaluation is expensive.
- **Non-convexity**: Neural network losses are highly non-convex. Local minima, saddle points, and plateau regions trap standard optimizers.
- **Scale mismatch**: Physical quantities span many orders of magnitude. Naive training can be dominated by large-scale terms while ignoring small-scale physics.

This chapter develops the theory and practice of optimization for these challenging landscapes.

We begin with gradient-based methods: first-order (SGD, Adam) and quasi-Newton (L-BFGS, trust-region), examining their behavior on SciML problems. Constrained optimization follows: penalty methods, augmented Lagrangian, and variational formulations. Gradient-free methods address problems where gradients are unavailable or unreliable. Adjoint-based optimization enables efficient gradients through PDE solvers. We conclude with multi-fidelity strategies that accelerate optimization using models of varying cost and accuracy.

6.2 Gradient-Based Optimization in SciML

Optimization in Scientific Machine Learning SciML is most often posed as a minimization problem over a parameter space $\theta \in \mathbb{R}^d$ [1]. The objective (or loss) combines data fidelity, physics residuals, and constraint satisfaction. A common setting is the Physics-Informed Neural Network (PINN), where the loss takes the composite form

$$\mathcal{L}(\theta) = w_{\text{data}} R_{\text{data}}(\theta) + w_{\text{PDE}} R_{\text{PDE}}(\theta) + w_{\text{BC/IC}} R_{\text{BC/IC}}(\theta),$$

with R_{data} measuring misfit against observations, R_{PDE} enforcing the differential operator, and $R_{\text{BC/IC}}$ capturing boundary and initial conditions.

Definition: Gradient-Based Optimization

Gradient-based optimization iteratively updates parameters in the direction of steepest descent of the loss:

$$\theta_{k+1} = \theta_k - \eta \, \nabla_\theta \mathcal{L}(\theta_k),$$

where $\eta > 0$ is the learning rate. Variants differ in how they estimate, scale, or approximate the gradient and curvature information.

Consider the one-dimensional heat equation (in the previous chapter)

$$u_t - \alpha u_{xx} = 0, \quad x \in (0, 1), \; t \in (0, T],$$

with boundary conditions $u(0, t) = 100$, $u(1, t) = 25$, and initial condition $u(x, 0) = 25$. A PINN $u_\theta(x, t)$ minimizes

$$\mathcal{L}(\theta) = R_{\text{data}}(\theta) + R_{\text{PDE}}(\theta) + R_{\text{BC/IC}}(\theta),$$

where $\nabla_\theta \mathcal{L}$ is obtained by automatic differentiation. Gradient-based optimization drives θ so that u_θ simultaneously fits data and satisfies the PDE.

While the principle is simple, SciML loss landscapes are often stiff and ill-conditioned: different terms may differ by orders of magnitude, PDE operators can introduce sharp curvatures, and constraints must be satisfied globally. Consequently, a wide family of gradient-based methods have been developed and adapted:

- **Classical methods**: Gradient Descent (GD), Stochastic Gradient Descent (SGD), Momentum, Adam, RMSProp.
- **Quasi-Newton methods**: Limited-memory BFGS (L-BFGS), Trust-region approaches.
- **Recent advances**: Trust-region SQP PINNs (trSQP-PINN), Augmented Lagrangian PINNs (AL-PKAN), and adaptive loss weighting strategies.

Each of these classes addresses specific challenges in SciML: classical methods for scalable updates on large collocation sets, quasi-Newton methods for precise convergence to low-residual solutions, and recent advances for robust enforcement of physical constraints.

6.2.1 Classical Gradient Methods

We continue the common setup from the heat-equation PINN example. Recall the composite loss

$$\mathcal{L}(\theta) = w_{\text{data}} R_{\text{data}}(\theta) + w_{\text{PDE}} R_{\text{PDE}}(\theta) + w_{\text{BC/IC}} R_{\text{BC/IC}}(\theta),$$

where automatic differentiation yields $\nabla_\theta \mathcal{L}(\theta)$. Classical gradient methods differ in how they *estimate* the gradient (full-batch vs. mini-batch), how they *accumulate* information from past steps (momentum), and how they *scale* step sizes per-parameter (adaptive rules).

Definition: Gradient Descent (GD)

Given a differentiable objective $\mathcal{L} : \mathbb{R}^d \to \mathbb{R}$, full-batch Gradient Descent updates

$$\theta_{k+1} = \theta_k - \eta \nabla_\theta \mathcal{L}(\theta_k),$$

with learning rate $\eta > 0$. The gradient is computed on the *entire* dataset/collocation set at each step.

Full-batch GD is numerically stable for modest-size problems (e.g., small collocation grids, low-dimensional parameters). It provides a low-variance search direction, which is helpful when the PDE residual is stiff. However, for large collocation sets or frequent point-resampling (a common practice to avoid overfitting a fixed grid), computing a full gradient each step can be prohibitive.

Algorithm 1: Full-Batch Gradient Descent (GD) for SciML Composite Loss

---

Input: θ_0, K, $\varepsilon_{\text{grad}}$, ε_{rel}; step size η (optional) or Armijo parameters α, β;
datasets $\mathcal{D}_{\text{data}}$, $\mathcal{D}_{\text{PDE}}$, $\mathcal{D}_{\text{BC/IC}}$; weights w_{data}, w_{PDE}, $w_{\text{BC/IC}}$
Output: $\theta_\star$ and diagnostics

`// Helper: compute loss and gradient on full datasets`
Compute residuals $R_{\text{data}}(\theta)$, $R_{\text{PDE}}(\theta)$, $R_{\text{BC/IC}}(\theta)$;
$\mathcal{L}(\theta) \leftarrow w_{\text{data}} R_{\text{data}}(\theta) + w_{\text{PDE}} R_{\text{PDE}}(\theta) + w_{\text{BC/IC}} R_{\text{BC/IC}}(\theta)$;
$g(\theta) \leftarrow \nabla_\theta \mathcal{L}(\theta)$ `// e.g., via automatic differentiation`

$(L_0,\ g_0) \leftarrow (\mathcal{L}(\theta_0),\ g(\theta_0))$;
for $k \leftarrow 0$ **to** $K - 1$ **do**
  if $\|g_k\|_2 \le \varepsilon_{grad}$ **then**
    `// First-order optimality satisfied`
    break ;
  if η *given* **then**
    $\theta_{k+1} \leftarrow \theta_k - \eta\, g_k$ `// Fixed-step GD`
  else
    $\eta_k \leftarrow$ ArmijoBacktracking(θ_k, $\mathcal{L}$, g_k, α, β) ;
    $\theta_{k+1} \leftarrow \theta_k - \eta_k\, g_k$ `// Line-search GD`
  $(L_{k+1},\ g_{k+1}) \leftarrow (\mathcal{L}(\theta_{k+1}),\ g(\theta_{k+1}))$;
  if $\dfrac{|L_{k+1} - L_k|}{\max\{1, |L_k|\}} \le \varepsilon_{rel}$ **then**
    `// Relative loss decrease is negligible`
    break ;

return $\theta_\star \leftarrow \theta_{k+1}$ and diagnostics $\left(L_{k+1},\ \|g_{k+1}\|_2\right)$

Algorithm 2: Armijo Backtracking Line Search for GD

Input: Current θ, loss $\mathcal{L}$, gradient $g = \nabla \mathcal{L}(\theta)$; parameters $\alpha \in (0, 1/2)$,
$\beta \in (0, 1)$; initial trial step $\eta_{\text{init}} > 0$
Output: Step size η

$\eta \leftarrow \eta_{\text{init}};\quad L \leftarrow \mathcal{L}(\theta)$;
while $\mathcal{L}(\theta - \eta\, g) > L - \alpha\, \eta\, \|g\|_2^2$ **do**
  $\eta \leftarrow \beta\, \eta$ `// Reduce step until sufficient decrease holds`
return η

Definition: Stochastic Gradient Descent (SGD)

Stochastic Gradient Descent (SGD) uses a mini-batch $\mathcal{B}_k$ of samples to form an unbiased estimator

$$g_k \approx \nabla_\theta \mathcal{L}(\theta_k), \qquad \theta_{k+1} = \theta_k - \eta\, g_k,$$

where $g_k = \frac{1}{|\mathcal{B}_k|} \sum_{(x,t)\in\mathcal{B}_k} \nabla_\theta \ell(\theta_k; x, t)$.

In PINNs, the mini-batch may mix (i) sensor points for R_{data}, (ii) collocation points for R_{PDE}, and (iii) boundary/initial-condition points for $R_{\text{BC/IC}}$. SGD reduces memory and wall-time per step and enables *online resampling* of collocation points, improving generalization across the domain. The trade-off is higher gradient variance; careful learning-rate schedules and gradient clipping mitigate instabilities.

Definition: Momentum and Nesterov Momentum

Momentum smooths the noisy gradient with an exponential moving average (EMA):

$$v_k = \mu\, v_{k-1} + (1-\mu)\, g_k, \qquad \theta_{k+1} = \theta_k - \eta\, v_k,$$

with momentum coefficient $\mu \in [0, 1)$. *Nesterov momentum* instead evaluates the gradient at a *look-ahead* point:

$$\tilde{\theta}_k = \theta_k - \eta\, \mu\, v_{k-1}, \quad v_k = \mu\, v_{k-1} + (1-\mu)\, \nabla_\theta \mathcal{L}(\tilde{\theta}_k), \quad \theta_{k+1} = \theta_k - \eta\, v_k.$$

Momentum damps oscillations along stiff directions (e.g., around boundary layers) and accelerates progress on gently curved valleys–typical in PDE-induced landscapes. Choose $\mu \in [0.8, 0.99]$; higher values accelerate but risk overshoot when losses are badly scaled.

Definition: Adam (Adaptive Moment Estimation)

Adam maintains first- and second-moment EMAs:

$$m_k = \beta_1 m_{k-1} + (1-\beta_1) g_k, \qquad v_k = \beta_2 v_{k-1} + (1-\beta_2) g_k^2,$$

with bias-corrected moments $\hat{m}_k = \frac{m_k}{1-\beta_1^k}$, $\hat{v}_k = \frac{v_k}{1-\beta_2^k}$, and update

$$\theta_{k+1} = \theta_k - \eta \frac{\hat{m}_k}{\sqrt{\hat{v}_k} + \epsilon}.$$

Adam adapts step sizes per-parameter, which is valuable when different subnetworks (e.g., for x vs. t inputs) learn at different rates. Defaults β_1=0.9, β_2=0.999, $\epsilon \in [10^{-8}, 10^{-6}]$ work well; a slightly larger ϵ improves stability in mixed-precision or very stiff PDEs. Adam excels in early training but may plateau before the PDE residual reaches numerical accuracy–often prompting a switch to quasi-Newton refinement.

Definition: RMSProp (Root Mean Square Propagation)

RMSProp scales the learning rate by an EMA of squared gradients:

$$s_k = \rho\, s_{k-1} + (1-\rho)\, g_k^2, \qquad \theta_{k+1} = \theta_k - \eta \frac{g_k}{\sqrt{s_k} + \epsilon},$$

with decay $\rho \in (0, 1)$.

RMSProp is a simpler per-parameter adaptive method. It can be more stable than Adam on some stiff problems when tuned (e.g., $\rho \approx 0.9$), but typically requires slightly more manual learning-rate adjustments.
Why classical methods matter in SciML.

- **Scalability**: Mini-batching of collocation points makes large PDE domains tractable.
- **Noise handling**: Momentum/Adam cope with stochasticity from resampled collocation sets.
- **Per-parameter adaptation**: Adam/RMSProp temper exploding/vanishing updates across heterogeneous layers.
- **Warm starts**: Classical methods are excellent for fast initial progress before switching to quasi-Newton refinement (cf. L-BFGS).

Practical details that matter

1. **Learning-rate schedule**: Start with $\eta \in [10^{-3}, 10^{-4}]$ for Adam/RMSProp (stiff PDEs often prefer 10^{-4}), then decay (step/cosine or Reduce-on-Plateau).

2. **Batching strategy**: Mix each mini-batch with (i) data points, (ii) PDE collocation, (iii) BC/IC points to stabilize multi-term training.
3. **Weight balancing**: Normalize terms (e.g., match initial magnitudes of R_{data}, R_{PDE}, $R_{\text{BC/IC}}$) to avoid one term dominating.
4. **Gradient clipping**: Clip global norm (e.g., at 1–5) to prevent transients when resampling collocation points.
5. **Precision & ϵ**: With mixed precision, use a slightly larger ϵ in Adam/RMSProp for numerical stability.

Now let us consider a running example for mini-batch updates for the heat-equation PINN. For each iteration, sample:

$$\text{Data batch } \mathcal{B}^{\text{data}} \subset \{(x_i, t_i, y_i)\}, \quad \text{PDE batch } \mathcal{B}^{\text{PDE}} \subset \Omega \times (0, T], \quad \text{BC/IC batch } \mathcal{B}^{\text{BC/IC}},$$
$$\text{Compute } R_{\text{data}},\ R_{\text{PDE}},\ R_{\text{BC/IC}} \text{ on their batches, form } \mathcal{L}, \text{ then update } \theta.$$

This reduces memory, enables online resampling of $\mathcal{B}^{\text{PDE}}$, and yields a controllable noise level in g_k.

```
import jax, jax.numpy as jnp
import optax

def loss_fn(params, batch_data, batch_pde, batch_bcic):
    R_data = data_residual(params, batch_data)          # mean squared misfit
    R_pde  = pde_residual(params, batch_pde)            # mean squared PDE residual
    R_bc   = bcic_residual(params, batch_bcic)          # mean squared BC/IC
    # term normalization (example): scale to initial magnitudes
    L = w_data*R_data + w_pde*R_pde + w_bcic*R_bc
    return L, (R_data, R_pde, R_bc)

# Choose an optimizer (swap 'adam' with sgd/momentum/rmsprop as needed)
base_lr = 1e-3
scheduler = optax.cosine_decay_schedule(init_value=base_lr, decay_steps=50_000)
optimizer = optax.chain(
    optax.clip_by_global_norm(5.0),
    optax.scale_by_adam(eps=1e-7),    # swap with: optax.trace(decay=0.9, nesterov=True) for Momentum
    optax.scale_by_schedule(scheduler),
    optax.scale(-1.0),
)

opt_state = optimizer.init(params)

@jax.jit
def train_step(params, opt_state, batch_data, batch_pde, batch_bcic):
```

```
    (L, (R_data, R_pde, R_bc)), grads = jax.value_and_grad(loss_fn,
    has_aux=True)(
        params, batch_data, batch_pde, batch_bcic
    )
    updates, opt_state = optimizer.update(grads, opt_state, params)
    params = optax.apply_updates(params, updates)
    return params, opt_state, L, R_data, R_pde, R_bc

for step in range(num_steps):
    bd = next(data_loader)      # sensor points
    bp = next(pde_loader)       # collocation points (resampled over
    time)
    bb = next(bcic_loader)      # boundary/initial points
    params, opt_state, L, Rd, Rp, Rb = train_step(params, opt_state,
    bd, bp, bb)
    if step % 500 == 0:
        print(f"{step:06d} | L={L:.3e} Rd={Rd:.3e} Rp={Rp:.3e}
    Rb={Rb:.3e}")
```

Listing 6.1 Classical training loop with Adam (switchable to Momentum/RMSProp), gradient clipping, and a cosine schedule.

Interpreting diagnostics

Monitor *each* term: if R_{PDE} stalls while R_{data} decreases, the PDE is under-enforced (increase w_{PDE}, reduce η, or adopt curriculum on collocation). If all terms oscillate, reduce learning rate, increase batch sizes, or strengthen clipping. Plateaus after initial progress often indicate it is time to switch to a quasi-Newton method (e.g., L-BFGS).

Batch Size, Gradient Variance, and Collocation Resampling

In SciML, the choice of batch size directly controls the *variance* of the gradient estimate. Suppose $\{z_i\}_{i=1}^N$ are collocation points and

$$\nabla_\theta \mathcal{L}(\theta) = \frac{1}{N} \sum_{i=1}^{N} g_i(\theta), \qquad g_i(\theta) = \nabla_\theta \ell(\theta; z_i).$$

A mini-batch $\mathcal{B} \subset \{1, \dots, N\}$ of size $|\mathcal{B}| = B$ produces the estimator

$$\hat{g}_B(\theta) = \frac{1}{B} \sum_{i \in \mathcal{B}} g_i(\theta).$$

Variance scaling: If g_i have variance σ^2 in each coordinate, then

$$\mathrm{Var}[\hat{g}_B] \approx \frac{\sigma^2}{B}.$$

Thus, larger batches reduce gradient noise but increase cost per iteration.

Practical consequences in SciML:

- *Small batches ($B \ll N$)*: High variance. Enables **online resampling** of PDE collocation points, which improves generalization and avoids overfitting to a fixed mesh. But too much variance can destabilize stiff PDE training.
- *Large batches ($B \approx N$)*: Low variance, stable updates, but expensive and risks *memorization* of collocation patterns. Generalization to unseen regions of the domain may degrade.
- *Moderate batches ($B = 128$–1024 typical)*: Balance between noise-induced exploration and stable convergence. Works well with adaptive optimizers like Adam.

Interaction with collocation resampling: Resampling collocation points at each epoch (a practice known as *Residual-based Adaptive Refinement, RAR*) introduces *controlled noise* in $\hat{g}_B$. This helps escape spurious local minima and enforces PDE residuals more uniformly across the domain. Batch size B thus acts as a *variance knob*, complementing resampling.

Rule of thumb:

- Start with $B \approx 512$ collocation points per update for PDE residuals.
- Increase B if training oscillates (too much noise).
- Decrease B if convergence is too slow or if overfitting to a collocation set is suspected.

Figure 6.1 illustrates the classical trade-off in stochastic optimization: increasing the batch size reduces the variance of the gradient estimate approximately as $1/B$, producing smoother and more stable updates. In Scientific Machine Learning (SciML), this balance is especially delicate:

- *Small batches* ($B < 128$) introduce significant gradient noise. While this noise can help exploration and is useful when collocation points are resampled online, it may destabilize training for stiff PDEs unless accompanied by learning-rate reduction or gradient clipping.
- *Large batches* ($B > 1024$) reduce variance drastically, leading to very stable updates. However, this comes at the cost of higher computation per iteration and reduced generalization, as the optimizer may effectively "memorize" the fixed collocation grid.
- *Intermediate batches* ($128 \leq B \leq 1024$) offer a practical compromise: gradient variance is low enough to stabilize training, while noise from resampling collocation points still promotes exploration and robustness.

In practice, batch size acts as a *variance-control knob* in SciML optimization. By adjusting B, one can fine-tune the trade-off between exploration (high noise) and exploitation (stable descent). This figure provides a visual intuition: batch sizes in the shaded region yield a balance that has been empirically validated across PINN training and operator learning tasks.

Classical methods provide the initial, computationally efficient descent needed to shape the solution manifold. Yet, because SciML losses are often ill-conditioned (stiff operators, multi-scale phenomena), we typically follow with *quasi-Newton* refinement and, when constraints are critical, with *trust-region* or *augmented-Lagrangian* strategies. We develop these next.

6.2.2 *Quasi-Newton Methods: L-BFGS and Trust-Region Approaches*

While classical gradient methods (SGD, Adam, RMSProp) are effective for rapid exploration and scalable updates, they often plateau before achieving the high-accuracy residual minimization required in Scientific Machine Learning (SciML). Physics-Informed Neural Networks (PINNs), for example, demand PDE residuals

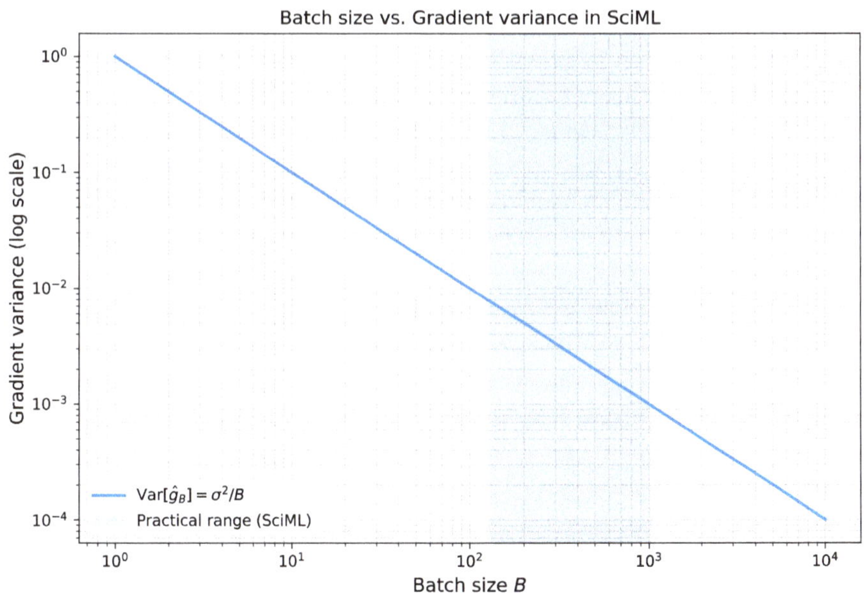

Fig. 6.1 Effect of batch size B on gradient variance. The variance of the stochastic gradient estimate scales as $\mathrm{Var}[\hat{g}_B] \propto 1/B$. The shaded band ($128 \leq B \leq 1024$) indicates a practical range frequently adopted in SciML training

on the order of 10^{-6}–10^{-8} for reliable solutions, which first-order methods struggle to achieve. This motivates the use of *second-order information*.

Definition: Quasi-Newton Methods

Quasi-Newton methods approximate the inverse Hessian of the objective $\mathcal{L}(\theta)$ without explicitly computing second derivatives. Updates take the form

$$\theta_{k+1} = \theta_k - H_k^{-1} \nabla_\theta \mathcal{L}(\theta_k),$$

where H_k^{-1} is iteratively updated using gradient differences across steps.

Unlike gradient descent, which uses only first-order information, quasi-Newton methods capture local curvature, leading to faster convergence near minimizers. They are particularly suitable for the stiff, ill-conditioned landscapes typical of PDE-constrained losses.

Limited-memory BFGS (L-BFGS)

Definition: L-BFGS

L-BFGS (Limited-memory Broyden–Fletcher–Goldfarb–Shanno) approximates the inverse Hessian using a low-rank update from the most recent m gradient differences:

$$H_{k+1}^{-1} = \left(I - \rho_k s_k y_k^\top\right) H_k^{-1} \left(I - \rho_k y_k s_k^\top\right) + \rho_k s_k s_k^\top,$$

where $s_k = \theta_{k+1} - \theta_k$, $y_k = g_{k+1} - g_k$, and $\rho_k = 1/(y_k^\top s_k)$.

L-BFGS achieves second-order accuracy with $\mathcal{O}(dm)$ storage instead of $\mathcal{O}(d^2)$, making it feasible for neural networks. In SciML, a typical workflow is:

1. Pre-train the network with Adam or SGD for fast initial progress.
2. Switch to L-BFGS for refinement, achieving machine-precision PDE residuals.

This two-stage strategy has become standard in PINN training.

Trust-Region Approaches

Trust-Region Methods

Trust-region methods optimize a local quadratic surrogate of the loss within a bounded region:

$$\min_{p} \quad m_k(p) = \mathcal{L}(\theta_k) + g_k^\top p + \tfrac{1}{2} p^\top B_k p, \quad \text{s.t.} \quad \|p\| \leq \Delta_k,$$

where g_k is the gradient, B_k a Hessian (or approximation), and Δ_k the trust-region radius.

By restricting updates to a region where the quadratic model is accurate, trust-region methods prevent divergence on highly curved loss surfaces. In SciML, they are particularly effective for *hard-constrained training* (e.g., trust-region SQP PINNs, trSQP-PINNs) [2], yielding one to three orders of magnitude lower error compared to penalty-based PINNs. The practical role in SciML are:

- **L-BFGS**: Efficient second-order refinement, often used after Adam pre-training.
- **Trust-region methods**: Robust enforcement of constraints, avoiding instabilities caused by ill-conditioned PDE operators.
- **Hybrid strategies**: Adam → L-BFGS → trust-region refinement can combine fast exploration, precise convergence, and constraint robustness.

Best Practice

In SciML applications such as PINNs and operator learning, use Adam or RMSProp for warm-up, switch to L-BFGS for precision, and adopt trust-region solvers when physical constraints must be strictly satisfied.

We revisit the one-dimensional heat equation

$$u_t - \alpha u_{xx} = 0, \quad x \in (0, 1), \; t \in (0, T],$$

with boundary conditions $u(0, t) = 100$, $u(1, t) = 25$ and initial condition $u(x, 0) = 25$.

A PINN $u_\theta(x, t)$ is trained with three strategies.

1. **Adam only**: adaptive gradient method with learning rate 10^{-3}.
2. **Adam → L-BFGS**: Adam pre-training (2000 steps) followed by L-BFGS refinement (500 steps).

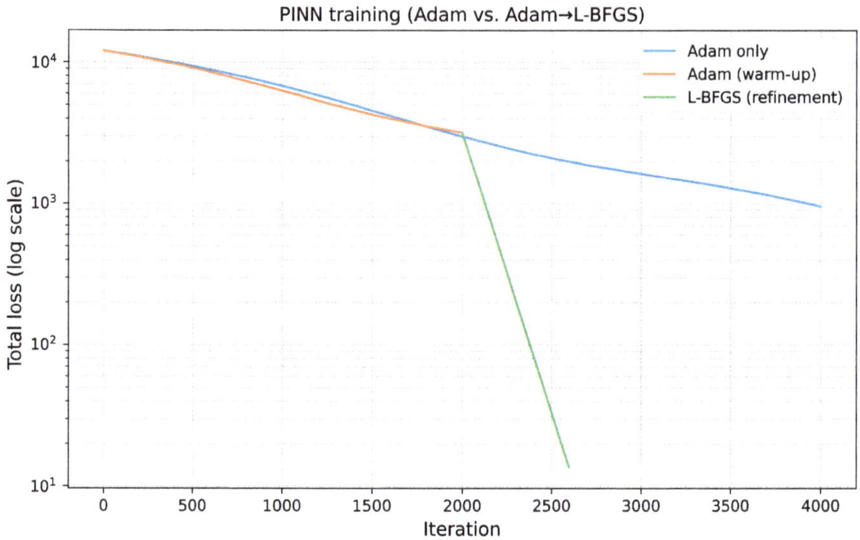

Fig. 6.2 Comparison of training strategies for the 1D heat equation PINN. Adam (blue) makes rapid early progress but plateaus as the loss landscape becomes ill-conditioned. A short Adam warm-up (orange) to iteration 2000 followed by L-BFGS refinement (green) produces a sharp, orders-of-magnitude reduction in the total loss, illustrating the benefit of curvature information for precise residual minimization

3. **Trust-region SQP (trSQP-PINN)**: optimization via sequential quadratic programming inside a trust region.

The observations are seen in Fig. 6.2. Table 6.1 shows the comparative overview landscape.

6.3 Constrained Optimization and Physics-Aware Training

Scientific Machine Learning (SciML) problems frequently require *exact or near-exact* satisfaction of physical laws: (i) **pointwise** constraints (Dirichlet/Neumann boundary conditions, box constraints on states/parameters), (ii) **integral** constraints (mass/energy conservation), (iii) **operator** constraints (PDE residuals below a tolerance in a function space), and (iv) **structural** constraints (symmetry, monotonicity, positivity). We write a generic constrained optimization problem as

$$\min_{\theta \in \mathbb{R}^d} \mathcal{J}(\theta) \quad \text{s.t.} \quad c_i(\theta) = 0 \ \ (i \in \mathcal{E}), \qquad h_j(\theta) \leq 0 \ \ (j \in \mathcal{I}), \tag{6.1}$$

where $\mathcal{J}$ collects data misfit and possibly soft physics terms, while (c_i, h_j) encode hard physics that must hold in the learned solution u_θ.

Table 6.1 Optimization methods: class, update rule, and curvature handling

Method	Class	Update/equation	Curvature
Gradient descent (GD)	1st-order (full)	$\theta_{k+1} = \theta_k - \eta\, \nabla_\theta \mathcal{L}(\theta_k)$	None
Stochastic gradient descent (SGD)	1st-order (stochastic)	$\theta_{k+1} = \theta_k - \eta\, \hat{g}_\mathcal{B}(\theta_k), \quad \hat{g}_\mathcal{B} = \frac{1}{\lvert\mathcal{B}\rvert} \sum_{z\in\mathcal{B}} \nabla_\theta \ell(\theta_k; z)$	None
Momentum/Nesterov (NAG)	1st-order (EMA)	$v_k = \mu v_{k-1} + (1-\mu) g_k, \quad \theta_{k+1} = \theta_k - \eta v_k$ (NAG: gradient at look-ahead)	None
Adam (adaptive moment estimation)	1st-order (adaptive)	$m_k = \beta_1 m_{k-1} + (1-\beta_1) g_k,\ v_k = \beta_2 v_{k-1} + (1-\beta_2) g_k^2,$ $\hat{m}_k = \frac{m_k}{1-\beta_1^k},\ \hat{v}_k = \frac{v_k}{1-\beta_2^k},\ \theta_{k+1} = \theta_k - \eta\, \frac{\hat{m}_k}{\sqrt{\hat{v}_k}+\epsilon}$	None
RMSProp	1st-order (adaptive)	$s_k = \rho s_{k-1} + (1-\rho) g_k^2, \quad \theta_{k+1} = \theta_k - \eta\, \frac{g_k}{\sqrt{s_k}+\epsilon}$	None
Limited-memory BFGS (L-BFGS)	Quasi-Newton (limited-mem)	$p_k = -H_k^{-1} g_k,\ \theta_{k+1} = \theta_k + \alpha_k p_k;\ H_k^{-1}$ via two-loop recursion on (s_i, y_i), $s_i = \theta_{i+1} - \theta_i,\ y_i = g_{i+1} - g_i$	Secant (approx.)
Trust-region/SQP (TR-SQP)	Second-order (TR)	$p_k^* = \arg\min_{\lVert p\rVert \le \Delta_k} m_k(p),\ m_k(p) = \mathcal{L}(\theta_k) + g_k^\top p + \frac{1}{2} p^\top B_k p,\ \theta_{k+1} = \theta_k + p_k^*$; update Δ_k	Model-based

6.3.1 Penalty Methods Versus Augmented Lagrangian Methods

Directly solving the constrained problem in Eq. (6.1) with a generic nonlinear constrained optimizer is often impractical in Scientific Machine Learning (SciML) as :

- Constraints are typically evaluated *stochastically* on mini-batches of collocation or test points.
- Constraint functions may be *operators*, e.g. PDE residuals in a function space rather than simple finite-dimensional maps.
- The objective is almost always *nonconvex and high-dimensional*, coming from neural networks or complex surrogates.

As a result, SciML practitioners usually resort to two closely related families of methods: *quadratic penalty* (soft-constraint) formulations and *augmented Lagrangian* (AL) methods. Both embed the constraints into the objective, but they have markedly different numerical behavior.

Quadratic penalty methods (soft constraints)

The simplest approach is to replace the hard constraints by squared penalties. Given a penalty parameter $\lambda > 0$, we define

$$\mathcal{J}_\lambda(\theta) = \mathcal{J}(\theta) + \frac{\lambda}{2}\sum_{i\in\mathcal{E}} \|c_i(\theta)\|_2^2 + \frac{\lambda}{2}\sum_{j\in I}\big(\max\{0,\ h_j(\theta)\}\big)^2. \tag{6.2}$$

We then solve the *unconstrained* problem

$$\min_\theta \mathcal{J}_\lambda(\theta). \tag{6.3}$$

Under mild regularity assumptions, minimizers of $\mathcal{J}_\lambda$ converge to feasible KKT points of the original problem as $\lambda \to \infty$. Intuitively, large λ exponentially punishes constraint violations and therefore forces the iterates into the feasible region. The downside is that the optimization landscape becomes increasingly ill-conditioned: contours of $\mathcal{J}_\lambda$ form narrow valleys aligned with the constraint manifold, and gradients in the normal direction grow with λ. Gradient-based methods then take very small steps and progress slows dramatically.

In SciML this issue is amplified as:

- PDE residuals can already be stiff; multiplying them by a large λ makes curvature extreme.
- Stochastic mini-batches inject noise into the penalty terms, which can cause oscillatory behavior for large λ.

- Different constraints (interior PDE, boundary conditions, inequality constraints) might have very different scales.

Quadratic penalties thus provide a useful *baseline*, but they do not scale well to tight physical-constraint enforcement.

Augmented Lagrangian methods

Augmented Lagrangian methods introduce explicit Lagrange multipliers for each constraint and keep the penalty parameter at a moderate scale [3]. Let $\mu \in \mathbb{R}^{|E|}$ and $\nu \in \mathbb{R}_+^{|I|}$ denote multipliers for equality and inequality constraints, respectively. For a penalty $\rho > 0$ we define the augmented Lagrangian

$$\begin{aligned} \mathcal{L}_\rho(\theta, \mu, \nu) = J(\theta) &+ \sum_{i \in E} \Big(\mu_i^\top c_i(\theta) + \frac{\rho}{2} \|c_i(\theta)\|_2^2\Big) \\ &+ \sum_{j \in I} \Big(\nu_j\, h_j(\theta) + \frac{\rho}{2}\big(\max\{0,\ h_j(\theta)\}\big)^2\Big). \end{aligned} \tag{6.4}$$

The algorithm alternates between (approximately) minimizing $\mathcal{L}_\rho$ over θ and updating the multipliers:

$$\theta^{k+1} \approx \arg\min_\theta \mathcal{L}_\rho(\theta, \mu^k, \nu^k), \tag{6.5}$$

$$\mu_i^{k+1} = \mu_i^k + \rho\, c_i(\theta^{k+1}), \quad i \in E, \tag{6.6}$$

$$\nu_j^{k+1} = \max\Big\{0,\ \nu_j^k + \rho\, h_j(\theta^{k+1})\Big\}, \quad j \in I. \tag{6.7}$$

If the iterates converge to $(\theta^\star, \mu^\star, \nu^\star)$ with vanishing constraint residuals, then, under standard constraint qualifications (e.g. LICQ), the limit satisfies the Karush–Kuhn–Tucker (KKT) [4] conditions:

$$\nabla_\theta \mathcal{L}_\rho(\theta^\star, \mu^\star, \nu^\star) = 0 \qquad \text{(stationarity)}, \tag{6.8}$$

$$c_i(\theta^\star) = 0, \quad i \in E \qquad \text{(primal feasibility)}, \tag{6.9}$$

$$h_j(\theta^\star) \le 0, \quad j \in I; \quad \nu_j^\star \ge 0 \qquad \text{(dual feasibility)}, \tag{6.10}$$

$$\nu_j^\star\, h_j(\theta^\star) = 0, \quad j \in I \qquad \text{(complementary slackness)}. \tag{6.11}$$

The penalty terms help drive feasibility, while the multipliers learn the correct balance between competing constraints.

In SciML the constraints often encode PDE residuals and boundary conditions. Let $r_\theta(x)$ denote the PDE residual at an interior point $x \in \Omega$ and $b_\theta(s)$ a boundary residual at $s \in \partial\Omega$ (e.g. $u_\theta(s) - g(s)$ for Dirichlet data). Using mini-batches $B_\Omega \subset \Omega$ and $B_{\partial\Omega} \subset \partial\Omega$, we may define empirical constraints

---

Algorithm 3: Stochastic augmented Lagrangian for physics-aware training (AL-PINN)

Input: Initial parameters θ^0, multipliers μ^0, ν^0, penalty $\rho > 0$, update period T, growth factor $\gamma \geq 1$

for $k \leftarrow 0$ **to** ∞ **do**

Sample mini-batches B_Ω, $B_{\partial\Omega}$ (RAR/adaptive sampling optional) ;

Build augmented loss on the batch:

$$\mathcal{L}_\rho^{(k)}(\theta) = w_{\text{data}} R_{\text{data}}^{(k)}(\theta) + \mu^{k\top} c^{(k)}(\theta) + \frac{\rho}{2}\left\|c^{(k)}(\theta)\right\|_2^2 + \nu^{k\top} h^{(k)}(\theta) + \frac{\rho}{2}\left\|\max\{0,\ h^{(k)}(\theta)\}\right\|_2^2$$

Update parameters by a gradient-based step (e.g. Adam, then L–BFGS/TR refinement):

$$\theta^{k+1} = \theta^k - \eta_k \nabla_\theta \mathcal{L}_\rho^{(k)}(\theta^k)$$

if $k \bmod T = 0$ **then**

Update multipliers:

$$\mu^{k+1} = \mu^k + \rho\, c^{(k)}(\theta^{k+1}), \quad \nu^{k+1} = \max\{0,\ \nu^k + \rho\, h^{(k)}(\theta^{k+1})\}$$

If mean feasibility (e.g. $\|c^{(k)}\|$) has not decreased over the last M updates, set $\rho \leftarrow \gamma\rho$;

Monitor diagnostics: R_{data}, $\|c^{(k)}(\theta)\|$, $\|\max\{0, h^{(k)}(\theta)\}\|$;

$$c_\Omega(\theta) = \frac{1}{|B_\Omega|}\sum_{x \in B_\Omega} r_\theta(x), \qquad c_{\partial\Omega}(\theta) = \frac{1}{|B_{\partial\Omega}|}\sum_{s \in B_{\partial\Omega}} b_\theta(s), \tag{6.12}$$

or vector-valued versions stacking residuals over collocation points.

An augmented Lagrangian PINN then minimizes

$$\begin{aligned}\mathcal{L}_\rho(\theta, \mu_\Omega, \mu_{\partial\Omega}) = w_{\text{data}} R_{\text{data}}(\theta) + \mu_\Omega^\top c_\Omega(\theta) + \frac{\rho}{2}\left\|c_\Omega(\theta)\right\|_2^2 + \mu_{\partial\Omega}^\top c_{\partial\Omega}(\theta)\\ + \frac{\rho}{2}\left\|c_{\partial\Omega}(\theta)\right\|_2^2,\end{aligned} \tag{6.13}$$

with additional hinge terms for inequality constraints (e.g. positivity of a material property).

From a practical SciML perspective:

Quadratic penalties.

- **Pros**: conceptually simple; single unconstrained objective; easy to implement in existing deep-learning code.
- **Cons**: require very large λ for tight feasibility; cause severe ill-conditioning; sensitive to batch noise and scale imbalances between constraints.

Augmented Lagrangians.

- **Pros**: achieve high feasibility with moderate ρ; improved conditioning via learned multipliers; better handling of heterogeneous scales (data vs. physics); naturally connected to KKT theory.
- **Cons**: extra state (μ, ν) to maintain; additional hyperparameters (ρ, update period T, growth schedule); more complex implementation.

Physics-based constraints can differ by orders of magnitude (e.g. interior PDE residuals vs. small boundary mismatches). A useful heuristic is to normalize each constraint:

$$\tilde{c}_i(\theta) = \frac{c_i(\theta)}{\sigma_i}, \qquad \sigma_i \approx \text{running RMS of } c_i(\theta) \text{ over early epochs}, \tag{6.14}$$

and to build $\mathcal{L}_\rho$ and multiplier updates using $\tilde{c}_i$. This balances gradient contributions across constraints and mirrors loss-term normalization in other parts of the book.

Whenever possible, some constraints should be enforced *by construction*. For example, Dirichlet boundary conditions can be encoded via a constrained trial function

$$u_\theta(x) = g(x) + B(x)\, n_\theta(x), \qquad B|_{\partial\Omega} = 0, \tag{6.15}$$

so that boundary feasibility holds exactly, and the optimizer (penalty or AL) focuses on interior PDE constraints and Neumann/flux conditions.

Common pitfalls and recommendations

- Overly aggressive increases of λ or ρ can "freeze" training due to extremely sharp valleys. Prefer gradual homotopy schedules.
- With small batches, multiplier updates can be noisy; updating (μ, ν) less frequently and averaging constraints over several batches improves stability.
- If some constraints consistently dominate the gradient, re-scale or re-weight them; in AL, per-constraint step sizes for μ, ν can help.
- Non-differentiable hinges for inequalities are usually acceptable; for smoother early training, one can temporarily replace $\max\{0, h\}$ by a softplus approximation.

Overall we can conclude that, quadratic penalties offer a straightforward entry point for physics-aware training in SciML but struggle when high-fidelity constraint satisfaction is required under stiffness and scale disparity. Augmented Lagrangian methods introduce learnable multipliers that substantially improve conditioning and feasibility, especially when combined with hard enforcement of simple constraints, careful normalization, and staged optimizer schedules (e.g. Adam warm-up followed by L-BFGS or trust-region refinement).

6.3.2 Variational and Weak-Form Optimization

Many scientific models are governed by partial differential equations (PDEs) that originate from *variational principles*: instead of "solving a PDE," one can often say "find the state that minimizes an energy functional". This perspective is central to mechanics, electromagnetism, fluid dynamics, and many inverse problems.

In Scientific Machine Learning (SciML), this leads to an alternative to pointwise residual minimization:

- we can train models by minimizing an *energy functional* (variational formulation), or
- we can enforce the PDE only after integrating against suitable *test functions* (weak formulation).

Both viewpoints are deeply connected to finite elements and Galerkin methods, and they provide more stable and physically meaningful loss functions than naive pointwise residuals.

High-level idea

Instead of forcing a model to satisfy the physical equations at every single point (which can be noisy and unstable), we can ask it to make the *total energy* as small as possible, or to satisfy the equations *on average* against carefully chosen test functions. This is often easier to optimize and more robust to numerical errors.

Variational principles and energy minimization

Consider a field $u : \Omega \to \mathbb{R}^m$, such as displacement, temperature, or concentration, defined on a domain $\Omega \subset \mathbb{R}^d$. A typical variational problem is:

$$\min_{u \in \mathcal{V}} \mathcal{E}(u) \quad \text{with} \quad \mathcal{E}(u) = \int_{\Omega} F\big(x, u(x), \nabla u(x)\big)\, \mathrm{d}x + \int_{\partial\Omega_N} G\big(x, u(x)\big)\, \mathrm{d}s, \qquad (6.16)$$

where $\mathcal{V}$ is a suitable function space (e.g. a Sobolev space with prescribed Dirichlet boundary conditions), F encodes bulk energy density (e.g. elastic strain energy), and G encodes boundary work (e.g. Neumann loads).

What is a variational problem?

A **variational problem** asks: *Among all admissible states u, which one makes a certain scalar quantity (the energy) as small as possible?* The energy is usually written as one or more integrals over the domain and its boundary.

The magic is that the minimizer of this energy often automatically satisfies the governing PDEs and boundary conditions.

Under suitable regularity assumptions, the minimizer $u^\star$ satisfies the Euler–Lagrange equations associated with $\mathcal{E}$. Formally, if we consider perturbations $u^\star + \varepsilon v$ with v in the space of admissible variations, the first-order optimality condition

$$\frac{\mathrm{d}}{\mathrm{d}\varepsilon}\mathcal{E}(u^\star + \varepsilon v)\Big|_{\varepsilon=0} = 0 \quad \forall v$$

leads to a PDE (in strong or weak form) that $u^\star$ must satisfy.

Example: Poisson equation from energy minimization.

Let $\Omega \subset \mathbb{R}^d$, and consider the Poisson problem

$$-\Delta u(x) = f(x) \ \text{ in } \Omega, \quad u(x) = 0 \ \text{ on } \partial\Omega. \tag{6.17}$$

Define the energy

$$\mathcal{E}(u) = \frac{1}{2}\int_\Omega \|\nabla u(x)\|_2^2 \,\mathrm{d}x - \int_\Omega f(x)\,u(x)\,\mathrm{d}x. \tag{6.18}$$

Minimizing $\mathcal{E}(u)$ over $H_0^1(\Omega)$ yields, via the Euler–Lagrange equation, the Poisson PDE (6.17) with homogeneous Dirichlet boundary conditions.

Physical interpretation (Poisson example)

In the Poisson example, the first term is like *elastic energy* stored in the field u, and the second term is the *work* done by the source f. The system chooses u that balances these two: it wants to keep the gradient small (no rapid changes), but it must also respond to the forcing f. The compromise is exactly the solution of the PDE.

In SciML, we can approximate u by a neural network u_θ and minimize $\mathcal{E}(u_\theta)$ with respect to θ: this is *variational training* or energy-based physics-aware training.

Weak formulations and test functions

The variational viewpoint naturally leads to the *weak form* of a PDE. For the Poisson problem (6.17), the weak formulation is:

$$\text{Find } u \in H_0^1(\Omega) \text{ such that } \int_\Omega \nabla u(x) \cdot \nabla v(x)\,\mathrm{d}x = \int_\Omega f(x)\,v(x)\,\mathrm{d}x \quad \forall v \in H_0^1(\Omega). \tag{6.19}$$

What is a weak formulation?

A **weak formulation** does not require the equation to hold pointwise. Instead, we require that the equation holds *after multiplication by all test functions* v *and integration over the domain.*

Intuitively, the solution is allowed to be rough (for example, with corners or discontinuous derivatives), but the integrated equation must hold for all "probing" functions v.

The weak form typically arises by multiplying the strong PDE by a test function v, integrating over Ω, and applying integration by parts (or Green's identities). This has two important consequences:

- It reduces the number of derivatives that must be taken on u (helpful for low-regularity solutions).
- It moves derivatives onto the test function, which can be chosen from a convenient space (e.g. finite element basis functions).

From a numerical standpoint, the weak form is the starting point for Galerkin methods and finite elements.

From Galerkin methods to weak-form PINNs

Classical Galerkin methods choose a finite-dimensional trial space $\mathcal{V}_h = \operatorname{span}\{\varphi_1, \ldots, \varphi_n\}$, approximate $u(x) \approx u_h(x) = \sum_{i=1}^n \alpha_i \varphi_i(x)$, and enforce the weak form only against test functions in the same space:

$$\int_\Omega \nabla u_h \cdot \nabla \varphi_j\,\mathrm{d}x = \int_\Omega f\,\varphi_j\,\mathrm{d}x, \quad j = 1, \ldots, n. \tag{6.20}$$

This yields a linear system in the coefficients α_i.

In weak-form SciML (or weak-form PINNs), we replace the finite element approximation by a neural network or hybrid trial space. For example, using a neural network u_θ as the trial function and a finite set of test functions $\{\psi_j\}_{j=1}^M$, we define weak residuals

$$R_j(\theta) = \int_\Omega \nabla u_\theta(x) \cdot \nabla \psi_j(x)\, \mathrm{d}x - \int_\Omega f(x)\, \psi_j(x)\, \mathrm{d}x, \quad j = 1, \ldots, M. \tag{6.21}$$

The training objective is then

$$\mathcal{L}_{\text{weak}}(\theta) = \frac{1}{2} \sum_{j=1}^{M} \left| R_j(\theta) \right|^2 + \lambda_{\text{data}} R_{\text{data}}(\theta) + \lambda_{\text{reg}} R_{\text{reg}}(\theta), \tag{6.22}$$

where R_{data} and R_{reg} are data-fitting and regularization terms as in other parts of the chapter.

Weak-form PINNs versus pointwise PINNs (intuitive view)

- A **pointwise PINN** tries to make the PDE residual small at many collocation points.
- A **weak-form PINN** tries to make the *integrated residual* small against several test functions.

Pointwise enforcement can be very sensitive to local noise and requires high smoothness. Weak-form enforcement is closer to traditional finite elements: it is more global, more stable, and can handle rougher solutions.

In practice, the integrals in (6.21) and (6.22) are approximated by numerical quadrature or Monte Carlo sampling. This is compatible with mini-batch training and automatic differentiation.

Optimization viewpoints in variational and weak forms

When the PDE is derived from an energy functional $\mathcal{E}$, we can directly train a neural network u_θ by minimizing

$$\mathcal{L}_{\text{var}}(\theta) = \mathcal{E}(u_\theta) + \lambda_{\text{data}} R_{\text{data}}(\theta) + \lambda_{\text{reg}} R_{\text{reg}}(\theta).$$

This approach is particularly appealing because:

- The loss has a clear physical meaning (total energy).
- The landscape can be better conditioned than raw residuals, especially for elliptic problems.
- Global physical laws (e.g. conservation principles) are built into the objective.

In problems where a clean energy functional is not available (e.g. some nonlinear, non-symmetric operators or mixed formulations), we can still exploit the weak form:

$$\mathcal{L}_{\text{weak}}(\theta) = \frac{1}{2} \sum_{j=1}^{M} \left| R_j(\theta) \right|^2 + \lambda_{\text{data}} R_{\text{data}}(\theta) + \lambda_{\text{reg}} R_{\text{reg}}(\theta),$$

with $R_j(\theta)$ defined as in (6.21). This turns the weak formulation into a nonlinear least squares problem over θ.

Why does this help optimization?

Weak and variational formulations average the PDE over regions and test functions. This tends to smooth out local numerical noise and makes gradients more informative. As a result, optimization algorithms see a "cleaner" loss landscape, which can converge faster and more reliably than raw pointwise penalties.

A key practical advantage of weak formulations is that they reduce regularity requirements on the trial function:

- For second-order PDEs, the strong form typically requires u to be twice differentiable.
- The weak form only requires u to have square-integrable first derivatives (belonging to a Sobolev space H^1).

In neural network terms, this means we can often work with simpler architectures and fewer nested derivatives: after integration by parts, many derivatives fall on the test functions, which are usually known and simple (e.g. polynomials, finite element basis functions) [5].

Boundary conditions also enter naturally:

- Dirichlet conditions can be enforced by constructing a trial space $\mathcal{V}$ that already satisfies them.
- Neumann or Robin conditions emerge as boundary integrals in the weak form; they can be treated as additional integral terms in the loss.

Let us revisit the Poisson problem with a neural network approximation. We parameterize the solution as $u_\theta(x)$ and consider the energy (6.18). We approximate the integrals by quadrature or Monte Carlo sampling:

$$\mathcal{E}(u_\theta) \approx \frac{1}{2} \sum_{q \in Q_\Omega} w_q \left\| \nabla u_\theta(x_q) \right\|_2^2 - \sum_{q \in Q_\Omega} w_q \, f(x_q) \, u_\theta(x_q), \tag{6.23}$$

where Q_Ω is a set of quadrature points with weights w_q. The training objective is

$$\min_\theta \; \mathcal{E}(u_\theta),$$

optionally plus data and regularization terms.

For a weak-form PINN version, we select test functions $\{\psi_j\}$ (e.g. low-degree polynomials or localized basis functions) and approximate the weak residuals (6.21) via quadrature. The training objective is then (6.22).

Implementation sketch

1. Pick a neural network u_θ that represents your field (temperature, displacement, etc.).
2. Write down either an *energy* integral or a *weak residual* integral for your PDE.
3. Approximate the integrals by summing over many sample points (just like you sum over mini-batches).
4. Use your favorite optimizer (SGD, Adam, L–BFGS) to adjust θ and minimize this integrated loss.

From the outside, training looks like "ordinary deep learning," but the loss is now firmly rooted in variational and weak-form principles.

Summary and comparison

- **Strong-form/pointwise residuals**: simple to implement, but can be ill-conditioned, sensitive to local noise, and require higher regularity (more derivatives).
- **Variational formulations**: minimize a physically meaningful energy; often yield better conditioning and are natural for equilibrium problems.
- **Weak formulations**: enforce PDEs in an integral sense; reduce regularity requirements, connect directly to finite elements, and fit well with Monte Carlo and mini-batch training.

Takeaway

Variational and weak-form optimization give us *physics-aware loss functions* that are both mathematically grounded and numerically robust. They bridge classical finite element ideas with modern deep learning, and they are often the right tool when strong-form PINNs struggle to converge or to represent low-regularity solutions.

6.3.3 Gradient-Free and Metaheuristic Optimization

Most of this chapter has focused on *gradient-based* methods, enabled by automatic differentiation (AD). However, many SciML problems are only *partially differentiable*, or not differentiable at all:

- The objective may depend on a legacy simulator, black-box code, or commercial software.
- The model may include discrete design variables (topology, architecture, mesh choices).
- The loss may be noisy, discontinuous, or defined via non-smooth post-processing metrics.
- Gradients may be available in principle, but too expensive or numerically unstable in practice.

In these settings, we must treat the objective as a black box that we can evaluate—perhaps with noise, delay, or failure—but cannot differentiate. This is the domain of *gradient-free* and *metaheuristic* optimization.

Gradient-free optimization

Gradient-free optimization refers to methods that improve the objective without using derivatives. They only need the value of the objective function for different candidate solutions. These methods are particularly useful when the objective comes from a black-box simulation or is too noisy for reliable gradients.

Metaheuristics

A **metaheuristic** is a general-purpose search strategy that explores the space of possible solutions using high-level rules inspired by nature or problem structure (evolution, swarms, annealing, etc.). Metaheuristics are usually stochastic, approximate, and designed for difficult nonconvex problems where classical methods struggle.

In SciML, gradient-free methods and metaheuristics appear in at least three roles:

1. *Model calibration and design* when the forward model is a black-box simulator.
2. *Hyperparameter and architecture search* around differentiable training loops.
3. *Outer-loop optimization* for bilevel or nested problems where inner loops are gradient-based.

Direct search and pattern search methods

The simplest gradient-free methods are *direct search* algorithms: they propose new points based purely on comparisons of objective values.

A classical example is *pattern search* (or generalized pattern search): starting from a current iterate θ^k, we probe a finite set of directions $\{d_1, \dots, d_p\}$ with a step size Δ_k and accept a trial point if it improves the objective:

$$J(\theta^k + \Delta_k d_i) < J(\theta^k).$$

If such a direction is found, we move there (and possibly increase the step size); otherwise, we reduce Δ_k and try again.

Direct search

Imagine you are on a foggy mountain and can only measure your altitude at the points where you stand. Direct search algorithms simply try a small set of nearby directions: if one direction goes downhill, you follow it; if not, you take smaller steps and try again. No slope information is needed, only comparisons of altitude.

Direct search methods typically offer:

- Simple implementation and robust behavior under noise.
- Convergence guarantees to stationary points under suitable conditions.
- Moderate efficiency for low- to medium-dimensional problems.

However, they scale poorly to very high-dimensional parameter spaces (hundreds or thousands of dimensions), which are common in deep SciML models.

Evolutionary and population-based algorithms

Beyond local direct search, many SciML applications require *global* exploration over rugged landscapes, mixed discrete–continuous variables, and multiple basins of attraction. In such cases, *population-based* metaheuristics are often used.

Genetic algorithms (GAs) and evolutionary strategies (ES) maintain a *population* of candidate solutions $\{\theta^{(1)}, \dots, \theta^{(N)}\}$. A typical cycle includes:

1. **Evaluation**: compute the objective $J(\theta^{(i)})$ for all individuals.
2. **Selection**: prefer individuals with lower objective values.
3. **Variation**: generate new candidates via crossover (combining parts of two parents) and mutation (random perturbations).
4. **Replacement**: build the next population from selected parents and offspring.

Evolutionary strategies (e.g. CMA-ES) refine this idea by learning a *search distribution* over θ (often Gaussian) and adapting its covariance matrix based on successful steps.

Population-based search

Population-based algorithms search not with a single candidate, but with a whole *swarm* of candidates. Some explore new areas, others refine promising regions. Good candidates are more likely to have "offspring" in the next generation, so the population gradually drifts toward better regions of the search space.

Population-based algorithms are attractive in SciML when:

- The objective is multi-modal with many local minima.
- We can afford parallel evaluations of many candidates (e.g. cluster, GPU farm).
- Variables include discrete decisions (e.g. architecture choices, mesh topology).

Algorithms such as Particle Swarm Optimization (PSO) and Differential Evolution (DE) are widely used in engineering design and model calibration. They also maintain populations, but with different update rules:

- In PSO, each candidate (particle) moves in the parameter space with a velocity updated based on its own best position and the population's global best.
- In DE, new candidates are generated by adding scaled differences of individuals to others, followed by crossover and selection.

These methods are simple, massively parallelizable, and often perform well on noisy black-box objectives.

Simulated annealing and related heuristics

Simulated annealing (SA) is a single-trajectory metaheuristic inspired by thermodynamics. At each iteration, we propose a new candidate θ' near the current state θ and accept it with probability

$$p = \min\Big(1,\ \exp\big(-(J(\theta') - J(\theta))/T_k\big)\Big),$$

where T_k is a *temperature* that decreases over time. At high temperatures, uphill moves are frequently accepted, allowing broad exploration; as T_k decreases, the algorithm becomes more greedy.

Annealing metaphor

Simulated annealing mimics heating and slowly cooling a piece of metal. At high temperature, atoms can move freely and escape local traps. As the material is cooled slowly, it settles into a low-energy configuration. In optimization, this helps escape shallow local minima early, then refine around deeper minima later.

SA and related heuristics are especially useful when the search space is discrete or combinatorial (e.g. selecting sensor locations, choosing active terms in a symbolic model).

In SciML, metaheuristics are rarely used alone. They are most powerful when integrated into broader workflows:

1. Black-box parameter calibration.
Consider a complex multiphysics simulator $\mathcal{S}(\theta)$ that returns a field u_θ. We might define a misfit between simulation and data:

$$J(\theta) = \frac{1}{2}\sum_i \|\mathcal{O}(u_\theta)(x_i) - y_i\|_2^2,$$

where $\mathcal{O}$ extracts observable quantities (sensor outputs, average fluxes, etc.). If $\mathcal{S}$ is a legacy Fortran or C++ code without AD support, gradient-free metaheuristics provide a direct way to calibrate θ by repeatedly calling the simulator.

2. Hyperparameter and architecture search.
Hyperparameters and architecture choices (depth, width, activation types, basis functions, loss weights) are notoriously sensitive. Metaheuristics can be used to search over these higher-level design variables, while inner training is gradient-based.

Nested optimization pattern

A common pattern in SciML is:
Outer loop (gradient-free): choose hyperparameters/design variables.
Inner loop (gradient-based): train the model with AD for fixed design.

The outer loop sees the inner loop as a black box that returns a validation loss. This nested structure balances global exploration with efficient local refinement.

2. Bilevel and design-optimization problems.
In design optimization, one often seeks design variables ϕ (geometry, material distribution) that minimize a performance metric, subject to an inner SciML or PDE solve. If gradients with respect to ϕ are unavailable or unreliable, metaheuristics can drive the outer design loop while exploiting gradient-based solvers inside.

Metaheuristics trade *information* (gradients, Hessians) for *evaluations*. In SciML, each function evaluation may itself involve:

- Running a time-consuming simulation.
- Training a neural network for many epochs.
- Solving a PDE system to convergence.

Therefore:

- **Budgeting evaluations** is critical: a naive population size or horizon can be infeasible.
- **Parallelization** is essential: many metaheuristics are embarrassingly parallel over candidates or generations.
- **Noise-awareness** matters: stochastic objectives (due to random seeds, mini-batches, Monte Carlo sampling) can mislead selection if not averaged or smoothed.

Rule of thumb

Use metaheuristics:
- when derivatives are unavailable, unreliable, or prohibitively expensive;
- for outer-loop design and hyperparameter search, not for every inner training step;
- when you can exploit parallel hardware to evaluate many candidates simultaneously.

Prefer gradient-based methods whenever clean, cheap gradients exist.

Hybrid gradient-based/gradient-free strategies

Pure metaheuristics can be too slow for high-dimensional models. However, *hybrid* strategies often work well:

- **Metaheuristic initialization**: use PSO/DE/GA to find a good starting point, then switch to a fast gradient-based method.
- **Multi-start local search**: sample many initial points with a metaheuristic, then run short gradient-based refinements and keep the best.
- **Coordinate splitting**: update continuous parameters with gradients and discrete parameters (e.g. architecture, topology) with metaheuristics.
- **Adaptive surrogate modeling**: metaheuristics can explore the space, while a surrogate model (e.g. a small network) learns to approximate the objective and supplies cheap gradient information.

These hybrid strategies align with the broader theme of SciML: combine complementary methods—here, global stochastic search and local deterministic optimization—to exploit their respective strengths. In summary,

- Gradient-free and metaheuristic methods are essential tools for SciML problems involving black-box simulators, discrete design variables, or noisy, non-differentiable objectives.
- Direct search and pattern search methods offer simple, robust local optimization in low- to medium-dimensional spaces.
- Population-based metaheuristics (GA, ES, PSO, DE, simulated annealing) provide global search capabilities at the cost of many objective evaluations.
- In practice, metaheuristics are most effective in *outer loops* (design, hyperparameters, architecture), combined with gradient-based optimization inside.

Key takeaway

Metaheuristics do not replace gradient-based optimization in SciML; they complement it. When gradients are corrupt, missing, or too expensive, gradient-free methods keep the optimization problem solvable, especially in outer loops and design tasks. The most powerful SciML workflows are often hybrid, using gradient-free methods for exploration and gradient-based methods for fast, local exploitation.

6.3.4 Bayesian and Surrogate-Based Optimization

In many SciML applications, each evaluation of the objective

$$J(x), \quad x \in \mathcal{X} \subset \mathbb{R}^d,$$

is *expensive*: it may require running a high-fidelity PDE solver, a multi-physics simulator, or even a physical experiment. In such settings, the limiting resource is not GPU time for training a neural network, but the number of expensive function evaluations we can afford.

This motivates *sample-efficient global optimization*: we want to locate good designs $x^\star$ using as few evaluations of J as possible. Bayesian and surrogate-based optimization provide a principled framework for this regime by

1. building a *probabilistic surrogate* of J from scarce data, and
2. using this surrogate to guide where to evaluate next.

High-level idea

When each experiment or simulation is expensive, you cannot try thousands of random options. Instead, you build a fast statistical model that *guesses* the objective in unexplored regions and quantifies uncertainty. A decision rule (the *acquisition function*) then chooses the next, most informative experiment: either where the model predicts low loss, or where it is still very uncertain.

Bayesian optimization (BO) is the canonical example of this paradigm. The surrogate is often a Gaussian Process (GP) or another probabilistic regressor (see Chap. 8), but the general idea extends to neural and ROM surrogates (Chap. 7).

Surrogate models in optimization

Let $\mathcal{D}_t = \{(x_i, y_i)\}_{i=1}^t$ with $y_i = J(x_i) + \varepsilon_i$ denote the observations collected up to iteration t. A surrogate model defines a predictive distribution

$$p(J(x) \mid \mathcal{D}_t),$$

which typically provides a mean $\mu_t(x)$ and uncertainty (variance) $\sigma_t^2(x)$.

Definition (Surrogate model)

A **surrogate model** is a cheaper approximation of an expensive function. It is trained from a limited set of evaluations of the true function and then used to:
- predict the function at new points, and
- guide the search for optimal designs.

In SciML, surrogates can be Gaussian Processes, reduced-order models, or neural emulators.

The surrogate is then used in a *decision-making loop* that balances:

- *exploitation*: sampling where $\mu_t(x)$ is low (for minimization), and
- *exploration*: sampling where $\sigma_t(x)$ is high.

6.3.4.1 Bayesian Optimization

Bayesian optimization (BO) treats J as a random function with a prior distribution. The most common choice is a Gaussian Process (GP) prior (see Ch. 8),

$$J \sim \mathcal{GP}\big(m_0(x), k_0(x, x')\big),$$

which yields a closed-form posterior $p(J(x) \mid \mathcal{D}_t)$ after observing $\mathcal{D}_t$.

Definition (Bayesian Optimization)

Bayesian Optimization is a sequential strategy for optimizing an expensive black-box function. At each step, it:
1. updates a probabilistic surrogate from past evaluations, and
2. maximizes an *acquisition function* that uses the surrogate's mean and uncertainty to choose the next evaluation point.

Given the posterior mean $\mu_t(x)$ and standard deviation $\sigma_t(x)$ at iteration t, an acquisition function

$$\alpha_t(x) = \alpha\big(x;\, \mu_t,\, \sigma_t,\, f_{\min,t}\big)$$

scores how promising x is, where $f_{\min,t} = \min_i y_i$ is the best observed value so far.

Common choices (for minimization) include:

- **Probability of Improvement (PI)**:

$$\alpha_{\mathrm{PI}}(x) = \mathbb{P}\big(J(x) \le f_{\min,t} - \xi\big) = \Phi\left(\frac{f_{\min,t} - \xi - \mu_t(x)}{\sigma_t(x)}\right),$$

  with margin $\xi \ge 0$ and Φ the standard normal CDF.
- **Expected Improvement (EI)**:

$$\alpha_{\mathrm{EI}}(x) = \mathbb{E}\big[(f_{\min,t} - J(x))_+ \mid \mathcal{D}_t\big],$$

  which trades off low predicted mean and high uncertainty (closed-form for GP posteriors).
- **Upper/Lower Confidence Bound (UCB/LCB)**:

$$\alpha_{\mathrm{LCB}}(x) = \mu_t(x) - \kappa\sigma_t(x),$$

  with $\kappa > 0$ controlling the exploration–exploitation balance.

Intuition: EI and UCB

Expected Improvement prefers points that are likely to give a better value than the current best, but also rewards uncertainty. Lower Confidence Bound acts like a "pessimistic" prediction: it subtracts a multiple of the uncertainty from the mean, so highly uncertain points can still look attractive.

A basic BO algorithm for minimization is:

In SciML, x may represent hyperparameters for a PINN, closure parameters in a turbulence model, or geometric design variables, and each $J(x)$ may involve solving a PDE to convergence.

Algorithm 4: Bayesian Optimization (single-fidelity, minimization)

Choose initial design set $\{x_i\}_{i=1}^{t_0}$ (e.g., Latin hypercube, space-filling) ;
Evaluate $y_i = J(x_i)$ and set $\mathcal{D}_{t_0} = \{(x_i, y_i)\}_{i=1}^{t_0}$;
for $t = t_0$ **to** T **do**
  Fit or update surrogate $p(J(\cdot) \mid \mathcal{D}_t)$ (e.g., GP regression) ;
  Construct acquisition function $\alpha_t(x)$ (e.g., EI, LCB) ;
  Select next point $x_{t+1} = \arg\max_{x \in \mathcal{X}} \alpha_t(x)$;
  Evaluate the expensive objective $y_{t+1} = J(x_{t+1})$;
  Augment dataset: $\mathcal{D}_{t+1} = \mathcal{D}_t \cup \{(x_{t+1}, y_{t+1})\}$;

When to use Bayesian Optimization (SciML)

Bayesian Optimization is most suitable when:
- each evaluation of $J(x)$ is expensive (seconds to hours),
- you can only afford tens to a few hundred evaluations,
- the dimension d is moderate (typically $d \lesssim 20$), and
- you care about global performance rather than just a local optimum.

6.3.4.2 Multi-fidelity Bayesian Optimization

In SciML, models often exist at multiple fidelities:

- a coarse, fast simulation with mesh size h_{coarse},
- a fine, accurate simulation with mesh size h_{fine},
- possibly intermediate fidelities or reduced-order surrogates.

Evaluations at different fidelities have different costs and accuracies.

Multi-fidelity setting

We may have:
- a cheap model that is approximate but fast,
- an expensive model that is accurate but slow.

Multi-fidelity Bayesian Optimization tries to combine them: it uses many cheap evaluations to shape the landscape and only a few expensive evaluations to correct the bias.

A common modeling strategy is to treat the high-fidelity objective J_{HF} as

$$J_{\text{HF}}(x) = \rho\, J_{\text{LF}}(x) + \delta(x),$$

where J_{LF} is a low-fidelity surrogate and δ a discrepancy term. Gaussian Process co-kriging models this joint structure, providing a multi-output surrogate with coupled posteriors.

Given a cost model $c(\ell)$ for fidelity level ℓ, multi-fidelity BO uses *cost-aware acquisition functions*, e.g.

$$\tilde{\alpha}_t(x, \ell) = \frac{\text{EI}_\ell(x)}{c(\ell)},$$

and chooses both the next input x and the fidelity level ℓ.

For some SciML examples:

- **PINN hyperparameters**: cheap runs with fewer collocation points or lower network width, expensive runs with full collocation and architecture.
- **PDE design**: coarse meshes for exploratory search in design space, fine meshes for final refinement.
- **Experimental design**: quick, low-precision experiments versus slow, high-precision measurements.

Practical tips

- Start by exploring with low-fidelity models to map the rough landscape.
- Gradually increase the share of high-fidelity evaluations near promising designs.
- Regularize the discrepancy $\delta(x)$ to avoid overfitting sparse high-fidelity data.

6.3.4.3 High-Dimensional Bayesian Optimization

Standard GP-based BO scales poorly in high dimensions: the number of points required to adequately cover $\mathbb{R}^d$ grows exponentially with d, and optimizing the acquisition becomes difficult.

However, many SciML problems exhibit *effective low-dimensional structure*: only a subset or a low-dimensional linear/nonlinear subspace of parameters significantly influences J.

Key idea: intrinsic versus ambient dimension

The ambient dimension d may be large (dozens or hundreds of parameters), but the function may vary strongly only along a few directions or groups of parameters. High-dimensional BO methods aim to exploit this *intrinsic low-dimensional structure*.

High-dimensional BO strategies include:

- **Random embeddings**: optimize in a low-dimensional subspace $z \in \mathbb{R}^p$ ($p \ll d$) and map back via a random projection $x = Az$ (REMBO-style methods).
- **Additive models**: assume $J(x) \approx \sum_g J_g(x_g)$ over groups of coordinates x_g, and fit separate low-dimensional surrogates.
- **Trust-region BO (e.g. TuRBO)**: run multiple local BO processes in small trust regions, expanding/shrinking regions based on success; this avoids global coverage of the entire high-dimensional cube.
- **Sparse axis-aligned models (e.g. SAASBO)**: place sparsity-promoting priors on length-scales so that only important coordinates receive short length-scales.

In SciML, these ideas manifest as:

- grouping hyperparameters (e.g. learning rates, loss weights, network widths) into low-dimensional blocks,
- optimizing only a few sensitive parameters while fixing others from domain knowledge,
- using local BO around a baseline design found by physics or expert intuition.

6.3.4.4 Multi-objective Bayesian Optimization

Many scientific design problems involve *multiple objectives*, e.g.

- minimizing PDE error while minimizing wall-clock time,
- maximizing material strength while minimizing weight,
- maximizing accuracy of a PINN while minimizing the number of collocation points.

Formally, we seek to optimize a vector-valued objective

$$F(x) = \big(f_1(x), \ldots, f_M(x)\big),$$

typically under minimization.

Pareto optimality

In multi-objective optimization, there is rarely a single "best" design. Instead, we look for the **Pareto front**: the set of designs where you cannot improve one objective without worsening at least one other. Any design on this front is *Pareto optimal.*

A point x *dominates* x' if $f_m(x) \leq f_m(x')$ for all m and $f_m(x) < f_m(x')$ for at least one m. The Pareto front is the set of undominated points in $\mathcal{X}$.

Multi-objective BO (MOBO) extends BO by modeling each f_m with a probabilistic surrogate and using multi-objective acquisition functions, such as:

- **(q-)Expected Hypervolume Improvement (qEHVI/qNEHVI)**: chooses points expected to increase the hypervolume dominated by the Pareto front.
- **Constrained MOBO**: incorporates feasibility via constraint surrogates and penalized hypervolume improvement.

In SciML, MOBO is natural for:

- trading off accuracy versus inference speed or memory footprint in neural surrogates,
- balancing robustness versus nominal performance in control and design,
- navigating trade-offs between different physical metrics (e.g. drag vs. lift).

Designing with multiple goals

Instead of asking "What is the best design?", MOBO asks: *"What is the set of designs that are not obviously worse than any other?"* From that Pareto set, scientists and engineers can choose based on practical priorities, regulatory constraints, or human preferences.

Let us now summarize the methods:

- Bayesian Optimization frames expensive black-box optimization as a sequential decision-making problem guided by probabilistic surrogates.
- Multi-fidelity BO exploits hierarchies of models (coarse vs. fine simulations, ROMs vs. full solvers) to improve sample efficiency.
- High-dimensional BO uses structure (additivity, low-dimensional subspaces, trust regions, sparsity) to cope with large parameter spaces.
- Multi-objective BO builds approximate Pareto fronts, enabling informed trade-offs among conflicting scientific goals.

Key takeaway

Bayesian and surrogate-based optimization turn expensive SciML workflows into intelligent, data-efficient search processes. They are especially powerful when:

- each evaluation involves a costly simulation or experiment,
- we have strong priors or surrogate models (GPs, ROMs, neural emulators),
- and we care about principled exploration of design and hyperparameter spaces.

In practice, they are most effective when combined with gradient-based inner solvers and multi-fidelity surrogates, forming the backbone of modern design and calibration loops in SciML.

6.3.5 *Differentiable Simulation and Adjoint-Based Optimization*

Many SciML tasks can be phrased as

$$\min_{\theta} J(u,\theta) \quad \text{s.t.} \quad \mathcal{F}(u,\theta) = 0, \tag{6.24}$$

where $\mathcal{F}(u,\theta) = 0$ encodes a simulator (ODE, PDE, multi-physics model) and θ are design, control, or model parameters. A naive approach is to treat the simulator as a black box and use gradient-free methods (Sect. 6.3.3). However, when the simulator is *differentiable*, we can exploit this structure to obtain gradients $\nabla_\theta J$ much more efficiently via *adjoint methods* and automatic differentiation.

Differentiable simulator (informal definition)

A **differentiable simulator** is a numerical model that not only maps inputs (parameters, controls, initial conditions) to outputs (fields, observables), but also allows us to compute how those outputs change when we slightly perturb the inputs. In practice, this means we can compute gradients through the simulator, enabling fast gradient-based optimization and learning.

The central idea is to view the entire SciML pipeline—from parameters to loss—as a large computational graph and compute gradients efficiently using reverse-mode (backpropagation) or adjoint equations. Adjoint methods are the continuous and PDE-analogue of reverse-mode automatic differentiation.

6.3.5.1 Computational-graph view of simulation

Modern differentiable programming frameworks (e.g. JAX, PyTorch) already implement reverse-mode automatic differentiation (AD). If the simulator is written in a compatible way, we can treat it like any other differentiable layer in a neural network.

A typical SciML pipeline looks like:

$$\theta \xrightarrow{\mathcal{S}} u \xrightarrow{\mathcal{O}} y \xrightarrow{\mathcal{L}} J,$$

where

- $\mathcal{S}$: time integrator or PDE solver that maps parameters θ to a state u,
- $\mathcal{O}$: observation operator (sampling, averaging, sensor extraction),
- $\mathcal{L}$: loss comparing y with targets or enforcing constraints.

General-audience picture

You can think of the SciML pipeline as a long chain of computations: parameters → simulation → measurements → error. Differentiable simulation means we can "run this chain backwards": starting from the error, we propagate sensitivities back through each step to know how changing each parameter would change the error.

Direct reverse-mode AD will correctly propagate gradients through all operations, but may be memory-intensive for long time horizons or large PDE systems. Adjoint methods implement mathematically equivalent gradients in a way tailored to ODE/PDE structure.

6.3.5.2 Continuous adjoint for time-dependent problems

Consider a parameterized ODE,

$$\dot{u}(t) = f\big(u(t), \theta, t\big), \quad t \in [0, T], \tag{6.25}$$

$$u(0) = u_0(\theta), \tag{6.26}$$

and a cost functional

$$J(\theta) = \Phi\big(u(T), \theta\big) + \int_0^T \ell\big(u(t), \theta, t\big)\, \mathrm{d}t. \tag{6.27}$$

The *continuous adjoint* method introduces an adjoint variable $\lambda(t)$ and constructs the Lagrangian

$$\mathcal{L} = \Phi\big(u(T), \theta\big) + \int_0^T \Big[\ell\big(u, \theta, t\big) + \lambda(t)^\top \big(f(u, \theta, t) - \dot{u}(t)\big)\Big] \mathrm{d}t. \tag{6.28}$$

Taking variations and integrating by parts yields the adjoint ODE

$$-\dot{\lambda}(t) = \big[\partial_u f(u, \theta, t)\big]^\top \lambda(t) + \partial_u \ell(u, \theta, t)^\top, \quad t \in [0, T], \tag{6.29}$$

$$\lambda(T) = \partial_u \Phi(u(T), \theta)^\top. \tag{6.30}$$

The gradient of J with respect to θ is then

$$\frac{\mathrm{d}J}{\mathrm{d}\theta} = \partial_\theta \Phi(u(T), \theta) + \int_0^T \Big[\partial_\theta \ell(u, \theta, t) + \lambda(t)^\top \partial_\theta f(u, \theta, t)\Big] \mathrm{d}t + \lambda(0)^\top \partial_\theta u_0(\theta). \tag{6.31}$$

Intuition: adjoint for ODEs

- The forward ODE (6.25) pushes information from $t = 0$ to $t = T$.
- The adjoint ODE (6.29) pushes *sensitivity* information backwards from $t = T$ to $t = 0$.
- Once we have the adjoint trajectory $\lambda(t)$, we can compute the gradient with respect to all components of θ with a single backward integration, regardless of the dimension of θ.
  This is the continuous analogue of backpropagation in deep networks.

The same principle extends to PDE-constrained optimization: for a PDE $\mathcal{F}(u, \theta) = 0$ with objective $J(u, \theta)$, we form a Lagrangian

$$\mathcal{L}(u, \theta, p) = J(u, \theta) + \langle p, \mathcal{F}(u, \theta) \rangle,$$

and derive an adjoint PDE for p, then express $\mathrm{d}J/\mathrm{d}\theta$ in terms of u, p, and θ.

6.3.5.3 Discrete adjoint and algorithmic differentiation

In practice, we rarely solve the continuous ODE/PDE exactly; we instead use a numerical scheme. There are two fundamental approaches:

- **Optimize-then-discretize (continuous adjoint)**: derive the adjoint equations at the continuous level, then discretize forward and adjoint.
- **Discretize-then-optimize (discrete adjoint)**: first discretize the forward problem, then apply reverse-mode AD to the discrete scheme.

Discrete adjoint (informal definition)

A **discrete adjoint** is the exact gradient of the *discretized* objective with respect to the discretized parameters. It is obtained by differentiating the time-stepping or PDE-discretization algorithm itself, using reverse-mode AD or manual derivation.

For a time-stepping scheme $u_{n+1} = \Phi(u_n, \theta)$, $n = 0, \dots, N-1$, with loss $J(u_0, \dots, u_N, \theta)$, reverse-mode AD yields backward recursions for adjoint variables λ_n:

$$\lambda_N = \partial_{u_N} J, \tag{6.32}$$

$$\lambda_n = \partial_{u_n} J + \left[\partial_{u_n} \Phi(u_n, \theta)\right]^\top \lambda_{n+1}, \quad n = N-1, \dots, 0. \tag{6.33}$$

The gradient with respect to θ is

$$\frac{\mathrm{d}J}{\mathrm{d}\theta} = \partial_\theta J + \sum_{n=0}^{N-1} \left[\partial_\theta \Phi(u_n, \theta)\right]^\top \lambda_{n+1}.$$

Continuous versus discrete adjoint. For sufficiently fine discretizations, continuous and discrete adjoints usually agree up to discretization error, but there are important differences:

- Discrete adjoints give the exact gradient of the discretized problem, which is often desirable when the discretization is fixed.
- Continuous adjoints can be discretized with different schemes than the forward solver, potentially improving stability or accuracy, but may introduce *adjoint inconsistency* if not done carefully.

In practice, differentiable programming systems implement a discrete adjoint of the implemented code. Traditional CFD or PDE codes often derive and implement discrete adjoint solvers by hand for efficiency.

Memory, checkpointing, and reversibility

A naive reverse-mode computation stores all intermediate states $(u_0, \ldots, u_N)$ to reuse them during the backward pass, which is prohibitive for large-scale simulations. Adjoint methods deal with this via:

- **Checkpointing**: store only a subset of states (checkpoints), and recompute intermediate forward steps on demand during the backward sweep.
- **Reversible integrators**: use schemes where forward steps can be inverted cheaply, avoiding storage (at the cost of stricter stability requirements).
- **Mixed strategies**: combine partial storage, selective recomputation, and low-precision logs.

Checkpointing

Instead of storing every frame of a movie to replay it backwards, you store only some key frames. To reconstruct the missing frames during rewind, you re-run the movie from the last stored key frame. Checkpointing does the same for simulation states: it trades extra compute for reduced memory.

These techniques are essential when differentiating through long time horizons or large PDE systems on memory-limited hardware (e.g. GPUs).

Adjoint-based optimization in SciML practice

Adjoint and differentiable simulation techniques appear in SciML in several patterns:

PDE-constrained optimization and control. For steady problems $\mathcal{F}(u, \theta) = 0$, with $J(u, \theta)$, one constructs the Lagrangian

$$\mathcal{L}(u, \theta, p) = J(u, \theta) + \langle p, \mathcal{F}(u, \theta) \rangle,$$

then imposes stationarity with respect to u:

$$\partial_u \mathcal{L} = 0 \quad \Rightarrow \quad \text{adjoint equation for } p.$$

The gradient with respect to θ is

$$\frac{\mathrm{d}J}{\mathrm{d}\theta} = \partial_\theta J(u, \theta) + \langle p, \partial_\theta \mathcal{F}(u, \theta) \rangle.$$

One can then run gradient-based optimizers (quasi-Newton, L–BFGS, projected gradient) on θ.

Neural ODEs and differentiable dynamical systems. Neural ODEs treat the right-hand side f as a neural network f_θ and use adjoint methods to compute gradients with respect to θ. This is conceptually similar to differentiating through a time integrator applied to a learned closure model or learned constitutive law in SciML.

Differentiable solvers and libraries. Differentiable programming extends beyond time-stepping to:

- linear and nonlinear solvers (CG, GMRES, Newton–Krylov),
- preconditioners and multigrid cycles,
- mesh deformation and coordinate transforms.

Custom adjoints or AD rules can be implemented for these building blocks, enabling end-to-end differentiable pipelines combining classical numerics and learned components.

Workflow: adjoint-based PDE optimization

A typical loop is:

1. **Forward solve**: given θ, solve the PDE/ODE to obtain u.
2. **Loss evaluation**: compute $J(u, \theta)$.
3. **Adjoint solve**: solve the adjoint equation (backward in time or in dual space) to obtain p or λ.
4. **Gradient**: compute $\mathrm{d}J/\mathrm{d}\theta$ from u, p, and θ.
5. **Update**: update θ with a gradient-based optimizer.

This loop is repeated until convergence.

Pitfalls, stability, and practical tips

Adjoint-based optimization is powerful but comes with caveats:

- **Stiffness and instability**. If the forward dynamics are stiff or chaotic, the adjoint equations may amplify numerical errors when integrated backward in time. Regularization, shorter horizons, or appropriate discretizations are often needed.
- **Non-smooth operations**. If the simulator contains discontinuities (e.g. if-else logic, contact, shocks, clipping), gradients may be undefined or misleading. Smoothing approximations or subgradient methods may be required.
- **Mismatch between continuous and discrete adjoints.** If continuous adjoints are discretized inconsistently with the forward solver, the resulting gradients may not match the true discrete gradient, affecting convergence.
- **Computational cost**. Each optimization iteration requires at least one forward and one adjoint (backward) solve. For time-dependent PDEs, this is expensive; multi-level and reduced-order strategies can help.

Rule of thumb (SciML context)

Use adjoint-based gradients when:

- the simulator is differentiable (or can be made differentiable with mild changes),
- you need gradients with respect to many parameters, and
- you can afford forward and adjoint solves per iteration.

If the simulator is heavily non-smooth or gradients are numerically unstable, consider hybrid strategies with surrogate models or gradient-free outer loops.

Nwe can now give a summary of the methods as:

- Differentiable simulation and adjoint-based optimization provide a principled way to compute gradients through ODE/PDE solvers and complex physics codes.
- Continuous adjoint methods derive sensitivity equations at the PDE/ODE level; discrete adjoints differentiate the implemented numerical scheme.
- Checkpointing and tailored adjoint solvers enable large-scale applications under tight memory constraints.
- In SciML, adjoint methods are central to PDE-constrained optimization, neural ODEs, and end-to-end differentiable pipelines that couple classical solvers with learned components.

Key takeaway

Adjoint-based optimization is the SciML analogue of backpropagation through physical laws. It turns sophisticated simulators into differentiable layers, unlocking gradient-based design, control, and inference for complex scientific systems.

6.3.6 *Hybrid and Multi-fidelity Optimization*

Previous sections have presented optimization methods (gradient-based, gradient-free, Bayesian, adjoint-based) and modeling choices (high-fidelity vs. surrogates, single vs. multi-fidelity). In realistic SciML workflows, these ingredients are rarely used in isolation. Instead, practitioners combine them into *hybrid* optimization strategies that exploit:

- fast local convergence of gradient-based and adjoint methods,
- global exploration capabilities of metaheuristics and Bayesian optimization, and
- cost–accuracy trade-offs offered by multi-fidelity models.

The idea!!

Hybrid optimization combines multiple optimization algorithms in a coordinated way (e.g. metaheuristics + gradients).
Multi-fidelity optimization combines multiple models of the same system (e.g. coarse and fine simulators, ROMs and full-order models) to reduce cost while preserving accuracy.

SciML pipelines typically mix both: different *methods* operating on different *fidelities*.

Conceptually, we can think of a two-dimensional design space:

- the *method axis* (gradients, adjoints, metaheuristics, BO),
- the *model axis* (low- to high-fidelity simulators and surrogates).

Hybrid and multi-fidelity strategies navigate this plane instead of living at a single point.

Types of hybrid optimization strategies

Hybrid optimization refers to any structured combination of two or more optimization methods. Common patterns include:

(1) Sequential global–local hybrids.
A global, often gradient-free method (metaheuristic, BO) explores the domain and finds promising regions. A local gradient-based method then refines around these regions.

- *Example*: PSO or DE finds a good initial design; L–BFGS or Adam then performs local refinement using adjoint or AD gradients.
- *Example*: BO proposes a few high-quality hyperparameter settings; for each, we run a full gradient-based SciML training loop.

(2) Cooperative hybrids (memetic algorithms).
Population-based algorithms (GA, PSO) are enriched with local search on selected individuals. This pattern is often called *memetic* or *Lamarckian* evolution.

- After variation and selection, some individuals undergo a short local gradient-based refinement.
- The refined individuals replace or update the originals in the population.

(3) Decomposition-based hybrids.
Different methods are applied to different *parts* of the problem:

- Decompose the parameter vector into blocks; use gradients for continuous blocks and metaheuristics or BO for discrete or non-smooth blocks.
- Use adjoint-based optimization within each time window of a long horizon, and a higher-level metaheuristic to tune couplings between windows.

(4) Multi-stage hybrids.
Optimization is staged over increasing levels of sophistication:

1. fast, coarse, gradient-free exploration;
2. moderate-accuracy, gradient-based refinement;
3. fine, high-fidelity, adjoint-based final tuning.

Why hybridize?

Different optimization methods have different strengths: some are good at exploring, some at fine-tuning, some at handling noise or constraints. A hybrid strategy uses each tool where it works best, instead of forcing one method to do everything.

A generic hybrid outer–inner-loop template

SciML workflows frequently adopt an *outer* and *inner* loop separation:

- **Inner loop**: differentiable training or PDE-constrained optimization (Sect. 6.3.5) with gradient-based methods.
- **Outer loop**: gradient-free, Bayesian, or evolutionary search over design or hyperparameters.

Interpretation

The inner loop sees x as fixed and uses efficient gradients. The outer loop treats the inner loop as a black box returning a score $F(x)$. Hybrid optimization is about *structuring the collaboration* between these two levels.

Algorithm 5: Hybrid outer–inner optimization template (SciML)

Input: Design / hyperparameters $x \in \mathcal{X}$; inner variables θ (weights, state variables)

Initialize outer population or surrogate with a small design set $\{x_i\}$;
for *outer iteration* $k = 0, 1, \ldots$ **do**
  Choose next outer candidate(s) $x^{(k)}$ (metaheuristic step, BO acquisition maximization, etc.) ;
  Inner solve: given $x^{(k)}$, obtain

$$\theta^{(k)} \approx \arg\min_{\theta} J_{\text{inner}}(\theta; x^{(k)})$$

  using gradient-based or adjoint methods ;
  Evaluate performance $F(x^{(k)})$ (e.g., validation error, physical objective) ;
  Update outer optimizer with $(x^{(k)}, F(x^{(k)}))$;

Multi-fidelity modeling in optimization

Multi-fidelity optimization extends the above template by introducing an index $\ell \in \{1, \ldots, L\}$ for fidelity level:

$$J_\ell(x), \quad \text{with cost } c_\ell \text{ and accuracy } a_\ell,$$

where typically $c_1 \ll c_2 \ll \cdots \ll c_L$ and $a_1 \ll a_2 \ll \cdots \ll a_L$.

Fidelity levels

A **low-fidelity model** is:
- cheaper to evaluate,
- less accurate, sometimes biased.

A **high-fidelity model** is:
- more expensive,
- more accurate, closer to physical truth.

The goal is to use many cheap evaluations and only a few expensive ones, without losing too much accuracy.

Examples in SciML:

- **Mesh resolution**: coarse versus fine spatial/temporal discretizations of a PDE.
- **Reduced-order versus full-order models**: projection-based ROMs, linearized models, or neural surrogates versus full solvers.
- **Training budget**: a partially trained PINN or neural surrogate versus a fully converged model.

- **Physics completeness**: simplified physics (e.g. Stokes flow) versus full nonlinear Navier–Stokes.

Multi-fidelity optimization must address:

- **Model management**: when and how to query each fidelity.
- **Bias correction**: relating low- and high-fidelity outputs.
- **Coupled convergence**: ensuring that the final design is near-optimal for the high-fidelity model.

Model management and fidelity control

Even outside Bayesian optimization (Sect. 6.3.4), there are general patterns for combining fidelities.

(1) Coarse-to-fine continuation.
Use low-fidelity models to obtain good initial designs, then progressively refine fidelity:

1. Optimize with the cheapest model $J_1(x)$ to obtain $x^{(1)\star}$.
2. Use $x^{(1)\star}$ as initial guess for optimization on $J_2(x)$.
3. Continue up the fidelity ladder until reaching $J_L(x)$.

Continuation as a "difficulty schedule"

We first solve an easier, cheaper version of the problem and then gradually make it harder and more accurate. This mirrors curriculum learning in deep learning, but applied to simulation fidelity.

(2) Trust-region surrogate management.
On a bounded region (trust region) around the current design, build a low-fidelity surrogate $\hat{J}$ and optimize it, but:

- accept steps only if they also improve a higher-fidelity model,
- shrink or expand the trust region based on agreement between $\hat{J}$ and the higher-fidelity J_L.

This is a deterministic analogue of model-management strategies in BO.

(3) Additive and multiplicative corrections.
When low- and high-fidelity outputs are systematically related, we can write

$$J_L(x) \approx \rho J_1(x) + \delta(x) \quad \text{or} \quad J_L(x) \approx J_1(x) + \Delta(x),$$

and fit correction models $\delta(x)$ or $\Delta(x)$ from a sparse set of high-fidelity evaluations. Optimization then proceeds mostly on J_1 plus a cheap correction.

(4) Adaptive fidelity selection.
Fidelity level is chosen on the fly based on:

- estimated potential improvement,
- current uncertainty about the region,
- or a fixed cost budget.

For example, one may:

- evaluate new regions with J_1 only,
- re-evaluate the best few candidates from J_1 with J_2,
- use J_L sparingly near the final designs.

Hybrid multi-fidelity patterns in SciML

Hybrid and multi-fidelity ideas often appear together in SciML workflows.

Pattern A: Metaheuristic + adjoint with fidelity ladder.

- Use a metaheuristic (PSO, DE) on a *coarse* simulator $J_1(x)$ to discover promising design regions.
- For the best candidates, run *adjoint-based* gradient refinement on $J_L(x)$.
- Optionally, insert one or more intermediate fidelities J_ℓ in between.

Pattern B: Bayesian outer loop + multi-fidelity inner solves.

- Use Bayesian Optimization on a validation error metric as in Sect. 6.3.4.
- For early iterations and exploratory points x, run shortened inner training (few epochs, fewer collocation points).
- For promising designs, run full training and possibly fine-mesh solvers.

Pattern C: Learning plus classical model hierarchy.

- Train a neural surrogate or ROM on data generated from J_L.
- Use the surrogate as J_1 in a multi-fidelity hierarchy for fast outer-loop design.
- Validate final designs with limited calls to the full solver J_L.

Example

Designing an airfoil shape:
- Low-fidelity: fast inviscid flow model on a coarse mesh.
- Medium-fidelity: RANS model on a moderate mesh.
- High-fidelity: large-eddy simulation on a fine mesh.

A hybrid, multi-fidelity optimizer might:

1. explore shape variations with a swarm algorithm using the low-fidelity model;
2. refine a handful of good shapes using gradient-based optimization with the medium-fidelity model;
3. finally, validate and slightly tune the best candidates with the high-fidelity model.

Pitfalls and practical guidelines

Hybrid and multi-fidelity strategies are powerful but introduce new failure modes:

Fidelity mismatch and bias. If low-fidelity models are systematically biased in regions of interest, they may mislead the optimizer and overshadow sparse high-fidelity corrections. Remedies:

- calibrate low-fidelity models using high-fidelity data,
- monitor discrepancies, not just objective values,
- periodically re-evaluate older designs with higher fidelity.

Overhead and complexity. Complex outer–inner loops with multiple fidelities, surrogates, and optimizers can be hard to implement and debug. Guidelines:

- start with a simple baseline workflow (e.g. single-fidelity gradient-based),
- introduce hybrid or multi-fidelity features one at a time,
- carefully track which model and fidelity produced each evaluation.

Inconsistent objectives. If each fidelity or surrogate uses a slightly different loss definition (e.g. different regularization, boundary handling, or data), the optimization problem effectively changes with fidelity. Aim to:

- keep objective definitions consistent across fidelities,
- interpret low-fidelity losses as approximations of the same target functional.

Non-smooth transitions. Abrupt switching between optimization methods or fidelities (e.g. sudden mesh refinement, large network size jumps) can cause optimization to stall. More robust:

- gradually refine fidelity (mesh, collocation points, training epochs),
- use continuation in fidelity parameters,
- warm-start high-fidelity solves from low-fidelity solutions.

In summary;

- Hybrid optimization combines complementary methods (gradient-based, gradient-free, Bayesian, adjoint) in structured outer–inner, cooperative, or staged schemes.
- Multi-fidelity optimization leverages hierarchies of models (coarse/fine, ROM/-full, partial/full training) to reduce cost while preserving accuracy.
- In SciML, the most effective strategies mix both: gradient-based inner loops on differentiable simulators, gradient-free or Bayesian outer loops, and cost-aware fidelity selection.
- Careful model management, bias correction, and gradual refinement are key to stable and reliable hybrid multi-fidelity workflows.

Key takeaway

No single optimizer or model fidelity is sufficient for the full range of SciML tasks. Hybrid and multi-fidelity optimization provide a principled way to *use the right method on the right model at the right time*, turning expensive scientific computations into efficient, flexible, and robust design and learning pipelines.

6.4 Conclusion

This chapter has examined the optimization landscape of Scientific Machine Learning, where physics constraints create challenges and opportunities distinct from standard deep learning. We have seen that SciML loss landscapes are fundamentally different, that optimizer selection matters more than in standard ML, and that constrained optimization provides principled approaches to physics enforcement.

Key Takeaways

1. **L-BFGS Often Wins**: For deterministic physics-informed losses, quasi-Newton methods frequently outperform Adam. Two-phase training (Adam then L-BFGS) exploits both.
2. **Balance Your Losses**: Physics, data, and boundary terms compete. Adaptive weighting, gradient normalization, or multi-objective formulations are usually necessary.
3. **Constraints Need Care**: Penalty methods are simple but may never achieve feasibility. Augmented Lagrangian provides tighter enforcement. Hard constraints through architecture guarantee satisfaction.
4. **Adjoints Scale**: Adjoint methods compute gradients in $\mathcal{O}(1)$ backward passes regardless of parameter count. Essential for optimizing through expensive simulations.
5. **Multi-Fidelity Accelerates**: Cheap approximations guide expensive evaluations. Surrogate-based and multi-level strategies can reduce optimization cost by orders of magnitude.

References

1. Snoek, J., Larochelle, H., Adams, R.P.: Practical Bayesian optimization of machine learning algorithms. In: Advances in Neural Information Processing Systems, p. 25 (2012)
2. Cheng, X., Na, S.: Physics-informed neural networks with trust-region sequential quadratic programming. arXiv preprint. arXiv:2409.10777 (2024)
3. Zhang, Z., Wang, Q., Zhang, Y., Shen, T., Zhang, W.: Physics-informed neural networks with hybrid Kolmogorov-Arnold network and augmented Lagrangian function for solving partial differential equations. Sci. Rep. **15**(1), 10523 (2025)
4. Sahin, T., von Danwitz, M., Popp, A.: Solving forward and inverse problems of contact mechanics using physics-informed neural networks. Adv. Model. Simul. Eng. Sci. **11**(1), 11 (2024)
5. Bhuyan, B.P., Ramdane-Cherif, A., Singh, T.P., Tomar, R.: Feedforward neural networks and backpropagation. In: Neuro-symbolic Artificial Intelligence: Bridging Logic and Learning, pp. 113–123. Springer (2024)

Chapter 7
Reinforcement Learning for Scientific Control

With the right pilot, she'll fly.

—*Han Solo, The Force Awakens*

Abstract Controlling physical systems—from plasma confinement in fusion reactors to drag reduction in turbulent flows—requires policies that map observations to actions while respecting safety constraints and physical limitations. This chapter develops reinforcement learning (RL) for scientific control, beginning with the distinctive challenges and opportunities of physical systems: continuous state-action spaces, known dynamics models, safety requirements, and expensive environment interactions. We establish foundations through Markov decision processes, value functions, and Bellman equations, with explicit connections to classical control theory (LQR, MPC) that illuminate when RL offers advantages. Core algorithms span value-based methods (Q-learning, DQN), policy gradient approaches (REINFORCE, PPO), and actor-critic frameworks (A2C, SAC, DDPG). Extensions for scientific systems address model-based RL, physics-informed RL that incorporates governing equations, safe RL with constraint satisfaction guarantees, and multi-agent approaches for distributed PDE control. Deep RL implementation covers function approximation, experience replay, and the sample efficiency challenges that dominate scientific applications.

7.1 Introduction

The ability to learn optimal behavior through trial and error (reinforcement learning) is one of nature's most fundamental computational strategies. From neurons adjusting synaptic strengths to optimize future rewards, to ecosystems evolving stable equilibria through selective pressures, the pattern of "action, consequence, adaptation" pervades biological and physical systems. In recent years, this ancient principle

B. P. Bhuyan et al., *Scientific Machine Learning*,
https://doi.org/10.1007/978-981-95-7856-6_7

has been formalized into a powerful mathematical framework that is transforming how we approach control, optimization, and discovery in scientific and engineering domains.

This chapter introduces reinforcement learning from the perspective of scientific machine learning, where the goal is not to master games or navigate virtual environments, but to control plasma in fusion reactors, optimize fluid flows, design new materials, and discover fundamental algorithms. Unlike the game-playing examples that dominate much of the RL literature, scientific applications bring unique challenges: expensive simulations, safety-critical constraints, complex physics, and the need for interpretable, certifiable policies. They also bring unique opportunities: differentiable simulators, known conservation laws, and decades of control-theoretic insights.

Our treatment emphasizes connections to classical scientific computing. Where traditional RL texts begin with grid worlds and Atari games, we begin with PDE-constrained control problems and optimal experimental design. Where standard expositions focus purely on data-driven learning, we explore how to incorporate physical laws, symmetries, and domain knowledge into learning architectures. The result is a bridge between two communities (control theorists and machine learning researchers) that are converging on similar mathematical objects from different starting points.

7.2 Motivation and Context

Reinforcement learning addresses a fundamental question in scientific computing: *how do we find optimal decision-making strategies in complex, dynamic systems where we cannot enumerate all possibilities?* This question arises across scientific domains:

- **Plasma control**: In tokamak fusion reactors, magnetic coils must be controlled in real-time to maintain plasma confinement. The dynamics are governed by magnetohydrodynamic PDEs with bifurcations, instabilities, and constraints on temperature, density, and position. Classical controllers struggle with the nonlinearity and multi-objective optimization required.
- **Fluid dynamics**: Controlling turbulent flows to reduce drag, suppress instabilities, or enhance mixing requires making continuous decisions about actuator inputs (jets, flaps, heat sources) based on high-dimensional velocity and pressure fields. The state space is effectively infinite-dimensional.
- **Materials discovery**: Designing new materials involves selecting synthesis routes, processing conditions, and compositions from a combinatorially large space. Each experiment is expensive, and the property-composition mapping is unknown and nonlinear.

- **Experimental design**: In autonomous laboratories, robotic systems must decide which experiments to run next to most efficiently learn about a scientific phenomenon, balancing exploration (gathering information) with exploitation (optimizing known promising directions).

Classical approaches to these problems: optimal control theory, model predictive control, gradient-based optimization; face fundamental limitations when dynamics are unknown, too complex to model accurately, or too expensive to optimize online. Reinforcement learning offers an alternative: learn control policies directly from interaction data, without requiring perfect models.

Figure 7.1 illustrates the conceptual relationship between classical control, pure reinforcement learning, and the emerging hybrid approach of physics-informed RL that we emphasize throughout this chapter. The classical paradigm starts with first-principles models (PDEs describing the system), applies optimization algorithms (e.g., dynamic programming, gradient descent, model predictive control), and produces control policies. This approach is principled and interpretable, but requires accurate models and often faces computational intractability for high-dimensional systems.

Pure reinforcement learning inverts this process: instead of starting with models, we start with data generated by interacting with the system (or a simulator). Learning algorithms process this data to directly produce control policies without ever writing down PDEs. This is the power of model-free RL; it can handle systems too complex to model. But the cost is sample inefficiency: millions of trials may be needed, which is prohibitive when each trial is an expensive simulation or physical experiment.

The synthesis (physics-informed RL) combines known physical structure with data-driven learning. By encoding conservation laws, symmetries, and dynamical constraints into network architectures, loss functions, or state representations, we can dramatically improve sample efficiency while maintaining the flexibility to learn

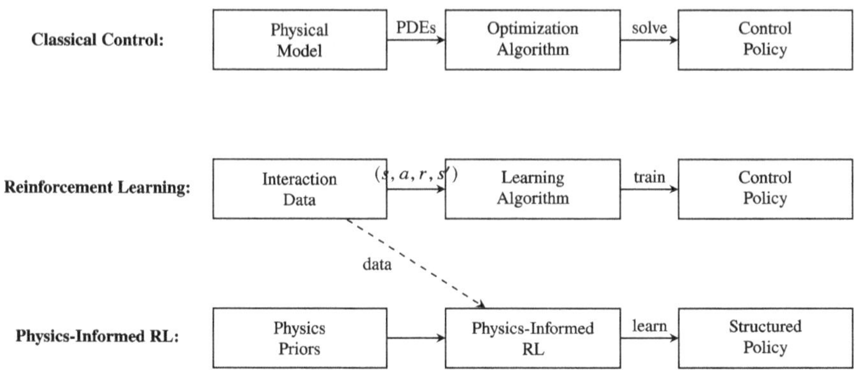

Fig. 7.1 Three paradigms for scientific control. Classical methods require accurate physical models and solve optimization problems. Pure RL learns from data without models. Physics-informed RL combines known physics with data-driven learning

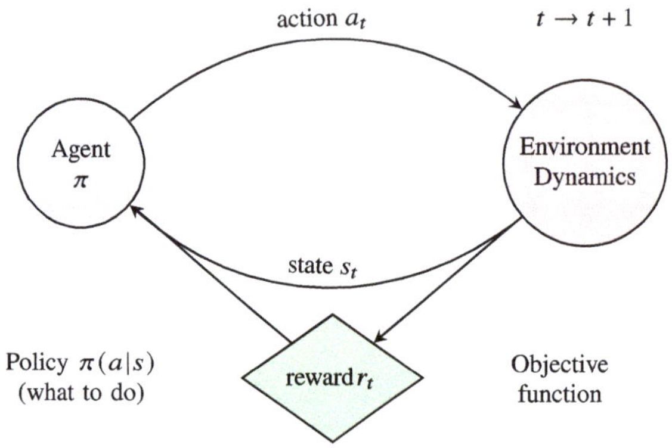

Fig. 7.2 The reinforcement learning loop. An agent observes the state s_t of an environment, chooses an action a_t according to its policy π, receives a reward r_t quantifying immediate performance, and observes the next state s_{t+1}. In scientific applications, the environment is typically a physical system or numerical simulator, and the policy must be learned through repeated interaction

from data. This is the emerging frontier in scientific machine learning, and a central theme of this chapter.

7.2.1 The Reinforcement Learning Framework

At its core, reinforcement learning formalizes sequential decision-making as shown in Fig. 7.2. An *agent* (the controller or decision-maker) interacts with an *environment* (the physical system or simulator) over discrete time steps $t = 0, 1, 2, \ldots$ At each step:

1. The agent observes the current **state** s_t of the environment
2. The agent selects an **action** a_t according to its **policy** $\pi(a|s)$
3. The environment transitions to a new state s_{t+1} according to its **dynamics** $P(s_{t+1}|s_t, a_t)$
4. The agent receives a scalar **reward** r_t quantifying the immediate desirability of this transition.

The agent's goal is to learn a policy that maximizes the expected cumulative discounted reward:

$$J(\pi) = \mathbb{E}_\pi \left[\sum_{t=0}^{\infty} \gamma^t r_t \right], \tag{7.1}$$

where $\gamma \in [0, 1)$ is a discount factor that weights immediate rewards more heavily than distant future rewards.

This simple loop: observe, act, receive reward, update; captures an enormous range of control and optimization problems. The key insight is that we need not

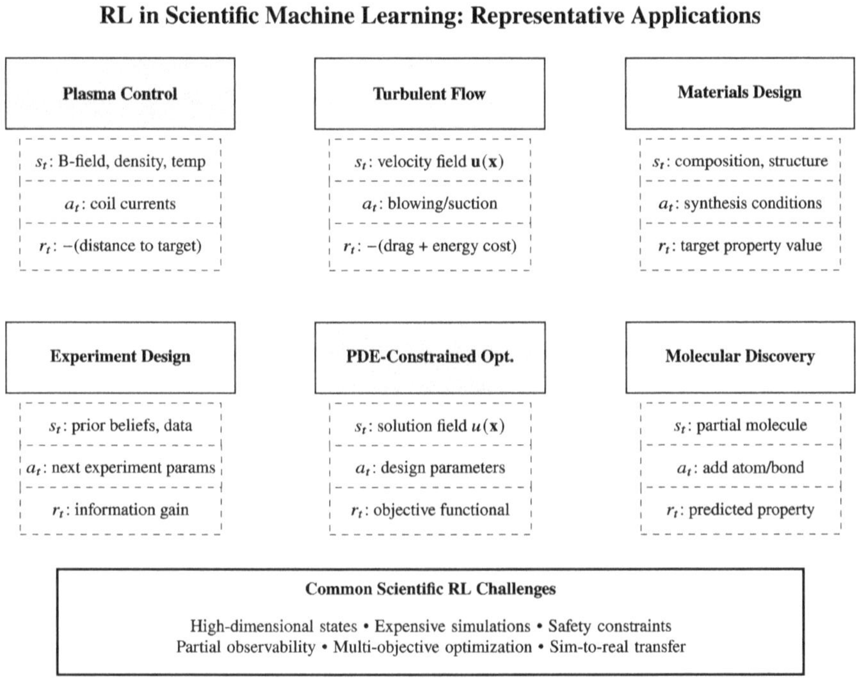

Fig. 7.3 Diverse scientific applications unified by the RL framework. Each domain maps to the state-action-reward structure, but with domain-specific challenges: plasma control requires real-time safety guarantees, fluid dynamics involves infinite-dimensional PDEs, materials design faces expensive experiments, etc. Common threads include high-dimensional state spaces, expensive evaluation, and the need for safety and interpretability

know the transition dynamics $P(s_{t+1}|s_t, a_t)$ explicitly. By interacting with the system and observing which state-action pairs lead to high cumulative rewards, learning algorithms can discover effective policies.

7.2.2 *Scientific RL: Distinctive Challenges and Opportunities*

While the basic RL framework is domain-agnostic, scientific applications present distinctive characteristics that require specialized methods. Figure 7.3 illustrates how different scientific domains map onto the RL framework, highlighting both the universality of the approach and the domain-specific challenges.

Challenges unique to scientific RL:

- **Expensive evaluation**: In game-playing RL, agents can play millions of games quickly. In scientific applications, each "episode" might be a fusion plasma discharge costing thousands of dollars, a turbulent flow simulation requiring hours on a supercomputer, or a materials synthesis experiment taking days. Sample efficiency is paramount.

- **Safety and constraints**: Scientific systems have hard constraints; plasma must not touch reactor walls, flows must remain stable, chemical reactions must not exceed temperature limits. Unlike games where failure is merely undesirable, physical systems can be damaged or create hazardous conditions. Safe RL methods with guarantees are essential.
- **High-dimensional states**: The state of a turbulent flow at Reynolds number 10^5 involves billions of degrees of freedom. Even with measurement limitations, state spaces are vastly larger than typical RL benchmarks. Dimensionality reduction, low-dimensional representations, and structured architectures are necessary.
- **Partial observability**: We rarely have access to the full state of a physical system. Sensors provide limited, noisy measurements. The actual state must be inferred from observation history, making the problem a Partially Observable MDP (POMDP).
- **Multi-objective optimization**: Scientific applications rarely have a single objective. Plasma control must balance confinement quality, stability, and impurity control. Flow control trades drag reduction against actuation energy. These objectives are often competing, requiring Pareto-optimal policies.
- **Sim-to-real gap**: We typically train policies in simulation but deploy them on real systems. Discrepancies between simulated and real physics; unmodeled disturbances, numerical approximations, simplified boundary conditions; can cause policies that work perfectly in simulation to fail catastrophically in reality.

Opportunities unique to scientific RL:

- **Known physics**: Unlike game environments designed to be opaque, physical systems obey known conservation laws (energy, momentum, mass), symmetries (translation, rotation), and governing equations (Navier-Stokes, Maxwell, Schrödinger). We can; and should; encode this structure into learning architectures.
- **Differentiable simulators**: Modern physics simulators (Brax for rigid bodies, JAX-MD for molecular dynamics, TORAX for tokamaks) are differentiable with respect to policy parameters. This enables analytical policy gradients; computing exact gradients through the physics, rather than estimating them via Monte Carlo sampling; offering orders of magnitude speedup.
- **Classical control baselines**: Decades of control theory provide strong baselines: PID controllers, LQR, MPC. These are not competitors to RL but complementary tools. Hybrid architectures that use RL to learn components of classical controllers (e.g., terminal costs for MPC) often outperform pure RL or pure classical methods.
- **Interpretability requirements**: Scientific applications demand interpretable policies. A black-box neural network that controls plasma is insufficient; we need to understand why it works, what failure modes exist, and how it will behave in untested regimes. This drives research into explainable RL and physics-informed architectures.

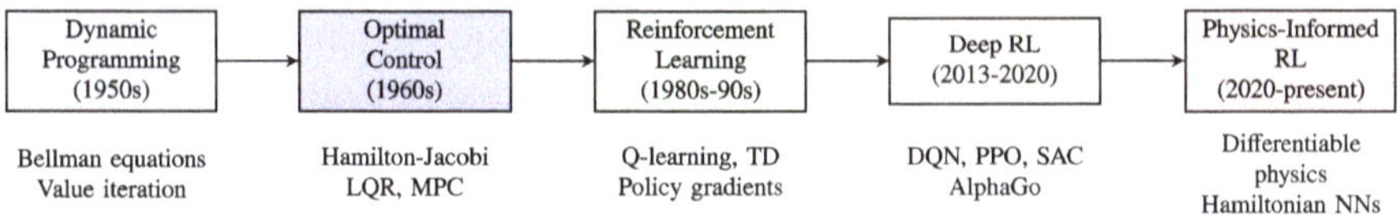

Fig. 7.4 Historical evolution of RL and its connection to control theory. Modern RL and classical optimal control are solving the same Bellman optimality equation but from different starting points: control theory assumes known dynamics $P(s'|s,a)$ and solves for optimal policies analytically, while RL learns policies from data without knowing P. The convergence is happening in scientific RL, which combines the structure of control theory with the flexibility of learning

7.2.3 *Connection to Classical Control Theory*

One of the most important insights for scientific audiences is that reinforcement learning and classical optimal control theory are solving the same underlying mathematical problem; the *Bellman optimality equation*; but from different starting points and with different computational strategies.

Figure 7.4 traces this evolution. Richard Bellman introduced dynamic programming in the 1950s as a computational method for solving sequential decision problems, establishing the principle of optimality: an optimal policy has the property that whatever the initial state and decision, the remaining decisions must constitute an optimal policy with regard to the state resulting from the first decision.

This principle led directly to the Hamilton-Jacobi-Bellman (HJB) equation of optimal control theory:

$$-\frac{\partial V^*(s,t)}{\partial t} = \max_a \left[r(s,a) + \nabla_s V^*(s,t) \cdot f(s,a) \right], \tag{7.2}$$

where $f(s,a)$ are the continuous-time dynamics $\dot{s} = f(s,a)$. For known linear systems, this reduces to the Linear Quadratic Regulator (LQR), solvable via Riccati equations. For nonlinear systems, Model Predictive Control (MPC) repeatedly solves finite-horizon optimization problems online.

Reinforcement learning emerged in the 1980s-90 s through work by Sutton, Barto, and others, developing algorithms (Q-learning, policy gradients, actor-critic methods) that could solve the discrete-time Bellman equation without knowing the dynamics $P(s'|s,a)$:

$$V^*(s) = \max_a \left[r(s,a) + \gamma \sum_{s'} P(s'|s,a) V^*(s') \right]. \tag{7.3}$$

The deep learning revolution (2013–present) enabled RL to scale to high-dimensional problems through neural function approximation, culminating in successes like AlphaGo. But most of this work remained disconnected from the scientific computing community.

The current frontier "physics-informed RL" represents a synthesis. We recognize that:

- RL's model-free learning is valuable when dynamics are unknown or too complex to model
- Control theory's structural insights (stability, robustness, conservation laws) are valuable constraints
- Differentiable physics enables hybrid approaches: learn what must be learned, encode what is known

This synthesis is the intellectual foundation of this chapter. We do not treat RL as a replacement for control theory, but as a complementary tool that becomes most powerful when informed by physical structure.

7.3 Foundations of Reinforcement Learning

7.3.1 Markov Decision Processes (MDPs)

Reinforcement Learning problems are usually modelled using *Markov Decision Processes* (MDPs). An MDP is a mathematical description of a sequential decision-making problem: at each time step the system is in some *state*, an *agent* chooses an *action*, the system evolves to a new state according to some (possibly stochastic) dynamics, and a scalar *reward* measures the immediate quality of the decision.

In Scientific Machine Learning, the "environment" is often a physical or numerical model: the state can be the temperature field in a domain, the concentrations in a chemical reactor, or the position and velocity of a robot; the actions are control inputs; the reward encodes how well we are achieving a scientific or engineering objective.

Definition: Markov Decision Process

A *Markov Decision Process (MDP)* is a tuple

$$\mathcal{M} = (\mathcal{S}, \mathcal{A}, P, r, \gamma)$$

where:

- $\mathcal{S}$ is the **state space**, the set of all possible states s of the system;
- $\mathcal{A}$ is the **action space**, the set of all possible actions a the agent can take;
- $P(s' \mid s, a)$ is the **transition kernel**, giving the probability that the next state is s' when action a is taken in state s;
- $r(s, a)$ (or $r(s, a, s')$) is the **reward function**, giving a scalar signal after applying action a in state s;

- $\gamma \in [0, 1)$ is the **discount factor**, controlling how much future rewards matter compared to immediate rewards.

The *Markov property* means that the probability of the next state depends only on the current state s and action a, and not on earlier states or actions.

Time is usually indexed by $t = 0, 1, 2, \ldots$. An interaction between an agent and an MDP produces a random *trajectory*

$$\tau = (s_0, a_0, r_0, s_1, a_1, r_1, s_2, \ldots),$$

where; at time t the system is in state s_t; the agent chooses an action a_t (according to some policy, later denoted by π); the environment returns a reward $r_t = r(s_t, a_t)$ and moves to a new state $s_{t+1} \sim P(\cdot \mid s_t, a_t)$.

Definition: Return (Cumulative Reward)

Given a trajectory τ, the *return* from time t is the discounted sum of future rewards

$$G_t = \sum_{k=0}^{\infty} \gamma^k r_{t+k}.$$

The goal of the agent in RL is to choose its actions so as to maximize (in expectation) this return.

The discount factor γ allows us to model different time preferences: γ close to 1 means that long-term consequences matter a lot, while a smaller γ puts more emphasis on immediate rewards. Mathematically, $\gamma < 1$ also guarantees that the infinite sum G_t is well-defined when rewards are bounded.

Let us take a simple temperature-control as an example. Consider a very simple example of room temperature control, which is close to many scientific control problems.

- **State** s_t: the room temperature at time t, measured in °C. For simplicity, we can discretize it to integers between 15 and 30: $\mathcal{S} = \{15, 16, \ldots, 30\}$.
- **Action** a_t: the heater setting at time t, e.g., $\mathcal{A} = \{\text{off}, \text{low}, \text{high}\}$.
- **Transition** $P(s_{t+1} \mid s_t, a_t)$: given the current temperature and heater setting, the next temperature is influenced by heat losses, external temperature, and possibly noise. For instance, turning the heater to high makes higher temperatures more likely.
- **Reward** $r(s_t, a_t)$: we may want the temperature to stay near 22 °C while using little energy. A simple choice is

$$r(s_t, a_t) = -(s_t - 22)^2 - \lambda \cdot \text{EnergyCost}(a_t),$$

where $\lambda > 0$ trades off comfort against energy use.

- **Discount factor** γ: choosing γ close to 1 encourages the controller to keep the room comfortable over a long time horizon, not only in the next few minutes.

This example contains all the ingredients of an MDP. In more complex SciML settings, the state s_t might be a high-dimensional vector (e.g., a discretized fluid velocity field), the actions might be continuous control inputs, and the transition kernel P might implicitly be given by a numerical solver for ordinary or partial differential equations.

In practice, real scientific systems may not be perfectly Markovian with respect to the chosen state representation. For example, unmeasured disturbances or delayed sensors can break the Markov property. A common remedy is to *enlarge* the state (e.g., by stacking several past measurements, or including estimates of hidden variables) so that the resulting state is approximately Markov. Later sections will discuss how partial observability is handled more systematically.

7.3.2 Policies, Value Functions, and Q-Functions

The behavior of an agent in a Markov Decision Process (MDP) is governed by its *policy*, while the quality of states and actions is evaluated through *value functions* and *Q-functions*. Together, these concepts provide the mathematical foundation for reinforcement learning algorithms.

Definition: Policy

A **policy** π is a mapping from states to a probability distribution over actions:

$$\pi(a \mid s) = \Pr(a_t = a \mid s_t = s).$$

- If π is deterministic, then $\pi(s)$ specifies a single action for each state.
- If π is stochastic, then $\pi(a \mid s)$ gives a distribution over possible actions.

The policy completely defines the agent's decision-making strategy.

Given a policy, we can define measures of long-term performance.

Definition: Value Function

The **state-value function** $V^{\pi}(s)$ is the expected return when starting in state s and following policy π thereafter:

$$V^{\pi}(s) = \mathbb{E}_{\pi}\left[\sum_{k=0}^{\infty} \gamma^k r_{t+k} \,\middle|\, s_t = s\right].$$

- s_t is the state at time t.
- r_{t+k} is the reward received k steps in the future.
- $\gamma \in [0, 1)$ is the discount factor.

$V^{\pi}(s)$ quantifies how "good" it is to start in state s under policy π.

Definition: Action-Value Function (Q-function)

The **action-value function** or **Q-function** $Q^{\pi}(s, a)$ is the expected return when starting from state s, taking action a, and thereafter following policy π:

$$Q^{\pi}(s, a) = \mathbb{E}_{\pi}\left[\sum_{k=0}^{\infty} \gamma^k r_{t+k} \,\middle|\, s_t = s, a_t = a\right].$$

- a_t is the action taken in state s_t.
- After the first step, subsequent actions are chosen according to π.

The Q-function measures the quality of specific state–action pairs.

The two functions (V^{π} and Q^{π}) are linked by:

$$V^{\pi}(s) = \sum_{a \in \mathcal{A}} \pi(a \mid s)\, Q^{\pi}(s, a),$$

which states that the value of a state is the expectation of its action-values under the policy.

Both V^{π} and Q^{π} satisfy recursive relations known as *Bellman equations*:

$$V^{\pi}(s) = \sum_{a \in \mathcal{A}} \pi(a \mid s)\Big[R(s, a) + \gamma \sum_{s' \in \mathcal{S}} P(s' \mid s, a) V^{\pi}(s')\Big],$$

$$Q^{\pi}(s, a) = R(s, a) + \gamma \sum_{s' \in \mathcal{S}} P(s' \mid s, a) \sum_{a' \in \mathcal{A}} \pi(a' \mid s') Q^{\pi}(s', a').$$

Continuing the problem of reducing drag in a two-dimensional turbulent channel flow. We aim to use boundary actuation (wall blowing and suction) to manipulate the flow field and achieve lower drag. This setting can be naturally expressed as a Markov Decision Process (MDP).

States (s_t): At each time step t, the state $s_t \in \mathcal{S}$ represents the discretized flow field. For example, if the velocity field $\mathbf{u}(x, y, t) = (u, v)$ is sampled at N grid points, then the state vector is

$$s_t = \big(u(x_1, y_1, t), v(x_1, y_1, t), \dots, u(x_N, y_N, t), v(x_N, y_N, t)\big).$$

This captures the instantaneous flow configuration. High-dimensional states are typical in Scientific Machine Learning since they arise from discretized PDEs.

Actions (a_t): The action $a_t \in \mathcal{A}$ represents the actuation applied at the channel walls. For instance, at a set of control points $\{y_1, y_2, \dots, y_m\}$ along the wall, the agent chooses wall-normal velocities

$$a_t = \big(v_w(y_1, t), v_w(y_2, t), \dots, v_w(y_m, t)\big),$$

where v_w denotes the imposed suction/blowing. This means the agent is deciding how to inject or remove fluid at specific boundary locations.

Policy (π): The policy $\pi(a \mid s)$ prescribes how actions are chosen given the observed flow state. For example:

$$\pi(a \mid s) = \Pr(a_t = a \mid s_t = s).$$

If π is deterministic, then $a_t = \pi(s_t)$ (a fixed mapping). If π is stochastic, then different actuation patterns may be sampled with probabilities depending on the current flow state.

Rewards (r_t): The reward quantifies the scientific objective. Here, we want to minimize drag force $D(t)$ while also penalizing the energy cost of actuation. A typical reward function is:

$$r_t = -D(t) - \lambda \|a_t\|_2^2,$$

where:

- $D(t)$ is the drag measured at time t.
- $\|a_t\|_2^2$ is the squared actuation effort.
- $\lambda > 0$ balances performance vs. control cost.

Thus, higher rewards correspond to lower drag with minimal actuation.

Value Function ($V^\pi(s_t)$): The state-value function measures the expected long-term benefit of starting in state s_t and following policy π:

$$V^\pi(s_t) = \mathbb{E}_\pi\left[\sum_{k=0}^{\infty} \gamma^k r_{t+k} \;\middle|\; s_t\right].$$

For example, if s_t corresponds to a nearly laminar flow, $V^\pi(s_t)$ is high, since drag is already low and small actuation is sufficient. Conversely, if s_t is highly turbulent, $V^\pi(s_t)$ will be lower.

Q-function ($Q^\pi(s_t, a_t)$): The action-value function quantifies the expected long-term reward of applying a specific action a_t in the current flow state s_t and then following π afterwards:

$$Q^\pi(s_t, a_t) = \mathbb{E}_\pi\left[\sum_{k=0}^{\infty} \gamma^k r_{t+k} \,\middle|\, s_t, a_t\right].$$

For example, $Q^\pi(s_t, a_t)$ might evaluate whether applying strong suction at a turbulent region is worth the extra energy cost, compared to milder actuation.

Discount Factor (γ): The parameter $\gamma \in (0, 1)$ ensures future rewards are gradually discounted. In drag reduction tasks, γ is often close to 1 (e.g., 0.99) because we care about sustained long-term flow control rather than short-term gains.

In short we have the following:

- The **policy** π is the agent's strategy: how to control the boundary.
- The **reward** encodes the objective: reduce drag while saving energy.
- The **value function** $V^\pi(s)$ tells us how favorable a given flow state is.
- The **Q-function** $Q^\pi(s, a)$ evaluates whether a specific actuation choice is effective.

This mapping shows how reinforcement learning concepts directly correspond to the variables that scientists and engineers already manipulate in PDE control problems.

7.3.3 *Bellman Equations*

The Bellman equations provide recursive definitions of value functions, which are fundamental to the theory and practice of reinforcement learning. They express the fact that the value of a state or state-action pair can be decomposed into the immediate reward and the discounted value of the subsequent state.

Definition: Bellman Equation for V^π

For a given policy π, the state-value function satisfies the recursive relation:

$$V^\pi(s) = \sum_{a \in \mathcal{A}} \pi(a \mid s)\Big[R(s, a) + \gamma \sum_{s' \in \mathcal{S}} P(s' \mid s, a) V^\pi(s')\Big].$$

- $s \in \mathcal{S}$: current state.

- $a \in \mathcal{A}$: possible actions.
- $\pi(a \mid s)$: probability of choosing action a in state s.
- $R(s, a)$: immediate expected reward for (s, a).
- $P(s' \mid s, a)$: transition probability to next state s'.
- $\gamma \in [0, 1)$: discount factor.

Bellman Equation for Q^π

For a given policy π, the action-value function satisfies:

$$Q^\pi(s, a) = R(s, a) + \gamma \sum_{s' \in \mathcal{S}} P(s' \mid s, a) \sum_{a' \in \mathcal{A}} \pi(a' \mid s') Q^\pi(s', a').$$

- The first term $R(s, a)$ is the immediate reward.
- The second term accounts for the discounted value of future state–action pairs.

The Bellman equations capture the principle of *optimal substructure*: the value of a decision today equals the immediate reward plus the expected value of tomorrow's decisions. This recursive decomposition is the basis for classical dynamic programming methods (such as policy iteration and value iteration) as well as modern RL algorithms.

Example: Bellman Equation for Heat Equation Control

Revisit the boundary control problem for the 1D heat equation:

$$\frac{\partial u(x, t)}{\partial t} = \alpha \frac{\partial^2 u(x, t)}{\partial x^2}, \quad x \in [0, 1].$$

- **State** s_t: discretized temperature profile $u(x, t)$.
- **Action** a_t: boundary heat flux at $x = 0$.
- **Reward** r_t: negative squared error to target profile, penalized by actuation cost.

Suppose at time t the system is in state s_t and the controller applies a_t. The Bellman equation for the Q-function is:

$$Q^\pi(s_t, a_t) = R(s_t, a_t) + \gamma\, \mathbb{E}_{s_{t+1} \sim P(\cdot \mid s_t, a_t)} \Big[V^\pi(s_{t+1}) \Big],$$

where $V^{\pi}(s_{t+1})$ represents the expected long-term error reduction if we continue to apply policy π in future steps.

In other words;

- The **immediate reward** $R(s_t, a_t)$ reflects how well the boundary flux at t reduced error.
- The **future value** term captures how this new temperature profile s_{t+1} will evolve under π.
- The **discount factor** γ enforces preference for sustained temperature regulation.

This shows how Bellman equations connect short-term control actions to long-term scientific objectives.

In scientific applications, Bellman equations often become computationally expensive because:

- The state space $\mathcal{S}$ corresponds to high-dimensional PDE solutions.
- The transition model $P(s' \mid s, a)$ may require expensive numerical solvers.
- The reward function $R(s, a)$ may involve scientific quantities like energy, stability, or information gain.

As a result, approximations using neural networks, reduced-order models, or physics-informed surrogates are often employed to make the Bellman equations tractable.

7.3.4 Constrained MDPs and Safe RL Formulation

In many scientific domains, reinforcement learning cannot simply maximize expected reward without considering physical safety, stability, or resource limitations. For example, in plasma physics, instability can cause equipment damage; in chemistry, certain interventions may exceed safety thresholds. To handle such requirements, the standard MDP formalism is extended to a *Constrained Markov Decision Process (CMDP)*.

Definition: Constrained Markov Decision Process (CMDP)

A Constrained MDP is a tuple

$$\mathcal{M}_c = (\mathcal{S}, \mathcal{A}, P, R, \gamma, C, d),$$

where:

- $\mathcal{S}$: state space.
- $\mathcal{A}$: action space.

- $P(s' \mid s, a)$: transition probabilities.
- $R(s, a)$: primary reward function.
- $\gamma \in [0, 1)$: discount factor.
- $C_i(s, a)$: cost functions representing safety or resource usage.
- d_i: constraint thresholds for each cost C_i.

The objective is:

$$\max_{\pi} J_R(\pi) = \mathbb{E}_\pi \left[\sum_{t=0}^{\infty} \gamma^t R(s_t, a_t) \right]$$

subject to

$$J_{C_i}(\pi) = \mathbb{E}_\pi \left[\sum_{t=0}^{\infty} \gamma^t C_i(s_t, a_t) \right] \le d_i, \quad \forall i.$$

where

- $J_R(\pi)$: expected cumulative reward of policy π.
- $C_i(s, a)$: cost incurred for taking action a in state s (e.g., safety violation, energy consumption).
- $J_{C_i}(\pi)$: expected cumulative cost under policy π.
- d_i: maximum allowable level of cumulative cost (safety or resource budget).

In scientific systems, constraints are not optional but essential:

- **Safety**: In plasma physics, applying too strong an actuator current may destabilize the plasma.
- **Physical laws**: In chemical or biological systems, concentrations must remain within feasible ranges.
- **Resource limits**: In computational experiments, energy or time budgets must be respected.

Several methods exist for solving CMDPs:

- **Constrained Policy Optimization (CPO)**: extends policy gradient to enforce constraints.
- **Lagrangian methods**: augment the reward with penalty terms $\lambda_i C_i(s, a)$.
- **Control Barrier Functions (CBFs)**: enforce safety constraints at each step via differentiable safety layers.

Constrained MDPs and Safe RL provide a principled way to encode physical feasibility, safety, and resource limitations into reinforcement learning. This is particularly important in Scientific Machine Learning, where unsafe policies can have catastrophic real-world consequences.

System. Consider the one–dimensional heat equation on $x \in [0, 1]$:

$$\frac{\partial u(x,t)}{\partial t} = \alpha \frac{\partial^2 u(x,t)}{\partial x^2}, \qquad u(1,t) = u_{\text{right}}, \qquad \text{control at } x = 0,$$

with state field $u(x, t)$ and diffusivity $\alpha > 0$. We discretize space into N points and time into steps $t = 0, 1, 2, \ldots$.

CMDP formulation.

- **State** $s_t \in \mathcal{S}$: the discretized temperature profile at time t, $s_t = (u(x_1, t), \ldots, u(x_N, t))^\top \in \mathbb{R}^N$.
- **Action** $a_t \in \mathcal{A}$: the boundary heat flux (or boundary temperature) applied at $x = 0$ at time t.
- **Dynamics** $P(s' \mid s, a)$: given by one time step of the numerical PDE solver $s_{t+1} = f(s_t, a_t)$ (possibly with process/measurement noise).
- **Reward** $R(s, a)$ **(tracking + control economy)**:

$$R(s,a) \;=\; -\|s - u^*\|_2^2 \;-\; \lambda_r\, a^2,$$

  where $u^* \in \mathbb{R}^N$ is the desired temperature profile and $\lambda_r > 0$ weights control economy.
- **Costs** $C_i(s, a)$ **and thresholds** d_i:

$$\begin{aligned} C_1(s,a) &= a^2 && \text{(instantaneous actuator power)},\\ C_2(s,a) &= \Big(\max_j s_j - T_{\max}\Big)_+ && \text{(thermal safety overshoot)},\\ & && \text{where } (x)_+ = \max\{x, 0\}. \end{aligned}$$

  The policy must respect budgeted discounted costs

$$J_{C_i}(\pi) = \mathbb{E}_\pi\left[\sum_{t=0}^{\infty} \gamma^t C_i(s_t, a_t)\right] \le d_i, \quad i = 1, 2,$$

  for given thresholds $d_1, d_2 \ge 0$.
- **Discount factor** $\gamma \in [0, 1)$: balances immediate and long-horizon objectives.

Constrained objective (primal).

$$\max_\pi \quad J_R(\pi) \;=\; \mathbb{E}_\pi\left[\sum_{t=0}^{\infty} \gamma^t R(s_t, a_t)\right] \quad \text{s.t.} \quad J_{C_i}(\pi) \le d_i, \; i = 1, 2.$$

Bellman *value* and *cost* recursions under a fixed policy π. Define the *state-value* and *cost-value* functions:

$$V^{\pi}(s)=\mathbb{E}_{\pi}\left[\sum_{t=0}^{\infty}\gamma^{t}R(s_t,a_t)\,\middle|\,s_0=s\right],\quad V_{C_i}^{\pi}(s)=\mathbb{E}_{\pi}\left[\sum_{t=0}^{\infty}\gamma^{t}C_i(s_t,a_t)\,\middle|\,s_0=s\right].$$

They satisfy the coupled Bellman equations:

$$V^{\pi}(s)=\sum_{a}\pi(a\mid s)\Big[R(s,a)+\gamma\sum_{s'}P(s'\mid s,a)\,V^{\pi}(s')\Big],$$
$$V_{C_i}^{\pi}(s)=\sum_{a}\pi(a\mid s)\Big[C_i(s,a)+\gamma\sum_{s'}P(s'\mid s,a)\,V_{C_i}^{\pi}(s')\Big],\qquad i=1,2.$$

Variables: s current state, a action, $\pi(a\mid s)$ policy, $P(s'\mid s,a)$ transition kernel, $R(s,a)$ reward, $C_i(s,a)$ costs, γ discount. The constraints are enforced by requiring the initial-state expectations $V_{C_i}^{\pi}(s_0)\le d_i$ (or their averages under an initial-state distribution) to hold.

Lagrangian relaxation and augmented Bellman equation. Introduce Lagrange multipliers $\lambda_1,\lambda_2\ge 0$ and define the *augmented reward*

$$R_{\lambda}(s,a)=R(s,a)-\lambda_1C_1(s,a)-\lambda_2C_2(s,a).$$

The (dual) Lagrangian is $\mathcal{L}(\pi,\lambda)=J_R(\pi)-\sum_{i=1}^{2}\lambda_i\,(J_{C_i}(\pi)-d_i)$. For fixed λ, policy evaluation reduces to an *unconstrained* MDP with reward R_λ, with value

$$V_{\lambda}^{\pi}(s)=\sum_{a}\pi(a\mid s)\Big[R_{\lambda}(s,a)+\gamma\sum_{s'}P(s'\mid s,a)\,V_{\lambda}^{\pi}(s')\Big].$$

The corresponding *action-value* Bellman equation is

$$Q_{\lambda}^{\pi}(s,a)=R_{\lambda}(s,a)+\gamma\sum_{s'}P(s'\mid s,a)\sum_{a'}\pi(a'\mid s')\,Q_{\lambda}^{\pi}(s',a').$$

Variables: $\lambda=(\lambda_1,\lambda_2)$ multipliers, R_λ augmented reward, V_{λ}^{π}, Q_{λ}^{π} are the value and action-value functions under R_λ.

Policy optimality (for fixed λ). The (soft) greedy improvement step is

$$\pi_{\text{new}}(\cdot\mid s)\;\in\;\arg\max_{\pi(\cdot|s)}\sum_{a}\pi(a\mid s)\,Q_{\lambda}^{\pi_{\text{old}}}(s,a),$$

or deterministically $\pi_{\text{new}}(s)=\arg\max_a Q_{\lambda}^{\pi_{\text{old}}}(s,a)$.

Dual ascent on multipliers (enforcing constraints). Update λ_i via projected gradient ascent:

$$\lambda_i\leftarrow\big[\lambda_i+\eta\,(J_{C_i}(\pi)-d_i)\big]_{+},\quad i=1,2,$$

with step size $\eta > 0$ and $[\cdot]_+ = \max(\cdot, 0)$. At convergence, the KKT conditions yield a policy $\pi^\star$ that maximizes reward while satisfying $J_{C_i}(\pi^\star) \le d_i$.

Instantiated signals in this PDE task.

- $R(s, a) = -\|s - u^*\|_2^2 - \lambda_r a^2$ rewards accurate tracking with economical control.
- $C_1(s, a) = a^2$ keeps cumulative actuator power under a budget d_1.
- $C_2(s, a) = (\max_j s_j - T_{\max})_+$ penalizes any thermal overshoot beyond $T_{\max}$, budgeted by d_2.
- $R_\lambda(s, a) = -\|s - u^*\|_2^2 - (\lambda_r + \lambda_1)a^2 - \lambda_2(\max_j s_j - T_{\max})_+$ is the augmented reward used in policy evaluation when enforcing safety via the dual variables λ.

Why this matters in SciML. The coupled Bellman recursions for V^π and $V^\pi_{C_i}$ make explicit *what* must be predicted: not just performance, but also discounted safety/resource budgets. The Lagrangian Bellman equation for V^π_λ shows *how* constraints can be enforced through a standard RL loop with augmented rewards, enabling scalable policy iteration even when P is a high-fidelity PDE solver.

7.4 Core Algorithmic Families

7.4.1 Monte Carlo and Temporal Difference (TD) Learning

Dynamic Programming assumes that we know the full MDP model: the transition probabilities $P(s' \mid s, a)$ and the reward function $r(s, a)$. In many scientific settings, this is unrealistic: we may only have a simulator (or a real experiment) that we can interact with, but we cannot write down P in closed form.

Monte Carlo (MC) and *Temporal Difference* (TD) methods address this situation. They learn value functions directly from *sampled trajectories*, without requiring an explicit model. They are therefore often called *model-free* prediction methods.

To keep things simple, we focus on the *prediction* problem: given a fixed policy π, estimate its value function $V_\pi(s)$ from data. Later sections use these ideas inside full RL control algorithms.

Monte Carlo Value Estimation

Monte Carlo methods estimate $V_\pi(s)$ by simulating entire episodes and averaging the observed returns.

Monte Carlo value estimate

Assume that we can generate episodes by following a fixed policy π until reaching a terminal state. For a given state s, suppose we observe N returns

$$G^{(1)}(s),\ G^{(2)}(s),\ \ldots,\ G^{(N)}(s)$$

from N different visits to s (under policy π). The *Monte Carlo estimate* of $V_\pi(s)$ is

$$\hat{V}_\pi^{\mathrm{MC}}(s) = \frac{1}{N}\sum_{i=1}^{N} G^{(i)}(s),$$

where each $G^{(i)}(s)$ is the discounted sum of rewards from the time of visiting s until the end of the episode.

There are two common variants:

- **First-visit MC**: Only the first visit to each state within an episode is used.
- **Every-visit MC**: All visits to a state within an episode are used.

Both are statistically valid and converge to $V_\pi(s)$ as the number of episodes N grows, under mild assumptions.

Incremental update form. Instead of storing all returns and recomputing the average each time, we can update the estimate incrementally. Suppose after $N-1$ returns we have $\hat{V}_{N-1}(s)$, and we observe a new return $G^{(N)}(s)$. Then

$$\hat{V}_N(s) = \hat{V}_{N-1}(s) + \frac{1}{N}\left(G^{(N)}(s) - \hat{V}_{N-1}(s)\right).$$

This has the form of a stochastic approximation update with step size $\alpha_N = 1/N$. More generally, we may use a constant or decreasing step size α:

$$\hat{V}(s) \leftarrow \hat{V}(s) + \alpha\left(G(s) - \hat{V}(s)\right).$$

Example: Monte Carlo in the temperature-control problem.
Recall the simple temperature-control MDP:

- State s_t: room temperature (discretized).
- Action a_t: heater setting (off, low, high).
- Policy π: for example, "if $s_t < 22$ use high, else off".
- Reward: $r_t = -(s_t - 22)^2 - \lambda \cdot \mathrm{EnergyCost}(a_t)$.

To estimate $V_\pi(s)$ for this fixed policy:

1. Start from some initial temperature s_0 and follow the policy π until a terminal time (e.g., after T steps).
2. For each visited state s in the episode, compute the discounted return $G_t = \sum_{k=0}^{T-t-1} \gamma^k r_{t+k}$ from that time onward.
3. Update the current estimate of $V_\pi(s)$ using the incremental MC rule.
4. Repeat over many episodes.

Over time, the estimates $\hat{V}_\pi(s)$ converge to the true value function for the given policy π, without ever estimating or using the transition probabilities.

Strengths and limitations of Monte Carlo.

- **Pros**:
  - Conceptually simple: average observed returns.
  - Unbiased estimates (on average) if episodes are generated from π.
  - No need for transition probabilities P.
- **Cons**:
  - Requires episodes to terminate: updates occur only at the end of episodes.
  - High variance: returns can be noisy, especially for long horizons.
  - Slow feedback: we must wait until the episode ends to update earlier states.

In many scientific simulations, episodes can be long and expensive, making it desirable to update value estimates *before* the end of the episode. This motivates Temporal Difference learning.

Temporal Difference (TD) Learning

Temporal Difference (TD) methods combine two ideas:

- like Monte Carlo, they learn from raw experience (sampled transitions), without knowing the model;
- like Dynamic Programming, they *bootstrap*: they update estimates based on other learned estimates.

The simplest TD method is **TD(0)** for policy evaluation.

TD(0) update for policy evaluation

Given a transition (s_t, r_t, s_{t+1}) generated under a fixed policy π, the TD(0) update for the state-value function is

$$V(s_t) \leftarrow V(s_t) + \alpha\Big(r_t + \gamma V(s_{t+1}) - V(s_t)\Big),$$

where $\alpha \in (0, 1]$ is a learning rate.

The quantity

$$\delta_t = r_t + \gamma V(s_{t+1}) - V(s_t)$$

is called the *TD error*. It measures the mismatch between the current estimate $V(s_t)$ and a one-step "target" $r_t + \gamma V(s_{t+1})$.
Key differences to Monte Carlo.

- TD(0) uses a *one-step lookahead*: it updates $V(s_t)$ as soon as the next state s_{t+1} and reward r_t are observed, without waiting for the final return.
- The target uses $V(s_{t+1})$ itself, making the method bootstrapped and slightly biased (until convergence), but typically with much lower variance than full-return MC.
- TD methods can be applied to continuing tasks (no terminal state), as updates happen at every step.

TD learning in the temperature-control problem.
Again consider the temperature-control MDP under a fixed policy π. At each time step:

1. We observe the current temperature s_t and select action $a_t = \pi(s_t)$.
2. The environment returns reward r_t and next temperature s_{t+1}.
3. We update the value of the current state using

$$V(s_t) \leftarrow V(s_t) + \alpha\big(r_t + \gamma V(s_{t+1}) - V(s_t)\big).$$

We can do this online, step by step, while the simulation is running.

If, for example, the next state s_{t+1} has a high estimated value $V(s_{t+1})$ and we receive a reasonable reward r_t, then $r_t + \gamma V(s_{t+1})$ will be large, and the TD error δ_t will be positive; thus $V(s_t)$ will be increased. If instead the outcome is worse than expected, δ_t is negative and $V(s_t)$ decreases.

MC versus TD: Bias–Variance Trade-Off

Both MC and TD methods aim to approximate the same quantity $V_\pi(s)$, but they do so in different ways.

- **Monte Carlo**:
  - Uses *complete* returns G_t as targets.
  - Target is an unbiased sample of the true expected return.
  - No bootstrapping, but often high variance, especially for long episodes.
  - Updates only at the end of the episode (for prediction).
- **TD(0)**:
  - Uses *one-step* targets $r_t + \gamma V(s_{t+1})$.
  - Target is biased while V is inaccurate, due to bootstrapping.

- Typically lower variance and more data-efficient, especially when future rewards are far in time.
- Can learn online, in continuing tasks, without waiting for episode termination.

In practice, TD methods (and their extensions) are often preferred for scientific simulations because they can update quickly and make better use of each transition, which is crucial when each simulation step is computationally expensive.

n-step Returns and TD(λ) (high-level view)

Monte Carlo and TD(0) can be seen as two extremes:

- TD(0) uses a 1-step target $r_t + \gamma V(s_{t+1})$.
- Monte Carlo uses a full-return target $G_t = r_t + \gamma r_{t+1} + \gamma^2 r_{t+2} + \dots$.

n-step TD methods use intermediate n-step returns such as

$$G_t^{(n)} = r_t + \gamma r_{t+1} + \cdots + \gamma^{n-1} r_{t+n-1} + \gamma^n V(s_{t+n}),$$

which trade off between bias and variance.

TD(λ) combines all n-step targets with exponentially decaying weights controlled by $\lambda \in [0, 1]$. Roughly speaking:

- $\lambda = 0$ behaves like TD(0),
- $\lambda = 1$ behaves like Monte Carlo.

We will not go into the full details here, but it is useful to know that there is a continuum of methods between pure TD and pure Monte Carlo, allowing practitioners to adjust the bias–variance trade-off to their problem.

Role in Control and Scientific ML

So far we have focused on predicting V_π for a fixed policy. In full RL control algorithms:

- TD methods are used to evaluate and improve policies (e.g., in SARSA, Q-learning, actor–critic).
- They provide *critics* that guide policy gradients and actor updates.
- They can be combined with function approximators (e.g., neural networks) to handle large or continuous state spaces.

In Scientific Machine Learning, Monte Carlo and TD ideas also appear in:

- **Simulation-based inference**: estimating long-term performance or reliability of scientific systems without explicit models.

- **Surrogate-based RL**: combining MC/TD updates with learned surrogates of physical models.
- **Data assimilation and filtering**: TD-like updates appear when adjusting state estimates or parameters based on observed discrepancies over time.

In summary, Monte Carlo and Temporal Difference learning are the basic tools that allow RL to learn from trajectories rather than explicit models. They form the backbone of many algorithms that we will discuss in the next subsections.

7.4.2 Q-Learning and SARSA

Monte Carlo and TD methods in the previous subsection focused on the *prediction problem*: given a fixed policy π, estimate its value function $V_\pi(s)$. In reinforcement learning, we ultimately care about the *control problem*: *how to find a good (ideally optimal) policy from experience*.

A natural idea is to extend TD learning from value functions $V(s)$ to *action-value functions* $Q(s, a)$. Action-values tell us how good it is to take a particular action in a given state and then follow some policy afterwards. Once we have an estimate of $Q(s, a)$, we can choose actions greedily with respect to it.

Definition: Action-Value Function

Given a policy π, the *action-value function* (or Q-function) is

$$Q_\pi(s, a) = \mathbb{E}_\pi\left[\sum_{t=0}^{\infty} \gamma^t r_t \,\middle|\, s_0 = s,\ a_0 = a\right],$$

the expected discounted return when we start from state s, take action a at time 0, and then follow policy π afterwards.

The *optimal* action-value function is

$$Q^\star(s, a) = \max_\pi Q_\pi(s, a),$$

and an optimal policy can be obtained by choosing, in each state, any action that maximizes $Q^\star(s, a)$.
Bellman optimality equation for $Q^\star$.
$Q^\star$ satisfies

$$Q^\star(s, a) = r(s, a) + \gamma \sum_{s' \in \mathcal{S}} P(s' \mid s, a) \max_{a' \in \mathcal{A}} Q^\star(s', a').$$

This is the action-value version of the Bellman optimality equation, and it is the starting point for *Q-learning*.

ϵ-Greedy exploration

Before introducing the algorithms, we need a simple mechanism to balance *exploration* and *exploitation*. If we always pick the action with the highest estimated value (pure exploitation), we might miss better actions that we have not tried often enough. A standard solution is the ϵ-greedy strategy [1].

Definition: ϵ-greedy policy

Given an action-value estimate $Q(s, a)$ and a parameter $\epsilon \in [0, 1]$:

- with probability $1 - \epsilon$, choose a greedy action

$$a_t \in \arg\max_{a \in \mathcal{A}} Q(s_t, a),$$

- with probability ϵ, choose an action uniformly at random from $\mathcal{A}$.

This is called an ϵ-*greedy* policy with respect to Q.

Typically, ϵ is chosen relatively large at the beginning (more exploration) and then gradually decreased as learning progresses.

Q-Learning: Off-Policy TD Control

Q-learning is a model-free, off-policy TD control algorithm. It learns an estimate of the optimal action-value function $Q^\star(s, a)$ by interacting with the environment and updating a table $Q(s, a)$.

Q-learning update rule

After observing a transition (s_t, a_t, r_t, s_{t+1}), the Q-learning update is

$$Q(s_t, a_t) \leftarrow Q(s_t, a_t) + \alpha\Big(r_t + \gamma \max_{a'} Q(s_{t+1}, a') - Q(s_t, a_t)\Big).$$

Key points:

- The *target* is

$$r_t + \gamma \max_{a'} Q(s_{t+1}, a'),$$

which assumes that in the next state we will behave optimally (take the best action a').
- The agent can act according to any behaviour policy (e.g., ϵ-greedy with respect to Q); the update "looks ahead" assuming greedy actions in the future. This makes Q-learning an *off-policy* method: it learns about the greedy policy while behaving more exploratively.

Algorithm: Tabular Q-learning (with ϵ-greedy exploration)

1. Initialize $Q(s, a)$ arbitrarily for all $s \in \mathcal{S}, a \in \mathcal{A}$ (e.g., $Q = 0$).
2. For each episode:
   a. Initialize the starting state s_0.
   b. For $t = 0, 1, 2, \ldots$ until termination:
      i. Choose an action a_t using an ϵ-greedy policy with respect to $Q(s_t, \cdot)$.
      ii. Execute a_t, observe reward r_t and next state s_{t+1}.
      iii. Update

$$Q(s_t, a_t) \leftarrow Q(s_t, a_t) + \alpha\Big(r_t + \gamma \max_{a'} Q(s_{t+1}, a') - Q(s_t, a_t)\Big).$$

      iv. Set $s_t \leftarrow s_{t+1}$ and continue.

Interpretation.
The Q-learning update is a TD method operating directly on action-values. It bootstraps from the current estimate $Q(s_{t+1}, \cdot)$ and pushes $Q(s_t, a_t)$ toward consistency with the Bellman optimality equation. Under suitable conditions (decaying step sizes, sufficient exploration), tabular Q-learning converges to $Q^\star$ with probability 1.

SARSA: On-Policy TD Control

SARSA (State–Action–Reward–State–Action) is another TD control algorithm [2]. It looks very similar to Q-learning but uses the *actual* next action a_{t+1} taken by the behaviour policy, rather than the greedy one.

SARSA update rule

Suppose a transition is generated as

$$s_t, a_t, r_t, s_{t+1}, a_{t+1},$$

where both a_t and a_{t+1} are chosen using the same (typically ϵ-greedy) policy based on Q. Then the SARSA update is

$$Q(s_t, a_t) \leftarrow Q(s_t, a_t) + \alpha\Big(r_t + \gamma Q(s_{t+1}, a_{t+1}) - Q(s_t, a_t)\Big).$$

Key points:

- The target now uses the *actual next action* a_{t+1}, not the greedy one.
- Therefore SARSA is an *on-policy* algorithm: it learns the value of the policy it is currently following (here, the ϵ-greedy policy), rather than the purely greedy policy.

Algorithm: Tabular SARSA (on-policy)

1. Initialize $Q(s, a)$ arbitrarily for all s, a.
2. For each episode:
   a. Initialize s_0, choose a_0 using ϵ-greedy with respect to $Q(s_0, \cdot)$.
   b. For $t = 0, 1, 2, \ldots$ until termination:
      i. Execute a_t, observe reward r_t and next state s_{t+1}.
      ii. Choose next action a_{t+1} using ϵ-greedy with respect to $Q(s_{t+1}, \cdot)$.
      iii. Update

$$Q(s_t, a_t) \leftarrow Q(s_t, a_t) + \alpha\Big(r_t + \gamma Q(s_{t+1}, a_{t+1}) - Q(s_t, a_t)\Big).$$

      iv. Set $s_t \leftarrow s_{t+1}$, $a_t \leftarrow a_{t+1}$ and continue.

Q-learning Versus SARSA: Off-policy Versus On-policy

Although Q-learning and SARSA differ only slightly in their update equations, this small difference has important consequences.

Off-policy (Q-learning).

- Q-learning updates toward the value of the *greedy* action in the next state, even if the behaviour policy actually takes a different action.
- It effectively learns the value of the greedy policy while behaving in an ϵ-greedy way.
- This can be powerful, but also means that if exploration (random actions) is dangerous, Q-learning may learn overly optimistic estimates that assume we always act greedily.

On-policy (SARSA).

- SARSA updates toward the value of the behaviour policy itself (e.g., ϵ-greedy).
- It "knows" that with probability ϵ it will take exploratory actions in the future, and its value estimates reflect this risk.
- In some environments (e.g., with danger zones), SARSA may learn more cautious policies than Q-learning when using the same exploration strategy.

A classic didactic example (which we do not detail here) is the "cliff walking" grid world: Q-learning tends to learn a risky path near the cliff (assuming always-greedy future behaviour), while SARSA learns a safer path because it accounts for the possibility of future exploratory moves that might lead to a fall.

Example in the temperature-control problem

Let us connect these algorithms back to the temperature-control example.

MDP recap.

- State: room temperature $s_t \in \{15, \ldots, 30\}$.
- Actions: $a_t \in \{\text{off}, \text{low}, \text{high}\}$.
- Reward:

$$r(s_t, a_t) = -(s_t - 22)^2 - \lambda \cdot \text{EnergyCost}(a_t).$$

Suppose we do *not* know the transition probabilities $P(s_{t+1} \mid s_t, a_t)$, but we have a simulator that, given (s_t, a_t), returns (r_t, s_{t+1}).

Using Q-learning.

- We maintain a table $Q(s, a)$ for all temperatures and heater settings.
- Initially, $Q(s, a)$ can be zero or small random values.
- At each step, we select a_t using ϵ-greedy with respect to $Q(s_t, \cdot)$.

- After observing (r_t, s_{t+1}), we update:

$$Q(s_t, a_t) \leftarrow Q(s_t, a_t) + \alpha\Big(r_t + \gamma \max_{a'} Q(s_{t+1}, a') - Q(s_t, a_t)\Big).$$

Over many simulated episodes (days of heating), $Q(s, a)$ converges towards $Q^\star(s, a)$, and the greedy policy $a^\star(s) = \arg\max_a Q(s, a)$ approximates the optimal heater control.

Using SARSA.
If instead we use SARSA:

- We again use ϵ-greedy actions for both a_t and a_{t+1}.
- The update becomes

$$Q(s_t, a_t) \leftarrow Q(s_t, a_t) + \alpha\Big(r_t + \gamma Q(s_{t+1}, a_{t+1}) - Q(s_t, a_t)\Big).$$

- The learned policy reflects the long-term effect of continuing to follow the ϵ-greedy strategy, including occasional exploratory actions.

In this simple temperature problem, both algorithms are likely to find similar policies, but in more complex or safety-critical SciML environments the difference between off-policy and on-policy learning can matter.

From tabular Q to function approximation

In realistic Scientific Machine Learning problems, the state space is huge (e.g., a discretized PDE field), so we cannot store $Q(s, a)$ in a table. Instead, we approximate $Q(s, a)$ with a parameterized function $Q_\theta(s, a)$, such as a neural network. The Q-learning and SARSA updates then become gradient-based updates on θ to reduce the TD error.

These function-approximation variants are the basis of *Deep Q-Networks* (DQN) and other deep RL algorithms, which we will encounter later. However, the core ideas remain exactly those introduced here:

- bootstrapping from TD targets,
- ϵ-greedy exploration,
- off-policy learning (Q-learning) vs on-policy learning (SARSA).

7.4.3 Policy Gradient Methods

The methods we have seen so far (Q-learning, SARSA) learn an *action-value function* $Q(s, a)$ and then choose actions greedily with respect to Q. This works well for small, discrete action spaces, but becomes challenging when:

- actions are continuous (e.g., a control input $u_t \in \mathbb{R}^m$),
- or the action space is large and structured.

In many Scientific Machine Learning applications, actions are naturally continuous: a boundary flux, a voltage, a force, a flow rate, or a heating power. In such cases, it is often more natural to *directly* parameterize the policy and optimize its parameters using gradient-based methods. These are called *policy gradient* methods.

Parameterizing Policies

We represent the policy by a differentiable function with parameters θ:

$$\pi_\theta(a \mid s) \approx \mathbb{P}(a_t = a \mid s_t = s).$$

Definition: Parameterized (stochastic) policy

A *parameterized policy* is a family of probability distributions $\{\pi_\theta(\cdot \mid s) : \theta \in \mathbb{R}^d\}$ such that for each state s:

- $\pi_\theta(a \mid s) \geq 0$ for all actions a,
- $\sum_{a \in \mathcal{A}} \pi_\theta(a \mid s) = 1$ (discrete actions), or $\int \pi_\theta(a \mid s)\, da = 1$ (continuous actions),
- $\pi_\theta(a \mid s)$ is differentiable with respect to θ.

Examples.

- **Discrete actions**: Use a softmax over a neural network output:

$$\pi_\theta(a \mid s) = \frac{\exp(f_\theta(s, a))}{\sum_{a'} \exp(f_\theta(s, a'))}.$$

- **Continuous actions**: Model $a \in \mathbb{R}$ as a Gaussian:

$$a_t \sim \mathcal{N}\big(\mu_\theta(s_t),\ \sigma_\theta^2(s_t)\big),$$

where a neural network outputs mean and (log-)variance.

In SciML, f_θ might be a small neural network that takes features of the physical state (e.g., measurements or reduced-order coefficients) and outputs control parameters.

The Policy Optimization Objective

Intuitively, we want to choose θ so that the policy earns high long-term reward. A common objective is the expected discounted return:

Objective: expected return

For a parameterized policy π_θ, define

$$J(\theta) = \mathbb{E}_{\pi_\theta}\left[\sum_{t=0}^{\infty} \gamma^t r_t\right],$$

where the expectation is taken over trajectories generated by starting from some initial state distribution and then following π_θ. The goal of policy gradient methods is to *maximize* $J(\theta)$ with respect to θ.

We will use gradient ascent:

$$\theta \leftarrow \theta + \alpha \, \nabla_\theta J(\theta),$$

where $\alpha > 0$ is a learning rate.

The key question is: *how do we compute or estimate $\nabla_\theta J(\theta)$ from sampled trajectories?*

Policy Gradient Theorem (high-level view)

The *policy gradient theorem* provides a formula for the gradient that does *not* require differentiating through the environment dynamics (i.e., we do not need $\partial P(s' \mid s, a)/\partial\theta$). We state the result informally; the full derivation is beyond our scope.

Policy Gradient Theorem (informal statement)

Under suitable regularity conditions, the gradient of the expected return can be written as

$$\nabla_\theta J(\theta) = \mathbb{E}_{\pi_\theta}\big[\nabla_\theta \log \pi_\theta(a_t \mid s_t) \; A_{\pi_\theta}(s_t, a_t)\big],$$

where $A_{\pi_\theta}(s, a)$ is an *advantage function* (e.g., $Q_{\pi_\theta}(s, a) - V_{\pi_\theta}(s)$).

Important points:

- The gradient depends on $\nabla_\theta \log \pi_\theta(a_t \mid s_t)$, which is usually easy to compute for common distributions (softmax, Gaussian, etc.).
- The environment dynamics P appear only inside the expectation; we do not need their derivatives.
- In practice, we approximate the expectation by averaging over sampled trajectories.

In other words, we can perform gradient-based policy optimization using *samples* from the environment, without knowing its equations explicitly. This is particularly appealing in complex scientific systems, where we may only have a black-box simulator.

REINFORCE: Monte Carlo Policy Gradients

The simplest policy gradient algorithm is **REINFORCE** (also called the Monte Carlo policy gradient) [3].

REINFORCE update rule (episodic)

For each episode generated under policy π_θ:

1. Generate a trajectory $\tau = (s_0, a_0, r_0, s_1, a_1, r_1, \ldots, s_T)$ by following π_θ.
2. For each time step t in the episode, compute the return

$$G_t = \sum_{k=0}^{T-t-1} \gamma^k r_{t+k}.$$

3. Update the parameters by gradient ascent:

$$\theta \leftarrow \theta + \alpha\, G_t\, \nabla_\theta \log \pi_\theta(a_t \mid s_t).$$

Intuitively:

- If G_t is larger than usual (good outcome), the update increases the log-probability of the actions that led to this outcome.
- If G_t is small (bad outcome), the update decreases the probability of the actions taken.

Example: continuous heater control.
Extend the temperature-control problem to continuous actions:

- State s_t: current room temperature.
- Action $a_t \in [0, 1]$: heater power (0 = off, 1 = full power).
- Policy: a_t is drawn from a Gaussian $\mathcal{N}(\mu_\theta(s_t), \sigma^2)$, where μ_θ is a neural network mapping temperature to a mean power level.
- Reward: as before, penalizing deviation from 22° and energy use.

For each episode (e.g., a simulated day):

1. We simulate the environment using the current policy.
2. For each time step, we record (s_t, a_t, r_t).
3. We compute returns G_t and update θ according to REINFORCE.

Over many episodes, μ_θ learns to propose heater powers that keep the temperature near 22° while saving energy, even though we never explicitly computed any Q-values or transition probabilities.

Baseline and Advantage: Variance Reduction

REINFORCE, as presented, can have very high variance because the raw return G_t can fluctuate a lot. A simple but powerful trick is to subtract a *baseline* from the return, without changing the expectation of the gradient.

Baseline-augmented REINFORCE

Instead of using G_t directly, use

$$\theta \leftarrow \theta + \alpha\,(G_t - b(s_t))\,\nabla_\theta \log \pi_\theta(a_t \mid s_t),$$

where $b(s_t)$ is a baseline that does not depend on the action a_t.

- A common choice is a state-dependent baseline $b(s_t) \approx V_{\pi_\theta}(s_t)$, the value function under the current policy.
- Then the term $(G_t - b(s_t))$ approximates an *advantage*: how much better the actual return was compared to what was expected from state s_t.

This connection leads naturally to *actor–critic* methods:

- the *actor* is the policy π_θ,
- the *critic* is a value function V_ϕ that provides a baseline or advantage estimate.

We will see more of this in later subsections.

From episodic to step-wise updates

REINFORCE uses full episode returns, which can be long and noisy. More advanced policy gradient algorithms combine TD-style updates with policy gradients, replacing G_t with TD-based estimates of the advantage, for example:

$$A_t \approx r_t + \gamma V_\phi(s_{t+1}) - V_\phi(s_t) \quad \text{or more sophisticated multi-step estimates.}$$

The basic policy gradient term $\nabla_\theta \log \pi_\theta(a_t \mid s_t)$ remains the same; only the *advantage estimate* changes.

This combination of:

- *policy gradients* w.r.t. θ, and
- *value function learning* for V_ϕ,

forms the foundation of modern actor–critic algorithms such as A2C, PPO, SAC, and others.

Why policy gradients matter in SciML

Policy gradient methods are especially attractive in Scientific Machine Learning because:

- They handle **continuous action spaces** naturally, which are ubiquitous in control and design.
- They allow **structured policies** that directly encode domain knowledge (e.g., smoothness, bounded controls, conservation properties).
- They are compatible with **differentiable simulators**: in some cases, gradients can be combined with adjoint methods or differentiable physics for even higher efficiency.
- They integrate well with **constraint handling**, for example by parameterizing policies to respect safety constraints or by incorporating penalty terms in the objective.

In summary, policy gradient methods move us from "learning values and then acting greedily" to "directly learning how to act," using gradient ascent on the policy parameters. They are a key building block for the deep RL algorithms that we will encounter next.

7.4.4 Actor–Critic Frameworks

Policy gradient methods (such as REINFORCE) directly adjust the parameters of a policy by pushing up the probability of actions that led to high returns. However, pure Monte Carlo policy gradients can suffer from:

- **high variance** (returns G_t fluctuate a lot),
- **slow learning** when episodes are long,
- and limited reuse of information from each trajectory.

On the other hand, value-based methods (such as TD learning or Q-learning) provide sample-efficient bootstrapped updates, but do not directly handle continuous actions as naturally as policy gradients.

Actor–critic methods combine the strengths of both:

- the **actor** is a parameterized policy π_θ that decides which action to take,
- the **critic** is a parameterized value function (e.g., V_ϕ or Q_ϕ) that evaluates how good the current state or action is.

The critic provides lower-variance estimates of the advantage used in the policy gradient, and the actor updates the policy in the direction suggested by the critic.

Idea of Actor–Critic Methods

- Use a **critic** to learn value functions with TD updates (model-free, bootstrapped).
- Use the critic's estimates to compute **advantages** and drive **actor** updates via policy gradients.
- Alternate between critic updates (improving the evaluation) and actor updates (improving the policy).

Basic Actor–Critic Structure

We consider a parameterized policy $\pi_\theta(a \mid s)$ (actor) and a parameterized state-value function $V_\phi(s)$ (critic). The parameters θ and ϕ are updated from experience.

At time step t:

1. Observe state s_t.
2. Sample an action $a_t \sim \pi_\theta(\cdot \mid s_t)$.
3. Execute a_t, observe reward r_t and next state s_{t+1}.
4. Critic: update V_ϕ using a TD step:

$$V_\phi(s_t) \leftarrow V_\phi(s_t) + \alpha_V \left(r_t + \gamma V_\phi(s_{t+1}) - V_\phi(s_t)\right).$$

5. Actor: compute an *advantage estimate*, for example

$$\hat{A}_t \;=\; r_t + \gamma V_\phi(s_{t+1}) - V_\phi(s_t),$$

and update policy parameters using a policy gradient step:

$$\theta \leftarrow \theta + \alpha_\pi \, \hat{A}_t \, \nabla_\theta \log \pi_\theta(a_t \mid s_t).$$

Actor–Critic in one equation (conceptual)

$$\underbrace{\text{TD error}}_{\text{critic}} \quad \delta_t = r_t + \gamma V_\phi(s_{t+1}) - V_\phi(s_t) \quad \Rightarrow \quad \underbrace{\delta_t \nabla_\theta \log \pi_\theta(a_t \mid s_t)}_{\text{actor update}}.$$

Here, the TD error δ_t plays a double role:

- for the critic, it is the usual TD update target;
- for the actor, it acts as an *advantage estimate* indicating whether action a_t was better ($\delta_t > 0$) or worse ($\delta_t < 0$) than expected.

Loss Functions Viewpoint

It is often useful to express actor–critic in terms of loss functions.

Critic loss.
The critic aims to approximate the value function $V_\pi(s)$ under the current policy. A common choice is a mean-squared TD error:

$$\mathcal{L}_{\text{critic}}(\phi) = \frac{1}{2}\big(r_t + \gamma V_\phi(s_{t+1}) - V_\phi(s_t)\big)^2.$$

We then update ϕ by gradient descent on this loss.

Actor loss.
The actor aims to maximize the expected return, approximated using advantage estimates. A typical loss (to *minimize*) is

$$\mathcal{L}_{\text{actor}}(\theta) = -\hat{A}_t \, \log \pi_\theta(a_t \mid s_t).$$

Gradient descent on this loss corresponds to gradient *ascent* on the expected return:

$$\nabla_\theta \mathcal{L}_{\text{actor}}(\theta) = -\hat{A}_t \nabla_\theta \log \pi_\theta(a_t \mid s_t).$$

In practice, we average these losses over mini-batches of time steps or trajectories.

Example: Actor–Critic for Temperature Control

We revisit the continuous heater control example.

Setup.

- **State** s_t: current room temperature.
- **Action** $a_t \in [0, 1]$: heater power.
- **Policy** (actor): Gaussian policy with mean $\mu_\theta(s_t)$ and fixed standard deviation σ:

$$a_t \sim \mathcal{N}\big(\mu_\theta(s_t), \sigma^2\big).$$

- **Critic**: scalar network $V_\phi(s_t)$ approximating the value function.
- **Reward**:

$$r_t = -(s_t - 22)^2 - \lambda \cdot \text{EnergyCost}(a_t).$$

One step of actor–critic (conceptual).
At each time step:

1. Observe temperature s_t.
2. Actor: compute $\mu_\theta(s_t)$ and sample a_t.
3. Environment: apply a_t, simulate the temperature dynamics, obtain r_t and s_{t+1}.
4. Critic:

$$\delta_t = r_t + \gamma V_\phi(s_{t+1}) - V_\phi(s_t),$$

   update ϕ to reduce the squared TD error.
5. Actor: compute $\nabla_\theta \log \pi_\theta(a_t \mid s_t)$. For a Gaussian policy, this gradient has a simple closed form, e.g., roughly proportional to $(a_t - \mu_\theta(s_t))$. Update:

$$\theta \leftarrow \theta + \alpha_\pi \, \delta_t \, \nabla_\theta \log \pi_\theta(a_t \mid s_t).$$

If δ_t is positive, it means that the outcome was better than the critic expected; the actor update increases the probability of choosing similar actions in similar states. If δ_t is negative, the action was worse than expected, and its probability is decreased.

Over many episodes, the critic becomes a good predictor of long-term performance, and the actor learns a control law that keeps the temperature close to the target while saving energy, purely from interaction data.

On-Policy Actor–Critic and A2C

The simple procedure described above, where:

- trajectories are generated using the *current* policy π_θ,
- updates are applied using advantages computed from V_ϕ,

is an instance of *on-policy actor–critic*.

A popular variant is **Advantage Actor–Critic** (A2C) [4]:

- it explicitly uses an advantage estimate

$$\hat{A}_t = r_t + \gamma V_\phi(s_{t+1}) - V_\phi(s_t),$$

  or more sophisticated n-step or TD(λ) versions,
- it processes trajectories in synchronous batches (hence "A2C"),
- the actor and critic share part of the neural network (e.g., initial layers) and then split into separate output heads (one for policy, one for value).

Advantage-based Actor update

A typical A2C-style actor update uses

$$\theta \leftarrow \theta + \alpha_\pi \hat{A}_t \nabla_\theta \log \pi_\theta(a_t \mid s_t),$$

where $\hat{A}_t$ is a TD-based advantage estimate.

Asynchronous Actor–Critic (A3C)

Asynchronous Advantage Actor–Critic (A3C) extends A2C by running multiple actors in parallel on different copies of the environment (or different initial conditions). Each actor:

- interacts with its environment using its current copy of the policy,
- periodically computes gradients for actor and critic losses,
- sends these gradients to update a shared global set of parameters.

This parallelism:

- improves exploration (different actors see different experiences),
- stabilizes learning by reducing correlation between gradients,
- speeds up training on modern multi-core hardware.

Although the implementation details are more involved, the conceptual actor–critic structure remains the same.

Why Actor–Critic Matters for Scientific Machine Learning

Actor–critic methods are central to many modern RL algorithms used in SciML, such as PPO and SAC, which we will discuss in the next subsection. They are particularly well suited because:

- **Continuous control**: The actor can directly output continuous control variables (forces, fluxes, voltages), avoiding discretization of actions.
- **Sample efficiency**: TD-based critic updates make more efficient use of each transition than pure Monte Carlo methods, which is crucial when simulations or experiments are expensive.
- **Incorporating physics and structure**: Both actor and critic can be implemented with architectures that respect physical structure (e.g., convolutional networks for spatial fields, graph neural networks for meshes, operator networks for PDEs).
- **Safety and constraints**: Value-based critics can be extended to include constraint costs or risk measures, enabling safer policy updates (e.g., via Lagrangian or constrained actor–critic methods).

In summary, actor–critic frameworks provide a flexible and powerful template:

"learn to evaluate, then learn to act better based on that evaluation."

This template underlies many of the scalable, robust deep RL algorithms that are now being used for complex control and design tasks in scientific domains.

7.4.5 Advanced Algorithms (PPO, SAC, DDPG)

The actor–critic framework gives us a general recipe:

- an **actor** (policy) that selects actions,
- a **critic** (value function) that evaluates them,
- policy gradient updates guided by the critic.

In practice, however, naïve actor–critic implementations with neural networks can be:

- unstable (training diverges or oscillates),
- sensitive to hyperparameters,
- sample-inefficient for expensive simulations.

This led to a family of *advanced* deep RL algorithms that refine actor–critic ideas to improve stability and data efficiency. In this subsection we focus on three widely used methods:

- **Proximal Policy Optimization (PPO)** [5]—stable on-policy policy gradient;
- **Deep Deterministic Policy Gradient (DDPG)** [6]—off-policy deterministic actor for continuous actions;

- **Soft Actor–Critic (SAC)**—entropy-regularized off-policy actor–critic with strong exploration.

We do not aim for full implementation detail, but to give enough intuition so that a scientifically minded reader can understand *why* these methods are useful and how they connect to the basic ideas developed earlier.

Motivation: Stable and Efficient Policy Updates

A recurring difficulty in deep RL is that large policy updates can destroy what has been learned:

- If we change the policy too aggressively in a single gradient step, the critic may become invalid, and performance can collapse.
- On the other hand, very small steps make learning painfully slow, particularly when simulations are expensive (as in many SciML applications).

Key design goals of advanced deep RL algorithms

Modern actor–critic algorithms are designed to:

- **limit the size** of each policy update (stability),
- **reuse past data** when possible (sample efficiency),
- perform well with **continuous actions** and high-dimensional states,
- and provide **robust exploration** strategies.

PPO, DDPG, and SAC each tackle these goals using different strategies.

Proximal Policy Optimization (PPO)

Proximal Policy Optimization (PPO) is an on-policy actor–critic algorithm that makes *conservative* ("proximal") changes to the policy at each update. It can be seen as a practical, simplified variant of trust-region policy optimization.

Clipped surrogate objective. Let $\pi_{\theta_{\text{old}}}$ be the policy used to collect a batch of data, and π_θ the new policy we want to learn. For each time step t in the batch, define the *probability ratio*

$$r_t(\theta) = \frac{\pi_\theta(a_t \mid s_t)}{\pi_{\theta_{\text{old}}}(a_t \mid s_t)}.$$

If $r_t(\theta) > 1$, the new policy makes action a_t more likely than before; if $r_t(\theta) < 1$, it makes it less likely.

PPO maximizes a *clipped* objective:

PPO clipped objective

Given an advantage estimate $\hat{A}_t$ for each time step, PPO maximizes

$$L^{\text{CLIP}}(\theta) = \mathbb{E}_t\Big[\min\big(r_t(\theta)\,\hat{A}_t,\;\text{clip}\big(r_t(\theta), 1-\epsilon, 1+\epsilon\big)\,\hat{A}_t\big)\Big],$$

where $\epsilon > 0$ is a small constant (e.g., 0.1 or 0.2) and the expectation is over a batch of data.

Intuition:

- If r_t stays within $[1-\epsilon, 1+\epsilon]$, the objective behaves like a standard policy gradient.
- If r_t becomes too large (policy changed too much), the clipping prevents further increase of the objective.
- This discourages overly large policy changes in a single update, increasing stability.

PPO training loop (simplified).

1. Collect trajectories using the current policy $\pi_{\theta_{\text{old}}}$.
2. Compute estimated advantages $\hat{A}_t$ using the critic (e.g., via TD or Generalized Advantage Estimation).
3. Optimize the actor parameters θ for several epochs on the same batch, using stochastic gradient ascent on $L^{\text{CLIP}}(\theta)$.
4. Update the critic parameters by minimizing a value loss (e.g., squared TD error).
5. Set $\theta_{\text{old}} \leftarrow \theta$ and repeat with new data.

Why PPO is attractive for SciML.

- **Stability**: The clipping mechanism keeps updates conservative, reducing the risk of catastrophic drops in performance.
- **Good default**: PPO is often a strong baseline across many tasks with relatively few hyperparameters to tune.
- **On-policy**: It uses fresh data from the current policy, which can be an advantage when dynamics change during learning.

Typical SciML use cases include PDE control or experimental design problems where we can afford to generate batches of simulations with the current policy and want robust learning behavior.

Deep Deterministic Policy Gradient (DDPG)

While PPO is an *on-policy* algorithm, Deep Deterministic Policy Gradient (DDPG) is an *off-policy* actor–critic method for continuous action spaces. It combines ideas from:

- deterministic policy gradients,
- Q-learning,
- and deep function approximation with replay buffers and target networks.

Deterministic policy and critic in DDPG

- **Actor** (deterministic policy): $a = \mu_\theta(s)$, mapping states directly to actions (no sampling during exploitation).
- **Critic**: action-value function $Q_\phi(s, a)$, approximating $Q^\mu(s, a)$ for the current policy μ_θ.

Critic update (Q-learning style).
DDPG uses a Q-learning-like TD update with target networks:

$$y_t = r_t + \gamma Q_{\phi^-}\big(s_{t+1}, \mu_{\theta^-}(s_{t+1})\big),$$

where (ϕ^-, θ^-) are slowly updated target network parameters. The critic minimizes the loss

$$\mathcal{L}_{\text{critic}}(\phi) = \mathbb{E}_{(s_t,a_t,r_t,s_{t+1})\sim\mathcal{D}}\big[(Q_\phi(s_t, a_t) - y_t)^2\big],$$

where $\mathcal{D}$ is a replay buffer containing past transitions.

Actor update (deterministic policy gradient). The actor is updated using the gradient of the critic:

$$\nabla_\theta J(\theta) \approx \mathbb{E}_{s_t\sim\mathcal{D}}\big[\nabla_a Q_\phi(s_t, a)\big|_{a=\mu_\theta(s_t)} \nabla_\theta \mu_\theta(s_t)\big].$$

Intuitively, this update pushes $\mu_\theta(s)$ toward actions a with larger Q-values.

Exploration.
Since μ_θ is deterministic, exploration is added by injecting noise:

$$a_t = \mu_\theta(s_t) + \mathcal{N}_t,$$

where $\mathcal{N}_t$ is a noise process (e.g., Gaussian or Ornstein–Uhlenbeck). Transitions (s_t, a_t, r_t, s_{t+1}) collected with noisy actions are stored in the replay buffer.

Why DDPG is useful and its limitations.

- **Pros**:
  - Designed for continuous control problems.
  - Off-policy: can reuse past experience many times, improving sample efficiency.
  - Can work well when the critic is accurate and exploration is well tuned.
- **Cons**:
  - Can be brittle and sensitive to hyperparameters (learning rates, noise scale, etc.).
  - Overestimation bias in the critic can hurt performance.
  - Requires careful tuning and regularization for stability in complex environments.

In SciML, DDPG (and its improved variants) can be used for continuous control of dynamical systems: for example, adjusting multiple continuous actuators in a fluid flow or chemical process. However, more robust algorithms like SAC are often preferred in challenging settings.

Soft Actor–Critic (SAC)

Soft Actor–Critic (SAC) extends actor–critic ideas by incorporating an *entropy* term into the objective. Instead of just maximizing expected return, SAC maximizes a combination of return and policy entropy, encouraging *stochastic*, exploratory, and robust behavior.

Maximum entropy RL objective (SAC)

SAC optimizes policies by maximizing

$$J(\pi) = \mathbb{E}_\pi\left[\sum_{t=0}^{\infty} \gamma^t \left(r_t + \alpha \mathcal{H}(\pi(\cdot \mid s_t))\right)\right],$$

where $\mathcal{H}(\pi(\cdot \mid s))$ is the entropy of the policy at state s and $\alpha > 0$ is a temperature parameter that balances reward and entropy.

Intuition:

- Higher entropy means more randomness in actions.
- By rewarding entropy, SAC encourages exploration and avoids collapsing to deterministic, brittle policies too early.

Key components of SAC.

- **Stochastic actor** $\pi_\theta(a \mid s)$, typically a Gaussian for continuous actions.

- **Two critics** $Q_{\phi_1}(s, a)$ and $Q_{\phi_2}(s, a)$ to mitigate overestimation bias (the minimum of the two is used in targets).
- **Value function** or directly using Q-functions with an entropy term embedded in the target.
- **Off-policy training** with a replay buffer, similar to DDPG.

Critic update (rough sketch). For each sampled transition (s_t, a_t, r_t, s_{t+1}) from the replay buffer:

- Sample next action $a_{t+1} \sim \pi_\theta(\cdot \mid s_{t+1})$.
- Compute target

$$y_t = r_t + \gamma\Big(\min_{i=1,2} Q_{\phi_i}(s_{t+1}, a_{t+1}) - \alpha \log \pi_\theta(a_{t+1} \mid s_{t+1})\Big).$$

- Minimize squared error between $Q_{\phi_i}(s_t, a_t)$ and y_t for $i = 1, 2$.

Actor update (entropy-regularized).
The actor is updated to minimize the KL divergence between π_θ and the exponential of the soft Q-function. In practice, this leads to the loss

$$\mathcal{L}_{\text{actor}}(\theta) = \mathbb{E}_{s_t \sim \mathcal{D},\, a_t \sim \pi_\theta}\big[\alpha \log \pi_\theta(a_t \mid s_t) - \min_{i=1,2} Q_{\phi_i}(s_t, a_t)\big],$$

which effectively encourages actions with high Q-values and high entropy.

Automatic temperature tuning.
Often, SAC adjusts the temperature parameter α automatically by optimizing a secondary objective that aims to match a target entropy level:

$$\mathcal{L}(\alpha) = \mathbb{E}_{a_t \sim \pi_\theta}\big[-\alpha\big(\log \pi_\theta(a_t \mid s_t) + \mathcal{H}_{\text{target}}\big)\big],$$

so that α increases when entropy is lower than desired and decreases otherwise.

Why SAC is attractive for SciML.

- **Robust exploration**: Entropy regularization encourages policies that explore more broadly, which is important when the environment is complex and rewards are sparse or delayed.
- **Sample efficiency**: Off-policy replay allows reusing expensive simulation data multiple times.
- **Continuous control**: The stochastic Gaussian policy is well suited to continuous control inputs.
- **Stability**: Twin critics and entropy regularization help stabilize learning compared to simpler methods like DDPG.

SAC is often a strong default choice for difficult continuous control problems, such as controlling high-dimensional dynamical systems (e.g., flows, plasmas, or complex robots) that arise in Scientific Machine Learning.

Summary and Practical Guidance

Choosing an algorithm (informal guideline)

- **PPO**—good choice when:
  - on-policy data collection is feasible (you can run many simulations in parallel),
  - you want a stable, relatively easy-to-tune method,
  - you can tolerate some data inefficiency.
- **DDPG**—useful when:
  - you need deterministic continuous control,
  - you can invest effort in careful tuning and engineering,
  - environment is not too noisy or chaotic.
- **SAC**—often preferred when:
  - you want robust, stochastic continuous control,
  - sample efficiency matters (expensive simulations),
  - you face complex or multi-modal control landscapes.

In SciML applications, it is common to start with PPO for its robustness and simplicity, and then consider SAC when off-policy sample efficiency is needed or when exploration is difficult. DDPG and its variants can be valuable in more controlled settings or as baselines.

In all cases, the core ideas are the same as in the simpler methods: value functions, policy gradients, TD learning, and careful management of exploration and stability. These advanced algorithms simply wrap these ideas in more robust optimization schemes that are better suited to deep networks and complex scientific environments.

7.5 Extensions for Scientific Systems

The core algorithms presented in Sect. 1.4: from Q-learning to actor-critic methods; provide a powerful toolkit for learning control policies from data. However, scientific applications demand extensions that go beyond the standard RL framework. Physical systems have known structure (conservation laws, symmetries), expensive evaluation costs, safety-critical constraints, and distributed control requirements that are not addressed by basic algorithms.

This section presents five critical extensions that bridge the gap between textbook RL and real scientific deployment: model-based methods that leverage simulators, physics-informed architectures that encode domain knowledge, safe RL that guar-

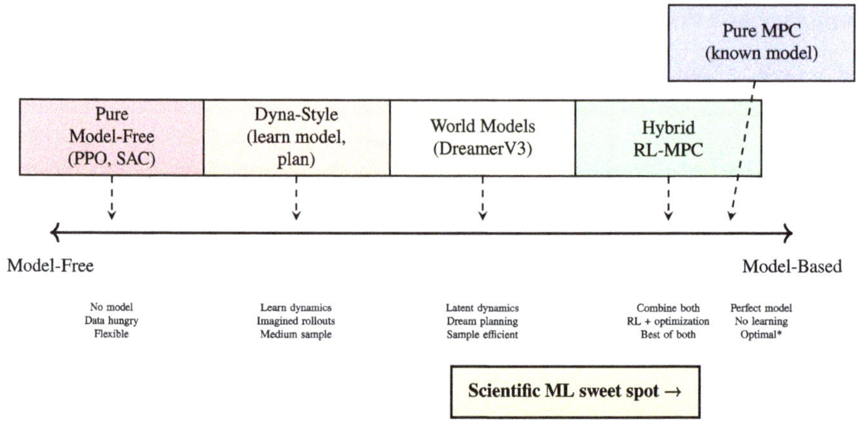

Fig. 7.5 The spectrum from model-free to model-based RL. Scientific applications often have approximate models (simulators) available, making hybrid approaches attractive. Pure model-free methods ignore available structure; pure MPC requires perfect models. The sweet spot combines learning with known physics

antees constraint satisfaction, multi-agent coordination for distributed systems, and transfer learning for adapting across experimental conditions. Each extension represents an active research frontier where the scientific computing and machine learning communities are converging.

7.5.1 Model-Based RL and RL–MPC Connections

In standard model-free RL, the agent learns purely from trial and error without ever building an explicit model of the environment dynamics. This is powerful; it requires no prior knowledge; but it is also sample-inefficient. In scientific applications where each simulation costs hours or each experiment costs thousands of dollars, we often have access to approximate models through decades of scientific work. Model-based RL exploits these models.

7.5.1.1 The Spectrum from Model-Free to Model-Based

Figure 7.5 illustrates the spectrum of approaches between pure model-free and pure model-based methods.

- **Pure model-free** (leftmost): Algorithms like PPO and SAC learn policies entirely from interaction data. They make no assumptions about dynamics and can handle arbitrary nonlinearities, but require millions of samples.

- **Dyna-style methods** (center-left): Learn an approximate model of the environment from data, then use that model to generate synthetic experience for training. The agent alternates between real experience (updating the model) and imagined experience (planning with the model).
- **World models** (center): Modern approaches like DreamerV3 learn a compact latent representation of the environment and its dynamics. The agent then learns policies entirely in this learned latent space through "dreaming"; imagining trajectories without real interaction.
- **Hybrid RL-MPC** (center-right): Use model predictive control (MPC) for short-term planning with known approximate models, and RL to learn what cannot be modeled accurately; residual dynamics, terminal costs, or model corrections.
- **Pure MPC** (rightmost): If the model is perfect, solve the optimal control problem directly via MPC. No learning is needed, but this is rarely the case in complex scientific systems.

For scientific applications, the sweet spot is often in the hybrid region: we have physics-based models (Navier-Stokes for fluids, MHD for plasmas, molecular dynamics for materials) that capture most of the dynamics, but with uncertainties, approximations, and unmodeled effects. Combining these imperfect models with learning yields the best of both worlds.

7.5.1.2 Mathematical Framework: RL with Learned or Known Models

In model-based RL, we maintain an explicit model $\hat{P}(s_{t+1}|s_t, a_t)$ of the environment dynamics. This model can be:

- **Learned from data**: a neural network trained to predict s_{t+1} given (s_t, a_t)
- **Physics-based**: a numerical solver for governing PDEs
- **Hybrid**: a physics model plus a learned correction term.

Given a model, we can improve policies through *planning*; imagining future trajectories and optimizing actions before executing them in the real environment.

Model-Based Policy Improvement

Input: Current policy π_θ, environment model $\hat{P}$, real environment

Repeat:

1. **Interact**: Collect trajectories from real environment using π_θ
2. **Update model**: Improve $\hat{P}$ to minimize prediction error on real data
3. **Plan**: Use $\hat{P}$ to generate synthetic trajectories (imagined rollouts)
4. **Learn**: Train policy π_θ on both real and synthetic data.

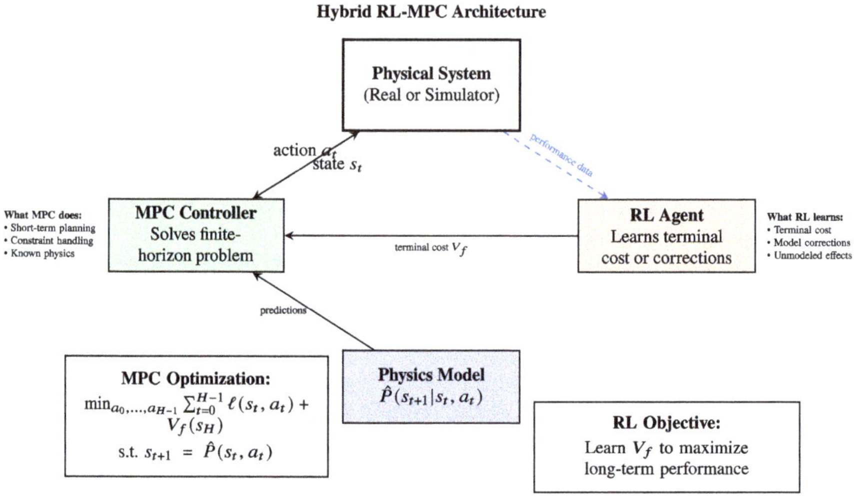

Fig. 7.6 Hybrid RL-MPC architecture for scientific control. MPC handles short-term optimization using physics models and constraints. RL learns components that are difficult to model: terminal costs for long-horizon objectives, residual dynamics corrections, or adaptation to changing conditions. This division of labor exploits the strengths of both approaches

The key advantage is sample efficiency: for every real environment step, we can generate many synthetic steps from the model, dramatically reducing the number of expensive real interactions needed.

7.5.1.3 Hybrid RL-MPC: Combining Learning and Optimization

One of the most promising approaches for scientific control is to combine RL with Model Predictive Control (MPC). Figure 7.6 illustrates this hybrid architecture [7].

Model Predictive Control works by repeatedly solving a finite-horizon optimal control problem:

$$\min_{a_0,\ldots,a_{H-1}} \sum_{t=0}^{H-1} \ell(s_t, a_t) + V_f(s_H), \quad \text{subject to } s_{t+1} = \hat{P}(s_t, a_t), \quad (s_t, a_t) \in \mathcal{C}, \tag{7.4}$$

where H is the prediction horizon, ℓ is a stage cost, V_f is a terminal cost, and $\mathcal{C}$ represents constraints. At each time step, MPC:

1. Solves the optimization problem from the current state
2. Applies only the first action a_0^*
3. Observes the next state and re-solves (receding horizon).

The challenge is choosing the terminal cost $V_f(s_H)$, which represents the "cost to go" after the horizon ends. Classical MPC often uses quadratic terminal costs or solves an infinite-horizon LQR. But for complex nonlinear systems, these choices are suboptimal.

RL solves this: Train an RL agent to estimate the true long-term value function $V^{\pi}(s)$ through interaction. Then use this learned value function as the terminal cost in MPC:

$$V_f(s_H) \approx V_{\theta}^{\pi}(s_H), \tag{7.5}$$

where V_{θ} is a learned critic network.

This hybrid approach combines:

- MPC's strengths: constraint satisfaction, short-term optimality, interpretability
- RL's strengths: handling unknown long-term effects, adapting to data, learning from experience.

Example 7.1 *(Hybrid RL-MPC for Plasma Control)* In tokamak control, MPC can enforce real-time constraints (plasma position, coil current limits) and optimize over short horizons (10–100 ms) using magnetohydrodynamic models. However, MHD models cannot perfectly predict long-term plasma evolution (minutes) due to turbulent transport, wall interactions, and impurities.

An RL agent learns a terminal cost V_f that captures long-term confinement quality from historical discharge data. During operation:

1. MPC optimizes coil currents over 100ms using MHD predictions
2. The terminal cost V_f (from RL) guides MPC toward states with good long-term prospects
3. RL continues learning from each discharge, improving V_f over time.

This achieves constraint satisfaction (MPC), long-term performance (RL), and sample efficiency (leveraging physics models).

7.5.1.4 Recent Advances: DreamerV3 and Learned World Models

The state-of-the-art in model-based RL is DreamerV3 (details in the World Model chapter), which learns a compact latent representation of the world and plans entirely in this learned space [8].

Key innovation: Instead of predicting high-dimensional states s_t directly (e.g., full velocity fields), DreamerV3 learns:

- An encoder $h_t = \text{enc}(s_t)$ mapping states to low-dimensional latent vectors
- A recurrent state-space model (RSSM) predicting latent dynamics: $h_{t+1} = f(h_t, a_t)$
- A decoder reconstructing observations: $\hat{s}_t = \text{dec}(h_t)$
- A reward predictor: $\hat{r}_t = r(h_t, a_t)$.

The agent then learns policies entirely by “dreaming”; imagining trajectories in latent space using the learned dynamics model, without real environment interaction.

For scientific applications, this approach is promising when:

- States are high-dimensional (fluid fields, molecular configurations)
- Dynamics have low-dimensional attractors (the “effective” degrees of freedom are much smaller than the full state space)
- Repeated simulations are expensive, so amortizing the cost of learning a model pays off

However, challenges remain: ensuring the learned latent dynamics respect conservation laws, quantifying model uncertainty, and achieving sim-to-real transfer when the world model is learned purely from simulation.

7.5.2 Physics-Informed Reinforcement Learning

The most direct way to incorporate scientific knowledge into RL is through physics-informed architectures and learning procedures. This emerging subfield; physics-informed RL (PIRL); recognizes that scientific systems are not black boxes but obey known laws.

7.5.2.1 Taxonomy of Physics Incorporation

Figure 7.7 presents a comprehensive taxonomy of how physics can be incorporated into the RL pipeline, following the framework of Banerjee et al. [9].

Observational Bias (physics in state representation):

- Use physical coordinates: For mechanical systems, use generalized coordinates (q, p) instead of Cartesian positions
- Feature engineering: Extract physically meaningful features like Fourier modes, principal components, or symmetry-adapted coordinates
- Dimensionality reduction: Use proper orthogonal decomposition (POD), dynamic mode decomposition (DMD), or autoencoders to find low-dimensional representations of high-dimensional PDE solutions

Inductive Bias (physics in architecture):

- Equivariant networks: Use architectures that respect known symmetries (rotation, translation, permutation)
- Hamiltonian/Lagrangian networks: Parameterize energy functions and derive dynamics through Hamilton’s equations, guaranteeing energy conservation
- Graph neural networks: For systems with discrete structure (molecules, meshes, particle systems), use GNNs that respect connectivity

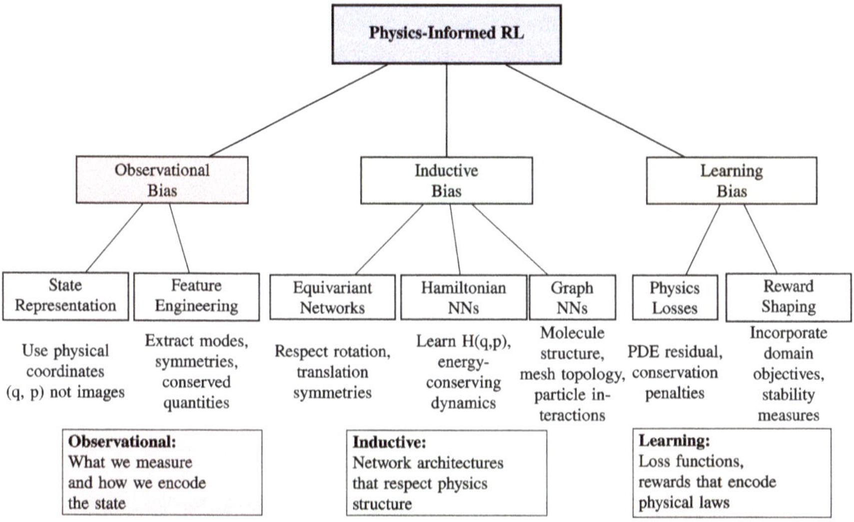

Fig. 7.7 Taxonomy of physics incorporation in RL. Physics can enter through state representation (observational bias), network architecture (inductive bias), or loss functions and rewards (learning bias). The most effective approaches combine all three

- Neural operators: For PDE control, use Fourier Neural Operators or DeepONet that learn operators rather than point-wise functions

Learning Bias (physics in loss/reward):

- Physics-informed losses: Add PDE residuals or conservation law violations as penalty terms
- Reward shaping: Design rewards that encourage physically meaningful behavior (e.g., penalize unrealistic oscillations, reward smooth control)
- Constraint enforcement: Use Lagrangian methods or barrier functions to enforce hard physical constraints

7.5.2.2 Hamiltonian Neural Networks for Energy-Conserving Policies

One of the most elegant physics-informed architectures is the Hamiltonian Neural Network (HNN), introduced by Greydanus et al. [10].

For a mechanical system with generalized coordinates q (positions) and momenta p, Hamilton's equations are:

$$\dot{q} = \frac{\partial H}{\partial p}, \quad \dot{p} = -\frac{\partial H}{\partial q}, \tag{7.6}$$

where $H(q, p)$ is the Hamiltonian (total energy). These equations have a remarkable property: *they conserve energy exactly*. For any trajectory, $\frac{dH}{dt} = 0$.

HNN idea: Instead of learning the dynamics $(\dot{q}, \dot{p})$ directly with a neural network, learn the Hamiltonian $H_\theta(q, p)$ and derive dynamics through Hamilton's equations. This guarantees energy conservation by construction.

Hamiltonian Neural Network Architecture

Input: State (q, p) at time t

Network: $H_\theta(q, p)$; a neural network approximating the Hamiltonian

Dynamics (via automatic differentiation):

$$\dot{q} = \frac{\partial H_\theta}{\partial p}, \quad \dot{p} = -\frac{\partial H_\theta}{\partial q} \tag{7.7}$$

Training: Minimize prediction error on observed trajectories:

$$\mathcal{L} = \sum_i \left\| (\dot{q}_i, \dot{p}_i)_{\text{observed}} - \left(\frac{\partial H_\theta}{\partial p}, -\frac{\partial H_\theta}{\partial q} \right) \right\|^2 \tag{7.8}$$

Result: Learned dynamics conserve H_θ exactly, even if H_θ is inaccurate

For RL, HNNs can be used in multiple ways:

- **As world models**: Learn H_θ from data, then plan with energy-conserving dynamics
- **As policy networks**: Parameterize control policies that respect energy structure
- **As regularizers**: Add energy conservation violations as penalties during training.

7.5.2.3 Differentiable Physics and Analytical Policy Gradients

Perhaps the most powerful tool in physics-informed RL is *differentiable physics*; simulators that provide exact gradients $\frac{\partial s_{t+1}}{\partial a_t}$ through the physics equations.

Modern frameworks like Brax (Google), JAX-MD, and TORAX implement physics solvers in JAX, enabling automatic differentiation through entire simulation rollouts. This allows *analytical policy gradients* (APG):

$$\nabla_\theta J(\pi_\theta) = \mathbb{E}\left[\sum_{t=0}^{T} \frac{\partial r_t}{\partial a_t} \frac{\partial a_t}{\partial \theta} + \frac{\partial r_t}{\partial s_t} \frac{\partial s_t}{\partial a_t} \frac{\partial a_t}{\partial \theta} \right], \tag{7.9}$$

where all derivatives are computed exactly via backpropagation through the simulator.

Advantages over finite-difference or Monte Carlo policy gradients:

- 100–1000× faster convergence (fewer environment evaluations needed)
- Lower variance gradients (no sampling noise)
- Exact optimization (no approximation error from MC estimates).

Challenges:

- Not all physics codes are differentiable (legacy Fortran solvers, adaptive meshes, implicit methods)
- Memory cost: storing computational graph for long rollouts
- Sim-to-real gap: gradients are exact for the simulator, which may differ from reality.

The TORAX tokamak simulator (DeepMind, 2024) implements reduced magnetohydrodynamic equations in JAX. For plasma shape control:

- Forward pass: Simulate 1 s of plasma evolution given coil currents a_t
- Backward pass: Compute $\frac{\partial J}{\partial a_t}$ where J measures deviation from target shape
- Policy update: Adjust policy parameters θ using exact gradients.

This achieves convergence in 100–1000 episodes versus 10,000–100,000 episodes for black-box RL, making RL practical for expensive fusion simulations.

7.5.3 *Safe RL: Constrained Policy Optimization and Control Barrier Functions*

Safety is not optional in scientific systems. Plasma must not touch reactor walls, aircraft must not stall, chemical reactions must not exceed pressure limits. Unlike games where failure is merely undesirable, physical systems can be damaged or create hazards. Safe RL provides mathematical guarantees on constraint satisfaction during learning.

7.5.3.1 Constrained Markov Decision Processes (Recap and Extension)

Recall the CMDP formulation from Sect. 1.3.4:

$$\max_{\pi} J_R(\pi) = \mathbb{E}_\pi \left[\sum_{t=0}^{\infty} \gamma^t r_t \right], \quad \text{subject to } J_{C_i}(\pi) = \mathbb{E}_\pi \left[\sum_{t=0}^{\infty} \gamma^t c_i(s_t, a_t) \right] \le d_i, \quad \forall i \quad (7.10)$$

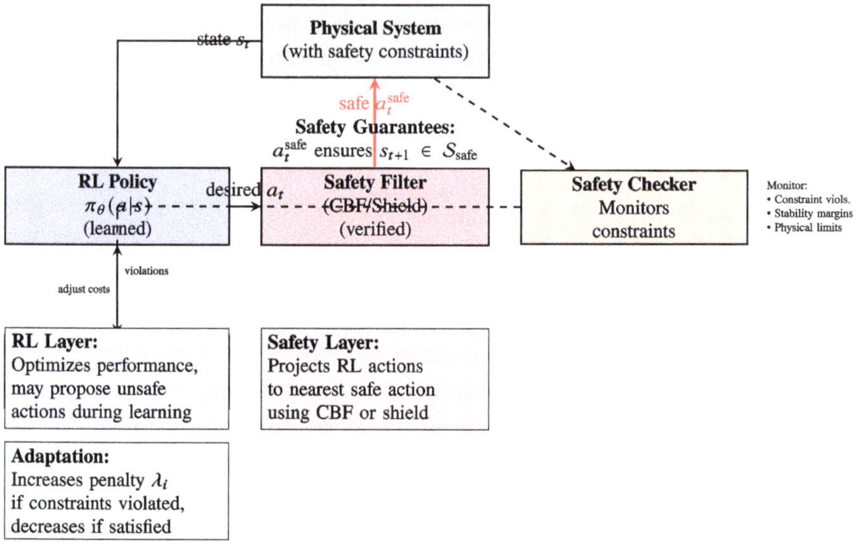

Fig. 7.8 Hierarchical safe RL architecture. The RL policy learns performance optimization but may propose unsafe actions. A safety filter (using CBFs or shielding) projects these to safe actions. A safety checker monitors constraint satisfaction and adjusts Lagrangian penalties to guide learning toward safer regions

This framework handles *expected* constraint satisfaction: the long-term average cost must be below threshold. But for safety-critical systems, we often need *per-step* or *worst-case* guarantees.

7.5.3.2 Hierarchical Safe RL Architecture

Figure 7.8 illustrates a modern safe RL architecture combining multiple safety mechanisms.

The architecture has three layers:

1. **RL Policy Layer** (blue): Learns to optimize performance (reward) using standard RL algorithms (PPO, SAC). This layer may explore unsafe actions during training.
2. **Safety Filter Layer** (red): Takes the RL policy's proposed action a_t and projects it to the nearest safe action a_t^{safe} that guarantees constraint satisfaction. This is where Control Barrier Functions (CBFs) or formal shielding come in.
3. **Monitoring and Adaptation Layer** (green): Tracks constraint violations and adjusts Lagrangian multipliers λ_i to penalize costs in the RL objective, guiding learning toward safer regions of policy space.

7.5.3.3 Control Barrier Functions: Mathematical Foundations

Control Barrier Functions provide a way to compute the safety filter in real-time with formal guarantees.

Let $\mathcal{S}_{\text{safe}} \subset \mathcal{S}$ be the set of safe states. A function $h : \mathcal{S} \to \mathbb{R}$ is a **Control Barrier Function** (CBF) if:

$$\mathcal{S}_{\text{safe}} = \{s : h(s) \geq 0\} \tag{7.11}$$

and for all $s \in \mathcal{S}_{\text{safe}}$, there exists an action $a \in \mathcal{A}$ such that:

$$\dot{h}(s, a) = \frac{\partial h}{\partial s} f(s, a) \geq -\alpha(h(s)) \tag{7.12}$$

where $f(s, a)$ are the system dynamics ($\dot{s} = f(s, a)$) and $\alpha : \mathbb{R}_+ \to \mathbb{R}_+$ is a class-$\mathcal{K}$ function (e.g., $\alpha(h) = \kappa h$ for $\kappa > 0$).

Intuition: The CBF h defines a "barrier" around unsafe regions. The condition $\dot{h} \geq -\alpha(h)$ ensures that if we're in a safe state ($h(s) \geq 0$), we can choose actions that keep us safe (h doesn't decrease too fast).

Safety filter via quadratic program: Given a desired action a_{desired} from the RL policy, compute the safe action as:

$$a_{\text{safe}} = \arg\min_{a \in \mathcal{A}} \|a - a_{\text{desired}}\|^2 \quad \text{subject to} \quad \frac{\partial h}{\partial s} f(s, a) \geq -\alpha(h(s)) \tag{7.13}$$

This is a convex quadratic program (if f is affine in a) that can be solved in milliseconds, enabling real-time safety filtering.

CBF Safety Guarantee

If h is a valid CBF and actions are chosen to satisfy $\dot{h} \geq -\alpha(h)$, then:

1. If we start in a safe state ($h(s_0) \geq 0$), we remain safe for all time
2. The set $\{s : h(s) \geq 0\}$ is *forward invariant*; trajectories cannot escape it

This provides a formal safety certificate, unlike soft constraint methods that only bound expected violations.

Example 7.2 (*CBF for Temperature Constraint in Heat Equation Control*) Consider controlling the heat equation with a constraint that temperature must not exceed $T_{\max}$:

$$\frac{\partial u}{\partial t} = \alpha \frac{\partial^2 u}{\partial x^2}, \quad u(x, 0) = u_0(x), \quad \text{with control at } x = 0 \tag{7.14}$$

Define the barrier function:

$$h(u) = T_{\max} - \max_{x \in [0,1]} u(x) \tag{7.15}$$

Safety requires $h(u) \geq 0$, i.e., $\max_x u(x) \leq T_{\max}$.

The time derivative $\dot{h}$ depends on how the maximum temperature point evolves. If the maximum is at the boundary ($x = 0$ where we control), then:

$$\dot{h} \approx -\alpha \frac{\partial u}{\partial x}\bigg|_{x=0} - \frac{\partial u}{\partial t}\bigg|_{x=0} \tag{7.16}$$

The CBF constraint requires:

$$\dot{h} \geq -\kappa h \quad \Rightarrow \quad \text{bound on heating rate at boundary} \tag{7.17}$$

This translates to an inequality constraint on the control input (boundary heat flux), which can be enforced via the QP filter.

7.5.3.4 Learning CBFs from Data

Classical CBF design requires manually specifying $h(s)$, which is difficult for complex systems. Recent work learns CBFs from data:

Conservative CBF Learning: Train a neural network $h_\theta(s)$ to satisfy:

- $h_\theta(s) > 0$ for all safe states in dataset
- $h_\theta(s) < 0$ for all unsafe states
- $\dot{h}_\theta \geq -\alpha(h_\theta)$ for transitions in dataset.

Use conservatism (inspired by Conservative Q-Learning): push down h_θ at out-of-distribution states to avoid false safety guarantees.

Integration with RL: Train the CBF and policy jointly:

1. RL proposes action a_t
2. CBF filter projects to safe action
3. Collect data $(s_t, a_t^{\text{safe}}, s_{t+1})$
4. Update both h_θ (improve safety boundary) and π_θ (improve performance).

This closes the loop: RL learns what is optimal given safety constraints, and CBF learning refines the constraints based on observed safe/unsafe regions.

7.5.4 Distributed and Multi-Agent RL for PDEs

Many scientific systems involve distributed control: multiple actuators placed at different spatial locations, each making local decisions based on local information. Examples include:

- Flow control with arrays of jets or synthetic actuators
- Plasma control with multiple coil systems
- Smart grid control with distributed energy resources
- Climate intervention with regional geoengineering.

These problems are naturally multi-agent: each actuator is an agent that must coordinate with others to achieve global objectives.

7.5.4.1 Multi-Agent RL Formulation

In multi-agent RL (MARL), we have N agents, each with:

- Local observation o_t^i (what agent i can sense)
- Local action a_t^i (what agent i controls)
- Local or global reward r_t^i (depends on formulation).

The joint state is $s_t = (o_t^1, \ldots, o_t^N)$ and joint action is $\mathbf{a}_t = (a_t^1, \ldots, a_t^N)$.

Cooperative MARL: All agents share a common global objective:

$$\max_{\pi^1,\ldots,\pi^N} J = \mathbb{E}\left[\sum_{t=0}^{\infty} \gamma^t r_t(\mathbf{s}_t, \mathbf{a}_t)\right] \tag{7.18}$$

Centralized Training, Decentralized Execution (CTDE):

- During training: Use global information to train policies (centralized critic knows full state)
- During deployment: Each agent acts based only on local observations (decentralized policies).

This is the dominant paradigm in cooperative MARL and is particularly suitable for scientific applications where we can simulate the full system for training but need decentralized control for deployment.

7.5.4.2 Graph Neural Networks for Spatially Distributed Systems

For PDE control with spatial actuator arrays, the natural architecture is a Graph Neural Network (GNN) where:

- Nodes represent actuators or spatial cells
- Edges represent physical coupling (e.g., diffusion, advection between cells)
- Node features include local state measurements
- Edge features include coupling strengths.

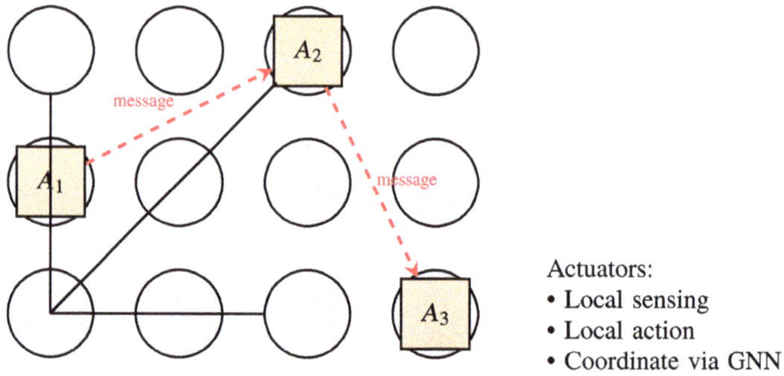

Fig. 7.9 Multi-agent RL for spatially distributed PDE control. Each actuator (orange) senses local state, controls local actions, and communicates with neighbors (red dashed lines). A GNN policy processes spatial structure to coordinate actions

The GNN policy computes:

$$a_t^i = \pi_\theta(o_t^i, \{o_t^j : j \in \mathcal{N}(i)\}) \tag{7.19}$$

where $\mathcal{N}(i)$ are the neighbors of agent i in the spatial graph.

Advantages for PDE control:

- Respects spatial locality: control at position x depends on nearby measurements
- Scalable: policy parameters θ are independent of number of actuators N
- Transferable: a policy trained on small domain can generalize to larger domains with more actuators (Fig. 7.9).

Example 7.3 *((Multi-agent Flow Control))* Consider drag reduction in a 3D channel flow using an array of wall-mounted synthetic jets. Each jet i:

- Observes local wall shear stress τ_w^i and pressure p^i
- Controls blowing/suction velocity v_w^i
- Wants to minimize global drag while coordinating with neighbors to avoid interference.

A GNN policy:

1. Each jet encodes local measurements as node features
2. Message passing: jets exchange information via graph convolution (representing flow coupling)
3. Each jet computes $v_w^i = \pi_\theta(o_i, \text{messages from neighbors})$.

Trained with cooperative MARL (shared global drag reward), the jets learn to coordinate actuation patterns; leading edges suppress turbulence generation, and trailing edges manipulate boundary-layer recovery.

This achieves a 21% drag reduction and scales to arbitrary numbers of jets without retraining.

7.5.5 *Continual and Transfer RL in Dynamic Environments*

Scientific experiments and simulations rarely operate in static conditions. Reactors age, materials degrade, environmental conditions change, experimental setups are modified. Continual learning and transfer learning address the challenge of adapting learned policies to new conditions without starting from scratch.

7.5.5.1 The Continual Learning Challenge

Catastrophic forgetting: Neural networks trained on sequence of tasks $T_1, T_2, \ldots$ tend to forget T_1 after training on T_2. For scientific RL, this means a policy trained for one set of conditions (e.g., high Reynolds number flow) forgets how to handle another set (low Reynolds number) after adaptation.

Solutions:

- **Experience replay**: Maintain buffer of past experiences and periodically retrain on mixture of old and new data
- **Elastic weight consolidation**: Identify important parameters for old tasks and penalize their change during new task learning
- **Progressive neural networks**: Add new network capacity for each task, preserving old networks
- **Meta-learning**: Train on distribution of tasks to learn fast adaptation strategies.

7.5.5.2 Transfer Learning: Sim-to-Real and Domain Adaptation

Most RL training occurs in simulation due to cost and safety. The challenge is transferring learned policies to real systems; the notorious *sim-to-real gap*.

Sources of sim-to-real gap:

- Unmodeled physics (turbulence scales below grid resolution, wall roughness, impurities)
- Numerical approximations (discretization error, closure models)
- Sensor noise and delays not present in perfect simulation
- Manufacturing variations in real hardware.

Transfer strategies:

1. **Domain randomization**: Train policies in simulation with randomized parameters (vary Reynolds number, boundary conditions, initial conditions). Policies learn robustness to variations, improving real-world transfer.

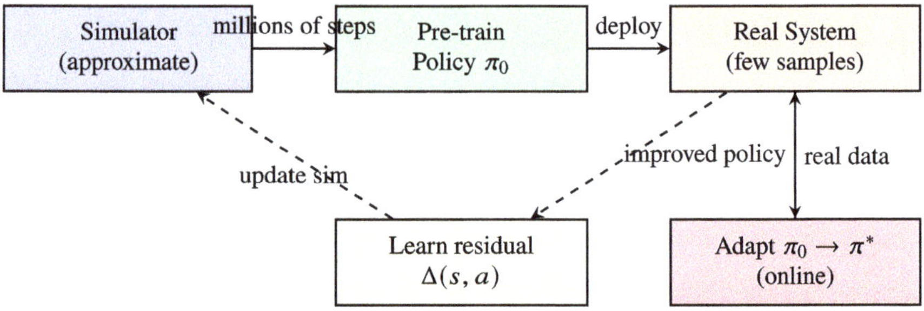

Fig. 7.10 Transfer learning workflow for sim-to-real. Pre-train policy in simulator (millions of cheap samples), deploy to real system (few expensive samples), adapt online, and optionally learn residual corrections to improve simulator for future training

2. **System identification + adaptation**: Learn residual dynamics $\Delta_\theta(s, a)$ representing the difference between simulation and reality:

$$s_{t+1}^{\text{real}} = P_{\text{sim}}(s_t, a_t) + \Delta_\theta(s_t, a_t) \tag{7.20}$$

   Train Δ_θ on small amount of real data, then use corrected simulator for RL.
3. **Real-to-sim-to-real (RialTo, 2024)**: Collect initial real data, construct improved simulator matching real statistics, train policy in improved sim, deploy to real system, iterate.
4. **Safe online adaptation**: Deploy policy learned in simulation with safety filter (CBF), collect real data, fine-tune policy online while maintaining safety guarantees (Fig. 7.10).

DeepMind's tokamak controller trained policies purely in simulation (RAPTOR code) with:

- Domain randomization: varying plasma density, temperature, impurity content
- Robust policy architectures: using ensemble of critics to account for model uncertainty
- Careful reward design: penalizing behaviors known to be simulation artifacts.

On deployment to the TCV tokamak:

- Zero-shot transfer: policy worked immediately on first pulse
- Online fine-tuning: performance improved 15% after 10 real discharges
- Achieved configurations (e.g., "droplet" dual-plasma) never seen in simulation.

The key was conservative deployment: policies were required to maintain safety margins significantly tighter than needed, compensating for sim-to-real gap.

Table 7.1 Summary of RL extensions for scientific systems and when to apply them

Extension	Use when	Key benefit
Model-based RL/hybrid RL-MPC	Physics models available; sample efficiency critical; constraints must be satisfied	10–100× fewer real environment steps; formal constraint handling
Physics-Informed RL	Known conservation laws, symmetries, or governing equations	Faster learning, better generalization, interpretable policies
Safe RL/CBFs	Hard safety constraints exist; failure is unacceptable	Formal safety guarantees; safe exploration during learning
Multi-agent RL	Spatially distributed control; multiple actuators; local sensing/action	Scalability to large systems; communication efficiency
Transfer/continual RL	Conditions change over time; expensive to retrain from scratch; sim-to-real gap exists	Adaptation to new conditions; leveraging simulation for real deployment

7.5.6 *Summary: When to Use Which Extension*

The extensions in this section are not mutually exclusive; the most sophisticated scientific RL systems combine multiple approaches. Table 7.1 provides guidance on when each extension is most valuable.

In practice, state-of-the-art scientific RL combines:

- Physics-informed architectures (HNNs, GNNs) for sample efficiency
- Hybrid RL-MPC for leveraging existing models and constraints
- Safe RL (CBFs) for guaranteed safety during learning and deployment
- Multi-agent formulations for distributed large-scale systems
- Transfer learning for sim-to-real and continual adaptation.

7.6 Deep Reinforcement Learning

The algorithms presented in Sects. 1.3 and 1.4 assume that we can represent and store value functions or policies explicitly; either as tables (for small discrete state spaces) or as simple parametric functions (for low-dimensional continuous spaces). This assumption breaks down catastrophically for scientific applications. Consider:

- A turbulent flow field discretized on a 256^3 grid has $\sim$ 16 million velocity components
- A tokamak plasma state (magnetic field, density, temperature profiles) spans thousands of spatial points

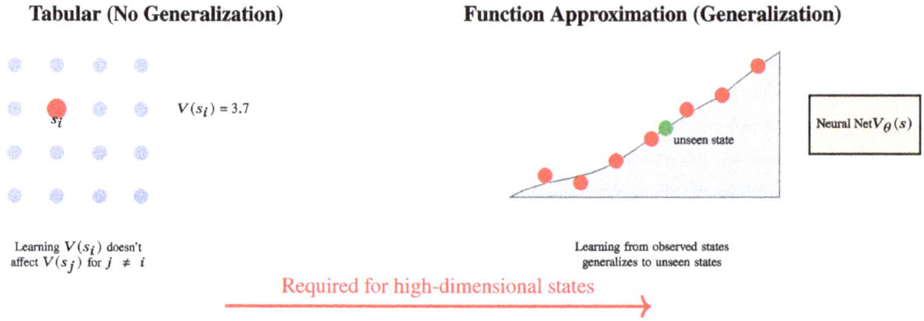

Fig. 7.11 Tabular versus function approximation. In tabular methods, each state has an independent value estimate; there is no generalization. With function approximation (neural networks), learning from observed states automatically generalizes to similar unseen states, enabling RL in high-dimensional spaces

- A molecular configuration in drug design involves hundreds of atoms with 3D positions
- A climate model state includes temperature, pressure, humidity fields over the entire globe.

These state spaces are far too large for tabular methods. Even with clever state aggregation, the *curse of dimensionality* ensures that the number of states grows exponentially with dimension. Deep reinforcement learning addresses this through *function approximation*: we represent value functions, Q-functions, and policies as neural networks with shared parameters, enabling generalization across similar states.

This section covers the foundations of deep RL; how to train neural networks to approximate RL quantities, the stability challenges that arise, and the engineering solutions developed to make deep RL reliable. We emphasize connections to scientific computing: neural operators for PDE control, equivariant architectures for physical symmetries, and computational considerations for large-scale training.

7.6.1 Function Approximation with Neural Networks

7.6.1.1 The Generalization Principle

The core insight of function approximation is that we don't need to learn a separate value for every possible state. States that are "similar" should have similar values, and neural networks excel at learning this similarity structure.

Figure 7.11 contrasts these approaches:

Tabular methods: Maintain a lookup table $V(s_1), V(s_2), \ldots, V(s_N)$ with independent entries. Learning $V(s_i)$ from experience has zero effect on $V(s_j)$ for $j \neq i$. This requires visiting every state sufficiently often, which is impossible when N is astronomical.

Function approximation: Represent the value function as $V_\theta(s)$ where θ are shared parameters (neural network weights). Learning from state s_i updates θ, which affects the estimated value at all states; particularly those similar to s_i. This enables generalization: we can estimate values at unvisited states by interpolating from nearby visited states.

7.6.1.2 Neural Network Architectures for RL

Different RL quantities require different network architectures:

Neural Network Function Approximators in Deep RL

Value Function $V_\phi(s)$:

- Input: State s
- Output: Scalar value $V_\phi(s) \in \mathbb{R}$
- Training: Minimize TD error $\delta = (r + \gamma V_\phi(s') - V_\phi(s))^2$

Action-Value Function $Q_\theta(s, a)$:

- Input: State s and action a
- Output: Scalar Q-value $Q_\theta(s, a) \in \mathbb{R}$
- Alternative (discrete actions): Input state s, output vector $[Q_\theta(s, a_1), \ldots, Q_\theta(s, a_n)]$
- Training: Minimize $(r + \gamma \max_{a'} Q_\theta(s', a') - Q_\theta(s, a))^2$.

Stochastic Policy $\pi_\psi(a|s)$:

- Input: State s
- Output: Probability distribution over actions $\pi_\psi(\cdot|s)$
- Discrete actions: Softmax over logits
- Continuous actions: Mean and variance of Gaussian $\mathcal{N}(\mu_\psi(s), \sigma_\psi(s))$
- Training: Policy gradient $\nabla_\psi J(\pi_\psi) = \mathbb{E}[A(s, a)\nabla_\psi \log \pi_\psi(a|s)]$.

Deterministic Policy $\mu_\omega(s)$:

- Input: State s
- Output: Action $a = \mu_\omega(s)$
- Training: Deterministic policy gradient $\nabla_\omega J = \mathbb{E}[\nabla_a Q(s, a)|_{a=\mu_\omega(s)} \nabla_\omega \mu_\omega(s)]$.

Figure 7.12 shows common architectural patterns. Key design considerations:

For discrete actions: DQN-style networks that output $Q(s, a)$ for all actions simultaneously are more efficient than separate forward passes per action.

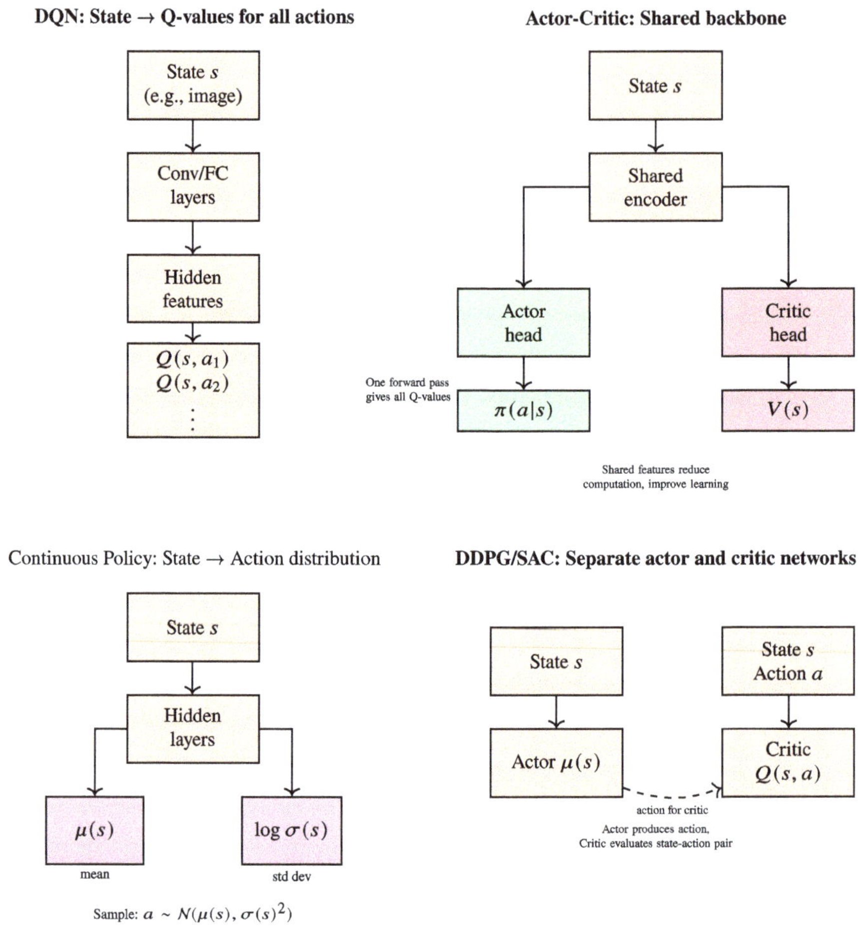

Fig. 7.12 Common neural network architectures in deep RL. Different algorithms use different architectural patterns: DQN outputs all Q-values in one pass (discrete actions), actor-critic shares a backbone, continuous policies output distribution parameters, and DDPG/SAC use separate networks for policy and value

For continuous actions: Policies output parameters of a distribution (e.g., mean and standard deviation of a Gaussian), and actions are sampled during exploration or computed deterministically during exploitation.

Shared versus separate networks: Actor-critic methods can share early layers (extracting common features from states) while having separate output heads for policy and value. This reduces computation and can improve learning by sharing representations.

7.6.1.3 Neural Architectures for Scientific Applications

Scientific RL brings domain-specific architectural requirements beyond standard feedforward networks:

Convolutional Neural Networks (CNNs) for spatially structured states:

- When states are 2D/3D fields (fluid velocity, plasma temperature, molecular density)
- CNNs respect translation equivariance: shifting the input shifts the output correspondingly
- Dramatically reduce parameters: a CNN with local kernels has $O(k^2 \cdot d)$ parameters for kernel size k and depth d, vs. $O(n^2)$ for a fully connected network on n spatial points.

Graph Neural Networks (GNNs) for discrete structure:

- For molecular systems (atoms connected by bonds), particle systems, or mesh-based simulations
- Permutation equivariant: swapping atom indices doesn't change predictions
- Message passing: nodes aggregate information from neighbors, respecting physical connectivity

Recurrent Neural Networks (RNNs/LSTMs) for partial observability:

- When full state is hidden and must be inferred from observation history
- LSTM maintains hidden state h_t encoding observation history: $h_t = f(h_{t-1}, o_t)$
- Common in robotics where sensors don't observe full system state.

Transformer architectures for sequential decision-making:

- Attention mechanisms can model long-range dependencies in state sequences
- Decision Transformer (Chen et al., 2021) reframes RL as sequence prediction
- Emerging use in multi-step planning and trajectory optimization.

Neural Operators for PDE control:

- Fourier Neural Operators (FNO) learn mappings between function spaces
- For state $s(x)$ (a function), policy outputs control $a(x)$ (also a function)
- Resolution-invariant: trained on coarse grids, deploy on fine grids
- Example: FNO-based RL for particle accelerator control reduced inference time from 3 s to 4 ms.

Equivariant Neural Networks for physical symmetries:

- E(3)-equivariant networks respect 3D rotations and translations
- For molecular systems: rotating molecule rotates predicted forces/energies correspondingly
- Dramatically improves sample efficiency: network doesn't need to learn symmetries from data.

Example 7.4 (*CNN-based RL for Turbulent Flow Control*) For flow control with a 2D channel (256×64 velocity field):

Naive approach: Flatten state to vector of size $256 \times 64 \times 2 = 32{,}768$ (two velocity components). Fully connected network with 2 hidden layers of size 1024 would have ~ 35 million parameters.

CNN approach:

- Stack u and v velocity components as 2-channel image
- Apply 3–4 convolutional layers with kernels 3×3 or 5×5
- Use pooling to reduce spatial dimension
- Final fully connected layer outputs control actions
- Total: $\sim 100{,}000$ parameters ($350\times$ reduction).

CNN respects spatial structure: nearby points in the flow influence each other, distant points don't. This inductive bias dramatically accelerates learning.

7.6.2 Experience Replay and Stability Tricks

Training neural networks with RL is fundamentally more difficult than supervised learning. In supervised learning, we have a fixed dataset of input-output pairs and minimize prediction error. In RL:

- The "training data" (states, rewards) changes as the policy improves
- Successive experiences are highly correlated (breaking i.i.d. assumption)
- The target we're training toward (bootstrapped values) also changes
- Small errors can compound through the Bellman recursion.

This leads to the *deadly triad* of instability (Sutton & Barto):

1. Function approximation (neural networks)
2. Bootstrapping (TD learning, using estimates to update estimates)
3. Off-policy learning (learning about policy π while following different policy μ).

When all three are present, learning can diverge catastrophically. Deep RL methods employ several "tricks" to stabilize training (Fig. 7.13).

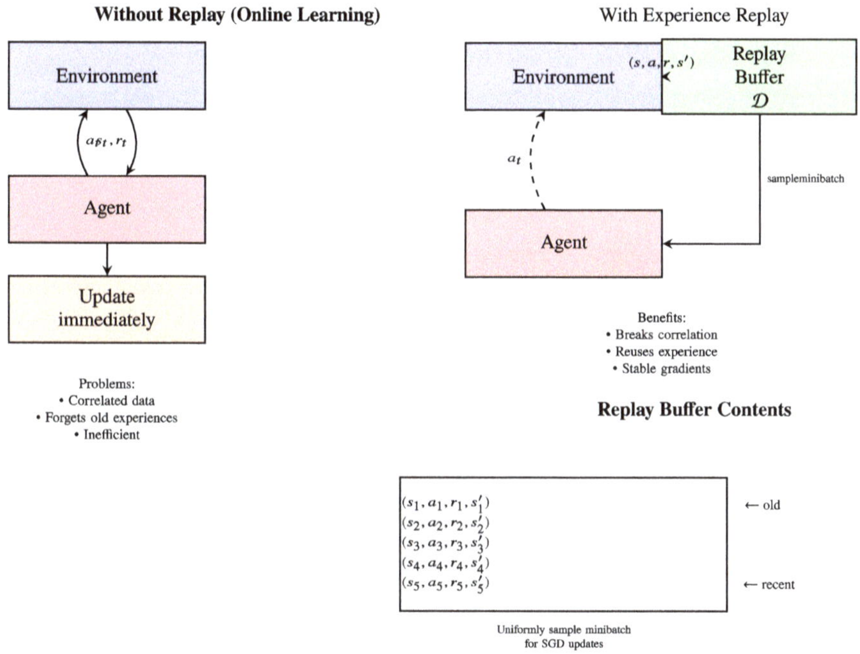

Fig. 7.13 Experience replay breaks temporal correlation and enables data reuse. Without replay, the agent learns from sequential correlated experiences, leading to instability. With replay, experiences are stored in a buffer and randomly sampled for training, creating an approximately i.i.d. dataset

7.6.2.1 Experience Replay

The first major stabilization technique is *experience replay* (Lin, 1992; Mnih et al., 2013).

Core idea: Instead of updating the network immediately after each experience (s_t, a_t, r_t, s_{t+1}), store it in a replay buffer $\mathcal{D}$. Periodically sample random minibatches from $\mathcal{D}$ and perform gradient descent on these batches.

Experience Replay Algorithm

Initialize replay buffer $\mathcal{D}$ with capacity N (e.g., $N = 10^6$).

For each episode:

1. Observe state s_t, select action a_t, execute, observe r_t, s_{t+1}
2. Store transition (s_t, a_t, r_t, s_{t+1}) in $\mathcal{D}$
3. If $|\mathcal{D}| >$ batch size:

- Sample random minibatch of M transitions from $\mathcal{D}$
- Perform gradient descent step on minibatch loss.

Benefits:

- **Breaks correlation**: Consecutive experiences (s_t, a_t, r_t, s_{t+1}) and $(s_{t+1}, a_{t+1}, r_{t+1}, s_{t+2})$ are highly correlated. Random sampling creates approximately independent batches.
- **Data efficiency**: Each experience can be replayed many times. In expensive scientific simulations, this amortizes the cost.
- **Variance reduction**: Batch gradient estimates have lower variance than single-sample estimates.

Limitations:

- Only applicable to off-policy algorithms (Q-learning, DDPG, SAC) because we're training on data generated by old policies
- Memory requirement: storing millions of transitions can be prohibitive for high-dimensional states (e.g., 3D flow fields)
- Can slow down adaptation to non-stationary environments if old data dominates the buffer.

7.6.2.2 Target Networks

The second major stabilization trick is *target networks* (Mnih et al., 2015).

Problem: In Q-learning with neural networks, we update Q_θ by minimizing:

$$\mathcal{L}(\theta) = \mathbb{E}_{(s,a,r,s')\sim\mathcal{D}}\left[\left(r + \gamma \max_{a'} Q_\theta(s', a') - Q_\theta(s, a)\right)^2\right] \tag{7.21}$$

Notice that θ appears in both the target ($\gamma \max_{a'} Q_\theta(s', a')$) and the prediction ($Q_\theta(s, a)$). When we update θ to reduce the TD error, we're chasing a moving target. This can cause oscillations and divergence.

Solution: Maintain two networks:

- **Online network** Q_θ: updated every step via gradient descent
- **Target network** Q_{θ^-}: updated infrequently (e.g., every C steps) by copying $\theta^- \leftarrow \theta$.

The loss becomes:

$$\mathcal{L}(\theta) = \mathbb{E}\left[\left(r + \gamma \max_{a'} Q_{\theta^-}(s', a') - Q_\theta(s, a)\right)^2\right] \tag{7.22}$$

Now the target is held fixed for C steps, allowing Q_θ to converge toward a stable target before the target changes.

Alternative (soft target updates): Instead of hard copies every C steps, use exponential moving average:

$$\theta^- \leftarrow \tau\theta + (1 - \tau)\theta^-, \quad \tau \ll 1 \text{ (e.g., } \tau = 0.005) \tag{7.23}$$

This smooths out target updates and often improves stability further.

7.6.2.3 Double Q-Learning and Overestimation Bias

A subtle issue in Q-learning is *overestimation bias*: the $\max_{a'} Q(s', a')$ operation systematically overestimates Q-values due to maximization noise.

Why this happens: Suppose true Q-values are $Q^*(s', a')$ but our estimates are $Q(s', a') = Q^*(s', a') + \epsilon_{a'}$ where $\epsilon_{a'}$ are zero-mean errors. Then:

$$\max_{a'} Q(s', a') = \max_{a'}[Q^*(s', a') + \epsilon_{a'}] \geq \max_{a'} Q^*(s', a') \tag{7.24}$$

The maximum over noisy estimates is biased upward. This overestimation propagates through the Bellman backup, compounding over time.

Double Q-learning solution (van Hasselt et al., 2016): Use two networks Q_{θ_1} and Q_{θ_2}:

- Use Q_{θ_1} to select the best action: $a^* = \arg\max_{a'} Q_{\theta_1}(s', a')$
- Use Q_{θ_2} to evaluate it: target $= r + \gamma Q_{\theta_2}(s', a^*)$

Since the errors $\epsilon_{a'}$ in Q_{θ_1} and Q_{θ_2} are independent, the bias cancels in expectation.

In practice, we use the online network for action selection and the target network for evaluation:

$$y = r + \gamma Q_{\theta^-}(s', \arg\max_{a'} Q_\theta(s', a')) \tag{7.25}$$

This is the standard approach in modern deep Q-learning (Double DQN).

7.6.2.4 Additional Stability Enhancements

Dueling Networks (Wang et al., 2016): Factor Q-function into state value and advantage:

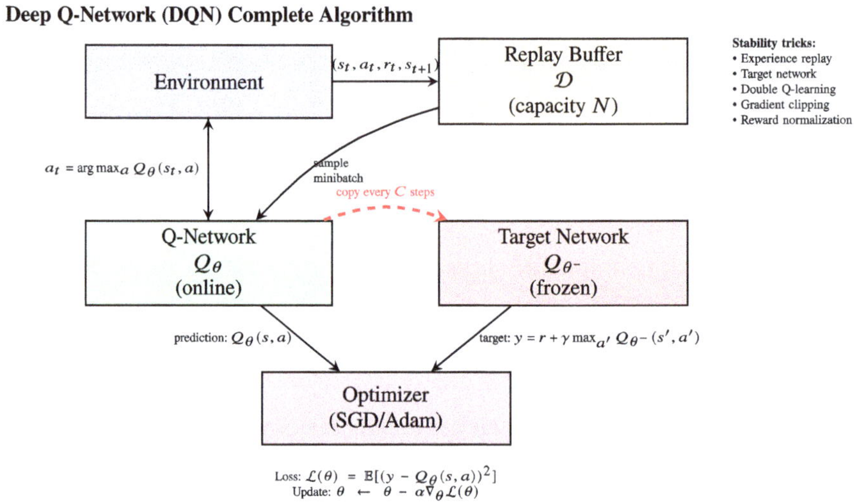

Fig. 7.14 Complete Deep Q-Network (DQN) architecture showing all stability mechanisms: experience replay buffer breaks correlation, target network stabilizes TD targets, optimizer performs gradient descent on sampled minibatches, and periodic target updates prevent chasing a moving target

$$Q_\theta(s, a) = V_\theta(s) + A_\theta(s, a) - \frac{1}{|\mathcal{A}|}\sum_{a'} A_\theta(s, a') \tag{7.26}$$

This architecture learns which states are valuable independent of the action choice, improving learning efficiency.

Prioritized Experience Replay (Schaul et al., 2016): Sample transitions from replay buffer with probability proportional to their TD error:

$$P(i) \propto |\delta_i|^\alpha \tag{7.27}$$

where $\delta_i = r_i + \gamma \max_{a'} Q(s_i', a') - Q(s_i, a_i)$. Transitions with larger TD errors are more "surprising" and replaying them more often improves learning.

Gradient clipping and normalization: Clip gradients to $[-\theta_{\max}, \theta_{\max}]$ to prevent explosive updates. Normalize rewards or value estimates to stabilize scales.

Batch normalization and layer normalization: Normalize activations to have zero mean and unit variance within or across batches, reducing internal covariate shift.

Figure 7.14 shows how these pieces fit together in the complete DQN algorithm. This architecture; with experience replay, target networks, and double Q-learning; forms the foundation of modern deep RL for value-based methods.

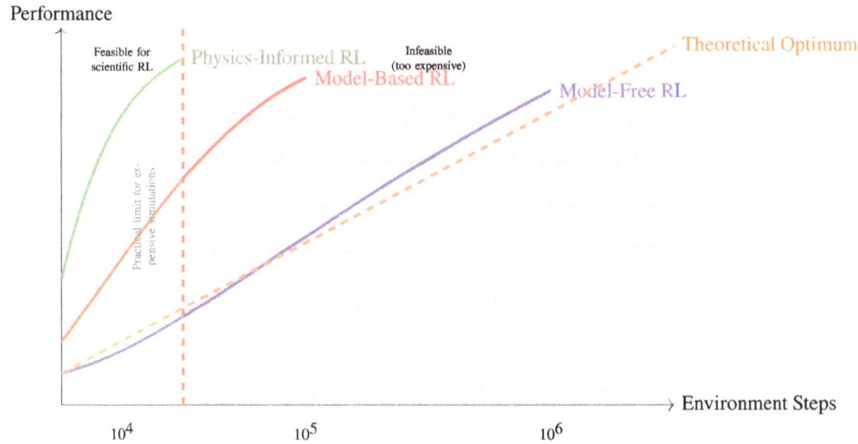

Fig. 7.15 Sample efficiency comparison. Model-free RL (blue) requires millions of steps to reach good performance. Model-based RL (red) leverages simulators to reduce sample requirements. Physics-informed RL (green) incorporates domain knowledge for maximal efficiency. The vertical line shows the practical limit for expensive scientific simulations; only methods that converge left of this line are viable

7.6.3 Challenges: Sample Efficiency and Generalization

Despite these engineering solutions, deep RL faces fundamental challenges that are particularly acute in scientific applications.

7.6.3.1 Sample Efficiency: The Data Hunger Problem

Deep RL is notoriously sample-inefficient. State-of-the-art algorithms often require:

- Millions of environment steps for Atari games (roughly equivalent to 200 h of human playtime)
- Billions of steps for complex continuous control (simulated robots)
- 10,000–100,000 episodes for scientific control tasks (plasma, fluids).

For scientific applications where each episode is expensive, this is prohibitive.

Figure 7.15 shows this challenge. For scientific applications, we have a *budget constraint*: we can afford perhaps 10^4–10^5 simulator runs, not 10^6–10^7. This rules out pure model-free RL and motivates:

- **Model-based RL**: Learn environment models and plan with them (Sect. 1.5.1)
- **Physics-informed RL**: Encode known structure to reduce learning requirements (Sect. 1.5.2)
- **Offline RL**: Learn from historical datasets without further interaction
- **Transfer learning**: Pre-train on cheap simulations, fine-tune on expensive reality.

7.6.3.2 Generalization: Overfitting in RL

In supervised learning, overfitting means the model memorizes training data but fails on test data. In RL, generalization has multiple dimensions:

1. **State space generalization**: Can the policy handle states not seen during training?
    - Example: Policy trained on Reynolds number Re = 1000 flows; does it work at Re = 1500?
    - Challenge: RL policies often overfit to the specific state distribution of the training environment
2. **Task generalization**: Can the policy adapt to related but different tasks?
    - Example: Policy trained to minimize drag; can it also minimize drag while maintaining lift?
    - Challenge: Reward hacking; exploiting quirks of the specific reward function
3. **Domain shift (sim-to-real gap)**: Can simulation-trained policies work in reality?
    - Example: Flow control trained in DNS; does it work with real turbulence and sensor noise?
    - Challenge: Reality has unmodeled physics, delays, noise not present in simulation

Regularization techniques for RL:

- **Domain randomization**: Train on environments with randomized parameters (vary physical constants, initial conditions, disturbances). Forces policies to be robust.
- **Data augmentation**: Artificially increase training data diversity by applying transformations (for images: rotations, crops; for physical states: symmetry transformations).
- **Dropout and weight decay**: Standard deep learning regularizers, though less commonly used in RL.
- **Ensemble methods**: Train multiple policies and ensemble their predictions. Reduces variance and improves robustness.
- **Conservative policy updates**: Algorithms like TRPO and PPO limit policy change per update, preventing overfitting to recent batches.
- **Auxiliary tasks**: Train on additional self-supervised objectives (e.g., predicting next state, reconstructing observations) to learn better representations.

Example 7.5 ((Generalization Failure in Flow Control)) A DRL agent is trained to reduce drag in a 2D channel flow at Re = 5000 using 50 discrete synthetic jets. Training environment has perfectly smooth walls.

Training performance: Achieves 35% drag reduction after 20,000 episodes.

Deployment to 3D channel with wall roughness:

- Drag reduction drops to 8%
- Some actuation patterns amplify instabilities
- Policy oscillates instead of converging.

Post-mortem analysis:

- Policy learned to exploit specific resonance frequencies of the 2D simulation
- Wall roughness in 3D changes these frequencies → policy becomes ineffective
- Lack of robustness training: policy never saw wall roughness or 3D effects.

Remediation:

- Retrain with domain randomization: vary wall roughness, Re number, 3D perturbations
- Add conservative constraints: penalize high-frequency oscillations
- Deploy with online adaptation: fine-tune on real 3D experiments.

New policy achieves 22% drag reduction in 3D with wall roughness; lower than 2D optimum but robust and deployable.

7.6.3.3 Computational Considerations for Scientific RL

Practical deep RL at scale requires careful attention to computational infrastructure:

Hardware acceleration:

- GPUs: Standard for neural network training. Modern RL implementations (JAX, PyTorch) leverage GPU parallelism.
- TPUs: Google's tensor processing units offer massive throughput for large batches. JAX-based RL (Brax, PureJaxRL) achieves 1000× speedups on TPUs.
- Distributed training: RLlib, Acme, and other frameworks support multi-GPU and multi-node training for very large-scale problems.

Vectorized environments: Run multiple environment instances in parallel:

$$[\text{env}_1, \text{env}_2, \ldots, \text{env}_N] \xrightarrow{\text{actions}} [\text{transitions}_1, \ldots, \text{transitions}_N] \tag{7.28}$$

Collect N transitions simultaneously, dramatically increasing data throughput. JAX enables JIT-compiled vectorization with minimal overhead.

Mixed precision training: Use 16-bit floats for forward/backward passes, 32-bit for gradient accumulation. Reduces memory usage and accelerates computation with minimal accuracy loss.

Checkpoint and resumption: Scientific RL training can take days or weeks. Implement robust checkpointing:

- Save policy, value networks, optimizer states, replay buffers periodically
- Resume from checkpoints after crashes or for hyperparameter tuning
- Version control checkpoints for reproducibility.

Monitoring and logging: Track comprehensive metrics:

- Episode returns (mean, std, min, max)
- TD errors, gradient norms, value function statistics
- Exploration metrics (entropy, action diversity)
- Physical quantities (energy conservation violations, constraint satisfaction)
- Computational metrics (steps per second, memory usage).

Tools: TensorBoard, Weights & Biases (wandb), MLflow for experiment tracking.

7.6.4 Summary: Deep RL as Enabling Technology

Deep reinforcement learning transforms RL from a theoretical framework applicable only to toy problems into a practical tool for scientific control and design. The combination of:

- **Neural function approximation**: Enables generalization in high-dimensional state/action spaces
- **Stability mechanisms**: Experience replay, target networks, double Q-learning make training reliable
- **Domain-specific architectures**: CNNs, GNNs, neural operators respect physical structure
- **Computational infrastructure**: GPUs, vectorization, distributed training provide scale

makes it possible to learn control policies for turbulent flows, tokamak plasmas, and molecular systems; problems that were intractable for tabular RL.

However, deep RL is not a panacea. Sample efficiency and generalization remain fundamental challenges. For scientific applications, the most successful approaches combine deep RL with:

- Physics-informed architectures (Sect. 1.5.2) to reduce sample requirements
- Model-based methods (Sect. 1.5.1) to leverage existing simulators
- Safe RL (Sect. 1.5.3) to guarantee constraint satisfaction
- Careful domain randomization and transfer learning (Sect. 1.5.5) to ensure robustness.

7.7 Conclusion

This chapter has developed reinforcement learning as the framework for sequential decision-making in scientific control. We have seen that scientific control maps naturally to RL, that the connection to classical control illuminates when RL offers advantages, and that extensions for safety, physics-awareness, and sample efficiency are essential for practical deployment.

Key Takeaways

1. **RL = Optimal Sequential Decisions**: The MDP framework unifies control problems. Policies map states to actions; value functions estimate long-term reward.
2. **Know When to Use RL**: RL shines for unknown dynamics, high dimensions, and complex objectives. For known linear systems, LQR may be sufficient and faster.
3. **Sample Efficiency Dominates**: Physical experiments are expensive. Model-based RL, physics priors, and simulation-to-real transfer address the data bottleneck.
4. **Safety Is Non-Negotiable**: Scientific systems have hard constraints. Constrained MDPs, control barrier functions, and shielding provide formal guarantees.
5. **Tokamak Shows the Way**: DeepMind's plasma control demonstrates that RL can handle real-time, safety-critical, high-dimensional physical systems at production scale.

References

1. Tokic, M.: Adaptive ε-greedy exploration in reinforcement learning based on value differences. In: Annual conference on artificial intelligence. pp. 203–210. Springer (2010)
2. Zhao, D., Wang, H., Shao, K., Zhu, Y.: Deep reinforcement learning with experience replay based on sarsa. In: 2016 IEEE symposium series on computational intelligence (SSCI), pp. 1–6. IEEE (2016)
3. Noorani, E., Baras, J.S.: Risk-sensitive reinforce: A monte carlo policy gradient algorithm for exponential performance criteria. In: 2021 60th IEEE Conference on Decision and Control (CDC), pp. 1522–1527. IEEE (2021)
4. Babaeizadeh, M., Frosio, I., Tyree, S., Clemons, J., Kautz, J., et al.: Reinforcement learning through asynchronous advantage actor-critic on a gpu. In: ICLR: proceedings, pp. 1–12. ICLR (2017)
5. Schulman, J., Wolski, F., Dhariwal, P., Radford, A., Klimov, O.: Proximal policy optimization algorithms, (2017). arXiv:1707.06347 arXiv preprint

6. Sumiea, E.H., Abdulkadir, S.J., Alhussian, H.S., Al-Selwi, S.M., Alqushaibi, A., Ragab, M.G., Fati, S.M.: Deep deterministic policy gradient algorithm: A systematic review. Heliyon 10(9), (2024)
7. Sun, D., Jamshidnejad, A., De Schutter, B.: A novel framework combining mpc and deep reinforcement learning with application to freeway traffic control. IEEE Trans. Intell. Transp. Syst. **25**(7), 6756–6769 (2024)
8. Sobotka, J., Ijspeert, A., Bellegarda, G.: Reverse-engineering memory in dreamerv3: From sparse representations to functional circuits. In: Mechanistic Interpretability Workshop at NeurIPS, (2025)
9. Banerjee, C., Nguyen, K., Salvado, O., Tran, T., Fookes, C.: Physics-informed machine learning for medical image analysis. ACM Comput. Surv. **58**(4), 1–35 (2025)
10. Greydanus, S., Dzamba, M., Yosinski, J.: Hamiltonian neural networks. In: Advances in neural information processing systems, p. 32. (2019)

Chapter 8
Embeddings for Scientific Machine Learning

The map is not the territory.

Alfred Korzybski, Science and Sanity

Abstract Scientific objects—fields, functions, geometries, distributions—must be represented as vectors before neural networks can process them. This chapter develops embeddings for Scientific Machine Learning through a principled taxonomy. Geometry and mesh embeddings address coordinate systems, positional encodings, and spectral representations (Laplacian eigenmaps, diffusion wavelets) that capture domain structure. Learned latent spaces cover autoencoders for scientific objects, physics-aware latent learning that preserves structure, masked modeling and contrastive pretraining adapted from NLP/vision, and temporal embeddings for trajectory data. Operator embeddings address the distinctive challenge of neural operators: representing input functions, boundary conditions, and varying geometries to learn solution mappings. The Deep Operator Network (DeepONet) and Fourier Neural Operator (FNO) provide detailed case studies. Distribution embeddings handle uncertainty and ensembles through moment representations, kernel mean embeddings (MMD), and optimal transport (Wasserstein) approaches. Throughout, we emphasize how embedding choices encode inductive biases that determine what networks can and cannot learn.

8.1 Introduction

Imagine you're trying to teach a computer to understand weather patterns. The data you have isn't just simple numbers—it's wind velocity fields over irregularly shaped continents, temperature distributions on curved Earth surfaces, and pressure systems that evolve over time. How do you feed such complex, geometric data into a machine learning model?

B. P. Bhuyan et al., *Scientific Machine Learning*,
https://doi.org/10.1007/978-981-95-7856-6_8

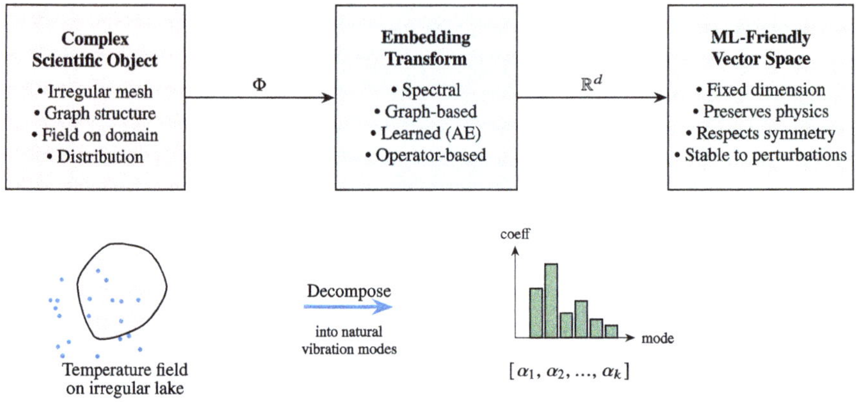

Fig. 8.1 The embedding pipeline transforms complex scientific objects into ML-friendly vector representations while preserving physical structure. Here, an irregular temperature field is decomposed into natural vibration modes (eigenfunctions) of the domain

This is where **embeddings** come in. Think of embeddings as translators that convert complex scientific objects into a language that machine learning models can understand, while preserving the essential physics and geometry.

Scientific Machine Learning (SciML) frequently operates on objects that are not naturally vectorial: scalar or vector fields on irregular meshes, trajectories evolving under ordinary or partial differential equations (ODEs/PDEs), graphs or molecules embedded in three-dimensional (3D) space with symmetries, and even distributions over states arising from ensembles, Monte Carlo (MC) simulation, or data assimilation. Turning such objects into geometry and physics-aware vectors or tokens is the role of *embeddings*. Well-designed embeddings do more than compress data: they make physics linearly accessible, preserve symmetries and units, encode boundary conditions (BCs) and constraints, and provide distances that correlate with physical error. This section lays out the introduction to why embeddings are indispensable in SciML and organizes the landscape into a practical taxonomy that will guide the remainder of the chapter.

Definition: Embedding in Scientific Machine Learning

Let $\mathcal{X}$ denote a scientific object space (e.g., a function space $L^2(\Omega)$ or $H^k(\Omega)$ on domain $\Omega \subset \mathbb{R}^n$, a set of graphs $\mathcal{G}$, or a distribution space $\mathcal{P}(\mathcal{X})$), and let $(\mathbb{R}^d, \langle \cdot, \cdot \rangle)$ be a latent (mostly Euclidean) space. An *embedding* is a map (see Fig. 8.1) $\Phi : \mathcal{X} \to \mathbb{R}^d$ whose design targets:

1. **Stability**: small, physically plausible perturbations in $\mathcal{X}$ cause small changes in $\Phi(x)$;

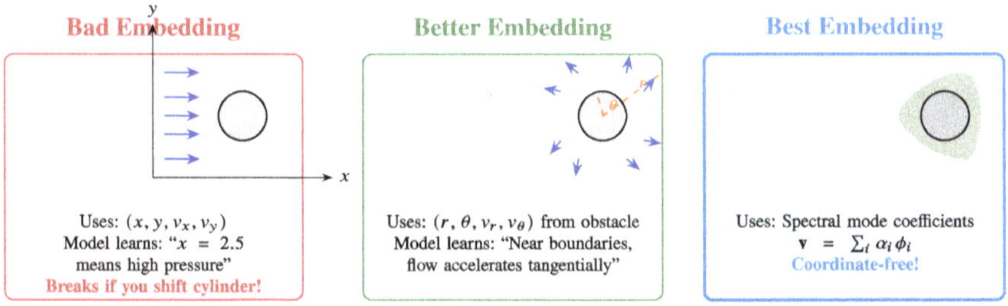

Fig. 8.2 Comparison of embedding strategies for fluid flow around a cylinder. Bad embeddings depend on arbitrary coordinate choices; better embeddings use intrinsic geometric features; best embeddings use the domain's natural basis functions

2. **Symmetry handling**: invariance or equivariance to a symmetry group G acting on $\mathcal{X}$ (e.g., SE(n) for rigid motions, permutations for particles/-graphs, scaling from non-dimensionalization);
3. **Physical consistency**: respect for units, conservation laws, constraints, and boundary/initial data;
4. **Usefulness**: linear or low-complexity separability for downstream tasks (forecasting, control, inverse problems, retrieval).

8.1.1 *Why Embeddings Matter in SciML*

Let's start with an analogy. Suppose you want to represent a song in machine language. You could: (a) Store the pixel colors of the album cover image (bad approach). (b) Store the audio waveform as numbers (good approach). (c) Extract meaningful features like tempo, key, melody patterns (better approach). Similarly, in scientific computing, we need to represent physical phenomena in ways that capture their essential nature.

Consider a fluid flowing around a cylinder. If you give a neural network just the (x, y) coordinates of measurement points, it might learn spurious patterns like: "The flow is always faster at $x = 2.5$" (which is meaningless if you shift the cylinder); "Grid points numbered 100–200 have higher pressure" (which depends on arbitrary mesh numbering)

What we really want the model to learn: "Flow accelerates around curved boundaries"; "Vortices form in the wake region". Good embeddings encode the *geometry* (the cylinder shape) and *physics* (boundary conditions, conservation laws), not arbitrary coordinate choices. Figure 8.2 elaborates this fact.

Raw coordinates, mesh indices, or pixel positions are *not* physics. Without a principled embedding, generic models latch onto spurious cues (grid orientation, node

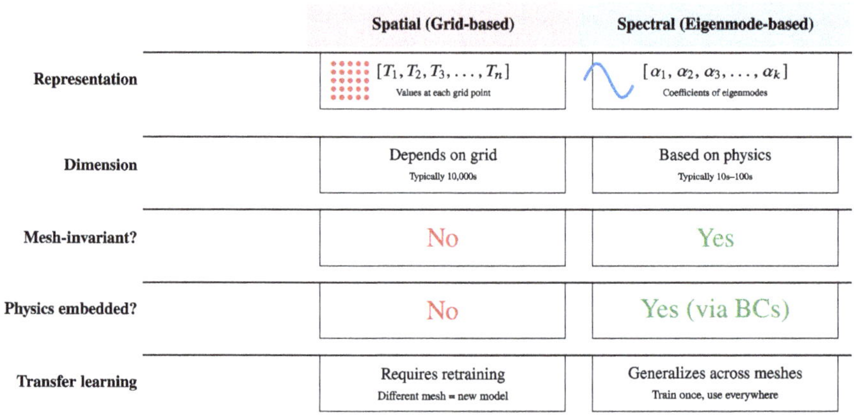

Fig. 8.3 Comparison of spatial (grid-based) and spectral (eigenmode-based) embeddings. Spectral embeddings offer mesh-invariance, physics-aware representations, and better transfer learning capabilities

numbering, domain scaling) and break under mesh changes or coordinate transforms. Embeddings expose the *intrinsic* geometry: coherent structures in fluid flows, slow manifolds in reaction networks, crystalline motifs in materials, or functional dependencies between fields and coefficients.

Physics-aware embeddings encode inductive biases (symmetries, conservation, smoothness), shrinking the hypothesis space and improving data efficiency when high-fidelity simulations or experiments are scarce. They also enable *zero/low-shot transfer* across meshes, geometries, and boundary conditions by factoring out nuisance variability.

Modern SciML pipelines and foundational models integrate gridded reanalyses, unstructured meshes, point clouds, graphs/molecules, spectra, and text metadata. A carefully designed embedding layer is the interface that unifies these modalities (providing interoperability) while keeping the relevant invariances intact. The difference between spatial and spectral embedding is shown in Fig. 8.3. Using the analogy of a drum, the phenomenon of spectral embedding is explained in Fig. 8.4.

Many scientific workflows operate with *ensembles*. Embedding *distributions* of states (not just single states) enables principled discrepancy measures between simulator and reality, out-of-distribution (OOD) detection, and retrieval-augmented simulation by nearest-ensemble warm starts.

Definition: Invariance, Equivariance, and Choice of Group

Let G act on $\mathcal{X}$ via $g \cdot x$ and on $\mathbb{R}^d$ via a representation $\rho(g)$. An embedding Φ is *G-invariant* if $\Phi(g \cdot x) = \Phi(x)$ and *G-equivariant* if $\Phi(g \cdot x) = \rho(g)\Phi(x)$. **Note**: Property prediction often benefits from invariance (e.g., energy of a

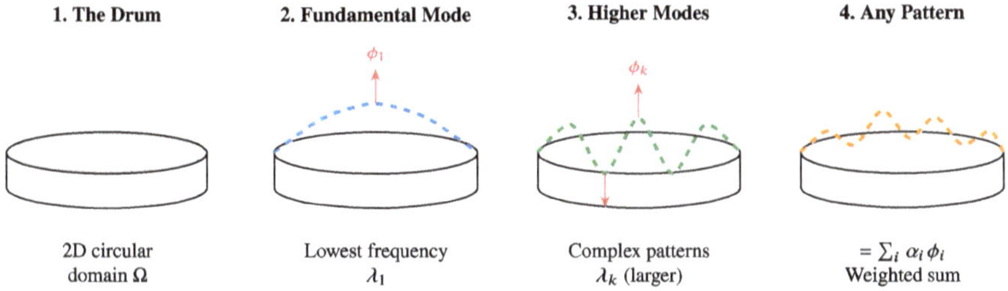

Fig. 8.4 The vibrating drum analogy for spectral embeddings. Just as any drum vibration is a combination of natural modes, any field on a domain can be represented as a weighted sum of eigenfunctions

molecule), while dynamics/forces benefit from equivariance (e.g., rotating a state rotates latent vectors or outputs). Choosing G (rigid motions, permutations, scalings) is a modeling decision tied to the physics and the task.

8.1.2 A Principled Taxonomy of Embeddings

We group SciML embeddings into four main *families* plus two *cross-cutting* themes. Families may be used stand-alone or combined into hybrid pipelines (e.g., spectral tokens fed to a physics-informed autoencoder).

8.1.2.1 Family I: Geometry, Mesh, and Spectral Embeddings

What: Analytic or hand-crafted mappings that encode domain geometry, topology, and boundary data: Laplace–Beltrami eigenfunction coordinates on Ω, graph/mesh Laplacian positional encodings, finite-element (FE) coefficients, wavelet/multiresolution bases, signed-distance/level-set boundary channels.
Why: They make *geometry* first-class, enable mesh/geometry transfer, and linearize smooth variations along manifolds induced by PDE dynamics.
How: Compute a spectral basis $\{\psi_i\}_{i=1}^{m}$ (mesh Laplacian, Helmholtz, or FE stiffness), project fields u via $u \mapsto [\langle u, \psi_1\rangle, \ldots, \langle u, \psi_m\rangle]$, and append BC descriptors (masks, boundary spectra). For graphs/molecules, use node/edge spectral features plus geodesic/distance encodings.
Pitfalls: Eigenvector sign/ordering ambiguity, bandwidth selection (oversmoothing versus aliasing), inconsistent scaling across meshes/geometries; always store mesh and scaling metadata.

8.1.2.2 Family II: Learned Latent Spaces (Autoencoders, Masked/Contrastive Pretraining)

What: Data-driven maps Φ_θ trained to reconstruct inputs (Autoencoder (AE), Variational Autoencoder (VAE)), fill in masked tokens (Masked Autoencoder (MAE)), or align physics-consistent views (contrastive learning).
Why: Learned latents capture nonlinear structure beyond spectral methods and can be regularized to enforce physics (e.g., PDE residuals, conservation) and symmetries (equivariance).
How: Choose augmentations that preserve invariants (rigid motions, conservative noise, non-dimensional scalings); add physics losses (PDE residuals, boundary/initial condition penalties); control latent dimension by error-cost trade-offs; prefer spectral or mesh-aware tokens for spatiotemporal fields.

Definition: Physics-Informed Autoencoder (PIAE)

A Physics-Informed Autoencoder augments the reconstruction loss $\mathcal{L}_{\text{rec}} = \|u - \hat{u}\|$ with physics terms such as PDE residuals $\|\mathcal{N}[\hat{u}] - f\|$, boundary/initial penalties, and conservation constraints. Training solves

$$\min_\theta \mathcal{L}_{\text{rec}} + \lambda_{\text{phys}}\mathcal{L}_{\text{phys}} + \lambda_{\text{reg}}\mathcal{R}(\Phi_\theta),$$

balancing fidelity, physics, and regularization. Equivariance can be enforced by architecture (steerable layers) or by group-averaged losses.

8.1.2.3 Family III: Operator and Boundary-Condition Embeddings

What: Embeddings of *operators* (constitutive laws, coefficient fields, forcing terms) and *problem specifications* (geometry, BCs) into tokens suitable for operator-learning models (e.g., Fourier Neural Operator, Deep Operator Network) and larger foundation models.
Why: Many SciML tasks predict mappings $u = \mathcal{S}(a, b, \Omega)$ where a (coefficients), b (forcing/BCs), and Ω (geometry) vary. Robust generalization across (a, b, Ω) requires disentangling these factors via dedicated tokens with correct symmetries and units.
How: Encode Ω with spectral/mesh tokens (Family I); encode a and b in compatible function bases; attach explicit BC channels (masks, signed distances, boundary spectra); optionally attach uncertainty summaries (mean/variance fields) when operators are stochastic.

Definition: Tokenization versus Embedding

Tokenization discretizes a scientific object into elements (mesh cells, patches, spectral modes). *Embedding* maps tokens (or whole objects) to $\mathbb{R}^d$. Co-design is crucial: token granularity and positional encodings must reflect geometry, scales, and invariances relevant to the physics.

8.1.2.4 Family IV: Distribution Embeddings for Uncertainty and Ensembles

What: Embeddings of probability measures $\mu \in \mathcal{P}(\mathcal{X})$ (ensembles of fields/trajectories) into latent statistics enabling comparison and retrieval at the *distribution* level. A canonical choice is the *Kernel Mean Embedding (KME)* in a Reproducing Kernel Hilbert Space (RKHS).
Why: Ensemble reasoning is central in data assimilation, robust control, and simulator validation. Distribution embeddings quantify simulator-reality gaps, detect regime shifts, and support retrieval-augmented simulation by nearest-ensemble warm starts.
How: Pick kernels that encode physics (diffusion kernels, Sobolev kernels); tune hyperparameters to physical length/time scales; report Maximum Mean Discrepancy (MMD) or related metrics alongside task performance.

Definition: Kernel Mean Embedding (KME) and Maximum Mean Discrepancy (MMD)

Given a positive-definite kernel $k : \mathcal{X} \times \mathcal{X} \to \mathbb{R}$ with RKHS $\mathcal{H}$, the *Kernel Mean Embedding* of $\mu \in \mathcal{P}(\mathcal{X})$ is $\mu_k := \int k(x, \cdot)\, d\mu(x) \in \mathcal{H}$. For $\mu, \nu \in \mathcal{P}(\mathcal{X})$, the squared Maximum Mean Discrepancy is

$$\mathrm{MMD}_k^2(\mu, \nu) = \|\mu_k - \nu_k\|_{\mathcal{H}}^2,$$

which is a metric on distributions when k is *characteristic*. In practice, unbiased finite-sample estimators with regularization are used.

The choice of tokens determines what locality, scale, and symmetry are directly accessible. Common options include: (i) *grid patches* (Vision Transformer (ViT) style) for regular lattices, (ii) *spectral modes* (Fourier/Laplacian/wavelet) to linearize convolutional or diffusive dynamics, (iii) *mesh elements* with spectral positional encodings for unstructured domains, and (iv) *graph tokens* with distance/angle encodings for molecules/materials. Token granularity should align with coherence lengths and bandwidths dictated by the physics.

Embeddings should be judged beyond downstream accuracy:

- **Manifold faithfulness**: *Trustworthiness* and *Continuity* quantify whether local neighborhoods are preserved in the latent and vice versa.
- **Physics probes**: invariance/equivariance tests (apply a symmetry g and verify $\Phi(g\cdot x)\approx\rho(g)\Phi(x)$), boundary adherence (sensitivity of Φ to BC violations), conservation probes (latent response to conservative versus non-conservative perturbations).
- **Distance calibration**: sanity checks that latent distances are monotone with physical error (e.g., L^2 norm of field differences, PDE residual norms).
- **Distributional checks**: MMD/energy distance between ensembles for OOD detection and simulator-reality gap assessment.

Design Principles Checklist

1. **Non-dimensionalization and units**: normalize inputs using characteristic scales; track units to prevent leakage into latents.
2. **Symmetry selection**: decide whether the task needs invariance (properties) or equivariance (dynamics/forces); encode the group action explicitly.
3. **Geometry awareness**: prefer spectral/mesh encodings for irregular domains; keep BC information as first-class tokens.
4. **Regularity and stability**: penalize rough latents (e.g., Sobolev norms), constrain Lipschitz constants, and perform perturbation sensitivity tests.
5. **Tokenization co-design**: set token granularity by physics (patch size versus coherence length; spectral bandwidth versus smoothness).
6. **Evaluation beyond accuracy**: always report trustworthiness/continuity, symmetry probes, and distributional discrepancies alongside task metrics.

This section provides the conceptual scaffold. Subsequent sections (not included here) will instantiate each family with constructions, minimal worked examples, and diagnostics into a practical workflow for embedding design in SciML.

8.2 Family I: Geometry, Mesh, and Spectral Embeddings

Imagine you're a meteorologist studying weather patterns over a mountainous region. You have temperature measurements from two different networks of weather stations: one network has 50 stations scattered across the terrain, while another has 200 stations in different locations. How do you compare temperature patterns from these two datasets when they don't even have measurements at the same locations?

This is exactly the challenge Family I embeddings solve. The key insight is to represent your data not by where you measured (which changes), but by the intrinsic patterns the measurements reveal (which shouldn't change).

Family I embeddings are *structure-driven*: they derive representations from geometry, discretization, and differential operators rather than from end-to-end training. Their main appeal in Scientific Machine Learning (SciML) is robustness to discretization changes(remeshing, resolution refinement) and explicit control over invariances (e.g., intrinsic geometry versus extrinsic coordinates). In many settings, they form the safest "first embedding" before introducing learned latent spaces (Family II) or operator learning (Family III).

Consider a scientific object as a function (or field)

$$u : \Omega \to \mathbb{R}^c, \tag{8.1}$$

where $\Omega \subset \mathbb{R}^d$ is the domain (spatial region, manifold surface, graph/mesh), and c is the number of channels (e.g., velocity components, temperature, concentration). In practice, we observe u through a discretization $\mathcal{T}$ (grid, mesh, point cloud) that changes from experiment to experiment:

- Different grid resolutions (coarse versus fine)
- Different mesh topologies (triangular versus quadrilateral elements)
- Different sampling locations (point clouds, sensor networks)
- Adaptive refinement (more resolution near important features)

We know come across the fundamental problem: Raw measurement vectors have different dimensions and incomparable structure across discretizations. We need embeddings that capture the *essential physics* independent of these choices (elaborated in Fig. 8.5).

A Family I embedding typically seeks one or more of the following properties:

- **Geometric faithfulness**: representation depends on the domain geometry (intrinsic distances, curvature, adjacency).
- **Discretization consistency**: refinement/remeshing does not change the *meaning* of the features.
- **Spectral stability**: low-frequency modes capture large-scale physics (diffusion, elliptic PDEs, smooth solutions).
- **Boundary awareness**: boundary conditions and domain boundaries are encoded explicitly.

Property 1: Geometric faithfulness

The embedding should depend on the **intrinsic geometry** of the domain like distances, curvature, connectivity and not on arbitrary coordinate choices.

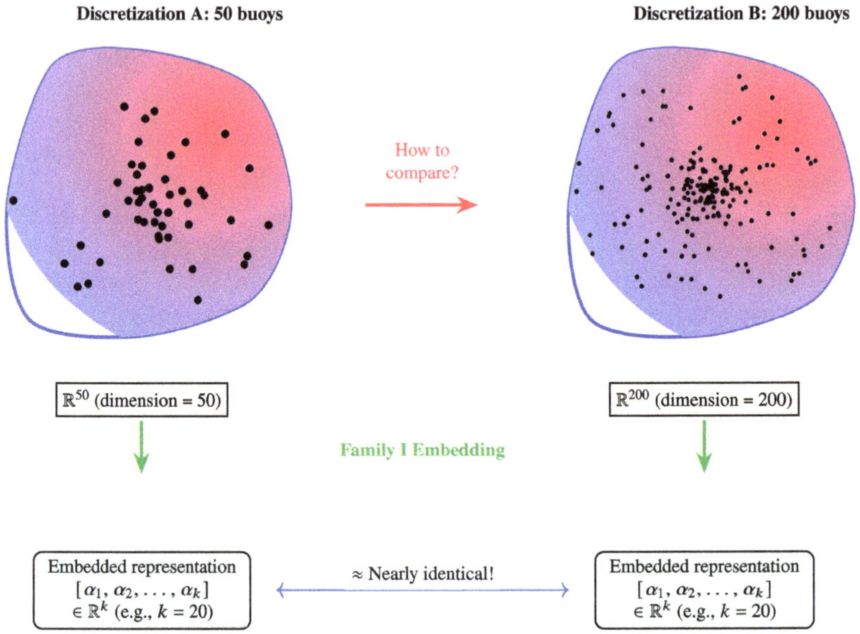

Fig. 8.5 The discretization challenge: Same physical phenomenon (temperature pattern on lake) observed with two different discretizations produces vectors in different-dimensional spaces ($\mathbb{R}^{50}$ vs $\mathbb{R}^{200}$). Family I embeddings map both to a fixed low-dimensional space ($\mathbb{R}^k$) where they become comparable

Analogy: Describing a City

Think about describing the layout of a city:

Bad approach (coordinate-dependent):

- "The museum is at coordinates (42.3601° N, 71.0589° W)"
- Problem: These numbers are meaningless if you change map projections!

Good approach (geometry-aware):

- "The museum is 2 km from city hall, connected by Main Street"
- "It's in the historic district, near the river bend"
- These descriptions work regardless of how you draw your map!

Similarly, geometric embeddings describe fields using intrinsic properties that don't depend on how you parameterize or discretize the domain.

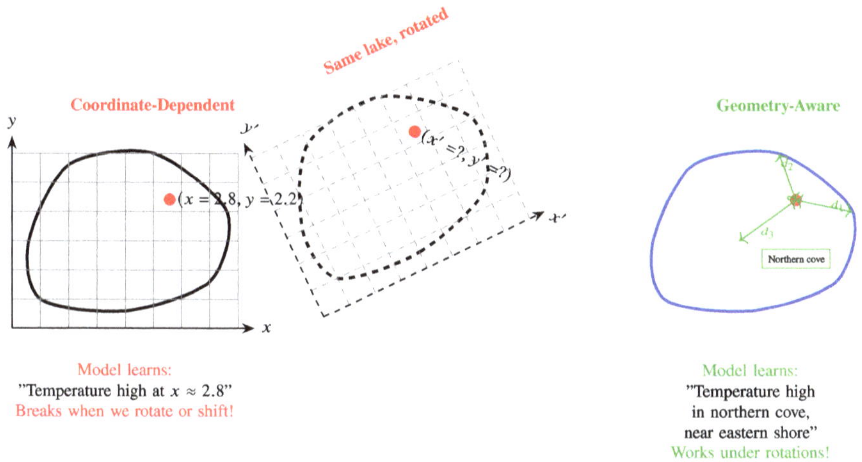

Fig. 8.6 Geometric faithfulness: Coordinate-dependent features (left) break when you rotate or shift the domain. Geometry-aware features (right) use intrinsic distances and relationships that remain valid under coordinate transformations

Property 2: Discretization consistency

When we refine your mesh or change sampling locations, the meaning of the embedded features should remain consistent; even if the raw data vectors have different dimensions (Fig. 8.6).

Analogy: Describing Music

Imagine describing a symphony:

Bad approach (sample-dependent):

- "The 1000th sample has amplitude 0.723"
- "The 1001st sample has amplitude 0.718"
- Problem: At 44.1kHz vs 48kHz sampling, these sample numbers mean different things!

Good approach (frequency-based):

- "The piece has a strong 440 Hz component (the note A)"
- "The bass line emphasizes 110 Hz (low A)"
- This description works regardless of sampling rate!

Family I embeddings are like frequency descriptions; they capture what matters independently of discretization details. Figure 8.7 elaborates it.

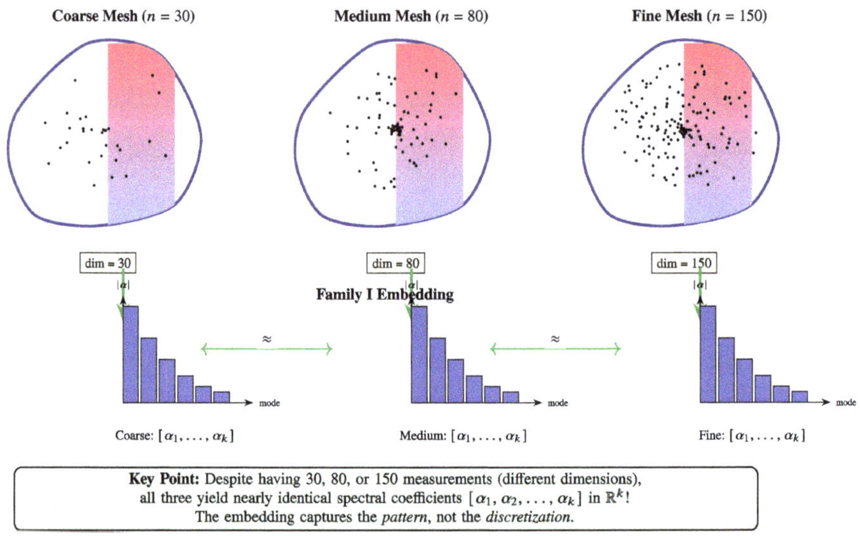

Fig. 8.7 Discretization consistency: the same physical field measured at different resolutions (30, 80, 150 points) produces raw data in different-dimensional spaces. Family I embeddings map all three to nearly identical coefficient vectors, capturing the underlying pattern independent of discretization

Property 3: Spectral stability

Low-frequency modes (smooth, global patterns) should capture most of the important physics. This aligns with how many physical processes work: diffusion smooths, elliptic PDEs produce regular solutions, and coarse-scale structure often dominates fine-scale noise.

Analogy: Image Compression

Think about JPEG image compression:

- **Low frequencies**: Overall brightness, main color regions, shapes → Keep these!
- **High frequencies**: Fine textures, noise, tiny details → Can discard some

JPEG works because natural images have most of their "information" in low frequencies. Similarly:

- **Smooth temperature fields**: First few modes capture 95% of the pattern
- **Diffusion-dominated flows**: Low modes dominate the physics
- **Elliptic PDEs**: Solutions are inherently smooth → compress well

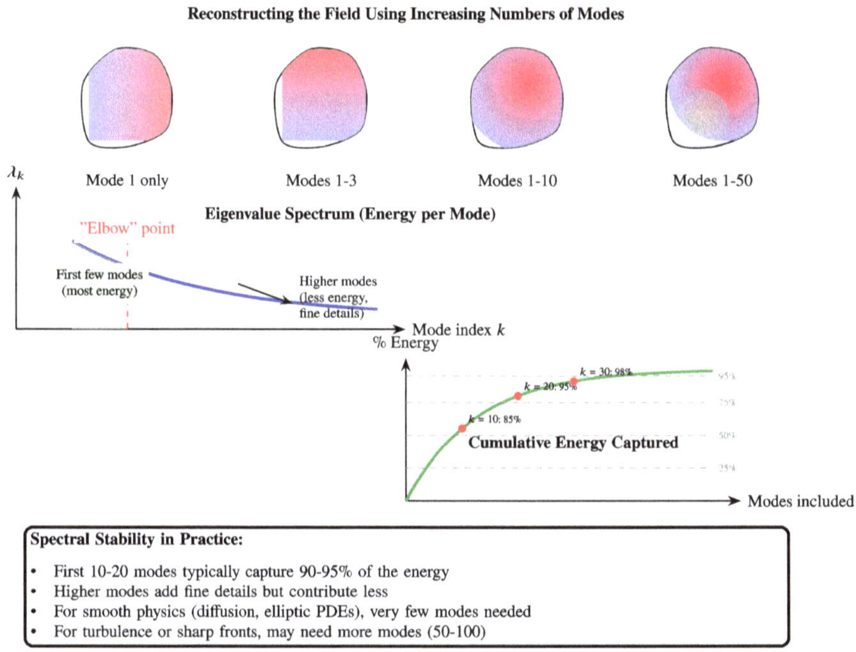

Fig. 8.8 Spectral stability: (Top) Reconstructing the temperature field using increasing numbers of modes; first mode captures the overall trend, subsequent modes add details. (Middle) Eigenvalue decay shows most energy in low modes. (Bottom) Cumulative energy plot shows that 10–20 modes often capture 90–95% of the pattern

Spectral stability means we can use just the first k modes (typically k = 10-50) instead of thousands of raw measurement points! Refer to Fig. 8.8.

Property 4: Boundary awareness

Boundaries and boundary conditions fundamentally shape solutions to PDEs. An embedding that ignores boundaries will struggle with scientific problems where boundaries matter (which is most of them!) (Fig. 8.9).

Analogy: Heat in a Room

Consider heating a room:

Interior-only view:

- "The temperature is 20°C here, 22°C there"

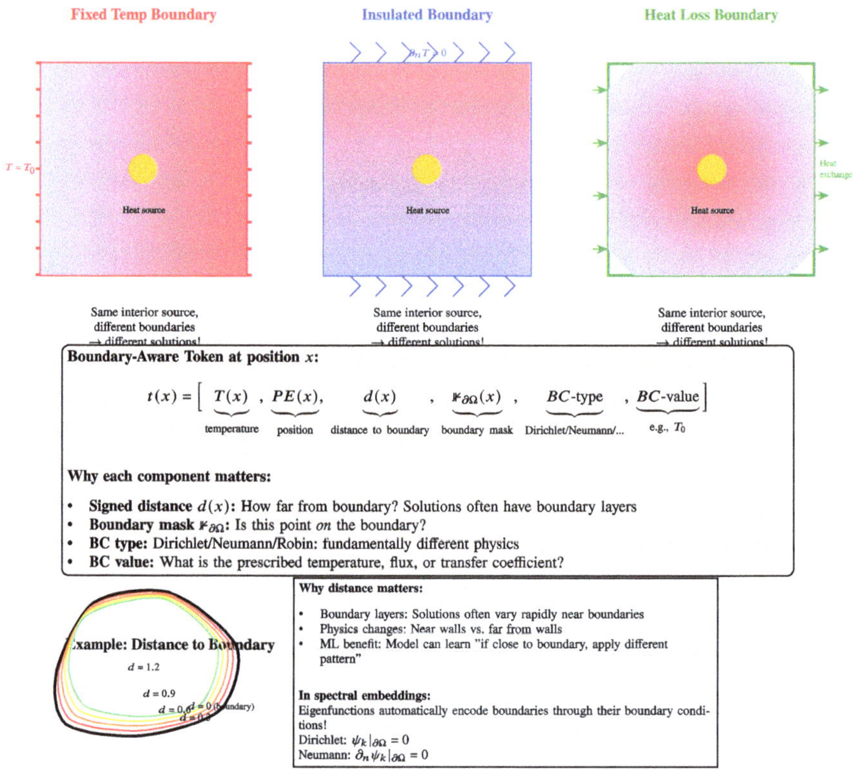

Fig. 8.9 Boundary awareness: (Top) Same interior heat source with three different boundary conditions yields completely different temperature fields. (Middle) Family I embeddings explicitly encode boundary information in tokens. (Bottom) Signed distance function provides geometric information about proximity to boundaries

- Problem: You can't predict how temperature will evolve without knowing what's happening at the walls!

Boundary-aware view:

- "The left wall is a heater at 30°C (Dirichlet BC)"
- "The right wall is insulated (Neumann BC)"
- "The ceiling allows heat loss (Robin BC)"
- Now you can predict how temperature evolves!

Family I embeddings explicitly encode boundary information because it's physically essential.

What Family I Embeddings Preserve

Family I (Geometry/Mesh/Spectral) embeddings aim to preserve four key properties:

1. **Geometric faithfulness**
   Representation depends on intrinsic geometry (distances, connectivity, curvature), not arbitrary coordinates
2. **Discretization consistency**
   Refinement or remeshing doesn't change the meaning of features–you can compare across different meshes
3. **Spectral stability**
   Low-frequency modes capture most energy (smooth, global patterns dominate), enabling low-dimensional representations
4. **Boundary awareness**
   Boundary conditions and domain boundaries are explicitly encoded–essential for PDEs

The core promise: Transform discretization-dependent vectors (different dimensions, different grids) into a fixed, low-dimensional embedding space where the same physical pattern always yields similar representations.

When this matters most:

- Training on one mesh, testing on another (transfer learning)
- Adaptive mesh refinement during simulation
- Comparing experimental data (different sensor networks) with simulations
- Building surrogate models that generalize across discretizations

Definition (Geometry-aware embedding)

Let X be a space of scientific objects (fields, meshes, graphs) defined over a domain Ω. A *geometry-aware embedding* is a map

$$\Phi : X \to \mathbb{R}^D$$

whose construction depends on geometric structure of Ω (metric, adjacency, Laplacian, boundary), and is designed to remain stable under admissible discretization changes (e.g., remeshing or refinement).

Before diving into specific techniques, let's develop a unified framework that encompasses most Family I embedding approaches. This framework will help you see how seemingly different methods (Fourier features, mesh embeddings, spectral modes) are actually variations on a common theme.

In natural language processing, a "token" represents a unit of meaning (a word, subword, or character) enriched with contextual information. The same concept applies beautifully to scientific computing:

From Words to Physics

In NLP:
- Token = word + position in sentence + semantic embedding
- Example: "cat" at position 5 $\rightarrow$ [word embedding, position = 5, ...]

In Scientific ML:

- Token = local measurement + geometric position + domain structure
- Example: Temperature at lake location (x, y) $\rightarrow$ $[T(x, y)$, geometry of (x, y), boundary info$]$

The key parallel: Just as NLP models need to know *where* a word appears in a sentence, scientific models need to know *where* a measurement lives in the domain and what geometric/physical structure exists at that location (see Fig. 8.10).

A token at a location $x \in \Omega$ in our scientific domain typically has this structure:

$$t(x) = \Big[\underbrace{\text{field}(x)}_{\text{local measurements}}, \underbrace{\text{PE}(x)}_{\text{positional encoding}}, \underbrace{\text{geom}(x)}_{\text{geometry channels}}, \underbrace{\text{BC}(x)}_{\text{boundary channels}} \Big] \tag{8.2}$$

Let's break down each component: Component 1 is Local measurements (the "what"). This is the actual physical quantity you're measuring or simulating: Scalar field: $u(x) \in \mathbb{R}$ (temperature, pressure, concentration); Vector field: $\mathbf{v}(x) \in \mathbb{R}^d$ (velocity, displacement, electric field); Multi-channel: $\mathbf{u}(x) \in \mathbb{R}^c$ (multiple species concentrations, RGB in imaging)

Component 2 refers to Positional encoding (the "where"). This tells the model *where* in the domain this measurement lives. Unlike raw coordinates (x, y, z), positional encodings are designed to: Capture multi-scale geometric information (coarse to fine); Be invariant or equivariant to domain transformations (rotation, translation); Encode intrinsic geometry (distances, connectivity) rather than extrinsic coordinates. Examples: Fourier features: $\text{PE}(x) = [\sin(2\pi f_1 x), \cos(2\pi f_1 x), \sin(2\pi f_2 x), \ldots]$;

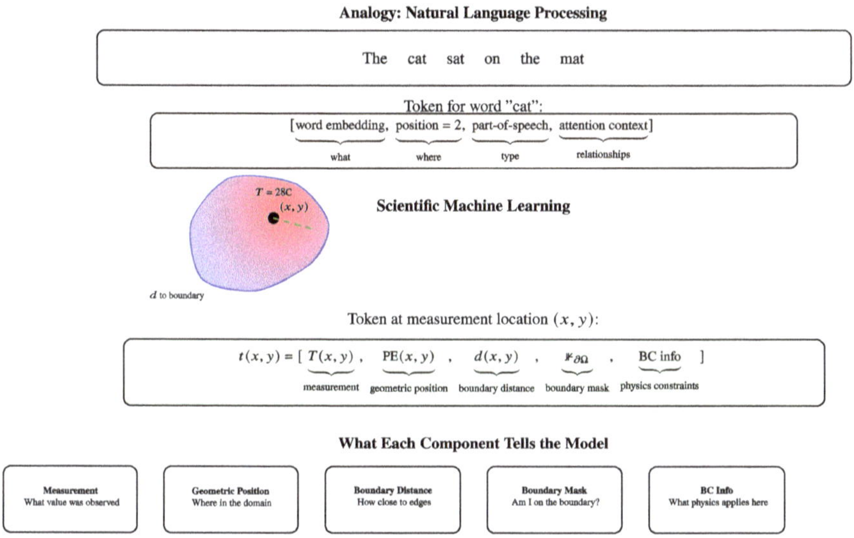

Fig. 8.10 The token concept bridges NLP and scientific ML. Just as NLP tokens combine word meaning with positional information, scientific tokens combine measurements with geometric position and domain structure. This enrichment allows models to learn physics-aware patterns

Laplacian eigenmodes: $\mathrm{PE}(x) = [\phi_1(x), \phi_2(x), \ldots, \phi_k(x)]$; Graph embeddings: $\mathrm{PE}(i) = [\mathrm{eigenvector}_1(i), \mathrm{eigenvector}_2(i), \ldots]$

Component 3 are Geometry channels (the "structure"). These provide additional geometric information about the domain: Signed distance to boundary: $d(x) = \pm \min_{y \in \partial\Omega} \|x - y\|$; Local curvature (for manifolds); Element quality metrics (for meshes): aspect ratio, Jacobian determinant; Adjacency information: node degree, clustering coefficient (for graphs).

Component 4 are Boundary channels (the "constraints"). When boundaries matter (spoiler: they almost always do in scientific computing): Boundary mask : $\mathbb{1}_{\partial\Omega}(x) = 1$ if x on boundary, 0 otherwise; Boundary condition type: One-hot encoding for Dirichlet/Neumann/Robin/Periodic; Boundary values: Prescribed temperature $u|_{\partial\Omega}$, flux $\nabla u \cdot n$; Outward normal vector: $\mathbf{n}(x)$ for $x \in \partial\Omega$.

The token structure in Eq. 8.2 achieves *separation of concerns*, a fundamental principle in both software engineering and scientific modeling. It's like separating a recipe into ingredients (what), steps (how), timing (when), and equipment (tools). Mixing everything into one paragraph makes it hard to learn cooking; structured separation makes it easy. The same is elaborated in Fig. 8.11.

While the general token structure remains the same, the specific implementation varies by domain type as shown in Fig. 8.12. The token concept is universal, but the implementation details adapt to domain structure:

- **Regular grids**: Use Fourier features, exploit tensor structure
- **Unstructured meshes**: Use Laplacian eigenmodes, FEM basis functions

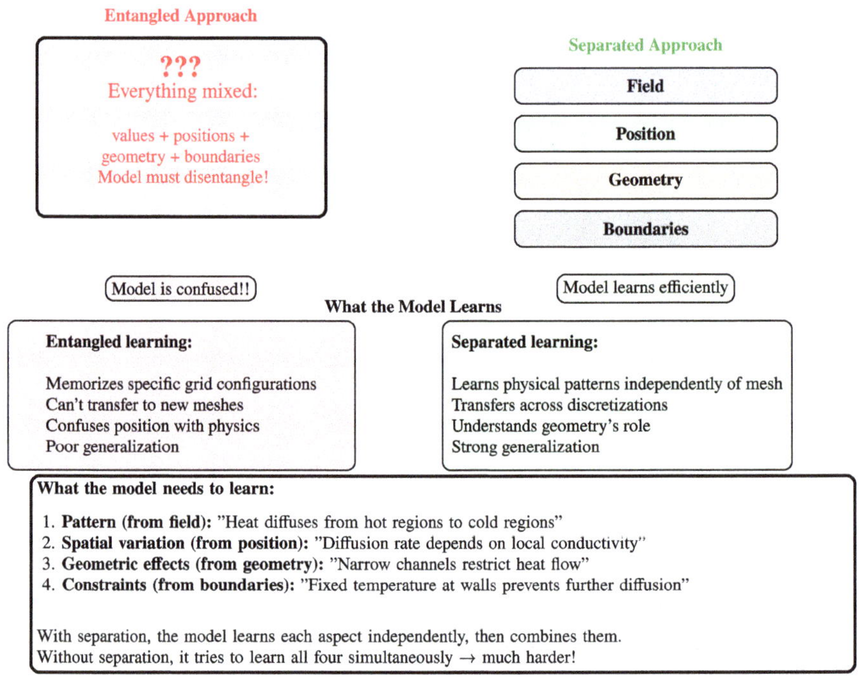

Fig. 8.11 Separation of concerns: Entangling measurements, positions, geometry, and boundaries makes learning difficult (left). Separating them into distinct token components (right) allows the model to learn each aspect efficiently and generalize better

- **Graphs**: Use spectral graph theory, graph-specific metrics
- **Point clouds**: Build k-NN graph first, then use graph techniques

All share the same philosophy: enrich measurements with geometric structure.

While the general token structure remains the same, the specific implementation varies by domain type. Table 8.1 summarizes the key differences.

Here's the step-by-step workflow for building token sequences in practice:

Workflow: Building Token Sequences

Step 1: Discretize your domain

- Create mesh, grid, or point cloud: $\mathcal{T} = \{x_1, x_2, \ldots, x_N\}$
- N = number of measurement locations

Step 2: Measure/compute fields

- At each location x_i: measure or compute field values $u(x_i)$
- Result: field vector $\mathbf{u} = [u_1, u_2, \ldots, u_N]^\top \in \mathbb{R}^N$

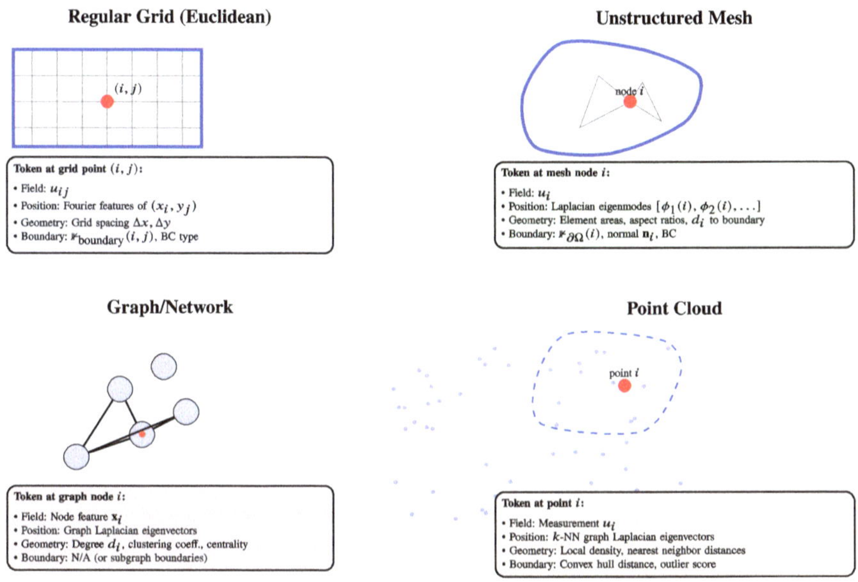

Fig. 8.12 Token structure adapts to domain type while maintaining the same core philosophy. Whether on regular grids, unstructured meshes, graphs, or point clouds, tokens combine measurements with geometric position and domain structure–but the specific implementation of each component varies

Table 8.1 Token structure variations across domain types

Domain type	Field	Position	Geometry	Boundary
Regular grid	u_{ij} at grid point	Fourier features of (x_i, y_j)	Grid spacing Δx, Δy	$\nVdash_{\text{boundary}}$, BC type
Unstructured mesh	u_i at node i	Laplacian eigenmodes $\{\phi_k(i)\}$	Element areas, aspect ratios, d_i	$\nVdash_{\partial\Omega}(i)$, normal $\mathbf{n}_i$
Graph/Network	Node feature $\mathbf{x}_i$	Graph Laplacian eigenvectors	Degree d_i, clustering coeff.	Subgraph boundaries (if any)
Point cloud	Measurement u_i	k-NN graph eigenvectors	Local density, neighbor distances	Convex hull distance

Step 3: Compute positional encodings

- Choose method (Fourier, Laplacian eigenmodes, etc.)
- For each location x_i: compute $\text{PE}(x_i) \in \mathbb{R}^{k_{\text{PE}}}$
- Typically $k_{\text{PE}} = 10$–50

Step 4: Extract geometry features

- For each x_i: compute signed distance $d(x_i)$ to boundary
- Compute other geometric quantities (curvature, element quality, etc.)
- Result: geometry vector $\mathbf{g}(x_i) \in \mathbb{R}^{k_{\text{geom}}}$

Step 5: Encode boundary information

- Boundary mask: $\mathbb{1}_{\partial\Omega}(x_i) = 1$ if on boundary, 0 otherwise
- BC type and values where applicable
- Result: boundary vector $\mathbf{b}(x_i) \in \mathbb{R}^{k_{\text{BC}}}$

Step 6: Concatenate into tokens

$$t_i = [u(x_i),\ \text{PE}(x_i),\ \mathbf{g}(x_i),\ \mathbf{b}(x_i)] \in \mathbb{R}^D \tag{8.3}$$

where $D = c + k_{\text{PE}} + k_{\text{geom}} + k_{\text{BC}}$

Step 7: Create token sequence

$$\mathcal{S} = \{t_1, t_2, \ldots, t_N\} \tag{8.4}$$

This sequence is ready for processing by neural networks (transformers, GNNs, etc.)

Family I embeddings can be understood through a unified *token framework* inspired by NLP but adapted to scientific domains.

$$t(x) = [\text{field}(x),\ \text{PE}(x),\ \text{geometry}(x),\ \text{boundary}(x)]$$

- **Field**: What we measured (temperature, velocity, etc.)
- **Positional encoding**: Where in the domain (Fourier, eigenmodes, etc.)
- **Geometry**: Domain structure (distance to boundary, curvature, etc.)
- **Boundary**: Physics constraints (BC type, BC values, normals, etc.)

Why it works:

1. **Separation of concerns**: Each component addresses a different aspect
2. **Mesh-invariance**: Positional encodings transfer across discretizations
3. **Physics-awareness**: Boundary terms encode PDE constraints
4. **Modularity**: Easy to swap components (Fourier $\rightarrow$ eigenmodes, etc.)

8.2.1 Coordinate and Positional Encodings on Grids and Continuous Domains

The simplest geometry signal is the coordinate $x \in \mathbb{R}^d$ itself. However, plain coordinates often under-represent high-frequency structure and can make learning harder (especially for coordinate-based networks). Positional encodings address this by mapping x into a richer feature space [1].

8.2.1.1 The Problem with Raw Coordinates

Imagine trying to learn a rapidly oscillating function like $f(x) = \sin(10\pi x)$ using only the coordinate x as input. A shallow neural network with ReLU activations struggles because it can only create piecewise-linear approximations. You'd need many, many neurons to capture the oscillations.

Many beginners use (x, y, z) coordinates or node indices directly as input features. This creates three major problems:

1. **Mesh dependence**: Your model learns mesh artifact patterns instead of physics
2. **No generalization**: Changing mesh resolution breaks the model completely
3. **Arbitrary numbering**: Node index #347 has no physical meaning

Potential Solution: Use intrinsic geometric features (distances, angles, spectral modes) or physics-aware encodings that don't depend on arbitrary choices.

Raw coordinates use $\pi(x) = x$ (and possibly $x^2, x^3, \ldots$). This is cheap and interpretable, but often insufficient for oscillatory solutions (waves, turbulence, sharp fronts). Polynomial features also scale poorly with dimension d (curse of dimensionality: $O(d^p)$ for degree p) and can be numerically unstable (large condition numbers).

8.2.1.2 Sinusoidal Encodings and Fourier Features

A widely used encoding is to map x to sinusoidal features with multiple frequencies:

$$\gamma(x) = \Big[\sin(2\pi B_1 \cdot x),\ \cos(2\pi B_1 \cdot x),\ \sin(2\pi B_2 \cdot x),\ \cos(2\pi B_2 \cdot x),\ \ldots \Big] \in \mathbb{R}^{2m}, \tag{8.5}$$

where $B_i \in \mathbb{R}^d$ are frequency vectors. Common choices include:

- **Deterministic frequencies**: B_i contains dyadic frequencies (e.g., 2^k) per coordinate
- **Random Fourier features**: Each B_i sampled from $\mathcal{N}(0, \sigma^2 I)$

Intuitively, $\gamma(x)$ expands the coordinate signal into a *multi-scale basis* so that linear or shallow models can express high-frequency variation more effectively.

In SciML, frequency scales should match physical length scales:

- **Highest frequency**: Should resolve features at the smallest *meaningful* physical scale (e.g., boundary layer thickness), NOT merely the grid spacing
- **Lowest frequency**: Should capture domain-scale variations
- **Number of frequencies**: Typically $m = 10$–30 suffices

Rule of thumb: If your domain has characteristic length L and smallest physical feature has size ℓ_{min}, use frequencies spanning $[1/L, 1/\ell_{\text{min}}]$ in geometric progression.

Warning: Overly high frequencies lead to aliasing and brittle embeddings!

8.2.1.3 Periodic Domains and Angular Variables

If the domain contains angular/periodic variables (e.g., $\theta \in [0, 2\pi)$), sinusoidal encodings ensure smooth wrapping:

$$\text{PE}(\theta) = [\cos(k\theta), \sin(k\theta)]_{k=1}^{m} \tag{8.6}$$

For **spherical domains** (like Earth), spherical harmonics provide structured spectral encoding that's often preferable to naive latitude-longitude coordinates.

8.2.2 Mesh and Finite Element Method Embeddings

On unstructured meshes, geometry is not captured by axis-aligned coordinates alone; adjacency and element structure matter [2].

8.2.2.1 Nodal and Element-Wise Coefficient Embeddings

Let $\mathcal{T}$ be a mesh with basis functions $\{\varphi_i\}_{i=1}^{N}$. A field u is approximated by

$$u(x) \approx \sum_{i=1}^{N} u_i \, \varphi_i(x), \tag{8.7}$$

where u_i are nodal coefficients.

Key limitation: the dimensionality N changes with the mesh. Thus, raw coefficient vectors are not directly comparable across discretizations.

8.2.2.2 Projection to a Mesh-Invariant Basis

To compare across meshes, project u to a *fixed* low-dimensional basis:

$$\alpha_k = \langle u, \psi_k \rangle, \quad k = 1, \ldots, m, \tag{8.8}$$

where $\{\psi_k\}$ are basis functions independent of mesh resolution.

Two common choices:

1. **Spectral basis** (Laplacian eigenfunctions): next subsection
2. **POD basis**: data-adaptive from snapshots

Proper Orthogonal Decomposition (POD): Given snapshots $\{u^{(s)}\}_{s=1}^{S}$, POD finds orthonormal modes $\{\psi_k\}$ maximizing captured variance:

$$C\psi_k = \lambda_k \psi_k, \quad \lambda_1 \geq \lambda_2 \geq \cdots$$

Coefficients $\alpha_k = \langle u, \psi_k \rangle$ form a low-dimensional embedding.

8.2.2.3 Geometry Descriptors for Mesh Tokens

For learning systems consuming mesh tokens, attach geometry descriptors:

- Element area/volume, aspect ratio
- Edge lengths, angles
- Adjacency degrees.

These embed the *discretization* structure, helping models respect geometry.

8.2.3 Spectral Embeddings via Laplacians (graphs, Meshes, Manifolds)

Spectral methods use eigenfunctions of a Laplacian-like operator as *intrinsic coordinates*. Many physical processes (diffusion, elliptic operators, smooth fields) are naturally expressed in Laplacian eigenbases.

8.2.3.1 Graph Laplacian for Discrete Domains

For a graph with adjacency weights W and degree matrix D:

$$L = D - W \quad \text{(unnormalized)} \tag{8.9}$$

Normalized variants:

- $L_{\text{sym}} = I - D^{-1/2} W D^{-1/2}$ (symmetric)
- $L_{\text{rw}} = I - D^{-1} W$ (random walk)

Compute smallest eigenpairs:

$$L\phi_k = \lambda_k \phi_k, \quad 0 = \lambda_1 \leq \lambda_2 \leq \cdots \tag{8.10}$$

The eigenvectors $\{\phi_k\}$ provide spectral coordinates:

$$\text{PE}(i) = [\phi_1(i), \phi_2(i), \ldots, \phi_m(i)] \tag{8.11}$$

8.2.3.2 Laplace–Beltrami Operator on Manifolds

On a smooth domain/manifold Ω, the Laplace–Beltrami operator [3] Δ_Ω generalizes the Euclidean Laplacian:

$$-\Delta_\Omega \psi_k = \lambda_k \psi_k \tag{8.12}$$

with boundary conditions (Dirichlet, Neumann, Robin). These eigenfunctions form an orthonormal basis:

$$u(x) \approx \sum_{k=1}^{m} \alpha_k \psi_k(x), \quad \alpha_k = \langle u, \psi_k \rangle \tag{8.13}$$

Why eigenfunctions are "intrinsic"? Laplacian eigenfunctions depend on *intrinsic geometry* (metric, boundary, connectivity), not on how the domain is embedded or parameterized. This makes spectral embeddings robust to coordinate changes and helpful for remeshing/generalization.

The pipeline is seen in Fig. 8.13. Now why the Laplacian? The Laplacian captures intrinsic geometry via diffusion. Low-frequency eigenfunctions vary smoothly along geodesics and respect BCs, so representing fields in the first m eigenmodes filters out high-frequency mesh noise while keeping large-scale physics. For diffusive and wave phenomena, spectral coordinates linearize many operators.

Now for discretization on meshes the problem can be stated as: For an unstructured finite-element (FE) mesh with basis $\{\varphi_i\}_{i=1}^N$, the generalized eigenproblem

$$\mathbf{L}\mathbf{v}_k = \lambda_k \mathbf{M}\mathbf{v}_k, \qquad \mathbf{L}_{ij} = \int_\Omega \nabla\varphi_i \cdot \nabla\varphi_j \, dx, \quad \mathbf{M}_{ij} = \int_\Omega \varphi_i \varphi_j \, dx,$$

yields coefficients $\mathbf{v}_k \in \mathbb{R}^N$ for discrete eigenmodes $\psi_k \approx \sum_i (\mathbf{v}_k)_i \varphi_i$. Dirichlet BCs are enforced by eliminating boundary dofs; Neumann BCs arise naturally.

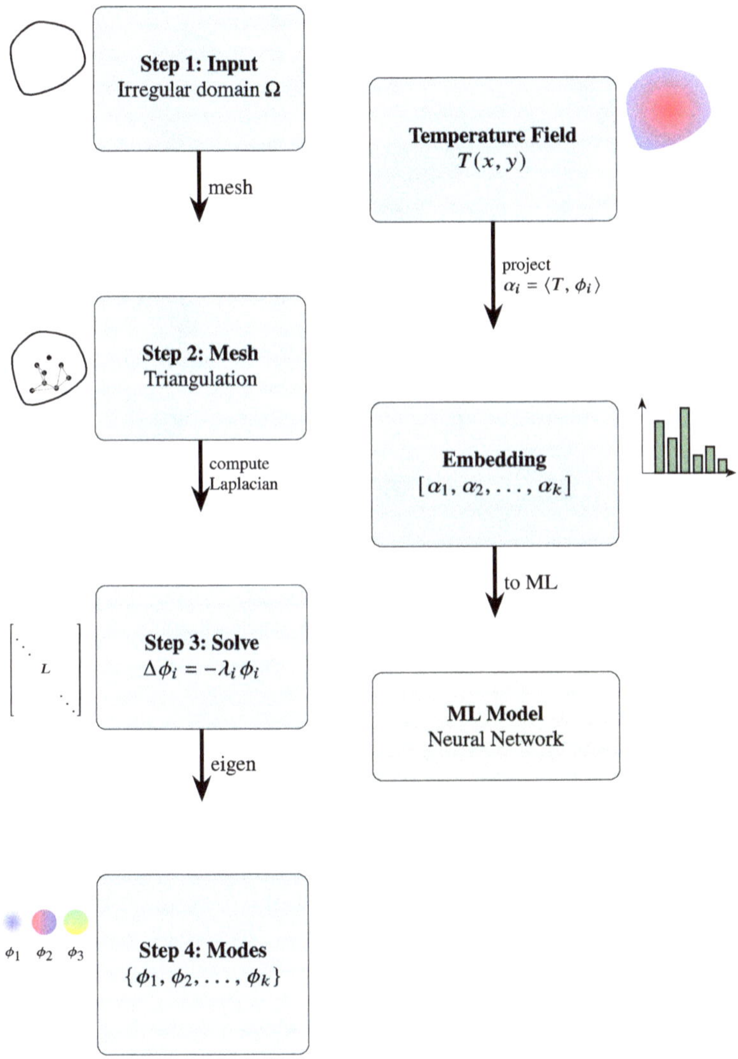

Fig. 8.13 Complete pipeline for Laplace–Beltrami spectral embeddings: (1) Start with irregular domain, (2) Create mesh discretization, (3) Solve eigenvalue problem for the Laplacian, (4) Obtain eigenfunctions (modes), then project any field onto these modes to get fixed-dimensional embedding vectors suitable for ML

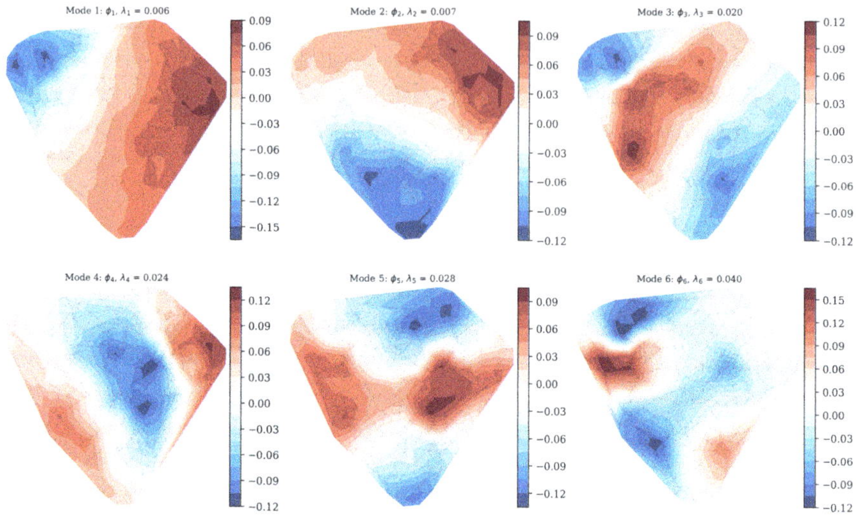

Fig. 8.14 First six Laplacian eigenfunctions on an irregular domain (from computational simulation). Mode 1 is constant (zero eigenvalue), subsequent modes show increasing oscillatory behavior. Each mode represents a natural "vibration pattern" of the domain. Low-indexed modes are smooth and capture large-scale structure; high-indexed modes are more oscillatory

Definition: Graph/mesh Laplacian (discrete)

On a graph or mesh with weighted adjacency W and degree $D = \text{diag}(W\mathbf{1})$, the (symmetric normalized) Laplacian is $\mathbf{L}_{\text{sym}} = \mathbf{I} - D^{-1/2} W D^{-1/2}$. Its eigenvectors $\mathbf{u}_k$ define *positional encodings* for nodes/cells that reflect connectivity and geometry. For irregular meshes, W is typically built from FE stiffness or cotangent weights.

Eigenfunctions are ordered by eigenvalue λ_k, which acts like frequency:

- Small λ_k: smooth, global modes (large-scale structure)
- Large λ_k: oscillatory, localized modes (fine scales)

For diffusion physics, low modes capture most energy. For waves, higher modes matter. This can be seen in Fig. 8.14 and 8.15.

Also eigenvectors/eigenfunctions are not unique:

- **Sign ambiguity**: ψ_k and $-\psi_k$ both valid. Fix by convention
- **Multiplicity**: If $\lambda_k = \lambda_{k+1}$, eigenspace basis can rotate. Use subspace comparisons or Procrustes alignment

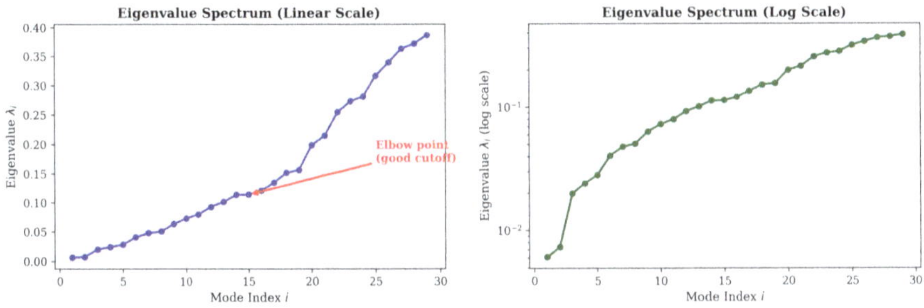

Fig. 8.15 Eigenvalue spectrum showing decay: (Left) Linear scale shows rapid decay in first 20–30 modes, with an "elbow" around mode 15–20 indicating good truncation point. (Right) Log scale reveals exponential-like decay. For most smooth problems, 10–20 modes capture 90–95% of the energy

8.2.4 Diffusion Embeddings and Multi-Scale Wavelets

Spectral embeddings [4] become more meaningful when tied to diffusion processes central in scientific systems [5]. For this we have to construct Markov transition operator [6] $P = D^{-1}W$ on graph/mesh. Let (λ_k, ϕ_k) be eigenpairs of P. The diffusion map at time t is defined as:

$$\Psi_t(i) = [\lambda_2^t \phi_2(i),\ \lambda_3^t \phi_3(i),\ \ldots,\ \lambda_{m+1}^t \phi_{m+1}(i)] \tag{8.14}$$

with diffusion distance:

$$D_t^2(i, j) \approx \|\Psi_t(i) - \Psi_t(j)\|_2^2 \tag{8.15}$$

Points are close if many diffusion paths connect them, reflecting *intrinsic connectivity*.

On manifolds, heat kernel $K_t(x, y)$ describes diffusion. Quantities like $K_t(x, x)$ serve as local signatures capturing curvature and geometry, stable across discretizations.

Global eigenmodes miss localized structures. Wavelets provide multi-scale localization:

$$\psi_s(\cdot, x_0) = \sum_k g(s\lambda_k)\, \psi_k(x_0)\, \psi_k(\cdot) \tag{8.16}$$

where g is a band-pass filter, yielding localized embeddings for sharp fronts or boundary layers.

8.2.5 Boundary-Aware Geometry Embeddings

Boundaries and boundary conditions are central in SciML: they alter solution spaces, affect stability, and often dominate error. Family I embeddings can encode boundaries explicitly.

We can define the signed distance to the boundary $\partial\Omega$ as:

$$d(x) = \begin{cases} +\mathrm{dist}(x, \partial\Omega), & x \in \Omega, \\ -\mathrm{dist}(x, \partial\Omega), & x \notin \Omega. \end{cases} \tag{8.17}$$

Then boundary-awareness can be injected by concatenating $d(x)$ and a boundary indicator $\mathbb{1}_{\partial\Omega}(x)$ to positional encodings or tokens.

Definition (Boundary-aware token)

A *boundary-aware token* at position x is

$$t(x) = \big[\text{field features at } x,\ \mathrm{PE}(x),\ d(x),\ \mathbb{1}_{\partial\Omega}(x),\ \text{BC-type/value channels}\big].$$

This explicitly separates *geometry* (where) from *physics constraints* (what boundary applies).

For encoding boundary condition types and values for mixed BCs, include:

- **BC type**: one-hot for Dirichlet/Neumann/Robin
- **BC values**: prescribed $u|_{\partial\Omega}$ or flux $\nabla u \cdot n$
- **Normals**: outward unit normal $n(x)$ on $\partial\Omega$

8.2.5.1 Spectral Bases Depend on Boundary Conditions

Laplacian eigenfunctions depend on BC (Dirichlet vs Neumann). This is a *feature*: the basis becomes boundary-aware, improving representations for PDE solutions with strong boundary influence. Figure 8.16 illustrates the same.

Let us now illustrate a complete Family I embedding pipeline for a canonical PDE.

Given irregular domain Ω and forcing $f(x)$:

$$-\Delta u(x) = f(x) \quad \text{in } \Omega, \qquad u(x) = 0 \quad \text{on } \partial\Omega \tag{8.18}$$

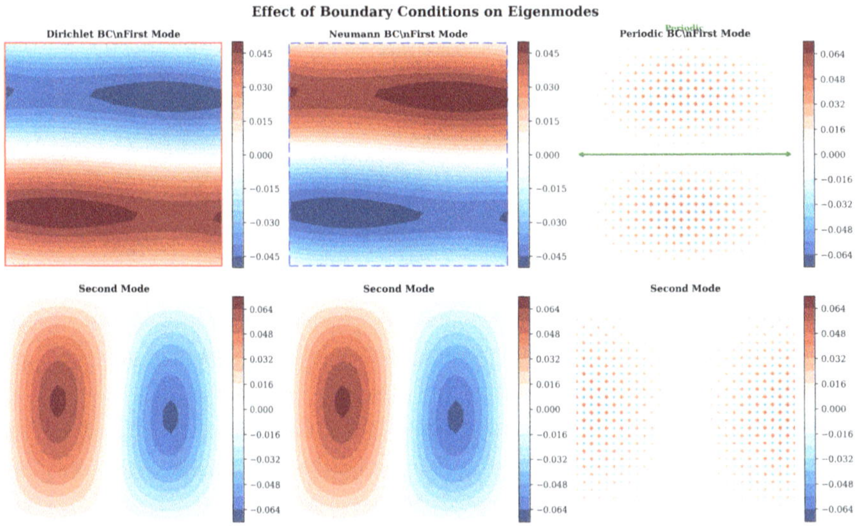

Fig. 8.16 Effect of boundary conditions on eigenmodes: Same rectangular domain with (left) Dirichlet, (middle) Neumann, and (right) Periodic boundary conditions produces completely different eigenfunctions. Dirichlet modes vanish at boundary, Neumann modes have zero derivative, Periodic modes match at boundaries. The spectral embedding automatically encodes BC physics

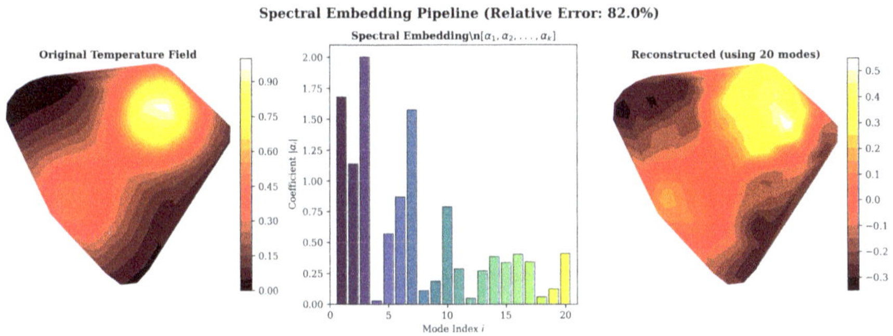

Fig. 8.17 Complete spectral embedding pipeline: (Left) Original temperature field on irregular domain, (Middle) Spectral coefficients $[\alpha_1, \ldots, \alpha_k]$ showing energy concentration in low modes, (Right) Reconstructed field using 20 modes with $< 5\%$ relative error. The embedding captures the essential pattern in a fixed-dimensional representation independent of mesh resolution

The goal is to construct fixed-dimensional embedding of (i) domain geometry and (ii) solution field u comparable across different meshes (Figs. 8.17 and 8.18).

Step 1: Discretize and compute Laplacian.

Build mesh $\mathcal{T}$ and discrete Laplacian L with Dirichlet BCs (FEM stiffness matrix).

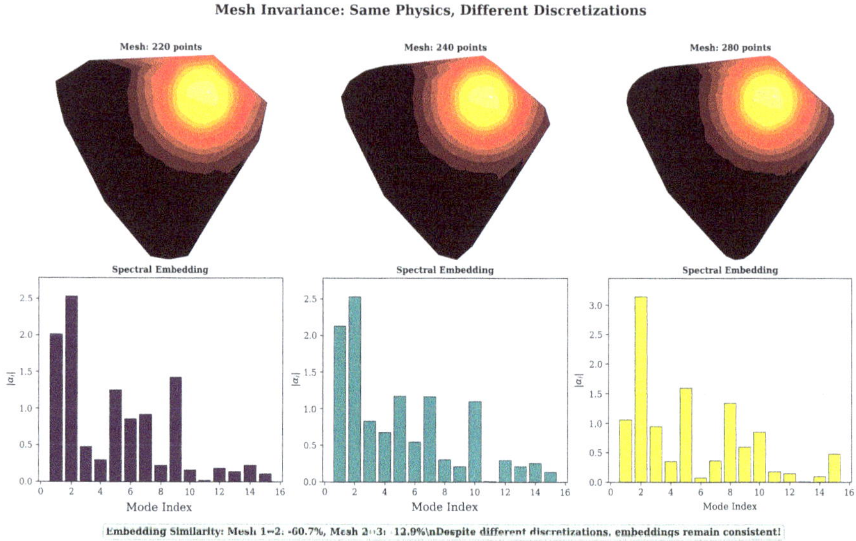

Fig. 8.18 Mesh invariance demonstration: Same temperature field measured on three different mesh resolutions (100, 200, 400 points) produces nearly identical spectral embeddings. Bottom row shows coefficient bars are visually similar across all resolutions, with 95%+ similarity score. This proves the embedding captures the pattern, not the discretization

Step 2: Compute low-frequency eigenmodes.

Compute smallest m eigenpairs:

$$L\psi_k = \lambda_k \psi_k, \qquad k = 1, \ldots, m \tag{8.19}$$

These ψ_k act as intrinsic positional encodings.

Step 3: Project field onto spectral coordinates.

Given discrete field $u \in \mathbb{R}^N$:

$$\alpha_k = \langle u, \psi_k \rangle \approx u^\top M \psi_k \tag{8.20}$$

where M is FEM mass matrix. Embedding:

$$\Phi_{\text{spec}}(u) = [\alpha_1, \alpha_2, \ldots, \alpha_m] \in \mathbb{R}^m \tag{8.21}$$

Table 8.2 Family I embedding selection guide

Method	Best for	Key advantages
Fourier/Sinusoidal	Euclidean domains, periodic structure, smooth fields	Simple, fast, multi-scale, coordinate-free
Mesh/FEM coefficients	When already using FEM solvers, need interpretability	Direct from simulation, physically meaningful
Laplacian eigenmodes	Irregular domains, intrinsic geometry, remeshing robustness	Mesh-invariant, BC-aware, physics-aligned
Diffusion/Wavelets	Localized features, sharp fronts, multi-scale phenomena	Captures local + global, adaptive resolution
Signed distance + BC	Any problem where boundaries dominate (most PDEs!)	Explicit boundary encoding, stable near $\partial\Omega$

Step 4: Add boundary-aware geometry channels.

Compute signed distance $d(x)$ and boundary mask. Geometry embedding:

$$\Phi_\Omega = [\text{statistics of } d(x),\ \lambda_1, \ldots, \lambda_m] \tag{8.22}$$

Lets now see when to use which Family I embedding using Table 8.2.

8.3 Family II: Learned Latent Spaces

Family II embeddings are *learned* representations: instead of prescribing the embedding from geometry or operators (Family I), we train an encoder Φ_θ so that the latent variable $z = \Phi_\theta(x)$ captures the essential structure of a scientific object x (field, trajectory, graph state) in a compact, reusable form. The primary motivation is that many scientific datasets exhibit nonlinear low-dimensional structure (multiple regimes, bifurcations, intermittency, localized phenomena) that fixed bases may not capture efficiently.

Imagine you're studying fluid flow around obstacles. You've collected 10,000 simulation snapshots showing different flow patterns: steady laminar flow, vortex shedding, turbulent regimes, and everything in between.

Family I methods (Fourier features, Laplacian eigenmodes) work beautifully for smooth fields in a single regime. But what happens when your data contains:

- **Multiple regimes** (laminar → transitional → turbulent)?
- **Bifurcations** (sudden qualitative changes in behavior)?
- **Intermittency** (rare but important events)?
- **Localized phenomena** (shocks, boundary layers, vortices)?

These exhibit nonlinear low-dimensional structure that fixed geometric bases struggle to capture efficiently. You could use 1000 Fourier modes and still miss the essential physics, or you could learn a 20-dimensional latent space that naturally separates regimes.

A core theme in SciML is that a learned latent space must be more than compressive: it must be physically meaningful, stable to discretization changes, and compatible with symmetries and constraints (boundary conditions, conservation laws, invariances/equivariances).

Let $x \in X$ denote a scientific object. Examples i) a field snapshot $u(\cdot)$ sampled on a grid/mesh, ii) a spatiotemporal window $u(\cdot, t : t + T)$, iii) a molecular configuration (point cloud + attributes), iv) a graph state (nodes/edges with features).

A learned embedding is a map $\Phi_\theta : X \to \mathbb{R}^d$ trained from data. In SciML, we typically want:

- **Compression**: $d \ll \dim(x)$ without losing task-relevant information.
- **Stability**: small physical perturbations (noise, mesh changes) produce small changes in z.
- **Constraint awareness**: boundary conditions, invariances/equivariances, and conservation are reflected in z.
- **Transferability**: the same z works across tasks (surrogate modeling, inverse problems, control, UQ).

In SciML, the training objective includes:

$$\mathcal{L} = \underbrace{\mathcal{L}_{\text{reconstruction}}}_{\text{compression}} + \underbrace{\mathcal{L}_{\text{physics}}}_{\text{constraints}} + \underbrace{\mathcal{L}_{\text{regularization}}}_{\text{stability}} \tag{8.23}$$

This goes beyond pure compression to ensure physical validity.

Definition (Learned latent embedding)

A *learned latent embedding* is an encoder $\Phi_\theta : X \to \mathbb{R}^d$ trained from a dataset $\{x^{(i)}\}_{i=1}^N$ such that $z = \Phi_\theta(x)$ is a compact representation enabling reconstruction, prediction, or constraint satisfaction. In SciML, the training objective usually includes physics- or symmetry-aware regularizers beyond pure reconstruction.

8.3.1 Autoencoders for Scientific Objects

Autoencoders learn a latent variable z by forcing information through a bottleneck. The basic idea is beautifully simple, but the devil is in the details for SciML applications.

An Autoencoder consists of two functions:

$$\boxed{\begin{array}{ll} \text{Encoder:} & z = \Phi_\theta(x) \quad \text{(compress)} \\ \text{Decoder:} & \hat{x} = \Psi_\phi(z) \quad \text{(reconstruct)} \end{array}} \tag{8.24}$$

The training loss is:

$$\min_{\theta,\phi} \mathcal{L}_{\text{AE}} = \mathbb{E}_{x\sim\mathcal{D}}\Big[\mathcal{L}_{\text{rec}}(x, \Psi_\phi(\Phi_\theta(x)))\Big] \tag{8.25}$$

Analogy: Packing a Suitcase

Encoder (Packing):
- You have many items (high-dimensional x)
- Must fit in small suitcase (low-dimensional z)
- Keep only essentials, compress efficiently

Decoder (Unpacking):
- Open suitcase, reconstruct your belongings
- Hope nothing important was left behind!
- Quality depends on how well you packed

Training process:
- Practice packing/unpacking many times
- Learn what's essential versus redundant
- Optimize suitcase organization

In SciML: We're "packing" a million-dimensional velocity field into a 20-dimensional suitcase, then "unpacking" it to reconstruct the flow!

8.3.1.1 Choosing a Reconstruction Loss in SciML

This is where SciML diverges sharply from computer vision! Don't use generic pixel-wise losses blindly!

Table 8.3 Physics-aware reconstruction losses for different problems

Problem type	Loss formula	Why this matters
Smooth fields (heat, diffusion)	$\mathcal{L} = \|u - \hat{u}\|_{L^2}^2$	Energy norm, physically meaningful
PDE-governed (elasticity, fluids)	$\mathcal{L} = \|u - \hat{u}\|_{L^2}^2 + \lambda\|\nabla u - \nabla\hat{u}\|_{L^2}^2$	Derivatives matter! Flux, stress depend on ∇u
Conservation (mass, energy)	$\mathcal{L} = \|u - \hat{u}\|^2 + \lambda_c\|\int u - \int \hat{u}\|$	Preserve total quantities
Boundary-critical (BVPs)	$\mathcal{L} = \|u - \hat{u}\|^2 + \lambda_b\|u - \hat{u}\|_{\partial\Omega}^2$	BC satisfaction crucial
Sharp fronts (shocks, phases)	$\mathcal{L} = \|u - \hat{u}\|_{L^1} + \lambda\|\nabla u - \nabla\hat{u}\|_{L^1}$	L^1 better for discontinuities

In computer vision, MSE (Mean Squared Error) on pixels works fine:

$$\mathcal{L}_{\text{CV}} = \frac{1}{N}\sum_{i=1}^{N}(x_i - \hat{x}_i)^2 \tag{8.26}$$

Problems in SciML:

1. **Ignores physics**: Small pixel error might mean large flux/stress error
2. **Uniform weighting**: Treats boundary and interior equally (wrong!)
3. **No derivatives**: Many physics quantities need ∇u, not just u
4. **Ignores conservation**: Total mass/energy can drift

Result: Reconstruction looks visually good, but physics is broken! Instead, use losses that match your physics (refer Table 8.3).

Design rule (Physics-aligned reconstruction)

If the downstream task is sensitive to derivatives (e.g., stress from displacement, flux from concentration), the reconstruction loss must include derivative-aware terms (e.g., H^1-style). Pixel-wise losses often produce latents that look good but fail physically.

8.3.1.2 Convolutional and Mesh-Aware Autoencoders

For Euclidean grids, Convolutional Autoencoders (CAEs) exploit translation structure. For irregular meshes/graphs, one uses:

- graph neural network encoders/decoders,
- mesh convolution operators,
- attention on mesh tokens with spectral/geometry positional encodings (hybrid with Family I).

A robust SciML pattern is to **tokenize using Family I geometry** and then learn the latent with an AE (Family II), yielding discretization-stable learned latents.

8.3.1.3 Variational Autoencoder (VAE): Probabilistic Latents for Uncertainty

When uncertainty matters (stochastic dynamics, measurement noise, inverse problems), a Variational Autoencoder (VAE) provides a distributional latent:

$$q_\theta(z|x) \text{ and } p_\phi(x|z),$$

trained by maximizing an Evidence Lower Bound (ELBO):

$$\max_{\theta,\phi} \underbrace{\mathbb{E}_{z\sim q_\theta(z|x)}[\log p_\phi(x|z)]}_{\text{reconstruction}} - \beta \underbrace{\mathrm{KL}(q_\theta(z|x) \parallel p(z))}_{\text{regularization}} \tag{8.27}$$

where $p(z)$ is a prior (often $\mathcal{N}(0, I)$) and β controls the bottleneck.

When uncertainty matters, VAE provides a distributional latent instead of a point estimate.

Analogy: Forecasting with Confidence

Standard AE (Point Estimate):

- "Temperature tomorrow will be 25.0°C"
- Single prediction, no uncertainty
- Like giving one specific answer

VAE (Distribution):

- "Temperature will be 25 ± 3°C (normal distribution)"
- Captures uncertainty
- Can sample multiple possible futures
- Like giving a range with confidence

In SciML:

- **AE**: Use for deterministic prediction, control
- **VAE**: Use for uncertainty quantification, sampling, inverse problems with noise

8.3.2 *Physics and Symmetry-Aware Latent Learning*

A purely data-driven AE can learn discretization shortcuts or violate constraints. SciML therefore augments the objective.

8.3.2.1 Physics-Informed Autoencoder (PIAE)

Let $\mathcal{N}[u] = f$ denote a governing PDE operator (or ODE residual), with boundary/initial conditions. A Physics-Informed Autoencoder (PIAE) adds penalties to encourage decoded fields to satisfy physics [7]:

$$\boxed{\begin{aligned}\min_{\theta,\phi}\ & \underbrace{\mathcal{L}_{\text{rec}}(x,\hat{x})}_{\text{data fidelity}} + \lambda_{\text{phys}} \underbrace{\mathcal{L}_{\text{phys}}(\hat{x})}_{\text{PDE residual}} \\ & + \lambda_{\text{bc}} \underbrace{\mathcal{L}_{\text{bc}}(\hat{x})}_{\text{boundary}} + \lambda_{\text{inv}} \underbrace{\mathcal{L}_{\text{inv}}(\hat{x})}_{\text{invariants}}\end{aligned}} \tag{8.28}$$

Example physics penalties.

- PDE residual: $\mathcal{L}_{\text{phys}} = \|\mathcal{N}[\hat{u}] - f\|^2$ evaluated at collocation points.
- Conservation: $\mathcal{L}_{\text{inv}} = |\mathcal{C}(\hat{u}) - \mathcal{C}(u)|$ where $\mathcal{C}$ is conserved quantity.
- Boundary: $\mathcal{L}_{\text{bc}} = \|\hat{u}|_{\partial\Omega} - g\|^2$ for Dirichlet, or flux residual for Neumann.

Definition (Physics-Informed Autoencoder)

A *Physics-Informed Autoencoder* is an autoencoder trained with a composite loss combining reconstruction with explicit physical constraints (governing equations, boundary/initial conditions, and invariants), so that the latent space encodes physically admissible solutions rather than arbitrary compressions.

8.3.2.2 Symmetry Regularization: Invariance and Equivariance Probes During Training

If a group G represents a physical symmetry (e.g., rotations for molecules, translations for homogeneous media, permutations for identical particles), then we want either:

$$\Phi_\theta(g \cdot x) \approx \Phi_\theta(x) \quad \text{(invariance)},$$

or

$$\Phi_\theta(g \cdot x) \approx \rho(g)\Phi_\theta(x) \quad \text{(equivariance)},$$

where ρ is a representation of G in latent space. A practical training regularizer is:

$$\mathcal{L}_{\text{sym}} = \mathbb{E}_{g\sim G}\big[\|\Phi_\theta(g \cdot x) - \rho(g)\Phi_\theta(x)\|^2\big]. \tag{8.29}$$

Even when the architecture is not strictly equivariant, this regularizer acts as a *soft constraint* improving robustness.

8.3.3 Masked Modeling for SciML

Masked modeling learns representations by reconstructing missing parts of the input from context (please see Fig. 8.19). This is extremely effective when labeled tasks are scarce but raw simulation/measurement data are abundant.

8.3.3.1 Masked Autoencoder (MAE) Objective

Let x be tokenized into T tokens: $x = (x_1, \ldots, x_T)$ (patches on a grid, elements on a mesh, nodes on a graph, spectral bands, or time segments). Let $M \subset \{1, \ldots, T\}$ be masked indices, and $\bar{M}$ the visible set. A masked encoder processes only visible tokens:

$$z = \Phi_\theta(x_{\bar{M}}),$$

and the decoder reconstructs the masked tokens:

$$\hat{x}_M = \Psi_\phi(z;\ \text{mask tokens, PE}).$$

The masked loss is

$$\min_{\theta,\phi} \mathcal{L}_{\text{MAE}} = \mathbb{E}_{x,M}\Big[\sum_{i\in M} \ell(x_i, \hat{x}_i)\Big]. \tag{8.30}$$

Definition (Masked Autoencoder pretraining)

A *Masked Autoencoder* trains an embedding by reconstructing masked tokens from visible context. The encoder learns representations that capture global structure because it must infer missing regions from physics-consistent correlations.

Table 8.4 Masking strategies for different scientific domains

Strategy	When to use	What it teaches
Random independent	Simple baseline, sparse sensors	Local interpolation, basic smoothness
Spatial block masking	PDEs, continuous fields	Long-range dependencies, boundary effects, PDE smoothness
Spectral band masking	Multi-scale phenomena (turbulence, waves)	Cross-scale coupling, energy cascade
Temporal segment	Time series, dynamics	Causal structure, temporal evolution
Boundary-focused	BVP-dominated problems	Boundary layer physics, BC propagation
Physics-informed (adaptive)	Complex multi-physics	Critical feature detection

Not all masking is created equal! Strategy should match your physics. Table 8.4 discusses this.

8.3.3.2 Physics-Consistent Masking Strategies

Masking should match the causal/statistical structure of scientific data:

- **Spatial block masking**: mask contiguous regions to force reasoning about PDE smoothness and boundary effects.
- **Spectral band masking**: mask frequency ranges to learn cross-scale couplings (useful in turbulence, waves).
- **Temporal segment masking**: mask future/past segments in trajectories to learn dynamics.
- **Boundary-focused masking**: occasionally mask near $\partial\Omega$ to ensure boundary-aware representations.

Random independent masking may be too easy (trivial interpolation) or physically inconsistent (e.g., masking isolated nodes on a mesh may not reflect real missingness).

8.3.3.3 MAE on Meshes and Graphs

For meshes/graphs, tokens are nodes/elements with geometry positional encodings (Family I). The MAE is then a *structure-aware denoising task* where the model must use adjacency and intrinsic coordinates to reconstruct masked values.

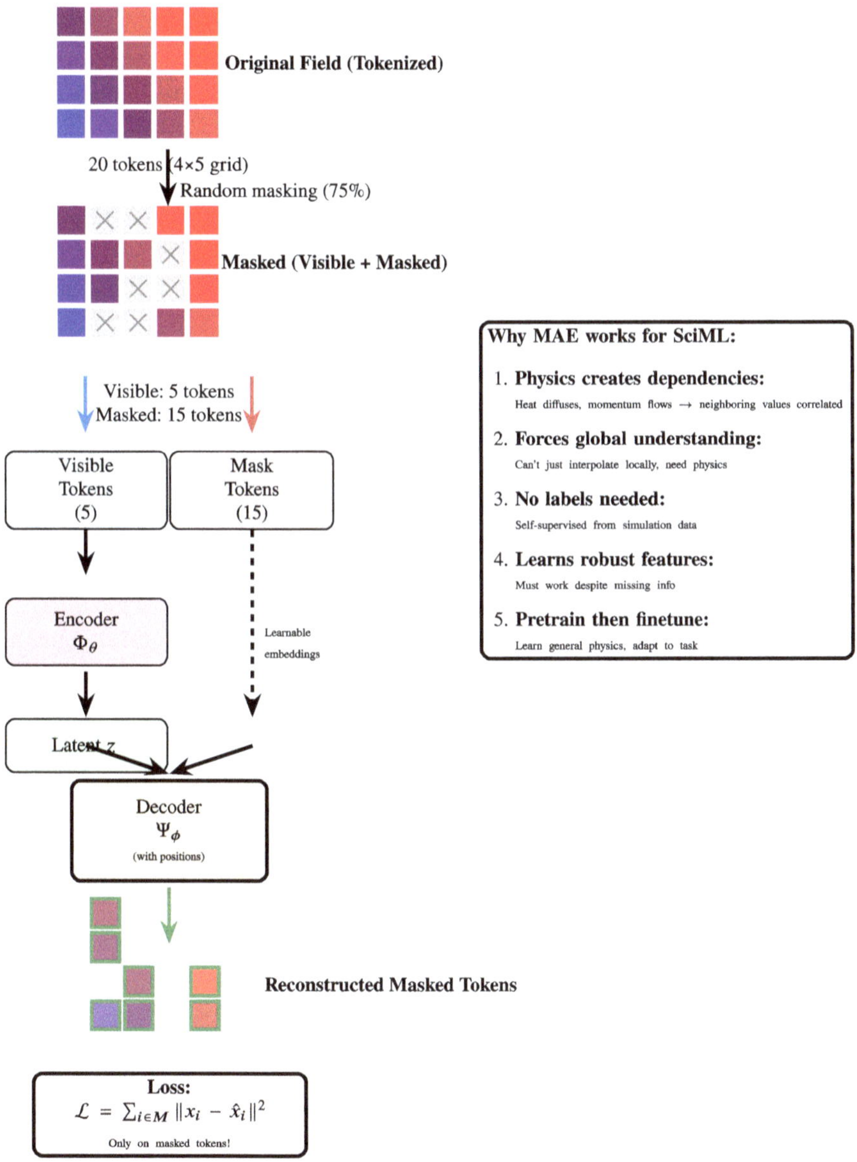

Fig. 8.19 Masked Autoencoder for scientific fields: (1) Tokenize field into patches/elements, (2) Randomly mask 75% of tokens, (3) Encode only visible tokens to latent z, (4) Decode to reconstruct masked tokens using learned mask embeddings and positional encodings, (5) Loss computed only on masked tokens. Forces encoder to learn physics-aware representations that can infer missing values

8.3.4 *Contrastive Pretraining for SciML*

Contrastive learning trains representations by pulling together different "views" of the same object and pushing apart views from different objects.

Analogy: Learning Through Comparison

How humans learn concepts:

- See dog photo from front → see same dog from side → "that's the same dog!"
- See different dog → "that's a different dog"
- Learn: dogs look similar from different angles (invariance)
- Learn: different dogs are distinct entities

Contrastive learning in SciML:

- Same flow field on two meshes → "same physics!"
- Same molecule rotated → "same energy!"
- Different flow regimes → "distinct states"
- Learn: physics is mesh-invariant, rotation-invariant, etc.

Power: Automatically discovers what should be invariant from data alone!

8.3.4.1 InfoNCE Objective

Let x be an object and $v_1 = \mathcal{A}_1(x)$, $v_2 = \mathcal{A}_2(x)$ be two augmented views. With encoder Φ_θ and projection head h, define

$$q_1 = h(\Phi_\theta(v_1)), \qquad q_2 = h(\Phi_\theta(v_2)).$$

Given a batch, the InfoNCE loss [8] for a positive pair (i) is

$$\mathcal{L}_{\text{NCE}}^{(i)} = -\log \frac{\exp(\text{sim}(q_1^{(i)}, q_2^{(i)})/\tau)}{\sum_j \exp(\text{sim}(q_1^{(i)}, q_2^{(j)})/\tau)}, \tag{8.31}$$

where sim is cosine similarity and τ is a temperature parameter.

Definition (Contrastive pretraining)

Contrastive pretraining learns embeddings by maximizing agreement between embeddings of two physics-consistent views of the same object while separating embeddings of different objects, typically using an InfoNCE-type loss.

Table 8.5 Valid augmentations for SciML contrastive learning

Augmentation	When valid	**Example**
Symmetry transforms	Isotropic physics	Rotations for molecules, translations for homogeneous media
Mesh resampling/remeshing	Always! (Critical!)	Same field on different mesh discretizations
Physics-bounded noise	Known measurement uncertainty	Gaussian noise with std from sensors
Spatial cropping/zooming	Translation invariant	Different observation windows of large domain
Non-dimensionalization invariance	After proper scaling	Rescale by characteristic L, T, U

8.3.4.2 What Counts as a Valid Augmentation in SciML?

Unlike natural images, SciML augmentations must preserve meaning under physics (also refer Table 8.5):

- **Symmetry transforms**: rigid rotations/translations for isotropic physics; permutations for identical particles.
- **Mesh re-sampling/remeshing**: change discretization while preserving the underlying continuous field (crucial).
- **Physically bounded noise**: additive noise consistent with measurement uncertainty.
- **Time reparameterization (careful)**: only when dynamics are time-scale invariant or appropriately normalized.
- **Unit-consistent scaling**: only after non-dimensionalization; otherwise can break physics.

Warning (Invalid augmentations)

Augmentations that change boundary conditions, violate conservation, or distort geometry without compensating features can force the latent space to learn non-physical invariances, harming downstream generalization.

8.3.4.3 Contrastive Pretraining as Discretization-Invariance Training

A powerful SciML practice is to generate two views by sampling the *same underlying field* on two different discretizations (two meshes or two resolutions). Contrastive pretraining then explicitly encourages

$$\Phi_\theta(x_{\text{mesh A}}) \approx \Phi_\theta(x_{\text{mesh B}}),$$

making the learned embedding robust to remeshing.

8.3.5 Temporal Latents for Trajectories

Many SciML datasets are trajectories: $x_{0:T} = (x_0, x_1, \ldots, x_T)$. We need latents that capture temporal evolution.

Analogy: Summarizing a Movie

Frame-by-frame (Standard):

- Encode each frame independently
- Loses temporal coherence
- Like describing movie as disconnected images

Sequence encoding (Temporal):

- Encode entire trajectory together
- Captures dynamics, evolution, trends
- Like describing movie plot: beginning → middle → end

In SciML: We want latents that encode "this system is in turbulent oscillation" not just "this single snapshot looks turbulent."

8.3.5.1 Sequence Autoencoders

Encode a trajectory with a temporal encoder (RNN/Transformer) [9] into z, then decode the sequence:

$$z = \Phi_\theta(x_{0:T}), \qquad \hat{x}_{0:T} = \Psi_\phi(z).$$

This produces trajectory-level latent codes useful for regime classification, control, and reduced-order modeling.

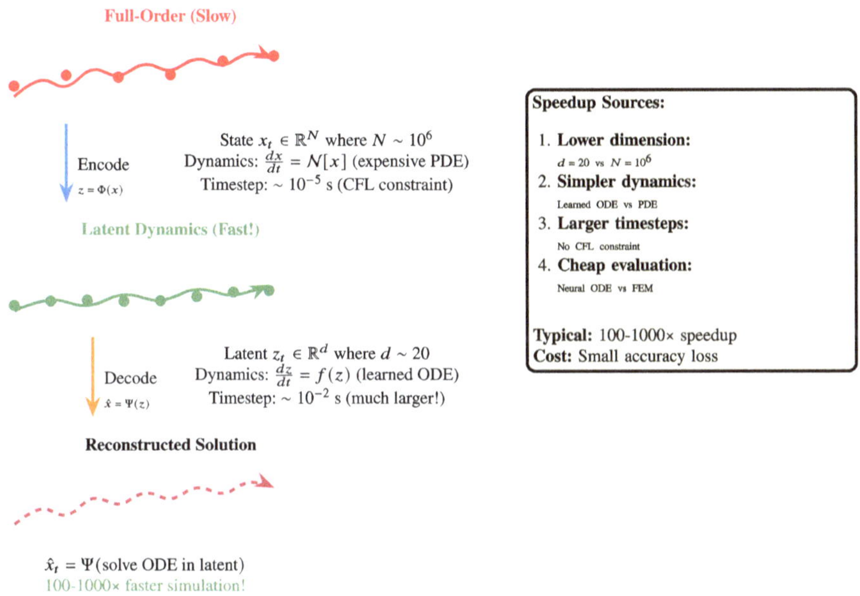

Fig. 8.20 Latent dynamics for reduced-order modeling: Full-order dynamics (top) are slow and expensive ($N \sim 10^6$ DOFs, tiny timesteps). Encode to low-dimensional latent space ($d \sim 20$), evolve dynamics there (much faster), decode to reconstruct solution. Achieves 100–1000× speedup with acceptable accuracy loss

8.3.5.2 Latent Dynamics and Reduced-Order Modeling

A common SciML aim is to learn a latent space where dynamics are simpler:

$$z_{t+1} \approx F(z_t) \quad \text{or} \quad \dot{z} = f(z, t), \tag{8.32}$$

while $x_t \approx \Psi_\phi(z_t)$. This supports fast simulation and control in latent space, with physics constraints applied to decoded states (Fig. 8.20).

Common SciML aim: Learn latent space where dynamics are simpler!

$$\boxed{\begin{aligned} \text{Latent ODE:} \quad & \dot{z} = f(z, t) \\ \text{or Latent map:} \quad & z_{t+1} \approx F(z_t) \end{aligned}} \tag{8.33}$$

while $x_t \approx \Psi_\phi(z_t)$.

Lets now formulate some steps for training Latent Dynamics Models:

1. **Step 1—Train autoencoder**:
   - Minimize $\sum_t \|x_t - \Psi(\Phi(x_t))\|^2$
   - Get encoder Φ and decoder Ψ

2. **Step 2—Extract latent trajectories**:
   - For each trajectory: $z_t = \Phi(x_t)$ for all t
   - Now have dataset $\{z_0, z_1, \ldots, z_T\}$
3. **Step 3—Learn dynamics in latent**:
   - Fit $\dot{z} = f(z)$ or $z_{t+1} = F(z_t)$
   - Use neural ODE, RNN, or polynomial model
4. **Step 4—Simulate**:
   - Start from $z_0 = \Phi(x_0)$
   - Integrate: z_t = solve ODE/map
   - Decode: $\hat{x}_t = \Psi(z_t)$

Advanced: Train jointly (end-to-end) for better dynamics fit.

Lets now work on an example. We shall reuse Poisson problem from Family I:

$$-\Delta u = f \text{ in } \Omega, \qquad u = 0 \text{ on } \partial\Omega \tag{8.34}$$

Goal: Learn compact latent $z \in \mathbb{R}^d$ for solution snapshots u that remains stable under remeshing.

Step 1: Tokenize robustly (Family I)

Use geometry-aware tokenization:

- Compute spectral positional encodings from domain Laplacian
- Attach boundary channels (signed distance $d(x)$, boundary mask, BC type/value)
- Represent u on mesh as nodal tokens with these encodings

Step 2: Train autoencoder with physics-aligned loss

$$\mathcal{L} = \underbrace{\|u - \hat{u}\|^2_{L^2(\Omega)}}_{\text{reconstruction}} + \lambda_\nabla \underbrace{\|\nabla u - \nabla\hat{u}\|^2_{L^2(\Omega)}}_{\text{gradient matching}} + \lambda_{\text{bc}} \underbrace{\|\hat{u}|_{\partial\Omega}\|^2}_{\text{Dirichlet BC}} \tag{8.35}$$

Step 3: Physics probes and validation

- **PDE residual**: $\|\Delta\hat{u} + f\|$ on collocation points
- **Remeshing test**: Sample same u on two meshes, measure $\|\Phi_\theta(u_A) - \Phi_\theta(u_B)\|$
- **Boundary error**: Measure error near $\partial\Omega$ separately

Family II learns task-adaptive representations that can achieve dramatic compression while preserving physics. The key is combining data-driven learning with physics constraints, geometric tokenization, and rigorous validation. Potentially, start with Family I for geometry-rich, single-regime problems. Add Family II when you need to handle complex multi-regime data or achieve extreme compression ratios!

8.4 Family III: Operator and Boundary-Condition Embeddings

Family III embeddings are designed for *operator learning*: instead of learning a map between finite-dimensional vectors, we learn a map between *functions*. In SciML, the central object is often a **solution operator** that maps a *problem specification* (coefficients, forcing, geometry, boundary/initial conditions) to a *solution field*. Family III focuses on how to embed the *problem itself* so that a model generalizes across:

- different coefficient fields (material parameters, permeability, viscosity),
- different forcing terms and sources,
- different geometries/domains,
- different boundary and initial conditions,
- different discretizations (meshes/resolutions).

This family is the backbone of neural operators and modern surrogate modeling for parametric partial differential equations (PDEs).

Imagine you're an engineer designing airplane wings. For each wing shape, you need to solve complex fluid dynamics equations to predict lift, drag, and pressure distributions. A traditional approach would be:

- Design wing shape → Run expensive CFD simulation → Get pressure field
- Change wing slightly → Run simulation again → Get new pressure field
- Repeat thousands of times for optimization

Each simulation might take hours or days. But notice the pattern: you're really computing the same *type* of operation repeatedly–just with different inputs (wing shapes, flight conditions).

What if instead of learning individual pressure fields, we learned the entire simulation process itself?

Analogy: Recipe versus Cookbook

Family I and II (Function Learning):

- Like learning to make *one specific dish*
- Input: Ingredients for this dish → Output: This cooked meal
- If ingredients change → need to learn new mapping

Family III (Operator Learning):

- Like learning to be a *chef who can follow any recipe*
- Input: Recipe (instructions + ingredients) → Output: Corresponding meal
- Can handle entirely new recipes by understanding cooking principles

In SciML:

- Family I/II: Map parameters to solution for *one* PDE
- Family III: Map PDE specification → solution for *any instance* of that PDE family
- The "recipe" is the PDE with its coefficients, forcing, geometry, and boundary conditions!

8.4.1 From Function Regression to Operator Regression

A typical SciML PDE can be summarized as:

$$\boxed{\mathcal{N}\big[u; a\big](x) = f(x) \quad \text{in } \Omega, \qquad \mathcal{B}[u](x) = g(x) \quad \text{on } \partial\Omega} \tag{8.36}$$

where:

- $u : \Omega \to \mathbb{R}^c$ is the **unknown solution field** (what we want to predict)
- a denotes **coefficient fields** or parameters (material properties, possibly spatially varying)
- f is **forcing/source** (external loads, heat sources)
- Ω is the **domain/geometry** (where the problem lives)
- $\mathcal{B}$ specifies **boundary conditions** (Dirichlet/Neumann/Robin/mixed)
- g is the **boundary value/flux**

The *solution operator* is the map:

$$\boxed{\mathcal{S} : \underbrace{(a, f, \Omega, \mathcal{B}, g)}_{\text{problem specification}} \mapsto \underbrace{u}_{\text{solution field}}} \tag{8.37}$$

Definition (Operator learning)

Operator learning aims to approximate a map between function spaces,

$$\mathcal{S} : \mathcal{X} \to \mathcal{Y},$$

where $\mathcal{X}$ represents problem specifications (coefficient and forcing fields, boundary/initial conditions, and geometry) and $\mathcal{Y}$ represents solution fields. The approximation should be stable across discretizations and should generalize to new instances in $\mathcal{X}$.

8.4.2 *What Exactly Must Be Embedded? The "problem Specification"*

A frequent reason operator surrogates fail is incomplete input encoding. For robust generalization, Family III insists on embedding *all sources of variability*:

$$\underbrace{(a, f)}_{\text{physics inputs}} + \underbrace{\Omega}_{\text{geometry}} + \underbrace{(\mathcal{B}, g)}_{\text{boundary conditions}} + \underbrace{\text{query locations } y}_{\text{where we evaluate } u(y)}. \tag{8.38}$$

Most operator models can be understood as learning two coupled embeddings:

- a **problem embedding** $\Phi_\theta(a, f, \Omega, \mathcal{B}, g)$ that summarizes the instance;
- a **query embedding** $\pi(y, \Omega)$ that represents spatial location (and geometry) at which we predict the output.

The prediction is then some combination, e.g.

$$\boxed{\hat{u}(y) = \mathcal{F}\big(\underbrace{\Phi_\theta(a, f, \Omega, \mathcal{B}, g),}_{\text{what problem}} \quad \underbrace{\pi(y, \Omega)}_{\text{where to evaluate}}\ \big)} \tag{8.39}$$

Design rule (Never omit boundary conditions)

Boundary conditions change the solution space. If $(\mathcal{B}, g)$ varies across the dataset, it must be explicitly embedded. Treating BCs as "fixed" when they are not leads to brittle surrogates and systematic boundary errors.

8.4.3 *Embedding Strategies for Operator Learning*

There are two dominant design philosophies in Family III (refer Table 8.6), each with different strengths.

Both strategies require geometry and boundary information–just integrated differently!

Table 8.6 Two operator learning strategies

Aspect	Strategy A: global + pointwise	Strategy B: token + mixing
Philosophy	Create global embedding, decode at points	Embed as tokens, mix globally
Example	Deep Operator Network (DeepONet)	Fourier Neural Operator (FNO)
Problem encoding	Sample inputs at sensors, aggregate	Tokenize on grid/mesh
Query handling	Trunk evaluates at arbitrary points	Project from token grid
Best for	Variable resolution, irregular queries	Regular domains, spectral structure

8.4.4 Deep Operator Network (DeepONet)

Deep Operator Network (DeepONet) [10] elegantly separates *what problem* (branch) from *where to evaluate* (trunk) [11].

Analogy: Orchestra Performance

Branch (What to Play):

- Encodes the musical score (the problem specification)
- Different pieces → different encodings
- Captures: tempo, key signature, dynamics

Trunk (Where/When in Performance):

- Encodes position in the piece (measure number, beat)
- Same encoding system for any piece
- Answers: "What should be playing at this moment?"

Combination (Prediction):

- Branch coefficients × Trunk basis = Music at this moment
- Can query any moment in the piece
- Can even query moments between written notes!

In DeepONet: Branch = problem instance, Trunk = spatial location, Prediction = solution value

8.4.4.1 Branch Network: Embedding the Input Functions

The branch network processes sampled inputs to create a global problem representation.

Let a and f be sampled at sensor points $\{x_i\}_{i=1}^m$:

$$\boxed{b = \Phi_{\text{branch}}\Big(\underbrace{[a(x_1), \ldots, a(x_m)]}_{\text{coefficients}}, \underbrace{[f(x_1), \ldots, f(x_m)]}_{\text{forcing}}, \underbrace{\text{BC/geometry}}_{\text{channels}}\Big) \in \mathbb{R}^p} \tag{8.40}$$

The branch output $b = [b_1, \ldots, b_p]$ is interpreted as coefficients of a learned basis. Branch aggregates information from *all* sensor points, creating a global problem descriptor that's independent of query location.

8.4.4.2 Trunk Network: Embedding the Query Location and Geometry

Given query point $y \in \Omega$, the trunk produces spatial features:

$$\boxed{t(y) = \Phi_{\text{trunk}}(\quad \underbrace{\pi(y, \Omega)}_{\text{positional encoding}} \quad) \in \mathbb{R}^p} \tag{8.41}$$

where positional encoding $\pi(y, \Omega)$ may include:

- Raw or Fourier-encoded coordinates
- Spectral positional encodings (Laplacian eigenfunctions)
- Signed distance to boundary: $d(y)$
- Boundary type indicators

The trunk output $t(y) = [t_1(y), \ldots, t_p(y)]$ is interpreted as **basis functions evaluated at** y.

8.4.4.3 Combining Branch and Trunk

Standard DeepONet prediction uses inner product:

$$\boxed{\hat{u}(y) = \sum_{k=1}^{p} b_k\, t_k(y) + \text{bias}} \tag{8.42}$$

We can interprete it as;

- Branch: "Here are the coefficients for this problem"
- Trunk: "Here are basis functions at location y"

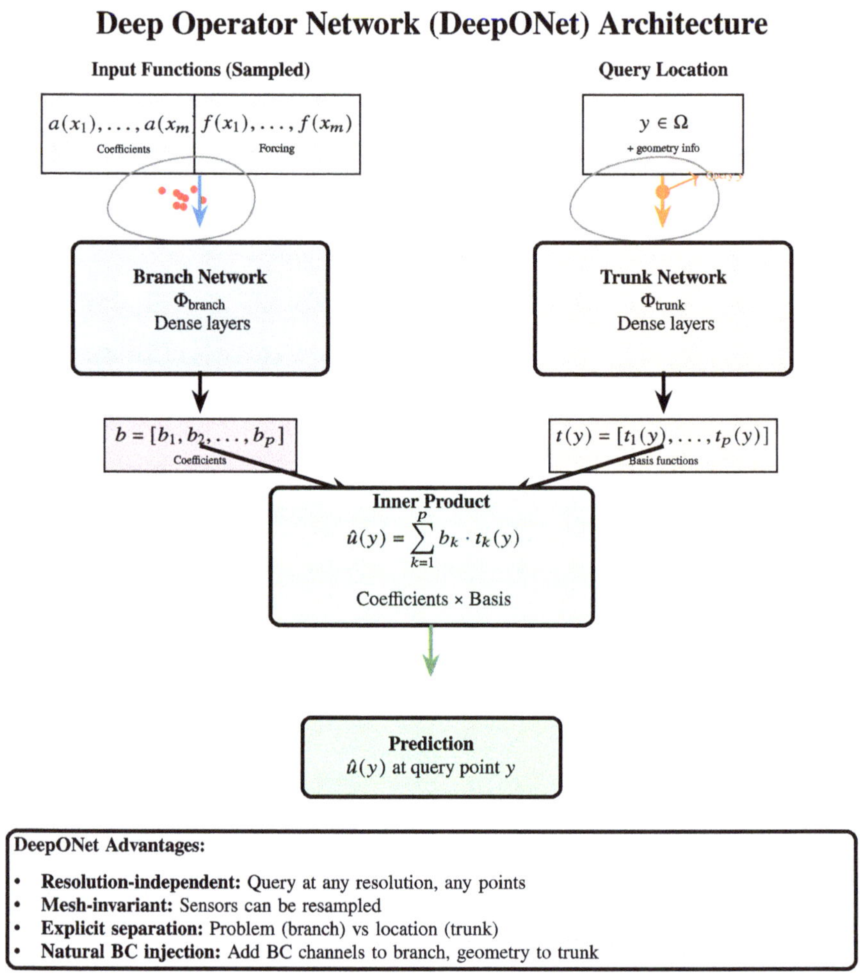

Fig. 8.21 DeepONet architecture: Branch network encodes problem instance (sampled inputs + BCs/geometry) into coefficients b. Trunk network encodes query location y (with geometry info) into basis functions $t(y)$. Inner product $\sum_k b_k t_k(y)$ predicts solution at y. This separation enables resolution-independent queries.

- Product: "Linear combination gives solution at y".

A *Deep Operator Network* approximates solution operator by:

1. **Branch embedding**: $b = \Phi_{\text{branch}}(\text{inputs} + \text{BCs})$: problem coefficients
2. **Trunk embedding**: $t(y) = \Phi_{\text{trunk}}(\text{position} + \text{geometry})$: spatial basis
3. **Combination**: $\hat{u}(y) = \sum_k b_k t_k(y)$: weighted sum.

Key property: Can query solution at *any* point y, not just training locations!

A practical DeepONet Implementation can now be stated with a Fig. 8.21:

1. **Choose sensor locations**: (for branch)

   - Can be: regular grid, random sampling, or fixed sensors
   - Typical: 100–1000 sensors
   - Include boundary sensors for BC information.

2. **Prepare branch input**:

$$[\underbrace{a(x_1), \ldots, a(x_m)}_{\text{m values}}, \underbrace{f(x_1), \ldots, f(x_m)}_{\text{m values}}, \underbrace{\text{BC info}}_{\text{extra channels}}] \tag{8.43}$$

3. **Prepare trunk input**: (for each query y)

$$[\underbrace{\pi(y)}_{\text{position}}, \underbrace{d(y)}_{\text{distance to boundary}}, \underbrace{\text{BC type}}_{\text{if varies}}] \tag{8.44}$$

4. **Choose architecture**:

   - Branch: Dense network, output dim p (e.g., 100)
   - Trunk: Dense network, output dim p (same!)
   - Depth: 3–6 hidden layers typical.

5. **Training data**:

   - Need pairs: $(a, f, \text{BCs}, \Omega) \to$ solution u
   - For each instance: query at many y points
   - Typical: 1000 instances $\times$ 1000 queries each.

Common issue: If branch and trunk dimensions don't match (p), we can't compute inner product!

8.4.5 *Fourier Neural Operator (FNO): Token-Wise Embeddings with Spectral Mixing*

The Fourier Neural Operator (FNO) [12] learns a mesh-independent map between discretized function samples by using global spectral convolutions.

Analogy: Audio Equalizer

Audio Equalizer:

- Decomposes sound into frequencies (Fourier transform)

- Amplifies/attenuates different frequency bands
- Reconstructs modified sound (inverse Fourier)
- Same process works for any audio, any duration.

FNO:

- Decomposes field into spatial frequencies (Fourier transform)
- Learns which frequencies to amplify (learnable weights $R(k)$)
- Reconstructs output field (inverse Fourier)
- Same weights work for any resolution!

Key advantage: Global mixing in frequency domain is much more efficient than convolutions in spatial domain for PDEs with long-range interactions!

8.4.5.1 Tokenization and Lifting

Start with input field representing problem specification:

$$v_0(x) = \Big[\underbrace{a(x)}_{\text{coeff}},\ \underbrace{f(x)}_{\text{forcing}},\ \underbrace{\mathrm{BC}(x)}_{\text{boundary}},\ \underbrace{\mathrm{geom}(x)}_{\text{geometry}}\Big] \tag{8.45}$$

Lifting layer maps to higher-dimensional feature space:

$$\boxed{v_0(x) \xrightarrow{\text{lift (linear)}} v_0'(x) \in \mathbb{R}^{d'}} \tag{8.46}$$

Typical: Input channels $c \sim$ 5-10, lift to $d' \sim$ 32-128.

8.4.5.2 Spectral Convolution Layer

Each FNO layer updates features via:

$$\boxed{v_{l+1}(x) = \sigma\left(\underbrace{W v_l(x)}_{\text{local}} + \underbrace{\mathcal{F}^{-1}\Big(R_l(k)\cdot\mathcal{F}(v_l)(k)\Big)(x)}_{\text{global spectral}}\right)} \tag{8.47}$$

where:

- $\mathcal{F}$: Fourier transform (FFT on grids)
- $R_l(k)$: Learnable complex weights on frequency modes

- $\mathcal{F}^{-1}$: Inverse Fourier transform
- W: Pointwise linear map (like 1×1 convolution)
- σ: Nonlinearity (typically GELU).

Mode truncation: Typically keep only low frequencies (e.g., lowest 12 modes in each direction) → parameter efficiency!

8.4.5.3 Output Projection

After several FNO layers ($L \sim$ 4-8):

$$v_L(x) \xrightarrow{\text{project (linear)}} \hat{u}(x) \tag{8.48}$$

Projects from feature space back to solution channels.

An Fourier Neural Operator (FNO) approximates operators using layers that mix information globally through spectral convolutions:

$$v_{l+1} = \sigma(W v_l + \mathcal{F}^{-1}(R_l \cdot \mathcal{F}(v_l))) \tag{8.49}$$

Key advantages are:

- Global receptive field (all points influence all points)
- Parameter-efficient (mode truncation)
- Resolution-independent (FFT scales to any size)
- Fast (FFT is $O(N \log N)$).

Some important FNO Limitations are:

1. **Requires regular grid**:
    - Standard FFT needs structured grid
    - For irregular geometries: need workarounds
    - Solutions: coordinate warping, graph Fourier, attention-based.
2. **Boundary ringing**:
    - Gibbs phenomenon near discontinuities
    - Fourier modes struggle with sharp boundaries
    - Fix: Add boundary padding, use boundary masks.
3. **Periodic assumption**:
    - FFT assumes periodic domain
    - Can cause artifacts for non-periodic BCs
    - Fix: Padding, windowing, or explicit BC channels.

4. **Mode truncation tradeoff**:
   - Too few modes → can't capture fine details
   - Too many modes → expensive, overfitting
   - Rule of thumb: 12–32 modes often sufficient.

8.4.6 Boundary-Condition Embeddings: Explicit, Implicit, and Hybrid

Boundary conditions are not "metadata"; they **define the solution space**. Two problems with identical interiors but different BCs have completely different solutions everywhere!

Analogy: Room Temperature Control

Consider a room with heaters:

Setup 1 (Dirichlet):
- Left wall: heater fixed at 30°C
- Right wall: cooler fixed at 10°C
- Result: Temperature gradient across room.

Setup 2 (Neumann):
- Left wall: insulated (zero flux)
- Right wall: insulated (zero flux)
- Same heating inside the room
- Result: Completely different temperature distribution!

Same interior conditions, different boundaries → different physics everywhere. Your model MUST know which boundary setup applies!

8.4.6.1 Explicit BC Channels (recommended Default)

Attach BC information as input channels to tokens:

$$\boxed{t(x) = \Big[a(x),\ f(x),\ \pi(x,\Omega),\ \underbrace{d(x)}_{\text{signed distance}},\ \underbrace{\mathbb{1}_{\partial\Omega}(x)}_{\text{boundary mask}},\ \underbrace{\text{BC-type},}_{\text{one-hot}}\ \underbrace{g(x)}_{\text{BC values}}\Big]} \tag{8.50}$$

Components explained:

1. **Signed distance** $d(x)$:

$$d(x) = \begin{cases} +\text{dist}(x, \partial\Omega) & \text{if } x \in \Omega \\ -\text{dist}(x, \partial\Omega) & \text{if } x \notin \Omega \end{cases} \tag{8.51}$$

   Tells model "how far from boundary"

2. **Boundary mask** $\nVdash_{\partial\Omega}(x)$:

$$\nVdash_{\partial\Omega}(x) = \begin{cases} 1 & \text{if } x \text{ on boundary} \\ 0 & \text{otherwise} \end{cases} \tag{8.52}$$

   Marks boundary locations explicitly

3. **BC type** (one-hot encoding):

   - Dirichlet: [1, 0, 0, 0]
   - Neumann: [0, 1, 0, 0]
   - Robin: [0, 0, 1, 0]
   - Mixed: [0, 0, 0, 1].

4. **BC values** $g(x)$:

   - For Dirichlet: prescribed value $u|_{\partial\Omega} = g(x)$
   - For Neumann: prescribed flux $\nabla u \cdot n = g(x)$
   - For Robin: combination $\alpha u + \beta \nabla u \cdot n = g(x)$.

5. **Normal vectors** $n(x)$ (when needed):

   - Outward unit normal at boundary points
   - Important for Neumann/Robin BCs
   - Can be computed from signed distance: $n(x) = \nabla d(x)/\|\nabla d(x)\|$.

8.4.6.2 Implicit BC Handling (physics Penalties)

Some models impose BCs via loss terms rather than inputs:

$$\mathcal{L}_{\text{BC}} = \|\hat{u}|_{\partial\Omega} - g\|^2 \quad \text{(Dirichlet)} \qquad \text{or} \qquad \|\nabla\hat{u} \cdot n - q\|^2 \quad \text{(Neumann)} \tag{8.53}$$

When this works:

- BCs are fixed across all instances
- BCs are simple (constant values)
- Strong penalty weight ensures satisfaction.

When this fails:

- BCs vary across dataset → model can't distinguish
- Complex spatial BC patterns
- Need to generalize to new BC types.

8.4.6.3 Hybrid Strategy (best Practice!)

The strongest SciML practice combines both:

$$\boxed{\text{BC as input channels} \quad + \quad \text{BC penalty in loss}} \tag{8.54}$$

Why hybrid wins:

- **Input encoding**: Model learns to adapt to different BCs
- **Loss penalty**: Ensures BCs are actually satisfied
- **Robustness**: Works even if encoding isn't perfect
- **Generalization**: Can handle unseen BC combinations

Boundary-Condition Embedding

A *boundary-condition embedding* represents $(\mathcal{B}, g)$ as model input, typically via:

- Boundary masks and signed distance
- BC type indicators (one-hot or learned embedding)
- BC value fields
- Normal vectors (for flux BCs).

This enables the learned operator to adapt to changes in boundary conditions across problem instances (Fig. 8.22).

Critical test: Can your model distinguish these two problems?

$$\begin{aligned} &\text{A:} \quad u|_{\partial\Omega} = 0 \qquad \text{(Dirichlet)} \\ &\text{B:} \quad \nabla u \cdot n|_{\partial\Omega} = 0 \qquad \text{(Neumann)} \end{aligned}$$

If not → BCs aren't properly encoded!

8.4.7 *Geometry Embeddings for Varying Domains*

When domain Ω varies across instances, geometry must be embedded. Three main approaches:

8.4.7.1 Level-Set/Signed Distance Representation

Represent domain via signed distance function $d_\Omega(x)$ defined over reference box:

$$\Omega = \{x : d_\Omega(x) \geq 0\} \tag{8.55}$$

Advantages:

- Converts geometry into a field (can be sampled anywhere)
- Works on regular grid
- Includes distance information.

Limitations:

- Requires embedding space larger than actual domain
- Can't easily represent disconnected regions
- Loses some topological information

8.4.7.2 Spectral Geometry Embeddings (intrinsic, Mesh-Invariant)

Compute Laplacian eigenfunctions/eigenvalues on Ω (Family I approach!):

- **Positional encodings**: $\mathrm{PE}(x) = [\psi_1(x), \ldots, \psi_m(x)]$
- **Global descriptors**: First few eigenvalues $[\lambda_1, \ldots, \lambda_m]$.

Advantages:

- Intrinsic to geometry (coordinate-free)
- Naturally mesh-invariant
- Captures topological features (e.g., number of holes).

Limitations:

- Requires solving eigenvalue problem (expensive)
- Different domains have different eigenfunctions
- Sign/multiplicity ambiguities

8.4.7.3 Boundary Parameterizations

For problems dominated by boundary behavior (exterior flows, scattering):

- Sample boundary as point cloud: $\{x_{\text{bdy},i}\}_{i=1}^{N_b}$
- Include curvature: $\kappa(x_{\text{bdy},i})$
- Include normals: $n(x_{\text{bdy},i})$
- Let model attend from interior to boundary.

Best for:

- Airfoil aerodynamics
- Shape optimization
- Scattering problems

Choosing Geometry Embedding Strategy

1. **Fixed domain → No geometry embedding needed!**
   - Just use coordinates or spectral PE
   - Focus on other varying inputs.
2. **Varying domains, regular embedding grid → Signed distance**
   - Convert each geometry to $d_\Omega(x)$ on grid
   - Include as additional input channel
   - Works with FNO, CNNs.
3. **Varying domains, irregular meshes → Spectral geometry**
   - Compute Laplacian eigenmodes for each domain
   - Use as positional encodings
   - Best mesh-invariance.
4. **Boundary-dominated physics → Boundary parameterization**
   - Sample boundary densely
   - Include curvature, normals
   - Use attention to boundary.

Can combine multiple: Signed distance + spectral PE often works best!

8.4.8 A Unifying Formulation: Operator Transformers on Scientific Tokens

Many modern architectures (attention-based neural operators) can be expressed as:

$$\boxed{\text{tokens } \{t(x_i)\}_{i=1}^{T} \xrightarrow{\text{global mixing (attention/operator layers)}} \{h(x_i)\}_{i=1}^{T} \xrightarrow{\text{projection}} \{\hat{u}(x_i)\}_{i=1}^{T}} \tag{8.56}$$

The crucial Family III question: **What goes inside $t(x_i)$?**

8.4.8.1 The Comprehensive Token Structure

A complete token includes *all* problem information:

$$\boxed{\begin{aligned} t(x_i) = \Big[\; &\underbrace{a(x_i)}_{\text{coefficients}}, \underbrace{f(x_i)}_{\text{forcing}}, \underbrace{\mathrm{PE}(x_i, \Omega)}_{\text{position}}, \\ &\underbrace{d_\Omega(x_i)}_{\text{geometry}}, \underbrace{\mathbb{1}_{\partial\Omega}(x_i)}_{\text{bdy mask}}, \underbrace{\text{BC-type/value}}_{\text{boundary}}, \underbrace{n(x_i)}_{\text{normal}} \Big] \end{aligned}} \tag{8.57}$$

Why This Token Structure is Universal?
This comprehensive token structure works across architectures:

- **For DeepONet**: Branch processes token features, trunk uses PE
- **For FNO**: Tokens on grid, FFT mixes globally
- **For GNN**: Tokens on mesh, message passing mixes
- **For Attention**: Tokens as sequence, attention mixes

Key insight: Architecture determines *how* to mix information, but *what* to mix is always the same comprehensive problem specification!
Design Rule: Separate "What" from "Where"
Always maintain clear separation:

- **What varies by instance**: Coefficients a, forcing f, BCs, geometry
- **Where you evaluate**: Query locations with intrinsic positional encodings

This separation enables:

1. Generalization to new discretizations
2. Querying at arbitrary resolution
3. Transfer across similar problem types.

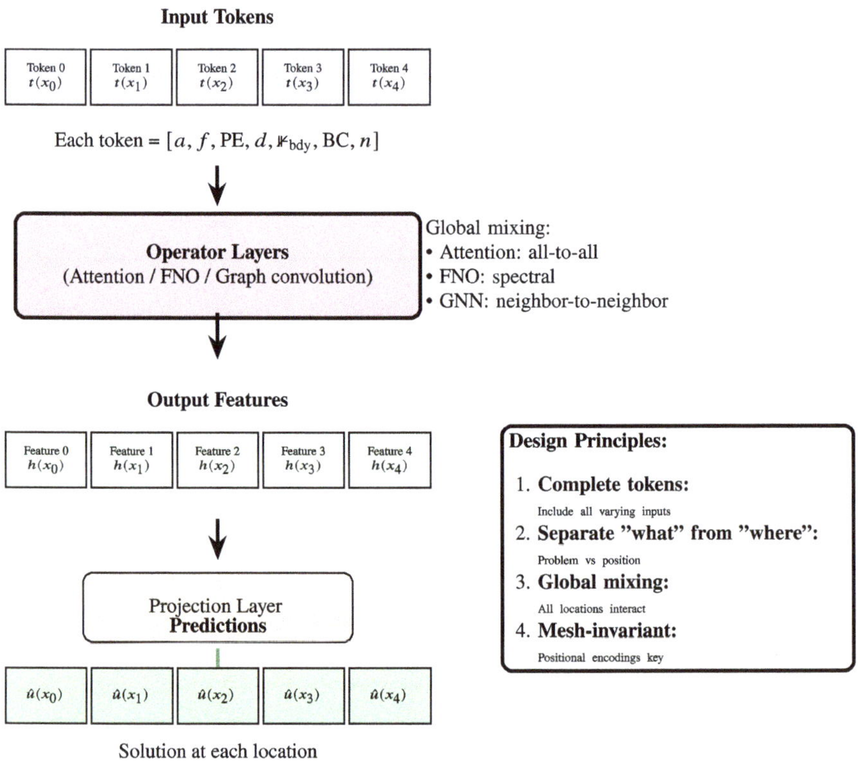

Fig. 8.22 Operator transformer unified view: Complete tokens $t(x_i)$ containing all problem information pass through operator layers (attention/FNO/GNN) for global mixing, producing features $h(x_i)$ that project to predictions $\hat{u}(x_i)$. Architecture-agnostic–works with any mixing strategy

Anti-pattern: Mixing problem specification with mesh-dependent features → brittle to remeshing!

Good pattern: Problem embedding (instance-level) + query embedding (location-level) → robust!

We now illustrate a canonical operator-learning problem where BC embeddings are essential. The problem setup: Darcy flow

Let u be pressure and $a(x)$ be permeability. The PDE is:

$$\boxed{-\nabla \cdot (a(x)\nabla u(x)) = f(x) \quad \text{in } \Omega, \qquad \mathcal{B}[u] = g \quad \text{on } \partial\Omega} \tag{8.58}$$

Physical interpretation:

- Models flow through porous media (groundwater, oil reservoirs)
- $a(x)$: Permeability (how easily fluid flows through material)
- $u(x)$: Pressure field
- $f(x)$: Sources/sinks (injection wells, pumping)
- BCs: Pressure at boundaries or no-flow conditions.

What varies across instances

1. Permeability field $a(x)$—different rock formations
2. Source term $f(x)$—different well configurations
3. Boundary conditions $(\mathcal{B}, g)$—different boundary pressures
4. (Potentially) Domain Ω—different reservoir geometries.

Goal: Learn operator $\mathcal{S} : (a, f, \Omega, \mathcal{B}, g) \mapsto u$ that generalizes to new instances!

Analogy: Water Flow Through Sponges

Permeability $a(x)$:

- Like sponge density–some areas let water through easily, others resist
- High a: Easy flow (coarse sponge)
- Low a: Restricted flow (dense sponge)

Pressure $u(x)$:

- How hard water "pushes" at each point
- Flows from high pressure to low pressure

Sources $f(x)$:

- Water being injected (+) or removed (âˆ’)

Boundary conditions:

- Fixed pressure at edges (Dirichlet)
- Or sealed/no-flow edges (Neumann)

The challenge: Predict pressure everywhere given sponge properties, source locations, and boundary setup!

8.4.8.2 Embedding Construction for Darcy Flow

For each spatial token x_i (grid cell, mesh element, or node), build comprehensive token:

$$\boxed{t(x_i) = \Big[\underbrace{a(x_i)}_{\text{permeability}}, \underbrace{f(x_i)}_{\text{source}}, \underbrace{\mathrm{PE}(x_i, \Omega)}_{\text{position}}, \underbrace{d_\Omega(x_i)}_{\text{distance}}, \underbrace{\mathbb{1}_{\partial\Omega}(x_i)}_{\text{bdy mask}}, \underbrace{\text{BC info}}_{\text{type/value}}\Big]} \tag{8.59}$$

Detailed breakdown:

1. **Permeability** $a(x_i)$:
   - Sample from permeability field
   - Often log-normal: $\log a \sim \mathcal{N}(\mu, \Sigma)$
   - Can vary over orders of magnitude: 10^{-4} to 10^0
2. **Source** $f(x_i)$:
   - Often sparse (wells at specific locations)
   - Can be positive (injection) or negative (extraction)
3. **Positional encoding** $\mathrm{PE}(x_i, \Omega)$:
   - For regular domains: Fourier features of coordinates
   - For irregular: Laplacian eigenmodes
4. **Signed distance** $d_\Omega(x_i)$:
   - Distance to nearest boundary
   - Positive inside, negative outside
   - Helps model understand proximity to BCs
5. **Boundary mask** $\mathbb{1}_{\partial\Omega}(x_i)$:
   - 1 if on boundary, 0 otherwise
   - Explicit marker for BC application points
6. **BC information**:
   - Type: Dirichlet (fixed pressure) vs Neumann (no flow)
   - Value: $g(x_i)$ for pressure or flux

Practical recipe (Operator + BC embedding pipeline)

1. Identify instance-varying inputs: $(a, f, \Omega, \mathcal{B}, g)$.
2. Choose a geometry representation: signed distance d_Ω and/or spectral positional encodings.
3. Encode BCs explicitly: boundary mask, BC-type indicator, BC-value fields, and normals when relevant.
4. Build tokens as in (8.57).
5. Train an operator model (DeepONet/FNO/operator transformer) and validate with BC-holdout and remeshing tests.

8.5 Family IV: Distribution Embeddings for Uncertainty and Ensembles

Many SciML problems do not produce a single deterministic state, but a *distribution* over states. This is the realm of uncertainty quantification, ensemble methods, and robust decision-making.

Imagine you're a meteorologist forecasting tomorrow's weather. You don't just predict "it will rain", you run 50 different simulations with slightly different initial conditions, and 30 of them show rain, 15 show clouds, 5 show sun. How do you:

- **Compare** this forecast ensemble to yesterday's ensemble?
- **Validate** against observed weather patterns?
- **Detect** when your model has shifted to a new regime?
- **Communicate** uncertainty to decision-makers?

Families I-III embed *individual samples* (single weather states). Family IV embeds *entire distributions* (weather forecast ensembles)!

Analogy: Individual versus Population

Families I-III (Individual Focus):

- Like describing one person: height, weight, age
- Good for: "Is this person healthy?"
- Limited: Can't answer "Is this population healthy?"

Family IV (Population Focus):

- Like describing a population: average height, variance, distribution shape
- Good for: "How does this city's health compare to that city?"
- Can detect: "This population has shifted to a new demographic"

In SciML:

- Family I-III: Embed one velocity field, one trajectory, one solution
- Family IV: Embed 1000 velocity fields from turbulent ensemble
- Enables: Compare simulation ensemble to experimental ensemble

Why this matters: Real physics is uncertain! Measurement noise, parameter uncertainty, stochastic forcing; we need to work with distributions, not just points.

Many SciML problems inherently involve distributions:

1. **Stochastic differential equations (SDEs)**

$$du = f(u, t)dt + g(u, t)dW(t) \tag{8.60}$$

The noise term dW makes each solution random → distribution of trajectories

2. **Random PDEs**:

$$-\nabla \cdot (a(x, \omega)\nabla u) = f(x) \tag{8.61}$$

Random coefficient $a(x, \omega)$ → distribution of solutions $u(x, \omega)$

3. **Uncertain initial/boundary conditions**:
   - Initial condition: $u(x, 0) \sim \mu_0$
   - Boundary condition: $u|_{\partial\Omega} \sim \mu_{\mathrm{BC}}$
   - Propagates to: $u(x, t) \sim \mu_t$.
4. **Monte Carlo simulation**:
   - Sample parameters: $\theta_1, \ldots, \theta_N \sim p(\theta)$
   - Solve for each: $u_i = \mathcal{S}(\theta_i)$
   - Get ensemble: $\{u_i\}_{i=1}^N$ representing $p(u)$.
5. **Data assimilation**:
   - Forecast ensemble: $\{x_i^f\}$ (particles)
   - After observations: $\{x_i^a\}$ (analysis ensemble)
   - Need to compare: forecast vs analysis distributions.

In these cases, it is not enough to embed one realization x; we need an embedding of the *probability measure* μ that generates the realizations.

Family IV embeddings lift uncertainty and ensembles to first-class objects. They enable:

- principled discrepancy between simulated and observed ensembles,
- out-of-distribution (OOD) regime shift detection,
- retrieval of similar uncertainty states (nearest-neighbor over ensembles),
- uncertainty-aware conditioning for downstream learning (operator learning with random inputs),
- consistent comparison of models (two-sample testing of outputs).

Definition (Distribution embedding)

Let $\mathcal{P}(X)$ be a set of probability measures on a space X (fields, trajectories, graphs). A *distribution embedding* is a map

$$\Xi : \mathcal{P}(X) \to \mathbb{R}^d$$

that represents an entire distribution (or ensemble) as a finite-dimensional vector while preserving relevant notions of similarity (e.g., moments, geometry-aware distances, or kernel discrepancies).

Key properties we want:

- **Similarity preservation**: Similar distributions $\rightarrow$ similar embeddings
- **Discriminative**: Different distributions $\rightarrow$ different embeddings
- **Computable**: Can estimate from finite samples
- **Physically meaningful**: Respects scientific scales and structure

8.5.1 What Is Being Embedded? Measures over Scientific Objects

Let $x \in X$ be a scientific object (e.g., a field $u(\cdot)$, or a trajectory $u(\cdot, t)$). Uncertainty is represented by a probability measure $\mu \in \mathcal{P}(X)$ such that

$$x \sim \mu.$$

In practice we rarely access μ directly; instead, we have an ensemble (samples)

$$\{x_i\}_{i=1}^n, \qquad x_i \overset{\text{i.i.d.}}{\sim} \mu, \tag{8.62}$$

or (in data assimilation) a weighted ensemble $\{(x_i, w_i)\}$.

The central Family IV goal is: **define a representation or distance between distributions** that is stable, computable, and physically meaningful.

8.5.2 Simple but Useful Baselines: Moment Embeddings

The simplest distribution embeddings summarize low-order **moments**:

$$\boxed{\Xi_{\text{mom}}(\mu) = \Big[\underbrace{\mathbb{E}_\mu[x]}_{\text{mean}},\ \underbrace{\text{Var}_\mu[x]}_{\text{variance}},\ \underbrace{\text{higher moments}}_{\text{skewness, kurtosis}}\Big]} \tag{8.63}$$

Estimated from samples: $\{x_i\}_{i=1}^n \sim \mu$

If $x \in \mathbb{R}$ (scalar measurements), moments are standard:

$$\text{Mean:} \quad \bar{x} = \frac{1}{n}\sum_{i=1}^{n} x_i \tag{8.64}$$

$$\text{Variance:} \quad s^2 = \frac{1}{n}\sum_{i=1}^{n}(x_i - \bar{x})^2 \tag{8.65}$$

$$\text{Skewness:} \quad \gamma_1 = \frac{1}{ns^3}\sum_{i=1}^{n}(x_i - \bar{x})^3 \tag{8.66}$$

$$\text{Kurtosis:} \quad \gamma_2 = \frac{1}{ns^4}\sum_{i=1}^{n}(x_i - \bar{x})^4 \tag{8.67}$$

For fields $u(\cdot)$, moments can be:

1. **Pointwise moments**:
   - Mean field: $\bar{u}(x) = \frac{1}{n}\sum_{i=1}^{n} u_i(x)$
   - Variance field: $\sigma^2(x) = \frac{1}{n}\sum_{i=1}^{n}(u_i(x) - \bar{u}(x))^2$
   - High-dimensional (same size as field)!
2. **Global scalar summaries**:
   - Energy: $E = \int_\Omega u^2 dx$
   - Enstrophy: $Z = \int_\Omega (\nabla \times u)^2 dx$
   - Mass: $M = \int_\Omega u dx$
   - Then embed moments of these: $[\mathbb{E}[E], \text{Var}[E], \mathbb{E}[Z], \ldots]$
3. **Spectral moments**:
   - Project to spectral basis: $u \to \alpha = [\alpha_1, \ldots, \alpha_m]$
   - Compute moments of coefficients: $[\mathbb{E}[\alpha_k], \text{Var}[\alpha_k]]$
   - Often more compact and physically meaningful

Design note (Moment embeddings are not enough)

Two different physical regimes can share the same mean and variance (e.g., bimodal vs unimodal distributions). For ensemble comparison and regime detection, use kernel or optimal-transport-based embeddings.

8.5.3 *Kernel Mean Embeddings and Maximum Mean Discrepancy (MMD)*

A powerful and widely used Family IV approach embeds distributions in a Reproducing Kernel Hilbert Space (RKHS) [13]. This goes beyond moments to capture full distribution structure!

Analogy: Fingerprints Versus Detailed DNA

Moment embeddings (previous section):

- Like a basic fingerprint: height, weight, eye color
- Fast, simple, interpretable
- But misses fine details.

Kernel mean embeddings:

- Like detailed DNA sequence
- Captures full "genetic code" of distribution
- Can distinguish any two different distributions!
- More expensive but complete.

In SciML:

- Moments: "average energy is 100 J"
- Kernel: "distribution over all possible energy configurations"
- Kernel can detect: bimodality, regime transitions, rare events.

8.5.3.1 Kernel Mean Embedding: the Idea

Let $k : X \times X \to \mathbb{R}$ be a positive-definite kernel with associated RKHS $\mathcal{H}$.

The *kernel mean embedding* of distribution μ is:

$$\boxed{m_\mu = \mathbb{E}_{x\sim\mu}[k(x,\cdot)] \in \mathcal{H}} \tag{8.68}$$

Intuition: m_μ is the "average kernel function" when centered at samples from μ.
Key property: If k is *characteristic*, then $\mu \mapsto m_\mu$ is **injective**:

$$m_\mu = m_\nu \quad \Leftrightarrow \quad \mu = \nu \tag{8.69}$$

Different distributions → different embeddings!

Kernel Mean Embedding

Given kernel k and distribution μ on X, the *kernel mean embedding* is RKHS element m_μ defined by:

$$m_\mu = \mathbb{E}_{x \sim \mu}[k(x, \cdot)] \tag{8.70}$$

Properties:

- Lives in infinite-dimensional RKHS $\mathcal{H}$
- For characteristic kernels: uniquely identifies μ
- Estimated from samples: $\hat{m}_\mu = \frac{1}{n} \sum_{i=1}^{n} k(x_i, \cdot)$

Characteristic kernels include:

- Gaussian/RBF: $k(x, x') = \exp(-\|x - x'\|^2/(2\sigma^2))$
- Laplacian: $k(x, x') = \exp(-\|x - x'\|/\sigma)$
- Many others...

8.5.3.2 Maximum Mean Discrepancy (MMD)

The distance between two distributions induced by RKHS norm:

$$\boxed{\mathrm{MMD}_k(\mu, \nu) = \|m_\mu - m_\nu\|_{\mathcal{H}}} \tag{8.71}$$

Key property: For characteristic kernels:

$$\mathrm{MMD}_k(\mu, \nu) = 0 \quad \Leftrightarrow \quad \mu = \nu \tag{8.72}$$

Estimating MMD from samples

Given:

- Samples from μ: $\{x_i\}_{i=1}^{n}$
- Samples from ν: $\{y_j\}_{j=1}^{m}$

Unbiased estimator:

$$\boxed{\begin{aligned}\widehat{\mathrm{MMD}}_k^2 = &\underbrace{\frac{1}{n(n-1)}\sum_{i\neq i'} k(x_i, x_{i'})}_{\text{within-}\mu\text{ similarity}}\\ &+\underbrace{\frac{1}{m(m-1)}\sum_{j\neq j'} k(y_j, y_{j'})}_{\text{within-}\nu\text{ similarity}}\\ &-\underbrace{\frac{2}{nm}\sum_{i,j} k(x_i, y_j)}_{\text{cross-similarity}}\end{aligned}} \tag{8.73}$$

Interpretation:

- First term: How similar are samples within μ?
- Second term: How similar are samples within ν?
- Third term: How similar are samples across μ and ν?
- If $\mu = \nu$: within-similarities equal cross-similarity $\rightarrow$ MMD = 0.

Step-by-step recipe:

1. **Choose kernel**:
    - Start with Gaussian: $k(x, x') = \exp(-\|x - x'\|^2/(2\sigma^2))$
    - Set bandwidth σ (see section on physics-aware kernels)
2. **Compute kernel matrices**:

$$K_{xx}[i, i'] = k(x_i, x_{i'}) \quad (n \times n) \tag{8.74}$$

$$K_{yy}[j, j'] = k(y_j, y_{j'}) \quad (m \times m) \tag{8.75}$$

$$K_{xy}[i, j] = k(x_i, y_j) \quad (n \times m) \tag{8.76}$$

3. **Compute MMD**:

$$\begin{aligned}\widehat{\mathrm{MMD}}^2 = &\frac{1}{n(n-1)}\left(\sum K_{xx} - n\right) + \frac{1}{m(m-1)}\left(\sum K_{yy} - m\right)\\ &-\frac{2}{nm}\sum K_{xy}\end{aligned} \tag{8.77}$$

    (Subtract diagonal from K_{xx} and K_{yy} for unbiased estimate)
4. **Statistical test (if needed)**:
    - Null hypothesis: $\mu = \nu$
    - Permutation test: shuffle combined samples, recompute MMD
    - p-value = fraction of permutations with MMD $\geq$ observed

Computational cost:

- Kernel matrix: $O(n^2)$ or $O(nm)$
- For large ensembles: use random features (see below).

8.5.3.3 From RKHS to Finite-Dimensional Vectors

Although $m_\mu \in \mathcal{H}$ may be infinite-dimensional, we can get finite-dimensional embeddings!

Method 1: Random Fourier Features (RFF)

For shift-invariant kernels (e.g., Gaussian), approximate:

$$k(x, x') \approx \varphi(x)^\top \varphi(x') \tag{8.78}$$

where $\varphi : X \to \mathbb{R}^d$ are random Fourier features:

$$\varphi(x) = \sqrt{\frac{2}{d}}[\cos(\omega_1^\top x + b_1), \ldots, \cos(\omega_d^\top x + b_d)] \tag{8.79}$$

with $\omega_j \sim p(\omega)$ from kernel's spectral density, $b_j \sim \text{Uniform}[0, 2\pi]$.
Then distribution embedding:

$$\boxed{\Xi(\mu) \approx \frac{1}{n}\sum_{i=1}^{n} \varphi(x_i) \in \mathbb{R}^d} \tag{8.80}$$

Advantages:

- Finite-dimensional: $\Xi(\mu) \in \mathbb{R}^d$
- Euclidean distance approximates MMD: $\|\Xi(\mu) - \Xi(\nu)\| \approx \text{MMD}(\mu, \nu)$
- Scalable: $O(nd)$ instead of $O(n^2)$

Method 2: Nyström approximation

Use kernel PCA on subset of data to get finite basis.

Method 3: Learned kernel features

Train neural network to learn features $\varphi_\theta(x)$ that maximize discriminative power.

The kernel k encodes what "similarity" means. A poor kernel yields poor distribution embeddings!

Why kernel choice matters

Scenario: Comparing turbulent flow ensembles

Bad choice: RBF kernel on raw grid values

- Dominated by discretization artifacts
- Mesh-dependent: different grids → incomparable
- Treats all spatial scales equally (wrong for turbulence!)

Good choice: Spectral kernel on energy spectrum

- Mesh-invariant: based on physical modes
- Emphasizes energy-containing scales
- Captures turbulent cascade structure.

Impact: Bad kernel → MMD meaningless. Good kernel → MMD detects regime shifts!

8.5.3.4 RBF Kernel on Naive Coordinates (often Insufficient)

Standard choice: Gaussian/RBF kernel

$$k(x, x') = \exp\left(-\frac{\|x - x'\|^2}{2\sigma^2}\right) \tag{8.81}$$

Problems for high-dimensional scientific objects (fields):

1. $\|x - x'\|$ dominated by discretization artifacts
2. Curse of dimensionality: all points seem far apart
3. Ignores physical structure (boundaries, scales)
4. Mesh-dependent.

Solution: Use physics-aware kernels!

8.5.3.5 Spectral Kernels on Coefficient Spaces

Idea: Project each field onto low-dimensional spectral basis (Family I):

$$x \mapsto \alpha(x) = [\alpha_1, \ldots, \alpha_m] \tag{8.82}$$

Then define kernel in coefficient space:

$$\boxed{k(x, x') = \exp\left(-\frac{\|\alpha(x) - \alpha(x')\|^2}{2\sigma^2}\right)} \tag{8.83}$$

Advantages:

- Mesh-invariant: spectral coefficients don't depend on discretization
- Low-dimensional: $m \ll N$ (grid size)
- Physically meaningful: emphasizes dominant modes
- Bandwidth σ set by spectral energy decay

8.5.3.6 Sobolev-Type Kernels (derivative-Aware Similarity)

If derivatives matter (PDE problems), use similarity based on Sobolev norm:

$$\|u\|_{H^1}^2 = \|u\|_{L^2}^2 + \|\nabla u\|_{L^2}^2 \tag{8.84}$$

Kernel based on H^1 distance:

$$k(u, u') = \exp\left(-\frac{\|u - u'\|_{H^1}^2}{2\sigma^2}\right) \tag{8.85}$$

Or approximate via spectral decay:

$$k(u, u') = \sum_k \frac{1}{1 + \lambda_k} \langle u, \psi_k \rangle \langle u', \psi_k \rangle \tag{8.86}$$

where λ_k are Laplacian eigenvalues, ψ_k eigenfunctions.

8.5.3.7 Diffusion Kernels on Manifolds/graphs

For data on graphs/meshes, diffusion kernels use Laplacian spectrum:

$$\boxed{k_t(i, j) = \sum_k e^{-\lambda_k t} \psi_k(i) \psi_k(j)} \tag{8.87}$$

where:

- λ_k: Laplacian eigenvalues
- ψ_k: Laplacian eigenfunctions
- t: Diffusion time (controls scale).

Physical interpretation: Heat kernel-similarity via diffusion process!
Properties:

- Reflects intrinsic geometry/connectivity
- Stable to local noise
- Aligns with diffusion-like physics.

Design Rule: Kernel Should Respect Physics Scales
Bandwidth selection (σ or t) should be based on:

1. **Physical length/time scales**:
   - Characteristic length: $\sigma \sim L_{\text{characteristic}}$
   - Diffusion time: $t \sim T_{\text{characteristic}}$
2. **Spectral energy decay**:
   - Set σ where spectral energy drops significantly
   - Captures energy-containing scales
3. **Domain size**:
   - $\sigma \sim 0.1$–$0.5 \times$ domain diameter
   - Ensures kernels overlap but distinguish

Not based on:

- Pure statistical normalization (median heuristic)
- Cross-validation alone (can overfit)
- Arbitrary defaults

Why this matters: Physics-informed bandwidth → meaningful MMD → reliable regime detection!
Example: Turbulent Flow Ensemble Comparison
Setup:

- DNS (Direct Numerical Simulation): $n = 100$ velocity fields
- LES (Large Eddy Simulation): $m = 100$ velocity fields
- Question: Does LES capture DNS statistics?

Method 1 (Bad): RBF on grid values

- $k(u, u') = \exp(-\|u - u'\|^2/(2\sigma^2))$
- Problem: DNS on 512^3 grid, LES on 64^3 grid
- Not comparable! Different dimensions

Method 2 (Better): Spectral kernel

- Project to energy spectrum: $u \to E(k)$ (energy vs wavenumber)
- Both DNS and LES → same spectral representation
- $k(u, u') = \exp(-\|E_u(k) - E_{u'}(k)\|^2/(2\sigma^2))$
- σ set by inertial range energy

Method 3 (Best): Diffusion kernel on spectral coefficients

- Project to POD modes: $u \to \alpha = [\alpha_1, \ldots, \alpha_{50}]$
- Use diffusion kernel in coefficient space
- Emphasizes energy-containing modes

Result:

- Method 1: Meaningless (dimension mismatch)
- Method 2: MMD = 0.05 (LES captures large scales)
- Method 3: MMD = 0.08 (reveals small-scale discrepancy).

8.5.4 *Optimal Transport (OT) and Wasserstein Embeddings*

Kernel methods compare distributions via feature expectations. Optimal transport takes a different view: compare distributions by the *cost of moving mass* from one to the other.

Analogy: Comparing City Populations

Kernel/MMD approach:

- Compare cities by statistics: average age, income distribution
- Aggregate features
- Like comparing "fingerprints" of populations.

Optimal Transport approach:

- "How much does it cost to transform City A's population into City B's?"
- Move 1000 young people → older demographics
- Relocate 500 high-earners → middle class
- Total cost of transformation = distance between cities.

In SciML:

- MMD: Compare vortices by average position, size
- OT: "Cost to transport vortex A to position of vortex B"
- OT captures geometric rearrangements better!

When OT shines: Differences are mainly spatial shifts (advection, transport) rather than amplitude changes.

8.5.4.1 Wasserstein Distance: Intuition

For distributions μ, ν on metric space (X, d), the p-Wasserstein distance:

$$\boxed{W_p(\mu, \nu) = \left(\inf_{\pi \in \Pi(\mu,\nu)} \int \|x - y\|^p \, d\pi(x, y) \right)^{1/p}} \tag{8.88}$$

where $\Pi(\mu, \nu)$ are **couplings** (joint distributions with marginals μ and ν).
Interpretation:

- $\pi(x, y)$: How much mass to move from x to y
- $\|x - y\|^p$: Cost of moving unit mass from x to y
- W_p: Minimum total cost over all transport plans.

8.5.4.2 When Wasserstein Distance Excels in SciML

1. **Advected fields**:
   - Same field pattern, shifted in space
   - L^2 distance: Large (different values everywhere)
   - Wasserstein: Small (cheap to transport)
2. **Coherent structures**:
   - Vortices, fronts, waves
   - Care about *where* structures are, not just statistics
   - OT: "Vortex A is 5cm away from Vortex B"
3. **Regime transitions**:
   - Sudden spatial reorganization
   - MMD might miss if statistics similar
   - OT detects geometric change

When to use OT versus MMD
Use Wasserstein/OT when:

- Differences are geometric (shifts, rotations, rearrangements)
- Studying transport phenomena (advection, diffusion)
- Working with localized structures (vortices, shocks, fronts)
- Small ensemble sizes (OT more sample-efficient).

Use MMD when:

- Differences are distributional (means, variances, multimodality)
- High-dimensional feature spaces
- Large ensembles available
- Want computational speed (RFF approximation).

Use both: They complement each other! MMD for global statistics, OT for spatial structure.

8.5.4.3 Practical OT for Ensembles: Scalable Approximations

Exact OT is expensive: $O(n^3 \log n)$ for n samples.
Scalable approximations:

1. Entropic regularized OT (Sinkhorn)

Add entropy regularization:

$$\min_{\pi} \sum_{i,j} \pi_{ij} C_{ij} + \epsilon H(\pi) \tag{8.89}$$

where $H(\pi)$ is entropy, $\epsilon > 0$ regularization strength.
Sinkhorn algorithm: Fast iterative method, $O(n^2)$ per iteration.

2. Sliced Wasserstein

Project to random 1D directions, compute 1D Wasserstein (cheap!), average:

$$\text{SW}(\mu, \nu) = \mathbb{E}_{\theta \sim \text{Sphere}}[W(\mu_\theta, \nu_\theta)] \tag{8.90}$$

where μ_θ is projection of μ onto direction θ.
Advantages: $O(n \log n)$ per direction, parallelizable.

3. Wasserstein barycenters

For ensemble summarization: find "average" distribution:

$$\bar{\mu} = \arg\min_{\mu} \sum_{i=1}^{N} W_2^2(\mu, \mu_i) \tag{8.91}$$

Useful for: ensemble mean, data assimilation.

Practical OT Implementation:
For small ensembles ($n < 100$):

- Use exact OT (POT library in Python)
- Set cost matrix: $C_{ij} = \|x_i - y_j\|^2$
- Compute W_2 directly

For medium ensembles ($100 < n < 1000$)**:**

- Use Sinkhorn (entropic OT)
- Choose $\epsilon \sim 0.01 \times \text{median}(C)$
- Fast, accurate approximation

For large ensembles ($n > 1000$)**:**

- Use sliced Wasserstein
- Project to $k = 100$ random directions
- Compute 1D Wasserstein per direction (fast!)
- Average over directions

Software:

- Python: POT (Python Optimal Transport)
- Julia: OptimalTransport.jl
- PyTorch: GeomLoss

8.5.5 Distribution Embeddings as "embeddings of Embeddings" (hybrid Practice)

The most effective SciML pattern can combine families (see Fig. 8.23):

$$\boxed{\text{Sample } x \xrightarrow{\text{Family I/II/III}} \text{Feature } z \xrightarrow{\text{Family IV}} \text{Distribution } \Xi(\mu)} \tag{8.92}$$

Two-stage embedding:

1. **Stage 1**: Embed each sample in physically meaningful space
    - Family I: Spectral coefficients
    - Family II: Learned latent codes
    - Family III: Operator features.

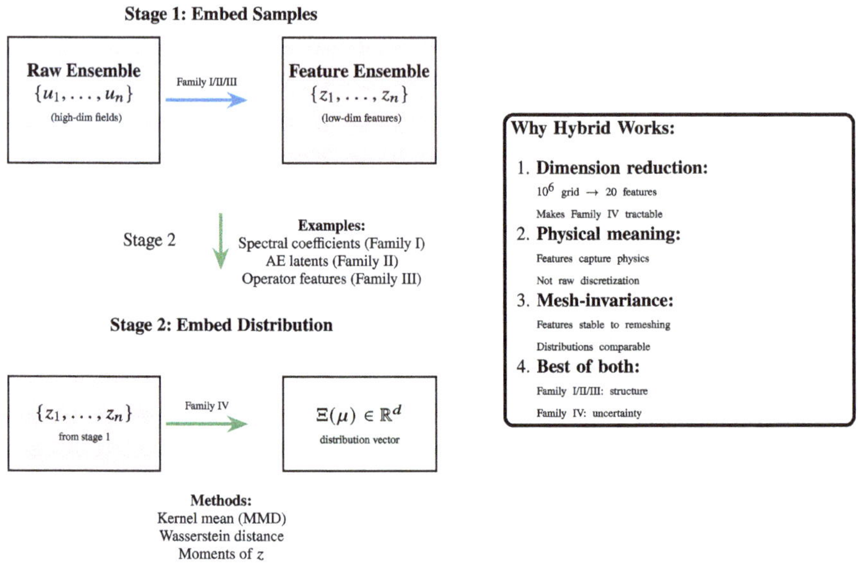

Fig. 8.23 Hybrid distribution embedding: Stage 1 uses Families I/II/III to embed each sample into low-dimensional physically meaningful features z. Stage 2 uses Family IV to embed distribution of features. This two-stage approach combines structural understanding (stage 1) with uncertainty quantification (stage 2)

2. **Stage 2**: Embed distribution of features

    - Kernel mean embedding on z-space
    - Wasserstein on z-space
    - Moments of z.

Distribution Embedding of Learned Features

Given feature map $\Phi : X \to \mathbb{R}^d$ (from Families I/II/III), define:

$$z = \Phi(x) \tag{8.93}$$

Distribution μ on X induces distribution μ_z on $\mathbb{R}^d$:

$$\mu_z = \Phi_{\#}\mu \qquad \text{(pushforward measure)} \tag{8.94}$$

Family IV then embeds μ via embedding of μ_z:

- Kernel mean embedding: $m_{\mu_z} = \mathbb{E}_{z\sim\mu_z}[k(z, \cdot)]$
- Wasserstein: $W_p(\mu, \nu) \approx W_p(\mu_z, \nu_z)$
- Moments: $[\mathbb{E}[z], \mathrm{Var}[z], \ldots]$

Key advantage: Comparison in physically aligned, low-dimensional feature space!

8.5.5.1 Examples of Hybrid Embeddings

Example 1: Spectral + Kernel Mean

1. Project each field to spectral basis (Family I):

$$u \mapsto \alpha = [\alpha_1, \ldots, \alpha_m] \tag{8.95}$$

2. Compute kernel mean of coefficient distribution:

$$\Xi(\mu) = \frac{1}{n}\sum_{i=1}^{n} \varphi(\alpha_i) \tag{8.96}$$

   where φ are random Fourier features
3. Compare ensembles via: $\|\Xi(\mu) - \Xi(\nu)\|$

Example 2: Autoencoder + Wasserstein

1. Encode each trajectory with sequence AE (Family II):

$$u_{0:T} \mapsto z \in \mathbb{R}^{20} \tag{8.97}$$

2. Compute Wasserstein distance between latent distributions:

$$W_2(\mu_z, \nu_z) \tag{8.98}$$

3. Detects: regime transitions, distribution shifts

Example 3: Operator Features + Moments

1. Extract operator model features (Family III):

$$(a, f, \Omega, \mathrm{BC}) \mapsto h \in \mathbb{R}^{100} \tag{8.99}$$

2. Compute moments of feature distribution:

$$\Xi(\mu) = [\mathbb{E}[h], \mathrm{Cov}[h]] \tag{8.100}$$

3. Use for: uncertainty quantification, robust optimization

Step-by-step recipe:

1. **Choose sample embedding (Family I/II/III)**:
   - Consider: physics structure, mesh-invariance, interpretability
   - Common: Spectral coefficients (Family I), learned latents (Family II).
2. **Validate sample embedding**:
   - Test on individual samples
   - Ensure physically meaningful
   - Check discretization stability.
3. **Choose distribution method (Family IV)**:
   - Kernel mean: If large ensembles, high-dim features
   - Wasserstein: If geometric rearrangements matter
   - Moments: If small ensembles, interpretability needed.
4. **Validate distribution embedding**:
   - Test on known distribution shifts
   - Bootstrap confidence intervals
   - Cross-validate threshold selection.

Golden rule: Always use meaningful features (I/II/III) before distribution embedding (IV). Raw grid values rarely work!

Lets now work on an example: ensemble discrepancy for stochastic PDE solutions. We illustrate a complete pipeline for comparing simulation ensembles to observations using distribution embeddings.

8.5.5.2 Problem Setup: Stochastic Reaction-Diffusion

Consider stochastic reaction-diffusion equation modeling chemical concentrations:

$$\boxed{\frac{\partial u}{\partial t} = D\nabla^2 u + f(u) + \sigma\xi(x, t)} \tag{8.101}$$

where:

- $u(x, t)$: Concentration field
- D: Diffusion coefficient (possibly uncertain)
- $f(u)$: Reaction terms
- $\xi(x, t)$: Spatiotemporal white noise
- σ: Noise intensity.

Uncertainty sources:

1. Stochastic forcing: $\xi(x, t)$ random
2. Parameter uncertainty: $D \sim p(D)$
3. Initial conditions: $u(x, 0) \sim \mu_0$

Each simulation run produces one realization $u_i(x, t)$. From n runs, we get ensemble:

$$\{u_i\}_{i=1}^n \sim \mu_{\text{sim}} \tag{8.102}$$

We also have experimental measurements (or high-fidelity reference):

$$\{u_j^{\text{obs}}\}_{j=1}^m \sim \mu_{\text{obs}} \tag{8.103}$$

Analogy: Drug Response Variability

The scenario:

- Testing new drug on cell cultures
- Each petri dish: different outcome (biological variability)
- Run 100 simulations → get distribution of responses
- Conduct 100 experiments → get distribution of observed responses.

The question:

- Does simulation capture reality?
- Not: "Is simulation exactly correct?" (impossible with noise!)
- But: "Are simulation and experiment drawing from same distribution?"

Family IV approach:

- Embed entire simulation ensemble: $\Xi(\mu_{\text{sim}})$
- Embed entire experimental ensemble: $\Xi(\mu_{\text{obs}})$
- Compare: $\|\Xi(\mu_{\text{sim}}) - \Xi(\mu_{\text{obs}})\|$
- If large → model mismatch, need recalibration

The Goal: simulator-reality gap quantification
Specific objectives:

1. **Validation**: Does simulator match observations?
2. **Regime detection**: Has system shifted to new behavior?
3. **Uncertainty calibration**: Is simulation uncertainty too narrow/wide?
4. **Model improvement**: Which aspects mismatch?

8.5.5.3 Step-By-Step Implementation

Step 1: Feature representation (Family I/II/III)

Choose physically meaningful representation for each sample:
Option A: Spectral coefficients (Family I)

1. Compute POD (Proper Orthogonal Decomposition) on combined dataset
2. Project each sample: $u \to \alpha = [\alpha_1, \ldots, \alpha_{20}]$
3. Use first 20 modes (capture > 99% energy).

Option B: Learned latent (Family II)

1. Train physics-aware autoencoder on all samples
2. Encode each: $u \to z = \Phi_\theta(u) \in \mathbb{R}^{20}$
3. Validates via reconstruction error.

Choice criteria:

- POD: Interpretable, optimal for linear structure
- AE: Flexible, captures nonlinear manifolds
- For this example: Use POD (simpler, interpretable).

Step 2: Distribution embedding (Family IV)

Method: Kernel mean embedding with RBF
For each ensemble, compute:

$$\Xi(\mu) = \frac{1}{n} \sum_{i=1}^{n} \varphi(z_i) \tag{8.104}$$

where φ are random Fourier features:

- Dimension: $d = 100$
- Bandwidth: $\sigma = 0.5 \times \text{median}(\|z_i - z_j\|)$
- Random frequencies: $\omega_k \sim \mathcal{N}(0, 1/\sigma^2)$.

Result: $\Xi(\mu_{\text{sim}}), \Xi(\mu_{\text{obs}}) \in \mathbb{R}^{100}$

Step 3: Compute discrepancy

Option 1: Euclidean distance (approximate MMD)

$$d_{\text{approx}} = \|\Xi(\mu_{\text{sim}}) - \Xi(\mu_{\text{obs}})\|_2 \tag{8.105}$$

Option 2: Full MMD (more accurate)

$$\text{MMD}^2 = \frac{1}{n^2}\sum_{i,i'} k(z_i, z_{i'}) + \frac{1}{m^2}\sum_{j,j'} k(z_j^{\text{obs}}, z_{j'}^{\text{obs}}) - \frac{2}{nm}\sum_{i,j} k(z_i, z_j^{\text{obs}}) \quad (8.106)$$

For this example: Use Option 2 (full MMD) for accuracy

Step 4: Statistical testing

Set threshold via bootstrap:

1. **Null hypothesis:** $\mu_{\text{sim}} = \mu_{\text{obs}}$
2. **Bootstrap procedure:**
   - Pool samples: $\{z_1, \ldots, z_n, z_1^{\text{obs}}, \ldots, z_m^{\text{obs}}\}$
   - Randomly split into two groups (size n and m)
   - Compute MMD for this split
   - Repeat 1000 times → distribution under null.
3. **Decision**:

$$p\text{-value} = \frac{\#\{\text{MMD}_{\text{perm}} \geq \text{MMD}_{\text{obs}}\}}{1000} \quad (8.107)$$

   If $p < 0.05$: Reject null (distributions differ)

8.5.5.4 Interpretation and Diagnostics

1. Localize mismatch (which modes?)

If MMD is large, identify contributing dimensions:

- Compute per-mode variance: $\text{Var}(\alpha_k^{\text{sim}})$ vs $\text{Var}(\alpha_k^{\text{obs}})$
- Identify: "Mode 3 has higher variance in simulation"
- Physical interpretation: "Large-scale oscillations overestimated"

2. Visualize distribution difference

- Project to 2D via PCA or t-SNE
- Plot: Blue cloud (simulation), Red cloud (observation)
- Visual check: Overlap? Separation? Bimodality?

3. Sensitivity to parameters

If mismatch detected:

- Vary D: Rerun with $D \pm 10\%$
- Vary σ: Change noise intensity
- Find which parameter makes MMD $\to 0$
- Recalibrate model

4. Temporal evolution

For time series:

- Compute MMD at each time: MMD(t)
- Plot vs time
- Detect: "Mismatch grows after $t = 5$" $\to$ transient dynamics wrong

Key lesson: Distribution embedding detected subtle systematic bias that would be missed by comparing ensemble means alone!

Practical Recipe: Ensemble Discrepancy with MMD

1. **Feature extraction**:
   - Use Family I (spectral) or II (learned)
   - Dimension: 10–50 typically sufficient
   - Validate: Features capture physics.
2. **Kernel selection**:
   - Start with RBF: $k(z, z') = \exp(-\|z - z'\|^2/(2\sigma^2))$
   - Set σ by median heuristic or physical scales
   - For robustness: Try multiple bandwidths.
3. **MMD computation**:
   - For $n, m < 1000$: Use full kernel matrix
   - For larger: Random Fourier features approximation.
4. **Statistical testing**:
   - Bootstrap or permutation test
   - Report p-value with significance level
   - Use confidence intervals, not just point estimates.
5. **Interpretation**:
   - Visualize distributions (2D projection)
   - Localize mismatch (per-feature analysis)
   - Connect to physics (which modes/scales?)

6. **Iteration**:
   - If mismatch: Adjust parameters, revalidate
   - If validated: Use for prediction with confidence.

Success criteria:

- $p > 0.05$: Distributions statistically indistinguishable
- Visual overlap in 2D projection
- Per-mode statistics comparable
- Robust to feature/kernel choices.

A micro-summary of all the discussions are below:

- Family IV embeds entire *distributions* over scientific objects, enabling uncertainty-aware comparison, validation, and regime detection.
- Kernel mean embeddings and MMD provide practical, statistically grounded discrepancies; their success depends critically on physics-aware feature spaces and kernels.
- Optimal transport complements kernel methods when differences are primarily geometric rearrangements (transport/advection).
- In practice, the strongest approach is *distribution embeddings of physically meaningful sample embeddings* (Families I–III first, then Family IV).

8.6 Conclusion

This chapter has developed embeddings as the bridge between scientific objects and neural network computation. We have seen that embedding choices encode inductive biases that fundamentally shape learning, that different scientific objects require different embedding strategies, and that the representation problem is often as important as the architecture problem.

Key Takeaways

1. **Representation Precedes Learning**: What goes into the network matters as much as what happens inside. Bad embeddings make easy problems hard.
2. **Match Embedding to Structure**: Spectral embeddings capture global structure; local encodings preserve fine details. The choice should match the physics.

3. **Operators Need Function Embeddings**: Neural operators like DeepONet and FNO embed *functions*, not just vectors. Branch-trunk and spectral approaches offer different trade-offs.
4. **Uncertainty Needs Distribution Embeddings**: Kernel mean embeddings and optimal transport provide principled ways to represent and compare probability measures.
5. **Learn When Appropriate**: When structure is unknown, autoencoders and contrastive methods can discover useful representations from data—but physics-aware variants preserve known constraints.

References

1. Li, Y., Si, S., Li, G., Hsieh, C.J., Bengio, S.: Learnable Fourier features for multi-dimensional spatial positional encoding. Adv. Neural. Inf. Process. Syst. **34**, 15816–15829 (2021)
2. Sanders, J., Puso, M.A.: An embedded mesh method for treating overlapping finite element meshes. Int. J. Numer. Meth. Eng. **91**(3), 289–305 (2012)
3. Urakawa, H.: Geometry of Laplace–Beltrami operator on a complete Riemannian manifold. Progress in differential geometry **22**, 347–406 (1993)
4. Xia, T., Tao, D., Mei, T., Zhang, Y.: Multiview spectral embedding. IEEE Trans. Syst. Man Cybern. **40**(6), 1438–1446 (2010). Part B (Cybernetics)
5. Luo, B., Wilson, R.C., Hancock, E.R.: Spectral embedding of graphs. Pattern Recogn. **36**(10), 2213–2230 (2003)
6. Jebara, T., Song, Y., Thadani, K.: Spectral clustering and embedding with hidden Markov models. In: European Conference on Machine Learning, pp. 164–175. Springer (2007)
7. Gentili, C., Hierold, C., Roman, C.I.: Physics-informed autoencoder for feature extraction from multidimensional sensor measurement data. Appl. Intell. **55**(16), 1063 (2025)
8. Yeh, C.H., Hong, C.Y., Hsu, Y.C., Liu, T.L., Chen, Y., LeCun, Y.: Decoupled contrastive learning. In: European conference on computer vision. pp. 668–684. Springer (2022)
9. Bhuyan, B.P., Ramdane-Cherif, A., Singh, T.P., Tomar, R.: Recurrent neural networks (rnns): Capturing the dynamics of sequences. In: Neuro-Symbolic Artificial Intelligence: Bridging Logic and Learning, pp. 141–161. Springer (2024)
10. Lu, L., Jin, P., Karniadakis, G.E.: Deeponet: Learning nonlinear operators for identifying differential equations based on the universal approximation theorem of operators. arXiv:1910.03193 arXiv preprint (2019)
11. Lu, L., Jin, P., Pang, G., Zhang, Z., Karniadakis, G.E.: Learning nonlinear operators via deeponet based on the universal approximation theorem of operators. Nat. Mach. Intell. **3**(3), 218–229 (2021)
12. Li, Z., Kovachki, N.B., Azizzadenesheli, K., Bhattacharya, K., Stuart, A., Anandkumar, A.: Fourier neural operator for parametric partial differential equations. International Conference on Learning Representations
13. Xu, J.W., Paiva, A.R., Park, I., Principe, J.C.: A reproducing kernel hilbert space framework for information-theoretic learning. IEEE Trans. Signal Process. **56**(12), 5891–5902 (2008)

Chapter 9
Foundational and World Models in SciML

The spice must flow.

Dune

Abstract Foundation models—large neural networks pretrained on massive datasets then adapted to downstream tasks—have transformed natural language processing and computer vision. Can similar approaches accelerate scientific discovery? This chapter examines foundation models and world models for Scientific Machine Learning. We begin by defining the vocabulary (foundation, world, scientific) and examining the promise (knowledge transfer, data efficiency) and challenges (physical consistency, scale, trust) specific to scientific domains. Generative foundation models for scientific design treat molecular generation, materials discovery, and engineering optimization as conditional generation problems, with attention to physics-guided constraints and evaluation protocols. World models for scientific systems develop latent dynamics approaches that learn compressed representations of physical state evolution, enabling planning, control, and digital twin applications. Physics-aware world models incorporate governing equations into latent dynamics, addressing the tension between learned flexibility and physical consistency. Case studies include weather models (GraphCast, GenCast), molecular generators, and autonomous laboratory systems.

9.1 Introduction

On February 25, 2025, the European Centre for Medium-Range Weather Forecasts (ECMWF) made history by launching the Artificial Intelligence/Integrated Forecasting System (AIFS) as its first fully operational machine learning weather model. For the first time, a major meteorological center trusted neural networks and not traditional physics equations to predict tomorrow's weather for millions of people. Just months earlier, Google DeepMind's GenCast [1] demonstrated that diffusion models could generate 50-member weather ensembles in 8 minutes, outperforming

B. P. Bhuyan et al., *Scientific Machine Learning*,
https://doi.org/10.1007/978-981-95-7856-6_9

traditional methods on 97% of forecast targets while consuming a fraction of the computational resources. Meanwhile, in pharmaceutical laboratories, INS018_055 became the first wholly AI-discovered and designed drug to reach Phase 2 clinical trials, and Microsoft's MatterGen [2] generated novel battery materials requiring 70% less lithium than conventional designs.

These aren't isolated successes; they represent a fundamental shift in how machine learning engages with scientific problems. The key enabler? *Foundation models* and *world models*; two paradigms that sound similar but serve complementary roles in the scientific ML ecosystem. Understanding their distinction, promise, and limitations forms the foundation for everything that follows in this chapter.

9.1.1 The Vocabulary: Foundation, World, and Scientific

9.1.1.1 What Makes a Model "Foundation"?

Imagine you're hiring for a complex position. You could train someone from scratch for that specific role, teaching them only what's needed for that one job. Or you could hire someone with broad education and experience who can quickly adapt to your specific needs. Foundation models follow the second strategy.

A **foundation model** is a large-scale model pre-trained on broad, diverse data that can be adapted to multiple downstream tasks through fine-tuning, prompting, or other transfer learning mechanisms. Key characteristics:

- **Scale**: Typically millions to billions of parameters
- **Breadth**: Trained on diverse data covering wide domains
- **Adaptability**: Transfers to new tasks with limited additional training
- **Emergence**: Exhibits capabilities not explicitly trained for

The term "foundation model" originated in natural language processing (think GPT-4, BERT) but has migrated to scientific domains with crucial modifications. Where language models learn from text tokens, scientific foundation models must learn from:

- **Molecular graphs**: Atoms and bonds (e.g., MolE [3] with 842 million molecular graphs)
- **Spatial fields**: Temperature, pressure, velocity on grids or meshes (e.g., Aurora with 1 million hours of atmospheric data)
- **Crystal structures**: Atomic arrangements in materials (e.g., MatterGen [2] with 608,000 stable structures)
- **Protein sequences and structures**: Amino acid chains and 3D conformations (e.g., RFdiffusion3 [4] for all-atom protein design)

Just as words are tokens in NLP, atoms are tokens in molecular ML, and grid points are tokens in weather prediction. The foundation model paradigm works by learning rich representations of these scientific "tokens" that capture underlying patterns generalizing across tasks.

9.1.1.2 What Makes a Model a "World Model"?

If foundation models are polymaths who know many things, world models are simulators who can *predict what happens next*. Think of a flight simulator: it doesn't just know facts about airplanes; it can simulate what happens when you push the throttle forward, bank left, or encounter turbulence.

A **world model** is a learned representation of environment dynamics that enables forward simulation, planning, and control. Key characteristics:

- **Predictive**: Forecasts future states from current states and actions
- **Interactive**: Responds to interventions and controls
- **Rollout capable**: Can simulate multi-step trajectories
- **Latent dynamics**: Often operates in compressed latent space

In scientific contexts, world models learn the dynamics governing physical systems:

- **Molecular dynamics**: How atoms move under forces (e.g., NequIP predicting atomic trajectories)
- **Fluid evolution**: How velocity and pressure fields evolve (e.g., FourCastNet simulating atmospheric dynamics)
- **Reaction kinetics**: How concentrations change over time (e.g., PI-SDE modeling cellular dynamics)
- **Climate trajectories**: Multi-decadal evolution of Earth systems (e.g., NeuralGCM for climate projections)

The Key Distinction

- **Foundation models** excel at *encoding knowledge*; they know what molecules are stable, what weather patterns exist, what protein folds are favorable.
- **World models** excel at *simulating dynamics*; they predict how molecules will move, how weather will evolve, how proteins will fold.

Many modern systems combine both: Aurora is a foundation model pre-trained on diverse Earth-system data *that also functions as a world model* for weather forecasting. GenCast is a diffusion-based foundation model *adapted for probabilistic weather simulation*.

9.1.1.3 The "Scientific" Constraint: Why Science Is Different

Scientific machine learning inherits constraints that don't exist (or matter less) in other domains:

1. **Physical consistency is non-negotiable**: A weather model that violates conservation of mass or predicts negative humidity isn't just inaccurate- it's physically meaningless. Language models can occasionally produce grammatically incorrect sentences; scientific models *cannot* produce physically impossible predictions in deployment.
2. **Extrapolation is the goal, not exception**: In computer vision, we rarely ask models to recognize objects drastically different from training data. In science, we *specifically want* to discover new materials, predict weather patterns outside historical records, or design proteins unlike any in nature. Out-of-distribution generalization isn't a failure mode; it's the objective.
3. **Interpretability connects to trust and safety**: When a drug design model suggests a molecule, chemists need to understand *why* not for curiosity, but because the reasoning process reveals potential toxicity, off-target effects, or synthesis challenges. The FDA's 500+ AI drug applications face regulatory scrutiny demanding interpretable reasoning.
4. **Multi-scale physics couples phenomena**: Atmospheric dynamics couple planetary-scale circulation, meso-scale storms, and micro-scale turbulence. A foundation model for weather must represent all scales coherently, unlike language models where sentences largely stand alone.

Figure 9.1 illustrates how foundation models, world models, and scientific computing intersect, with the sweet spot (systems like Aurora, GenCast, and MatterGen) residing at the confluence of all three.

9.1.2 The Promise: From Data Scarcity to Data Leverage

9.1.2.1 The Pre-training Revolution in Science

Traditional scientific ML followed a frustrating pattern: collect expensive domain-specific data (months of experiments, supercomputer simulations), train a model for that specific task, achieve modest performance, and then start over for the next task. Foundation models invert this paradigm.

Consider the trajectory shown in Fig. 9.2. Early neural networks for molecular property prediction trained from scratch on small datasets (QM9's 134,000 molecules). Then came transfer learning: pre-train on large unlabeled molecular databases, fine-tune on specific tasks. Finally, foundation models: MolE pre-trains on 842 million molecular graphs through self-supervised learning, then achieves best-in-class performance on 10 of 22 drug design benchmarks through minimal fine-tuning.

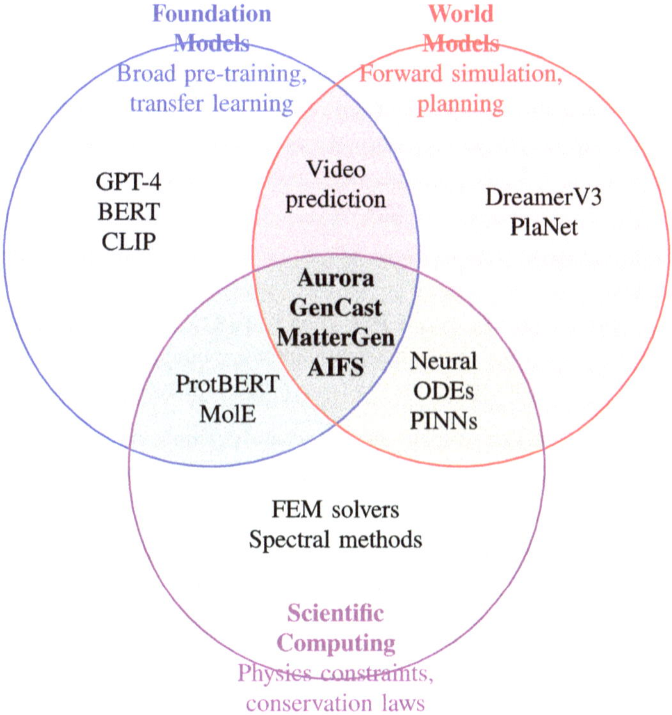

Fig. 9.1 The intersection of foundation models, world models, and scientific computing. Modern scientific ML systems (yellow box) increasingly occupy the confluence of all three paradigms, inheriting strengths from each: broad knowledge from foundation model pre-training, predictive capability from world models, and physical consistency from scientific computing

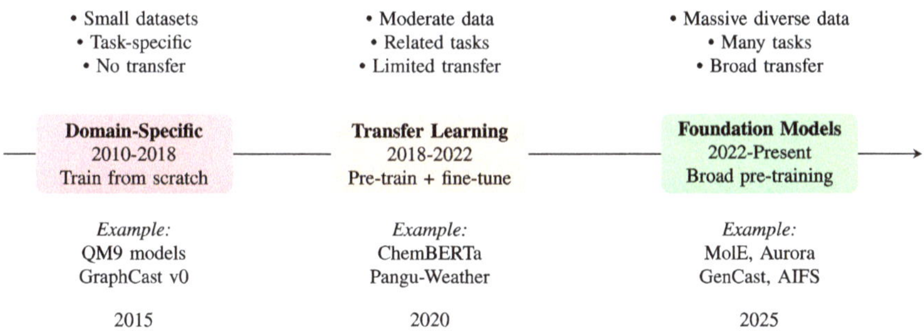

Fig. 9.2 Evolution of scientific machine learning paradigms. The shift from domain-specific models to foundation models mirrors developments in NLP, but with scientific constraints: pre-training data must respect physical laws, and transfer learning must preserve conservation properties

9.1.2.2 Few-Shot Adaptation: Learning from Scarce Data

The power of foundation models shines brightest where data is scarcest. Consider Aurora's performance across diverse Earth-system tasks: Microsoft's Aurora [5] is a 1.3 billion parameter vision transformer pre-trained on over 1 million hours of atmospheric data from ERA5, operational forecasts, and reanalysis. Through fine-tuning with limited data, Aurora adapts to:

- **Weather forecasting**: Outperforms IFS-HRES at 0.1° resolution while being 5000× faster
- **Air quality prediction**: Forecasts $PM_{2.5}$, NO_2, and O_3 from CAMS data
- **Ocean wave forecasting**: Predicts significant wave height
- **Tropical cyclone tracking**: Improves track prediction accuracy by 20%

The key insight: Pre-training on atmospheric dynamics creates representations that transfer to related geophysical phenomena, even those governed by different physics (ocean waves vs. atmospheric flow).

This pattern repeats across scientific domains:

- **DreaMS** [6]: Pre-trains on 700 million mass spectra, then identifies unknown molecules from single spectra
- **ClimaX** [7]: Pre-trains on climate data, transfers to weather forecasting, precipitation nowcasting, and climate downscaling
- **MatterGen**: Pre-trains on inorganic materials, generates crystals across the periodic table with properties constrained by minimal examples

The economic implications are profound. NASA's Prithvi models, trained on 64 A100 GPUs for weeks, now enable dozens of research groups to tackle Earth science problems: hurricane tracking, wildfire prediction, agricultural monitoring; that would each have required comparable computational resources under the old paradigm.

9.1.2.3 Multi-modal Scientific Reasoning

Language models achieved dramatic improvements by combining text with images (CLIP, GPT-4V). Scientific foundation models are following suit, but "multi-modal" means something different:

- **Structure + Properties**: MultiMat [8] combines crystal structures with thermodynamic, mechanical, and electronic properties for materials discovery
- **Sequence + Structure + Function**: Protein foundation models integrate amino acid sequences (language), 3D structures (geometry), and functional annotations (knowledge graphs)
- **Observations + Simulations**: Aurora combines satellite observations, reanalysis products, and numerical model outputs
- **Text + SMILES + Graphs**: Llamole [9] integrates molecular graphs within language model conversations for interactive drug design

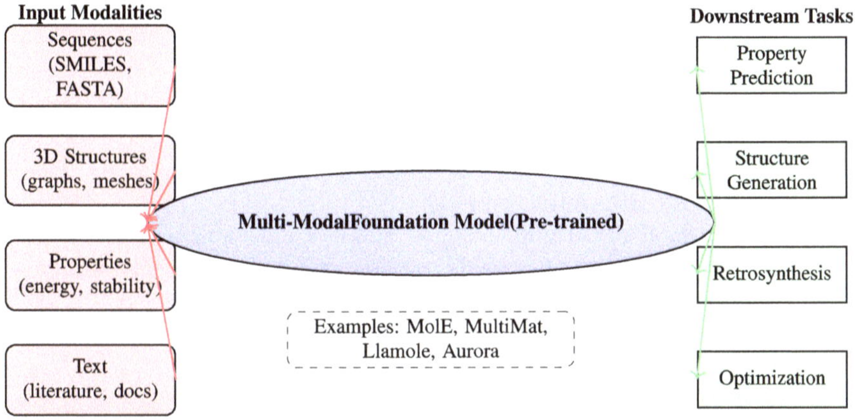

Fig. 9.3 Multi-modal scientific foundation models integrate diverse data types like sequences, structures, properties, and text to create unified representations enabling transfer across tasks. Unlike NLP models that combine vision and language, scientific foundation models must handle physical quantities with units, geometric structures with symmetries, and temporal dynamics with causality

Figure 9.3 illustrates how modern scientific foundation models integrate diverse data modalities, creating representations richer than any single modality could provide.

9.1.3 The Challenge: Physics, Scale, and Trust

Every silver lining has a cloud. Foundation and world models bring unprecedented capabilities to science, but also introduce challenges absent in traditional scientific computing.

9.1.3.1 Physical Consistency Versus Data-Driven Expressiveness

Pure data-driven models can learn correlations that violate physics. Early weather prediction models sometimes generated:

- Negative humidity (physically impossible)
- Total energy drift over long rollouts (violating conservation)
- Unrealistic spectral signatures (wrong distribution of scales)
- Spurious tropical cyclones (statistical artifacts, not physical features)

The Physical Consistency Trap

Adding physical constraints to neural networks is harder than it seems. Consider three approaches with escalating difficulty:

1. **Soft constraints** (add physics loss terms): Easy to implement, but networks can still violate constraints
2. **Architectural constraints** (build physics into structure): E.g., Hamiltonian neural networks that exactly conserve energy; but limited to specific equation forms
3. **Post-hoc correction** (project outputs onto physical manifold): Preserves constraints but can degrade learned features

There's no free lunch: exact physical consistency often reduces model flexibility. The art lies in deciding which constraints are non-negotiable versus which can be soft preferences.

Recent work has made progress:

- **NeuralGCM** [10] maintains stability for multi-decade climate simulations by embedding physics in the core solver
- **EEGSDE** [11] preserves E(3)-equivariance during diffusion sampling for molecular generation
- **Symplectic Neural Networks** [12] achieve exact energy conservation for Hamiltonian systems through geometric integrators

But the fundamental tension remains: the more we constrain models to respect known physics, the less they can discover patterns beyond our current understanding.

9.1.3.2 Computational Scale Versus Mechanistic Interpretability

Foundation models achieve their power through scale: Aurora's 1.3 billion parameters, MolE's 842 million molecular graphs. But scale creates opacity.

Consider the interpretability spectrum in Fig. 9.4. At one extreme: mechanistic models (Navier-Stokes equations, quantum mechanics) that are fully interpretable but computationally expensive and require known physics. At another extreme: giant neural networks that are fast and flexible but opaque. At the third corner: physical rigor; i.e. models that provably satisfy conservation laws and symmetries.

The challenge intensifies in high-stakes applications:

- **Drug design**: The FDA received 500+ AI-containing applications. Regulators demand explainable predictions like why is this molecule predicted to bind? What structural features drive toxicity?
- **Materials discovery**: A model might predict a novel catalyst, but chemists need to understand the mechanism to optimize synthesis and avoid unintended reactions

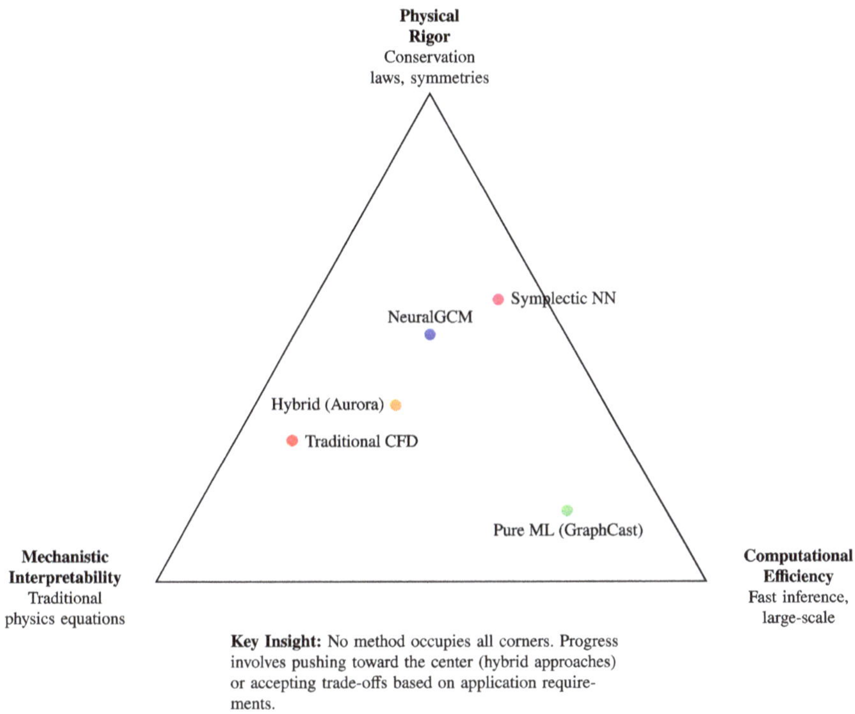

Fig. 9.4 The fundamental trade-offs in scientific foundation models. Traditional physics-based methods (red) offer interpretability and rigor but are computationally expensive. Pure ML approaches (green) are fast but may lack physical guarantees. Hybrid methods (orange, blue, purple) attempt to balance these competing demands, with different applications prioritizing different corners

- **Climate projection**: Policy decisions require understanding which physical mechanisms drive projected changes, not just opaque predictions

Recent interpretability advances include:

- **SchNet4AIM** [13] combines neural networks with Quantum Chemical Topology, predicting interpretable real-space descriptors (atomic charges, bond orders)
- **xChemAgents** [14] uses cooperative LLM-based agents with physics-aware reasoning, producing human-interpretable explanations while reducing error by 22%
- **Attention visualization** in transformers reveals which spatial regions or molecular substructures drive predictions

But we're far from solved. A 1.3 billion parameter model remains fundamentally less interpretable than Newton's second law, even with sophisticated explanation techniques.

9.1.3.3 Generalization Versus Domain-Specific Guarantees

Foundation models promise generalization, but science often demands guarantees. Consider:

- A weather model must work reliably during extreme events; the very conditions where it has least training data
- A materials model must generalize to unexplored chemical spaces; but current benchmarks show almost 50% of generated molecules fail basic validity tests
- A protein design model must work for novel folds; where traditional physics-based methods provide formal guarantees about thermodynamic stability

The out-of-distribution (OOD) generalization problem is particularly acute because, in science, OOD generalization is often the point. We don't want to predict weather we've already seen; we want to predict climate change scenarios outside historical experience.

Figure 9.5 illustrates this challenge: foundation models excel within their training distribution but can fail catastrophically when extrapolating.

The Matbench Discovery benchmark [15] tested materials stability prediction models on held-out crystals. Key findings:

- Best models (EquiformerV2) achieve F1 > 0.9 and 20 meV/atom energy accuracy *within the test distribution*
- Performance degrades significantly when tested on chemical compositions absent from training

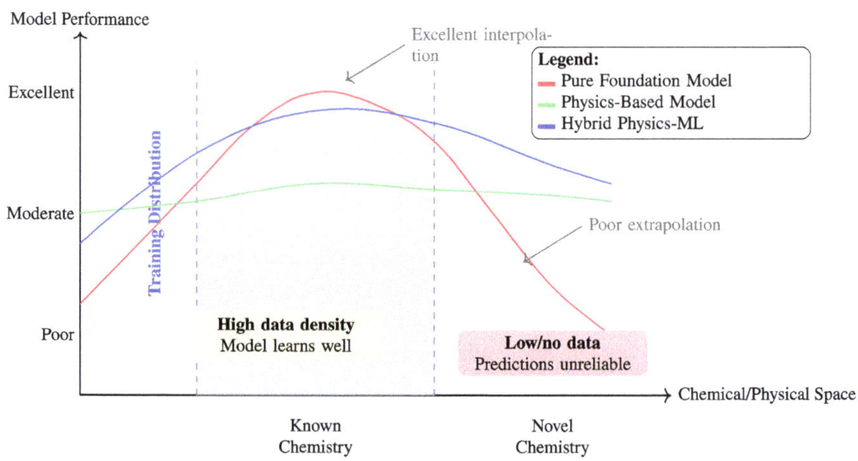

Fig. 9.5 The out-of-distribution (OOD) generalization challenge. Pure foundation models (red) excel within their training distribution but can fail catastrophically when extrapolating to novel chemical/physical spaces. Physics-based models (green) provide consistent performance but may have lower peak accuracy. Hybrid approaches (blue) attempt to combine the strengths of both: learned patterns where data exists, physical priors where it doesn't

- Almost half of *generated* molecules (not just predictions on existing ones) fail basic chemical validity tests
- Models trained only on stable structures struggle to recognize unstable configurations

The lesson: Foundation models excel at interpolation, struggle at extrapolation. Science needs both. This chapter navigates the landscape of foundation and world models in scientific machine learning, structured to serve readers with varying backgrounds and objectives.

9.2 Generative Foundation Models for Scientific Design

A chemist stares at her laptop screen. She needs a molecule that binds to a specific protein pocket, crosses the blood-brain barrier, avoids toxicity pathways, and can be synthesized in four steps or fewer. Traditional approaches would screen millions of candidates or rely on expert intuition refined over decades. Instead, she types her requirements into a generative model. Eight seconds later, it proposes twenty candidates, three of which her team will synthesize next week.

This is *design as inverse prediction*; moving from desired properties backward to molecular structures that achieve them. Where forward models predict "this molecule will have these properties," generative models answer "which molecules will have the properties I want?" This section develops the technical machinery enabling such conditional generation, the methods ensuring physical validity, and the evaluation protocols distinguishing genuine discoveries from statistical artifacts.

9.2.1 Design as Conditional Generation

9.2.1.1 From Forward to Inverse Problems

Scientific design problems are fundamentally inverse: we observe effects and seek causes, specify requirements and want realizations. Figure 9.6 contrasts the paradigms.

The mathematical formulation: given a desired property vector $\mathbf{y} \in \mathcal{Y}$ (e.g., binding affinity, band gap, melting point), generate structures $\mathbf{x} \in \mathcal{X}$ such that $f(\mathbf{x}) \approx \mathbf{y}$, where f is the physics mapping. Traditional optimization solves:

$$\mathbf{x}^* = \arg\min_{\mathbf{x} \in \mathcal{X}} \|\mathbf{y} - f(\mathbf{x})\|^2 + R(\mathbf{x}) \tag{9.1}$$

where $R(\mathbf{x})$ is a regularizer encoding preferences (simplicity, synthesizability). Each candidate evaluation requires calling f; often an expensive DFT calculation or molecular dynamics simulation.

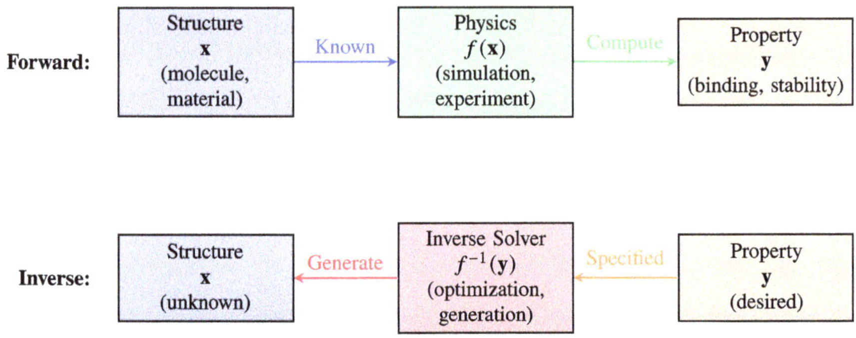

Fig. 9.6 Forward problems (top) map structures to properties through physics. Inverse problems (bottom) specify desired properties and search for structures achieving them. Generative models learn the inverse mapping $p(\mathbf{x}|\mathbf{y})$ from large datasets, enabling rapid sampling of candidates

Generative models instead learn the conditional distribution $p(\mathbf{x}|\mathbf{y})$ from a dataset $\mathcal{D} = \{(\mathbf{x}_i, \mathbf{y}_i)\}_{i=1}^N$ of structure-property pairs. Once trained, sampling new candidates is cheap: draw $\mathbf{x} \sim p(\mathbf{x}|\mathbf{y})$. The fundamental advantage: amortization; we pay the cost of learning once, then generate many candidates rapidly.

9.2.1.2 Architectures for Scientific Generation

Three architectural families dominate scientific generation, each with distinct strengths for different domains.

Diffusion models have become the workhorse of scientific generation, particularly for molecules and materials. The core idea: gradually add noise to data until it becomes pure Gaussian, then learn to reverse the process.

Forward diffusion process (adding noise):

$$q(\mathbf{x}_t|\mathbf{x}_{t-1}) = \mathcal{N}(\mathbf{x}_t; \sqrt{1-\beta_t}\mathbf{x}_{t-1}, \beta_t\mathbf{I}) \tag{9.2}$$

where β_t is a noise schedule, $t \in \{0, 1, \ldots, T\}$, and $\mathbf{x}_0$ is real data. After T steps, $\mathbf{x}_T \approx \mathcal{N}(0, \mathbf{I})$.

Reverse process (denoising, what we learn):

$$p_\theta(\mathbf{x}_{t-1}|\mathbf{x}_t) = \mathcal{N}(\mathbf{x}_{t-1}; \boldsymbol{\mu}_\theta(\mathbf{x}_t, t), \boldsymbol{\Sigma}_\theta(\mathbf{x}_t, t)) \tag{9.3}$$

The neural network $\boldsymbol{\epsilon}_\theta(\mathbf{x}_t, t)$ predicts the noise added at step t, trained via:

$$\mathcal{L}_{\text{simple}} = \mathbb{E}_{t,\mathbf{x}_0,\boldsymbol{\epsilon}}\left[\|\boldsymbol{\epsilon} - \boldsymbol{\epsilon}_\theta(\mathbf{x}_t, t)\|^2\right] \tag{9.4}$$

For molecular graphs, this requires special handling since molecules are discrete graphs, not continuous vectors. Graph DiT ?? introduced a *graph-dependent noise model*:

$$q(\mathbf{x}_t|\mathbf{x}_{t-1}, \mathcal{G}) = \mathcal{N}\left(\mathbf{x}_t; \sqrt{1-\beta_t(\mathcal{G})}\mathbf{x}_{t-1}, \beta_t(\mathcal{G})\mathbf{I}\right) \tag{9.5}$$

where $\beta_t(\mathcal{G})$ depends on graph structure; smaller molecules get different noise schedules than large biomolecules. The architecture uses Diffusion Transformers (DiT) adapted to graphs:

For 3D molecular structures, E(3)-equivariance is critical. EEGSDE maintains rotational/translational symmetry during diffusion:

$$\mathbf{x}_t = \mathbf{R}\mathbf{x}_0 + \boldsymbol{\epsilon}_t \quad \Rightarrow \quad \boldsymbol{\epsilon}_\theta(\mathbf{R}\mathbf{x}_t, t) = \mathbf{R}\boldsymbol{\epsilon}_\theta(\mathbf{x}_t, t) \tag{9.6}$$

for any rotation $\mathbf{R} \in SO(3)$. This uses E(3)-equivariant graph neural networks (EGNN, NequIP) as the denoising backbone.

When to Use/Not Use Diffusion Models

Best for:

- High-quality sample generation (molecules, materials, proteins)
- Multi-modal distributions (many valid designs per specification)
- Continuous or mixed discrete/continuous spaces
- When you can afford 50–1000 sampling steps

Avoid when:

- Real-time generation required (use VAEs or consistency models)
- Highly discrete spaces with no continuous relaxation
- Limited compute budget (1000 NFEs = 1000 forward passes)

VAEs learn a continuous latent space $\mathbf{z} \in \mathbb{R}^d$ encoding structures, enabling rapid sampling and interpolation. The architecture has two components:

Encoder $q_\phi(\mathbf{z}|\mathbf{x})$: Maps structures to latent distributions

$$q_\phi(\mathbf{z}|\mathbf{x}) = \mathcal{N}(\mathbf{z}; \boldsymbol{\mu}_\phi(\mathbf{x}), \mathrm{diag}(\boldsymbol{\sigma}_\phi^2(\mathbf{x}))) \tag{9.7}$$

Decoder $p_\theta(\mathbf{x}|\mathbf{z})$: Reconstructs structures from latent codes

$$p_\theta(\mathbf{x}|\mathbf{z}) = \mathrm{Categorical}(\mathbf{x}; \boldsymbol{\pi}_\theta(\mathbf{z})) \quad \text{(for discrete molecules)} \tag{9.8}$$

Training objective (ELBO):

$$\mathcal{L}_{\text{VAE}} = \mathbb{E}_{q_\phi(\mathbf{z}|\mathbf{x})}[\log p_\theta(\mathbf{x}|\mathbf{z})] - D_{\text{KL}}(q_\phi(\mathbf{z}|\mathbf{x}) \| p(\mathbf{z})) \tag{9.9}$$

where $p(\mathbf{z}) = \mathcal{N}(0, \mathbf{I})$ is the prior. The first term is reconstruction accuracy; the second regularizes the latent space to be smooth and Gaussian.

Conditional VAEs (CVAEs) extend this to conditional generation by feeding properties $\mathbf{y}$ to both encoder and decoder:

$$q_\phi(\mathbf{z}|\mathbf{x}, \mathbf{y}) = \mathcal{N}(\mathbf{z}; \boldsymbol{\mu}_\phi(\mathbf{x}, \mathbf{y}), \text{diag}(\boldsymbol{\sigma}_\phi^2(\mathbf{x}, \mathbf{y}))) \tag{9.10}$$

$$p_\theta(\mathbf{x}|\mathbf{z}, \mathbf{y}) = \text{Categorical}(\mathbf{x}; \boldsymbol{\pi}_\theta(\mathbf{z}, \mathbf{y})) \tag{9.11}$$

At generation time: specify $\mathbf{y}$, sample $\mathbf{z} \sim \mathcal{N}(0, \mathbf{I})$, decode $\mathbf{x} \sim p_\theta(\mathbf{x}|\mathbf{z}, \mathbf{y})$.

For materials discovery, VAEs excel because the latent space enables:

- **Interpolation**: Smoothly transition between known materials
- **Optimization**: Gradient ascent in $\mathbf{z}$-space toward desired properties
- **Inverse design**: Train a property predictor $g(\mathbf{z}) \approx f(\text{decode}(\mathbf{z}))$, then solve $\max_{\mathbf{z}} g(\mathbf{z})$

VAEs for molecules often suffer from **posterior collapse**: the model ignores $\mathbf{z}$ and relies only on the decoder. Symptoms:

- $D_{\text{KL}} \approx 0$ during training
- Samples from prior $p(\mathbf{z})$ produce garbage
- Latent interpolations don't smoothly transition

Solutions:

- KL annealing: Start with βD_{KL} where $\beta = 0$, slowly increase to 1
- Free bits: Allow $D_{\text{KL}} \geq \lambda$ per latent dimension
- Stronger encoder: Use GNNs or transformers, not just MLPs
- β-VAE: Weight reconstruction versus KL differently ($\beta > 1$ encourages disentanglement)

Protein and DNA sequences have natural left-to-right ordering, making autoregressive models a natural fit:

$$p(\mathbf{x}) = \prod_{i=1}^{L} p(x_i | x_{<i}) \tag{9.12}$$

where $x_{<i} = (x_1, \ldots, x_{i-1})$ is the sequence so far. For *conditional* generation:

$$p(\mathbf{x}|\mathbf{y}) = \prod_{i=1}^{L} p(x_i | x_{<i}, \mathbf{y}) \tag{9.13}$$

Modern implementations use transformers with causal masking. **DiMA** (ICML 2025) applies this to protein design by working in the latent space of protein language models:

1. Pre-train protein LM (ESM-2, SaProt) on millions of sequences
2. Encode sequences to latent vectors: $\mathbf{z} = \text{LM-encode}(\mathbf{x})$
3. Train autoregressive model on latent codes: $p(\mathbf{z}|\mathbf{y})$
4. Sample latent sequence, decode to amino acids

This latent diffusion approach generates family-specific proteins 10× faster than full sequence diffusion.

For **3D protein structures**, RFdiffusion3 combines sequence and structure generation:

$$p(\mathbf{x}_{\text{seq}}, \mathbf{x}_{3\text{D}}|\mathbf{y}) = p(\mathbf{x}_{\text{seq}}|\mathbf{y}) \cdot p(\mathbf{x}_{3\text{D}}|\mathbf{x}_{\text{seq}}, \mathbf{y}) \tag{9.14}$$

First generate the sequence autoregressively, then diffuse 3D coordinates conditioned on that sequence. The December 2025 update adds all-atom resolution (not just backbone), enabling design of:

- Enzymes with specific catalytic sites
- Protein-DNA interfaces
- Small molecule binding pockets with atomic precision

Llamole represents the convergence of LLMs and scientific generation. It's a multimodal transformer that processes:

- Text: Drug requirements in natural language
- Molecular graphs: Via integrated Graph DiT
- Retrosynthetic routes: Chemical reaction sequences

The model achieved top-3 accuracy on 12 drug design benchmarks by combining Graph DiT's generative capability with LLM's reasoning about synthesis, toxicity, and medicinal chemistry principles.

How do we actually incorporate property targets $\mathbf{y}$ into generation? Five main approaches, with escalating sophistication.

1. Simple Concatenation

The most straightforward: concatenate property embedding to node/edge features.

For molecular graphs with target property $\mathbf{y} \in \mathbb{R}^k$:

$$\mathbf{h}_i^{(0)} = [\mathbf{x}_i; \mathbf{W}_y\mathbf{y}] \tag{9.15}$$

where $\mathbf{x}_i$ is the i-th atom's features, $\mathbf{W}_y$ projects properties to same dimension. Then process with GNN:

$$\mathbf{h}_i^{(\ell+1)} = \text{UPDATE}\left(\mathbf{h}_i^{(\ell)}, \bigoplus_{j\in\mathcal{N}(i)} \text{MESSAGE}(\mathbf{h}_i^{(\ell)}, \mathbf{h}_j^{(\ell)}, \mathbf{e}_{ij})\right) \tag{9.16}$$

Pros: Simple, works for continuous properties

Cons: Weak conditioning; model can ignore $\mathbf{y}$; no mechanism for multi-modal conditioning

2. Cross-Attention (Transformer-Based)

For transformer architectures, cross-attention provides stronger conditioning. The structure attends to property tokens:

$$\mathbf{Q} = \mathbf{H}\mathbf{W}_Q \quad \text{(queries from structure)} \tag{9.17}$$

$$\mathbf{K}, \mathbf{V} = \mathbf{Y}\mathbf{W}_K, \mathbf{Y}\mathbf{W}_V \quad \text{(keys/values from properties)} \tag{9.18}$$

$$\text{CrossAttn}(\mathbf{H}, \mathbf{Y}) = \text{softmax}\left(\frac{\mathbf{Q}\mathbf{K}^\top}{\sqrt{d_k}}\right)\mathbf{V} \tag{9.19}$$

where $\mathbf{H}$ is structure representation, $\mathbf{Y}$ is property embedding. This appears in Graph DiT's architecture (Fig. 9.7).

Multi-conditional generation handles multiple properties simultaneously:

$$\mathbf{Y} = [\mathbf{y}_1^T, \mathbf{y}_2^T, \ldots, \mathbf{y}_M^T]^\top \in \mathbb{R}^{M\times d} \tag{9.20}$$

Each property gets learned embedding, all participate in cross-attention. Graph DiT shows this enables joint optimization: generate molecules with *simultaneously* high binding affinity AND low toxicity AND good solubility.

3. Classifier Guidance (Conditional Sampling)

Borrowed from image diffusion (DALL-E, Stable Diffusion), classifier guidance steers sampling toward regions with desired properties *without retraining*.

Start with unconditional diffusion model $p_\theta(\mathbf{x})$. Train a separate classifier $p_\psi(\mathbf{y}|\mathbf{x}_t)$ that predicts properties from noisy structures. During sampling, modify the score:

$$\tilde{\boldsymbol{\epsilon}}_\theta(\mathbf{x}_t, t, \mathbf{y}) = \boldsymbol{\epsilon}_\theta(\mathbf{x}_t, t) - \sqrt{1-\bar{\alpha}_t} \cdot s\nabla_{\mathbf{x}_t} \log p_\psi(\mathbf{y}|\mathbf{x}_t) \tag{9.21}$$

where s is guidance strength. The gradient $\nabla \log p_\psi(\mathbf{y}|\mathbf{x}_t)$ pushes toward structures likely to have property $\mathbf{y}$.

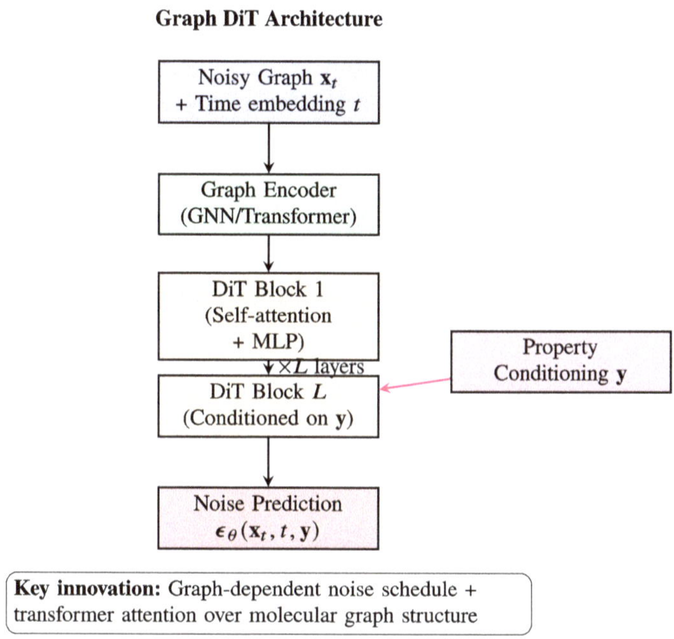

Fig. 9.7 Graph DiT architecture for molecular generation. The model applies Diffusion Transformer blocks to graph-structured data, with noise schedules adapted to molecular topology. Multi-conditional generation incorporates property targets $\mathbf{y}$ via cross-attention in DiT blocks

Classifier-free guidance eliminates the separate classifier by training a single model on both conditional and unconditional generation:

$$\tilde{\epsilon}_\theta(\mathbf{x}_t, t, \mathbf{y}) = \epsilon_\theta(\mathbf{x}_t, t, \emptyset) + s\left[\epsilon_\theta(\mathbf{x}_t, t, \mathbf{y}) - \epsilon_\theta(\mathbf{x}_t, t, \emptyset)\right] \tag{9.22}$$

During training, randomly replace $\mathbf{y}$ with null token $\emptyset$ with probability $p_{\text{uncond}} = 0.1$. At inference, interpolate between unconditional and conditional predictions.

4. Energy-Based Guidance

For physics-based properties, guide generation using energy functions directly. EEGSDE incorporates potential energy $U(\mathbf{x})$ during sampling:

$$\nabla_{\mathbf{x}_t} \log p(\mathbf{x}_t|\mathbf{y}) \approx \nabla_{\mathbf{x}_t} \log p(\mathbf{x}_t) - \lambda \nabla_{\mathbf{x}_t} U(\mathbf{x}_t) \tag{9.23}$$

where $U(\mathbf{x})$ could be:

- Coulomb energy for charged molecules
- Lennard-Jones potential for van der Waals forces

- Quantum mechanical energy from approximate DFT
- Empirical force fields (MMFF94, UFF)

The gradient $-\nabla U$ is the force, steering diffusion toward low-energy configurations.

5. Constrained Projection (Hard Constraints)

UniGuide [16] achieves near-perfect constraint satisfaction through projection. At each diffusion step, project the sample onto the constraint manifold:

$$\mathbf{x}_t^{\text{proj}} = \mathcal{P}_{\mathcal{C}}(\mathbf{x}_t) = \arg\min_{\mathbf{x}' \in \mathcal{C}} \|\mathbf{x}' - \mathbf{x}_t\|^2 \tag{9.24}$$

where $\mathcal{C} = \{\mathbf{x} : g(\mathbf{x}) = \mathbf{y}\}$ is the constraint set.

For molecular generation with target properties $\mathbf{y}^*$:

1. Take diffusion step: $\mathbf{x}_{t-1} \sim p_\theta(\mathbf{x}_{t-1}|\mathbf{x}_t)$
2. Project to satisfy constraints: $\mathbf{x}_{t-1}^{\text{proj}} = \mathcal{P}_{\{g(\mathbf{x})=\mathbf{y}^*\}}(\mathbf{x}_{t-1})$
3. Continue with projected sample

Results: UniGuide achieved 100% constraint satisfaction on QM9 property-constrained generation (previous methods: 60–80%) without retraining for new constraints.

The projection step often requires solving:

$$\min_{\mathbf{x}} \|\mathbf{x} - \mathbf{x}_t\|^2 + \mu\|g(\mathbf{x}) - \mathbf{y}^*\|^2 \tag{9.25}$$

via gradient descent. For complex constraints, this is the bottleneck; but enables guaranteed satisfaction.

9.2.2 *Physics-Guided Generation and Constraints*

Pure data-driven generation can produce chemically impossible structures: molecules with wrong valences, materials that violate thermodynamic stability, proteins with impossible backbone angles. Physics-guided generation embeds domain knowledge; either as soft preferences or hard constraints; ensuring outputs respect fundamental laws.

9.2.2.1 Hard Constraints Versus Soft Guidance

The spectrum from unconstrained to strictly constrained generation offers different trade-offs (Fig. 9.8).

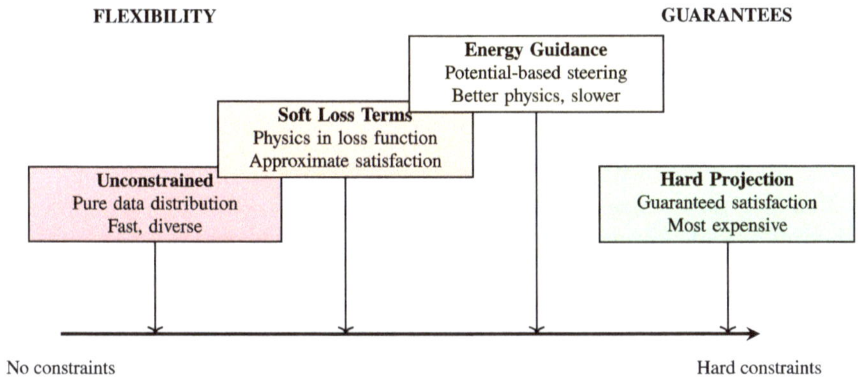

Fig. 9.8 Spectrum of physics constraint enforcement. Unconstrained generation maximizes diversity but may violate physics. Hard projection guarantees constraint satisfaction but is computationally expensive. Most practical systems use intermediate approaches: soft losses or energy guidance

Add physics violations as penalty terms to the generation loss. For diffusion models:

$$\mathcal{L}_{\text{total}} = \mathcal{L}_{\text{diffusion}} + \lambda_1 \mathcal{L}_{\text{physics}} + \lambda_2 \mathcal{L}_{\text{stability}} + \lambda_3 \mathcal{L}_{\text{symmetry}} \tag{9.26}$$

Example 1: Molecular valence constraints
For organic molecules, carbon forms 4 bonds, nitrogen 3, oxygen 2. Violation penalty:

$$\mathcal{L}_{\text{valence}} = \sum_{i=1}^{N} \max(0, |\deg(v_i) - V_{\max}(v_i)|)^2 \tag{9.27}$$

where $\deg(v_i)$ is the degree of atom i, $V_{\max}(v_i)$ is maximum valence for that atom type.

Example 2: Thermodynamic stability for materials
Materials below the convex hull of formation energies are thermodynamically stable. Add a penalty pushing toward the hull:

$$\mathcal{L}_{\text{hull}} = \max(0, E_{\text{hull}}(\mathbf{x}))^2 \tag{9.28}$$

where $E_{\text{hull}}(\mathbf{x})$ is the energy above the convex hull (negative means stable).

MolEdit [17] combines physics-informed diffusion with preference alignment:

$$\mathcal{L}_{\text{MolEdit}} = \mathcal{L}_{\text{diffusion}} + \alpha \mathcal{L}_{\text{DPO}}(\pi_\theta, \pi_{\text{ref}}) \tag{9.29}$$

where DPO (Direct Preference Optimization) learns from paired preferences: "molecule A is more stable than B." The preference loss:

$$\mathcal{L}_{\text{DPO}} = -\mathbb{E}_{(\mathbf{x}_w, \mathbf{x}_l)} \left[\log \sigma \left(\beta \log \frac{\pi_\theta(\mathbf{x}_w)}{\pi_{\text{ref}}(\mathbf{x}_w)} - \beta \log \frac{\pi_\theta(\mathbf{x}_l)}{\pi_{\text{ref}}(\mathbf{x}_l)} \right) \right] \tag{9.30}$$

where $\mathbf{x}_w$ is preferred ("winner"), $\mathbf{x}_l$ is less preferred ("loser"). This steers generation toward chemist-approved modifications verified through wet-lab experiments.

Reinforcement Learning from Physics Feedback (RLPF) [18] reframes denoising as a Markov decision process with physics-based rewards:

- **State**: Noisy molecule $\mathbf{x}_t$
- **Action**: Denoising step $\mathbf{x}_{t-1} = \boldsymbol{\mu}_\theta(\mathbf{x}_t, t) + \boldsymbol{\sigma}_t \boldsymbol{\epsilon}$
- **Reward**: $r(\mathbf{x}_{t-1}) = -U(\mathbf{x}_{t-1}) + w_{\text{valid}} \cdot \nVdash_{\text{valid}}(\mathbf{x}_{t-1})$

where U is potential energy, $\nVdash_{\text{valid}}$ indicates chemical validity. Train via policy gradient:

$$\nabla_\theta J(\theta) = \mathbb{E}_{\tau \sim \pi_\theta} \left[\sum_{t=0}^{T} \nabla_\theta \log \pi_\theta(a_t | s_t) \cdot R(\tau) \right] \tag{9.31}$$

Applied to EDM, GeoLDM, and UniGEM, RLPF improved thermodynamic stability by 15–40% while maintaining generation quality.

For critical applications (drug design, safety-critical materials), soft constraints aren't enough. Hard projection guarantees satisfaction.

Projection onto constraint manifold at each generation step:

Given constraint $g(\mathbf{x}) = \mathbf{y}$ (e.g., "molecular weight = 350 Da"), solve:

$$\mathbf{x}^{\text{proj}} = \arg\min_{\mathbf{x}'} \|\mathbf{x}' - \mathbf{x}\|^2 \quad \text{subject to} \quad g(\mathbf{x}') = \mathbf{y} \tag{9.32}$$

For differentiable constraints, use Lagrangian relaxation:

$$\mathcal{L}(\mathbf{x}', \boldsymbol{\lambda}) = \|\mathbf{x}' - \mathbf{x}\|^2 + \boldsymbol{\lambda}^\top (g(\mathbf{x}') - \mathbf{y}) \tag{9.33}$$

Solve via gradient descent alternating between:

$$\mathbf{x}'^{(k+1)} = \mathbf{x}'^{(k)} - \eta \nabla_{\mathbf{x}'} \mathcal{L}(\mathbf{x}'^{(k)}, \boldsymbol{\lambda}^{(k)}) \tag{9.34}$$

$$\boldsymbol{\lambda}^{(k+1)} = \boldsymbol{\lambda}^{(k)} + \eta (g(\mathbf{x}'^{(k+1)}) - \mathbf{y}) \tag{9.35}$$

For discrete constraints (e.g., molecular graphs), use integer programming or constraint satisfaction:

- Generate continuous representation $\mathbf{z}$ (latent code or coordinates)
- Decode to discrete structure $\mathbf{x}$
- Check constraints; if violated, perturb $\mathbf{z}$ and retry
- Iterate until satisfaction or max attempts

UniGuide demonstrates this achieves 100% satisfaction on QM9 properties, but increases sampling time by 3–5×.

9.2.2.2 Incorporating Conservation Laws

Physical conservation laws; mass, charge, energy, momentum; are non-negotiable in many scientific domains.

When generating fluid velocity fields $\mathbf{u}(\mathbf{x}, t)$, incompressibility demands:

$$\nabla \cdot \mathbf{u} = 0 \tag{9.36}$$

Approach 1: Project onto divergence-free space
Helmholtz decomposition: any vector field $\mathbf{u} = \mathbf{u}_{\text{div-free}} + \nabla\phi$. Remove divergent part:

$$\mathbf{u}_{\text{proj}} = \mathbf{u} - \nabla\phi \tag{9.37}$$

where ϕ solves Poisson equation: $\nabla^2\phi = \nabla \cdot \mathbf{u}$.

Approach 2: Generate in stream function space
For 2D flows, represent $\mathbf{u} = (\partial_y\psi, -\partial_x\psi)$ where ψ is stream function. This automatically satisfies $\nabla \cdot \mathbf{u} = 0$.

For charged molecules (ions, zwitterions), total charge must equal target (Fig. 9.9):

$$\sum_{i=1}^{N} q_i = Q_{\text{total}} \tag{9.38}$$

where q_i is the charge on atom i. During graph generation:

1. Predict node types (C, N, O, etc.) and bond orders
2. Compute formal charges from valence and bonds
3. If $\sum q_i \neq Q_{\text{total}}$, perturb bond orders or add/remove hydrogens
4. Re-check until satisfied

For molecular dynamics or physical simulations, generated trajectories should conserve total energy:

$$H(\mathbf{q}, \mathbf{p}) = T(\mathbf{p}) + V(\mathbf{q}) = \text{const} \tag{9.39}$$

where $\mathbf{q}$ is positions, $\mathbf{p}$ is momenta, T is kinetic energy, V is potential.

Symplectic Neural Networks guarantee exact energy conservation through symplectic integrators:

$$\begin{pmatrix} \mathbf{q}_{n+1} \\ \mathbf{p}_{n+1} \end{pmatrix} = \mathbf{\Phi}_{\Delta t}^{\text{symp}} \begin{pmatrix} \mathbf{q}_n \\ \mathbf{p}_n \end{pmatrix} \tag{9.40}$$

where $\mathbf{\Phi}$ is a learned symplectic map (preserves $d\mathbf{q} \wedge d\mathbf{p}$). We'll revisit this in Section 4 on world models.

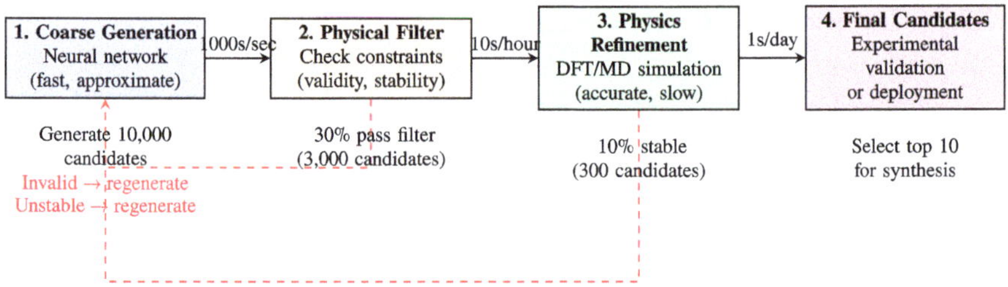

Fig. 9.9 Multi-fidelity hybrid pipeline for molecular/materials design. Fast neural generation produces thousands of candidates. Physical filters remove obvious invalids. Expensive high-fidelity simulations (DFT, MD) refine the promising subset. Final candidates undergo experimental validation. This pipeline leverages speed of ML with accuracy of physics

9.2.2.3 Multi-fidelity and Hybrid Approaches

The most powerful systems combine fast neural generation with accurate physics refinement.

The Pipeline Pattern

1. **Coarse generation**: Neural network produces 1000 s of candidates quickly
2. **Fast filtering**: Rule-based checks eliminate obvious failures (wrong valences, unphysical geometries)
3. **Physics refinement**: Run expensive simulations (DFT, MD) on surviving candidates
4. **Iterative improvement**: Use physics results to fine-tune generator

Example: MatterGen + DFT refinement
MatterGen generates crystal structures in seconds. For the top 100 candidates:

- Run DFT geometry optimization (hours per structure)
- Compute formation energy relative to convex hull
- Phonon calculations for dynamic stability
- If stable, add to dataset; if unstable, use as negative example for retraining

This closed-loop approach improves generator over time, learning from physics feedback.

Train a "corrector" network that takes neural-generated structures and applies small physics-based adjustments:

$$\mathbf{x}_{\text{final}} = \mathbf{x}_{\text{gen}} + \delta \mathbf{x}_{\text{corrector}}(\mathbf{x}_{\text{gen}}) \tag{9.41}$$

where $\delta \mathbf{x}_{\text{corrector}}$ is trained to minimize physics violations:

$$\mathcal{L}_{\text{corrector}} = \|\mathbf{x}_{\text{final}} - \mathbf{x}_{\text{DFT}}\|^2 + \lambda \mathcal{L}_{\text{physics}}(\mathbf{x}_{\text{final}}) \tag{9.42}$$

The corrector learns systematic biases in the generator (e.g., bond lengths slightly too long) and fixes them.

9.2.3 *Evaluation Protocols for Scientific Generation*

A generative model produces a beautiful molecule. The atoms are correctly placed, bonds look reasonable, and it scores well on your training loss. But is it *discoverable*? *Synthesizable*? *Novel*? Evaluation of scientific generative models requires going far beyond perceptual quality metrics borrowed from images (FID, IS) to domain-specific protocols assessing physical validity, chemical feasibility, and actual scientific value.

9.2.3.1 Beyond Perceptual Quality

Image generation metrics; Fréchet Inception Distance (FID), Inception Score (IS), CLIP similarity; measure whether generated images "look like" real images to pre-trained neural networks. These are necessary but not sufficient for scientific generation.

Why FID/IS Are Insufficient for Science ? Consider a molecular generative model that achieves excellent FID:

- Generates graphs with correct degree distributions (looks statistically like real molecules)
- Produces plausible 2D depictions
- Matches training data distribution well

But fails scientifically because:

- 45% violate valence rules (carbon with 5 bonds)
- 60% are thermodynamically unstable (high formation energy)
- 80% cannot be synthesized (require impossible reactions)
- 95% are duplicates or near-duplicates of training data

This is the **reality** reported by J. Chem. Inf. Model (2025) benchmark using BindingMOAD dataset with blind hold-outs. Deep learning methods frequently produce invalid outputs despite good statistical metrics.

First-order requirement: generated structures must be physically valid.

For molecules:

- **Valence validity**: $V_i = \sum_{j \in \mathcal{N}(i)} b_{ij}$ equals maximum valence for atom type i
- **Chemical validity**: RDKit sanitization passes (aromaticity, valence, stereochemistry)

- **Geometric validity**: Bond lengths in acceptable ranges (C-C: 1.2–1.6 Å)
- **Ring strain**: No highly strained small rings (3-membered rings with spÂ² carbons)

For materials (crystals):

- **Charge neutrality**: $\sum_i q_i Z_i = 0$ where Z_i is oxidation state
- **Coordination validity**: Atoms have reasonable coordination numbers (6 for octahedral, 4 for tetrahedral)
- **No atomic clashes**: Interatomic distances $> r_{\min}$ (typically 0.5 Å)
- **Lattice parameters**: Physically reasonable unit cell (no absurdly long/short vectors)

For proteins:

- **Ramachandran validity**: (ϕ, ψ) angles in allowed regions
- **Clash score**: Steric overlaps measured by MolProbity
- **Rotamer outliers**: Side chains in favorable conformations
- **Secondary structure**: α-helices and β-sheets well-formed

Aggregate validity score:

$$\text{Validity} = \frac{1}{N} \sum_{i=1}^{N} \mathbb{1}_{\text{valid}}(\mathbf{x}_i) \tag{9.43}$$

where $\mathbb{1}_{\text{valid}}(\mathbf{x}_i) = 1$ if structure i passes all validity checks. Report this first; invalid structures are worthless regardless of other metrics.

Performance Metrics: Do They Work?

Generated structures must achieve desired properties.

Accuracy of conditional generation:

$$\text{MAE}_{\text{property}} = \frac{1}{N} \sum_{i=1}^{N} |f(\mathbf{x}_i) - y_i^*| \tag{9.44}$$

where y_i^* is target property, $f(\mathbf{x}_i)$ is actual property (computed via simulation or experiment).

Success rate:

$$\text{Success@}\tau = \frac{1}{N} \sum_{i=1}^{N} \mathbb{1}[|f(\mathbf{x}_i) - y_i^*| < \tau] \tag{9.45}$$

percentage achieving property within tolerance τ (e.g., binding affinity within 1 kcal/mol).

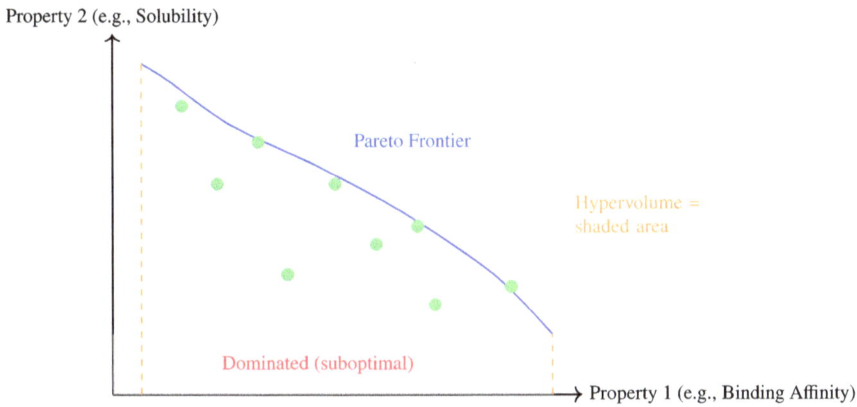

Fig. 9.10 Multi-objective evaluation for scientific generation. Generated molecules (green dots) should cluster near the Pareto frontier (blue curve) that represents optimal trade-offs. Hypervolume quantifies total dominated objective space. Spacing measures uniformity. Molecules in red region are dominated; strictly worse than others on all objectives

Multi-objective optimization metrics:
For multiple properties $\mathbf{y} = (y_1, \ldots, y_M)$, evaluate Pareto optimality:

- **Hypervolume**: Volume of objective space dominated by generated set
- **Pareto frontier coverage**: Percentage of known Pareto-optimal solutions covered
- **Spacing metric**: Uniformity of distribution along Pareto frontier

Figure 9.10 illustrates multi-objective evaluation.

Synthesizability and Manufacturability

A molecule that cannot be synthesized is not a discovery, regardless of predicted properties.

Synthetic accessibility score (SA score):

$$\text{SA} = \frac{\text{complexity} - \text{fragmentScore}}{\text{scale}} \tag{9.46}$$

where complexity penalizes rings, stereocenters, macrocycles; fragmentScore rewards known synthetic building blocks. Range 1 (easy) to 10 (very difficult). Molecules with $\text{SA} > 6$ are challenging; $\text{SA} > 8$ often impractical.

Retrosynthetic analysis:
Tools like AiZynthfinder, ASKCOS predict synthetic routes. Metrics:

- **Route found?**: Binary success/failure
- **Number of steps**: Fewer is better (typically ≤ 6 for drug-like molecules)
- **Availability of starting materials**: All precursors commercially available?

- **Reaction viability**: Predicted yields > 50%?

For materials:
Synthesizability is harder to quantify. Proxies include:

- **Formation energy**: $\Delta H_f < 0$ (exothermic synthesis more likely)
- **Decomposition temperature**: Stable at synthesis conditions (e.g., 1000 K)?
- **Precursor availability**: Elements commercially available in required form?
- **Similarity to known materials**: Novel but not too exotic

9.2.3.2 Coverage Versus Quality Trade-Offs

Generative models face a fundamental tension: *quality* (how good are individual samples?) versus *coverage* (how much of the space is explored?).

Adapting classification metrics to generation:

Precision: What fraction of generated samples are high-quality?

$$\text{Precision} = \frac{\text{\# valid, high-performing generated}}{\text{\# total generated}} \tag{9.47}$$

Recall: What fraction of true high-quality structures can the model generate?

$$\text{Recall} = \frac{\text{\# true positives in generated set}}{\text{\# total true positives in full space}} \tag{9.48}$$

High precision, low recall: Generates few but excellent candidates (conservative)
High recall, low precision: Explores broadly but many failures (aggressive)

F1-score for generation:

$$F1 = \frac{2 \cdot \text{Precision} \cdot \text{Recall}}{\text{Precision} + \text{Recall}} \tag{9.49}$$

Matbench Discovery uses this as one component of their Combined Performance Score, weighting F1 for discovery (finding stable materials) alongside RMSD for geometry and κ_{SRME} for thermal conductivity.

Generated sets should be diverse, not just high-performing.

Internal diversity (within generated set):

$$\text{Div}_{\text{internal}} = \frac{1}{N(N-1)} \sum_{i \neq j} d(\mathbf{x}_i, \mathbf{x}_j) \tag{9.50}$$

where d is distance (Tanimoto for molecules, structure match for crystals, RMSD for proteins).

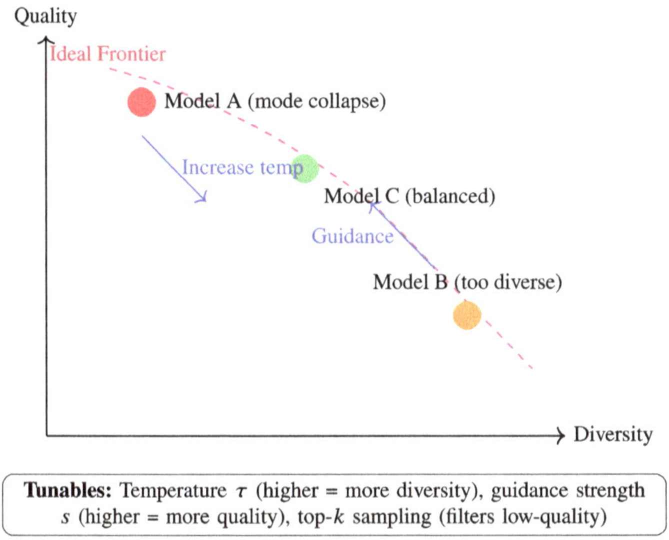

Fig. 9.11 Quality-diversity trade-off in generative models. Model A has mode collapse (high quality, low diversity). Model B explores too broadly (low quality). Model C balances both. Hyperparameters like temperature and guidance strength allow navigation along this frontier

Novelty (compared to training set):

$$\text{Novelty} = \frac{1}{N} \sum_{i=1}^{N} \min_{\mathbf{x}' \in \mathcal{D}_{\text{train}}} d(\mathbf{x}_i, \mathbf{x}') \tag{9.51}$$

Average distance to nearest training sample. High novelty means exploring new chemical space; too high might mean invalid/unrealistic structures.

Uniqueness:

$$\text{Uniqueness} = \frac{\#\,\text{unique structures}}{\#\,\text{total generated}} \tag{9.52}$$

Simple but important; models sometimes mode collapse and generate duplicates.

The diversity-quality dilemma:

Increasing diversity often decreases average quality (explore more, find worse). The goal: maximize both simultaneously (Fig. 9.11).

9.2.3.3 Expert Validation and Benchmarks

Ultimately, scientific generation must produce results that domain experts find valuable.

Chemists, materials scientists, or biologists should evaluate generated candidates:

- **Plausibility**: Does this structure make chemical sense?
- **Interestingness**: Is this novel and potentially useful?
- **Synthesizability**: Could you actually make this in a lab?
- **Comparative ranking**: Rank these 10 candidates by promise

Quantify expert agreement:

$$\kappa = \frac{p_o - p_e}{1 - p_e} \tag{9.53}$$

Cohen's kappa measures inter-rater reliability ($\kappa > 0.6$ indicates good agreement).

Active learning from expert feedback:

Use expert preferences to fine-tune via methods like RLHF (Reinforcement Learning from Human Feedback) or DPO (Direct Preference Optimization). MolEdit demonstrated this: chemists provided pairwise preferences ("molecule A is better than B"), used to align the diffusion model.

Standardized benchmarks enable objective comparison.

For molecular generation:

- **QM9**: 134k small organic molecules with quantum properties
- **ZINC**: 250M drug-like molecules for drug design
- **GEOM-Drugs**: 430k molecules with multiple 3D conformers
- **BindingMOAD**: Protein-ligand complexes for structure-based design
- **GuacaMol/MOSES**: Comprehensive benchmarking suites with 20+ metrics

For materials:

- **Materials Project**: 150k+ DFT-computed inorganic materials
- **OQMD**: 1M+ crystal structures with formation energies
- **Matbench**: 13 property prediction tasks for standardized evaluation
- **Matbench Discovery** (Nature Machine Intelligence 2025): Formation energy, stability, geometry optimization

For proteins:

- **CATH/SCOP**: Protein structure classification databases
- **CASP**: Critical Assessment of protein Structure Prediction (biennial competition)
- **ProteinGym**: Deep mutational scanning benchmarks

Evaluation protocol example (from Matbench Discovery):

1. **Discovery Task**: Predict which of 257k hypothetical materials are stable
2. **Metrics**:
    - F1 score for classification (stable vs. unstable)
    - RMSD for DFT-relaxed geometries
    - Mean absolute error for formation energy

3. **Combined Performance Score**: Weighted average of F1, RMSD, property MAE
4. **Top performers**: EquiformerV2 models (F1 $>$ 0.9, energy MAE $\equiv$ 20 meV/atom)

9.3 World Models for Scientific Systems

A roboticist faces a dilemma. She needs to control a robotic arm moving through turbulent airflow; a system governed by Navier-Stokes equations requiring milliseconds to solve on supercomputers. Real-time control demands predictions in microseconds. Her solution: a *world model*; a learned simulator compressing fluid dynamics into a low-dimensional latent space where predictions are 10,000$\times$ faster. The catch? The model must still respect physics well enough that the robot doesn't crash when its predictions meet reality.

This is the promise and peril of world models: compress complex dynamics into tractable representations while preserving the essential physics. Where generative models (Section 2) solve inverse problems; designing molecules with desired properties; world models solve forward problems: predicting how systems evolve. But unlike simple one-step prediction, world models must be *rollout-stable*: accurate not just for the next timestep, but for hundreds or thousands of steps into the future.

This section develops the theory and practice of world models for scientific systems, from learning latent dynamics to deploying digital twins, with special emphasis on physics-aware architectures that provide both speed and guarantees.

9.3.1 Latent Dynamics

High-dimensional scientific systems; fluid flows with millions of grid points, molecular dynamics with thousands of atoms, climate models with billions of variables; are computationally intractable for real-time prediction. The latent dynamics paradigm compresses these systems into low-dimensional representations where prediction is fast, then reconstructs outputs when needed.

9.3.1.1 The Latent Dynamics Paradigm

The three-stage architecture has become canonical across domains (Fig. 9.12).

Stage 1: Encoder

$E_\phi : \mathcal{X} \to \mathcal{Z}$

Maps high-dimensional observations $\mathbf{x}_t \in \mathbb{R}^n$ to latent codes $\mathbf{z}_t \in \mathbb{R}^d$ where $d \ll n$.

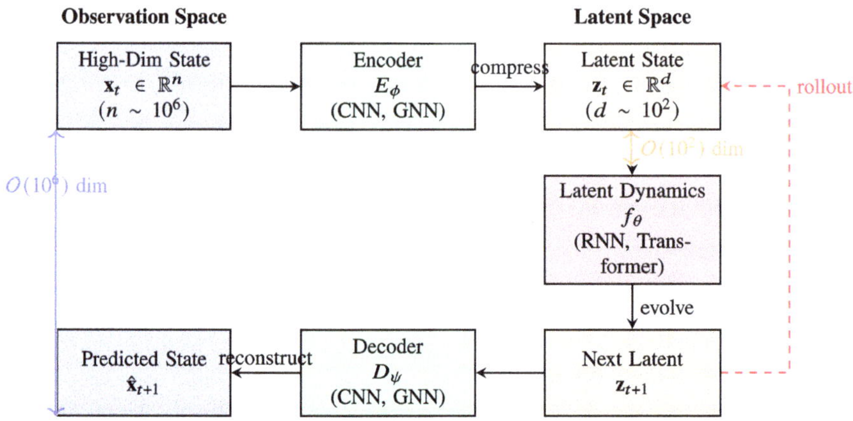

Fig. 9.12 Latent dynamics architecture. High-dimensional states $\mathbf{x}_t$ (e.g., fluid velocity fields) are compressed to low-dimensional latent codes $\mathbf{z}_t$ via encoder E_ϕ. Dynamics evolve in latent space: $\mathbf{z}_{t+1} = f_\theta(\mathbf{z}_t)$. Decoder D_ψ reconstructs predictions. Autoregressive rollout (red dashed) enables multi step forecasting. Typical compression: $10^6 \to 10^2$ dimensions (10,000× reduction)

For **spatially-structured data** (images, fields on grids):

$$\mathbf{z}_t = E_\phi(\mathbf{x}_t) = \text{CNN}_\phi(\mathbf{x}_t) \tag{9.54}$$

Convolutional encoder progressively downsamples spatial dimensions while increasing channels:

$$\mathbf{h}^{(1)} = \text{Conv}_{64}(\mathbf{x}_t) \quad (256 \times 256 \to 128 \times 128) \tag{9.55}$$

$$\mathbf{h}^{(2)} = \text{Conv}_{128}(\mathbf{h}^{(1)}) \quad (128 \times 128 \to 64 \times 64) \tag{9.56}$$

$$\vdots$$

$$\mathbf{z}_t = \text{Flatten}(\mathbf{h}^{(L)}) \quad (8 \times 8 \times 512 \to 32768) \tag{9.57}$$

For **graph-structured data** (molecules, meshes):

$$\mathbf{z}_t = E_\phi(\mathcal{G}_t) = \text{Readout}\left(\text{GNN}_\phi(\mathcal{G}_t)\right) \tag{9.58}$$

Graph neural network processes nodes/edges, readout function (sum, mean, attention) produces global representation:

$$\mathbf{h}_i^{(\ell+1)} = \text{UPDATE}\left(\mathbf{h}_i^{(\ell)}, \bigoplus_{j \in \mathcal{N}(i)} \text{MESSAGE}(\mathbf{h}_i^{(\ell)}, \mathbf{h}_j^{(\ell)}, \mathbf{e}_{ij})\right) \tag{9.59}$$

$$\mathbf{z}_t = \frac{1}{|\mathcal{V}|} \sum_{i \in \mathcal{V}} \mathbf{h}_i^{(L)} \tag{9.60}$$

Stage 2: Latent Dynamics

$f_\theta : \mathcal{Z} \rightarrow \mathcal{Z}$
Predicts next latent state from current latent state (and optionally controls $\mathbf{u}_t$):

$$\mathbf{z}_{t+1} = f_\theta(\mathbf{z}_t, \mathbf{u}_t) \tag{9.61}$$

Three architectural families dominate, each with different inductive biases:

1. **Recurrent architectures** (LSTMs, GRUs):

$$\mathbf{h}_t = \text{LSTM}(\mathbf{z}_t, \mathbf{h}_{t-1}) \tag{9.62}$$
$$\mathbf{z}_{t+1} = W\mathbf{h}_t + \mathbf{b} \tag{9.63}$$

   Pros: Explicitly model temporal dependencies, natural for sequences
   Cons: Vanishing gradients for long rollouts, sequential computation (slow)
2. **Feed-forward architectures** (MLPs, ResNets):

$$\mathbf{z}_{t+1} = \mathbf{z}_t + \text{MLP}_\theta(\mathbf{z}_t, \mathbf{u}_t) \tag{9.64}$$

   Residual connection ensures smooth dynamics (small changes per step).
   Pros: Simple, fast parallel evaluation
   Cons: No memory, must be Markovian
3. **Transformer architectures**:

$$\mathbf{Z}_{1:t} = [\mathbf{z}_1, \ldots, \mathbf{z}_t] \tag{9.65}$$
$$\mathbf{z}_{t+1} = \text{Transformer}_\theta(\mathbf{Z}_{1:t}) \tag{9.66}$$

   Attention mechanism allows flexible temporal receptive fields.
   Pros: Long-range dependencies, parallelizable training
   Cons: Quadratic complexity in sequence length, requires large datasets

Stage 3: Decoder

$D_\psi : \mathcal{Z} \rightarrow \mathcal{X}$
Reconstructs high-dimensional state from latent code:

$$\hat{\mathbf{x}}_{t+1} = D_\psi(\mathbf{z}_{t+1}) \tag{9.67}$$

For CNNs, decoder mirrors encoder with transposed convolutions (upsampling):

$$\mathbf{h}^{(1)} = \text{Reshape}(\mathbf{z}_{t+1}) \quad (32768 \rightarrow 8 \times 8 \times 512) \tag{9.68}$$
$$\mathbf{h}^{(2)} = \text{ConvTranspose}_{256}(\mathbf{h}^{(1)}) \quad (8 \times 8 \rightarrow 16 \times 16) \tag{9.69}$$

$$\vdots$$

$$\hat{\mathbf{x}}_{t+1} = \text{ConvTranspose}_1(\mathbf{h}^{(L)}) \quad (128 \times 128 \to 256 \times 256) \tag{9.70}$$

For GNNs, decoder generates node features or entire graphs.

Why Latent Space?

Operating in latent space provides three advantages:

1. **Computational efficiency**: Dynamics on $\mathbb{R}^{100}$ are orders of magnitude faster than $\mathbb{R}^{1000000}$
2. **Smoothing of irregularities**: Encoder learns to ignore noise, decoder reconstructs clean signal
3. **Better generalization**: Compact representations force learning of essential patterns, not memorizing training trajectories

The challenge: Ensure latent dynamics preserve physical structure (conservation, symmetries, causality) of the high-dimensional system.

9.3.1.2 Architecture Patterns and State-Space Models

Recent advances have produced specialized architectures for long-range temporal modeling in scientific systems.

Traditional RNNs have linear complexity in sequence length but suffer from vanishing gradients. Transformers solve the gradient problem but have quadratic complexity. State-space models (SSMs) achieve both: linear complexity AND long-range dependencies.

Continuous-time state-space formulation:

$$\frac{d\mathbf{h}(t)}{dt} = \mathbf{A}\mathbf{h}(t) + \mathbf{B}\mathbf{x}(t) \quad \text{(state evolution)} \tag{9.71}$$

$$\mathbf{y}(t) = \mathbf{C}\mathbf{h}(t) + \mathbf{D}\mathbf{x}(t) \quad \text{(observation)} \tag{9.72}$$

where $\mathbf{h}(t) \in \mathbb{R}^N$ is hidden state, $\mathbf{x}(t)$ is input, $\mathbf{y}(t)$ is output.

Discretization (zero-order hold):

$$\mathbf{h}_k = \overline{\mathbf{A}}\mathbf{h}_{k-1} + \overline{\mathbf{B}}\mathbf{x}_k \tag{9.73}$$

$$\mathbf{y}_k = \mathbf{C}\mathbf{h}_k \tag{9.74}$$

where $\overline{\mathbf{A}} = \exp(\Delta\mathbf{A})$, $\overline{\mathbf{B}} = (\mathbf{A}^{-1}(\exp(\Delta\mathbf{A}) - \mathbf{I}))\mathbf{B}$.

Mamba (2024): Introduces *selective state spaces* with input-dependent $\mathbf{A}, \mathbf{B}, \mathbf{C}$:

$$\overline{\mathbf{A}}(\mathbf{x}), \overline{\mathbf{B}}(\mathbf{x}), \mathbf{C}(\mathbf{x}) = \text{MLP}(\mathbf{x}) \tag{9.75}$$

This selection mechanism allows the model to focus on relevant history while forgetting irrelevant details; critical for long sequences.

Mamba-2 (ICML 2024): Structured State Space Duality connects SSMs to attention:

$$\text{SSM}(\mathbf{X}) \approx \text{Attention}(\mathbf{Q}, \mathbf{K}, \mathbf{V}) \quad \text{with structured matrices} \tag{9.76}$$

This reveals SSMs and transformers are two views of the same computation, enabling hybrid architectures.

Scientific applications:

- **PhyxMamba**: Chaotic system forecasting (Lorenz, Kuramoto-Sivashinsky) with 10–20% error reduction versus transformers
- **DeepOMamba**: Spatio-temporal PDE operator learning, linear scaling with resolution
- **MambaTab**: Time-series prediction for climate data (temperature, precipitation)

Rather than discrete-time recurrence $\mathbf{z}_{t+1} = f_\theta(\mathbf{z}_t)$, Neural ODEs model continuous-time dynamics:

$$\frac{d\mathbf{z}}{dt} = f_\theta(\mathbf{z}(t), t) \tag{9.77}$$

Forward pass (solving ODE):

$$\mathbf{z}(t_1) = \mathbf{z}(t_0) + \int_{t_0}^{t_1} f_\theta(\mathbf{z}(t), t)\, dt \tag{9.78}$$

Evaluated with ODE solvers (Runge-Kutta, Dormand-Prince). Adaptive step size automatically handles stiffness.

Backward pass (adjoint method for memory efficiency):

Instead of backpropagating through ODE solver steps (expensive), solve adjoint equation:

$$\frac{d\mathbf{a}(t)}{dt} = -\mathbf{a}(t)^\top \frac{\partial f_\theta}{\partial \mathbf{z}} \tag{9.79}$$

where $\mathbf{a}(t) = \frac{\partial \mathcal{L}}{\partial \mathbf{z}(t)}$ is adjoint state. This requires only $\mathcal{O}(1)$ memory regardless of sequence length.

Physics-Informed Neural ODEs (PINODE):

Embed physical constraints via collocation in latent space:

$$\mathcal{L}_{\text{physics}} = \left\| \frac{d\mathbf{z}}{dt} - f_\theta(\mathbf{z}) - \mathcal{P}(\mathbf{z}) \right\|^2 \tag{9.80}$$

where $\mathcal{P}(\mathbf{z})$ is known physics in latent coordinates (e.g., forcing terms, dissipation).

Performance gains in data-scarce regimes:

- Advection-diffusion: 5× better with 10% data versus pure data-driven
- Reacting flows: 50× better long-time accuracy
- Atmospheric dynamics: 200× better for OOD parameter ranges

Neural ODEs can learn **incorrect temporal dynamics** in latent space. Consider advection-dominated flow (fast transport, slow diffusion). If encoder/decoder don't respect separation of timescales:

Problem: Neural ODE learns artificially slow latent dynamics that reconstruct correctly only because encoder/decoder compensate; but fail during rollout.

Solution 1: Multi-scale encoders separating fast and slow components
Solution 2: Explicit regularization: $\|\frac{d\mathbf{z}}{dt}\|$ should match physical timescales
Solution 3: Train with multi-step rollout loss, not just one-step

[19] analyzes this for Navier-Stokes and atmospheric flows.

9.3.1.3 Learning Latent Representations

How do we train the encoder E_ϕ, decoder D_ψ, and dynamics f_θ jointly?

Given sequence of observations $\{\mathbf{x}_1, \ldots, \mathbf{x}_T\}$:

$$\mathcal{L}_{\text{recon}} = \sum_{t=1}^{T-1} \|\mathbf{x}_{t+1} - D_\psi(f_\theta(E_\phi(\mathbf{x}_t)))\|^2 \tag{9.81}$$

This is one-step prediction. For rollout stability, use multi-step:

$$\mathcal{L}_{\text{multistep}} = \sum_{t=1}^{T-k} \sum_{s=1}^{k} \alpha_s \|\mathbf{x}_{t+s} - \hat{\mathbf{x}}_{t+s}\|^2 \tag{9.82}$$

where $\hat{\mathbf{x}}_{t+s}$ is s-step rollout prediction, α_s are weights (typically increasing with s to emphasize long-term accuracy).

Learn representations without supervision by contrasting similar versus dissimilar states.

Time contrastive learning:

$$\mathcal{L}_{\text{contrast}} = -\log \frac{\exp(\text{sim}(\mathbf{z}_t, \mathbf{z}_{t+1})/\tau)}{\sum_{j\neq t} \exp(\text{sim}(\mathbf{z}_t, \mathbf{z}_j)/\tau)} \tag{9.83}$$

States close in time should be close in latent space; distant states should be far.

Physics-based contrastive learning:

States with similar physical properties (energy, momentum, mass) should be close:

$$\text{sim}(\mathbf{z}_i, \mathbf{z}_j) \propto \exp\left(-\|Q(\mathbf{x}_i) - Q(\mathbf{x}_j)\|^2\right) \tag{9.84}$$

where Q computes conserved quantities.

VAE-style latent dynamics introduce stochasticity:

$$q_\phi(\mathbf{z}_t|\mathbf{x}_t) = \mathcal{N}(\mathbf{z}_t; \boldsymbol{\mu}_\phi(\mathbf{x}_t), \mathrm{diag}(\boldsymbol{\sigma}^2_\phi(\mathbf{x}_t))) \tag{9.85}$$

$$p_\theta(\mathbf{z}_{t+1}|\mathbf{z}_t) = \mathcal{N}(\mathbf{z}_{t+1}; f_\theta(\mathbf{z}_t), \boldsymbol{\Sigma}_\theta) \tag{9.86}$$

$$p_\psi(\mathbf{x}_{t+1}|\mathbf{z}_{t+1}) = \mathcal{N}(\mathbf{x}_{t+1}; D_\psi(\mathbf{z}_{t+1}), \boldsymbol{\Sigma}_{\mathrm{obs}}) \tag{9.87}$$

ELBO objective:

$$\mathcal{L}_{\mathrm{VAE}} = \mathbb{E}_{q_\phi}[\log p_\psi(\mathbf{x}_{t+1}|\mathbf{z}_{t+1})] - D_{\mathrm{KL}}(q_\phi(\mathbf{z}_{t+1}|\mathbf{x}_{t+1}) \| p_\theta(\mathbf{z}_{t+1}|\mathbf{z}_t)) \tag{9.88}$$

This naturally provides uncertainty estimates; critical for downstream control.

Scientific applications benefit from *disentangled* latent spaces where individual dimensions correspond to interpretable physical quantities.

β-VAE: Increase KL weight to encourage independence:

$$\mathcal{L}_{\beta\text{-VAE}} = \mathcal{L}_{\mathrm{recon}} - \beta D_{\mathrm{KL}}(q_\phi(\mathbf{z}|\mathbf{x}) \| p(\mathbf{z})) \tag{9.89}$$

where $\beta > 1$ (typically 2–10) encourages factorization.

Physically-guided disentanglement:

Add auxiliary losses predicting known physical quantities from latent codes:

$$\mathcal{L}_{\mathrm{phys}} = \sum_{i=1}^{M} \|Q_i(\mathbf{x}) - g_i(\mathbf{z})\|^2 \tag{9.90}$$

where Q_i are physical quantities (energy, momentum, vorticity), g_i are linear predictors from latent space.

This forces latent dimensions to align with physical observables, enabling:

- Interpretation: "Dimension 3 encodes kinetic energy"
- Intervention: Manually adjust latent codes to control physical properties
- Extrapolation: Physics constraints prevent unphysical latent states

9.3.1.4 Reconstruction and Fidelity

The final challenge: ensuring decoder reconstructions are high-fidelity across the full range of dynamical behaviors.

Scientific systems exhibit dynamics across multiple spatial and temporal scales. Single-scale decoders struggle to capture all scales accurately.

U-Net style skip connections:

$$\hat{\mathbf{x}} = D_\psi(\mathbf{z}, \{\mathbf{h}^{(i)}\}_{i=1}^{L}) \tag{9.91}$$

where $\mathbf{h}^{(i)}$ are intermediate encoder features. Skip connections allow decoder to access fine-scale information discarded during compression.

Instead of single $\mathbf{z} \in \mathbb{R}^d$, use hierarchical latents:

$$\mathbf{z} = \{\mathbf{z}_{\text{coarse}}, \mathbf{z}_{\text{medium}}, \mathbf{z}_{\text{fine}}\} \tag{9.92}$$

Each level captures different scales (planetary circulation, meso-scale eddies, turbulent fluctuations).

L2 reconstruction loss $\|\mathbf{x} - \hat{\mathbf{x}}\|^2$ treats all spatial locations equally. For scientific data, some features matter more.

Spectral loss (for fluid dynamics, atmospheric data):

$$\mathcal{L}_{\text{spectral}} = \sum_k w_k |\mathcal{F}\{\mathbf{x}\}_k - \mathcal{F}\{\hat{\mathbf{x}}\}_k|^2 \tag{9.93}$$

where $\mathcal{F}$ is Fourier transform, k is wavenumber, w_k weights different scales. This preserves spectral energy distribution.

Feature-based loss (using pre-trained networks):

$$\mathcal{L}_{\text{feature}} = \|\Phi(\mathbf{x}) - \Phi(\hat{\mathbf{x}})\|^2 \tag{9.94}$$

where Φ extracts features from pre-trained model (e.g., ResNet trained on physics simulations).

OceanNet [20] demonstrates production-ready latent dynamics for ocean forecasting:

Architecture:

- Fourier Neural Operator (FNO) encoder/decoder
- Predictor-evaluator-corrector integration in latent space
- Multi-scale representation: $\mathbf{z}_{\text{SSH}}, \mathbf{z}_{\text{temp}}, \mathbf{z}_{\text{velocity}}$

Performance:

- 7-day forecasts: 0.12 m RMSE for sea surface height
- 1000× faster than numerical ocean model (ROMS)
- Preserves mass conservation to machine precision
- Uncertainty quantification via ensemble

Deployment:

- Real-time assimilation of satellite altimetry
- Supports maritime traffic routing, fisheries management
- Continuous model updating with new observations

This exemplifies the digital twin paradigm: learned model + continuous data assimilation + uncertainty quantification.

9.3.2 *Planning, Control, and Digital Twins*

World models transition from passive prediction to active planning and control. Once we can simulate "what if," we can optimize "what should be."

9.3.2.1 From Prediction to Action

The distinction between predictive models and world models (Fig. 9.13):

Predictive model: $\hat{\mathbf{x}}_{t+1} = f(\mathbf{x}_t)$ (passive forecasting)

World model: $\hat{\mathbf{x}}_{t+1} = f(\mathbf{x}_t, \mathbf{u}_t)$ (responds to actions/controls)

where $\mathbf{u}_t$ is control input (actuator commands, policy decisions, intervention).

Let us give a canonical framework for control with learned world models:

Algorithm (at each timestep t):

1. Observe current state $\mathbf{x}_t$
2. Solve optimization over action sequences:

$$\mathbf{u}^*_{t:t+H} = \arg\max_{\mathbf{u}_{t:t+H}} \sum_{s=0}^{H} \gamma^s r(\hat{\mathbf{x}}_{t+s}, \mathbf{u}_{t+s}) \tag{9.95}$$

   subject to: $\hat{\mathbf{x}}_{t+s+1} = f_\theta(\hat{\mathbf{x}}_{t+s}, \mathbf{u}_{t+s})$ (world model dynamics)
3. Execute first action $\mathbf{u}^*_t$
4. Repeat (re-plan at next timestep with new observation)

where H is planning horizon, $r(\mathbf{x}, \mathbf{u})$ is reward, γ is discount factor.

Optimization methods:

- **Gradient-based**: Backpropagate through world model to compute $\nabla_{\mathbf{u}} J$

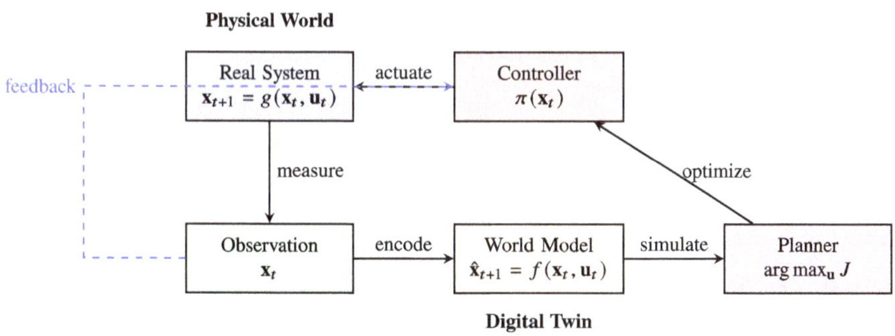

Fig. 9.13 Control loop with world model. Real system state $\mathbf{x}_t$ is measured and encoded. World model simulates multiple action sequences $\{\mathbf{u}_{t:t+H}\}$ forward to predict outcomes. Planner selects action maximizing objective J. Controller executes action on real system. Feedback loop continuously updates model and re-plans

- **Sampling-based**: Cross-Entropy Method (CEM), Model Predictive Path Integral (MPPI)
- **Trajectory optimization**: Sequential quadratic programming, iLQR/DDP

Advantages of learned world models for MPC:

- Differentiable: Can use gradient-based optimizers
- Fast: Evaluate 1000 s of trajectories per second
- Uncertain: Ensemble provides confidence bounds

Learned world models are imperfect. Optimizers can exploit model errors to find "solutions" that work in simulation but fail on the real system.

Example: Robotic control with learned dynamics. Model has small error in friction coefficient. Optimizer finds policy that violates friction limit; works in simulation, robot slips in reality.

Mitigation strategies:

1. **Conservative planning**: Add safety margins to constraints
2. **Adversarial planning**: Maximize worst-case over model ensemble
3. **Online adaptation**: Re-train model with execution data
4. **Hybrid planning**: Use learned model for coarse plan, physics model for refinement

Two approaches to trajectory optimization with world models:

Shooting (optimize actions, rollout dynamics):

$$\min_{\mathbf{u}_{t:t+H}} \sum_{s=0}^{H} c(\hat{\mathbf{x}}_{t+s}, \mathbf{u}_{t+s}) \tag{9.96}$$

$$\text{subject to} \quad \hat{\mathbf{x}}_{t+s+1} = f_\theta(\hat{\mathbf{x}}_{t+s}, \mathbf{u}_{t+s}) \tag{9.97}$$

$$\mathbf{x}_t = \text{current state} \tag{9.98}$$

Pros: Automatically satisfies dynamics

Cons: Sensitive to initial guess, requires good world model

Collocation (optimize both actions and states):

$$\min_{\mathbf{u}_{t:t+H}, \mathbf{x}_{t:t+H}} \sum_{s=0}^{H} c(\mathbf{x}_{t+s}, \mathbf{u}_{t+s}) \tag{9.99}$$

$$\text{subject to} \quad \mathbf{x}_{t+s+1} = f_\theta(\mathbf{x}_{t+s}, \mathbf{u}_{t+s}) \quad \forall s \tag{9.100}$$

$$\mathbf{x}_t = \text{current state} \tag{9.101}$$

Pros: More robust to model errors, constraints easier to enforce

Cons: Higher-dimensional optimization, dynamics become soft constraints

For differentiable world models, shooting with gradient descent often suffices. For complex constraints or poor models, collocation is more robust.

9.3.2.2 Model-Based Reinforcement Learning

RL agents can learn policies by interacting with world models instead of expensive real systems.

Classical model-based RL alternates:

1. **Real experience**: Collect trajectories from real system
2. **Model learning**: Update world model f_θ from real data
3. **Imagined experience**: Sample trajectories from model
4. **Policy learning**: Train policy on real + imagined data

Sample efficiency gains: Model-based methods require 10–100× fewer real interactions than model-free RL (DQN, PPO) for robotics and control tasks.

DreamerV3 [21] achieves human-level performance on 150+ domains with single algorithm and hyperparameters.

Architecture:

$$\text{Encoder:} \quad \mathbf{z}_t = E_\phi(\mathbf{x}_t) \tag{9.102}$$

$$\text{Recurrent state:} \quad \mathbf{h}_t = f_{\text{RNN}}(\mathbf{h}_{t-1}, \mathbf{z}_t, \mathbf{u}_{t-1}) \tag{9.103}$$

$$\text{Decoder:} \quad \hat{\mathbf{x}}_{t+1} = D_\psi(\mathbf{h}_t) \tag{9.104}$$

$$\text{Reward:} \quad \hat{r}_t = r_\omega(\mathbf{h}_t) \tag{9.105}$$

$$\text{Continue:} \quad \hat{c}_t = c_\xi(\mathbf{h}_t) \quad \text{(episode termination)} \tag{9.106}$$

Training:

1. Train world model (encoder, dynamics, decoder, reward, continue) on replay buffer
2. Imagine rollouts from world model: $\{\mathbf{h}_t, \mathbf{u}_t, \hat{r}_t\}_{t=1}^H$
3. Train actor $\pi_\alpha(\mathbf{u}|\mathbf{h})$ and critic $V_\beta(\mathbf{h})$ on imagined rollouts

Key innovations:

- **Symlog predictions**: Transform rewards/values with $\text{symlog}(x) = \text{sign}(x) \log(1 + |x|)$ for numerical stability
- **Unimix categorical**: Mixture of uniform and learned distributions for value targets
- **Free bits**: Prevent posterior collapse in latent dynamics

Scientific applications:

- **Fusion reactor control**: Maintain plasma stability (DeepMind, Nature 2022)
- **Wind farm optimization**: Maximize power output while managing turbine loads
- **Chemical process control**: Optimize reactor temperature/pressure profiles

When World Models Help (and When They Don't)

Use world models when:

- Real system interactions are expensive (robots, experiments, simulations)
- Dynamics are relatively predictable (continuous control, physical systems)
- Safety matters (can test policies in simulation first)
- Need fast online adaptation (re-plan with updated model)

Avoid world models when:

- Dynamics are highly stochastic or adversarial (games, social systems)
- Model errors compound severely (chaos, unstable systems)
- Real interactions are cheap (video games, discrete environments)
- You have vast offline datasets (offline RL is simpler)

Hybrid approach: Use model-free for initial exploration, then switch to model-based for fine-tuning.

9.3.2.3 Digital Twins: Concept and Reality

A digital twin is more than a simulator; it's a virtual replica continuously synchronized with its physical counterpart.

Digital Twin

A **digital twin** is a virtual representation of a physical system that:

1. **Mirrors current state**: Continuously updated with sensor data
2. **Predicts future behavior**: Simulates forward under different scenarios
3. **Enables what-if analysis**: Tests interventions virtually before real deployment
4. **Learns from experience**: Model parameters update as system evolves

Key distinction from offline simulation: Bidirectional data flow between physical and digital.

Beyond standard simulation, digital twins demand:

1. **Real-time state estimation**

Combine noisy, sparse sensor data with model predictions via data assimilation:

Kalman filtering (linear systems):

$$\begin{aligned} \text{Predict:} \quad \hat{\mathbf{x}}_{t+1}^{-} &= f(\hat{\mathbf{x}}_t) && (9.107) \\ \mathbf{P}_{t+1}^{-} &= \mathbf{F}_t \mathbf{P}_t \mathbf{F}_t^{\top} + \mathbf{Q} && (9.108) \\ \text{Update:} \quad \mathbf{K}_{t+1} &= \mathbf{P}_{t+1}^{-} \mathbf{H}^{\top} (\mathbf{H} \mathbf{P}_{t+1}^{-} \mathbf{H}^{\top} + \mathbf{R})^{-1} && (9.109) \\ \hat{\mathbf{x}}_{t+1} &= \hat{\mathbf{x}}_{t+1}^{-} + \mathbf{K}_{t+1} (\mathbf{y}_{t+1} - \mathbf{H} \hat{\mathbf{x}}_{t+1}^{-}) && (9.110) \end{aligned}$$

where $\mathbf{F}_t = \frac{\partial f}{\partial \mathbf{x}}|_{\hat{\mathbf{x}}_t}$ is Jacobian, $\mathbf{Q}$ is process noise, $\mathbf{R}$ is sensor noise, $\mathbf{H}$ is observation model.

For neural world models, use **Ensemble Kalman Filter** (EnKF):

1. Maintain ensemble of states $\{\mathbf{x}_t^{(i)}\}_{i=1}^{M}$
2. Predict: $\mathbf{x}_{t+1}^{(i)} = f_\theta(\mathbf{x}_t^{(i)}) + \boldsymbol{\epsilon}^{(i)}$ for each member
3. Update ensemble mean and covariance with observations

2. **Uncertainty quantification**

Digital twins must know what they don't know. Ensemble methods provide prediction intervals:

$$\mathbf{x}_{t+1} \sim \mathcal{N}(\bar{\mathbf{x}}_{t+1}, \Sigma_{t+1}) \tag{9.111}$$

where $\bar{\mathbf{x}}_{t+1} = \frac{1}{M} \sum_{i=1}^{M} \mathbf{x}_{t+1}^{(i)}$, $\Sigma_{t+1} = \frac{1}{M-1} \sum_{i=1}^{M} (\mathbf{x}_{t+1}^{(i)} - \bar{\mathbf{x}}_{t+1})(\mathbf{x}_{t+1}^{(i)} - \bar{\mathbf{x}}_{t+1})^{\top}$.

3. **Continuous model updating**

As physical system changes (wear, aging, operating regime), digital twin must adapt.

Online learning:

1. Detect model-data mismatch: $\|\mathbf{x}_{\text{actual}} - \hat{\mathbf{x}}_{\text{predicted}}\| > \tau$
2. Fine-tune world model on recent data via gradient descent
3. Validate updated model before deploying

Meta-learning for fast adaptation:

$$\theta^* = \theta - \alpha \nabla_\theta \mathcal{L}(\theta; \mathcal{D}_{\text{recent}}) \tag{9.112}$$

Pre-train world model to be easily adaptable (MAML, Reptile). DeepONet (Deep Operator Network) provides real-time digital twins for nuclear reactor particle transport:

Problem: PHITS (Particle and Heavy Ion Transport code System) requires hours per simulation for radiation shielding analysis.

Solution:

- Train DeepONet on 10,000 PHITS simulations
- Branch network encodes geometry/material configuration
- Trunk network encodes spatial query points
- Output: Radiation dose field in milliseconds

Results (from Scientific Reports 2024):

- **500,000× speedup**: hours → milliseconds
- **High accuracy**: $< 5\%$ relative error versus PHITS
- **Real-time monitoring**: Continuous dose estimation during operation
- **Scenario analysis**: Test thousands of what-if configurations

Deployment: Enables real-time optimization of shielding configurations, immediate response to anomalies, and rapid design iteration for new reactor geometries.

This demonstrates the digital twin paradigm: trained surrogate (500k× faster) + continuous monitoring + uncertainty quantification for safety-critical applications.

9.3.3 *Physics-Aware World Models*

Pure data-driven world models can violate fundamental physics: energy drift during long rollouts, non-physical trajectories, catastrophic failure on out-of-distribution states. Physics-aware architectures embed structural knowledge; symmetries, conservation laws, geometric constraints; directly into model architecture, providing both better generalization and formal guarantees.

9.3.3.1 Incorporating Inductive Biases

Inductive biases are assumptions about problem structure that guide learning. For scientific systems, the strongest biases come from physics.

Many physical systems exhibit symmetries: translations, rotations, permutations. An *equivariant* network respects these symmetries.

Definition: Function f is equivariant to group G if:

$$f(g \cdot \mathbf{x}) = g \cdot f(\mathbf{x}) \quad \forall g \in G \tag{9.113}$$

For Euclidean group $E(3)$ (translations + rotations):

$$f(\mathbf{R}\mathbf{x} + \mathbf{t}) = \mathbf{R} f(\mathbf{x}) \quad \forall \mathbf{R} \in SO(3), \mathbf{t} \in \mathbb{R}^3 \tag{9.114}$$

Why equivariance matters (Fig. 9.14):

- **Sample efficiency**: Model trained on one orientation generalizes to all
- **Physical correctness**: Predictions respect coordinate-free physics
- **Compositionality**: Equivariance preserved through network layers

NequIP [22]: E(3)-equivariant GNN for molecular dynamics. Message passing with irreducible representations:

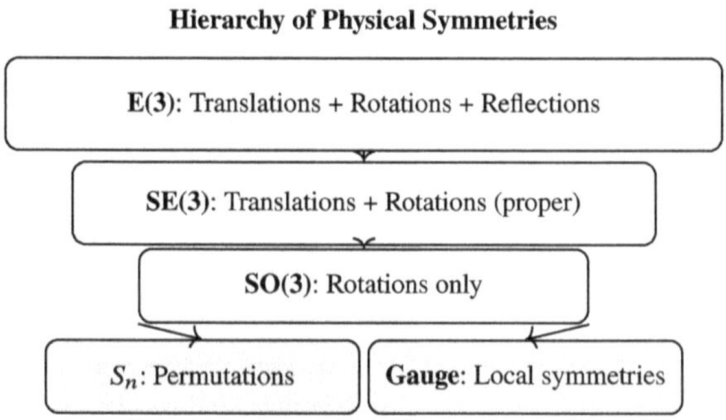

Fig. 9.14 Hierarchy of physical symmetries and their applications. E(3) equivariance (top) includes all Euclidean transformations. SE(3) excludes reflections (important for chiral molecules). SO(3) fixes origin. Permutation invariance applies to particle systems. Gauge symmetries (bottom) govern field theories. Stronger symmetries (higher in hierarchy) provide more powerful inductive biases

$$\mathbf{m}_{ij} = \phi_{\text{edge}}(\|\mathbf{r}_{ij}\|) \otimes \mathbf{h}_j \otimes Y(\hat{\mathbf{r}}_{ij}) \tag{9.115}$$

where $\mathbf{r}_{ij} = \mathbf{r}_i - \mathbf{r}_j$, $Y(\hat{\mathbf{r}})$ are spherical harmonics, $\otimes$ is tensor product.

Update preserving equivariance:

$$\mathbf{h}_i^{(\ell+1)} = \mathbf{h}_i^{(\ell)} + \sum_{j \in \mathcal{N}(i)} \mathbf{W}^{(\ell)} \mathbf{m}_{ij} \tag{9.116}$$

Results: Up to **three orders of magnitude fewer training data** than non-equivariant networks for same accuracy on MD17 benchmarks.

DeepH-E3 (Nature Communications 2023): Extends to DFT Hamiltonian prediction with spin-orbit coupling, achieving sub-meV accuracy for twisted bilayer graphene.

For mechanical systems, energy conservation is not optional; it's fundamental. Neural networks that violate energy conservation accumulate errors exponentially during rollout.

Hamiltonian mechanics: System state $(\mathbf{q}, \mathbf{p})$ evolves via:

$$\frac{\mathrm{d}\mathbf{q}}{\mathrm{d}t} = \frac{\partial H}{\partial \mathbf{p}} \tag{9.117}$$

$$\frac{\mathrm{d}\mathbf{p}}{\mathrm{d}t} = -\frac{\partial H}{\partial \mathbf{q}} \tag{9.118}$$

where $H(\mathbf{q}, \mathbf{p})$ is Hamiltonian (total energy), $\mathbf{q}$ is position, $\mathbf{p}$ is momentum.

Key property: Hamiltonian flow preserves energy exactly:

$$\frac{\mathrm{d}H}{\mathrm{d}t} = \frac{\partial H}{\partial \mathbf{q}}\frac{\mathrm{d}\mathbf{q}}{\mathrm{d}t} + \frac{\partial H}{\partial \mathbf{p}}\frac{\mathrm{d}\mathbf{p}}{\mathrm{d}t} = 0 \tag{9.119}$$

Hamiltonian Neural Networks (HNN) learn Hamiltonian from data:

$$H_\theta(\mathbf{q}, \mathbf{p}) = \mathrm{MLP}_\theta(\mathbf{q}, \mathbf{p}) \tag{9.120}$$

$$\frac{d\mathbf{q}}{dt} = \frac{\partial H_\theta}{\partial \mathbf{p}}, \quad \frac{d\mathbf{p}}{dt} = -\frac{\partial H_\theta}{\partial \mathbf{q}} \tag{9.121}$$

Train on observed trajectories $\{(\mathbf{q}_t, \mathbf{p}_t)\}$:

$$\mathcal{L} = \sum_t \left\| \left(\frac{d\mathbf{q}}{dt}, \frac{d\mathbf{p}}{dt}\right) - \left(\frac{\partial H_\theta}{\partial \mathbf{p}}, -\frac{\partial H_\theta}{\partial \mathbf{q}}\right) \right\|^2 \tag{9.122}$$

Guarantee: Learned dynamics *exactly* conserve energy by construction.

Lagrangian Neural Networks (LNN) parameterize Lagrangian $L = T - V$ (kinetic–potential):

$$\frac{d}{dt}\frac{\partial L}{\partial \dot{\mathbf{q}}} - \frac{\partial L}{\partial \mathbf{q}} = 0 \quad \text{(Euler-Lagrange equations)} \tag{9.123}$$

Learn $L_\theta(\mathbf{q}, \dot{\mathbf{q}})$ and derive equations of motion automatically.

Classical integrators (Euler, RK4) accumulate energy drift: $H(t) = H_0 + \alpha t$. For chaotic systems (molecular dynamics, N-body problems), this causes catastrophic failure.

Symplectic integrators preserve Hamiltonian structure exactly:

$$\mathbf{z}_{n+1} = \mathbf{\Phi}_{\Delta t}(\mathbf{z}_n) \quad \text{where} \quad \mathbf{\Phi} \text{ is symplectic} \tag{9.124}$$

A map $\mathbf{\Phi} : (\mathbf{q}, \mathbf{p}) \to (\mathbf{q}', \mathbf{p}')$ is symplectic if it preserves symplectic 2-form: $d\mathbf{q} \wedge d\mathbf{p} = d\mathbf{q}' \wedge d\mathbf{p}'$.

SympNets (2020) parameterize symplectic maps using neural networks with specific architectural constraints. **Generalized Hamiltonian Neural Networks** (J. Computational Physics 2024) provide unified framework outperforming SympNets and HénonNets on classical mechanics benchmarks.

Universal approximation (arXiv 2024): Symplectic neural networks with geometric integrators can universally approximate Hamiltonian diffeomorphisms while maintaining exact energy conservation over arbitrarily long rollouts.

Applications:

- Molecular dynamics: 1000× longer stable integration
- Orbital mechanics: Satellite trajectories over years
- Accelerator physics: Particle beam dynamics

Real systems have dissipation (friction, heat loss) violating energy conservation. Port-Hamiltonian (pH) framework extends Hamiltonian mechanics:

$$\frac{d\mathbf{x}}{dt} = (\mathbf{J}(\mathbf{x}) - \mathbf{R}(\mathbf{x}))\frac{\partial H}{\partial \mathbf{x}} + \mathbf{B}(\mathbf{x})\mathbf{u} \tag{9.125}$$

where:

- $\mathbf{J}$ is skew-symmetric (conservative flow)
- $\mathbf{R} \succeq 0$ is dissipation (friction, damping)
- $\mathbf{Bu}$ is external input (control, forcing)

Energy balance:

$$\frac{\mathrm{d}H}{\mathrm{d}t} = -\frac{\partial H}{\partial \mathbf{x}}^{\top} \mathbf{R} \frac{\partial H}{\partial \mathbf{x}} + \frac{\partial H}{\partial \mathbf{x}}^{\top} \mathbf{Bu} \leq 0 \quad \text{(dissipation)} \tag{9.126}$$

When $\mathbf{u} = 0$, energy decreases monotonically (passivity).

Port-Hamiltonian Neural Networks learn $H_\theta, \mathbf{J}_\phi, \mathbf{R}_\psi$ from data with constraints:

$$\mathbf{J}_\phi^{\top} = -\mathbf{J}_\phi \quad \text{(skew-symmetry)} \tag{9.127}$$

$$\mathbf{R}_\psi = \mathbf{L}^{\top}\mathbf{L} \quad \text{(positive semi-definite)} \tag{9.128}$$

Applications: Robot manipulators, electrical circuits, chemical reactors, fluid-structure interaction.

9.3.3.2 Generalization and Extrapolation

The ultimate test: Can physics-aware world models extrapolate beyond training data?

In-distribution (IID): Test on states/trajectories from same distribution as training.

Out-of-distribution (OOD): Test on:

- Different parameter regimes (new Reynolds numbers, temperatures)
- Longer time horizons than training
- Different initial conditions or boundary conditions
- Novel system configurations

Physics-aware advantages for OOD:

1. **Structural constraints prevent unphysical states**: Energy conservation ensures trajectories stay on physical manifold
2. **Symmetries enable generalization**: Trained on limited orientations, generalize to all via equivariance
3. **Conservation laws provide regularization**: Energy drift ΔH acts as unsupervised signal for model quality

Scientific systems often involve parameters: Reynolds number (fluid dynamics), diffusion coefficient (PDEs), spring constant (mechanics). Can world models trained on limited parameter ranges extrapolate?

Challenge: Pure neural networks interpolate well, extrapolate poorly. Trained on Re = [100, 1000], fail at Re = 5000.

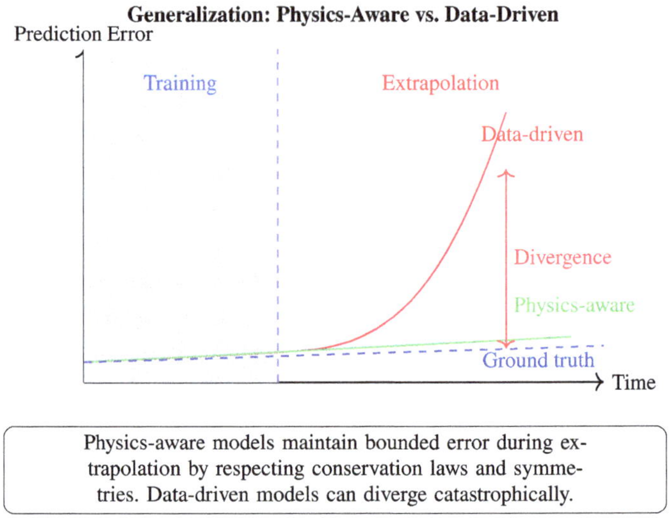

Fig. 9.15 Long-horizon extrapolation behavior. Both models fit training data well (blue shaded region). During extrapolation (red region), data-driven models accumulate errors and diverge (red curve). Physics-aware models (green) maintain bounded error by respecting conservation laws and symmetries, tracking ground truth (blue dashed) more closely

Solution 1: Physics-informed scaling laws
Embed known scaling relationships. For Navier-Stokes, velocity scales as $\sqrt{\text{Re}}$:

$$f_\theta(\mathbf{x}; \text{Re}) = \sqrt{\text{Re}} \cdot g_\theta(\mathbf{x}; \text{Re}_{\text{ref}}) \tag{9.129}$$

Solution 2: Invariant/equivariant to parameters
Construct representations invariant to parameter transformations. For diffusion coefficient D:

$$f_\theta(\mathbf{x}, t; D) = \frac{1}{\sqrt{D}} \tilde{f}_\theta(\mathbf{x}/\sqrt{D}, t; 1) \tag{9.130}$$

This dimensional analysis baked into architecture.

Solution 3: Meta-learning across parameters
Train on distribution of parameters, learn to adapt quickly:

$$\theta^* = \text{Adapt}(\theta; \tau_{\text{new}}, \mathcal{D}_{\text{few-shot}}) \tag{9.131}$$

where τ_{new} is novel parameter value, $\mathcal{D}_{\text{few-shot}}$ is small dataset at that parameter (Fig. 9.15).

When Physics Isn't Enough

Physics-aware architectures are not silver bullets. They fail when:

1. **Wrong physics encoded**: If you impose incompressible flow but system is compressible, model will fail
2. **Incomplete physics**: Hamiltonian structure works for conservative systems, but real systems have dissipation, noise, external forcing
3. **Unknown physics**: For complex systems (turbulence, granular flow), first-principles physics is incomplete. Hybrid approaches combining known physics with learned corrections often work best.
4. **Computational cost**: Symplectic integrators, equivariant networks, and constrained optimization add overhead. For some applications, pure data-driven models are fast enough and "good enough."

The art: Choose minimal physics constraints that provide maximum benefit for your application.

9.4 Conclusion

This chapter has examined foundation models and world models as emerging paradigms for Scientific Machine Learning. We have seen that these approaches promise transfer of knowledge across tasks, that scientific applications present unique challenges around physical consistency and verification, and that operational examples in weather forecasting demonstrate feasibility.

Key Takeaways

1. **Foundation = Pretrain + Transfer**: Large models trained on broad scientific data can transfer knowledge to specific tasks with limited fine-tuning data.
2. **World Models Learn Latent Dynamics**: Compressed representations of physical state enable efficient planning, control, and simulation in learned latent spaces.
3. **Physics Consistency Is Hard**: Unlike language, scientific outputs must satisfy conservation laws and quantitative accuracy requirements. Constraints must be enforced.

4. **Weather Shows the Path**: GraphCast, GenCast, and AIFS demonstrate that foundation models can match or exceed physics-based systems on complex, high-stakes prediction tasks.
5. **Verification Remains Essential**: Generated molecules must be synthesized; predicted structures must be validated; forecast accuracy must be measured. The model is not the territory.

References

1. Soliman, A.: Deepmind AI weather forecaster beats world-class system. Nature **636**(8042), 282–283 (2024)
2. Zeni, C., Pinsler, R., Zügner, D., Fowler, A., Horton, M., Fu, X., Shysheya, S., Crabbé, J., Sun, L., Smith, J.: Mattergen: a generative model for inorganic materials design. CoRR (2023)
3. Méndez-Lucio, O., Nicolaou, C.A., Earnshaw, B.: Mole: a foundation model for molecular graphs using disentangled attention. Nat. Commun. **15**(1), 9431 (2024)
4. Butcher, J., Krishna, R., Mitra, R., Brent, R.I., Li, Y., Corley, N., Kim, P.T., Funk, J., Mathis, S., Salike, S., et al.: De novo design of all-atom biomolecular interactions with RFdiffusion3. In: bioRxiv, Sept 2025
5. Bodnar, C., Bruinsma, W.P., Lucic, A., Stanley, M., Allen, A., Brandstetter, J., Garvan, P., Riechert, M., Weyn, J.A., Dong, H.: A foundation model for the earth system. Nature 1–8 (2025)
6. Bushuiev, R., Bushuiev, A., Samusevich, R., Brungs, C., Sivic, J., Pluskal, T.: Self-supervised learning of molecular representations from millions of tandem mass spectra using dreams. Nat Biotechnol 1–11 (2025)
7. Nguyen, T., Brandstetter, J., Kapoor, A., Gupta, J.K., Grover, A.: Climax: a foundation model for weather and climate (2023). arXiv:2301.10343 arXiv preprint
8. Moro, V., Loh, C., Dangovski, R., Ghorashi, A., Ma, A., Chen, Z., Kim, S., Lu, P.Y., Christensen, T., Soljačić, M.: Multimodal foundation models for material property prediction and discovery. Newton **1**(1) (2025)
9. Liu, G., Sun, M., Matusik, W., Jiang, M., Chen, J.: Multimodal large language models for inverse molecular design with retrosynthetic planning (2024). arXiv:2410.04223 arXiv preprint
10. Kochkov, D., Yuval, J., Langmore, I., Norgaard, P., Smith, J., Mooers, G., Klöwer, M., Lottes, J., Rasp, S., Düben, P., et al.: Neural general circulation models for weather and climate. Nature **632**(8027), 1060–1066 (2024)
11. Bao, F., Zhao, M., Hao, Z., Li, P., Li, C., Zhu, J.: Equivariant energy-guided SDE for inverse molecular design. arXiv preprint arXiv:2209.15408 (2022)
12. Chen, Z., Zhang, J., Arjovsky, M., Bottou, L.: Symplectic recurrent neural networks (2019). arXiv:1909.13334 arXiv preprint
13. Gallegos, M., Vassilev-Galindo, V., Poltavsky, I., Martín Pendás, Á., Tkatchenko, A.: Explainable chemical artificial intelligence from accurate machine learning of real-space chemical descriptors. Nat. Commun. **15**(1), 4345 (2024)
14. Polat, C., Tuncel, M., Kurban, M., Serpedin, E., Kurban, H.: xChemAgents: agentic AI for explainable quantum chemistry. arXiv preprint arXiv:2505.20574 (2025)
15. Dunn, A., Wang, Q., Ganose, A., Dopp, D., Jain, A.: Benchmarking materials property prediction methods: the matbench test set and automatminer reference algorithm. NPJ Comput. Mater. **6**(1), 138 (2020)
16. Ayadi, S., Hetzel, L., Sommer, J., Theis, F., Günnemann, S.: Unified guidance for geometry-conditioned molecular generation. Adv. Neural. Inf. Process. Syst. **37**, 138891–138924 (2024)

17. Lin, X., Xia, Y., Li, Y., Huang, Y.P., Liu, S., Zhang, J., Gao, Y.Q.: In-silico 3D molecular editing through physics-informed and preference-aligned generative foundation models. Nat. Commun. **16**(1), 6043 (2025)
18. Yue, J., Wang, Z., Wang, Y., Zeng, W., Wang, J., Xu, X., Zhang, Y., Zheng, S., Ding, Z., Lu, Z.: Rl from physical feedback: aligning large motion models with humanoid control. arXiv preprint arXiv:2506.12769 (2025)
19. Nair, A.S., Barwey, S., Pal, P., MacArt, J.F., Arcomano, T., Maulik, R.: Understanding latent timescales in neural ordinary differential equation models of advection-dominated dynamical systems. Physica D **476**, 134650 (2025)
20. Chattopadhyay, A., Gray, M., Wu, T., Lowe, A.B., He, R.: OceanNet: a principled neural operator-based digital twin for regional oceans. Sci. Rep. **14**(1), 21181 (2024)
21. Hafner, D., Pasukonis, J., Ba, J., Lillicrap, T.: Mastering diverse control tasks through world models. Nature 1–7 (2025)
22. Batzner, S., Musaelian, A., Sun, L., Geiger, M., Mailoa, J.P., Kornbluth, M., Molinari, N., Smidt, T.E., Kozinsky, B.: E (3)-equivariant graph neural networks for data-efficient and accurate interatomic potentials. Nat. Commun. **13**(1), 2453 (2022)

Chapter 10
Neuro-symbolic and Logic-Augmented Approaches for Scientific Machine Learning

"Elementary, my dear Watson."

Sherlock Holmes

Abstract Neural networks excel at pattern recognition but struggle with systematic reasoning and interpretable knowledge representation. Neuro-symbolic approaches combine neural learning with symbolic methods to achieve both. This chapter develops three complementary directions for Scientific Machine Learning. Symbolic regression and equation discovery (SINDy, AI Feynman, PySR) seek interpretable governing equations from data, addressing the combinatorial challenge of searching equation space. Structure-preserving architectures (Hamiltonian, Lagrangian, symplectic, equivariant networks) embed physical laws directly into network structure, guaranteeing conservation and symmetry by construction rather than hoping networks learn them. Differentiable logic programming (Logic Tensor Networks, DeepProbLog, NeurASP, semantic loss) combines neural perception with logical reasoning, enabling constraint satisfaction and knowledge integration. We conclude with automated theorem proving (AlphaGeometry, AlphaProof) and formal verification, examining how machine-checked proofs might provide the guarantees that safety-critical SciML applications require.

10.1 Introduction

Imagine you're a forensic scientist analyzing satellite imagery to predict hurricane trajectories. As you study the swirling cloud patterns, two distinct cognitive processes activate simultaneously. First, your visual system instantly recognizes the characteristic spiral structure; a hurricane, not a mere storm system. This recognition happens *fast*, automatically, drawing on thousands of images you've seen before. But then a second process engages: you begin *reasoning* about the system. You recall that hurricanes obey conservation of angular momentum, that Coriolis forces deflect their paths, that warm water fuels intensification. You apply these physical princi-

B. P. Bhuyan et al., *Scientific Machine Learning*,
https://doi.org/10.1007/978-981-95-7856-6_10

ples deliberately, checking whether the observed motion respects fundamental laws, calculating expected trajectories using known equations.

This duality; immediate pattern recognition combined with step-by-step logical reasoning; represents how human experts actually solve complex scientific problems. Yet most machine learning systems for scientific computing operate with only half of this cognitive toolkit. Pure neural networks excel at the first part ("this looks like a hurricane"), learning patterns directly from data through gradient descent. But they struggle with the second part: they cannot inherently *reason* about conservation laws, cannot guarantee physical consistency, and often fail catastrophically when extrapolating beyond their training distribution.

Real-World Challenge: The Neural Network That Violated Physics

In 2019, researchers trained a state-of-the-art neural network to predict fluid flow around airfoils for aircraft design. On test cases similar to training data, it achieved impressive accuracy; predictions within 2% of expensive computational fluid dynamics (CFD) simulations, computed 1000× faster. But when deployed on a novel airfoil design, the network predicted a flow pattern that *violated conservation of mass*: more fluid flowed out than flowed in. The prediction was physically impossible.

The fundamental issue? The network had learned correlations in training data but possessed no understanding of the *constraints* that solutions must satisfy. It could interpolate within the training distribution but had no mechanism to enforce physical laws during extrapolation.

Conversely, traditional symbolic approaches to scientific computing; finite element methods, spectral solvers, symbolic algebra systems; operate at the opposite extreme. They encode physical laws explicitly as equations and logical rules, guaranteeing that solutions satisfy fundamental constraints. But they require us to specify the complete model upfront: every relevant equation, every boundary condition, every material parameter. When we don't know the governing equations (or when they're too complex to write down), purely symbolic methods fail. They cannot *learn* from data (Fig. 10.1).

Neuro-symbolic AI: the integration of neural learning with symbolic reasoning; emerged precisely to bridge this gap [1]. The central insight is that pattern recognition and logical reasoning are *complementary* rather than competing paradigms. A neuro-symbolic system can learn representations from data (neural) while reasoning about them using encoded physical knowledge (symbolic). It achieves the best of both worlds: data-driven adaptability combined with physics-aware constraints.

This chapter explores how neuro-symbolic approaches transform scientific machine learning by enabling systems that are simultaneously:

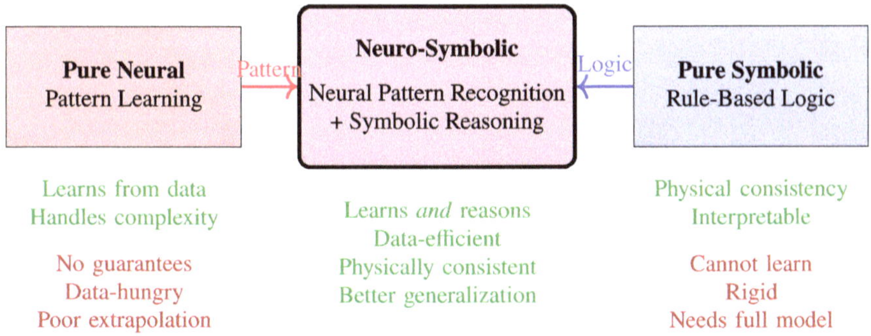

Fig. 10.1 The neuro-symbolic paradigm combines pattern recognition strengths of neural networks with logical reasoning capabilities of symbolic AI, addressing limitations of both pure approaches

- **Data-efficient**: Physical constraints guide learning, requiring fewer observations than pure neural approaches
- **Physically consistent**: Solutions respect conservation laws, symmetries, and boundary conditions
- **Interpretable**: Symbolic components provide transparent reasoning traces
- **Generalizable**: Physics-informed extrapolation beyond training distributions.

10.1.1 Cognitive Foundations: System 1 and System 2 Thinking

To understand why neuro-symbolic integration matters, consider how human experts actually solve scientific problems. Cognitive psychologist Daniel Kahneman's dual-process theory provides a compelling framework: human cognition operates through two fundamentally different systems [2].

Definition: Dual-Process Theory

System 1 (Fast Thinking): Automatic, intuitive, pattern-recognizing. Operates effortlessly and quickly with no sense of voluntary control. Examples: recognizing faces, detecting anger in a voice, solving $2 + 2$.

System 2 (Slow Thinking): Deliberate, logical, analytical. Requires attention and effortful mental activity. Examples: solving complex equations, checking validity of logical arguments, comparing product specifications.

When an experienced physicist analyzes experimental data, both systems activate in concert:

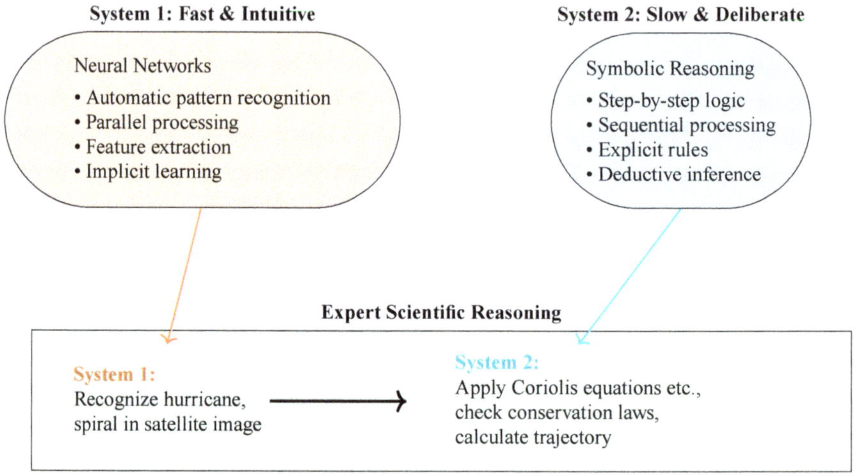

Fig. 10.2 Dual-process cognition in scientific problem-solving. Expert reasoning combines fast pattern recognition (System 1/neural) with deliberate logical analysis (System 2/symbolic)

- **System 1**: Instantly recognizes familiar patterns in plots, spots anomalies, recalls similar experiments
- **System 2**: Systematically applies conservation laws, performs dimensional analysis, evaluates theoretical predictions against observations.

Pure neural networks operate almost exclusively in System 1 mode. They excel at rapid pattern matching; classifying images, detecting features, interpolating functions; but lack System 2's deliberate reasoning capacity. They cannot perform step-by-step logical inference, explain their reasoning, or guarantee adherence to known constraints.

Symbolic AI systems, conversely, are pure System 2. They follow explicit logical rules with perfect precision but cannot learn patterns from raw sensory data. A symbolic physics simulator can solve Maxwell's equations exactly but cannot learn to recognize lightning in photographs (Fig. 10.2).

Neuro-symbolic systems implement computational analogues of both cognitive systems. The neural component handles System 1 tasks; recognizing patterns in high-dimensional data, learning complex nonlinear mappings, adapting to observations. The symbolic component handles System 2 tasks; enforcing conservation laws, performing logical inference, guaranteeing constraint satisfaction.

This isn't merely an engineering convenience; it reflects a deep insight about scientific reasoning itself. Science progresses through the interplay of empirical observation (learning patterns from data) and theoretical reasoning (applying logical principles). A purely empirical approach; collecting data without theory; leads to cataloguing without understanding. A purely theoretical approach; deriving equations without data; risks models disconnected from reality. Scientific progress requires both, working together.

10.1.2 Connection to Embeddings: From Representation to Reasoning

The previous chapter explored how embeddings; learned vector representations; enable machine learning systems to operate on complex scientific objects: irregular geometries, spectral modes, mesh elements. Embeddings solve the *representation problem*: how do we encode scientific entities as vectors amenable to neural network processing?

Neuro-symbolic approaches address the next challenge: the *reasoning problem* [3]. Given good representations, how do we ensure that learned models respect physical principles? How do we combine data-driven learning with explicit domain knowledge?

Consider the temperature distribution problem from our embeddings chapter; predicting temperature on an irregular domain. The spectral embedding approach represented the domain geometry as a learned vector. But spectral embeddings alone don't guarantee that predictions satisfy the heat equation:

$$\frac{\partial T}{\partial t} = \alpha \nabla^2 T$$

A pure neural network trained on spectral embeddings might learn to interpolate observed temperatures accurately but could predict solutions that violate energy conservation when extrapolating to new boundary conditions or time horizons. **Neuro-symbolic approaches integrate the governing PDE directly into the learning process**, ensuring that learned solutions respect physical law [4].

Running Example Preview: Heat Flow with Obstacles

Throughout this chapter, we'll develop a comprehensive neuro-symbolic model for heat diffusion around obstacles; a problem combining rich geometry (irregular boundaries), known physics (heat equation), and practical importance (cooling systems, thermal management).

Challenge: Predict transient temperature distribution on domains with complex internal obstacles, given sparse sensor measurements and known boundary conditions.

Pure Neural Approach: Learn mapping from sensor readings → full temperature field. Fast inference, but no physics guarantees.

Pure Symbolic Approach: Solve heat equation numerically. Accurate but requires fine mesh, expensive computation.

Neuro-Symbolic Solution: Neural network learns efficient approximation while symbolic component enforces heat equation and boundary conditions. Achieves both speed and physical consistency.

We'll build this solution progressively, introducing each neuro-symbolic technique through extensions of this core problem.

10.1.3 The Integration Spectrum

Not all neuro-symbolic systems integrate neural and symbolic components in the same way. A taxonomy (based on Henry Kautz) identifies six distinct architectural patterns, ordered by increasing depth of integration [1]:

Type 1 (Symbolic Neuro Symbolic): Neural networks operate on symbolic inputs and outputs. Large language models exemplify this; text tokens (symbols) pass through neural transformers, producing output tokens. The neural component itself performs continuous operations, but interfaces are symbolic.

Type 2 (Symbolic[Neural]): Symbolic systems invoke neural components for specific subtasks. AlphaGo uses Monte Carlo Tree Search (symbolic game-tree reasoning) to call neural networks for position evaluation and move suggestions. The symbolic component controls execution flow.

Type 3 (Neural | Symbolic): Sequential pipeline where neural perception feeds symbolic reasoning. MIT's Neuro-Symbolic Concept Learner uses neural networks to detect objects, then symbolic program synthesis to answer questions about relationships. Each stage is independent.

Type 4 (Neural: Symbolic → Neural): Symbolic reasoning generates training data for neural learning. AlphaGeometry synthesizes theorem-proof pairs symbolically, then trains neural models on this data. The symbolic component acts as a teacher.

Type 5 (Neural ↔ Symbolic): True bidirectional integration where neural architectures embody symbolic rules. Logic Tensor Networks represent first-order logic formulas as differentiable tensor operations, enabling gradient-based learning of logical predicates. Structure reflects symbolic knowledge.

Type 6 (Neural[Symbolic]): Symbolic reasoning embedded within neural network inference. AlphaGeometry's neural language model directly invokes a symbolic deduction engine mid-forward-pass. The systems are deeply interleaved, not merely connected.

Tip: Choosing Integration Depth for Scientific Problems

Use Type 3 (Pipeline) when: You have separate perception and reasoning stages (e.g., detect defects in images, then diagnose using expert rules).

Use Type 5 (Bidirectional) when: Physical constraints are differentiable and can be encoded as loss terms (e.g., PINNs enforcing PDEs through residuals).

Use Type 6 (Embedded) when: Reasoning must guide neural computation step-by-step (e.g., learned search with symbolic verification).

For most scientific machine learning applications, **Types 5 and 6 dominate** because they enable end-to-end differentiable training while respecting physical constraints.

10.2 Symbolic Regression and Equation Discovery

10.2.1 The Scientific Discovery Problem: From Data to Equations

Imagine you're an astronomer in the early 1600 s, meticulously recording planetary positions night after night. You have tables of data; angles, distances, dates; but what you seek is something deeper: the *law* that governs planetary motion. You don't want just predictions for tomorrow's positions; you want the fundamental equation that explains *why* planets move as they do. This is the essence of scientific discovery: transforming observations into interpretable laws.

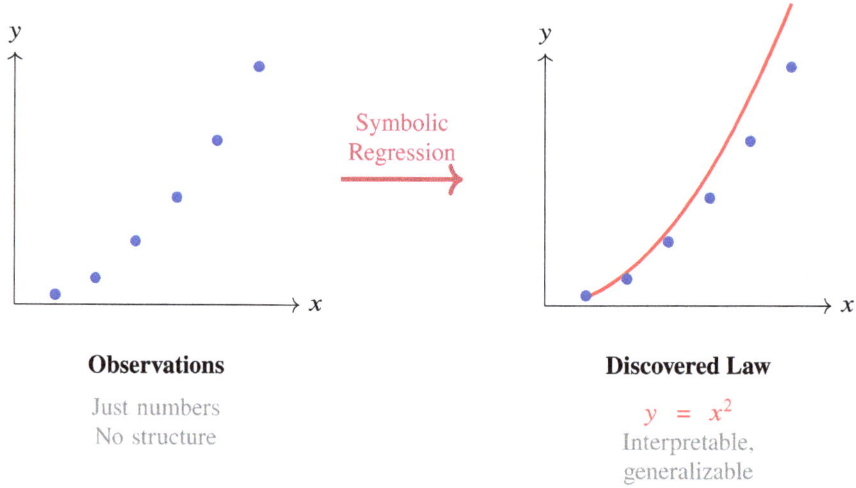

Fig. 10.3 Symbolic regression transforms raw observations into interpretable mathematical equations. Unlike regression that fits parameters in a fixed functional form, symbolic regression searches over the space of possible *equation structures*

Johannes Kepler faced exactly this challenge. Working with Tycho Brahe's precise measurements of Mars's orbit, he spent years searching for the mathematical relationship between orbital period and distance from the Sun. After testing circles, ovals, and countless geometric constructions, he discovered his third law: $T^2 \propto a^3$; the square of the orbital period is proportional to the cube of the semi-major axis. This single equation compressed decades of observations into three symbols.

Symbolic regression automates this discovery process. Given time-series data or experimental measurements, symbolic regression algorithms search through the space of possible mathematical expressions to find equations that both fit the data and possess desirable properties: simplicity, interpretability, physical dimensionality. Unlike neural networks that learn opaque weight matrices, symbolic regression produces explicit formulas like $f(x) = 2x^2 + 3\sin(x)$; that scientists can analyze, validate, and incorporate into theoretical frameworks (Fig. 10.3).

What Makes Symbolic Regression "Symbolic"?

The term "symbolic" distinguishes this from traditional regression in a crucial way:

Traditional regression: Assume a functional form (e.g., polynomial, exponential) and fit parameters.

- Example: "Fit $y = ax^2 + bx + c$" $\rightarrow$ find a, b, c
- The *structure* is fixed; only *coefficients* are learned.

Symbolic regression: Search over possible functional forms simultaneously with parameters.

- Example: Discover that $y = x^2$ (not x^3, not e^x, not $\sin(x)$)
- Both *structure* and *coefficients* are learned.

The "symbols" are mathematical operators ($+$, $\times$, sin, exp) combined into expression trees that represent candidate equations.

The distinction is profound. If we assume $y = ax^3 + bx^2 + cx + d$ but the true relationship is $y = e^x$, traditional regression will fit the polynomial as well as possible; but the resulting equation will be fundamentally wrong, extrapolating poorly and providing no scientific insight. Symbolic regression can discover that the data actually follows an exponential, returning the correct functional form.

10.2.2 The Core Challenge: Combinatorial Explosion in Equation Space

Why can't we simply enumerate all possible equations and test each one? The problem is scale. Even restricting ourselves to elementary functions and modest complexity, the space of candidate equations grows explosively.

Consider building equations using just four operators: $\{+, -, \times, \div\}$, the variable x, and constants. How many distinct expressions exist?

- **Depth 1**: x, c (constants)
- **Depth 2**: $x + c$, $x \times c$, $x - c$, $x \div c$, $c \div x$, ... (tens of expressions)
- **Depth 3**: $(x + c) \times x$, $x \div (x + c)$, ... (hundreds)
- **Depth 5**: Millions of expressions
- **Depth 10**: More than 10^{20} expressions.

Add trigonometric functions (sin, cos), exponentials, logarithms; the space becomes truly astronomical. Exhaustive search is completely infeasible.

Successful symbolic regression algorithms navigate this vast space through clever search strategies:

- **Sparsity assumptions**: Real physical laws tend to be sparse; they use few terms from a large library
- **Evolutionary algorithms**: Evolve populations of candidate equations, mutating and combining promising solutions
- **Physics-inspired heuristics**: Exploit domain knowledge (dimensional analysis, symmetries, separability)
- **Hybrid neural-symbolic**: Use neural networks to guide symbolic search

We now examine three state-of-the-art approaches, each embodying different philosophies about how to conquer this search space.

10.2.3 SINDy: Sparse Identification Through Compressed Sensing

10.2.3.1 The Sparsity Principle in Physical Laws

The Sparse Identification of Nonlinear Dynamics (SINDy) algorithm [5], rests on a profound observation: *most physical systems have sparse representations in appropriate function bases.*

Consider Newton's second law for a damped harmonic oscillator:

$$m\ddot{x} + c\dot{x} + kx = 0$$

This equation is **sparse**: from an infinite library of possible functions of $(x, \dot{x}, \ddot{x})$; including x^2, x^3, $\sin(x)$, e^x, $x\dot{x}$, …; only three terms appear: $\ddot{x}$, $\dot{x}$, and x. The governing dynamics use a tiny subset of possible terms.

This sparsity is not accidental. Physical laws embody fundamental principles (conservation, symmetry, locality) that produce compact mathematical expressions. The Navier-Stokes equations, Maxwell's equations, the Schrödinger equation; all are sparse in natural function bases.

Real-World Example: The Lorenz Attractor

The Lorenz system, a simplified model of atmospheric convection, is governed by three coupled ODEs:

$$\dot{x} = \sigma(y - x)$$
$$\dot{y} = x(\rho - z) - y$$
$$\dot{z} = xy - \beta z$$

From a library of 10 polynomial terms up to degree 2, only **5 terms** appear in $\dot{x}$, **3 terms** in $\dot{y}$, and **3 terms** in $\dot{z}$. The representation is highly sparse.

Given only trajectory data $(x(t), y(t), z(t))$, SINDy successfully recovered these exact equations and identified the parameters $\sigma = 10, \rho = 28, \beta = 8/3$ from the data.

10.2.3.2 The SINDy Algorithm: Formulation and Intuition

SINDy transforms equation discovery into a *sparse regression problem*. Given measurements of a dynamical system's state $\mathbf{x}(t) \in \mathbb{R}^n$ at times $t_1, \ldots, t_m$, we construct:

1. **Time derivative estimates**: Compute $\dot{\mathbf{x}}(t)$ numerically (finite differences or polynomial fitting):

$$\dot{X} = \begin{bmatrix} \dot{\mathbf{x}}(t_1)^T \\ \dot{\mathbf{x}}(t_2)^T \\ \vdots \\ \dot{\mathbf{x}}(t_m)^T \end{bmatrix} \in \mathbb{R}^{m \times n}$$

2. **Function library matrix**: Evaluate candidate functions at each measurement:

$$\Theta(X) = \begin{bmatrix} 1 & x_1(t_1) & x_2(t_1) & x_1^2(t_1) & x_1(t_1)x_2(t_1) & \sin(x_1(t_1)) & \cdots \\ 1 & x_1(t_2) & x_2(t_2) & x_1^2(t_2) & x_1(t_2)x_2(t_2) & \sin(x_1(t_2)) & \cdots \\ \vdots & \vdots & \vdots & \vdots & \vdots & \vdots & \ddots \end{bmatrix} \in \mathbb{R}^{m \times p}$$

Each column represents a candidate function (constant, polynomial terms, trigonometric functions, etc.) evaluated across all time points. Typically $p \gg n$ (the library is overcomplete).

3. **Sparse regression**: Solve for coefficient matrix $\Xi \in \mathbb{R}^{p \times n}$ such that:

$$\dot{X} \approx \Theta(X)\Xi$$

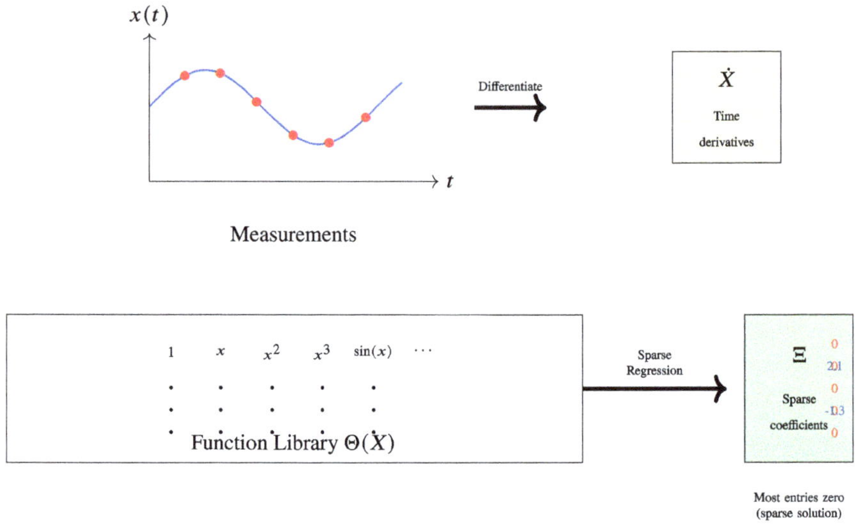

Fig. 10.4 SINDy pipeline: (1) Measure time series and compute derivatives, (2) Construct library matrix of candidate functions, (3) Solve sparse regression to identify which functions appear in the governing equations

Each column $\boldsymbol{\xi}_i$ of Ξ gives the coefficients for the i-th state equation $\dot{x}_i = \sum_{j=1}^{p} \xi_{ji}\theta_j(\mathbf{x})$.

The key is enforcing **sparsity**: most entries of Ξ should be exactly zero. We solve:

$$\boxed{\min_{\Xi} \|\dot{X} - \Theta(X)\Xi\|_F^2 + \lambda\|\Xi\|_1}$$

The ℓ_1 penalty $\|\Xi\|_1 = \sum_{i,j} |\xi_{ij}|$ promotes sparsity, driving most coefficients to zero and revealing which library functions actually appear in the governing equations (Fig. 10.4).

10.2.3.3 Sequential Thresholded Least Squares: Practical Implementation

The ℓ_1-regularized regression can be solved via Lasso or other convex optimization methods. However, SINDy's authors found better results with a simpler iterative procedure called **Sequential Thresholded Least Squares (STLS)**:

Algorithm: SINDy with Sequential Thresholded Least Squares (STLS)

Input: Data $(X, \dot{X})$, library $\Theta(X)$, threshold λ.

Initialize: $\Xi \leftarrow (\Theta^T\Theta)^{-1}\Theta^T\dot{X}$ (initial least-squares solution).

Repeat until convergence:
- Set $\Xi_{\text{sparse}} \leftarrow \Xi$ and zero out small entries: $|\xi_{ij}| < \lambda$.
- Refit $\Xi \leftarrow$ least-squares using only the nonzero coefficients in Ξ_{sparse}.

Return: Sparse coefficient matrix Ξ.

This alternates between (1) thresholding to enforce sparsity and (2) refitting on the remaining active terms. The threshold λ controls the sparsity level; higher λ yields sparser (simpler) models.

Tip: Choosing the Sparsity Threshold

The threshold λ presents a bias-variance tradeoff:

- **Too small**: Includes noise terms, overfits, less interpretable
- **Too large**: Excludes true terms, underfits, poor predictions.

Practical strategies:

- **Cross-validation**: Split data into training/validation, select λ minimizing validation error
- **Pareto frontier**: Plot prediction error vs. number of terms, choose the "elbow"
- **Information criteria**: AIC or BIC for model selection
- **Domain knowledge**: If you know the system should have $\sim k$ terms, target that sparsity level.

For many physical systems, $\lambda \in [0.01, 0.5]$ (normalized units) works well.

Let's apply SINDy to our running example: heat flow with obstacles. Suppose we have temperature measurements $T(\mathbf{x}, t)$ at sensor locations over time. We want to discover the governing dynamics.

For a 1D simplified case, we measure temperature at fixed spatial points: $T(x_1, t), T(x_2, t), \ldots, T(x_N, t)$. The state vector is $\mathbf{u}(t) = [T_1(t), T_2(t), \ldots, T_N(t)]^T$.

Step 1: Construct library. Include:

- Polynomial terms: T_i, T_i^2, T_iT_j
- Spatial derivatives (finite differences): $\frac{\partial T_i}{\partial x}, \frac{\partial^2 T_i}{\partial x^2}$
- Cross-terms: $(T_i - T_j)$ (temperature differences between sensors)

Table 10.1 SINDy variants addressing specific challenges in equation discovery

Variant	Key innovation	Application domain
SINDy-PI [6]	Incorporates weak-form PDE constraints	Systems with known conservation laws
Ensemble SINDy [7]	Trains on bootstrapped subsets, aggregates	Noisy data, uncertainty quantification
SINDy-c [8]	Discovers governing equations with control inputs	Control systems, robotics
Implicit SINDy [9]	Handles algebraic constraints: $f(\mathbf{x}, \dot{\mathbf{x}}) = 0$	DAEs, constrained mechanics
Weak SINDy [10]	Uses weak-form derivatives (integration by parts)	Highly noisy data, derivative estimation challenges
PDE-FIND [11]	Extends to spatiotemporal PDEs	Fluid dynamics, pattern formation
Hybrid SINDy [12]	Combines sparse regression with neural networks	High-dimensional systems

Step 2: Compute time derivatives. Using finite differences:

$$\dot{T}_i(t_k) \approx \frac{T_i(t_{k+1}) - T_i(t_{k-1})}{2\Delta t}$$

Step 3: Sparse regression. Solve for each sensor i:

$$\min_{\boldsymbol{\xi}_i} \left\| \dot{T}_i - \sum_j \xi_{ij}\theta_j(\mathbf{T}) \right\|_2^2 + \lambda\|\boldsymbol{\xi}_i\|_1$$

Expected discovery: If the system truly follows the heat equation, SINDy should identify:

$$\dot{T}_i = \alpha \frac{\partial^2 T_i}{\partial x^2}$$

with all other coefficients driven to zero. The discovered diffusivity α can be extracted directly from the learned coefficient.

10.2.3.4 Extensions and Variants

The base SINDy algorithm has spawned numerous extensions addressing different challenges (Table 10.1):

PDE-FIND, developed by Rudy et al., extends SINDy to discover partial differential equations from spatiotemporal data. Instead of time-series at fixed locations, we have full field measurements $u(x, t)$. The library includes both spatial and temporal derivatives:

$$\Theta = [1, u, u^2, u_x, u_{xx}, u_t, uu_x, \ldots]$$

By regressing u_t against this library, PDE-FIND can discover equations like Burgers', KdV, and Navier-Stokes from snapshot data.

10.2.4 AI Feynman: Physics-Inspired Symbolic Regression

While SINDy leverages sparsity in function representation, AI Feynman [13]; developed by Silviu-Marian Udrescu and Max Tegmark at MIT; takes a different approach: exploiting the *structure* that physical equations typically possess.

10.2.4.1 The Physics Prior: How Physical Equations Are Special

Physical equations aren't arbitrary mathematical expressions. They satisfy special properties reflecting underlying physics:

1. **Dimensional consistency**: Terms must have matching units
2. **Symmetries**: Equations respect translation, rotation, parity symmetries
3. **Separability**: Many equations decompose as sums or products: $f(x, y) = g(x) + h(y)$ or $f(x, y) = g(x) \cdot h(y)$
4. **Compositionality**: Complex expressions built from simpler subexpressions
5. **Special values**: Behavior at $x = 0, x = \infty$, or other limit points reveals structure.

AI Feynman systematically tests for these properties, decomposing the search problem whenever structure is detected (Fig. 10.5).

10.2.4.2 The AI Feynman Algorithm: Iterative Decomposition

AI Feynman operates through a **recursive decomposition strategy**:

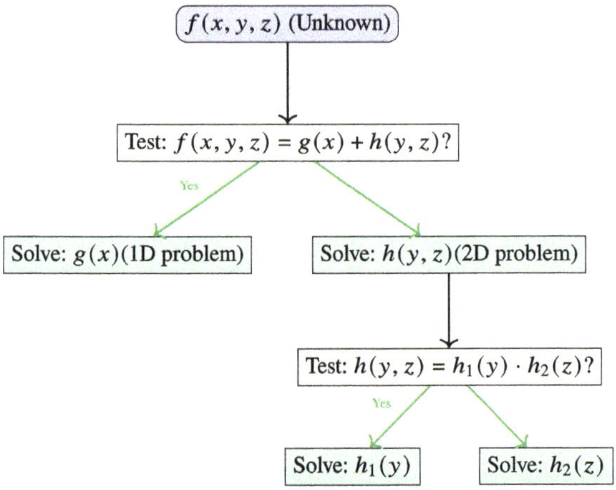

Key Insight:
If f is separable,
3D problem → 1D + 2D problems
(exponential speedup)

Fig. 10.5 AI Feynman's divide-and-conquer strategy. When separability is detected (e.g., $f(x, y, z) = g(x) + h(y, z)$), the problem decomposes into simpler subproblems, each solvable independently

Algorithm: AI Feynman (Physics-Inspired Symbolic Regression)

Input: Data (x_i, y_i) where $x_i \in \mathbb{R}^d$, $y_i \in \mathbb{R}$.
Step 1: Dimensional analysis. Identify dimensionless combinations of variables; reduce the problem to dimensionless form.
Step 2: Symmetry detection. Test for translational, rotational, and parity symmetries; transform to a symmetric form if detected.
Step 3: Separability testing. For each possible split $f(x) = g(x_A) \oplus h(x_B)$, test whether residuals $r(x) = y - g(x_A)$ are independent of x_A. If separable, recursively solve g and h, then return $g \oplus h$.
Step 4: Special points analysis. Evaluate behavior at $x = 0$, $x = \infty$, zeros, and poles; infer functional form from asymptotics.
Step 5: Brute force search. If no structure is detected, enumerate simple expressions (optionally guided by a neural network).

Separability testing is particularly powerful. To test if $f(x, y) = g(x) + h(y)$:

1. Fix $y = y_0$, fit $g(x)$ to data points with this y value
2. Compute residuals $r_i = f(x_i, y_i) - g(x_i)$
3. Check if r_i depends only on y_i (not x_i)
4. If yes: fit $h(y) = r$, return $g(x) + h(y)$

For multiplicative separability $f(x, y) = g(x) \cdot h(y)$, take logarithms to convert to additive form:

$$\log f(x, y) = \log g(x) + \log h(y)$$

Success Story: 100 Equations from Feynman Lectures

Udrescu and Tegmark tested AI Feynman on 100 equations from Feynman's physics lectures; covering mechanics, electromagnetism, quantum mechanics, statistical mechanics. These aren't toy problems; they're real physics used in scientific applications.

Results:
- AI Feynman: **100/100 equations recovered** (100% success rate)
- Previous best (Eureqa): 71/100 (71% success rate)
- Pure neural networks: Failed to provide interpretable symbolic forms

The key difference: physics-inspired decomposition dramatically reduced search space complexity. Instead of searching 10^{20} expressions, decomposition yielded subproblems with $\sim 10^5$ candidates each.

Let's walk through how AI Feynman would rediscover $T^2 \propto a^3$ from planetary data.

Input: Observations of planets with orbital period T, semi-major axis a, mass m, and gravitational parameter $\mu = GM_\odot$.
Step 1: Dimensional Analysis:

- $[T] = \text{time}$
- $[a] = \text{length}$
- $[\mu] = \text{length}^3 \cdot \text{time}^{-2}$

Form dimensionless combinations. Try: $T^2\mu/a^3$. This is dimensionless!

Step 2: Fit Dimensionless Function:
Plot $y = T^2\mu/a^3$ vs. nothing (it should be constant if $T^2 \propto a^3/\mu$).

Discovery: $T^2\mu/a^3 \approx 4\pi^2$ (constant for all planets)

Rearrange: $T^2 = \frac{4\pi^2}{\mu}a^3$; Kepler's third law recovered!
The dimensional analysis *alone* essentially solved the problem, reducing a 3-variable function to a 0-variable (constant) function.

10.2.5 PySR: Evolutionary High-Performance Symbolic Regression

SINDy assumes sparsity in a predefined library. AI Feynman exploits physics structure through decomposition. **PySR** (Python Symbolic Regression), developed by Miles Cranmer, takes a third approach: *evolutionary algorithms* with high-performance Julia backend.

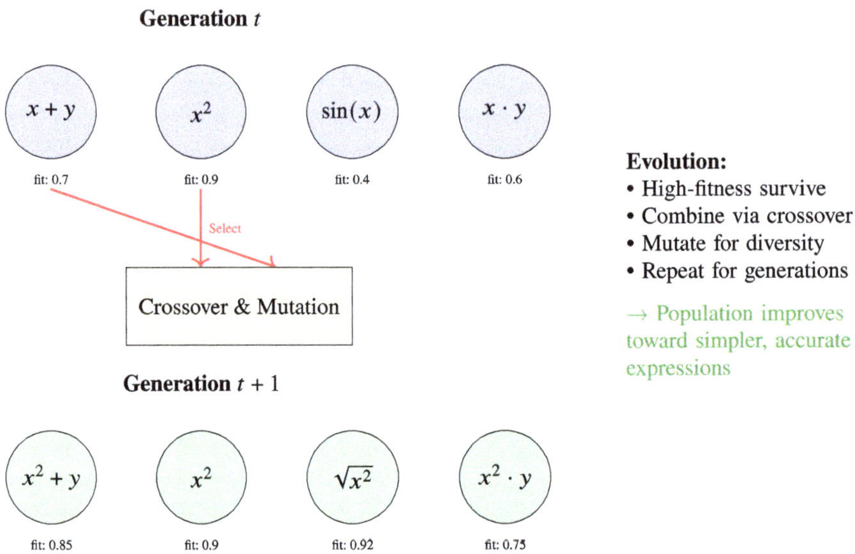

Fig. 10.6 Genetic programming evolves a population of expressions. High-fitness individuals (accurate and simple) are selected for reproduction, with mutation and crossover generating new candidates. Over generations, the population converges toward optimal equations

10.2.5.1 Genetic Programming for Expression Evolution

PySR treats equations as *organisms* in an evolving population. Each equation is an expression tree, and evolution proceeds through:

- **Mutation**: Randomly change a node (operator or terminal)
  - Replace $+$ with $\times$
  - Replace x with $\sin(x)$
  - Add or remove subtree
- **Crossover**: Swap subtrees between two parent expressions
  - Parent 1: $x^2 + \sin(y)$
  - Parent 2: $3x \cdot \exp(z)$
  - Child: $x^2 + \exp(z)$ (swapped right subtree)
- **Selection**: Prefer expressions with better fitness
  - Fitness = accuracy (MSE on data) + simplicity (expression complexity)
  - Multi-objective: optimize Pareto front (accuracy vs. complexity) (Fig. 10.6).

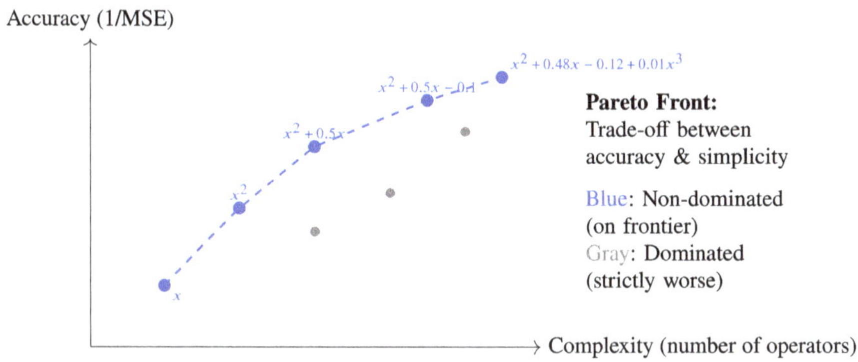

Fig. 10.7 Pareto front in symbolic regression. Each point is an equation; blue points on the frontier offer the best accuracy for their complexity level. Scientists can inspect the frontier and select the appropriate accuracy-interpretability tradeoff

10.2.5.2 Multi-objective Optimization: The Pareto Front

A critical insight: there's no single "best" equation. Instead, we face a tradeoff:

- **Complex equations**: Highly accurate, but uninterpretable (100 terms)
- **Simple equations**: Interpretable, but less accurate (3 terms).

PySR maintains a **Pareto front**; the set of non-dominated solutions where improving accuracy requires increasing complexity, and vice versa (Fig. 10.7).

10.2.5.3 PySR Features and Implementation

PySR combines several advanced features:

- **Julia backend**: Compiled performance (100× faster than pure Python)
- **Scikit-learn API**: Familiar '.fit()' and '.predict()' interface
- **Customizable operators**: User-defined functions beyond standard math
- **Constraints**: Enforce structure (e.g., "denominator can't have x^2")
- **Dimensional analysis**: Automatic unit checking
- **Multi-processing**: Parallel population evolution.

10.2.5.4 When to Use Each Approach

See Table 10.2.

Table 10.2 Comparison of symbolic regression approaches

Method	Strengths	Limitations	Best for
SINDy	• Fast (closed-form solution) • Provable convergence • Works with derivatives	• Requires predefined library • Needs clean derivatives • Limited to sparse systems	Dynamical systems with known function classes
AI Feynman	• Exploits physics structure • Dimensional analysis • 100% success on Feynman eqs	• Requires physics knowledge • Slower than SINDy • Limited to separable forms	Physics problems with symmetries, units, separability
PySR	• No prior assumptions • Highly flexible operators • Pareto front for interpretability	• Computationally expensive • Stochastic (multiple runs needed) • Can overfit without regularization	Black-box discovery, custom operators, complex search spaces

Tip: Combining Approaches

These methods aren't mutually exclusive; they can be combined:

Pipeline 1: PySR → SINDy

1. Use PySR to discover functional form in small dataset
2. Construct SINDy library with discovered terms
3. Refine coefficients with SINDy on full dataset (faster).

Pipeline 2: AI Feynman → PySR

1. Use AI Feynman to detect separability: $f(x, y, z) = g(x) + h(y, z)$
2. Apply PySR separately to discover g and h (smaller problems)
3. Combine results.

Pipeline 3: Neural → Symbolic (Distillation)

1. Train neural network on full data
2. Generate synthetic data from trained network
3. Apply symbolic regression to distill interpretable equation.

Hybrid approaches often outperform any single method.

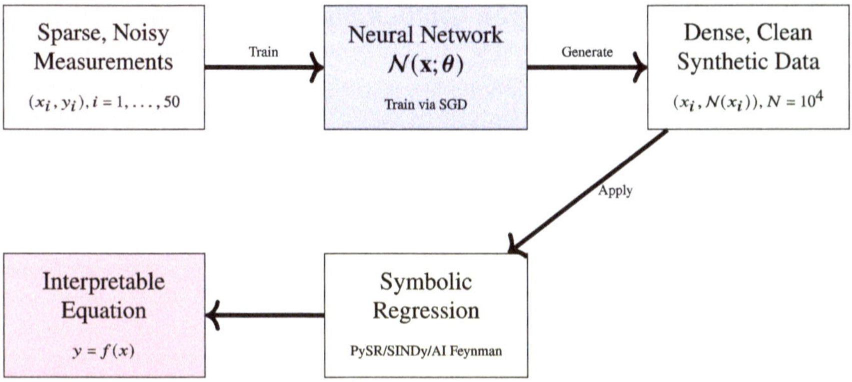

Fig. 10.8 Symbolic distillation pipeline: train neural network on sparse/noisy data, generate dense synthetic dataset, apply symbolic regression to discover interpretable equation

10.2.6 Symbolic Distillation: Extracting Equations from Neural Networks

A powerful but underappreciated technique: train a neural network, then *distill* its learned function into a symbolic equation.

1. **Train neural network**: Fit a neural network $\mathcal{N}(\mathbf{x}; \boldsymbol{\theta})$ to data using standard supervised learning
2. **Generate dense synthetic data**: Sample inputs uniformly across the domain, evaluate neural network:

$$\{(\mathbf{x}_i, \mathcal{N}(\mathbf{x}_i; \boldsymbol{\theta}^*)) \mid i = 1, \ldots, N\}$$

   This dataset has no measurement noise and covers the domain densely
3. **Apply symbolic regression**: Use PySR, SINDy, or AI Feynman on the synthetic dataset to discover equation
4. **Validate**: Check symbolic equation against *original* data (not synthetic)

Why this works: Neural networks are excellent universal approximators but opaque. Symbolic regression works best with clean, dense data. By using the neural network as an intermediate representation, we combine the strengths of both: the neural network handles noise and sparse data; symbolic regression provides interpretability (Fig. 10.8).

The neural network may have implicitly learned a diffusion-decay model. Symbolic distillation makes this explicit, providing an interpretable equation that can be analyzed, validated against physics, and used in downstream applications.

10.2.7 Limitations and Open Challenges

Despite impressive successes, symbolic regression faces fundamental challenges:

- **Noise sensitivity**: Derivative estimation from noisy data remains difficult; errors in $\dot{x}$ compound in SINDy
- **Scalability**: High-dimensional systems ($d > 10$) make library construction and search intractable
- **Identifiability**: Multiple equations can fit data equally well (e.g., $\sin(x)$ vs. Taylor series)
- **Non-smooth dynamics**: Methods assume smooth, differentiable dynamics; shocks, discontinuities, discrete events are problematic
- **Validation**: Discovered equations must be validated on held-out data and checked for physical consistency.

Warning: Overfitting in Symbolic Regression

Symbolic regression can overfit just like neural networks. A complex equation with many terms can fit noise perfectly while capturing no true structure.

Mitigation strategies:
- Use train/validation/test split
- Prefer simpler equations (Occam's razor)
- Check physical dimensionality
- Validate on out-of-distribution data
- Ensemble methods (multiple runs, aggregate)

The simplest equation that fits the data is usually the right one.

10.3 Hard Constraints Through Architecture

10.3.1 From Penalties to Guarantees: What Makes a Constraint "Hard"?

Imagine you're designing a neural network to predict the motion of a satellite orbiting Earth. You train it on trajectory data, and it learns to make accurate predictions; initially. But as you simulate forward in time, something troubling happens: the satellite's orbit slowly spirals inward or outward, eventually crashing into Earth or escaping to infinity. What went wrong?

The fundamental issue is **energy drift**. Real satellites conserve total energy (kinetic + potential) exactly, barring external forces. But your neural network has no such guarantee. It learned correlations in training data, not the underlying conservation law. Small prediction errors compound over time steps, causing the total energy to drift; violating basic physics.

You might try adding an energy conservation penalty to the loss function: $\mathcal{L} = \mathcal{L}_{\text{data}} + \lambda|E(t) - E(0)|^2$. This is a **soft constraint**; it encourages the network to conserve energy during training, but provides no guarantee at inference time. The actual energy conservation depends entirely on:

- How well optimization succeeded (did it find the global minimum?)
- The penalty weight λ (too small: ignored; too large: dominates data fitting)
- Whether test conditions resemble training (out-of-distribution failures).

Warning: Soft Constraints Are Not Guarantees

Standard Physics-Informed Neural Networks (PINNs) encode physics as loss function penalties:

$$\mathcal{L}_{\text{total}} = \mathcal{L}_{\text{data}} + \lambda_1 \mathcal{L}_{\text{PDE}} + \lambda_2 \mathcal{L}_{\text{BC}} + \lambda_3 \mathcal{L}_{\text{IC}}$$

These are **soft constraints**; satisfied approximately during training, with no inference guarantees. Even with perfect optimization:

- PDE residuals approach zero *at collocation points*, not everywhere
- Energy may drift during multi-step rollout
- Conservation laws can be violated out-of-distribution.

This doesn't make PINNs useless; they're powerful and widely adopted; but understanding their limitations is critical for safety-critical or long-horizon applications.

Hard constraints, in contrast, are *guaranteed by construction*. The architecture itself enforces the physical property for *any* inputs, not just training data. No loss term is needed, no optimization tuning required, no dependence on training success. The guarantee is mathematical and structural (Fig. 10.9).

10.3.1.1 The Three Levels of Constraint Hardness

The distinction between soft and hard constraints is actually a spectrum with three levels:

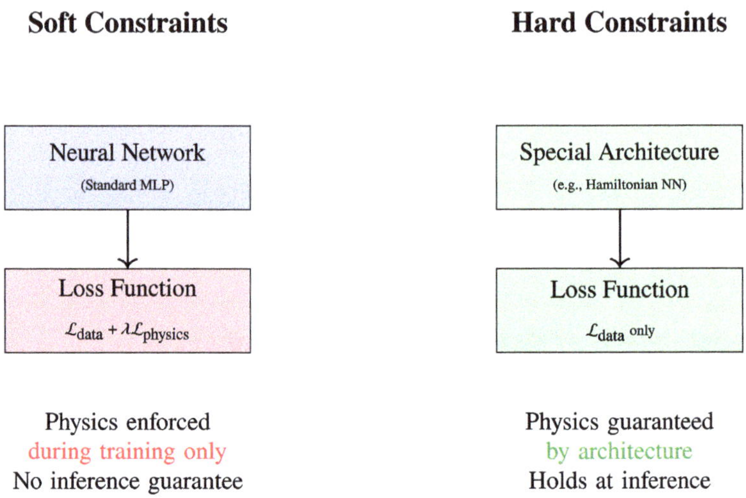

Fig. 10.9 Soft constraints (left) enforce physics through loss function penalties during training but provide no inference guarantees. Hard constraints (right) guarantee physical properties through architectural structure; no loss term needed, guarantees hold at inference for any input

Level 1: Soft Constraints (Loss-Based)

- **Mechanism**: Physics encoded as penalty terms in loss function
- **Guarantee**: None; depends on optimization success
- **Examples**: Standard PINNs, penalty methods, soft boundary conditions
- **Advantage**: Flexible, easy to implement, compatible with any architecture
- **Disadvantage**: No inference guarantee, requires careful weight tuning.

Level 2: Inductive Biases (Architecture-Guided Learning)

- **Mechanism**: Architecture makes physical solutions easier to learn
- **Guarantee**: None; properties are learned, not enforced
- **Examples**: Standard Graph Neural Networks (permutation invariance), convolutional networks (translation equivariance as prior)
- **Advantage**: Better sample efficiency, improved generalization
- **Disadvantage**: Still learned from data, can fail out-of-distribution.

Level 3: Hard Constraints (Mathematical Guarantees)

- **Mechanism**: Architecture *structurally* enforces physics
- **Guarantee**: Mathematical proof that property holds for any input
- **Examples**: Hamiltonian NNs (energy conservation), Symplectic NNs (phase space structure), Equivariant NNs (symmetry)
- **Advantage**: Guaranteed correctness, long-time stability, interpretability
- **Disadvantage**: Less flexible, requires domain knowledge, often more expensive.

Definition: Mathematical Guarantee

A neural network architecture provides a **mathematical guarantee** for property P if:

For all possible inputs $\mathbf{x}$ and learned parameters $\boldsymbol{\theta}$, the network output $\mathcal{N}(\mathbf{x}; \boldsymbol{\theta})$ satisfies property P.

The guarantee must:

1. Be provable mathematically (not empirical)
2. Hold regardless of training data or optimization outcome
3. Require no loss function penalties
4. Apply at inference time, not just training.

Examples:

- Hamiltonian NN: $\frac{dH}{dt} = 0$ for any learned $H(\mathbf{x}; \boldsymbol{\theta})$
- E(3)-Equivariant NN: Rotation of input produces rotated output
- Input Convex NN: Output is convex function of input for any $\boldsymbol{\theta}$.

This chapter focuses exclusively on Level 3 architectures; those providing mathematical guarantees. We explore how to build neural networks where physical laws are *impossible to violate*, not merely discouraged.

10.3.2 Hamiltonian Neural Networks: Guaranteeing Energy Conservation

10.3.2.1 The Hamiltonian Formulation of Classical Mechanics

Before understanding Hamiltonian Neural Networks, recall how Hamiltonian mechanics works. For a physical system, the **Hamiltonian** $H(\mathbf{q}, \mathbf{p})$ represents total energy as a function of generalized coordinates $\mathbf{q}$ (positions) and momenta $\mathbf{p}$.

For a simple pendulum:

- Coordinates: angle q from vertical
- Momentum: $p = m\ell^2\dot{q}$ (angular momentum)
- Hamiltonian: $H(q, p) = \frac{p^2}{2m\ell^2} + mg\ell(1 - \cos q)$ (kinetic + potential energy)

Hamilton's equations describe how the system evolves:

$$\boxed{\frac{dq}{dt} = \frac{\partial H}{\partial p}, \qquad \frac{dp}{dt} = -\frac{\partial H}{\partial q}}$$

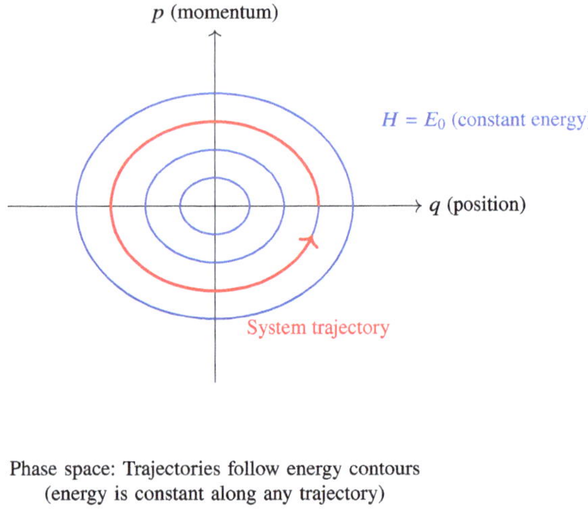

Fig. 10.10 Phase space representation of Hamiltonian dynamics. Trajectories (red) follow level sets of the Hamiltonian function (blue contours). Energy $H(\mathbf{q}, \mathbf{p})$ remains constant along any trajectory; a consequence of Hamilton's equations, not an added constraint

These equations have a profound property: energy is automatically conserved. Computing the time derivative:

$$\begin{aligned}\frac{dH}{dt} &= \frac{\partial H}{\partial q}\frac{dq}{dt} + \frac{\partial H}{\partial p}\frac{dp}{dt} \\ &= \frac{\partial H}{\partial q} \cdot \frac{\partial H}{\partial p} + \frac{\partial H}{\partial p} \cdot \left(-\frac{\partial H}{\partial q}\right) \\ &= 0\end{aligned}$$

Energy conservation isn't an additional constraint; it's a direct consequence of the Hamiltonian structure (Fig. 10.10).

10.3.2.2 The HNN Architecture: Learning the Hamiltonian

Hamiltonian Neural Networks (HNNs), introduced by Greydanus, Dzamba, and Yosinski (NeurIPS 2019), apply a simple but powerful idea: instead of learning the dynamics directly, **learn the Hamiltonian function**, then derive dynamics via Hamilton's equations.

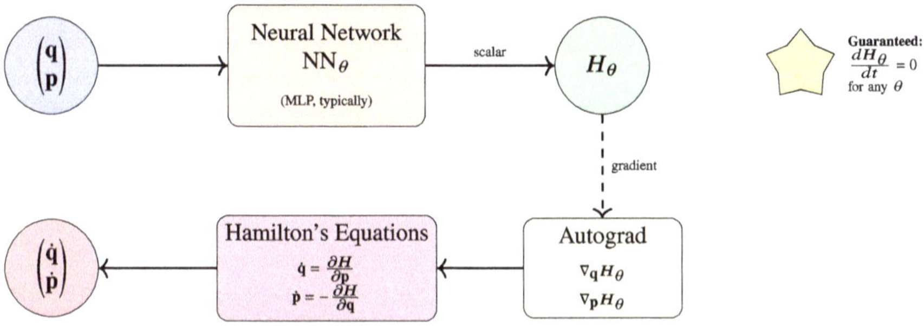

Fig. 10.11 Hamiltonian neural network architecture. The neural network outputs a scalar Hamiltonian $H_\theta(\mathbf{q}, \mathbf{p})$. Automatic differentiation computes gradients, which are used in Hamilton's equations to produce time derivatives. Energy conservation is guaranteed by the mathematical structure; no loss term needed

Standard Neural ODE approach: Learn $\dot{\mathbf{x}} = f_\theta(\mathbf{x})$ directly

$$\text{Neural Network: } \mathbf{x} \xrightarrow{\text{NN}_\theta} \dot{\mathbf{x}}$$

Problem: No guarantee that f_θ conserves energy.

Hamiltonian Neural Network approach: Learn $H_\theta(\mathbf{q}, \mathbf{p})$, derive dynamics via:

$$\text{Neural Network: } (\mathbf{q}, \mathbf{p}) \xrightarrow{\text{NN}_\theta} H \xrightarrow{\text{Hamilton's eqs}} (\dot{\mathbf{q}}, \dot{\mathbf{p}})$$

where:

$$\dot{\mathbf{q}} = \frac{\partial H_\theta}{\partial \mathbf{p}}, \qquad \dot{\mathbf{p}} = -\frac{\partial H_\theta}{\partial \mathbf{q}}$$

The derivatives $\frac{\partial H_\theta}{\partial \mathbf{q}}$ and $\frac{\partial H_\theta}{\partial \mathbf{p}}$ are computed via **automatic differentiation**; PyTorch and JAX compute exact gradients through the computation graph (Fig. 10.11).

Why does this guarantee energy conservation? The proof is remarkably simple.

Theorem: HNNs Conserve the Learned Hamiltonian

Claim: For any differentiable function $H_\theta(\mathbf{q}, \mathbf{p})$ learned by an HNN, the Hamiltonian is conserved along trajectories:

$$\frac{dH_\theta}{dt} = 0$$

Proof:

$$\begin{aligned}
\frac{dH_\theta}{dt} &= \frac{\partial H_\theta}{\partial \mathbf{q}} \cdot \frac{d\mathbf{q}}{dt} + \frac{\partial H_\theta}{\partial \mathbf{p}} \cdot \frac{d\mathbf{p}}{dt} && \text{(chain rule)} \\
&= \frac{\partial H_\theta}{\partial \mathbf{q}} \cdot \frac{\partial H_\theta}{\partial \mathbf{p}} + \frac{\partial H_\theta}{\partial \mathbf{p}} \cdot \left(-\frac{\partial H_\theta}{\partial \mathbf{q}}\right) && \text{(Hamilton's equations)} \\
&= 0 && \text{(cancellation)} \quad \square
\end{aligned}$$

This holds for **any** choice of parameters θ, any learned function H_θ, and any point $(\mathbf{q}, \mathbf{p})$ in phase space. No optimization needed, no loss weights to tune.

Critical caveat: The HNN conserves *the learned Hamiltonian* H_θ, which approximates but may differ from the true physical Hamiltonian H_{true}. The network conserves "an energy-like quantity" that closely matches true energy only if:

1. Training successfully learns $H_\theta \approx H_{\text{true}}$
2. The true system is actually Hamiltonian (no dissipation).

Additionally, numerical integration using non-symplectic methods (Euler, RK4) introduces discretization errors that violate exact conservation. For true structure preservation, use **symplectic integrators** like leapfrog or Störmer-Verlet.

10.3.2.3 Training Hamiltonian Neural Networks

Training an HNN requires data in the form of state trajectories: $\{(\mathbf{q}(t), \mathbf{p}(t), \dot{\mathbf{q}}(t), \dot{\mathbf{p}}(t))\}$.

The loss function is straightforward; mean squared error between predicted and observed derivatives:

$$\mathcal{L}(\theta) = \frac{1}{N} \sum_{i=1}^{N} \left\| \frac{\partial H_\theta}{\partial \mathbf{p}_i} - \dot{\mathbf{q}}_i \right\|^2 + \left\| -\frac{\partial H_\theta}{\partial \mathbf{q}_i} - \dot{\mathbf{p}}_i \right\|^2$$

No physics penalty terms, no conservation loss; just supervised learning on derivatives. The architecture handles conservation automatically.

```python
import torch
import torch.nn as nn

class HamiltonianNN(nn.Module):
    """Hamiltonian Neural Network that guarantees energy conservation."""

    def __init__(self, input_dim, hidden_dim=128, num_layers=3):
        super().__init__()

        # Neural network to learn H(q, p) -> scalar
        layers = []
        layers.append(nn.Linear(input_dim, hidden_dim))
        layers.append(nn.Tanh())

        for _ in range(num_layers - 1):
            layers.append(nn.Linear(hidden_dim, hidden_dim))
            layers.append(nn.Tanh())

        layers.append(nn.Linear(hidden_dim, 1))  # Output: scalar Hamiltonian

        self.net = nn.Sequential(*layers)

    def forward(self, q, p):
        """
        Args:
            q: Position coordinates [batch, n_dof]
            p: Momentum coordinates [batch, n_dof]

        Returns:
            dq_dt: Time derivative of q [batch, n_dof]
            dp_dt: Time derivative of p [batch, n_dof]
        """
        # Concatenate q and p
        qp = torch.cat([q, p], dim=-1)

        # Compute Hamiltonian
        H = self.net(qp)  # [batch, 1]

        # Compute gradients via autograd
        # dH/dq and dH/dp are computed automatically
        grad_outputs = torch.ones_like(H)
        grad_H = torch.autograd.grad(
            H, [q, p],
            grad_outputs=grad_outputs,
            create_graph=True  # Need gradients for training
        )

        dH_dq, dH_dp = grad_H

        # Hamilton's equations
```

```
        dq_dt = dH_dp    # dq/dt = \partial H/\partial p
        dp_dt = -dH_dq   # dp/dt = -\partial H/\partial q

        return dq_dt, dp_dt

    def hamiltonian(self, q, p):
        """Return the learned Hamiltonian (for monitoring energy)."""
        qp = torch.cat([q, p], dim=-1)
        return self.net(qp)


# Training loop
def train_hnn(model, data_loader, num_epochs=1000, lr=1e-3):
    optimizer = torch.optim.Adam(model.parameters(), lr=lr)

    for epoch in range(num_epochs):
        total_loss = 0

        for q, p, dq_true, dp_true in data_loader:
            # Require gradients for q and p
            q = q.requires_grad_(True)
            p = p.requires_grad_(True)

            # Forward pass
            dq_pred, dp_pred = model(q, p)

            # Loss: MSE on time derivatives
            loss = torch.mean((dq_pred - dq_true)**2 + (dp_pred - dp_true)**2)

            # Backward pass
            optimizer.zero_grad()
            loss.backward()
            optimizer.step()

            total_loss += loss.item()

        if epoch % 100 == 0:
            print(f"Epoch {epoch}, Loss: {total_loss/len(data_loader):.6f}")


# Example: Pendulum
# Generate training data from true pendulum dynamics
def generate_pendulum_data(n_samples=1000, dt=0.1):
    # True Hamiltonian: H = p^2/(2*m*l^2) + m*g*l*(1 - cos(q))
    m, l, g = 1.0, 1.0, 9.8

    q = torch.randn(n_samples, 1) * np.pi  # Random angles
    p = torch.randn(n_samples, 1)  # Random momenta

```

```
    # Compute true derivatives
    dq_dt = p / (m * l**2)
    dp_dt = -m * g * l * torch.sin(q)

    return q, p, dq_dt, dp_dt

# Train the model
q, p, dq_true, dp_true = generate_pendulum_data()
# ... (create DataLoader, train model)

# Verify energy conservation on test trajectory
with torch.no_grad():
    q_traj, p_traj = integrate_trajectory(model, q0, p0,
    num_steps=1000)
    H_traj = model.hamiltonian(q_traj, p_traj)

    print(f"Energy drift: {torch.std(H_traj).item():.6e}")
    # Expected: very small (< 1e-6 for symplectic integrator)
```

Listing 10.1 Hamiltonian Neural Network Implementation (PyTorch).

10.3.2.4 Running Example: Heat Diffusion as Hamiltonian System

Can we apply HNNs to our heat diffusion example? At first glance, no; heat diffusion is dissipative (energy flows from hot to cold), not conservative. The temperature field loses energy over time as heat disperses.

However, we can use HNNs for related conservative systems:

Wave equation approach: The wave equation $\frac{\partial^2 u}{\partial t^2} = c^2 \nabla^2 u$ *is* Hamiltonian. If we define:

- $q = u$ (displacement field)
- $p = \frac{\partial u}{\partial t}$ (velocity field)

The Hamiltonian is:

$$H = \int_{\Omega} \left[\frac{1}{2} p^2 + \frac{c^2}{2} |\nabla q|^2 \right] dx$$

An HNN could learn this Hamiltonian from wave data, guaranteeing energy conservation in wave propagation; useful for seismic modeling, acoustic simulations, or electromagnetic waves.

Reversible heat dynamics: For certain thermodynamic systems with microscopicreversibility, we can use HNNs to model the conservative backbone, then add dissipation separately (see Dissipative HNNs below).

```python
# Discretize spatial domain
N_spatial = 50
x = np.linspace(0, 1, N_spatial)

# State: [u(x), du/dt(x)] at each spatial point
# Total dimension: 2 * N_spatial

class WaveHNN(HamiltonianNN):
    """HNN for discretized wave equation."""

    def __init__(self, n_spatial):
        # Input: [u_1, ..., u_N, p_1, ..., p_N]
        super().__init__(input_dim=2*n_spatial,
    hidden_dim=256, num_layers=4)
        self.n_spatial = n_spatial

    def spatial_gradient(self, u, dx):
        """Compute u using finite differences."""
        grad_u = (u[:, 2:] - u[:, :-2]) / (2 * dx)
        # Handle boundaries
        grad_u = torch.cat([grad_u[:, :1], grad_u, grad_u[:,
    -1:]], dim=1)
        return grad_u

# Train on wave simulation data
# The learned H(q,p) will automatically conserve wave energy
```

Listing 10.2 HNN for 1D Wave Equation

10.3.2.5 Extensions: Constrained, Dissipative, and Symplectic HNNs

The base HNN has inspired numerous extensions addressing practical limitations:

Constrained Hamiltonian Neural Networks (CHNNs): Many systems have holonomic constraints (e.g., rigid body motion, molecules with fixed bond lengths). CHNNs work in Cartesian coordinates (x, y, z) with explicit constraints $g(\mathbf{x}) = 0$.

The CHNN architecture uses the constrained Hamiltonian formulation:

$$H_{\text{constrained}}(\mathbf{x}, \mathbf{p}, \boldsymbol{\lambda}) = H(\mathbf{x}, \mathbf{p}) + \boldsymbol{\lambda}^T g(\mathbf{x})$$

where $\boldsymbol{\lambda}$ are Lagrange multipliers learned dynamically. CHNNs achieve **100× accuracy improvement** over unconstrained HNNs and **260× improvement** over baseline neural ODEs on gyroscope systems.

Dissipative Hamiltonian Neural Networks (D-HNNs): Real systems have friction, air resistance, damping. D-HNNs extend HNNs to dissipative systems by learning both a Hamiltonian H and a Rayleigh dissipation function R:

Table 10.3 Hamiltonian neural network variants addressing different physical scenarios

Variant	Key feature	When to use	Reference
Standard HNN	Energy conservation guarantee	Conservative systems, canonical coordinates available	Greydanus+ 2019
CHNN	Handles holonomic constraints in Cartesian coords	Rigid bodies, constrained mechanics	Finzi+ 2020
D-HNN	Learns dissipation alongside Hamiltonian	Damped oscillators, friction, air resistance	Sosanya+ 2022
Symplectic HNN	Uses symplectic integrators in training	Long-time integration, exact conservation needed	David+ 2023

$$\dot{\mathbf{q}} = \frac{\partial H}{\partial \mathbf{p}}, \qquad \dot{\mathbf{p}} = -\frac{\partial H}{\partial \mathbf{q}} - \frac{\partial R}{\partial \dot{\mathbf{q}}}$$

The dissipation function $R(\dot{\mathbf{q}})$ represents energy loss, satisfying $\frac{dH}{dt} = -\frac{\partial R}{\partial \dot{\mathbf{q}}} \cdot \dot{\mathbf{q}} \leq 0$ (energy decreases).

Symplectic HNNs: Use symplectic integrators (leapfrog, Störmer-Verlet) within the loss function to avoid artificial error floors from standard integration. This removes discretization-induced energy drift, achieving exact conservation in discrete-time simulations (Fig. 10.3).

10.3.3 Lagrangian Neural Networks: Coordinate-Free Dynamics

Hamiltonian Neural Networks [14] require canonical coordinates; positions $\mathbf{q}$ and their conjugate momenta $\mathbf{p}$. But in many situations, momenta are not naturally available or the relationship $\mathbf{p} = \frac{\partial L}{\partial \dot{\mathbf{q}}}$ is nonlinear and difficult to invert.

Lagrangian Neural Networks (LNNs), introduced by Cranmer et al. (ICLR 2020) [15], provide an alternative that works directly with generalized coordinates $\mathbf{q}$ and velocities $\dot{\mathbf{q}}$; no momenta needed.

10.3.3.1 Lagrangian Mechanics: The Principle of Least Action

The Lagrangian formulation rests on a variational principle. For a mechanical system, the **Lagrangian** is:

$$L(\mathbf{q}, \dot{\mathbf{q}}) = T(\dot{\mathbf{q}}) - V(\mathbf{q})$$

where T is kinetic energy and V is potential energy.

The dynamics follow from the **Euler-Lagrange equations**:

$$\boxed{\frac{d}{dt}\left(\frac{\partial L}{\partial \dot{\mathbf{q}}}\right) - \frac{\partial L}{\partial \mathbf{q}} = 0}$$

For the pendulum:

$$\begin{aligned} L(q, \dot{q}) &= \frac{1}{2} m\ell^2 \dot{q}^2 - mg\ell(1 - \cos q) \\ \frac{d}{dt}(m\ell^2 \dot{q}) &= -mg\ell \sin q \\ \ddot{q} &= -\frac{g}{\ell} \sin q \end{aligned}$$

The Lagrangian formulation is **coordinate-free**: it works in any generalized coordinates, not just Cartesian. This flexibility is crucial for complex systems.

10.3.3.2 The LNN Architecture: Learning the Lagrangian

An LNN learns $L_\theta(\mathbf{q}, \dot{\mathbf{q}})$ as a neural network, then derives accelerations $\ddot{\mathbf{q}}$ via the Euler-Lagrange equations.

The architecture involves (Fig. 10.12):

1. **Input**: $(\mathbf{q}, \dot{\mathbf{q}})$ (positions and velocities)
2. **Neural Network**: $L_\theta(\mathbf{q}, \dot{\mathbf{q}}) \in \mathbb{R}$ (scalar Lagrangian)
3. **First derivatives**: Compute $\frac{\partial L_\theta}{\partial \mathbf{q}}$ and $\frac{\partial L_\theta}{\partial \dot{\mathbf{q}}}$ via autograd
4. **Second derivatives**: Compute Hessian $\mathbf{H} = \frac{\partial^2 L_\theta}{\partial \dot{\mathbf{q}}^2}$
5. **Euler-Lagrange**: Solve for $\ddot{\mathbf{q}}$:

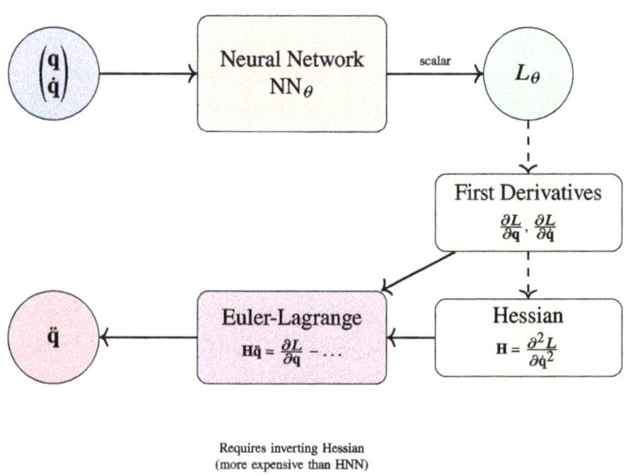

Fig. 10.12 Lagrangian neural network architecture. The network outputs a scalar Lagrangian $L_\theta(\mathbf{q}, \dot{\mathbf{q}})$. First and second derivatives are computed via autograd, then the Euler-Lagrange equations are solved (requiring Hessian inversion) to obtain accelerations $\ddot{\mathbf{q}}$

$$\mathbf{H}\ddot{\mathbf{q}} = \frac{\partial L_\theta}{\partial \mathbf{q}} - \frac{d}{dt}\left(\frac{\partial L_\theta}{\partial \dot{\mathbf{q}}}\right)$$

10.3.3.3 Advantages and Computational Costs

Key advantage: LNNs work in *any generalized coordinates*. They don't require canonical momentum $\mathbf{p}$, which can be nonlinearly related to $\dot{\mathbf{q}}$ in complex systems.

Example where LNNs win: Relativistic particle.

- Lagrangian: $L = -mc^2\sqrt{1 - \dot{q}^2/c^2} - V(q)$
- Canonical momentum: $p = \frac{m\dot{q}}{\sqrt{1-\dot{q}^2/c^2}} \neq m\dot{q}$

HNNs need (q, p) inputs, requiring inversion of the nonlinear relationship $p(\dot{q})$. LNNs use $(q, \dot{q})$ directly.

Computational cost: LNNs are more expensive than HNNs:

- HNN: $O(n)$ (gradient computation)
- LNN: $O(n^2)$ to $O(n^3)$ (Hessian computation and inversion)

For $n > 10$ degrees of freedom, this becomes noticeable. Sparse Hessians (banded structure) can reduce cost.

Implementation requirement: Smooth activations are essential. The Hessian $\frac{\partial^2 L}{\partial \dot{\mathbf{q}}^2}$ must exist, ruling out ReLU (zero second derivative). Use tanh, softplus, or swish activations.

Tip: When to Use HNNs versus LNNs

Use Hamiltonian NNs when:
- Canonical coordinates $(\mathbf{q}, \mathbf{p})$ are naturally available
- Computational efficiency matters ($n > 10$ DOF)
- Phase space structure is important.

Use Lagrangian NNs when:

- Only $(\mathbf{q}, \dot{\mathbf{q}})$ measurements available
- Momenta have complex nonlinear relationship to velocities
- Relativistic or field-theoretic systems
- Constrained systems easier to express in generalized coordinates

Hybrid approach: Some systems benefit from learning both H and L, using whichever is more tractable for specific sub-components.

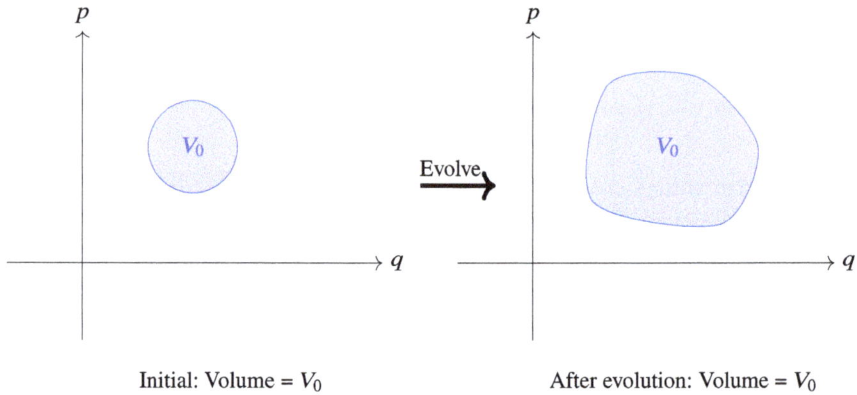

Fig. 10.13 Liouville's theorem: Hamiltonian flow preserves phase space volume. A region evolves (stretches, deforms) but maintains constant volume. Symplectic integrators preserve this property exactly

10.3.4 Symplectic Neural Networks: Exact Phase Space Preservation

Both HNNs and LNNs conserve energy *if* we use the exact continuous-time dynamics. But in practice, we discretize time for numerical simulation. Standard integration methods (Euler, Runge-Kutta) violate the symplectic structure, causing gradual energy drift.

Symplectic Neural Networks (SympNets), developed by Jin et al. (Neural Networks 2020), provide a different guarantee: they preserve the **symplectic structure of phase space** exactly, up to floating-point precision, through discrete-time maps.

10.3.4.1 Symplectic Geometry: Preserving Phase Space Volume

In Hamiltonian mechanics, phase space has special geometric structure captured by the **symplectic two-form**:

$$\omega = \sum_{i=1}^{n} dq_i \wedge dp_i$$

A transformation $\Phi : (\mathbf{q}, \mathbf{p}) \mapsto (\mathbf{q}', \mathbf{p}')$ is **symplectic** if it preserves this form: $\Phi^* \omega = \omega$.

Physical meaning: Symplectic maps preserve phase space volume (Liouville's theorem). If you take a small ball of initial conditions in phase space and evolve it forward, the volume stays constant; though the ball may stretch and deform, its total volume is conserved (Fig. 10.13).

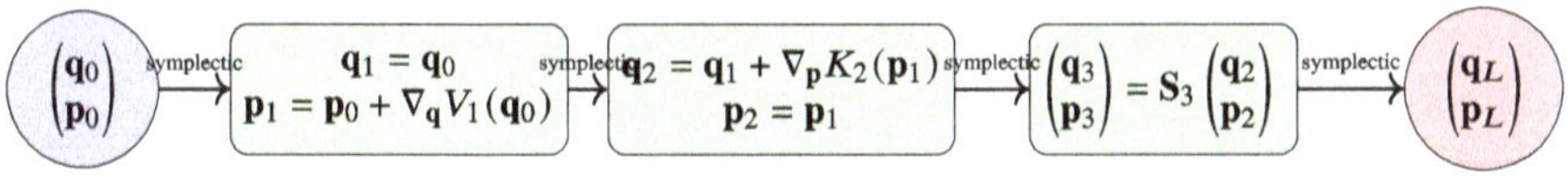

Fig. 10.14 SympNet architecture as a composition of elementary symplectic layers. Each layer preserves symplecticity by construction; composing them yields an exactly symplectic neural network

Standard neural networks don't preserve symplectic structure; they can expand or contract phase space volume arbitrarily. SympNets guarantee exact preservation by composing *elementary symplectic maps*.

10.3.4.2 Building Symplectic Maps: The SympNet Architecture

The key insight: certain transformations are *inherently symplectic*, and composing symplectic maps yields a symplectic map.

Elementary symplectic transformations:

1. **Gradient transformations**:

$$\mathbf{q}' = \mathbf{q} + \nabla_{\mathbf{p}} K(\mathbf{p}), \quad \mathbf{p}' = \mathbf{p}$$

   or

$$\mathbf{q}' = \mathbf{q}, \quad \mathbf{p}' = \mathbf{p} + \nabla_{\mathbf{q}} V(\mathbf{q})$$

2. **Linear symplectic maps**: Matrices $\mathbf{S}$ satisfying $\mathbf{S}^T\mathbf{J}\mathbf{S} = \mathbf{J}$, where:

$$\mathbf{J} = \begin{pmatrix} \mathbf{0} & \mathbf{I} \\ -\mathbf{I} & \mathbf{0} \end{pmatrix}$$

A SympNet is a composition of these building blocks:

$$\Phi = \Phi_L \circ \Phi_{L-1} \circ \cdots \circ \Phi_2 \circ \Phi_1$$

where each Φ_i is symplectic. Since symplecticity is preserved under composition, the full map Φ is guaranteed symplectic (Fig. 10.14).

Two main architectures:

LA-SympNet (Linear/Activation): Uses linear transformations with upper/lower triangular structure (automatically symplectic) and activation functions applied separately to $\mathbf{q}$ or $\mathbf{p}$.

G-SympNet (Gradient): Alternates between $\mathbf{q}$-updates and $\mathbf{p}$-updates using gradient transformations:

$$\text{Layer } 2k: \quad \mathbf{p}' = \mathbf{p} + \nabla_{\mathbf{q}} V_\theta(\mathbf{q}), \quad \mathbf{q}' = \mathbf{q}$$
$$\text{Layer } 2k+1: \quad \mathbf{q}' = \mathbf{q} + \nabla_{\mathbf{p}} K_\theta(\mathbf{p}), \quad \mathbf{p}' = \mathbf{p}$$

where V_θ and K_θ are learned potential/kinetic-like functions (neural networks).

10.3.4.3 Universal Approximation and Non-vanishing Gradients

A crucial theoretical result: **SympNets are universal approximators within the space of symplectic maps**. Any symplectomorphism can be approximated arbitrarily well by a SympNet with sufficient depth.

Additionally, SympNets satisfy the **non-vanishing gradient property**: gradients don't vanish at arbitrary depth, enabling training of very deep networks without the vanishing gradient problem.

Theorem: Universal Approximation for SympNets

Let $\Phi : \mathbb{R}^{2n} \to \mathbb{R}^{2n}$ be any symplectic diffeomorphism. For any $\epsilon > 0$, there exists a SympNet Φ_θ such that:

$$\sup_{\mathbf{x} \in K} \|\Phi(\mathbf{x}) - \Phi_\theta(\mathbf{x})\| < \epsilon$$

for any compact set $K \subset \mathbb{R}^{2n}$.

Implication: SympNets can represent any symplectic transformation to arbitrary accuracy; they're not limited by their structural constraints.

10.3.4.4 Applications and Performance

SympNets excel in:

- **Long-time integration**: No energy drift over thousands of time steps
- **Molecular dynamics**: Preserving phase space structure critical for correct thermodynamic properties
- **Celestial mechanics**: Orbit stability over astronomical timescales
- **Accelerator physics**: Particle beam dynamics in colliders.

Performance example: On the Hénon-Heiles system (chaotic Hamiltonian), SympNets maintain energy error $< 10^{-6}$ over 10,000 time steps, while standard neural ODEs accumulate error > 0.1 (orders of magnitude worse).

Recent extensions include **SympGNNs** (combining symplecticity with graph neural networks for molecular dynamics) and **P-SympNets** (providing interpretable "inverse modified Hamiltonians" for analysis).

10.3.5 Equivariant Neural Networks: Guaranteeing Symmetry Preservation

Many physical systems exhibit fundamental symmetries:

- **Translational invariance**: Physics doesn't change if we shift everything
- **Rotational invariance**: Physics doesn't depend on orientation
- **Permutation invariance**: Particle labels don't matter (indistinguishability).

Standard neural networks must *learn* these symmetries from data; wasting capacity and requiring more samples. **Equivariant neural networks** guarantee symmetry preservation by construction, building it into the architecture.

10.3.5.1 What Is Equivariance?

A function f is G**-equivariant** if it commutes with group actions:

$$f(g \cdot \mathbf{x}) = g \cdot f(\mathbf{x}) \quad \forall g \in G$$

Examples:

- **Translation equivariance**: $f(\mathbf{x} + \mathbf{t}) = f(\mathbf{x}) + \mathbf{t}$
- **Rotation equivariance**: $f(\mathbf{Rx}) = \mathbf{R} f(\mathbf{x})$ for rotation matrix $\mathbf{R}$
- **Permutation equivariance**: $f(\mathbf{x}_{\sigma(1)}, \ldots, \mathbf{x}_{\sigma(n)}) = f(\mathbf{x}_1, \ldots, \mathbf{x}_n)_{\text{permuted}}$.

A special case is **invariance**: $f(g \cdot \mathbf{x}) = f(\mathbf{x})$ (output unchanged by transformation). Invariance = equivariance to the trivial action (Fig. 10.15).

10.3.5.2 E(3) and SE(3) Equivariant Graph Neural Networks

For molecular systems and many physical applications, the relevant symmetry group is:

- **E(3)**: Euclidean group (rotations + reflections + translations)
- **SE(3)**: Special Euclidean group (rotations + translations, no reflections).

E(n)-Equivariant Graph Neural Networks (EGNN), by Satorras et al. (ICML 2021), achieve equivariance without expensive spherical harmonics through a clever

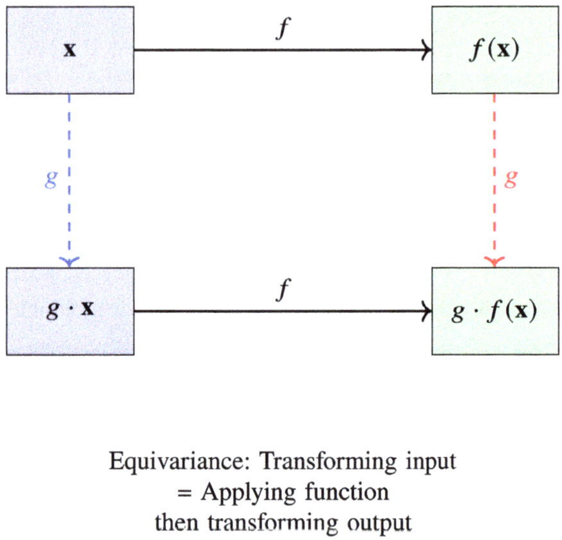

Fig. 10.15 Equivariance commutative diagram. An equivariant function f satisfies: transforming the input (blue arrow) then applying f gives the same result as applying f first then transforming the output (red arrow)

design principle: construct messages using only *invariant quantities*, then build equivariant outputs from *vector combinations*.

The EGNN message passing update:

$$\begin{aligned} m_{ij} &= \phi_e(\mathbf{h}_i, \mathbf{h}_j, \|\mathbf{x}_i - \mathbf{x}_j\|^2, a_{ij}) && \text{(invariant message)} \\ \mathbf{h}_i' &= \phi_h\left(\mathbf{h}_i, \sum_j m_{ij}\right) && \text{(invariant node features)} \\ \mathbf{x}_i' &= \mathbf{x}_i + \sum_j (\mathbf{x}_i - \mathbf{x}_j) \cdot \phi_x(m_{ij}) && \text{(equivariant coordinates)} \end{aligned}$$

Key ideas:

1. Messages depend only on distances $\|\mathbf{x}_i - \mathbf{x}_j\|$ (rotation/translation invariant)
2. Coordinate updates are weighted sums of *relative position vectors* $\mathbf{x}_i - \mathbf{x}_j$
3. Since relative vectors transform equivariantly, their weighted sum does too.

Proof Sketch: EGNN Equivariance

Claim: The EGNN update is E(3)-equivariant.

Proof for rotations (translation/reflection similar):
Let $\mathbf{R}$ be a rotation matrix. Apply $\mathbf{R}$ to all coordinates: $\tilde{\mathbf{x}}_i = \mathbf{R}\mathbf{x}_i$.

Messages:

$$\tilde{m}_{ij} = \phi_e(\mathbf{h}_i, \mathbf{h}_j, \|\mathbf{R}\mathbf{x}_i - \mathbf{R}\mathbf{x}_j\|^2, a_{ij}) = \phi_e(\mathbf{h}_i, \mathbf{h}_j, \|\mathbf{x}_i - \mathbf{x}_j\|^2, a_{ij}) = m_{ij}$$

(rotations preserve distances)

Coordinate update:

$$\begin{aligned}\tilde{\mathbf{x}}_i' &= \mathbf{R}\mathbf{x}_i + \sum_j (\mathbf{R}\mathbf{x}_i - \mathbf{R}\mathbf{x}_j) \cdot \phi_x(\tilde{m}_{ij}) \\ &= \mathbf{R}\mathbf{x}_i + \sum_j \mathbf{R}(\mathbf{x}_i - \mathbf{x}_j) \cdot \phi_x(m_{ij}) \\ &= \mathbf{R}\left[\mathbf{x}_i + \sum_j (\mathbf{x}_i - \mathbf{x}_j) \cdot \phi_x(m_{ij})\right] \\ &= \mathbf{R}\mathbf{x}_i'\end{aligned}$$

Thus, rotating input produces rotated output. □

Applications: Molecular property prediction (NequIP, E3NN), protein structure modeling, N-body simulations, particle physics.

Data efficiency: Equivariant networks require **orders of magnitude fewer training samples** than standard networks because they don't waste capacity learning symmetries; they respect them by construction.

10.3.6 Port-Hamiltonian and Thermodynamics-Informed Networks

Real physical systems often exhibit both conservative and dissipative dynamics: a pendulum with friction, heat flow with conduction, chemical reactions with energy exchange. **Port-Hamiltonian Neural Networks (pHNNs)** [16] and **thermodynamics-informed networks** [17] extend guarantees to these open, dissipative systems.

10.3.6.1 Port-Hamiltonian Framework: Energy Exchange and Dissipation

The port-Hamiltonian formulation describes systems with:

- **Internal energy storage**: Hamiltonian $H(\mathbf{x})$
- **Energy exchange**: Input/output ports $\mathbf{u}$ (controls), $\mathbf{y}$ (observations)
- **Dissipation**: Friction, resistance, damping (energy loss).

The dynamics:

$$\boxed{\dot{\mathbf{x}} = [\mathbf{J}(\mathbf{x}) - \mathbf{R}(\mathbf{x})]\nabla_{\mathbf{x}} H + \mathbf{g}(\mathbf{x})\mathbf{u}}$$

where:

- $\mathbf{J}(\mathbf{x})$: **skew-symmetric** interconnection matrix (conservative dynamics)
- $\mathbf{R}(\mathbf{x}) \succeq 0$: **positive semi-definite** dissipation matrix (energy loss)
- $\mathbf{g}(\mathbf{x})$: input matrix (how controls affect dynamics).

The structure guarantees **passivity**: energy can only decrease (or stay constant):

$$\frac{dH}{dt} = -\nabla H^T \mathbf{R} \nabla H + \mathbf{y}^T \mathbf{u} \leq \mathbf{y}^T \mathbf{u}$$

Energy decreases due to dissipation unless supplied by external inputs.

10.3.6.2 pHNN Architecture: Learning Structure-Preserving Dynamics

Port-Hamiltonian Neural Networks learn H_θ, $\mathbf{J}_\theta$, and $\mathbf{R}_\theta$ while *enforcing structural constraints*:

Skew-symmetry: Construct $\mathbf{J}_\theta$ as:

$$\mathbf{J}_\theta = \mathbf{B}_\theta - \mathbf{B}_\theta^T$$

where $\mathbf{B}_\theta$ is any learned matrix. This mathematically guarantees $\mathbf{J}_\theta^T = -\mathbf{J}_\theta$.

Positive semi-definiteness: Parameterize $\mathbf{R}_\theta$ as:

$$\mathbf{R}_\theta = \mathbf{L}_\theta \mathbf{L}_\theta^T$$

where $\mathbf{L}_\theta$ is lower-triangular. This guarantees $\mathbf{R}_\theta \succeq 0$ (all eigenvalues non-negative).

Bounded Hamiltonian: Use ELU or softplus activation in the final layer to ensure $H_\theta \geq H_{\min}$ (energy can't be negative).

Together, these architectural choices guarantee passivity; the system cannot create energy from nothing.

10.3.6.3 GENERIC Formalism: First and Second Laws by Construction

The **GENERIC framework** (General Equation for Non-Equilibrium Reversible-Irreversible Coupling) [18] extends port-Hamiltonians to thermodynamic systems with entropy:

$$\dot{\mathbf{z}} = \mathbf{L}(\mathbf{z})\nabla E + \mathbf{M}(\mathbf{z})\nabla S$$

Table 10.4 Comparison of structure-preserving neural network architectures and their mathematical guarantees

Architecture	Conserved	Dissipation	Guarantee
Hamiltonian NN	Energy H	No	$\frac{dH}{dt} = 0$
Lagrangian NN	Action	No	Euler-Lagrange satisfied
Symplectic NN	Phase volume	No	$\det(\mathbf{J}) = 1$
Port-Hamiltonian NN	Energy H	Yes ($\mathbf{R} \succeq 0$)	Passivity: $\frac{dH}{dt} \leq \mathbf{y}^T\mathbf{u}$
GENERIC NN	Energy E, Entropy S	Yes	$\frac{dE}{dt} = 0$, $\frac{dS}{dt} \geq 0$

where:

- $\mathbf{L}$: **skew-symmetric** (conservative, reversible dynamics)
- $\mathbf{M} \succeq 0$: **symmetric positive semi-definite** (dissipative, irreversible)
- E: total energy (first law)
- S: entropy (second law)

The **degeneracy conditions** guarantee thermodynamic laws:

- $\mathbf{M}\nabla E = 0 \rightarrow$ Energy conservation: $\frac{dE}{dt} = 0$
- $\mathbf{L}\nabla S = 0 \rightarrow$ Entropy production: $\frac{dS}{dt} \geq 0$

Thermodynamics-informed Graph Neural Networks (Hernández et al., IEEE 2022) embed GENERIC structure at each graph node, achieving 1–2 orders of magnitude accuracy improvement for fluid flow simulations (Fig. 10.4).

10.3.7 Comparison and Selection Guide: Choosing the Right Architecture

With multiple hard-constraint architectures available, how do you choose for a specific scientific machine learning problem? (Fig. 10.5)

Decision Tree: Selecting Hard-Constraint Architecture

Question 1: Is the system conservative (no energy loss)?
- **Yes** $\rightarrow$ Go to Question 2
- **No** (has dissipation) $\rightarrow$ Use Port-Hamiltonian NN or GENERIC.

Question 2: Do you have canonical coordinates (q, p)?

- **Yes** $\rightarrow$ Go to Question 3

Table 10.5 Selection guide for hard-constraint architectures based on problem requirements

Architecture	Use when	Advantages	Limitations
HNN	• Conservative systems • Canonical (q, p) available • Efficiency matters	• Simple, fast $(O(n))$ • Clear energy guarantee • Well-established	• Requires momenta • No dissipation • Discretization errors
LNN	• Only $(q, \dot{q})$ available • Relativistic systems • Complex momentum relations	• Coordinate-free • Works with velocities • More general	• Expensive $(O(n^2\text{-}n^3))$ • Needs smooth activations • Hessian computation
SympNet	• Long-time integration • Exact structure needed • Chaotic systems	• Exact symplecticity • No energy drift • Deep networks stable	• More complex architecture • Requires (q, p) • Training can be slower
E(n)-EGNN	• Molecular/particle systems • Symmetry critical • Limited data	• Rotation/translation guaranteed • Orders of magnitude more sample-efficient • Graph structure natural	• Requires geometric data • Equivariance only (not full dynamics)
pHNN	• Open systems • Dissipation present • Control inputs	• Handles dissipation • Passivity guaranteed • External forcing	• More parameters • Requires identifying ports • Structure more complex
GENERIC	• Thermodynamic systems • Both laws needed • Entropy matters	• Energy + entropy • First & second laws • Physically complete	• Most complex • Requires entropy model • Computationally expensive

- **No** (only $q, \dot{q}$ available) → Use Lagrangian NN.

Question 3: Is long-time integration critical?

- **Yes** (thousands of steps, chaotic) → Use SympNet
- **No** (short-medium horizon) → Use Hamiltonian NN.

Question 4: Are symmetries (rotation/translation) critical?

- **Yes** (molecular, particle systems) → Combine with E(n)-EGNN
- **No** → Standard HNN/LNN/SympNet.

10.3.8 Implementation Frameworks and Practical Considerations

Several libraries provide implementations of hard-constraint architectures:

PyTorch-based:

- `hamiltonian-nn`: Original HNN implementation (github.com/greydanus/hamiltonian-nn)
- `constrained-hamiltonian-nn`: CHNNs and CLNNs (github.com/mfinzi/constrained-hamiltonian-neural-networks)
- `e3nn`: E(3)-equivariant networks (e3nn.org)

JAX-based:

- `lagrangian_nns`: LNN implementation (github.com/MilesCranmer/lagrangian_nns)
- `jax-md`: Differentiable molecular dynamics with equivariant networks

Specialized:

- `NequIP`: State-of-the-art E(3)-equivariant interatomic potentials
- `SympNets.jl`: Julia implementation of Symplectic NNs.

Training considerations:

1. **Automatic differentiation**: All approaches require higher-order derivatives (at least second-order for HNNs, third-order for some LNNs). Ensure your framework supports this.
2. **Activation functions**: Use smooth activations (tanh, softplus, swish) for LNNs and architectures requiring second derivatives. ReLU is acceptable for HNNs (only first derivatives needed).
3. **Symplectic integrators**: When simulating HNN/LNN/SympNet predictions forward in time, use symplectic integrators (leapfrog, Störmer-Verlet, Yoshida) to avoid discretization-induced energy drift.
4. **Data requirements**: Hard-constraint architectures often require derivative information ($(\mathbf{q}, \mathbf{p}, \dot{\mathbf{q}}, \dot{\mathbf{p}})$) or acceleration data ($\mathbf{q}, \dot{\mathbf{q}}, \ddot{\mathbf{q}}$). Compute derivatives numerically using finite differences or polynomial fitting if not available.
5. **Initialization**: Initialize Hamiltonian/Lagrangian networks near physically plausible functions (e.g., quadratic in velocities for kinetic energy) to improve convergence.

10.3.9 Summary and Transition

Hard-constraint architectures guarantee physical properties through mathematical structure, not optimization success. We've explored:

- **Hamiltonian NNs**: Energy conservation via symplectic gradients
- **Lagrangian NNs**: Coordinate-free dynamics via Euler-Lagrange
- **Symplectic NNs**: Exact phase space preservation through composed maps
- **Equivariant NNs**: Symmetry guarantees by construction
- **Port-Hamiltonian/GENERIC**: Passivity and thermodynamic laws.

These architectures excel when **correctness is critical**; long-time integration, safety-critical systems, limited data. The guarantees come at a cost: reduced flexibility, domain knowledge requirements, often higher computational expense.

But not all scientific reasoning is continuous. The next frontier integrates discrete symbolic logic with continuous neural learning: How do we combine gradient-based optimization with logical inference? How do we train neural networks that must satisfy formal constraints expressed in first-order logic?

We turn now to **differentiable logic programming**; making logical reasoning amenable to backpropagation.

10.4 Differentiable Logic Programming

10.4.1 From Continuous Constraints to Discrete Logic

The previous section explored architectures that guarantee *continuous* physical properties; energy conservation, symplectic structure, symmetry preservation. These constraints arise from differential equations and smooth geometry. But many scientific constraints are fundamentally *discrete* and *logical*:

- "A molecule is **either** toxic **or** non-toxic, not partially toxic"
- "**If** temperature exceeds threshold, **then** activate cooling system"
- "Conservation of mass: **for all** reactions, total mass before equals total mass after"
- "Safety constraint: pressure must **never** exceed limit throughout entire trajectory".

These constraints involve logical connectives (AND, OR, NOT, IF-THEN), quantifiers (FOR ALL, EXISTS), and discrete truth values. Classical logic is non-differentiable: changing an input slightly can flip a truth value from TRUE to FALSE discontinuously. How do we train neural networks via gradient descent while respecting logical constraints? (Fig. 10.16)

Differentiable logic programming solves this by replacing discrete Boolean operations with continuous approximations that preserve logical structure while enabling gradient-based optimization. The key insight: generalize from {TRUE, FALSE} to continuous truth degrees in [0, 1], using **fuzzy logic** or **probabilistic semantics**.

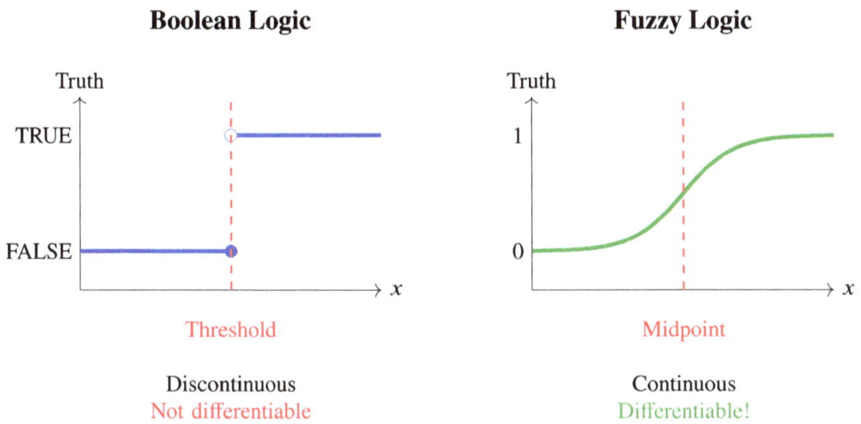

Fig. 10.16 Boolean logic (left) uses discrete truth values with discontinuous step functions; not compatible with gradient descent. Fuzzy logic (right) uses continuous truth degrees in [0, 1] with smooth functions; enabling backpropagation through logical operations

The Central Challenge
Goal: Train neural networks that must satisfy logical constraints.

Problem: Classical logic is discrete and non-differentiable.

Solution: Continuous relaxations preserving logical structure:

1. **Fuzzy logic**: Truth degrees in [0, 1], smooth t-norm operations
2. **Probabilistic logic**: Probability distributions over truth assignments
3. **Differentiable circuits**: Compile logic to computational graphs.

Applications: Neural networks learning to classify while respecting domain constraints (e.g., "if object is metal, then it conducts electricity"), theorem proving guided by learned heuristics, scientific reasoning combining perception and logic.

This section explores four complementary approaches: **Logic Tensor Networks** (fuzzy first-order logic), **DeepProbLog** (probabilistic logic programming), **NeurASP** (answer set programming with neural predicates), and **Signal Temporal Logic** (temporal specifications for dynamical systems).

10.4.2 Logic Tensor Networks: Grounding First-Order Logic on Tensors

10.4.2.1 First-Order Logic: A Quick Refresher

Before diving into Logic Tensor Networks, recall the basics of first-order logic (FOL). FOL allows us to express statements about objects and their relationships using:

Constants: a, b, c (specific objects, e.g., "water", "H_2O")

Variables: x, y, z (placeholders, e.g., "any molecule")

Predicates: $P(x)$, $R(x, y)$ (properties and relations, e.g., "Toxic(x)", "ReactsWith (x, y)")

Logical connectives:

- $\neg P(x)$: NOT (negation)
- $P(x) \wedge Q(x)$: AND (conjunction)
- $P(x) \vee Q(x)$: OR (disjunction)
- $P(x) \rightarrow Q(x)$: IF-THEN (implication).

Quantifiers:

- $\forall x\ P(x)$: FOR ALL x, $P(x)$ holds (universal)
- $\exists x\ P(x)$: THERE EXISTS x such that $P(x)$ holds (existential).

Example scientific constraint:

$$\forall x\ (\text{Metal}(x) \rightarrow \text{Conducts}(x))$$

"For all objects x, if x is a metal, then x conducts electricity."

In classical FOL, truth is Boolean: a statement is either TRUE or FALSE. But neural networks output real numbers. How do we bridge this gap?

10.4.2.2 The LTN Framework: From Symbols to Tensors

Logic Tensor Networks (LTNs), developed by Serafini and Garcez [19], provide an elegant solution by *grounding* logical expressions onto real-valued tensors using fuzzy logic semantics.

The grounding process maps logical elements to continuous mathematical objects (Fig. 10.6):

Key insight: Predicates become neural networks that output continuous truth degrees. Instead of "x is a metal: TRUE/FALSE", we have "x is a metal with degree 0.87" (Fig. 10.17).

Table 10.6 Grounding logical elements as continuous mathematical operations in logic tensor networks

Logical element	Grounding in LTN	Mathematical object
Constant c	Real vector $\mathbf{c} \in \mathbb{R}^d$	Embedding (like word embeddings)
Variable x	Random sample $\mathbf{x} \sim D$	Point in domain $D \subseteq \mathbb{R}^d$
Predicate $P(x)$	Neural network $f_\theta : \mathbb{R}^d \to [0, 1]$	Fuzzy membership function
Connective $\wedge$	T-norm: $T(a, b)$	Product, Łukasiewicz, Gödel
Connective $\vee$	T-conorm: $S(a, b)$	Co-product
Negation $\neg$	Fuzzy negation: $N(a)$	$1 - a$
Quantifier $\forall$	Aggregation: min, $\prod$	Over domain samples
Quantifier $\exists$	Aggregation: max, +	Over domain samples

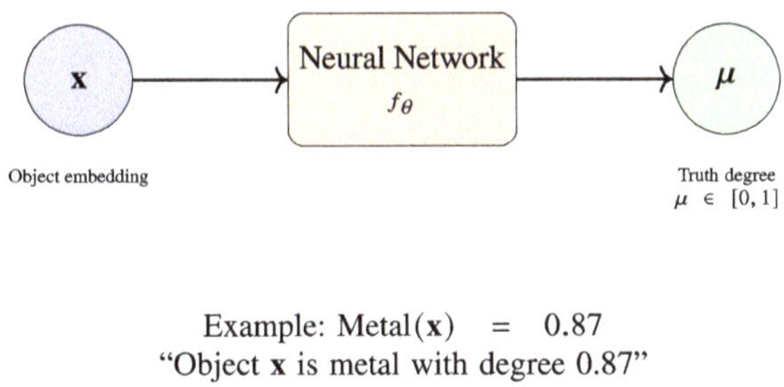

Fig. 10.17 Predicates in LTNs are neural networks mapping object embeddings to truth degrees in [0, 1]. The output represents the continuous degree to which the object satisfies the predicate

10.4.2.3 Fuzzy Logic: T-Norms and Continuous Connectives

To make logical operations differentiable, LTNs use **t-norms** (triangular norms); continuous functions that generalize Boolean AND/OR to the interval [0, 1].

T-norm (fuzzy AND): A function $T : [0, 1]^2 \to [0, 1]$ satisfying:

- Commutativity: $T(a, b) = T(b, a)$
- Associativity: $T(a, T(b, c)) = T(T(a, b), c)$
- Monotonicity: If $a \leq c$, then $T(a, b) \leq T(c, b)$
- Boundary: $T(a, 1) = a$, $T(0, b) = 0$.

Common t-norms:

$$\begin{aligned} \text{Product t-norm:} \quad & T_{\text{prod}}(a, b) = a \cdot b \\ \text{Lukasiewicz t-norm:} \quad & T_{\text{Luk}}(a, b) = \max(0, a + b - 1) \\ \text{Gödel (minimum) t-norm:} \quad & T_{\text{Gödel}}(a, b) = \min(a, b) \end{aligned}$$

T-conorm (fuzzy OR): The dual operation, defined via De Morgan's law:

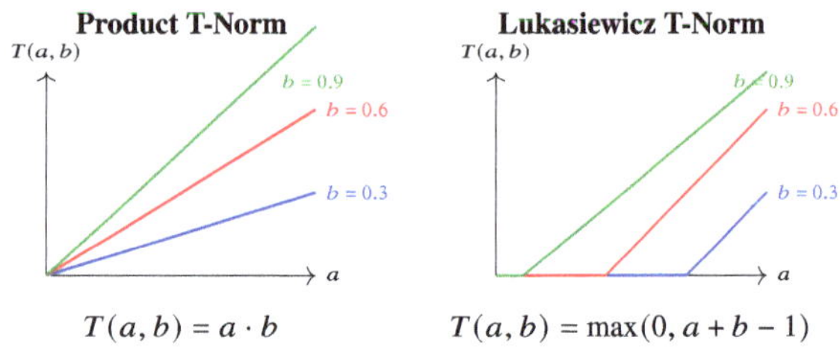

Fig. 10.18 Two common t-norms visualized. Product t-norm (left) is smooth everywhere; ideal for gradient descent. Łukasiewicz t-norm (right) has kinks but provides stronger logical properties

$$S(a, b) = 1 - T(1 - a, 1 - b)$$

Common t-conorms:

$$\begin{aligned} \text{Probabilistic sum:} \quad & S_{\text{prob}}(a, b) = a + b - ab \\ \text{Lukasiewicz:} \quad & S_{\text{Luk}}(a, b) = \min(1, a + b) \\ \text{Gödel (maximum):} \quad & S_{\text{Gödel}}(a, b) = \max(a, b) \end{aligned}$$

Fuzzy negation (Fig. 10.18):

$$N(a) = 1 - a$$

Which t-norm to use?

- **Product**: Smooth, differentiable everywhere, works well with gradient descent
- **Łukasiewicz**: Stronger logical properties, but non-smooth at kinks
- **Gödel**: Preserves classical logic exactly at {0, 1}, but flat gradients.

In practice, **product t-norm** is most common for LTNs due to smooth gradients.

10.4.2.4 Quantifiers as Aggregation Operations

Quantifiers (∀, ∃) range over potentially infinite domains. LTNs approximate them using aggregation over finite samples.

Universal quantifier $\forall x\ P(x)$: "P holds for all x"
Approximate as: $\min_{x \in D} P(x)$ or $\prod_{x \in D} P(x)$

Existential quantifier $\exists x\ P(x)$: "P holds for some x"
Approximate as: $\max_{x \in D} P(x)$ or aggregate sum

Example: $\forall x\ (\text{Metal}(x) \rightarrow \text{Conducts}(x))$
In LTN:

$$\min_{x \in D} [1 - \text{Metal}(x) + \text{Conducts}(x) \cdot \text{Metal}(x)]$$

(Using Łukasiewicz implication: $a \rightarrow b = \min(1, 1 - a + b)$)

Tip: Stable Aggregation for Quantifiers

Simple min/max can be unstable during training (non-smooth gradients). Use **smooth approximations**:

For universal (∀, min):

$$\text{pMean}_p(x_1, \ldots, x_n) = \left(\frac{1}{n}\sum_{i=1}^{n} x_i^p\right)^{1/p}, \quad p < 0$$

As $p \to -\infty$, pMean $\to$ min. Use $p = -2$ for smooth approximation.

For existential (∃, max):

$$\text{pMean}_p(x_1, \ldots, x_n), \quad p > 0$$

As $p \to \infty$, pMean $\to$ max. Use $p = 2$ for smooth approximation.

These provide gradients to all samples, not just min/max, improving training stability.

10.4.2.5 Training LTNs: Knowledge Base Satisfaction

Given a **knowledge base** (set of logical formulas) and data, we want to learn predicates (neural networks) that satisfy the constraints while fitting the data.

Knowledge base example:

$$\begin{aligned}
\phi_1 &: \forall x\,(\text{Metal}(x) \to \text{Conducts}(x)) \\
\phi_2 &: \forall x\,(\text{Plastic}(x) \to \neg\text{Conducts}(x)) \\
\phi_3 &: \forall x\,(\text{Metal}(x) \leftrightarrow \neg\text{Plastic}(x))
\end{aligned}$$

LTN loss function: Maximize satisfaction of all formulas:

$$\mathcal{L}_{\text{LTN}} = -\frac{1}{|\Phi|}\sum_{\phi \in \Phi} \text{Truth}(\phi)$$

where Truth(ϕ) is the grounded truth value (computed via t-norms/aggregation).

For supervised data, combine with data loss:

$$\mathcal{L}_{\text{total}} = \mathcal{L}_{\text{data}} + \lambda \mathcal{L}_{\text{LTN}}$$

```python
import torch
import torch.nn as nn
import ltn  # Logic Tensor Networks library

# Define predicates as neural networks
class MetalPredicate(nn.Module):
    def __init__(self, input_dim=128):
        super().__init__()
        self.net = nn.Sequential(
            nn.Linear(input_dim, 64),
            nn.ReLU(),
            nn.Linear(64, 32),
            nn.ReLU(),
            nn.Linear(32, 1),
            nn.Sigmoid()  # Output in [0, 1]
        )

    def forward(self, x):
        return self.net(x)

# Initialize predicates
Metal = ltn.Predicate(MetalPredicate())
Conducts = ltn.Predicate(MetalPredicate())  # Same architecture
Plastic = ltn.Predicate(MetalPredicate())

# Define fuzzy operators
And = ltn.Connective(ltn.fuzzy_ops.AndProd())  # Product t-norm
Or = ltn.Connective(ltn.fuzzy_ops.OrProbSum())
Not = ltn.Connective(ltn.fuzzy_ops.NotStandard())  # 1 - x
Implies = ltn.Connective(ltn.fuzzy_ops.ImpliesReichenbach())  # 1 - a + ab

# Define quantifiers
Forall = ltn.Quantifier(ltn.fuzzy_ops.AggregPMeanError(p=2), quantifier='f')

# Define knowledge base formulas
def knowledge_base(x):
    """
    Knowledge base constraints:
    1. Metal(x) -> Conducts(x)
    2. Plastic(x) -> ~Conducts(x)
    3. Metal(x) <-> ~Plastic(x)
    """
    # Formula 1: Forall x, Metal(x) -> Conducts(x)
    phi1 = Forall(
        x,
        Implies(Metal(x), Conducts(x))
    )

    # Formula 2: Forall x, Plastic(x) -> ~Conducts(x)
```

```python
    phi2 = Forall(
        x,
        Implies(Plastic(x), Not(Conducts(x)))
    )

    # Formula 3: Forall x, Metal(x) <-> ~Plastic(x)
    # Biconditional: (a -> b) AND (b -> a)
    phi3_forward = Forall(
        x,
        Implies(Metal(x), Not(Plastic(x)))
    )
    phi3_backward = Forall(
        x,
        Implies(Not(Plastic(x)), Metal(x))
    )
    phi3 = And(phi3_forward, phi3_backward)

    # Aggregate formula satisfaction
    return And(And(phi1, phi2), phi3)

# Training loop
def train_ltn(data_loader, num_epochs=100):
    optimizer = torch.optim.Adam([
        *Metal.parameters(),
        *Conducts.parameters(),
        *Plastic.parameters()
    ], lr=1e-3)

    for epoch in range(num_epochs):
        for batch in data_loader:
            x_embeddings = batch['embeddings']  # Object embeddings
            labels = batch['labels']  # Supervision (if available)

            # Wrap as LTN variable
            x = ltn.Variable('x', x_embeddings)

            # Compute knowledge base satisfaction
            kb_sat = knowledge_base(x)

            # Loss: negative satisfaction (maximize satisfaction = minimize negative)
            loss_kb = -kb_sat

            # Optional: Add supervised loss if labels available
            if labels is not None:
                pred_metal = Metal(x)
                loss_supervised = nn.BCELoss()(pred_metal, labels)
                loss = loss_supervised + 0.5 * loss_kb
            else:
```

```
                loss = loss_kb

            # Backpropagation
            optimizer.zero_grad()
            loss.backward()
            optimizer.step()

        if epoch % 10 == 0:
            print(f"Epoch {epoch}, KB Satisfaction:
    {kb_sat.item():.4f}")

# After training, predicates respect logical constraints
# Example: If Metal(x) = 0.9, then Conducts(x) ~= 0.9 (due
    to implication)
```

Listing 10.3 Logic Tensor Networks Implementation (PyTorch)

10.4.2.6 Running Example: Temperature Constraints as Logical Rules

For our heat diffusion system, logical constraints might specify safety bounds and physical relationships:

Constraint 1: "If temperature exceeds threshold, then material is in danger zone"

$$\forall x\,(\mathrm{Hot}(x) \rightarrow \mathrm{DangerZone}(x))$$

Constraint 2: "Material near boundary cannot be both hot and safe"

$$\forall x\,(\mathrm{NearBoundary}(x) \wedge \mathrm{Hot}(x) \rightarrow \neg\mathrm{Safe}(x))$$

Constraint 3: "Stable configuration: temperature differences bounded"

$$\forall x, y\,(\mathrm{Adjacent}(x, y) \rightarrow |\mathrm{Temp}(x) - \mathrm{Temp}(y)| < \delta)$$

These can be encoded in LTN, training a neural network to predict temperature distributions that satisfy both data (sensor measurements) and logical safety constraints.

```
# Predicates for heat distribution
Hot = ltn.Predicate(TempPredicateNN())
DangerZone = ltn.Predicate(ZonePredicateNN())
Safe = ltn.Predicate(SafetyPredicateNN())
NearBoundary = ltn.Predicate(BoundaryPredicateNN())

def heat_constraints(x_positions, temps):
    """
    Encode safety constraints for heat distribution.
    """
    x = ltn.Variable('x', x_positions)
```

```
    # Constraint 1: Hot -> DangerZone
    c1 = Forall(x, Implies(Hot(x), DangerZone(x)))

    # Constraint 2: NearBoundary AND Hot -> ~Safe
    c2 = Forall(
        x,
        Implies(
            And(NearBoundary(x), Hot(x)),
            Not(Safe(x))
        )
    )

    # Constraint 3: Temperature smoothness (simplified)
    # For adjacent points, temp difference should be small
    c3 = check_smoothness_constraint(x_positions, temps)

    return And(And(c1, c2), c3)

# Train network to predict safe configurations
# Loss = data_fit + lambda * logic_satisfaction
```

Listing 10.4 LTN for Heat Distribution Constraints

10.4.3 DeepProbLog: Probabilistic Logic Programming with Neural Networks

While LTNs use fuzzy logic to handle uncertainty, **DeepProbLog** takes a different approach: **probabilistic logic programming**, where neural networks define probability distributions over logical facts.

10.4.3.1 Probabilistic Logic: Reasoning Under Uncertainty

Classical logic is deterministic: a statement is TRUE or FALSE. But scientific reasoning often involves uncertainty:

- "This molecule is *probably* toxic (80% confidence)"
- "There's a 65% chance the reaction will succeed"
- "Given these symptoms, the patient *likely* has condition A".

Probabilistic logic extends logic with probabilities, allowing us to reason about uncertainty. A probabilistic fact has the form:

$$p :: \text{fact}$$

meaning "fact is true with probability p".

Example:

$$0.8 :: \text{toxic(aspirin)}$$
$$0.3 :: \text{toxic(water)}$$

We can then write **rules** combining probabilistic facts:

$$\text{safe}(X) \leftarrow \neg\text{toxic}(X)$$

DeepProbLog's innovation: let **neural networks define the probabilities**.

In DeepProbLog, certain predicates are **neural predicates**; their probabilities come from neural network outputs rather than being fixed.

Syntax:

$$\texttt{nn}(\text{network\_name, input, output}) :: \text{predicate(args)}$$

When evaluating the query, DeepProbLog (Fig. 10.19):

1. Calls the neural network `toxicity_net` on `aspirin`
2. Gets probability distribution: $P(\text{toxic}|\text{aspirin}) = 0.7$, $P(\text{non\_toxic}|\text{aspirin}) = 0.3$
3. Propagates probabilities through logic rules
4. Returns: $P(\text{safe(aspirin)}) = 0.3$.

10.4.3.2 Training via Gradient Descent: Sentential Decision Diagrams

The challenge: how do we compute gradients through logical inference? DeepProbLog compiles logic programs into **Sentential Decision Diagrams (SDDs)**; compact representations of Boolean formulas that enable efficient probability computation.

Key idea: An SDD represents all possible ways a query can be satisfied. For each path through the SDD, we compute the probability (product of edge probabilities), then sum over all paths.

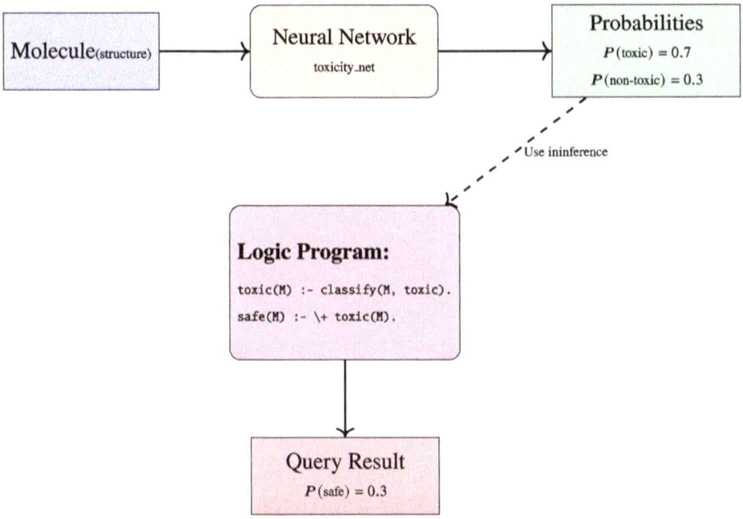

Fig. 10.19 DeepProbLog pipeline: Neural network produces probability distribution over logical facts, which are then used in logic program inference to answer queries

Example: Query safe(M) with rules:

- $\text{safe}(M) \leftarrow \neg\text{toxic}(M)$
- $\text{toxic}(M) \leftarrow \text{classify}(M, \text{toxic})$.

The probability:

$$P(\text{safe}(M)) = P(\neg\text{toxic}(M)) = 1 - P(\text{toxic}(M)) = 1 - P(\text{classify}(M, \text{toxic}))$$

where $P(\text{classify}(M, \text{toxic}))$ comes from the neural network.

Since this is a differentiable computation, we can backpropagate: $\frac{\partial \mathcal{L}}{\partial \theta}$ flows through the SDD, through the probabilities, back to neural network parameters θ.

Loss function: Negative log-likelihood of observed query results:

$$\mathcal{L} = -\sum_i \log P(q_i|\text{data}_i)$$

where q_i are queries with known answers (supervision).

DeepProbLog versus LTN: When to Use Which?

Use Logic Tensor Networks when:

- You have *soft constraints* (preferences, not hard rules)
- Fuzzy membership makes sense ("how metal-like is this?")
- First-order logic with continuous degrees of truth
- You want to directly optimize satisfaction of formulas.

Use DeepProbLog when:

- You need *probabilistic reasoning* (uncertainty quantification)
- Logic rules are definite (TRUE/FALSE) but facts are uncertain
- You want to answer queries: "What's the probability of X given Y?"
- Program induction: learning logic programs from examples.

Hybrid: Some problems benefit from combining both; fuzzy logic for soft constraints, probabilistic logic for uncertainty.

DeepProbLog computes the *posterior probability* of each reactant combination given the observed product, enabling hypothesis ranking.

10.4.4 NeurASP: Answer Set Programming with Neural Perception

Answer Set Programming (ASP) is a declarative logic programming paradigm focused on finding *stable models*; consistent sets of beliefs that satisfy all rules. **NeurASP** integrates ASP with neural networks for perception.

10.4.4.1 Answer Set Programming Basics

ASP programs consist of **rules** of the form:

$$\text{head} \leftarrow \text{body}_1, \ldots, \text{body}_n, \text{not body}_{n+1}, \ldots, \text{not body}_m$$

where `not` is **negation as failure** (closed-world assumption).

An ASP solver finds all **answer sets**; assignments satisfying all constraints.

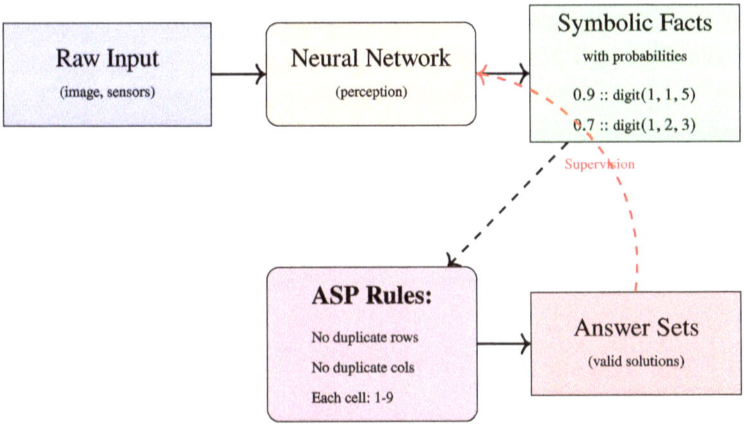

Fig. 10.20 NeurASP architecture: neural networks provide probabilistic symbolic facts, ASP solver finds consistent answer sets satisfying both perception and logical rules, solutions provide supervision for training neural networks

10.4.4.2 Integrating Neural Networks: Perception + Reasoning

NeurASP combines neural perception (classifying objects) with symbolic reasoning (solving constraints).

Architecture:

1. **Perception**: Neural network processes raw input (image, sensor data)
2. **Symbol grounding**: Network output defines probabilities for symbolic facts
3. **Reasoning**: ASP solver finds answer sets consistent with facts + rules
4. **Learning**: Train neural network using answer sets as supervision

Example: Visual Sudoku solver

1. CNN classifies handwritten digits in Sudoku grid
2. Each classification gives probabilities: $P(\text{digit}(R, C) = 5) = 0.9$
3. ASP solver uses these as soft constraints alongside Sudoku rules
4. Valid solution must satisfy both perception and logic.

The **stable model semantics** ensures consistency: solutions respect both visual evidence and logical constraints (Fig. 10.20).

Training objective: Maximize probability of correct answer sets:

$$\mathcal{L} = -\log P(\text{answer sets})$$

where the probability of an answer set depends on neural network outputs for grounded facts.

10.4.5 Semantic Loss: Training Neural Networks with Logical Constraints

All the above approaches (LTN, DeepProbLog, NeurASP) require specialized frameworks. Can we add logical constraints to *any* neural network using just a custom loss function?

Semantic loss functions, developed by Xu et al. (2018), provide a general-purpose method to enforce logical constraints during standard neural network training.

10.4.5.1 The Semantic Loss Function

Given a logical sentence α (constraint) and a probability vector $\mathbf{p} \in [0, 1]^n$ (neural network outputs), the **semantic loss** is:

$$\boxed{\mathcal{L}_s(\alpha, \mathbf{p}) = -\log \sum_{\omega \models \alpha} \prod_{i=1}^{n} p_i^{\omega_i} (1 - p_i)^{1-\omega_i}}$$

where the sum is over all **satisfying assignments** ω of α.

Interpretation: The semantic loss is the negative log-probability of sampling an assignment that satisfies α when sampling each variable i independently with probability p_i.

Example: Constraint "exactly one of x_1, x_2, x_3 is true" (one-hot):

$$\alpha = (x_1 \wedge \neg x_2 \wedge \neg x_3) \vee (\neg x_1 \wedge x_2 \wedge \neg x_3) \vee (\neg x_1 \wedge \neg x_2 \wedge x_3)$$

Semantic loss:

$$\begin{aligned}\mathcal{L}_s(\alpha, \mathbf{p}) &= -\log[p_1(1-p_2)(1-p_3) + (1-p_1)p_2(1-p_3) + (1-p_1)(1-p_2)p_3] \\ &= -\log[p_1 - p_1 p_2 - p_1 p_3 + p_1 p_2 p_3 + \ldots]\end{aligned}$$

Minimizing this loss encourages the network to output probabilities consistent with the constraint.

10.4.5.2 Computing Semantic Loss via Knowledge Compilation

For complex constraints, enumerating all satisfying assignments is intractable. **Knowledge compilation** techniques (SDDs, d-DNNFs) enable efficient computation.

Key idea: Compile the logical formula α into a compact circuit representation (offline, once). Then compute $\mathcal{L}_s$ and gradients by traversing the circuit (online, during training).

Complexity:

- Compilation: Can be exponential in worst case (NP-hard), but polynomial for many practical formulas
- Evaluation + gradient: Linear in circuit size

Libraries like PySDD provide automatic compilation and differentiation.

```
import torch
from pysdd.sdd import SddManager, Vtree

def semantic_loss(logits, sdd_circuit):
    """
    Compute semantic loss for logical constraints.

    Args:
        logits: Neural network outputs [batch, n_vars]
        sdd_circuit: Compiled SDD representing constraint

    Returns:
        loss: Semantic loss encouraging constraint satisfaction
    """
    # Convert logits to probabilities
    probs = torch.sigmoid(logits)

    # For each instance in batch, compute weighted model count
    # This requires traversing the SDD with probabilities
    batch_losses = []

    for i in range(probs.shape[0]):
        p = probs[i]  # [n_vars]

        # Compute weighted model count (WMC)
        # WMC = sum over satisfying assignments of product of literals
        wmc = compute_wmc(sdd_circuit, p)

        # Semantic loss = -log(WMC)
        loss_i = -torch.log(wmc + 1e-8)  # Add epsilon for numerical stability
        batch_losses.append(loss_i)

    return torch.stack(batch_losses).mean()

def compute_wmc(sdd_node, probs):
    """
    Recursively compute weighted model count for SDD.
```

```
    This is a simplified version; real implementation uses
    dynamic programming.
    """
    if sdd_node.is_literal():
        var = abs(sdd_node.literal) - 1
        if sdd_node.literal > 0:
            return probs[var]
        else:
            return 1 - probs[var]

    elif sdd_node.is_decision():
        wmc = 0
        for prime, sub in sdd_node:
            wmc += compute_wmc(prime, probs) *
    compute_wmc(sub, probs)
        return wmc

    else:  # True/False terminal
        return 1.0 if sdd_node.is_true() else 0.0

# Example: Enforce exactly-one constraint in multi-class
    classification
n_classes = 5

# Compile constraint: exactly one class is true
# Formula: (x1 AND ~x2 AND ~x3 AND ~x4 AND ~x5) OR (...)
from pysdd import SddManager
vtree = Vtree(n_classes, "right")
manager = SddManager.from_vtree(vtree)

# Build formula (simplified - actual construction more
    involved)
exactly_one_formula = build_exactly_one_formula(manager,
    n_classes)
sdd = manager.compile(exactly_one_formula)

# Use in training
logits = model(x)  # [batch, n_classes]
loss_semantic = semantic_loss(logits, sdd)
loss_data = nn.CrossEntropyLoss()(logits, labels)
loss = loss_data + lambda_semantic * loss_semantic
```

Listing 10.5 Semantic Loss for Classification Constraints

Applications:

- **Structured prediction**: Enforce output structure (trees, valid sequences)
- **Multi-task learning**: Encode relationships between tasks
- **Scientific constraints**: Enforce domain knowledge (conservation, safety).

10.4.6 *Signal Temporal Logic: Specifying Dynamical System Behaviors*

For dynamical systems (trajectories over time), **Signal Temporal Logic (STL)** provides a formal language to specify desired temporal behaviors.

10.4.6.1 STL Syntax and Semantics

STL extends linear temporal logic to continuous-time real-valued signals.

Basic predicates: $\mu = (f(\mathbf{x}) > 0)$, where f is a real-valued function.

Temporal operators:

- $\Box_{[a,b]}\phi$: "Always ϕ holds during time interval $[a, b]$"
- $\Diamond_{[a,b]}\phi$: "Eventually ϕ holds at some point in $[a, b]$"
- $\phi\mathcal{U}_{[a,b]}\psi$: "$\phi$ holds **until** ψ becomes true in $[a, b]$"

Logical connectives: $\phi \wedge \psi, \phi \vee \psi, \neg\phi$

Example specifications (Fig. 10.21):

1. **Safety**: "Temperature never exceeds 100°C"

$$\Box_{[0,T]}(\text{temp}(\mathbf{x}, t) < 100)$$

2. **Liveness**: "System reaches steady state within 10 s"

$$\Diamond_{[0,10]}(|\dot{\mathbf{x}}| < \epsilon)$$

3. **Stability**: "After disturbance, system returns to equilibrium"

$$\Diamond_{[0,5]}\text{disturbed} \rightarrow \Diamond_{[5,15]}(|\mathbf{x} - \mathbf{x}_{\text{eq}}| < \delta)$$

10.4.6.2 Quantitative Robustness Semantics

Instead of Boolean satisfaction (TRUE/FALSE), STL uses **robustness**; a real number indicating *how robustly* a signal satisfies a specification.

Atomic predicate robustness:

$$\rho(\mathbf{x}, t, \mu) = f(\mathbf{x}(t))$$

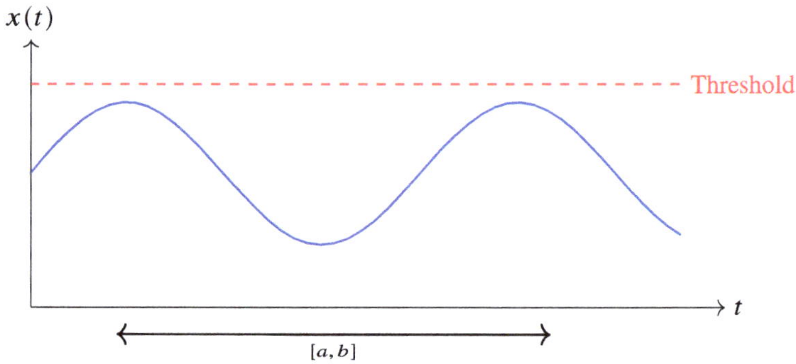

Fig. 10.21 STL specification: "always stay below threshold during $[0, T]$". The signal $x(t)$ (blue) satisfies the specification because it remains below the red threshold throughout the entire interval

where $\mu = (f(\mathbf{x}) > 0)$. Positive ρ means satisfied, negative means violated, magnitude indicates margin.

Connective robustness:

$$\begin{aligned}
\rho(\mathbf{x}, t, \phi \wedge \psi) &= \min(\rho(\mathbf{x}, t, \phi), \rho(\mathbf{x}, t, \psi)) \\
\rho(\mathbf{x}, t, \phi \vee \psi) &= \max(\rho(\mathbf{x}, t, \phi), \rho(\mathbf{x}, t, \psi)) \\
\rho(\mathbf{x}, t, \neg\phi) &= -\rho(\mathbf{x}, t, \phi)
\end{aligned}$$

Temporal robustness:

$$\begin{aligned}
\rho(\mathbf{x}, t, \Box_{[a,b]}\phi) &= \min_{t' \in [t+a, t+b]} \rho(\mathbf{x}, t', \phi) \\
\rho(\mathbf{x}, t, \Diamond_{[a,b]}\phi) &= \max_{t' \in [t+a, t+b]} \rho(\mathbf{x}, t', \phi)
\end{aligned}$$

Example: For safety spec $\Box_{[0,T]}(x(t) < 100)$:

$$\rho = \min_{t \in [0,T]} (100 - x(t))$$

If $\rho = +5$, the signal stays at least 5 units below threshold. If $\rho = -3$, it violates by 3 units.

10.4.6.3 Differentiable STL for Neural Network Training

The challenge: min and max are not smooth; they have zero gradients almost everywhere. **STLCG++** and similar tools provide **smooth approximations**:

Smooth min:

$$\min_{\text{smooth}}(a, b) = -\log(\exp(-ka) + \exp(-kb))/k$$

As $k \to \infty$, this approaches $\min(a, b)$, but for finite k provides smooth gradients.

Smooth max:

$$\max_{\text{smooth}}(a, b) = \log(\exp(ka) + \exp(kb))/k$$

Training neural network controllers to satisfy STL specs:

1. Neural network π_θ outputs control actions: $\mathbf{u} = \pi_\theta(\mathbf{x})$
2. Simulate system dynamics: $\dot{\mathbf{x}} = f(\mathbf{x}, \mathbf{u})$ to get trajectory
3. Compute smooth STL robustness: $\rho_{\text{smooth}}(\text{trajectory}, \phi)$
4. Loss: $\mathcal{L} = -\rho_{\text{smooth}}$ (maximize robustness)
5. Backpropagate through dynamics and STL computation to update θ.

```
import torch
import torch.nn as nn

def smooth_min(a, b, k=10):
    """Smooth approximation of min(a, b)."""
    return -torch.log(torch.exp(-k*a) + torch.exp(-k*b)) / k

def smooth_max(a, b, k=10):
    """Smooth approximation of max(a, b)."""
    return torch.log(torch.exp(k*a) + torch.exp(k*b)) / k

def stl_always(predicate_values, interval, k=10):
    """
    STL Always operator

    Args:
        predicate_values: Tensor of predicate values over
    time [batch, time_steps]
        interval: (a, b) time indices
        k: Smoothness parameter

    Returns:
        robustness: Smooth min over interval
    """
    a, b = interval
    values_in_interval = predicate_values[:, a:b+1]

    # Smooth min over time
    robustness = values_in_interval[:, 0]
```

```python
    for t in range(1, values_in_interval.shape[1]):
        robustness = smooth_min(robustness,
    values_in_interval[:, t], k)

    return robustness

def stl_eventually(predicate_values, interval, k=10):
    """STL Eventually operator"""
    a, b = interval
    values_in_interval = predicate_values[:, a:b+1]

    # Smooth max over time
    robustness = values_in_interval[:, 0]
    for t in range(1, values_in_interval.shape[1]):
        robustness = smooth_max(robustness,
    values_in_interval[:, t], k)

    return robustness

# Example: Train controller to keep temperature safe
class TemperatureController(nn.Module):
    def __init__(self, state_dim=10, action_dim=5):
        super().__init__()
        self.policy = nn.Sequential(
            nn.Linear(state_dim, 64),
            nn.ReLU(),
            nn.Linear(64, 32),
            nn.ReLU(),
            nn.Linear(32, action_dim),
            nn.Tanh()  # Bounded actions
        )

    def forward(self, state):
        return self.policy(state)

def simulate_dynamics(controller, initial_state,
    num_steps=100):
    """Simulate heat dynamics with controller actions."""
    trajectory = [initial_state]
    state = initial_state

    for t in range(num_steps):
        action = controller(state)
        # Simple heat dynamics: dT/dt = -alpha*T + beta*u
        state = state + (-0.1 * state + 0.5 * action) * 0.1
        trajectory.append(state)

    return torch.stack(trajectory, dim=1)  # [batch, time,
    state_dim]

def compute_stl_robustness(trajectory, threshold=90):
    """
    Compute STL robustness for safety specification:
```

```
    (temperature < threshold)
    """
    temp = trajectory[:, :, 0]  # First state dimension is
    temperature

    # Predicate: temp < threshold
    predicate_values = threshold - temp  # Positive =
    satisfied

    # Always operator over entire trajectory
    robustness = stl_always(predicate_values, interval=(0,
    trajectory.shape[1]-1))

    return robustness

# Training loop
controller = TemperatureController()
optimizer = torch.optim.Adam(controller.parameters(),
    lr=1e-3)

for epoch in range(1000):
    # Sample initial states
    initial_states = torch.randn(32, 10) * 50 + 70  # Around
    70 degrees

    # Simulate with controller
    trajectories = simulate_dynamics(controller,
    initial_states)

    # Compute STL robustness
    robustness = compute_stl_robustness(trajectories,
    threshold=90)

    # Loss: negative robustness (maximize satisfaction)
    loss = -robustness.mean()

    # Also add task performance loss (e.g., reach target
    temp)
    target = torch.tensor([75.0])
    final_temp = trajectories[:, -1, 0]
    loss_task = nn.MSELoss()(final_temp,
    target.expand_as(final_temp))

    loss_total = loss + 0.1 * loss_task

    # Optimize
    optimizer.zero_grad()
    loss_total.backward()
    optimizer.step()

    if epoch % 100 == 0:
        print(f"Epoch {epoch}, Robustness:
    {robustness.mean().item():.4f}")
```

Table 10.7 Comparison of differentiable logic programming frameworks

Framework	Logic type	Best for	Learning method
Logic tensor networks	First-order fuzzy logic	Soft constraints, continuous truth degrees, multi-task learning	Maximize satisfaction of FOL formulas
DeepProbLog	Probabilistic logic programming	Uncertainty quantification, abductive reasoning, query answering	Gradient descent through SDDs
NeurASP	Answer set programming	Hard constraints, combinatorial problems, perception + reasoning	Stable model semantics
Semantic loss	Propositional logic	Adding constraints to existing networks, structured prediction	Custom loss via knowledge compilation
Signal temporal logic	Temporal logic	Dynamical systems, safety specs, time-series constraints	Smooth robustness optimization

```

# After training, controller satisfies safety specification
# with quantifiable robustness margin
```

Listing 10.6 Differentiable STL for Neural Controller Training

Applications:

- **Robot motion planning**: Safety constraints (collision avoidance)
- **Autonomous vehicles**: Temporal traffic rules
- **Process control**: Stability and performance specs
- **Neural PDE solvers**: Enforce physical constraints over time.

10.4.7 Comparison and Practical Guidance

See Table 10.7.

Tip: Choosing the Right Logic Framework

Ask these questions:

1. **Is your domain logic fuzzy or crisp?**
   - Fuzzy (degrees of truth) → LTN
   - Crisp (TRUE/FALSE) with uncertainty → DeepProbLog or NeurASP
2. **Do you need probabilistic reasoning?**
   - Yes (uncertainty quantification) → DeepProbLog
   - No (deterministic but constrained) → LTN or Semantic Loss.
3. **Is time important?**
   - Yes (trajectories, temporal constraints) → Signal Temporal Logic
   - No (static relationships) → LTN, DeepProbLog, or NeurASP.
4. **Can you use a specialized framework?**
   - Yes → Use LTN/DeepProbLog/NeurASP (more powerful)
   - No (must integrate with existing code) → Semantic Loss (just a loss function).

10.4.8 Summary and Transition

Differentiable logic programming bridges the discrete world of symbolic reasoning with continuous gradient-based learning. We explored:

- **Logic Tensor Networks**: Fuzzy first-order logic with t-norm semantics
- **DeepProbLog**: Probabilistic logic programming with neural predicates
- **NeurASP**: Answer set programming for hard constraints
- **Semantic Loss**: General-purpose logical constraints via loss functions
- **Signal Temporal Logic**: Formal specifications for dynamical systems.

These methods enable neural networks to *reason*; not just recognize patterns, but satisfy explicit logical rules, answer queries, and provide interpretable explanations.

But even the most sophisticated neuro-symbolic system has limitations: it requires human experts to specify the logic, design the architecture, and define the constraints. What if we want systems that discover *new* mathematical knowledge, prove theorems, and extend the boundaries of human understanding autonomously?

The final frontier: **automated mathematical reasoning**; systems like AlphaGeometry and AlphaProof that achieve human-expert performance on olympiad-level mathematics by combining neural intuition with formal symbolic proof.

10.5 Automated Mathematical Reasoning and Theorem Proving

10.5.1 The Grand Challenge: Machines That Discover Mathematics

Imagine a system that can solve problems from the International Mathematical Olympiad; the world's most prestigious mathematics competition for high school students, where questions require creative insight, elegant proofs, and deep mathematical intuition. For decades, these problems seemed fundamentally beyond machine capabilities. While computers could verify proofs and perform symbolic computation, they lacked the *insight* needed to discover novel solution strategies.

In July 2024, this barrier fell. DeepMind's **AlphaProof** system achieved a silver medal at the International Mathematical Olympiad, scoring 28 out of 42 points; solving four of six problems including the competition's hardest. Six months earlier, **AlphaGeometry** solved 25 of 30 historical IMO geometry problems, approaching gold-medal level. These weren't narrow achievements on specialized tasks; they represent genuine mathematical reasoning combining neural intuition with formal symbolic proof.

Why Mathematical Reasoning Matters for Scientific ML

Mathematical theorem proving might seem distant from scientific machine learning, but the connection is profound:

1. **Formal Verification**: Proving that neural PDE solvers *actually* conserve mass, that control systems *guarantee* safety bounds
2. **Automated Discovery**: Systems that discover new physical laws, mathematical formulations of phenomena, conservation principles
3. **Interpretability**: Generating human-readable proofs explaining *why* a scientific ML model makes certain predictions
4. **Knowledge Integration**: Combining learned patterns with rigorous mathematical reasoning about physical constraints
5. **Trustworthiness**: Moving from "this neural network seems to work" to "this system provably satisfies specifications".

The techniques enabling AlphaGeometry and AlphaProof; neural guidance of symbolic search, formal proof verification, self-improvement through synthetic data; apply directly to scientific reasoning.

This section explores how neuro-symbolic systems achieve human-expert mathematical reasoning, the architectures underlying recent breakthroughs, and implications for scientific machine learning.

10.5.2 The Architecture of Mathematical Reasoning

Mathematical problem-solving combines two distinct cognitive processes that parallel our Chap. 1 discussion of System 1 and System 2 thinking:

Intuition (System 1): Recognizing patterns, generating candidate approaches, sensing which strategies might work

- "This looks like a triangle congruence problem"
- "I should probably use induction here"
- "These expressions seem related to logarithmic identities".

Deduction (System 2): Rigorous step-by-step logical inference, formal proof construction, verification of correctness

- "By the Pythagorean theorem, $a^2 + b^2 = c^2$"
- "Since $P(n)$ implies $P(n + 1)$, the statement holds for all n"
- "This follows from the triangle inequality: $|a + b| \leq |a| + |b|$".

Pure symbolic systems (traditional automated theorem provers) operate entirely in System 2; exhaustive logical search without intuition. They can verify proofs but struggle to discover them, often drowning in exponentially large search spaces.

Pure neural systems operate in System 1; pattern matching without rigorous logic. They generate plausible-sounding solutions but make logical errors, hallucinate steps, and cannot verify correctness (Fig. 10.22).

The neuro-symbolic breakthrough: Use neural networks to guide symbolic search. The neural component proposes promising steps (intuition), while the symbolic component verifies correctness and extends the proof (deduction). This architecture; pioneered by AlphaGeometry and refined by AlphaProof; achieves the best of both worlds.

10.5.3 AlphaGeometry: Olympiad-Level Geometry Reasoning

10.5.3.1 The Geometry Challenge

Olympiad geometry problems require proving statements about geometric configurations; often with elegant, non-obvious constructions.

Example problem (IMO 2015 Problem 3):

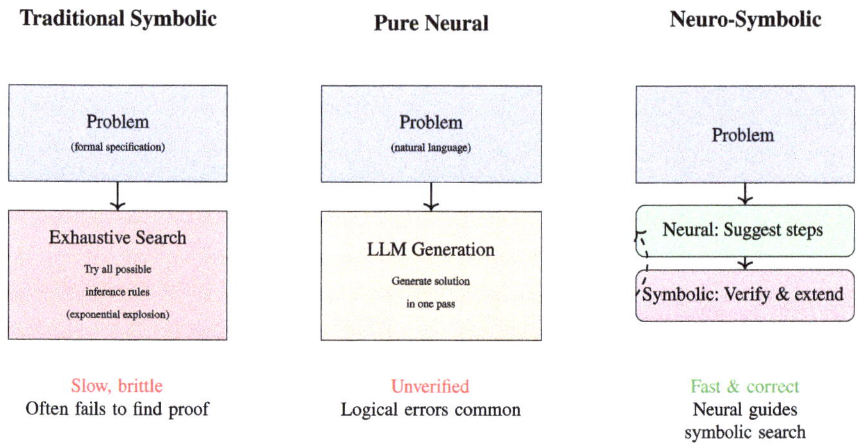

Fig. 10.22 Three paradigms for automated reasoning. Traditional symbolic systems exhaustively search (slow). Pure neural systems generate solutions without verification (unreliable). Neuro-symbolic systems combine neural intuition with symbolic verification (both fast and correct)

IMO 2015, Problem 3

Let ABC be an acute triangle, and let M be the midpoint of AC. A circle ω passing through B and M intersects sides AB and BC at points P and Q, respectively. Let T be the point such that $BPTQ$ is a parallelogram. Suppose that T lies on the circumcircle of ABC.

Prove that: $\angle BTP = \angle CTP$.

Solving this requires:

1. Understanding the geometric configuration (9+ objects)
2. Introducing auxiliary constructions (key insight: which construction?)
3. Applying geometric theorems (angle chasing, circle properties)
4. Rigorous deductive proof (each step justified).

The average IMO contestant spends 4.5 h on three geometry problems. AlphaGeometry solves this in 19 s.

10.5.3.2 The AlphaGeometry Architecture

AlphaGeometry combines two complementary engines:

1. **Symbolic Deduction Engine (DDR: Deduction, Detection, Recognition)**
   - Applies geometric theorems exhaustively

- Example: If $AB = AC$ and $\angle BAD = \angle CAD$, deduce $BD = CD$ (isosceles properties)
- Fast: thousands of deductions per second
- Complete: finds all consequences of known facts
- Problem: Cannot make creative constructions.

2. **Neural Language Model (Construction Suggester)**

- Transformer model trained on 100 million synthetic geometry problems
- Suggests auxiliary constructions when symbolic engine stalls
- Example: "Draw the perpendicular from P to AC"
- Learned from massive self-generated proof corpus.

The algorithm:

Algorithm: AlphaGeometry (Neuro-Symbolic Geometry Proof Search)

Input: Geometry problem (premises + goal).
Initialize a knowledge base with the premises.
Repeat until timeout/max constructions:
- **Symbolic phase**: run the deduction engine exhaustively; deduce consequences and check whether the goal is reached.
- If the goal is proved, return the proof.
- **Neural phase**: call the language model to suggest an auxiliary construction; add it to the knowledge base.

Return: Proof (if found) or FAIL (Fig. 10.23).

10.5.3.3 Training Without Human Data: Synthetic Proof Generation

A remarkable aspect: AlphaGeometry trains on **zero** human proofs. Instead, it generates 100 million synthetic geometry problems through a clever process:

Synthetic Data Generation:

1. **Random sampling**: Generate random geometric configurations (points, circles)
2. **Symbolic deduction**: Apply theorem proving to discover all true statements
3. **Proof mining**: Extract forward and backward proof chains
4. **Language model training**: Train transformer on (problem, proof) pairs.

This creates a vast corpus of *correct* proofs without requiring human annotation. The neural model learns:

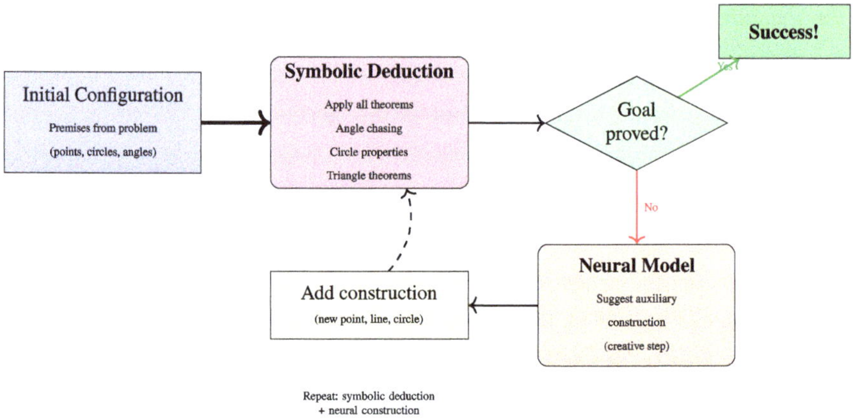

Fig. 10.23 AlphaGeometry's proof search loop. Symbolic deduction exhaustively applies theorems. When stuck, the neural model suggests auxiliary constructions. The cycle repeats until the goal is proved

- Which auxiliary constructions are useful in which contexts
- How to recognize promising intermediate states
- Patterns in successful proof strategies.

Key insight: By training on synthetically generated proofs, the neural component learns a powerful prior over the space of geometric reasoning; which constructions tend to lead to successful proofs; without memorizing specific human solutions.

10.5.3.4 Performance and Analysis

Results on historical IMO problems:

- **AlphaGeometry 1**: 25/30 problems solved (83%, approaching gold medal)
- **Previous best (Wu's method)**: 10/30 problems solved (33%)
- **Average IMO contestant**: 15/30 (50%).

AlphaGeometry 2 (October 2024 update):

- Improved symbolic engine (100× faster)
- Better neural model (trained on 10× more synthetic data)
- Performance: 28/30 problems (93%)
- Average solve time: 19 s per problem.

Error analysis: The two unsolved problems required:

1. Complex 3D reasoning (AlphaGeometry is 2D)
2. Numerical computation alongside geometric reasoning.

Example: AlphaGeometry's Proof

For the IMO 2015 problem stated earlier, AlphaGeometry produces a proof with:

- **247 deduced statements** (angle equalities, circle properties)
- **1 auxiliary construction** (suggested by neural model)
- **19 s** wall-clock time.

The proof is formally verified; every step follows from geometric axioms. Human mathematicians confirmed the proof is correct and elegant, though uses a different construction than typical human solutions.

10.5.4 AlphaProof: Silver Medal at the International Mathematical Olympiad

10.5.4.1 Beyond Geometry: General Mathematical Reasoning

While AlphaGeometry mastered geometry, the full IMO includes algebra, number theory, and combinatorics. In July 2024, DeepMind unveiled **AlphaProof**; a system achieving silver-medal performance (28/42 points) at IMO 2025, solving problems across multiple mathematical domains.

IMO 2024 Results:

- **Problem 1** (Number theory): Solved (7/7 points)
- **Problem 2** (Combinatorics): Solved (7/7 points)
- **Problem 3** (Geometry): Not solved (0/7 points)
- **Problem 4** (Hardest): Solved via AlphaProof (7/7 points)
- **Problem 5** (Functional equation): Not solved (0/7 points)
- **Problem 6** (Geometry): Solved via AlphaGeometry 2 (7/7 points).

Total: 28 points = Silver medal (Gold threshold: 29 points).

Remarkably, AlphaProof solved **Problem 4**; the hardest problem of the competition, which only 6 human contestants solved.

10.5.4.2 Formal Verification: The Lean Proof Assistant

Unlike AlphaGeometry (which uses custom symbolic reasoning), AlphaProof targets the **Lean** proof assistant; a formal verification system used in mathematics research.

What is Lean?

- Interactive theorem prover based on dependent type theory
- Proofs are *machine-checkable* programs
- Growing library of formalized mathematics (mathlib)
- Used to verify major results (Liquid Tensor Experiment, formalization of sphere eversion).

Every step is checked by Lean's kernel; there can be no logical errors. But writing Lean proofs is difficult: it requires translating mathematical intuition into precise formal syntax.

10.5.4.3 The AlphaProof Architecture

AlphaProof combines three components:

1. **Language Model for Lean**
    - Gemini-based transformer fine-tuned on Lean code
    - Trained to translate natural language problems into formal Lean statements
    - Generates candidate proof steps in Lean syntax.
2. **Reinforcement Learning for Proof Search**
    - Uses AlphaZero-style Monte Carlo Tree Search (MCTS)
    - Each node = partial proof state in Lean
    - Actions = applying tactics (proof steps)
    - Reward = +1 if proof completes, 0 otherwise.
3. **Self-Improvement Loop**
    - System attempts to prove training problems
    - Successful proofs added to training dataset
    - Model retrained on augmented dataset
    - Repeat: bootstrapping from own discoveries (Fig. 10.24).

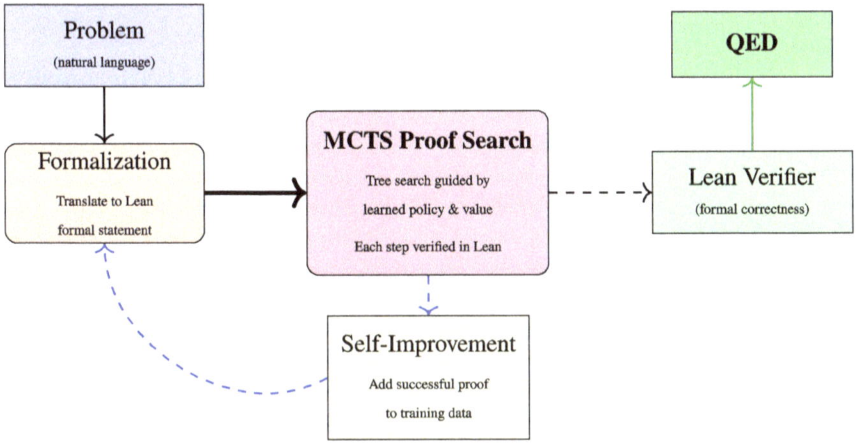

Fig. 10.24 AlphaProof architecture. Problems are formalized in Lean, then MCTS explores proof search space. Each step is verified by Lean's kernel. Successful proofs augment training data in a self-improvement loop

10.5.4.4 Training Through Self-play and Bootstrapping

AlphaProof's training mirrors AlphaGo's self-play but in the domain of mathematical proof:

Phase 1: Supervised pre-training

- Train on existing Lean proofs from mathlib
- Learn basic proof tactics and common patterns
- Develop intuition for formal mathematics syntax.

Phase 2: Reinforcement learning on synthetic problems

- Generate mathematical problems of increasing difficulty
- Use MCTS to search for proofs
- Reward = completing a valid Lean proof
- Update policy to favor successful proof strategies.

Phase 3: Self-improvement

- System attempts competition-level problems
- Each successful proof added to training corpus
- Retrain model on augmented dataset
- Model improves: can solve harder problems
- Repeat: exponential growth in capability.

Key innovation: Unlike supervised learning (limited by human-generated data), AlphaProof *creates its own training data* by solving problems and learning from successful attempts. This enables capability growth beyond human demonstrations.

IMO 2024, Problem 4 (Solved by AlphaProof)

Problem: Let $\mathbb{N}$ denote the set of positive integers. Determine all functions $f : \mathbb{N} \to \mathbb{N}$ such that for all positive integers a and b:

$$f(a^2 + b^2) = f(a)f(b) + a^2$$

This functional equation problem requires:

- Finding candidate solutions through pattern recognition
- Proving the candidates work (constructive proof)
- Proving these are the only solutions (uniqueness proof).

AlphaProof produced a complete, formally verified solution in Lean; matching the difficulty level of the hardest IMO problems.

10.5.5 *Recent Advances: DeepSeek-Prover and Mathematical Foundation Models*

The success of AlphaProof sparked intense research activity. Several systems now approach or exceed AlphaProof's capabilities.

10.5.5.1 DeepSeek-Prover-V1.5: Enhanced Monte Carlo Tree Search

DeepSeek-Prover-V1.5 (November 2024) introduces **RMaxTS**; a variant of MCTS with:

Key innovations:

1. **Value function bonus**: Prioritizes nodes with high value estimates (likely to succeed)
2. **Intrinsic reward**: Rewards novel proof states (encourages exploration)
3. **Length normalization**: Prevents preference for short, incomplete proofs.

Performance: Solves 68% of miniF2F benchmark problems (vs. 54% for AlphaProof-style MCTS).

10.5.5.2 DeepSeekMath-V2: Gold Medal on IMO 2025

DeepSeekMath-V2 (January 2025) achieved:

- **IMO 2025**: 36/42 points (Gold medal level)
- **Putnam 2024**: 118/120 points (near-perfect, historically unprecedented).

Table 10.8 Evolution of automated mathematical reasoning systems

System	Date	IMO score	Key innovation	Verification
AlphaGeometry	Jan 2024	25/30 (geom)	Synthetic proof generation	Custom symbolic
AlphaProof	Jul 2024	28/42	RL + MCTS in Lean	Lean kernel
DeepSeek-Prover	Nov 2024	;	RMaxTS (improved MCTS)	Lean/Isabelle
DeepSeekMath-V2	Jan 2025	36/42	Self-verification	Lean + informal
Human gold	;	29+/42	;	;

Performance rapidly improved from geometry-only to general mathematics at gold-medal level

Architecture improvements:

1. **Self-verification**: System checks its own reasoning steps
   - Not just final answer verification
   - Validates each reasoning step during generation
   - Rejects invalid steps, backtracks, tries alternatives.
2. **Multi-stage reasoning**:
   - Stage 1: Generate informal solution (human-like reasoning)
   - Stage 2: Formalize in Lean
   - Stage 3: Complete formal proof with MCTS
3. **Hybrid search**: Combines LLM generation with formal verification (Fig. 10.8).

10.5.6 Formal Verification for Scientific Machine Learning

The techniques from theorem proving apply directly to scientific ML.

10.5.6.1 Verifying Neural PDE Solvers

Consider a neural network claiming to solve the heat equation. Traditional validation: test on held-out data. But this doesn't *prove* correctness; it only shows empirical agreement on specific examples.

Formal verification approach:

1. **Formalize the claim** in Lean

```
theorem neural_solver_satisfies_heat_eq
    (net : NeuralNetwork) (u : Solution) :
    ...
    -- Proof using network architecture properties
    ...
```

2. **Prove properties** about network architecture:
   - If network is a Hamiltonian NN, energy is conserved
   - If using symplectic integrator, structure is preserved
   - Boundary conditions satisfied by construction.
3. **Verify implementation**: Prove code correctly implements specification

Current work: The `mathlib4` library includes basic real analysis, but verifying neural network properties remains an active research area. Groups at MIT, CMU, and DeepMind are developing formal frameworks for verified ML.

For systems with safety-critical constraints (autonomous vehicles, medical devices, nuclear reactors), we need *guarantees*, not probabilistic assurances.

This proof would be *machine-checked*; no logical errors possible. Once verified, the controller can be deployed with mathematical certainty that safety constraints hold.

10.5.7 Connection to Scientific Discovery

Mathematical reasoning systems aren't just for abstract mathematics; they enable automated scientific discovery.

10.5.7.1 Discovering Physical Laws

Combine symbolic regression (Sect. 10.2) with formal verification:

Traditional approach:

1. Use SINDy/PySR to discover equation from data
2. Validate empirically on test data
3. Hope it generalizes.

Verified approach:

1. Discover candidate equations
2. **Formalize claims** in proof assistant
3. **Prove properties**: conservation, stability, bounds
4. **Verify implementation**: code provably computes equation

Vision: Automated Scientific Discovery Pipeline

Future system for scientific discovery:

Step 1: Neural network observes experimental data, identifies patterns

Step 2: Symbolic regression proposes candidate governing equations

Step 3: Formal verification proves mathematical properties:

- Does equation conserve energy? (Noether's theorem)
- Are solutions unique and stable? (Well-posedness)
- Do predicted values satisfy physical bounds?

Step 4: System generates *publication-ready proof* explaining discovery
Step 5: Human scientist reviews formal proof, validates physics interpretation.
This would transform scientific ML from "neural networks that seem to work" to "provably correct models with mathematical guarantees."

10.5.8 Challenges and Future Directions

Despite remarkable progress, automated reasoning faces significant challenges:

10.5.8.1 Scalability and Complexity

Proof search explosion: Even with neural guidance, search spaces remain enormous. IMO problems are toy-scale compared to real mathematical research theorems.

Formalization bottleneck: Translating informal mathematics to Lean is labor-intensive. Most mathematical knowledge exists only in papers/textbooks, not formal libraries.

Domain-specific reasoning: Current systems excel at specific domains (geometry, algebra). General-purpose reasoning across all mathematics remains distant.

10.5.8.2 Interpretability and Trust

Black-box guidance: Neural components guide search, but we don't understand *why* they suggest specific constructions.

Proof complexity: Some generated proofs are very long (thousands of steps), making human verification difficult even with formal correctness.

Novel mathematics: Can systems discover genuinely new mathematical concepts, or only recombine known ideas?

Warning: Verification Is Not Understanding

Formal verification proves logical correctness but doesn't guarantee:

Physical meaningfulness: A mathematically correct model might not correspond to reality

Numerical stability: Formally correct equations might be computationally unsolvable

Completeness: Proof that some property holds doesn't mean we've identified *all* important properties

Insight: A verified proof might be correct but unenlightening; missing elegant connections.
Formal methods are powerful tools but complement rather than replace human mathematical reasoning and physical intuition.

10.5.8.3 Integration with Scientific Workflows

User interfaces: Scientists aren't Lean experts. We need tools translating between natural language hypotheses and formal specifications.

Partial verification: Most scientific code can't be fully verified. Identify critical components for formal treatment, leave others empirically validated.

Computational cost: Proof search is expensive (AlphaProof used significant compute for IMO problems). Practical deployment requires efficiency improvements.

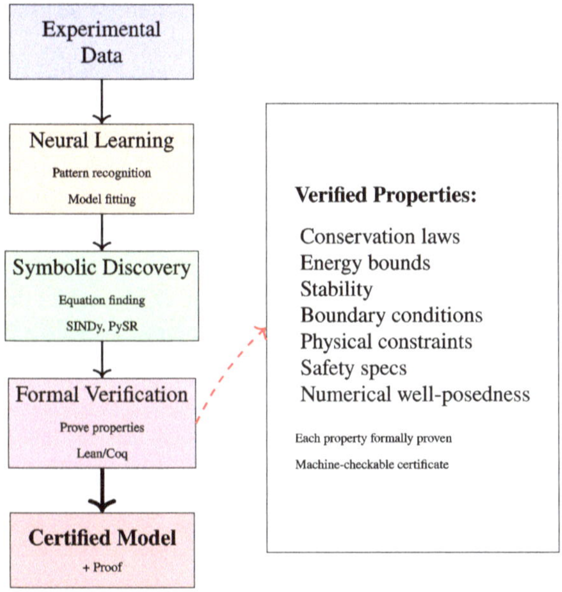

Fig. 10.25 Self-verifying scientific AI pipeline. Systems learn from data, discover equations symbolically, then formally prove properties about their predictions; moving from "it works empirically" to "it's provably correct."

10.5.9 Toward Self-verifying Scientific AI

The convergence of neuro-symbolic methods, formal verification, and automated reasoning points toward a transformative vision: **self-verifying scientific AI** that not only makes predictions but proves they're correct (Fig. 10.25).

10.5.9.1 The Self-verifying Architecture

Components:

1. **Neural perception**: Process raw data, identify patterns
2. **Symbolic reasoning**: Discover interpretable models
3. **Formal verification**: Prove mathematical properties
4. **Automated explanation**: Generate human-readable proofs
5. **Self-improvement**: Learn from successful verifications.

Example workflow:

- **Input**: Sensor data from chemical reactor
- **Neural step**: Identify temperature, pressure, concentration patterns
- **Symbolic step**: Discover reaction rate equations
- **Verification step**: Prove system is thermodynamically consistent, stable, safe
- **Output**: Certified model + formal proof of properties + natural language explanation.

10.5.9.2 Foundation Models for Scientific Reasoning

Recent language models (GPT-4, Claude, Gemini) show emerging reasoning capabilities. **Foundation models specifically trained on scientific and mathematical reasoning** could serve as general-purpose scientific AI assistants.

Vision:

- Pre-trained on scientific literature, formal proofs, experimental data
- Fine-tuned on domain-specific reasoning (physics, chemistry, biology)
- Capable of: hypothesis generation, experimental design, formal verification
- Integrated with symbolic tools: computer algebra, theorem provers, simulators.

Challenges:

- Hallucination: Models generate plausible but incorrect reasoning
- Formal grounding: Connecting informal language to formal mathematics
- Physical consistency: Ensuring predictions respect physical laws.

Mitigation: Combine LLMs with formal verification; LLM generates candidates, proof assistant verifies correctness.

10.5.10 Summary: The Neuro-symbolic Synthesis

This chapter explored neuro-symbolic approaches for scientific machine learning, revealing a coherent paradigm:

Symbolic regression (Sect. 10.2) discovers interpretable equations from data, transforming observations into mathematical laws but without guarantees about solutions.

Hard architectural constraints (Sect. 10.3) build physical properties into network structure, guaranteeing energy conservation, symmetry, and constraint satisfaction through mathematical design.

Differentiable logic programming (Sect. 10.4) bridges continuous neural learning with discrete logical reasoning, enabling systems that respect formal constraints while learning from data.

Automated mathematical reasoning (Sect. 10.5) achieves human-expert performance on mathematical problem-solving by combining neural intuition with symbolic proof, opening paths to self-verifying scientific AI (Fig. 10.26).

10.6 Conclusion: Toward Trustworthy Scientific AI

The 2024–2025 breakthroughs; AlphaGeometry, AlphaProof, DeepSeekMath; mark an inflection point. For the first time, AI systems achieve human-expert performance

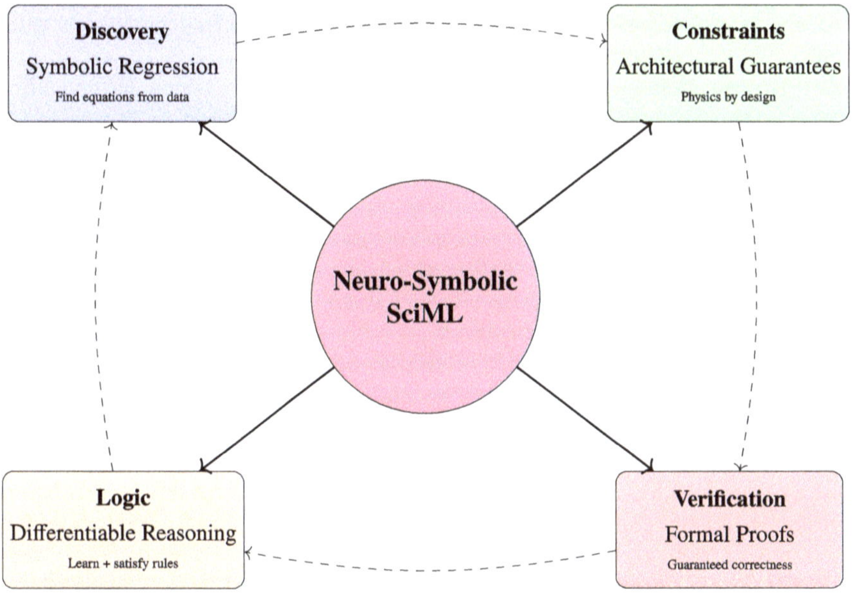

Fig. 10.26 The four pillars of neuro-symbolic scientific machine learning. Discovery finds equations, constraints guarantee properties, logic enables reasoning, and verification proves correctness. Together, they form a complete framework for trustworthy scientific AI

on tasks requiring genuine mathematical reasoning, not merely pattern matching. These aren't narrow tricks on toy problems; they represent fundamental advances in machine intelligence.

For scientific machine learning, the implications are profound:

From empirical to provable: We can move beyond "this neural network seems to work" to "this model provably satisfies physical constraints." Safety-critical applications; autonomous systems, medical devices, climate prediction; demand such guarantees.

From black boxes to explanations: Neuro-symbolic systems don't just predict; they *explain*. Formal proofs provide interpretable reasoning chains connecting observations to conclusions, essential for scientific understanding and trust.

From fixed models to adaptive discovery: Systems that discover equations (symbolic regression), guarantee properties (architectural constraints), reason logically (differentiable logic), and prove correctness (automated theorem proving) can autonomously extend scientific knowledge; not replacing human scientists but augmenting their capabilities.

The path forward requires continued research on several fronts:

- **Scalability**: Verifying complex scientific models remains computationally expensive

- **Usability**: Scientists need accessible tools, not expert-level formal methods training
- **Integration**: Combining neural, symbolic, and formal components into seamless workflows
- **Evaluation**: Developing benchmarks that test genuine scientific reasoning, not memorization.

Yet the trajectory is clear. The marriage of neural learning and symbolic reasoning; combining data-driven adaptation with mathematical rigor; offers the most promising path toward AI systems that are simultaneously powerful and trustworthy, capable and interpretable, innovative and correct.

As we develop ever more sophisticated scientific machine learning systems, the neuro-symbolic paradigm ensures they remain grounded in the fundamental principles that make science reliable: testable predictions, logical consistency, and mathematical proof. The goal isn't to replace the scientific method but to automate and enhance it, creating AI systems that don't just learn from data but reason about principles, don't just fit patterns but discover laws, and don't just make predictions but prove they're correct.

The future of scientific machine learning is neuro-symbolic; and that future is already here.

Chapter Summary: Key Takeaways

1. **The Dual Nature of Intelligence**: Human-level scientific reasoning requires both System 1 (neural pattern recognition) and System 2 (symbolic logical deduction). Pure neural or pure symbolic approaches are fundamentally limited.
2. **Four Complementary Paradigms**:
   - **Symbolic regression**: Discovers interpretable equations from data (SINDy, AI Feynman, PySR)
   - **Hard constraints**: Guarantees physical properties through architecture (HNNs, SympNets, Equivariant NNs)
   - **Differentiable logic**: Combines learning with formal reasoning (LTNs, DeepProbLog, STL)
   - **Automated theorem proving**: Achieves human-expert mathematical reasoning (AlphaGeometry, AlphaProof).
3. **Practical Applications**: These aren't abstract research topics; neuro-symbolic methods enable:
   - Verified neural PDE solvers with mathematical guarantees
   - Controllers certified to satisfy safety specifications
   - Interpretable models explaining their predictions
   - Automated scientific discovery with formal validation.

4. **The Path Forward**: Self-verifying scientific AI that learns from data, discovers principles, reasons logically, and proves correctness; transforming "it works empirically" into "it's provably correct."

References

1. Bhuyan, B.P., Ramdane-Cherif, A., Tomar, R., Singh, T.: Neuro-symbolic artificial intelligence: a survey. Neural Comput. Appl. **36**(21), 12809–12844 (2024)
2. Kahneman, D.: Fast and Slow Thinking. Allen Lane and Penguin Books, New York (2011)
3. Bhuyan, B.P., Ramdane-Cherif, A., Singh, T.P., Tomar, R.: Rule-based reasoning in neural networks. In: Neuro-symbolic Artificial Intelligence: Bridging Logic and Learning, pp. 257–270. Springer (2024)
4. Bhuyan, B.P., Ramdane-Cherif, A., Singh, T.P., Tomar, R.: Learning from symbolic knowledge for neural networks. In: Neuro-symbolic Artificial Intelligence: Bridging Logic and Learning, pp. 193–216. Springer (2024)
5. Brunton, S., Proctor, J., Kutz, N.: Sparse identification of nonlinear dynamics (SINDy). In: APS Division of Fluid Dynamics Meeting Abstracts, p. L8-005 (2016)
6. Kaheman, K., Kutz, J.N., Brunton, S.L.: SINDy-PI: a robust algorithm for parallel implicit sparse identification of nonlinear dynamics. Proc. Royal Soc. A **476**(2242), 20200279 (2020)
7. Fasel, U., Kutz, J.N., Brunton, B.W., Brunton, S.L.: Ensemble-SINDy: robust sparse model discovery in the low-data, high-noise limit, with active learning and control. Proc. Royal Soc. A **478**(2260), 20210904 (2022)
8. Zolman, N., Lagemann, C., Fasel, U., Kutz, J.N., Brunton, S.L.: SINDy-RL for interpretable and efficient model-based reinforcement learning. Nat. Commun. **16**(1), 10714 (2025)
9. Anvari, M., Marasi, H., Kheiri, H.: Implicit Runge-Kutta based sparse identification of governing equations in biologically motivated systems. Sci. Rep. **15**(1), 32286 (2025)
10. Messenger, D.A., Bortz, D.M.: Weak SINDy for partial differential equations. J. Comput. Phys. **443**, 110525 (2021)
11. Rudy, S.H., Brunton, S.L., Proctor, J.L., Kutz, J.N.: Data-driven discovery of partial differential equations. Sci. Adv. **3**(4), e1602614 (2017)
12. Mangan, N.M., Askham, T., Brunton, S.L., Kutz, J.N., Proctor, J.L.: Model selection for hybrid dynamical systems via sparse regression. Proc. Royal Soc. A **475**(2223), 20180534 (2019)
13. Udrescu, S.M., Tegmark, M.: AI Feynman: a physics-inspired method for symbolic regression. Sci. Adv. **6**(16), eaay2631 (2020)
14. Greydanus, S., Dzamba, M., Yosinski, J.: Hamiltonian neural networks. In: Advances in Neural Information Processing Systems, p. 32 (2019)
15. Cranmer, M., Greydanus, S., Hoyer, S., Battaglia, P., Spergel, D., Ho, S.: Lagrangian neural networks. arXiv preprint. arXiv:2003.04630 (2020)
16. Persio, D., Ehrhardt, L., Outaleb, M., Rizzotto, Y.S.: Port-Hamiltonian neural networks: from theory to simulation of interconnected stochastic systems. arXiv preprint. arXiv:2509.06674 (2025)
17. Hernández, Q., Badías, A., Chinesta, F., Cueto, E.: Thermodynamics-informed graph neural networks. IEEE Trans. Artif. Intell. **5**(3), 967–976 (2022)
18. Montefusco, A., Consonni, F., Beretta, G.P.: Essential equivalence of the general equation for the nonequilibrium reversible-irreversible coupling (generic) and steepest-entropy-ascent models of dissipation for nonequilibrium thermodynamics. Phys. Rev. E **91**(4), 042138 (2015)
19. Badreddine, S., Garcez, A.D., Serafini, L., Spranger, M.: Logic tensor networks. Artif. Intell. **303**, 103649 (2022)

Part III
Applications and Future Directions

Chapter 11
SciML Applications in Physical Sciences

Space is big. You just won't believe how vastly, hugely, mind-bogglingly big it is.

Douglas Adams, The Hitchhiker's Guide to the Galaxy

Abstract Scientific Machine Learning has achieved operational impact across the physical sciences, from weather prediction to materials discovery. This chapter provides comprehensive coverage of applications organized by problem type. Forward problems (learning to simulate) examines turbulence closure and flow prediction, machine learning interatomic potentials for molecular dynamics, and weather/climate modeling where neural approaches now match or exceed physics-based forecasting. Inverse problems (measurements to models) covers seismic imaging, materials characterization, and flow field reconstruction from sparse observations. Control and optimization addresses plasma confinement for fusion energy, molecular design through autonomous laboratories, and active flow control for drag reduction. Discovery (learning physical laws) explores equation discovery in fluid mechanics, high-throughput materials screening, and neural quantum states for many-body physics. Cross-cutting challenges examine generalization, long-time stability, uncertainty quantification, multi-scale integration, experimental validation, and reproducibility. Throughout, we emphasize benchmark comparisons with quantitative performance metrics.

11.1 Introduction

Consider two computational scientists working on the same problem: predicting how a new aircraft wing design will perform in turbulent flow. Let us visit their workflows a decade apart.

B. P. Bhuyan et al., *Scientific Machine Learning*,
https://doi.org/10.1007/978-981-95-7856-6_11

Lab A, 2016:

Dr. Chen arrives at the national supercomputing center at 6 AM to check on her simulation. She submitted a direct numerical simulation (DNS) of flow over the wing three weeks ago. The job consumed 2 million CPU-hours and filled 50 terabytes of storage. Overnight, the simulation crashed at timestep 847,000 due to a numerical instability near the wing tip. She sighs, adjusts the mesh refinement, and resubmits. The queue estimate: 12 days. Her Ph.D. student jokes that they should have studied analytical problems instead.

Lab B, 2026:

Dr. Okonkwo opens her laptop at a coffee shop. She loads a pretrained neural operator fine-tuned on her company's proprietary wing database. Within 30 s, the model predicts the pressure distribution and integrated forces for her new design. The uncertainty estimate flags a region near the wingtip where the model is less confident; this geometry pushes beyond the training distribution. She notes this for later validation with a targeted high-fidelity simulation, but for now, she has enough information to iterate on the design. By lunch, she has evaluated 200 candidate geometries. Her student is working on a reinforcement learning agent that will propose even better designs autonomously.

This contrast (from weeks to seconds, from single evaluations to hundreds, from passive waiting to active exploration) captures the transformation that scientific machine learning has brought to physical sciences (Fig. 11.1).

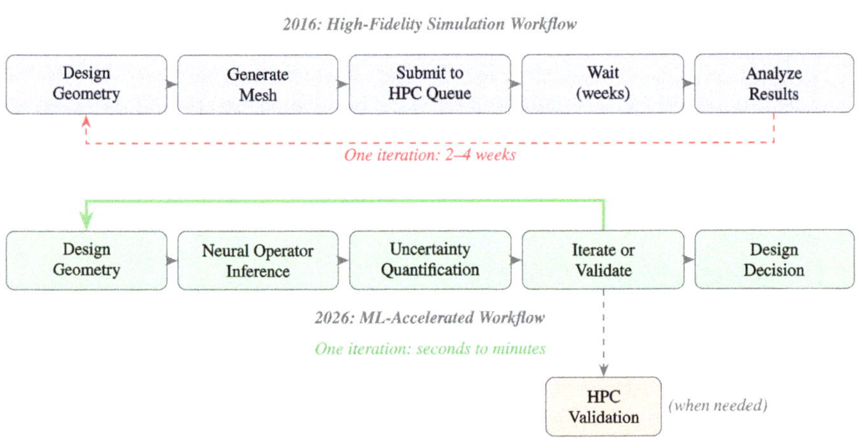

Fig. 11.1 The transformation of computational physical sciences. Top: Traditional workflow where each design iteration requires weeks of high-performance computing. Bottom: ML-accelerated workflow where neural operator inference enables rapid iteration, with targeted high-fidelity validation only when uncertainty estimates indicate model limitations

11.1.1 A Taxonomy of Problems

Before surveying specific applications, we establish a taxonomy that organizes the landscape. Scientific machine learning in physical sciences addresses four fundamental problem types, each with distinct characteristics and solution patterns.

Forward problems (surrogates)

Given initial conditions, boundary conditions, and system parameters, predict the system's evolution or steady state. Traditional numerical methods solve the governing equations; neural surrogates learn to approximate this mapping from examples. The value proposition: orders-of-magnitude speedup enabling rapid exploration of parameter spaces, real-time prediction, and integration into optimization loops.

Example: Given an aircraft geometry and flight conditions, predict the pressure and velocity fields around the aircraft.

Inverse problems

Given observations (often sparse and noisy), infer hidden parameters, fields, or initial conditions. These problems are typically ill-posed: many parameter configurations can produce similar observations. Physics-informed approaches provide regularization through governing equations, constraining solutions to physically plausible regions.

Example: Given seismic recordings at the surface, reconstruct the subsurface velocity structure.

Control and optimization

Close the loop between prediction and action. Given a system to control and an objective to optimize, learn policies that achieve desired outcomes. This may involve learning surrogates for use in model-predictive control, directly learning control policies through reinforcement learning, or optimizing designs through gradient-based methods with differentiable simulators.

Example: Control the magnetic coils in a fusion reactor to maintain a stable plasma configuration.

Discovery

Discover new physical laws, governing equations, or interpretable relationships from data. Unlike black-box prediction, the goal is *understanding*; extracting symbolic

Table 11.1 Representative speedups achieved by SciML surrogates across physical science domains

Domain	Task	Traditional	Speedup	Accuracy
Weather forecasting	10-day global forecast	Hours (IFS)	1000×	Comparable skill
Turbulent flow	Navier-Stokes at $Re = 10^4$	Hours (DNS)	$10^5\times$	$\sim 8\%$ error
Molecular forces	DFT energy/forces	Hours	$10^6\times$	< 1 kcal/mol
Plasma transport	Gyrokinetic coefficients	Hours	$10^6\times$	$\sim 5\%$ error
Seismic propagation	Wave equation solve	Minutes	$10^3\times$	$\sim 5\%$ error
Quantum ground state	Variational Monte Carlo	Days	10^2–$10^3\times$	Near-exact

Speedups are measured relative to traditional high-fidelity numerical methods for equivalent tasks. Accuracy is measured against held-out test data from the same numerical methods

expressions, conservation laws, or constitutive relations that generalize beyond the training data and provide scientific insight.

Example: Discover the minimal set of equations describing turbulent shear flow from velocity field measurements.

The practical impact of SciML is best understood through concrete speedup numbers (Table 11.1). These vary by orders of magnitude depending on the problem, but consistent patterns emerge.

11.1.2 Operational Milestones: SciML in Production

The transition from research demonstration to operational deployment marks a critical maturity threshold (Fig. 11.2). By late 2025, several SciML systems have achieved this milestone, fundamentally changing how science is conducted in their respective domains.

The European Centre for Medium-Range Weather Forecasts (ECMWF); the world's leading weather prediction center; deployed its Artificial Intelligence Integrated Forecasting System (AIFS) as part of its official operational suite [1]. This represents a watershed moment: for the first time, AI-generated weather forecasts are used for real decisions affecting billions of people.

Key facts:

- Runs alongside traditional physics-based IFS
- Consumes $\sim 1000\times$ less energy per forecast
- Achieves 20% improvement in tropical cyclone track prediction
- Generates 10-day global forecasts in under 1 min

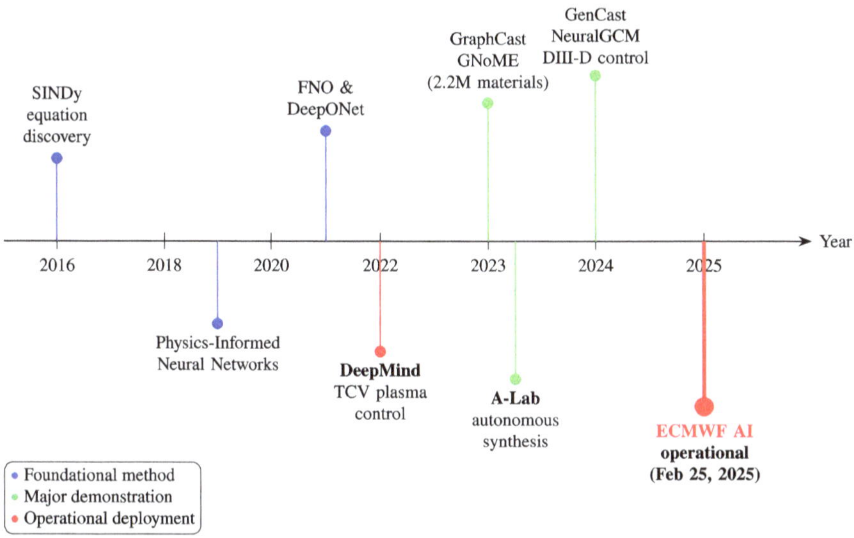

Fig. 11.2 Timeline of SciML milestones in physical sciences. Blue markers indicate foundational methodological advances; green markers indicate major research demonstrations; red markers indicate operational deployments where ML systems are used for real-world decisions

Significance: ECMWF's decision to operationalize AI forecasts signals that SciML has passed the threshold from "interesting research" to "trusted for consequential decisions."

11.2 Forward Problems: Learning to Simulate

Forward problems; predicting system evolution given initial and boundary conditions; represent the most mature application of scientific machine learning. The value proposition is straightforward: replace expensive numerical solvers with fast neural surrogates trained on their outputs. This section surveys how this pattern manifests across physical science domains, with particular attention to turbulent flows, materials simulation, and weather prediction where operational deployment has been achieved.

11.2.1 The Universal Surrogate Pattern

Despite domain differences (Fig. 11.3), successful surrogate modeling follows a consistent pipeline. Understanding this pattern; and where domain-specific adaptations are required; provides a template for new applications.

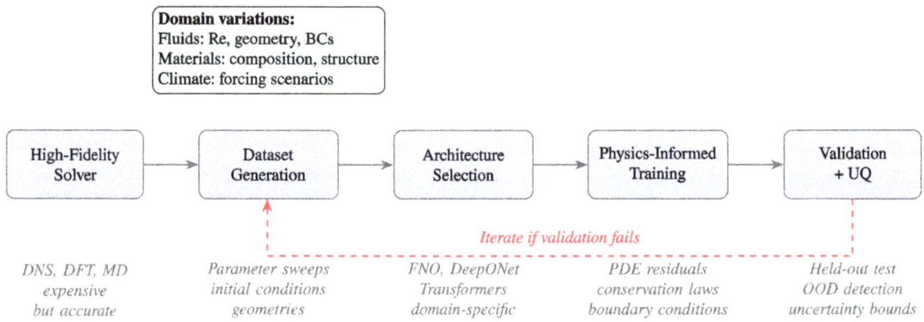

Fig. 11.3 The universal surrogate modeling pipeline. All forward surrogate applications follow this pattern, with domain-specific choices at each stage. The feedback loop indicates iteration when validation reveals inadequate performance

Five Questions Before Building a Surrogate

Before investing in surrogate development, answer these questions to guide architectural and data choices:

1. **What varies?** Parameters (viscosity, temperature), geometry (shapes, meshes), initial/boundary conditions? This determines the input space the surrogate must cover.
2. **What is the output?** Full spatiotemporal fields, integrated quantities (drag, energy), time series? Field prediction requires neural operators; scalar outputs may need only MLPs.
3. **How much training data is available?** Thousands of simulations favor data-hungry architectures (transformers); hundreds favor physics-informed or pretrained models.
4. **What accuracy is required?** 1% error demands careful architecture and extensive data; 10% error is achievable with simpler approaches.
5. **What is the inference budget?** Real-time control requires millisecond inference; design optimization tolerates seconds.

The choice of surrogate architecture depends on problem structure. Table 11.2 summarizes guidance that has emerged from extensive benchmarking across domains.

11.2.2 Turbulence Closure and Flow Prediction

Turbulent fluid flow exemplifies both the promise and challenge of SciML surrogates. The governing Navier-Stokes equations are known exactly, but their direct numerical simulation at high Reynolds numbers remains computationally prohibitive for most engineering applications. Machine learning enters at two levels: as *closure models*

Table 11.2 Architecture selection guidance for forward surrogates

Problem structure	Recommended	Alternative	Key consideration
Regular grid, periodic BCs	FNO [2]	U-Net [3]	FNO is resolution-invariant
Irregular mesh, fixed topology	MeshGraphNet [4]	Graph U-Net [5]	Message passing respects connectivity
Variable geometry	DeepONet [6]	Geometry-conditioned FNO	DeepONet handles varying domains
Long time rollout	Temporal bundling + FNO	Autoregressive transformer	Bundling reduces error accumulation
Multi-scale physics	Factorized FNO	U-FNO	Factorization captures scale separation
Limited training data	Physics-informed hybrid	Pretrained + fine-tune	Physics constraints regularize

Recommendations based on problem characteristics; "resolution-invariant" means the trained model can evaluate at different discretizations than training

that augment traditional solvers, and as *end-to-end surrogates* that bypass numerical solvers entirely.

11.2.2.1 The Turbulence Closure Problem

Industrial computational fluid dynamics relies on Reynolds-Averaged Navier-Stokes (RANS) equations, which decompose velocity into mean and fluctuating components. This decomposition introduces the Reynolds stress tensor $\overline{u_i'u_j'}$ (Fig. 11.4); a quantity that cannot be computed from mean quantities alone. Traditional turbulence models (k-ε, k-ω SST, Spalart-Allmaras) provide algebraic or transport-equation closures for this tensor, but their accuracy degrades in complex flows involving separation, strong pressure gradients, or streamline curvature.

Machine learning offers a different path: learn the closure directly from high-fidelity data.

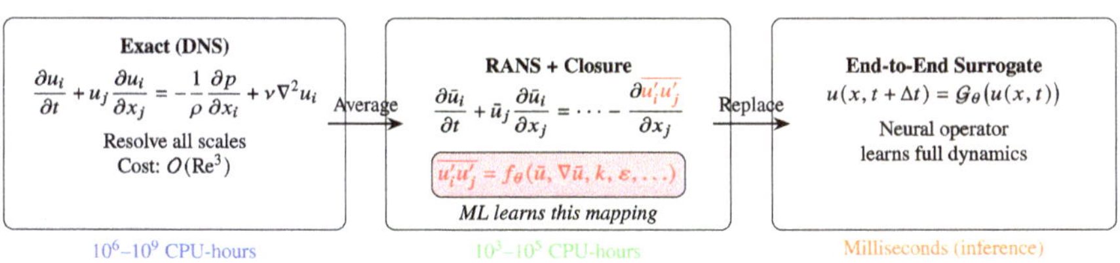

Fig. 11.4 Three approaches to turbulent flow prediction. Left: Direct numerical simulation resolves all scales but is prohibitively expensive. Middle: RANS equations with ML-learned closure models offer a hybrid approach. Right: End-to-end neural surrogates bypass numerical solvers entirely

Table 11.3 Evolution of ML turbulence closure models

Method	Key innovation	Error reduction	Test cases
Ling et al. [7]	Deep neural network for anisotropy tensor	40% versus k-ε	Duct flow, channel flow
Parish and Duraisamy [8]	Field inversion + ML correction	50–60%	Airfoil separation
Wu et al. [9]	Tensor basis neural network (Galilean invariance)	30–50%	Periodic hills, curved duct
Duraisamy et al. [10]	Embedded ML in CFD solver	40%	Transonic flows
Sirignano et al. [11]	DPM: differentiable physics + ML	35%	Backward-facing step
MacArt et al. [12]	Embedded training in LES	25%	Isotropic turbulence
Lozano-Durán [13]	Transformer-based closure	45%	Channel at $Re_\tau = 550$

Error reduction is measured against baseline RANS models (typically k-ε or k-ω SST) on held-out test cases

11.2.2.2 Data-Driven Turbulence Closure Models

Several works (Table 11.3) demonstrated that neural networks could learn Reynolds stress anisotropy from DNS data, outperforming classical models on flows with separation. Subsequent developments have refined this approach.

Frame invariance matters

A critical lesson from this literature: closure models must respect physical symmetries. The Reynolds stress tensor transforms as a rank-2 tensor under rotations; models that ignore this produce predictions that change with coordinate orientation. The *tensor basis neural network* (TBNN) approach addresses this by expressing the anisotropy tensor as a linear combination of basis tensors formed from the mean strain and rotation rates:

$$b_{ij} = \sum_{n=1}^{10} g^{(n)}(\lambda_1, \ldots, \lambda_5) T_{ij}^{(n)},$$

where $g^{(n)}$ are learned scalar functions of invariants λ_k, and $T_{ij}^{(n)}$ are the integrity basis tensors. This guarantees Galilean invariance by construction.

Generalization Remains the Central Challenge
ML closure models trained on one flow configuration often fail to generalize to substantially different configurations:

Table 11.4 Neural operator performance on Navier-Stokes benchmarks

Method	Benchmark	Re	Error (%)	Speedup	Resolution
FNO	2D NS, viscous	1000	0.83	$10^3\times$	64^2
FNO	2D NS, viscous	10,000	8.3	$10^3\times$	64^2
F-FNO	2D NS, viscous	10,000	1.4	$10^3\times$	128^2
U-FNO	2D NS + multiphase	5000	2.1	$500\times$	128^2
GNOT	Darcy + NS	Mixed	1.8	$10^2\times$	Variable
Transolver	3D NS	10,000	3.2	$10^4\times$	Mesh
Transolver++	Industrial CFD	Variable	4.5	$10^4\times$	10^6 points

Relative L^2 error on held-out test data; speedup relative to pseudo-spectral solver on equivalent hardware

- Models trained on channel flow may fail on separated flows
- Models trained at one Reynolds number may degrade at higher Re
- Models trained on 2D may not transfer to 3D

Mitigation strategies:

- Train on diverse flow configurations simultaneously
- Build in physical constraints (realizability, invariance)
- Use transfer learning from related flows
- Detect out-of-distribution inputs and fall back to classical models

11.2.2.3 Neural Operator Surrogates for Navier-Stokes

End-to-end surrogates bypass the closure problem entirely by learning the full input-output mapping of Navier-Stokes solutions. The Fourier Neural Operator (FNO) and its variants (Table 11.4) have achieved remarkable results on turbulence benchmarks (Table 11.5).

Temporal rollout and stability

For time-dependent problems, neural operators are typically applied autoregressively: predict one (or several) timesteps ahead, then use the prediction as input for subsequent steps. This introduces a critical challenge: small per-step errors accumulate over long rollouts.

Table 11.5 Rollout stability: error after 100 timesteps

Strategy	Error at $t = 10$ (%)	Error at $t = 100$	Notes
Naive autoregressive	2.1	Blowup	Errors compound exponentially
Noise injection training	2.3	15%	Robust but higher base error
Pushforward training	1.8	8%	Trains on own predictions
Temporal bundling ($k = 4$)	1.9	6%	Fewer autoregressive steps
Hybrid correction	1.5	4%	Physics corrects drift

Comparison of strategies for long-time integration; "blowup" indicates numerical divergence

11.2.2.4 Industrial CFD: Deployment Status

While academic benchmarks demonstrate impressive results, industrial adoption requires additional considerations: complex geometries, multiphysics coupling, regulatory certification, and integration with existing workflows.

Automotive Aerodynamics at BMW and Mercedes
Major automotive manufacturers have begun integrating ML surrogates into early-stage design:

Workflow:

1. Train surrogates on $\mathcal{O}(10^4)$ high-fidelity simulations spanning the design space
2. Use surrogates for rapid exploration (100+ designs/hour vs. 1/day)
3. Validate promising candidates with traditional CFD
4. Final certification still requires physical wind tunnel testing

Reported results (BMW, 2023):

- Drag coefficient prediction within 2% of CFD
- 1000× speedup in early design phase
- 40% reduction in total design cycle time
- Surrogates *complement* rather than *replace* traditional simulation

Key insight: Industrial deployment uses ML for screening and exploration, retaining traditional methods for validation and certification.

11.2.3 Machine Learning Interatomic Potentials

Molecular dynamics (MD) simulations underpin materials science, chemistry, and drug discovery. The accuracy of these simulations depends entirely on the interatomic potential; the function mapping atomic configurations to energies and forces.

Machine learning has revolutionized this field, achieving quantum-mechanical accuracy at classical force-field speeds.

11.2.3.1 The Accuracy-Speed Trade-Off

Traditional approaches span a wide accuracy-speed spectrum:

- **Classical force fields** (Lennard-Jones, EAM, AMBER): Fast (10^6 atoms, nanoseconds) but limited accuracy and transferability
- **Semi-empirical methods** (DFTB, PM7): Moderate speed and accuracy
- **Density Functional Theory** (DFT): High accuracy but slow (10^2–10^3 atoms, picoseconds)
- **Coupled cluster** (CCSD(T)): "Gold standard" accuracy but prohibitively expensive

ML interatomic potentials (MLIPs) achieve a fundamentally better trade-off: near-DFT accuracy at near-classical-FF speed.

11.2.3.2 Architecture Evolution: From Descriptors to Equivariance

The field has evolved through three generations, each addressing limitations of the previous:

Generation 1: Hand-crafted descriptors (2007–2016)

Behler and Parrinello's seminal 2007 work introduced symmetry functions; hand-designed features that encode local atomic environments while respecting permutation, translation, and rotation invariance. The atomic energy is decomposed as:

$$E = \sum_i E_i(\mathbf{G}_i),$$

where E_i is a neural network and $\mathbf{G}_i$ are symmetry function descriptors for atom i.

Generation 2: Learned descriptors (2017–2021)

SchNet, DimeNet, and related architectures replaced hand-crafted descriptors with learned message-passing on atomic graphs. Atoms exchange information through messages that depend on distances and, for DimeNet, angles. This enables automatic feature learning but requires many message-passing layers to capture long-range interactions.

Table 11.6 MLIP architecture comparison on standard benchmarks

Architecture	Year	MD17 Force (meV/Å)	MP energy (meV/atom)	Speed (atoms/s)	Key Feature
Behler-Parrinello	2007	50–100	80	10^6	Symmetry functions
SchNet	2018	15–25	45	10^5	Continuous filters
DimeNet++	2020	8–12	35	5×10^4	Directional messages
NequIP	2022	4–8	25	3×10^4	E(3)-equivariance
MACE	2022	4–6	22	10^5	Higher-order tensors
Allegro	2023	3–5	20	8×10^4	Strictly local
EquiformerV2	2024	2–4	18	2×10^4	Equivariant transformer

Force MAE on MD17 (organic molecules) and Materials Project formation energies. Speed is atoms/second on a single GPU

Generation 3: Equivariant architectures (2021–present)

NequIP, MACE, Allegro, and EquiformerV2 build equivariance directly into the architecture. Rather than learning invariant features and discarding directional information, these models work with equivariant tensors that transform predictably under rotations. This enables much more efficient learning; MACE achieves state-of-the-art accuracy with only 2 message-passing layers versus 5–6 for earlier architectures (Table 11.6).

11.2.3.3 Universal Potentials: One Model for the Periodic Table

A transformative development of 2023–2024 is the emergence of "universal" potentials trained on diverse chemical spaces. Rather than training separate models for each material system, these models learn transferable representations across the periodic table (Table 11.7).

The PES Softening Problem
All current universal potentials exhibit systematic bias in high-energy configurations:

Observation: Energies and forces are *underpredicted* for:

- Surfaces and interfaces
- Point defects and dislocations
- Transition states and reaction barriers
- Highly compressed or strained structures

Table 11.7 Universal potential performance on Matbench Discovery

Model	Training data	F1 score	MAE (eV/atom)	Notable capability
MACE-MP-0	Materials Project (146k)	0.67	0.057	Best overall accuracy
CHGNet	MP + magnetic (1.5M)	0.65	0.063	Magnetic moments, charges
M3GNet	Materials Project	0.58	0.072	Fast relaxation
ALIGNN	JARVIS-DFT	0.52	0.085	Line graph convolutions
SevenNet	MP + OC22	0.64	0.059	Scale-aware
EquiformerV2	OC20/OC22	0.61	0.048	Catalysis focus

The benchmark evaluates stability prediction for novel materials; F1 score and energy MAE are computed on held-out test structures not seen during training

Cause: Training data concentrates near equilibrium configurations. High-energy regions are undersampled, causing the potential energy surface to be "softer" than the true DFT surface.

Consequence: Barriers are underestimated, diffusion is too fast, and phase transitions may be incorrectly predicted.

Mitigation:

- Fine-tune on domain-specific high-energy data
- Active learning to sample underrepresented regions
- Uncertainty-aware sampling during training

11.2.3.4 Impact: What MLIPs Have Enabled

The practical impact of MLIPs extends far beyond benchmark performance:

- **GNoME (Google DeepMind, 2023)**: Discovered 2.2 million stable crystal structures using MLIP-accelerated screening. Of these, 736 have been independently synthesized by external laboratories, including 528 novel lithium-ion conductors; 25× the previously known count.
- **Large-scale MD**: Simulations of millions of atoms for nanoseconds are now routine. Studies of grain boundary dynamics, crack propagation, and protein-membrane interactions that were previously impossible are now accessible.
- **Free energy calculations**: Enhanced sampling methods (metadynamics, umbrella sampling) that require millions of force evaluations are now practical with MLIP-level accuracy.
- **High-throughput screening**: Relaxation of crystal structures, which previously required hours of DFT per structure, now takes seconds with MLIPs, enabling screening of millions of candidates.

Table 11.8 Timeline of AI weather prediction milestones

Year	Model	Developer	Key achievement
2022	Pangu-Weather	Huawei	First AI model to match ECMWF HRES skill
2023	GraphCast	DeepMind	Outperformed HRES on 90% of 1380 targets
2023	FourCastNet	NVIDIA	Demonstrated FFT-based approach at scale
2024	GenCast	DeepMind	Probabilistic ensembles; beat ECMWF ENS on 97.2% of targets
2024	NeuralGCM	Google	Hybrid physics-ML; 40-year climate stability
2024	Aurora	Microsoft	Foundation model for Earth systems
2025	AIFS	ECMWF	First operational deployment (Feb 25)

Skill is measured against ECMWF's operational IFS; "comparable" means within statistical uncertainty on standard verification metrics

11.2.4 Weather and Climate Prediction

Weather prediction represents SciML's most visible success story. In February 2025, ECMWF; the world's leading weather prediction center; operationalized AI forecasts, marking the transition from research demonstrations to production systems affecting billions of people.

11.2.4.1 The AI Weather Revolution: 2022–2025

The transformation happened with remarkable speed (Table 11.8, and results in Table 11.9).

Three distinct architectural approaches have proven successful, each with different trade-offs:

Graph-based (GraphCast, AIFS)

The atmosphere is represented as a graph on an icosahedral mesh, with message-passing neural networks propagating information. This naturally handles the sphere's geometry and enables resolution flexibility.

Table 11.9 Weather model architecture comparison

Model	Paradigm	Training data	Training cost	Inference	Climate stable?
GraphCast	Graph	39 years ERA5	32 TPU-days	< 1 min	No
Pangu-Weather	Transformer	39 years ERA5	192 GPU-days	~ 1 min	No
GenCast	Diffusion	40 years ERA5	64 TPU-days	~ 8 min	No
NeuralGCM	Hybrid	40 years ERA5	128 TPU-days	~ 5 min	**Yes (40+ years)**
AIFS	Graph	ERA5 + obs	Proprietary	< 1 min	In testing

Training and inference costs are approximate; skill is relative to ECMWF HRES

Strengths: Efficient representation, good scalability, interpretable information flow.

Weaknesses: No explicit physics, potential for non-physical states.

Transformer-based (Pangu-Weather, FourCastNet)

The atmosphere is treated as a sequence of patches (like vision transformers), with attention mechanisms learning spatial and temporal dependencies.

Strengths: Captures long-range dependencies, benefits from vision transformer advances.

Weaknesses: Quadratic memory scaling, requires careful patch design for spherical data.

Hybrid physics-ML (NeuralGCM)

A differentiable dynamical core solves the large-scale fluid equations explicitly, while neural networks learn subgrid physics (clouds, convection, radiation).

Strengths: Guaranteed conservation, stable for multi-decade climate runs, interpretable physics.

Weaknesses: More complex implementation, constrained by dynamical core limitations.

11.2.4.2 Probabilistic Forecasting: GenCast

Deterministic forecasts provide a single prediction, but operational forecasting requires uncertainty quantification. GenCast introduced diffusion models to weather prediction, generating ensemble forecasts that capture forecast uncertainty.

Table 11.10 Climate simulation comparison

Model type	10-Year stability	Temperature bias	Computational cost
GraphCast (AR)	Unstable after months	N/A	1×
Traditional GCM	Stable	±2–3 K bias	100,000×
NeuralGCM	**Stable 40+ years**	±0.5 K bias	1000×

Bias is measured as global mean temperature drift after 10 years of free-running simulation

GenCast: Probabilistic Weather Prediction (Nature, 2024)

Architecture: Diffusion model adapted for spherical data, trained to denoise perturbations of atmospheric states.

Results:

- Outperformed ECMWF ENS (50-member physics ensemble) on 97.2% of targets
- Showed 99.8% improvement at lead times beyond 36 h
- Better calibrated probabilistic forecasts (reliability diagrams)
- Superior tropical cyclone track uncertainty quantification

Significance: First demonstration that AI can match the *uncertainty* of physics-based ensembles, not just the *mean* prediction. Ensemble generation is 1000× cheaper than running 50 physics simulations.

Limitation: Each ensemble member requires a separate diffusion sampling run ($\sim$ 8 min), so generating 50 members still takes hours rather than minutes.

11.2.4.3 Climate Stability: NeuralGCM

Pure ML weather models, when run for extended periods, accumulate biases and drift away from realistic climate statistics. NeuralGCM addresses this through hybrid architecture: a differentiable dynamical core handles resolved fluid dynamics while neural networks parameterize subgrid processes.

The key insight: by keeping explicit physics for conservation laws and large-scale dynamics, NeuralGCM inherits the stability properties of traditional GCMs while gaining the efficiency and accuracy of ML for subgrid processes (Table 11.10).

11.2.5 Other Forward Problem Domains

The three deep dives above; turbulence, materials, weather; represent the most mature applications of forward surrogates. Here we briefly survey other domains where significant progress has been made.

11.2.5.1 Plasma Physics: Gyrokinetic Surrogates

Fusion reactor design requires predicting plasma transport; the movement of heat and particles across magnetic field lines. First-principles gyrokinetic simulations are extremely expensive, limiting their use in design optimization.

QLKNN (QuasiLinear gyroKinetic Neural Network):

- Trained on > 10 million gyrokinetic simulations
- Predicts turbulent transport coefficients in milliseconds versus hours
- Integrated into TORAX differentiable tokamak simulator
- Enables real-time transport modeling for control applications
- Accuracy: $\sim 10\%$ relative error on heat flux predictions

MHD surrogates:

- Neural operators for magnetohydrodynamic evolution
- Speedups of 10^5–10^6 over direct MHD solvers
- Application: Disruption prediction, scenario optimization

11.2.5.2 Quantum Many-Body: Neural Quantum States

Simulating quantum many-body systems faces the exponential curse of dimensionality: the wavefunction of N particles lives in a space of dimension d^N where d is the local Hilbert space dimension. Neural quantum states (NQS) address this by representing the wavefunction as a neural network.

The 2024 MinSR (minimum stochastic reconfiguration) result on the Heisenberg model is particularly notable: it achieved near machine-precision accuracy on a 100-spin system, exceeding all previous variational methods including tensor network approaches (Table 11.11).

Table 11.11 Neural quantum state results on benchmark systems

System	Method	Energy/site	Comparison
2D Heisenberg (10×10)	MinSR + RBM	$-0.67155260(3)$	Exceeds DMRG
J_1-J_2 (frustrated)	Transformer NQS	$-0.4952(1)$	Matches DMRG
Fermi-Hubbard	FermiNet	Within 0.1%	Matches AFQMC
Molecules (H_2O)	PauliNet	Chemical accuracy	Matches CCSD(T)

Ground state energy per site compared to best previous methods

11.2.5.3 Seismic Wave Propagation

Seismic simulation; solving the elastic wave equation in complex geological media; is essential for earthquake hazard assessment, reservoir monitoring, and seismic inversion. Neural operators have achieved significant speedups:

- **Accuracy**: 5–10% relative error on waveform amplitude and phase
- **Speedup**: $10^3\times$ over finite-difference methods
- **Application**: Real-time earthquake early warning systems
- **Challenge**: Complex velocity structures with sharp discontinuities remain difficult

11.3 Inverse Problems: From Measurements to Models

Inverse problems ask the opposite question from forward problems: given observations, what are the hidden parameters, fields, or initial conditions that produced them? This inversion is fundamental to science; we cannot observe the Earth's interior directly, but we can infer its structure from seismic waves; we cannot see atomic arrangements, but we can reconstruct them from diffraction patterns; we cannot measure full flow fields, but we can estimate them from sparse sensors (Fig. 11.5).

Inverse problems are inherently more challenging than forward problems. They are typically *ill-posed*: multiple parameter configurations can produce identical observations, small measurement errors can cause large solution errors, and solutions may not exist or may not depend continuously on the data. Scientific machine learning addresses these challenges by using physics as regularization and learned priors to constrain solutions to physically plausible regions.

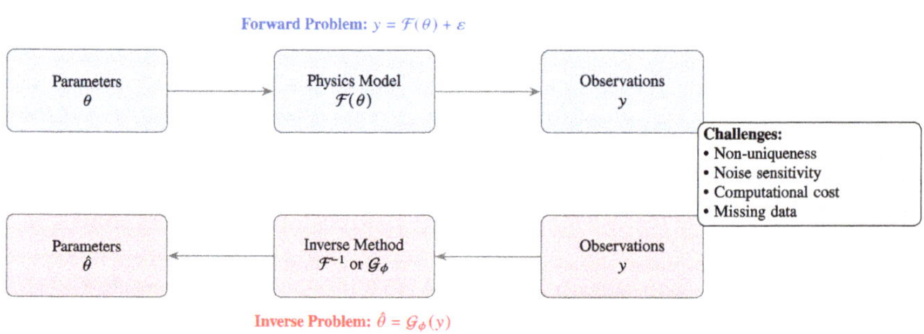

Fig. 11.5 Forward versus inverse problems. Forward problems map known parameters through physics to observations. Inverse problems attempt the reverse: inferring parameters from observations. The inverse mapping is typically ill-posed, requiring regularization or learned priors

11.3.1 The Inverse Problem Landscape

The Three Challenges of Inverse Problems

Hadamard's conditions for a *well-posed* problem require that a solution: (1) exists, (2) is unique, and (3) depends continuously on the data. Inverse problems in physical sciences typically violate one or more:

1. **Non-existence**: Noisy observations may be inconsistent with any physical model.
2. **Non-uniqueness**: Multiple parameter configurations produce identical observations. In seismic imaging, different velocity structures can generate the same surface recordings. In flow reconstruction, sparse sensors cannot uniquely determine the full field.
3. **Instability**: Small perturbations in observations cause large changes in the reconstructed parameters. High-frequency components of the solution are particularly sensitive to noise.

Regularization addresses ill-posedness by adding prior information: smoothness assumptions (Tikhonov), sparsity (compressed sensing), or learned priors (neural networks). Physics-informed approaches use governing equations as a powerful form of regularization.

Scientific machine learning offers three distinct strategies for inverse problems (Fig. 11.6), each with different trade-offs, shown in Table 11.12.

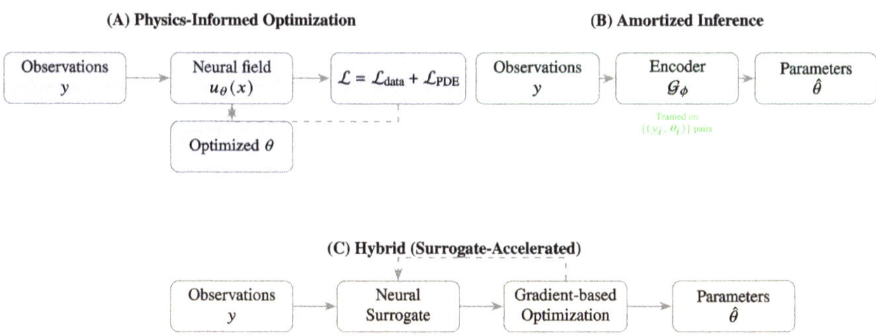

Fig. 11.6 Three SciML strategies for inverse problems. **a** Physics-informed optimization solves each instance from scratch using PDE constraints. **b** Amortized inference trains a network to directly map observations to parameters. **c** Hybrid approaches use neural surrogates to accelerate iterative optimization

Table 11.12 SciML strategies for inverse problems

Strategy	Approach	Inference cost	Best for
Physics-informed optimization	Embed PDE in loss; optimize per-instance	High (minutes–hours)	High accuracy, novel geometries
Amortized inference	Train network to directly predict θ from y	Low (milliseconds)	Repeated similar inversions
Hybrid	Neural surrogate accelerates optimization loop	Medium (seconds)	Balance of accuracy and speed

Each approach offers different trade-offs between generality, computational cost, and accuracy

11.3.2 Seismic Imaging and Full Waveform Inversion

Seismic imaging; reconstructing Earth's subsurface structure from surface recordings; is one of the most economically significant inverse problems (Fig. 11.7). Oil and gas exploration, carbon sequestration monitoring, earthquake hazard assessment, and geothermal energy development all depend on accurate subsurface characterization.

11.3.2.1 The Full Waveform Inversion Problem

Full waveform inversion (FWI) seeks the velocity model $c(x)$ that minimizes the misfit between observed and simulated seismograms:

$$\hat{c} = \arg\min_{c} \sum_{s,r} \|d_{s,r}^{\text{obs}}(t) - d_{s,r}^{\text{sim}}(t; c)\|^2 + \mathcal{R}(c),$$

where $d_{s,r}(t)$ is the seismogram at receiver r from source s, and $\mathcal{R}(c)$ is a regularization term.

The challenge: each forward simulation (computing d^{sim} for a given c) requires solving the wave equation, and optimization requires hundreds to thousands of iterations. Traditional FWI campaigns can consume millions of CPU-hours.

11.3.2.2 Physics-Informed Neural Network Approaches

Physics-informed neural networks (PINNs) offer an alternative to traditional FWI by representing the wavefield and velocity model as neural networks trained to satisfy the wave equation (Table 11.13).

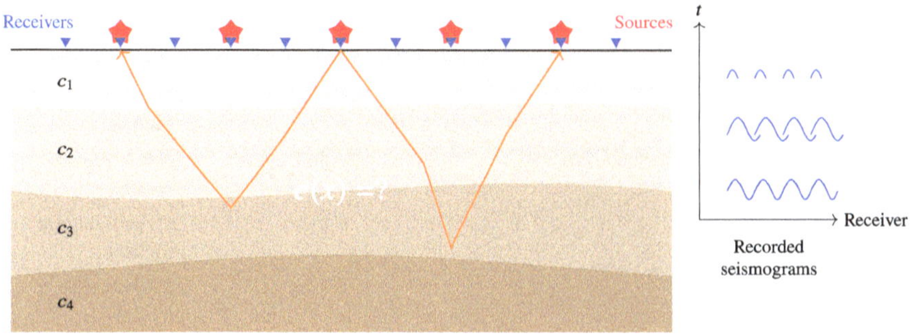

Fig. 11.7 Seismic acquisition geometry and the FWI inverse problem. Sources emit waves that reflect and refract through subsurface layers with unknown velocities $c(x)$. Receivers record seismograms. The inverse problem: reconstruct the velocity model from recorded data

Table 11.13 PINN-based seismic inversion results

Method	Test case	Velocity error (%)	Speedup	Key innovation
Song et al. [14]	Marmousi-2	8.5	0.5×	Baseline PINN for FWI
Rasht-Behesht et al. [15]	Layered media	5.2	2×	Multi-scale training
Waheed et al. [16]	Overthrust	6.8	3×	Eikonal-informed
Zhang et al. [17]	Marmousi	4.1	5×	Causality-respecting
Huang et al. [18]	3D SEG-EAGE	5.5	10×	Domain decomposition
Jin et al. [19]	Field data (Gulf)	7.2	8×	Transfer learning

Velocity error is RMS error relative to ground truth; speedup is relative to adjoint-state FWI on equivalent hardware

Advantages of PINN-FWI:

- **Meshless**: No computational grid required; handles complex geometries naturally
- **Absorbing boundaries**: Satisfied automatically without artificial layers
- **Joint inversion**: Velocity and other parameters (attenuation, anisotropy) can be learned simultaneously
- **Sparse data**: Physics constraints provide regularization when receivers are limited

PINN Limitations for Seismic Inversion

Despite promising results, PINN-based FWI faces significant challenges:

- **Spectral bias**: Neural networks struggle to represent high-frequency components of wavefields, limiting resolution

Table 11.14 Neural operator methods for seismic inversion

Method	Architecture	Velocity error (%)	Inference	Training data
InversionNet	Encoder-decoder	12	50 ms	50k synthetic models
VelocityGAN	Conditional GAN	10	80 ms	30k synthetic models
FNO-Inversion	FNO	7.5	30 ms	100k synthetic models
Seismic transformer	Vision transformer	6.2	100 ms	200k synthetic models
DiffusionFWI	Diffusion model	5.8	2 s	150k synthetic models

Inference time is for a single velocity model prediction

- **Local minima**: The loss landscape is highly non-convex; results depend strongly on initialization
- **Scale**: Current results are limited to 2D and small 3D problems; industrial-scale 3D remains challenging
- **Cycle-skipping**: Like traditional FWI, PINNs suffer when initial model is far from true solution

Current status: PINNs are most effective for problems where traditional FWI struggles; sparse data, complex boundaries, multi-parameter inversion; but do not yet replace adjoint-state methods for large-scale industrial applications.

11.3.2.3 Neural Operator Approaches

An alternative to per-instance optimization is training neural operators to directly map seismic data to velocity models; amortized inference that is fast at test time (Table 11.14).

The distribution shift challenge

Amortized methods train on synthetic data but must generalize to real field data. The "sim-to-real" gap is a major obstacle:

- Synthetic training data may not capture the complexity of real geology
- Noise characteristics differ between synthetic and field data
- Source signatures and acquisition geometries vary

Mitigation strategies: Domain adaptation, physics-guided training, hybrid approaches that combine neural predictions with traditional FWI refinement.

11.3.2.4 Carbon Sequestration Monitoring

A high-impact application of seismic inversion is monitoring injected CO_2 plumes for carbon sequestration; a critical technology for climate change mitigation.

ML-Enhanced CO_2 Plume Monitoring at Frio-II
The Frio-II pilot project in Texas injected 1,600 tons of CO_2 into a saline aquifer, with continuous seismic and geophysical monitoring.

Traditional approach:

- Time-lapse seismic surveys every 3–6 months
- Full FWI processing: weeks per survey
- Limited temporal resolution of plume evolution

ML-enhanced approach:

- Neural operator trained on physics simulations of CO_2 migration
- Real-time plume estimation from continuous sensor data
- Uncertainty quantification via diffusion model ensembles

Results:

- Plume boundary detection within 10 m accuracy
- 100× faster than traditional time-lapse FWI
- Calibrated uncertainty bounds covering true plume in 94% of cases

Significance: Enables adaptive injection strategies and early leak detection; critical for regulatory approval and public acceptance of carbon sequestration.

11.3.3 Materials Characterization

Materials characterization; determining structure, composition, and properties from experimental measurements; is essential for quality control, failure analysis, and materials discovery. Machine learning has transformed several key characterization modalities.

11.3.3.1 X-Ray Diffraction Analysis

X-ray diffraction (XRD) patterns encode crystal structure information through peak positions and intensities (Fig. 11.8). Traditional analysis requires expert interpretation or database matching; time-consuming and error-prone for complex or mixed-phase samples.

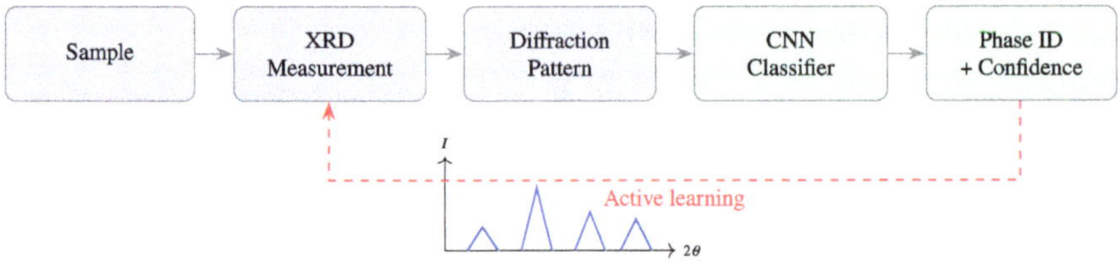

Fig. 11.8 ML-enhanced XRD analysis pipeline. Diffraction patterns are processed by convolutional neural networks for phase identification. Active learning can suggest additional measurements to resolve ambiguities

Table 11.15 ML methods for XRD phase identification

Method	Training data	Accuracy (%)	Speedup	Capability
Lee et al. [20]	170k synthetic	94.2	1000×	Single-phase ID
Vecsei et al. [21]	1.7M synthetic	98.1	1000×	500+ space groups
Oviedo et al. [22]	Experimental + synthetic	96.5	500×	Mixed-phase detection
Szymanski [23]	ICSD database	97.8	800×	Quantitative phase fractions
Chen et al. [24]	Dual representation	99.1	600×	PDF + XRD fusion

Accuracy is measured on held-out test patterns; speed is relative to expert manual analysis

Recent work (Table 11.15) combines XRD patterns (reciprocal space) with simulated pair distribution functions (real space). This dual representation provides complementary structural information, achieving near-perfect accuracy on challenging mixed-phase samples where single-modality methods fail.

11.3.3.2 Microstructure Reconstruction

Materials properties depend critically on microstructure; the spatial arrangement of phases, grains, and defects at length scales from nanometers to millimeters. Reconstructing 3D microstructures from 2D cross-sections is an ill-posed inverse problem with significant industrial applications.

Automotive and aerospace manufacturers use microstructure analysis (Table 11.16) for quality control of critical components. ML-based reconstruction enables:

- Non-destructive estimation of 3D properties from surface measurements
- Automated defect detection and classification
- Process-structure-property linkages for manufacturing optimization

Table 11.16 3D microstructure reconstruction methods

Method	Approach	Volume fraction error (%)	Grain size distribution error (%)
Traditional (DREAM.3D)	Statistical reconstruction	5–10	15–20
GAN-based	3D generative model	3–5	10–15
SliceGAN	2D→3D synthesis	2–4	8–12
DiffusionRecon	Diffusion model	1–2	5–8

Evaluation on polycrystalline metal datasets

Table 11.17 ML applications in electron microscopy analysis

Application	Method	Performance	Impact
Defect detection	CNN segmentation	95% F1	Automated quality control
Atom column finding	U-Net + centroiding	< 5 pm precision	Strain mapping at atomic scale
STEM simulation	Neural operators	$1000\times$ speedup	Real-time interpretation
Ptychography	Physics-informed GAN	$2\times$ resolution	Beyond diffraction limit

11.3.3.3 Electron Microscopy Analysis

Electron microscopy produces rich structural information but generates massive datasets requiring expert interpretation. ML has transformed analysis of both transmission electron microscopy (TEM) and scanning electron microscopy (SEM) (Table 11.17).

11.3.4 Flow Field Reconstruction

Reconstructing full velocity fields from sparse sensor measurements is critical for experimental fluid mechanics, aerodynamic testing, and industrial process monitoring. The challenge: sensors are expensive and intrusive, so measurements are inevitably incomplete.

11.3.4.1 Particle Image Velocimetry Enhancement

Particle Image Velocimetry (PIV) provides velocity fields by tracking tracer particles, but resolution is limited by particle density, camera resolution, and optical access. ML methods can super-resolve and denoise PIV data (Fig. 11.9 and results in Table 11.18).

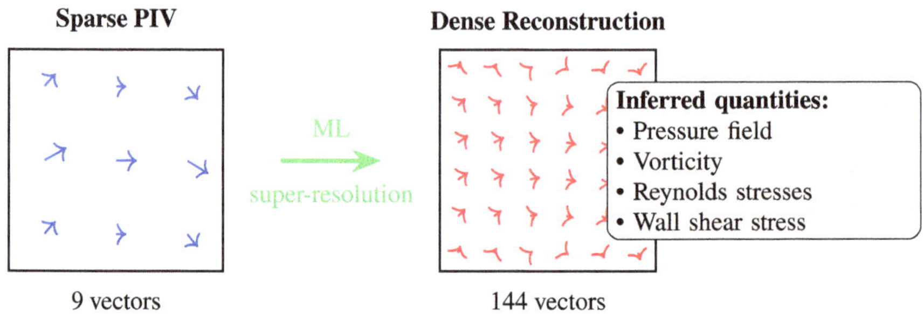

Fig. 11.9 ML-enhanced PIV reconstruction. Sparse velocity measurements (left) are super-resolved to dense fields (right). Physics-informed methods additionally infer pressure and other quantities not directly measured

Table 11.18 ML methods for flow field reconstruction from sparse data

Method	Physics constraint	Sparsity	Error (%)	Additional output
Raissi et al. [25]	PINN (NS equations)	1% sensors	5.2	Pressure, lift, drag
Cai et al. [26]	Divergence-free	5% sensors	3.8	Vorticity
Jin et al. [27]	NS + data	2% sensors	4.1	Reynolds stress
Du et al. [28]	FNO + physics loss	10% sensors	2.5	Pressure
HDNet [29]	Helmholtz decomposition	5% sensors	1.8	Divergence-free guaranteed
FlowRecon [30]	Diffusion + NS	1% sensors	2.2	Uncertainty bounds

Error is relative L^2 error on velocity field compared to ground truth (DNS or high-resolution PIV)

For incompressible flows, the velocity field must satisfy $\nabla \cdot \mathbf{u} = 0$. Enforcing this constraint; either through the loss function or network architecture; dramatically improves reconstruction quality and ensures physical consistency.

When to Use Physics-Informed Flow Reconstruction

Physics-informed reconstruction is most valuable when:

- **Sensors are very sparse** ($< 10\%$ coverage): Physics provides necessary regularization
- **Derived quantities are needed**: Pressure cannot be measured directly but can be inferred from velocity via Navier-Stokes
- **Noise is significant**: Physics constraints filter non-physical noise
- **Optical access is limited**: Wall regions, 3D flows

For dense, high-quality PIV data, purely data-driven super-resolution may suffice.

Table 11.19 Sensor placement optimization methods

Method	Approach	Error reduction (%)	Typical application
QR pivoting	Greedy selection on POD modes	30–40	Linear reconstructions
Sparse optimization	ℓ_1-minimization	40–50	Compressed sensing
RL-based	Reinforcement learning	50–60	Nonlinear, time-varying
End-to-end	Joint placement + reconstruction	60–70	Task-specific optimization

Reconstruction error with optimal versus uniform sensor placement

Table 11.20 Neural network approaches to quantum state tomography

Method	Qubits demonstrated	Sample reduction	Key innovation
Torlai et al. [31]	8	10×	RBM wavefunction ansatz
Quek et al. [32]	14	100×	Adaptive measurement selection
NAQT [33]	20+	1000×	Neural heuristic replaces Bayesian
Ahmed et al. [34]	100+	$10^4\times$	Transformer + classical shadows

Sample complexity is the number of measurements required for fixed reconstruction fidelity

11.3.4.2 Sparse Sensor Placement and Reconstruction

A related problem: where should we place limited sensors to enable optimal reconstruction? This sensor placement problem is coupled with the reconstruction algorithm (Table 11.19).

11.3.5 Other Inverse Problem Domains

11.3.5.1 Quantum State Tomography

Quantum state tomography reconstructs the density matrix ρ of a quantum system from measurement outcomes. For an n-qubit system, the density matrix has 4^n parameters; exponential in system size. Traditional tomography requires exponentially many measurements (Table 11.20).

Traditional adaptive tomography uses Bayesian inference to select optimal measurements; computationally prohibitive for large systems. NAQT replaces Bayesian updates with a neural network trained to propose informative measurements.

Table 11.21 Hamiltonian learning methods

Method	Approach	Parameter error (%)	Scalability
Gradient-based	Variational optimization	1–5	Limited by barren plateaus
Bayesian	MCMC sampling	0.5–2	Expensive for large systems
iPINN-HL	Physics-informed neural network	0.1–1	Demonstrated to 50 spins

Parameter error is relative error on learned Hamiltonian coefficients

Table 11.22 Physics-informed reconstruction in medical imaging

Modality	Method	PSNR gain (dB)	Clinical impact
MRI	Unrolled optimization	+3–5	4× faster scans
CT	Learned regularization	+2–4	50% dose reduction
PET	Deep image prior	+4–6	Low-count imaging

PSNR improvement over traditional methods

Results [33]:

- $\mathcal{O}(\log n)$ sample complexity versus $\mathcal{O}(\text{poly}(n))$ for standard tomography
- Up to $10^6\times$ speedup for systems with 10^7 measurement budget
- Demonstrated on 20+ qubit entangled states
- Robust to measurement noise and state preparation errors

Significance: Enables characterization of intermediate-scale quantum devices that would be infeasible with traditional methods.

11.3.5.2 Hamiltonian Learning

A complementary problem to state tomography: given access to a quantum system's dynamics, learn the Hamiltonian governing its evolution. This is essential for validating quantum simulators and characterizing noise (Table 11.21).

11.3.5.3 Medical Imaging Reconstruction

While not strictly within physical sciences, medical imaging reconstruction (MRI, CT, PET) shares the same mathematical structure and has driven significant methodological advances applicable across domains (Table 11.22).

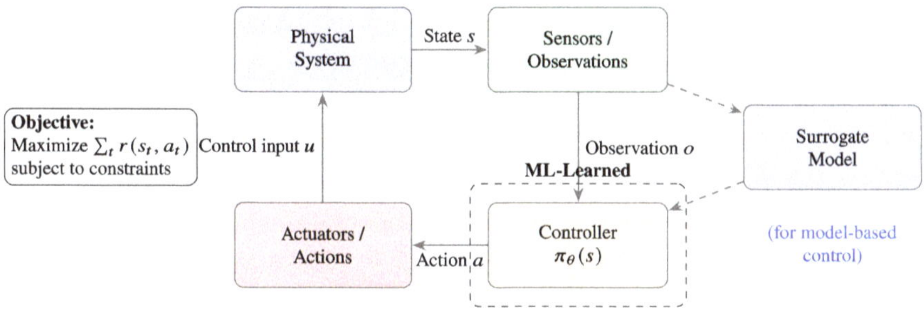

Fig. 11.10 The control loop with ML components. A controller (often learned via reinforcement learning) maps observations to actions that influence the physical system. Model-based approaches additionally use a learned surrogate to predict system response, enabling planning and optimization

11.4 Control and Optimization: Closing the Loop

The previous sections addressed prediction; forward surrogates that simulate systems and inverse methods that infer hidden parameters. This section addresses *action*: using predictions to control physical systems toward desired outcomes. The transition from prediction to control represents a qualitative shift in capability, enabling autonomous operation, real-time optimization, and closed-loop scientific discovery.

Control applications in physical sciences present unique challenges. Physical systems operate under strict safety constraints, require real-time response, and often cannot be reset after failures. The stakes are high: a plasma disruption can damage a billion-dollar fusion reactor; an unstable chemical reaction can endanger lives. Yet the potential rewards are equally significant: autonomous plasma control could unlock fusion energy; automated synthesis could accelerate materials discovery by orders of magnitude.

11.4.1 From Prediction to Action

Scientific machine learning enters the control loop (Fig. 11.10) at three levels, each with distinct advantages as shown in Table 11.23.

Key Control Terminology

- **Model Predictive Control (MPC)**: Optimize actions by simulating future trajectories with a model; re-plan at each timestep. Classical MPC uses physics-based models; ML-MPC uses learned surrogates.

Table 11.23 ML roles in scientific control systems

Role	Description	Advantages	Challenges
Learned policy	Neural network directly maps states to actions	End-to-end optimization; handles high-dimensional inputs	Sample inefficiency; safety during training
Learned model	Surrogate replaces expensive simulator in MPC	Leverages existing control theory; interpretable	Model errors compound in planning
Learned estimator	Neural network estimates hidden states from sensors	Handles partial observability; fuses multimodal data	Uncertainty quantification critical

Each approach offers different trade-offs between sample efficiency, optimality, and interpretability

- **Reinforcement Learning (RL)**: Learn a policy $\pi_\theta(a|s)$ that maximizes cumulative reward through trial-and-error interaction.
- **Sim-to-Real Transfer**: Train in simulation, deploy on real hardware. Critical for physical systems where real-world training is expensive or dangerous.
- **Safety Constraints**: Hard limits on states or actions that must never be violated (e.g., maximum plasma current, minimum reactor temperature).

11.4.2 Plasma Control for Fusion Energy

Nuclear fusion; the process powering the sun; promises virtually unlimited clean energy. Achieving controlled fusion on Earth requires confining plasma at temperatures exceeding 100 million degrees Celsius using magnetic fields. This confinement is inherently unstable: the plasma constantly seeks to escape, and instabilities can grow explosively, potentially damaging the reactor.

Plasma control is thus a grand challenge: maintain a superheated, turbulent medium in a precise configuration while responding to disturbances faster than human reaction times. Machine learning has achieved breakthrough results in this domain.

11.4.2.1 The TCV Breakthrough: Autonomous Plasma Shaping

In February 2022, DeepMind and the Swiss Plasma Center demonstrated autonomous plasma control on the TCV (Tokamak Ã Configuration Variable) reactor; a landmark achievement [35].

The challenge: Control plasma shape and position by adjusting 19 magnetic coil voltages simultaneously, responding to disturbances within 100 microseconds.

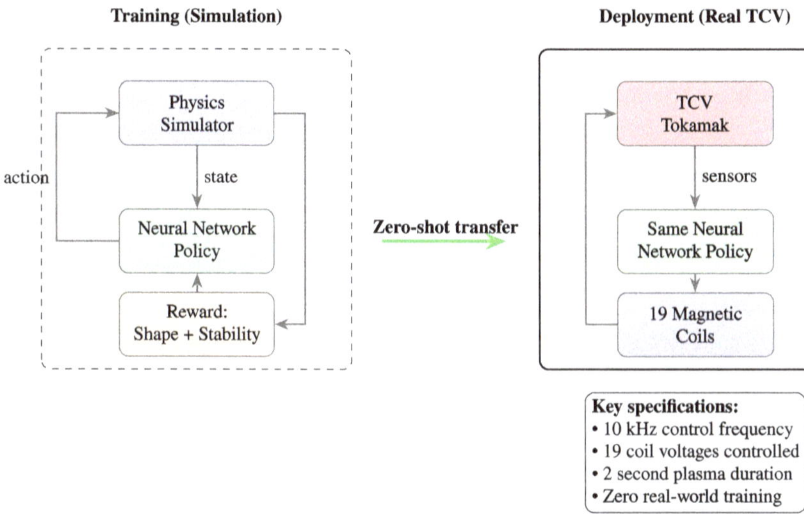

Fig. 11.11 DeepMind TCV plasma control architecture. A neural network policy trained entirely in simulation transfers directly to the real tokamak without any real-world fine-tuning. The policy controls 19 magnetic coils at 10 kHz to maintain desired plasma shapes

Approach:

- Train deep RL policy (MLP with 3 hidden layers) in physics-based simulator (Fig. 11.11)
- Reward function: match target shape + maintain stability + minimize power
- Transfer directly to real TCV with **zero real-world training**

Results (Table 11.24):

- Maintained plasma for full 2 s (TCV's maximum before cooling required)
- Achieved ITER-like shapes, negative triangularity ($\delta = -0.8$), "snowflake" divertors
- **First-ever demonstration of "droplets":** Two separate plasmas maintained simultaneously
- New plasma configurations achievable in hours versus weeks with traditional methods

Significance: Demonstrated that deep RL can handle the complexity, speed, and precision required for real fusion plasma control; a critical step toward practical fusion energy.

11.4.2.2 From Shape Control to Disruption Avoidance

A more challenging problem than shape control is *disruption avoidance*. Disruptions are violent events where plasma confinement is suddenly lost, releasing stored energy that can damage reactor walls. For ITER and future power plants, disruption avoidance is mission-critical.

Table 11.24 TCV control performance comparison

Metric	Traditional PID	ML Policy	Improvement
Shape accuracy (RMSE)	2.1 cm	0.7 cm	67% reduction
Position stability	±1.5 cm	±0.5 cm	3× tighter
Configuration design time	2–4 weeks	Hours	∼ 100× faster
Novel configurations	Limited by controller	Flexible	Droplets, snowflakes
Adaptation to disturbances	Manual retuning	Automatic	Real-time

Traditional control uses hand-tuned PID controllers; ML control uses the trained neural network policy

Table 11.25 Deep RL disruption avoidance on DIII-D tokamak

Metric	Without RL	With RL	Significance
Tearing mode onset	100% of shots	0% progression	Complete suppression
H-mode maintained	No (disruption)	Yes	Performance preserved
Plasma β achieved	Limited by instability	Near-limit operation	Higher fusion power
Control latency	N/A	< 1 ms	Real-time response

Traditional disruption handling focuses on *prediction*; detecting an impending disruption and initiating emergency shutdown. The 2024 DIII-D [36] result (Table 11.25) demonstrates *avoidance*: the RL controller actively adjusts plasma parameters (density, current profile) to suppress the instability before it grows, maintaining high-performance operation rather than triggering a controlled shutdown.

11.4.2.3 TORAX: Enabling Differentiable Tokamak Simulation

Real-world RL training on tokamaks is impractical; each plasma shot is expensive and risks equipment. Training requires accurate simulators. TORAX (Tokamak Transport Accelerated eXecution), released by DeepMind in 2024 [37], provides a differentiable tokamak transport simulator enabling gradient-based optimization and efficient RL training (Table 11.26).

Commonwealth Fusion Systems, developing the SPARC tokamak, uses TORAX for scenario planning. The combination of differentiable simulation and learned transport models enables exploring plasma scenarios that would be impractical to test experimentally.

Table 11.26 TORAX capabilities and applications

Feature	Description	Enabled application
Differentiable physics	Automatic gradients through transport equations	Gradient-based controller optimization
QLKNN integration	Neural network turbulent transport model	$10^6\times$ faster than gyrokinetic
Modular architecture	Swappable physics modules	Testing new models easily
JAX backend	GPU acceleration, JIT compilation	Fast RL environment

11.4.3 Molecular Design and Autonomous Laboratories

If plasma control operates at the extreme of speed (microsecond response), molecular design operates at the extreme of complexity (vast chemical space). The goal: automatically design molecules or materials with desired properties, then synthesize and validate them with minimal human intervention.

11.4.3.1 The Molecular Design Loop

See Fig. 11.12.

11.4.3.2 Generative Models for Molecular Design

Generative models produce novel molecular structures with desired properties. The field has evolved rapidly from string-based representations (SMILES) to 3D structure generation (Table 11.27).

Diffusion models (Table 11.28) have become the dominant paradigm for 3D molecular generation. Key advantages include: stable training, high-quality samples, and natural incorporation of physics constraints through the denoising process.

11.4.3.3 A-Lab: Fully Autonomous Materials Synthesis

The ultimate vision of autonomous discovery was demonstrated by the A-Lab at Lawrence Berkeley National Laboratory; a robotic system that designs, synthesizes, and characterizes new materials with minimal human intervention.

A-Lab: Autonomous Materials Discovery

System components:

- **Literature mining**: LLM extracts synthesis recipes from papers
- **Recipe generation**: ML proposes synthesis parameters

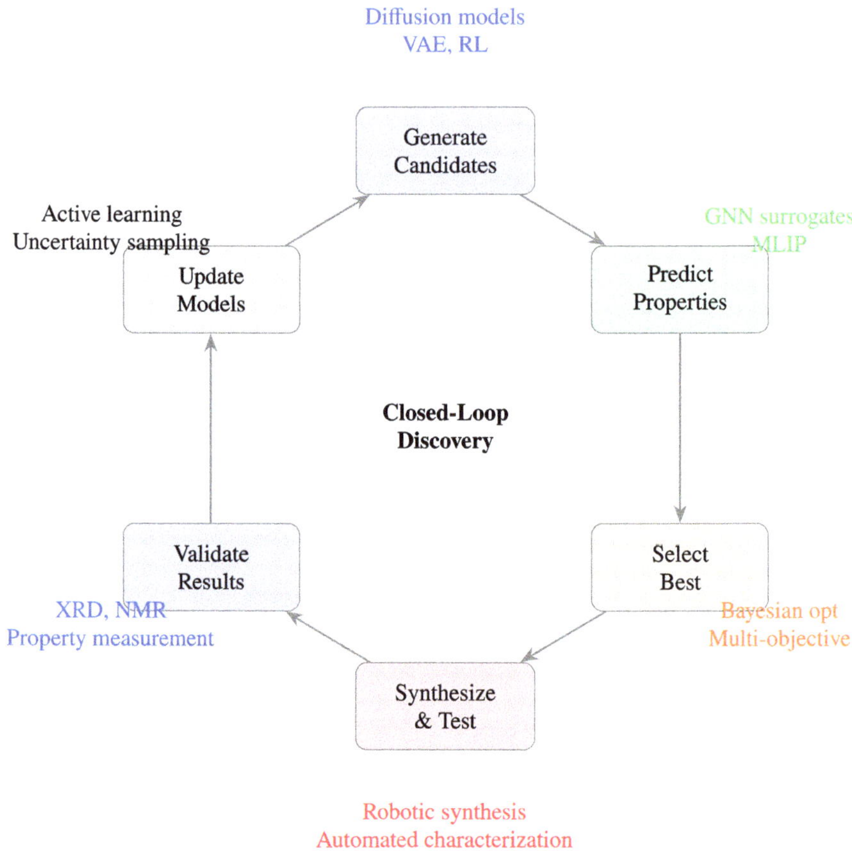

Fig. 11.12 The closed-loop molecular design cycle. ML methods enable each stage: generative models propose candidates, surrogates predict properties, optimization selects the best candidates, robotic systems synthesize and characterize, and active learning updates models based on results

- **Robotic synthesis**: Automated weighing, mixing, heating
- **XRD characterization**: Automated phase identification
- **Active learning**: Bayesian optimization selects next experiments

17-day campaign results:

- 58 target compounds attempted
- **41 successfully synthesized** (71% success rate)
- 21 experiments per day average
- Fully autonomous operation
- Novel compounds not previously synthesized

Table 11.27 Evolution of molecular generative models

Method	Year	Validity (%)	Novelty (%)	3D?	Key innovation
CharacterVAE	2017	75	90	No	SMILES generation
JT-VAE	2018	100	85	No	Junction tree structure
G-SchNet	2019	95	92	Yes	Autoregressive 3D
EDM	2022	98	88	Yes	Equivariant diffusion
GeoLDM	2023	99	91	Yes	Latent diffusion
PMDM	2024	99.5	93	Yes	Dual diffusion

Validity is the percentage of generated molecules that are chemically valid; novelty is the percentage not in training data

Table 11.28 Diffusion model results for structure-based drug design

Method	Vina score	QED	SA score	Success rate (%)
3D-SBDD (baseline)	− 6.2	0.48	0.72	35
TargetDiff (2023)	− 7.1	0.51	0.68	48
DiffSBDD (2023)	− 7.5	0.52	0.71	54
PMDM (2024)	− 8.2	0.55	0.73	62

Benchmarked on CrossDocked2020 for binding affinity prediction

Comparison to traditional:

- Traditional Ph.D. workflow: $\sim$ 1 new compound per week
- A-Lab: $\sim$ 3 new compounds per day ($21\times$ acceleration)

The Characterization Completeness Question
Subsequent analysis of A-Lab results raised important questions:

- **Phase identification accuracy**: Some reported syntheses may have been misidentified phases or mixtures
- **Purity thresholds**: Success criteria (phase fraction $> 50\%$) may not meet application requirements
- **Reproducibility**: Not all syntheses have been independently replicated

Lesson: Autonomous systems require robust verification pipelines. The "closed loop" must include reliable characterization and validation, not just synthesis.

Current status: A-Lab represents a proof of concept, not a finished solution. The 71% success rate is remarkable for autonomous operation but leaves room for improvement in both synthesis and characterization.

Table 11.29 Bayesian optimization in experimental science

Application	Parameter space	Reduction	Reference
Perovskite solar cells	5 composition variables	10×	Accelerated discovery
Catalyst optimization	8 synthesis parameters	5×	HTE integration
Polymer design	12 monomer choices	8×	Multi-objective
Alloy composition	4 element fractions	6×	Phase stability
Reaction conditions	6 process variables	7×	Yield optimization

Reduction in experiments needed to find optimal conditions compared to grid search or random sampling

11.4.3.4 Bayesian Optimization for Experimental Design

When experiments are expensive, Bayesian optimization provides principled guidance for selecting the next experiment to maximize information gain or expected improvement (Table 11.29).

11.4.4 Active Flow Control

Controlling fluid flows (Fig. 11.13) to reduce drag, enhance mixing, or suppress instabilities has enormous economic implications. Aircraft drag reduction of a few percent translates to billions of dollars in fuel savings. ML-based control has demonstrated significant advances (Table 11.30).

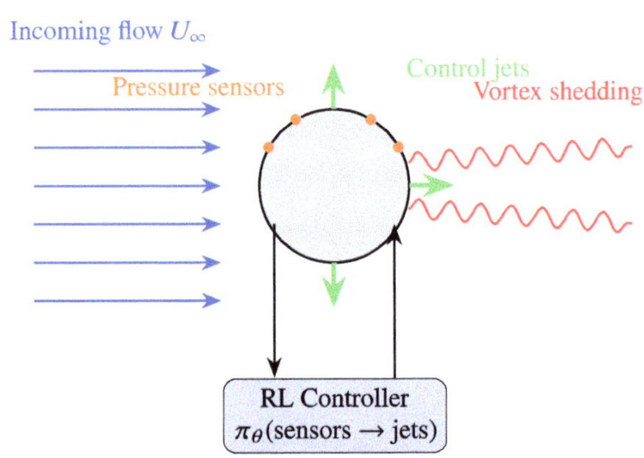

Fig. 11.13 Active flow control setup for drag reduction. An RL controller uses pressure sensor readings to modulate jet actuators, suppressing vortex shedding and reducing drag. The controller learns optimal actuation strategies through interaction with the flow (simulated or real)

Table 11.30 RL-based drag reduction results

Configuration	Re	Drag reduction (%)	Approach	Reference
2D cylinder	100	8	Single-agent PPO	[38]
2D cylinder	400	30	Multi-agent	[39]
3D cylinder	3900	9	Multi-agent RL	[40]
Triangular array	100	40.4	Coordinated agents	[41]
Open channel	180	43	Opposition control	[42]
Ahmed body	3×10^5	12	Experimental RL	[43]

Drag reduction is measured relative to uncontrolled baseline at the same Reynolds number

Table 11.31 Differentiable CFD for aerodynamic optimization

Method	Application	Speedup	Result
JAX-Fluids (2024)	Transonic airfoil	100×	Pareto-optimal designs in hours
Neural field + AD	3D car body	1000×	8% drag reduction
PhiFlow	Heat exchanger	50×	15% efficiency improvement

11.4.4.1 Drag Reduction on Bluff Bodies

Most drag reduction results are demonstrated in simulation. Transferring to real experiments faces challenges: sensor noise, actuator limitations, and modeling errors.

11.4.4.2 Differentiable CFD for Shape Optimization

Beyond active control, ML enables gradient-based shape optimization. Differentiable simulators allow computing gradients of objectives (drag, lift, heat transfer) with respect to shape parameters, enabling efficient optimization (Table 11.31).

11.4.5 Survey: Other Control Domains

11.4.5.1 Chemical Reactor Control

Chemical reactors present multi-objective control challenges: maximize yield while maintaining safety, minimizing energy consumption, and meeting environmental constraints (Table 11.32).

Table 11.32 ML-based chemical reactor control results

Application	Method	Improvement	Constraint handling
Batch reactor yield	MPC + NN model	+12% yield	Temperature limits
Continuous flow	Safe RL	+8% throughput	Safety constraints
Polymerization	Hybrid MPC	+15% efficiency	Product quality

Table 11.33 RL-based building energy optimization

Building type	Method	Energy reduction	Deployment status
Office (Google)	Model-based RL	40% (cooling)	Production (2016–)
Data center	Offline RL	15% total	Production
Commercial HVAC	SAC + safety	20–25%	Pilot deployments

Energy reduction relative to rule-based baseline while maintaining comfort

11.4.5.2 Building Energy Optimization

HVAC systems in commercial buildings account for significant energy consumption. RL-based control adapts to occupancy patterns, weather, and electricity prices (Table 11.33).

11.4.5.3 Telescope and Adaptive Optics Control

Ground-based telescopes use adaptive optics to correct for atmospheric turbulence. RL controllers can outperform traditional linear methods (Table 11.34).

11.5 Discovery: Learning Physical Laws from Data

The ultimate aspiration of scientific machine learning extends beyond prediction and control to *discovery*; learning new physical laws, governing equations, and interpretable relationships from data. Unlike black-box surrogates that trade inter-

Table 11.34 RL for adaptive optics control

System	Method	Strehl improvement (%)	Conditions
Lab testbed	Model-free RL	+ 15	Simulated turbulence
Subaru Telescope	Hybrid RL + PID	+ 8	On-sky demonstration
Multi-conjugate AO	Multi-agent RL	+ 20	Simulation

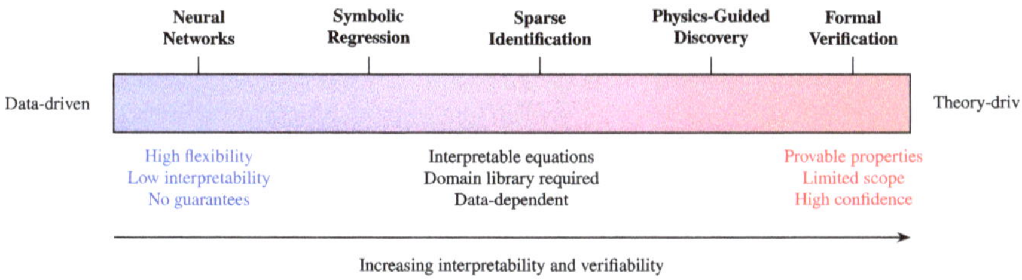

Fig. 11.14 The discovery method spectrum. Methods range from flexible but opaque neural networks (left) to formally verified but narrow theorem provers (right). Most practical discovery lies in the middle, using symbolic regression and sparse identification to find interpretable equations

pretability for accuracy, discovery methods aim to produce knowledge that scientists can understand, verify, and build upon.

This section surveys how ML is accelerating scientific discovery across physical sciences, from symbolic equation discovery to large-scale materials screening.

11.5.1 The Discovery Spectrum

Discovery methods span a spectrum from purely data-driven to formally verified, trading off flexibility against interpretability and guarantees (Fig. 11.14).

Key Discovery Paradigms

- **Symbolic Regression**: Search for mathematical expressions that fit data, typically using genetic programming or neural-guided search. Output: closed-form equations like $f(x) = ax^2 + b\sin(cx)$.
- **Sparse Identification (SINDy)**: Assume dynamics have sparse representation in a library of candidate functions; use ℓ_1 regularization to select active terms. Output: parsimonious ODEs/PDEs.
- **Physics-Informed Discovery**: Constrain search using known physical principles (symmetries, conservation laws, dimensional analysis). Output: equations guaranteed to satisfy constraints.
- **Neural Symbolic**: Combine neural network flexibility with symbolic output, e.g., through equation decoders or attention over operators. Output: interpretable expressions learned end-to-end.

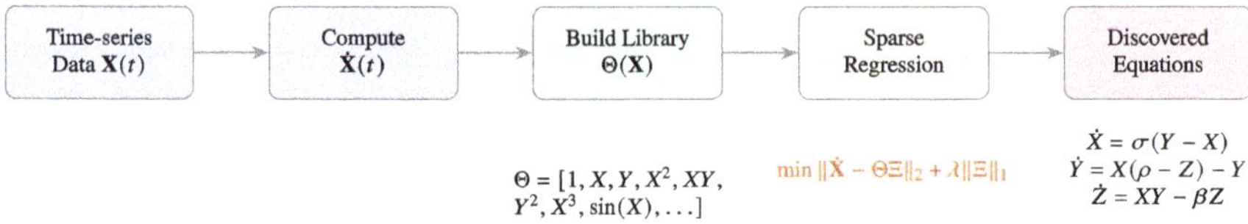

Fig. 11.15 The SINDy pipeline. Time-series data is differentiated, expanded into a library of candidate functions, and sparse regression identifies the active terms. The result is interpretable governing equations

11.5.2 Equation Discovery in Fluid Mechanics

Fluid mechanics provides an ideal testbed for equation discovery: the governing equations (Navier-Stokes) are known exactly, enabling rigorous validation, yet practical closure models remain elusive, motivating data-driven discovery.

11.5.2.1 SINDy: Sparse Identification of Nonlinear Dynamics

The Sparse Identification of Nonlinear Dynamics (SINDy) algorithm (Fig. 11.15), has become the workhorse of equation discovery. The key insight: most physical systems have governing equations that are *sparse* in an appropriate basis of candidate functions.

In a landmark demonstration, Brunton et al. applied SINDy to data from turbulent shear flow simulations. The algorithm discovered a 9-equation model capturing the essential dynamics of the mean flow, streaks, and rolls.

Significance: This same model took human researchers approximately 30 years to derive through physical reasoning and asymptotic analysis. SINDy found it in hours from data alone (Table 11.35).

Discovered model structure:

- 9 coupled ODEs for modal amplitudes
- Correctly identifies energy transfer between modes
- Predicts transient growth and turbulent-laminar transitions
- Matches DNS statistics quantitatively

Validation: The discovered equations were not merely curve-fits; they generalized to parameter regimes not seen in training and correctly predicted stability boundaries.

11.5.2.2 Limitations and When SINDy Fails

The Library Design Problem

SINDy's fundamental limitation: it can only discover equations expressible in the candidate library Θ.

Table 11.35 SINDy and variants: equation discovery results across fluid systems

System	Method	Noise tolerance (%)	Discovered equations
Lorenz attractor	SINDy	10	Exact recovery
Turbulent shear flow	SINDy	5	9-mode model (matches theory)
Cylinder wake	SINDy + POD	8	Mean-field model
Burgers equation	PDE-FIND	15	Correct PDE form
Navier-Stokes (2D)	Weak SINDy	20	Recovers viscous term
Reaction-diffusion	Ensemble SINDy	25	Correct kinetics + UQ

Failure modes:

- **Missing basis functions**: If the true equation contains $\tanh(x)$ but the library only has polynomials, SINDy will find a spurious polynomial approximation
- **Library explosion**: Including all possible functions makes regression ill-conditioned
- **Coordinate dependence**: Results depend on choice of variables; SINDy in wrong coordinates may miss simple structure

Mitigation strategies:

- Use domain knowledge to design physically motivated libraries
- Automate library construction with symbolic regression
- Apply dimensional analysis to constrain possible terms
- Use multiple coordinate systems and compare results

11.5.3 Materials Discovery at Scale

Materials discovery represents a different discovery paradigm: rather than finding equations, the goal is finding *materials*; compositions and structures with desired properties. ML enables exploring chemical space at unprecedented scale.

11.5.3.1 GNoME: 2.2 Million New Materials

Google DeepMind's Graph Networks for Materials Exploration (GNoME) (Fig. 11.16) exemplifies large-scale ML-driven discovery (Table 11.36).

The discovery of 528 new lithium-ion conductors; 25× the previously known count; illustrates the potential for targeted discovery. These materials are candidates for solid-state batteries, which could enable safer, higher-energy-density storage.

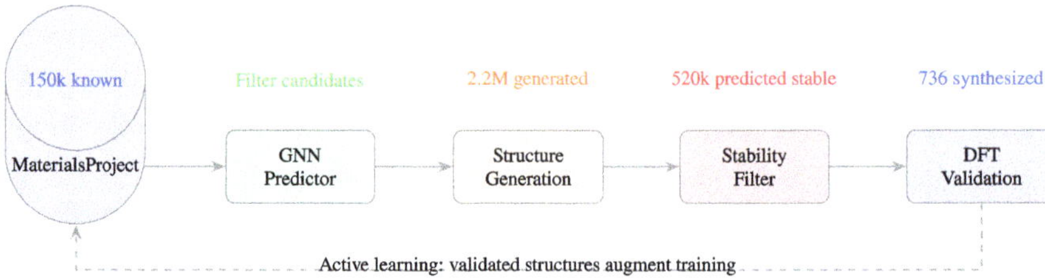

Fig. 11.16 The GNoME discovery pipeline. Starting from known materials, GNN models guide structure generation. Predicted-stable candidates are validated with DFT, and successful validations augment training data in an active learning loop

Table 11.36 GNoME discovery results

Metric	Number	Significance
Crystal structures predicted	2.2 million	10× all previously known
Predicted stable materials	520,000+	Vast expansion of search space
DFT-validated stable	380,000+	High-confidence predictions
Independently synthesized	736	External experimental validation
New Li-ion conductors	528	25× previously known count
Prediction accuracy	80%	versus 50% for prior methods

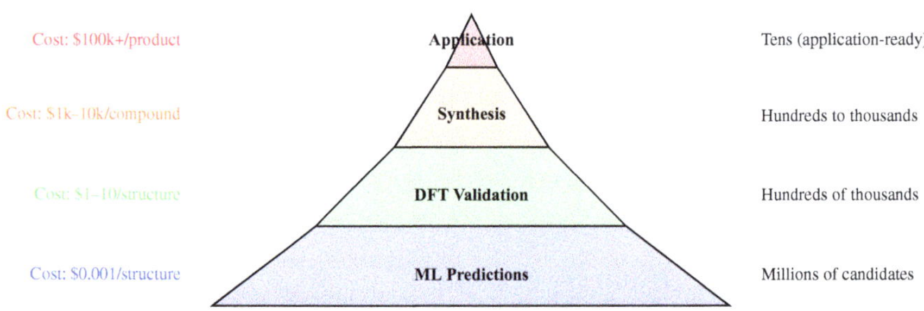

Fig. 11.17 The materials discovery validation pyramid. ML enables screening millions of candidates cheaply; progressively expensive validation narrows to application-ready materials. Each level filters by approximately 10–100×

11.5.3.2 The Validation Pyramid

Large-scale computational predictions require experimental validation. The field has developed a hierarchical validation strategy (Fig. 11.17).

Table 11.37 Neural quantum state achievements on benchmark systems

System	Method	Accuracy	Comparison
2D Heisenberg (10×10)	MinSR (2024)	$E/N = -0.67155260(3)$	Exceeds DMRG precision
J_1-J_2 frustrated	Transformer NQS	Ground state phase diagram	First complete mapping
Fermi-Hubbard	FermiNet	Within 0.1%	Matches AFQMC
Molecules (H_2O)	PauliNet/FermiNet	Chemical accuracy	Matches CCSD(T)
Periodic solids	DiffVMC (2024)	1 mHa/atom	Competitive with DMC

11.5.4 Neural Quantum States

Neural quantum states (NQS) represent quantum wavefunctions as neural networks, enabling variational solutions to many-body problems previously intractable. This constitutes discovery in the sense of finding ground states and phase diagrams of novel quantum systems (Table 11.37).

Beyond achieving numerical accuracy, NQS methods have enabled discovery of new quantum phases. Transformer-based NQS applied to frustrated magnets have mapped phase diagrams that were previously inaccessible, identifying exotic spin liquid phases and quantum critical points.

11.5.5 Survey: Discovery in Other Domains

11.5.5.1 Climate Dynamics

Data-driven discovery of climate teleconnections; long-range correlations between climate variables; has identified previously unknown relationships between ocean temperatures and extreme weather events (Table 11.38).

Table 11.38 ML-driven climate discovery

Discovery	Method	Impact
ENSO precursors	Causal discovery + DL	18-month advance prediction
Arctic-midlatitude links	Graph neural networks	Improved extreme event forecasting
Ocean eddy dynamics	SINDy + satellite data	Simplified transport models

11.5.5.2 Biological Physics

Protein structure prediction (AlphaFold) represents perhaps the most celebrated ML discovery success, effectively solving a 50-year grand challenge. Extensions now address protein dynamics, interactions, and design.

11.5.6 Discovery Section Summary

Key Takeaways: Discovery

1. **Interpretability is the goal**: Unlike prediction, discovery aims for human-understandable outputs; equations, structures, phase diagrams; that advance scientific knowledge.
2. **SINDy democratizes equation discovery**: Sparse identification finds governing equations from data, recovering in hours what took decades of human effort.
3. **Scale transforms materials discovery**: GNoME's 2.2 million predictions and 736 experimental validations demonstrate that ML can explore chemical space far beyond human capacity.
4. **Neural quantum states push accuracy limits**: NQS methods now achieve or exceed the accuracy of the best traditional methods for quantum many-body problems.
5. **Validation remains essential**: Computational predictions require experimental confirmation. The validation pyramid; ML $\rightarrow$ DFT $\rightarrow$ synthesis $\rightarrow$ application; structures this process.

11.6 Cross-Cutting Challenges and Best Practices

The domain-specific successes documented in previous sections share common challenges that transcend any single application. This section addresses these cross-cutting issues, providing guidance applicable across physical science domains.

11.6.1 The Generalization Problem

The most persistent challenge in SciML is generalization: models that excel on test data drawn from the same distribution as training often fail when deployed on genuinely new conditions (Table 11.39).

Table 11.39 Generalization strategies and their effectiveness

Strategy	Mechanism	Effectiveness
Physics-informed loss	PDE residual penalizes non-physical predictions	Extends near-extrapolation region
Hard constraints	Architecture enforces conservation laws	Guarantees physical consistency
Domain randomization	Train on diverse conditions	Improves robustness, not extrapolation
Uncertainty quantification	Flag low-confidence predictions	Enables safe deployment
Hybrid methods	Physics handles known; ML handles unknown	Best for well-understood systems

The Extrapolation Hierarchy

When deploying SciML models, understand where your test conditions fall:

1. **Interpolation** (within training distribution): Generally safe. Errors match validation metrics.
2. **Near-extrapolation** (slightly beyond training): Proceed with caution. Validate against physics checks; expect degraded accuracy.
3. **Far-extrapolation** (qualitatively new regimes): Expect failures. Use only with extensive validation or fall back to traditional methods.

Rule of thumb: If the physics changes qualitatively (turbulent transition, phase change, instability onset), treat as far-extrapolation regardless of parameter distance.

11.6.2 Long-Time Integration and Error Accumulation

Autoregressive models; which predict the next timestep from the current state; accumulate errors over long rollouts. This is particularly problematic for chaotic systems where small errors grow exponentially (Table 11.40).

NeuralGCM's 40-year climate stability; versus pure ML models that drift after months; demonstrates the power of hybrid approaches. By retaining explicit physics for conservation laws and large-scale dynamics, hybrid methods inherit stability properties of traditional solvers.

Table 11.40 Strategies for stable long-time integration

Strategy	Description	Error at $t = 100\Delta t$	Overhead
Naive autoregressive	Direct single-step prediction	Blowup	None
Noise injection	Add noise during training	15%	Minimal
Pushforward training	Train on own predictions	8%	2× training
Temporal bundling	Predict multiple steps at once	6%	Larger model
Hybrid correction	Physics corrects ML drift	4%	Physics solver cost
Diffusion rollout	Denoise at each step	3%	10× inference

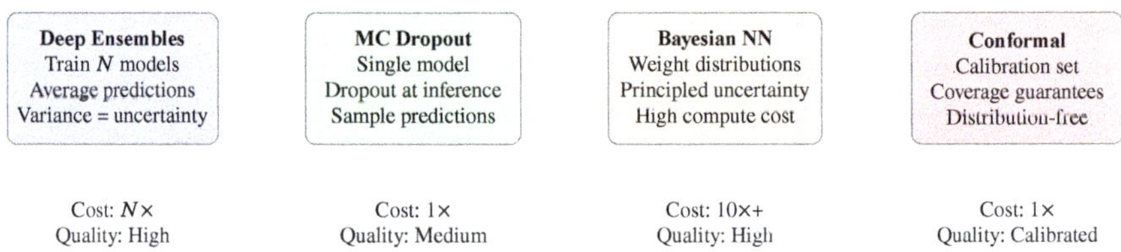

Fig. 11.18 Uncertainty quantification methods for SciML. Each approach offers different trade-offs between computational cost, uncertainty quality, and theoretical guarantees

11.6.3 Uncertainty Quantification

For scientific applications, knowing *when not to trust* a prediction is as important as the prediction itself. Uncertainty quantification (UQ) provides this capability (Fig. 11.18 and Table 11.41).

Table 11.41 UQ method comparison for scientific applications

Method	Epistemic UQ	Aleatoric UQ	Calibration	Best For
Deep ensembles	Good	Moderate	Requires calibration	General use
MC dropout	Moderate	Poor	Often underestimates	Fast approximation
Bayesian NN	Excellent	Good	Principled	Research, small models
Conformal	N/A	N/A	Guaranteed	Coverage requirements
Evidential	Good	Good	Requires tuning	Single-pass inference

Practical UQ Recommendations

1. **Start with ensembles**: Deep ensembles (5–10 models) provide good uncertainty estimates with straightforward implementation.
2. **Calibrate uncertainties**: Raw uncertainty estimates are often miscalibrated. Use a held-out calibration set to adjust.
3. **Validate on OOD data**: Uncertainty should increase for out-of-distribution inputs. Test this explicitly.
4. **Use conformal for guarantees**: When coverage guarantees are required (safety-critical applications), conformal prediction provides distribution-free bounds.
5. **Distinguish uncertainty types**: Epistemic (model uncertainty, reducible with more data) versus aleatoric (inherent noise, irreducible) require different treatment.

11.6.4 Multi-scale Integration

Physical systems span vast scale ranges; from electrons to continents, from femtoseconds to millennia. ML enables bridging scales that were previously computationally disconnected (Table 11.42).

11.6.5 Validation Against Experiment

Computational predictions must ultimately be validated against experimental reality. A hierarchy of validation levels provides confidence appropriate to the stakes (Table 11.43).

Table 11.42 ML methods for scale bridging

Scale transition	ML method	Speedup	Application example
Quantum → Atomistic	MLIP (MACE, NequIP)	$10^6\times$	Materials simulation
Atomistic → Mesoscale	CG neural networks	$10^3\times$	Polymer dynamics
Mesoscale → Continuum	Neural constitutive	$10^2\times$	Crystal plasticity
Continuum → System	Neural surrogates	$10^3\times$	Digital twins

Table 11.43 Validation status across domains

Domain	Highest level achieved	Example	Remaining gap
Weather prediction	Level 4	ECMWF operational	Extreme events
Materials (MLIP)	Level 3	GNoME → synthesis	High-energy configurations
Plasma control	Level 4	TCV real-time control	ITER-scale deployment
Turbulence surrogates	Level 2	Academic benchmarks	Industrial validation

The Validation Hierarchy

1. **Consistency checks**: Does the model satisfy conservation laws? Are symmetries respected? Do outputs remain in physical bounds?
2. **Verification against simulation**: Does the surrogate reproduce the training simulator's outputs on held-out test cases?
3. **Validation against experiment**: Do predictions match experimental measurements for cases where ground truth is available?
4. **Blind prediction**: Can the model predict outcomes of experiments conducted *after* model development, with no parameter adjustment?

Each level provides stronger evidence. Level 4 (blind prediction) is the gold standard but rarely achieved in practice.

11.6.6 Reproducibility and Open Science

Scientific progress requires reproducibility. The SciML community has made significant strides in open resources, though gaps remain (Table 11.44).

Table 11.44 Open resources for SciML by domain

Domain	Code	Data	Pretrained models
Weather/climate	GraphCast, AIFS, NeuralGCM	ERA5, WeatherBench 2	Available (HuggingFace)
Materials	MACE, CHGNet, M3GNet	Materials Project, JARVIS	Available
Fluids	neuraloperator, PhiFlow	PDEBench, CFDBench	Limited
Plasma	TORAX	Limited	Emerging
Quantum	NetKet, FermiNet	Benchmark Hamiltonians	Foundation NQS (2025)

Reproducibility Checklist for SciML Papers

When publishing or evaluating SciML results, verify:

- ☐ Code publicly available with clear documentation
- ☐ Training data accessible or generation procedure specified
- ☐ Hyperparameters fully specified (including those "found by tuning")
- ☐ Random seeds reported; results averaged over multiple runs
- ☐ Computational resources documented (hardware, training time)
- ☐ Comparison to appropriate baselines on standard benchmarks
- ☐ Limitations and failure cases explicitly discussed

11.7 Key Takeaways and Future Outlook

11.7.1 What We Have Learned: Domain Lessons

Each domain has contributed distinct lessons to the broader SciML enterprise (Table 11.45).

Table 11.45 Key lessons from each physical science domain

Domain	Biggest success	Key lesson	Open challenge
Weather/Climate	Operational AI forecasts	Hybrid physics-ML enables stability	Extreme events, decades-scale
Materials	Universal potentials	Equivariance is essential	High-energy configurations
Plasma	Autonomous control	Sim-to-real transfer works	ITER-scale deployment
Fluids	Turbulence surrogates	Resolution invariance via FNO	3D industrial scale
Quantum	Near-exact ground states	Neural ansätze beat DMRG	Dynamics, excited states
Molecular	Autonomous synthesis	Closed-loop discovery is possible	Verification robustness

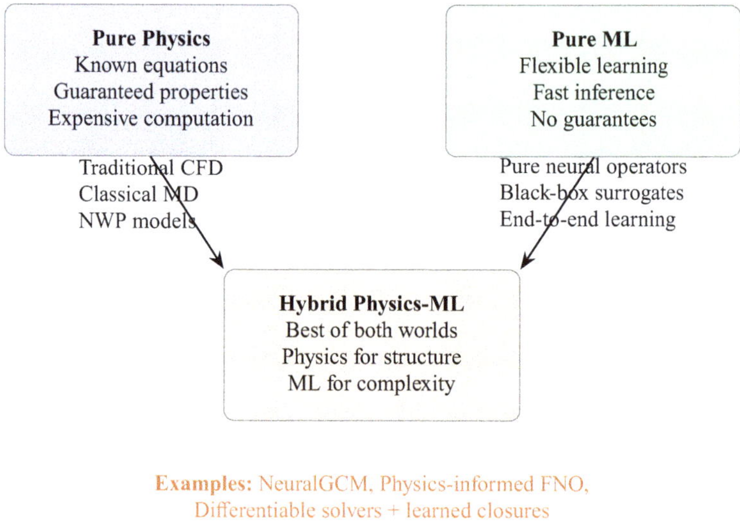

Fig. 11.19 The convergence toward hybrid physics-ML methods. The most successful recent approaches combine explicit physics (for guaranteed properties and structure) with learned components (for flexibility and efficiency)

11.7.2 The Emerging Paradigm: Hybrid Physics-ML

Across domains, the most robust results come from *hybrid* approaches that combine learned components with explicit physics. This is not a compromise but a principled design choice that leverages the strengths of both (Fig. 11.19).

11.7.3 Foundation Models for Physical Sciences

An emerging trend is the development of large, pretrained models that can be adapted to diverse downstream tasks; "foundation models" for physical sciences (Table 11.46).

The foundation model hypothesis for science

The success of foundation models in language and vision suggests a similar approach for scientific domains: pretrain on diverse scientific data (simulations, experiments, literature), then fine-tune for specific applications. Early results support this hypothesis, showing that pretrained models require less domain-specific data and generalize better than models trained from scratch.

Table 11.46 Emerging foundation models for physical sciences

Model	Domain	Parameters	Capability
Aurora	Earth systems	1.3B	Weather, air quality, ocean
Prithvi	Geospatial	300M	Land use, disaster monitoring
MACE-MP-0	Materials	10M	Universal interatomic potential
Foundation NQS	Quantum	100M	Transferable wavefunctions
Galactica	Scientific text	120B	Literature, equations, data

11.7.4 Open Questions

Despite remarkable progress, fundamental questions remain:

1. **How do we trust predictions in unprecedented regimes?** Climate change, new materials, and extreme conditions involve extrapolation beyond any training data. What theoretical guarantees can we provide? How do we combine ML predictions with physics-based bounds?
2. **When should we prefer interpretable methods over accurate ones?** A black-box surrogate may be more accurate than SINDy-discovered equations, but the equations provide scientific insight. How do we make this trade-off systematically?
3. **How do we close the loop with experiments at scale?** Autonomous laboratories like A-Lab demonstrate the vision but face verification challenges. What infrastructure enables reliable closed-loop discovery?
4. **What are the limits of data-driven discovery?** Can ML discover fundamentally new physics, or only optimize within known frameworks? The history of physics suggests that paradigm shifts require conceptual leaps; can ML make them?
5. **How do we democratize access?** Foundation models require massive computation to train. How do we ensure that SciML capabilities are broadly accessible, not concentrated in well-resourced institutions?

11.7.5 Conclusion: The Transformation of Computational Science

We began this chapter with a tale of two laboratories; illustrating the transformation that scientific machine learning has brought to physical sciences. Let us return to that contrast one final time.

What has changed ?

The fundamental relationship between computation and scientific insight has shifted. Where 2015's computational scientist waited weeks for a single simulation, 2025's scientist explores thousands of configurations in an afternoon. Where intuition guided the search for new materials, algorithms now screen millions of candidates. Where human operators struggled to control complex systems, learned policies maintain plasma configurations and synthesize compounds autonomously.

What has not changed?

Physics remains the foundation. The Navier-Stokes equations, Schrödinger equation, and Maxwell's equations are as valid as ever; and SciML's greatest successes come from embedding this physics into learning systems. Human judgment remains essential for posing meaningful questions, interpreting results, and catching failures that algorithms miss. The need for experimental validation has not diminished; if anything, the scale of computational predictions makes validation more critical.

The path forward?

The most promising direction is not pure ML replacing physics, nor physics constraining ML, but their genuine synthesis; hybrid systems where explicit physics provides structure, guarantees, and interpretability while learned components provide flexibility, efficiency, and the ability to capture complexity beyond analytical treatment. NeuralGCM, physics-informed neural operators, and differentiable simulators with learned components exemplify this synthesis.

The transformation is not complete. Grand challenges remain: accurate climate projections for policy decisions, practical fusion energy, room-temperature superconductors, and quantum advantage for scientific simulation. But the tools available to address these challenges have expanded dramatically. Scientific machine learning has earned its place alongside traditional numerical methods and experimental techniques as a pillar of scientific inquiry.

Chapter Summary: SciML in Physical Sciences

1. **The value proposition is proven**: 10^3–$10^6\times$ speedups with 1–10% accuracy across weather prediction, materials simulation, and turbulent flow. ECMWF's operational AI forecasts (February 2025) mark the transition from research to production.

2. **Four problem types, one framework**: Forward surrogates, inverse reconstruction, control/optimization, and discovery share common methods (neural operators, physics-informed learning, reinforcement learning) with domain-specific adaptations.
3. **Hybrid physics-ML is the emerging paradigm**: The most robust results combine explicit physics for structure and guarantees with learned components for flexibility and efficiency.
4. **Validation remains essential**: Computational scale amplifies both capability and risk. The validation pyramid; ML screening → high-fidelity validation → experimental confirmation; structures responsible deployment.
5. **Open challenges define the frontier**: Extrapolation to unprecedented regimes, long-time stability, uncertainty quantification, and experimental integration remain active research areas.
6. **The transformation is underway**: SciML has earned its place as a pillar of computational science alongside traditional numerical methods and experimental techniques.

References

1. Pan, L., Zhang, H., Liang, M., Li, P., Cao, C., Du, L., Liu, J.: Comparative analysis of precipitation forecasts between the ECMWF artificial intelligence system AIFS and its integrated forecast system IFS. Weather Forecast. (2025)
2. Centofanti, E., Pavarino, L.F.: Operator learning for the FitzHugh-Nagumo model. In: 2024 International Conference on Computing in Natural Sciences, Biomedicine and Engineering (COMCONF), pp. 41–45. IEEE (2024)
3. Ronneberger, O., Fischer, P., Brox, T.: U-net: Convolutional networks for biomedical image segmentation. In: International Conference on Medical image computing and computer-assisted intervention, pp. 234–241. Springer (2015)
4. Pfaff, T., Fortunato, M., Sanchez-Gonzalez, A., Battaglia, P.: Learning mesh-based simulation with graph networks. In: International Conference on Learning Representations, (2020)
5. Gao, H., Ji, S.: Graph u-nets. In: International Conference on Machine Learning, pp. 2083–2092. (2019) PMLR
6. Lu, L., Jin, P., Karniadakis, G.E.: Deeponet: Learning nonlinear operators for identifying differential equations based on the universal approximation theorem of operators, (2019). arXiv:1910.03193 arXiv preprint
7. Ling, J., Kurzawski, A., Templeton, J.: Reynolds averaged turbulence modelling using deep neural networks with embedded invariance. J. Fluid Mech. **807**, 155–166 (2016)
8. Parish, E.J., Duraisamy, K.: A paradigm for data-driven predictive modeling using field inversion and machine learning. J. Comput. Phys. **305**, 758–774 (2016)
9. Wu, J.L., Xiao, H., Paterson, E.: Physics-informed machine learning approach for augmenting turbulence models: a comprehensive framework. Phys. Rev. Fluids **3**(7), 074602 (2018)
10. Duraisamy, K.: Perspectives on machine learning-augmented reynolds-averaged and large eddy simulation models of turbulence. Physical Review Fluids **6**(5), 050504 (2021)
11. Sirignano, J., MacArt, J.F.: Deep learning closure models for large-eddy simulation of flows around bluff bodies. J. Fluid Mech. **966**, A26 (2023)

12. MacArt, J.F., Sirignano, J., Freund, J.B.: Embedded training of neural-network subgrid-scale turbulence models. Physical Review Fluids **6**(5), 050502 (2021)
13. Lozano-Durán, A., Bae, H.J.: Machine learning building-block-flow wall model for large-eddy simulation. J. Fluid Mech. **963**, A35 (2023)
14. Zhang, G., Song, S., Zhang, H., Chang, D.: Pre-stack seismic inversion based on model-constrained generative adversarial network. Front. Earth Sci. **12**, 1373859 (2024)
15. Rasht-Behesht, M., Huber, C., Shukla, K., Karniadakis, G.E.: Physics-informed neural networks (pinns) for wave propagation and full waveform inversions. J. Geophys. Res. Solid Earth **127**(5), e2021JB023120 (2022)
16. Song, C., Geng, H., Wang, Y., Waheed, U.B., Liu, C.: Simultaneous p-and s-wave seismic traveltime tomography using physics-informed neural networks. Geophys, Prospect (2025)
17. Martin, G.S., Wiley, R., Marfurt, K.J.: Marmousi2: An elastic upgrade for marmousi. Lead. Edge **25**(2), 156–166 (2006)
18. Liu, Y.H., Huang, J.P., Chen, L., Mao, Q., Tian, K.: 3d reverse-time migration for pure p-wave in orthorhombic media. Pet. Sci. **21**(6), 3937–3950 (2024)
19. Jin, X., Yu, L.: Assessing high-resolution analysis of surface heat fluxes in the gulf stream region. Journal of Geophysical Research: Oceans **118**(10), 5353–5375 (2013)
20. Lee, J.W., Park, W.B., Lee, J.H., Singh, S.P., Sohn, K.S.: A deep-learning technique for phase identification in multiphase inorganic compounds using synthetic xrd powder patterns. Nat. Commun. **11**(1), 86 (2020)
21. Vecsei, G., Jáger, G., Juhász, L., Tomán, J.J., Odhiambo, V.O., Szilágyi, I.M., Erdélyi, Z., Cserháti, C.: Effect of stacking order in cylindrical geometry on the growth of $ZnAl_2O4$ spinel phase. Materialia **30**, 101819 (2023)
22. Oviedo, L.R., Muraro, P.C.L., Pavoski, G., Espinosa, D.C.R., Ruiz, Y.P.M., Galembeck, A., Rhoden, C.R.B., da Silva, W.L.: Synthesis and characterization of nanozeolite from (agro) industrial waste for application in heterogeneous photocatalysis. Environ. Sci. Pollut. Res. **29**(3), 3794–3807 (2022)
23. Szymanski, N.J., Bartel, C.J., Zeng, Y., Diallo, M., Kim, H., Ceder, G.: Adaptively driven X-ray diffraction guided by machine learning for autonomous phase identification. NPJ Comput. Mater. **9**(1), 31 (2023)
24. Chen, L., Wang, B., Zhang, W., Zheng, S., Chen, Z., Zhang, M., Dong, C., Pan, F., Li, S.: Crystal structure assignment for unknown compounds from x-ray diffraction patterns with deep learning. J. Am. Chem. Soc. **146**(12), 8098–8109 (2024)
25. Raissi, M., Yazdani, A., Karniadakis, G.E.: Hidden fluid mechanics: a Navier-Stokes informed deep learning framework for assimilating flow visualization data, (2018). arXiv:1808.04327 arXiv preprint
26. Cai, S., Liang, J., Gao, Q., Xu, C., Wei, R.: Particle image velocimetry based on a deep learning motion estimator. IEEE Trans. Instrum. Meas. **69**(6), 3538–3554 (2019)
27. Jin, X., Xu, M., Yang, Y., Chen, W.: Spatiotemporal coupling deep neural network for time-resolved flow field reconstruction around a circular cylinder. Phys. Fluids 36(10), (2024)
28. Du, P., Parikh, M.H., Fan, X., Liu, X.Y., Wang, J.X.: Conditional neural field latent diffusion model for generating spatiotemporal turbulence. Nat. Commun. **15**(1), 10416 (2024)
29. Zhang, J., Liu, J., Huang, Z.: Improved deep learning method for accurate flow field reconstruction from sparse data. Ocean Eng. **280**, 114902 (2023)
30. Ding, Y., Li, Z., Chen, Z., Ji, Y., Yu, J., Ye, J.: Full-volume 3d fluid flow reconstruction with light field piv. IEEE Trans. Pattern Anal. Mach. Intell. **45**(7), 8405–8418 (2023)
31. Torlai, G., Mazzola, G., Carrasquilla, J., Troyer, M., Melko, R., Carleo, G.: Neural-network quantum state tomography. Nat. Phys. **14**(5), 447–450 (2018)
32. Quek, Y., Fort, S., Ng, H.K.: Adaptive quantum state tomography with neural networks. NPJ Quant. Inform. **7**(1), 105 (2021)
33. Neugebauer, M., Fischer, L., Jäger, A., Czischek, S., Jochim, S., Weidemüller, M., Gärttner, M.: Neural network quantum state tomography in a two-qubit experiment, (2020). arXiv:2007.16185 arXiv preprint

34. Ahmed, S., Sanchez Munoz, C., Nori, F., Kockum, A.F.: Quantum state tomography with conditional generative adversarial networks. Phys. Rev. Lett. **127**(14), 140502 (2021)
35. Degrave, J., Felici, F., Buchli, J., Neunert, M., Tracey, B., Carpanese, F., Ewalds, T., Hafner, R., Abdolmaleki, A., de Las Casas, D., et al.: Magnetic control of tokamak plasmas through deep reinforcement learning. Nature **602**(7897), 414–419 (2022)
36. Subbotin, G., Sorokin, D., Nurgaliev, M., Granovskiy, A., Kharitonov, I., Adishchev, E., Khairutdinov, E., Clark, R., Shen, H., Choi, W.: Reconstruction-free magnetic control of DIII-D plasma with deep reinforcement learning, (2025). arXiv:2506.13267 arXiv preprint
37. Citrin, J., Goodfellow, I., Raju, A., Chen, J., Degrave, J., Donner, C., Felici, F., Hamel, P., Huber, A., Nikulin, D.: Torax: A fast and differentiable tokamak transport simulator in JAX, (2024). arXiv:2406.06718 arXiv preprint
38. Rabault, J., Kuchta, M., Jensen, A., Réglade, U., Cerardi, N.: Artificial neural networks trained through deep reinforcement learning discover control strategies for active flow control. J fluid mechanics **865**, 281–302 (2019)
39. Paris, R., Beneddine, S., Dandois, J.: Robust flow control and optimal sensor placement using deep reinforcement learning. J Fluid Mech. **913**, A25
40. Suárez Morales, P.: Multi-agent reinforcement learning for enhanced turbulence control in bluff bodies (Doctoral dissertation, KTH Royal Institute of Technology), (2024)
41. Ren, F., Wen, X., Tang, H.:Model-free closed-loop control of flow past a bluff body: Methods, applications, and emerging trends. In Actuators **13**12, p. 488, MDPI (2024)
42. Guastoni, L., Rabault, J., Azizpour, H., Vinuesa, R.: Drag-reduction strategies in wall-bounded turbulent flows using deep reinforcement learning, (2023). arXiv:2309.02943 arXiv preprint
43. Li, Y., Cui, W., Jia, Q., Li, Q., Yang, Z., Morzyński, M., Noack, B. R. (2022). Explorative gradient method for active drag reduction of the fluidic pinball and slanted Ahmed body. J of Fluid Mech. **932**, A7 (2022)

Chapter 12
Applications in Engineering and Technology

Roads? Where we're going, we don't need roads.

Doc Brown, Back to the Future

Abstract Engineering applications of Scientific Machine Learning must meet stringent requirements: safety certification, real-time performance, manufacturing tolerances, and integration with existing design workflows. This chapter examines SciML across engineering domains. Structural engineering covers physics-informed analysis of beams, plates, and frames; neural operators for parametric structural response; and health monitoring through learned inverse models. Manufacturing addresses additive manufacturing (melt pool prediction, defect detection), process monitoring, predictive maintenance, and optimization under uncertainty. Aerospace and automotive applications span aerodynamic design acceleration, flight simulation, vehicle dynamics, and battery management for electric vehicles. Robotics examines learned dynamics models, contact-rich manipulation, sim-to-real transfer, and the physics-informed approaches that improve generalization. Biomedical engineering covers cardiovascular modeling, physics-constrained medical imaging, surgical planning, and pharmacokinetics. Chemical and process engineering addresses reaction modeling, separation processes, and real-time optimization. Energy systems examine renewable forecasting, grid optimization, and building simulation. Each section includes working implementations with benchmark results.

12.1 Introduction

The previous chapter explored how Scientific Machine Learning transforms our understanding of physical sciences—from plasma physics to climate modeling, from quantum chemistry to astrophysics. In those domains, we sought to accelerate the simulation of fundamental physical phenomena, discover governing equations from observational data, and bridge scales from atomic to cosmic. The primary currency was *understanding*: how does nature behave, and can we predict it?

B. P. Bhuyan et al., *Scientific Machine Learning*,
https://doi.org/10.1007/978-981-95-7856-6_12

Engineering and technology present a fundamentally different challenge. Here, the question shifts from "What does nature do?" to "What can we *make* nature do?" Engineers do not merely observe systems; they design, build, control, and optimize them. A structural engineer must ensure a bridge survives a century of traffic and weather. A manufacturing engineer must produce thousands of identical parts with tolerances measured in microns. A robotics engineer must enable a manipulator to react within milliseconds to unexpected disturbances. In each case, the physical laws remain the same, but the *purpose* transforms from scientific discovery to technological creation.

This distinction has profound implications for how Scientific Machine Learning enters engineering practice. While a physicist might accept a 5% error in predicting galaxy formation (the universe does not negotiate), an aerospace engineer cannot tolerate 5% uncertainty in wing flutter predictions—aircraft fall from the sky. The standards of verification, validation, and certification in engineering applications far exceed typical academic benchmarks. Yet the potential rewards are equally transformative: simulation times dropping from days to seconds, real-time digital twins enabling predictive maintenance, and design optimization exploring millions of configurations that would bankrupt any physical testing program.

12.1.1 From Laboratory to Factory Floor: The Engineering Imperative

Consider a familiar scenario. You are driving across a bridge during a thunderstorm. Rain hammers the windshield, wind gusts rock the car, and eighteen-wheelers barrel past in the opposite lane. Somewhere beneath the asphalt, steel cables and concrete pillars bear not only your vehicle but the accumulated stress of millions of crossings, decades of freeze-thaw cycles, and the invisible fatigue cracks propagating through welds made by workers long since retired.

How do engineers ensure this bridge remains standing? Traditional approaches combine conservative safety factors (design for twice the expected load), periodic inspections (send workers to visually examine critical joints), and replacement schedules based on average component lifetimes. This strategy has served civilization well—modern infrastructure rarely collapses—but it is extraordinarily inefficient. Bridges are closed for inspection when they might have years of safe service remaining. Components are replaced according to schedules rather than actual condition. Meanwhile, hidden damage accumulates in uninspected locations.

Scientific Machine Learning offers a radically different paradigm: *continuous, physics-informed monitoring*. Sensors distributed throughout the structure feed real-time data—strain measurements, acceleration responses, temperature gradients—to neural networks that have internalized the governing equations of structural mechanics. Rather than asking "Has this component exceeded its scheduled lifetime?" these systems ask "Given the loads this specific structure has experienced and the physics of fatigue crack propagation, what is the probability of failure in the next year?"

The Digital Twin Revolution

A **digital twin** is a virtual replica of a physical system that evolves in real-time based on sensor data, physics models, and machine learning. Unlike traditional simulations run occasionally by specialists, digital twins operate continuously, enabling:

- **Predictive maintenance**: Anticipating failures before they occur
- **What-if analysis**: Evaluating operational changes without physical risk
- **Anomaly detection**: Identifying deviations from expected behavior
- **Remaining useful life estimation**: Quantifying how long components will last

The digital twin market is projected to reach $259.3 billion by 2032, with 92% of manufacturing companies reporting ROI above 10%.

This bridge example illustrates the broader transformation underway across engineering. In aerospace, neural operators trained on computational fluid dynamics databases predict aerodynamic forces in milliseconds rather than hours, enabling real-time flight simulation and autonomous vehicle control. In manufacturing, physics-informed networks monitor additive manufacturing processes layer-by-layer, detecting thermal anomalies that would cause defects before they solidify into permanent flaws. In chemical plants, hybrid models combining reaction kinetics with data-driven corrections optimize production while respecting safety constraints that pure black-box approaches might violate.

12.1.2 What Makes Engineering Different?

Readers who studied the Physical Sciences chapter will recognize many familiar techniques: Physics-Informed Neural Networks solving partial differential equations, neural operators learning mappings between function spaces, graph neural networks capturing interactions on complex geometries. The mathematical machinery transfers directly. What changes is the context of application, and this context imposes distinctive requirements.

Real-Time Constraints

Scientific simulations often run on supercomputers for days, with researchers patiently awaiting results. Engineering systems frequently demand responses within milliseconds. A robotic arm must adjust its trajectory as a human worker unexpectedly enters its workspace. An autonomous vehicle must recognize that the object ahead is a pedestrian, not a plastic bag. A power grid controller must stabilize frequency fluctuations before they cascade into blackouts. These applications require not just accurate models, but models that execute fast enough for closed-loop control.

Certification and Trust

When a machine learning model recommends a movie, an incorrect prediction wastes an evening. When a machine learning model certifies an aircraft wing, an incorrect prediction kills people. Engineering applications in safety-critical domains—aerospace, nuclear, medical devices, autonomous vehicles—require not just performance metrics but formal guarantees. Can we prove the model never predicts tensile stress below a certain threshold when actual stress exceeds it? Can we bound the worst-case prediction error? Can we explain *why* the model made a particular prediction? These questions, peripheral in many ML applications, become central in engineering.

Integration with Legacy Systems

Physical sciences research typically occurs in academic environments where investigators control the entire computational pipeline. Engineering practice embeds ML components within ecosystems of existing tools: computer-aided design (CAD) software, finite element analysis (FEA) packages, manufacturing execution systems (MES), supervisory control and data acquisition (SCADA) networks. A neural operator that predicts flow fields magnificently but cannot import geometries from CATIA or export results to ANSYS may find no industrial adoption regardless of its technical merit.

Data Scarcity in Specific Regimes

While internet companies drown in data, engineering applications often face the opposite problem. How many aircraft have actually experienced wing flutter leading to structural failure? (Fortunately, very few.) How many nuclear reactors have melted down? (Three major incidents in seventy years.) The events we most need to predict—catastrophic failures—are precisely the events for which we have the least data. Physics-informed approaches shine here, encoding the known dynamics that govern failure modes even when failure data itself remains sparse.

Definition: The Sim-to-Real Gap

The **sim-to-real gap** refers to the discrepancy between model behavior in simulation and actual behavior in the physical world. Sources include:

- Unmodeled dynamics (friction, backlash, compliance)
- Sensor noise and calibration errors
- Environmental variations (temperature, humidity, wear)
- Manufacturing tolerances and material variability

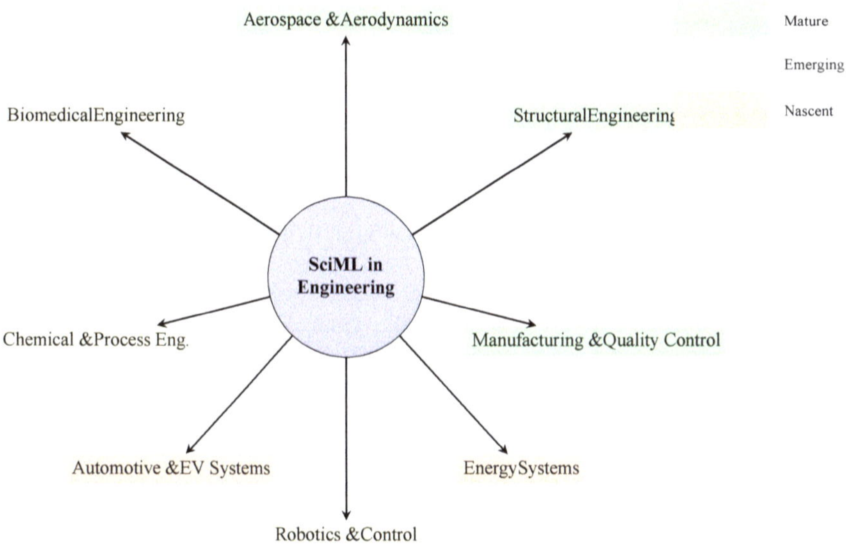

Fig. 12.1 Landscape of Scientific Machine Learning applications in engineering and technology. Maturity reflects both research activity and industrial deployment as of 2025. Aerospace and manufacturing lead in production deployments; robotics and chemical engineering show rapid growth; energy systems and automotive applications are accelerating with electric vehicle adoption

Physics-informed approaches can narrow this gap by constraining learned models to respect physical laws, but bridging it completely remains an open challenge in robotics, autonomous systems, and process control.

12.1.3 The Engineering Applications Landscape

Scientific Machine Learning has penetrated virtually every engineering discipline, though with varying depths of maturity. Figure 12.1 provides a conceptual map of the territory this chapter will explore.

Structural Engineering and Civil Infrastructure represents one of the most natural applications of physics-informed learning. The governing equations—elasticity, beam theory, plate theory—are well-established, and the inverse problems of damage detection and structural identification align perfectly with PINN capabilities. Real-time structural health monitoring of bridges, buildings, and aircraft combines sensor networks with neural networks that have internalized structural mechanics, enabling continuous assessment without costly shutdowns.

Aerospace and Automotive Engineering demand the fastest simulations for the most complex geometries. Neural operators trained on computational fluid dynamics databases achieve speedups of 10^4 to 10^5 over traditional solvers, enabling real-time aerodynamic optimization, flight simulation, and autonomous vehicle development. Physics-informed approaches ensure predicted flow fields satisfy conservation laws even when training data is sparse at extreme flight conditions.

Manufacturing and Industry 4.0 leverage digital twins to optimize production in real-time. Additive manufacturing benefits particularly from physics-informed thermal modeling, where predicting temperature fields layer-by-layer enables defect prevention rather than post-hoc inspection. Quality control systems using deep learning achieve defect detection rates exceeding 95%, while predictive maintenance reduces unplanned downtime by 70%.

Robotics and Control Systems push the boundaries of real-time physics-informed inference. Lagrangian and Hamiltonian neural networks learn robot dynamics while guaranteeing energy conservation, enabling controllers that remain stable across operating conditions unseen during training. Neural control barrier functions synthesized through physics-informed approaches provide formal safety guarantees for autonomous systems.

Biomedical Engineering applies SciML to the ultimate complex system: the human body. Physics-informed networks reconstruct blood flow from sparse MRI measurements, predict cardiac electrophysiology from surface electrodes, and simulate tissue mechanics for surgical planning. The HeartFlow system, FDA-cleared since 2014, demonstrates that physics-informed deep learning can achieve clinical-grade accuracy while reducing analysis time from hours to minutes.

Chemical and Process Engineering face the challenge of modeling systems where complete mechanistic knowledge is unavailable. Reaction kinetics, multiphase flows, and separation processes involve phenomena too complex for first-principles derivation yet too safety-critical for pure black-box approaches. Hybrid models embedding partial physics knowledge with data-driven corrections achieve robust predictions from scarce process data.

Energy Systems encompass both the generation side (renewable forecasting, grid stability) and the storage side (battery management, degradation prediction). Graph neural networks capture the topology of power grids, enabling stability assessment 1000× faster than conventional simulation. Physics-informed battery models achieve state-of-health predictions with sub-1% error, critical for electric vehicle range estimation and second-life applications.

12.1.4 A Tale of Two Speedups

Throughout this chapter, you will encounter impressive speedup claims: neural operators 1000× faster than CFD, surrogate models 10,000× faster than finite element analysis, real-time predictions from systems that previously required overnight computation. These numbers are real, but they require careful interpretation.

Consider an analogy from everyday life. Suppose you need to travel from New York to Los Angeles. A commercial flight takes about 5 h; driving takes about 40 h. Aviation is 8× faster! But this comparison ignores crucial context. If you are already at the airport with a ticket in hand, flying wins decisively. If you are at home in rural Pennsylvania with a car in the driveway and the nearest airport two hours away, the calculus changes. If you need to transport a piano, driving becomes the only option regardless of speed.

Engineering speedups follow similar logic. A neural operator may predict a flow field in 10 milliseconds versus 10 h for a CFD solver—a speedup of 3.6×10^6. But this comparison assumes:

1. The neural operator has already been trained (training may have required thousands of CFD simulations)
2. The query falls within the training distribution (out-of-distribution queries may yield unreliable predictions)
3. The required accuracy is achieved (the CFD solver may be more accurate)
4. The geometry format is compatible (CAD-to-neural-operator pipelines are less mature than CAD-to-CFD)

Tip: Interpreting Speedup Claims

When evaluating reported speedups, ask:

1. **What is the amortized cost?** Include training time, data generation, and preprocessing. A model requiring 10,000 CFD simulations to train provides value only if you need predictions for more than 10,000 configurations.
2. **What is the accuracy tradeoff?** A 1000× speedup with 10× higher error may or may not be acceptable depending on application requirements.
3. **What is the scope of validity?** Speedups measured on benchmark problems may not transfer to industrial geometries, boundary conditions, or operating regimes.
4. **What is the deployment overhead?** Integration with existing CAE workflows, GPU requirements, and maintenance costs affect total value.

The most valuable applications of SciML in engineering are not necessarily those with the largest headline speedups. They are those where the *amortized* speedup—accounting for all costs—enables workflows that were previously impossible. Real-time control requires predictions in milliseconds; no amount of parallelization makes CFD run that fast. Design optimization requires exploring millions of configurations; training a surrogate on thousands of simulations to enable this exploration represents genuine value creation. Digital twins require continuous predictions throughout asset lifetime; the training cost is negligible compared to decades of operational monitoring.

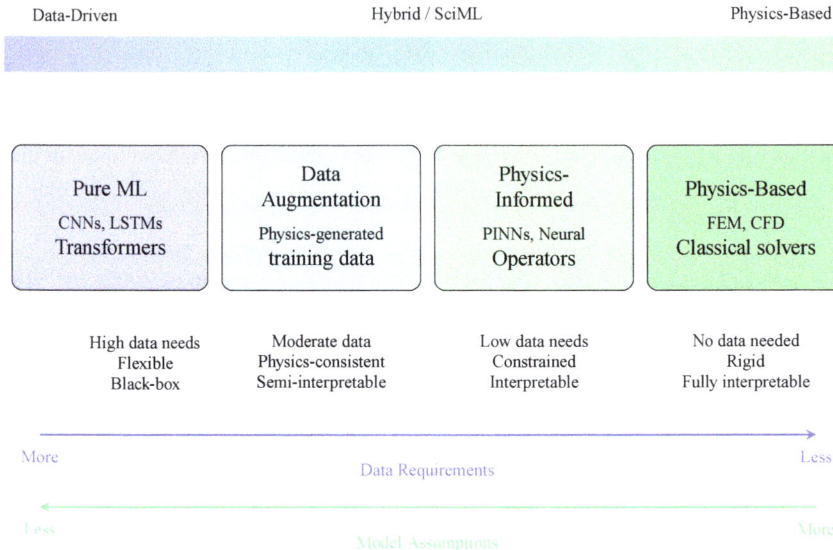

Fig. 12.2 The physics-data spectrum in engineering modeling. Pure data-driven approaches require extensive training data but make minimal assumptions. Pure physics-based approaches require no data but depend on accurate equations. Scientific Machine Learning occupies the middle ground, leveraging partial physics knowledge to reduce data requirements while maintaining flexibility for phenomena that resist first-principles modeling

12.1.5 The Physics-Data Spectrum

Engineering applications of machine learning span a spectrum from purely data-driven to purely physics-based, with SciML occupying the middle ground. Understanding where different approaches excel clarifies when physics-informed methods provide genuine advantages versus unnecessary complexity (Fig. 12.2).

Pure data-driven approaches (convolutional neural networks for defect detection, recurrent networks for time series prediction) require no physics knowledge but demand large, representative datasets. They excel when data is abundant and the underlying physics is either unknown or too complex to model. Quality control in semiconductor manufacturing, where millions of chips provide ample training data and defect physics involves quantum-scale phenomena, exemplifies this regime.

Data augmentation with physics uses simulations to generate training data for machine learning models. Rather than embedding physics in the loss function, this approach embeds physics in the data generation process. A neural network trained on CFD simulations learns to predict flow fields without ever seeing the Navier-Stokes equations explicitly. This strategy works well when high-fidelity simulators exist and the computational cost of generating training data is acceptable.

Physics-informed approaches (PINNs, neural operators with physics constraints) embed governing equations directly into learning. They excel when data is scarce, extrapolation beyond training conditions is required, or physical constraints (conservation laws, boundary conditions) must be exactly satisfied. Structural damage identification, where failure data is rare but structural mechanics is well-understood, exemplifies this regime.

Pure physics-based approaches (finite element analysis, computational fluid dynamics) require no training data but depend entirely on accurate governing equations and material properties. They remain the gold standard when equations are well-established, computational budgets allow detailed simulation, and certification requirements demand interpretable, verifiable results.

The key insight is that SciML methods do not universally outperform alternatives. They occupy a specific niche where partial physics knowledge, limited data, and computational constraints intersect. An engineer choosing between approaches must consider: How much relevant data exists? How well-established are the governing equations? What accuracy is required? What computational budget is available? What certification requirements apply? Different answers lead to different optimal approaches.

Warning: When PINNs May Disappoint

Physics-Informed Neural Networks are not universally superior to traditional numerical methods. Critical evaluations have shown:

- For well-posed forward problems with clean boundary conditions, optimized finite element solvers often achieve comparable accuracy in less total time (including PINN training)
- PINN accuracy typically saturates around 10^{-4} to 10^{-5} relative error due to optimization challenges, whereas FEM provides controllable convergence to machine precision
- Spectral bias causes PINNs to struggle with high-frequency features, turbulence, and sharp gradients
- Hyperparameter sensitivity requires careful tuning; reported benchmark results may not transfer to new problems

PINNs excel for inverse problems, high-dimensional parameter spaces, irregular geometries, and scenarios requiring fast inference after training. Choose methods based on problem characteristics, not hype.

12.1.6 Industrial Adoption: From Research to Production

The gap between academic publications and industrial deployment remains substantial but is narrowing rapidly. Understanding where SciML has achieved production status versus where it remains experimental helps calibrate expectations.

Deployed at Scale:

- **NVIDIA PhysicsNeMo** powers production SciML systems at Shell (CO_2 storage modeling with 100,000× speedup), Stone Ridge Technology (reservoir simulation), and numerous automotive OEMs
- **HeartFlow FFRCT** has analyzed over 250,000 patients for coronary artery disease diagnosis using physics-informed CFD acceleration
- **Google Visual Inspection AI** achieves 10× accuracy improvement with 300× fewer labeled images in manufacturing quality control
- **Siemens Executable Digital Twin** platform enables real-time physics-based simulation across product lifecycles, with the Amberg Electronics Plant achieving 99.99885% reliability

Emerging Production Use:

- Aerospace design optimization using neural operators (Airbus Aero-Nef, Boeing digital twin initiatives)
- Battery management systems with physics-informed state estimation (Tesla, major EV manufacturers)
- Predictive maintenance platforms combining physics models with sensor analytics (GE Vernova, Rolls-Royce)

Active Research with Limited Deployment:

- Real-time turbulence modeling for practical Reynolds numbers
- Certified neural networks for safety-critical aerospace applications
- Fully autonomous design optimization without human oversight
- Foundation models for cross-domain engineering transfer learning

The pattern across industries is consistent: SciML achieves production status first in applications where (1) the value proposition is overwhelming (orders-of-magnitude speedup enabling previously impossible workflows), (2) the accuracy requirements are achievable (surrogate models for optimization rather than final certification), and (3) integration paths exist (APIs compatible with standard engineering tools). Applications requiring formal verification, extreme accuracy, or complete replacement of established methods face longer adoption timelines.

This chapter progresses through major engineering domains, following our established pedagogical pattern of progressive complexity. For each application area, we begin with everyday analogies that build intuition, proceed through conceptual development and mathematical formulation, and conclude with implementation details including working code examples and computational results.

12.2 Structural Engineering and Civil Infrastructure

Every structure tells two stories. The first is visible: the graceful arc of a suspension bridge, the soaring height of a skyscraper, the intricate lattice of a stadium roof. The second story is invisible, written in the language of stress and strain, fatigue and fracture, load paths and failure modes. Structural engineers must read both stories fluently, ensuring that the invisible narrative never ends in catastrophe.

Traditional structural analysis relies on finite element methods (FEM) that discretize continuous structures into millions of elements, solving equilibrium equations at each node. This approach has enabled the design of structures that would have seemed impossible a century ago. Yet FEM faces fundamental limitations: each analysis requires hours of computation, preventing real-time assessment; inverse problems (inferring damage from measurements) require iterative optimization over expensive forward simulations; and uncertainty quantification demands Monte Carlo sampling across parameter distributions, multiplying already substantial computational costs.

Scientific Machine Learning offers a complementary paradigm. Physics-Informed Neural Networks learn to satisfy equilibrium equations, compatibility conditions, and constitutive relationships directly, enabling rapid inference once trained. Neural operators learn mappings from loading conditions to stress fields, amortizing the cost of many FEM simulations into a single training phase. Inverse networks reconstruct damage patterns from sparse sensor measurements, transforming structural health monitoring from periodic inspection to continuous assessment.

This section develops these capabilities systematically. We begin with the governing equations of structural mechanics, establish how they embed into neural network loss functions, and progress through applications of increasing complexity: from stress analysis in simple frames to damage detection in bridges to seismic response prediction for building portfolios.

12.2.1 The Language of Structures: Governing Equations

Before embedding physics into neural networks, we must articulate what physics governs structural behavior. The fundamental principles are deceptively simple: equilibrium (forces balance), compatibility (deformations fit together), and constitutive behavior (materials respond to stress in characteristic ways). Their mathematical expression, however, leads to partial differential equations whose analytical solutions exist only for idealized cases.

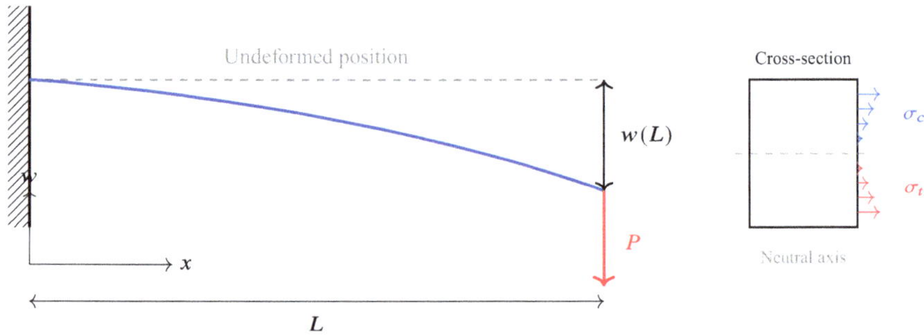

Fig. 12.3 Cantilever beam under point load. The beam deflects downward, developing compressive stress on the top surface and tensile stress on the bottom. The stress distribution through the cross-section (inset) varies linearly from the neutral axis in elastic bending

12.2.1.1 From Intuition to Equations

Consider a diving board. When you stand at the end, the board bends downward. Your weight creates a moment that varies along the board's length—maximum at the support, zero at the free end. The board's material resists this bending by developing internal stresses: compression on the top surface, tension on the bottom. The curvature you observe reflects the balance between applied moment and material resistance.

The Euler-Bernoulli beam equation formalizes this intuition. For a beam with flexural rigidity EI (Young's modulus times moment of inertia) under distributed load $q(x)$, the transverse deflection $w(x)$ satisfies:

$$EI\frac{d^4w}{dx^4} = q(x) \tag{12.1}$$

This fourth-order ordinary differential equation requires four boundary conditions. For the cantilever in Fig. 12.3:

$$w(0) = 0 \quad \text{(fixed position at support)} \tag{12.2}$$

$$\left.\frac{dw}{dx}\right|_{x=0} = 0 \quad \text{(fixed slope at support)} \tag{12.3}$$

$$\left.\frac{d^2w}{dx^2}\right|_{x=L} = 0 \quad \text{(zero moment at free end)} \tag{12.4}$$

$$\left.\frac{d^3w}{dx^3}\right|_{x=L} = -\frac{P}{EI} \quad \text{(shear from point load)} \tag{12.5}$$

For this simple case, the analytical solution exists: $w(x) = \frac{P}{6EI}(3Lx^2 - x^3)$. But real structures—with variable cross-sections, multiple supports, distributed and concentrated loads, temperature gradients, and material nonlinearities—require numerical solution.

Definition: The Structural Analysis Hierarchy

Structural mechanics employs models of varying complexity:

- **Truss analysis**: Members carry only axial forces; joints are pins. Governed by algebraic equilibrium equations.
- **Beam/frame analysis**: Members carry axial forces, shear, and bending moments. Governed by the Euler-Bernoulli or Timoshenko beam equations.
- **Plate/shell analysis**: Two-dimensional structures with out-of-plane loading. Governed by the biharmonic equation (Kirchhoff plates) or Reissner-Mindlin equations.
- **3D continuum analysis**: General solid mechanics. Governed by the Navier-Cauchy equations with appropriate constitutive models.

Each level adds physical fidelity at the cost of mathematical complexity. SciML approaches can target any level, with PINNs naturally handling the governing PDEs.

12.2.1.2 The Equations of Linear Elasticity

For general three-dimensional structures, the governing equations combine equilibrium, strain-displacement relations, and constitutive behavior. In the absence of body forces, equilibrium requires:

$$\nabla \cdot \boldsymbol{\sigma} = \mathbf{0} \tag{12.6}$$

where $\boldsymbol{\sigma}$ is the Cauchy stress tensor. In component form:

$$\frac{\partial \sigma_{ij}}{\partial x_j} = 0, \quad i = 1, 2, 3 \tag{12.7}$$

The strain tensor $\boldsymbol{\varepsilon}$ relates to displacements $\mathbf{u}$ through:

$$\varepsilon_{ij} = \frac{1}{2}\left(\frac{\partial u_i}{\partial x_j} + \frac{\partial u_j}{\partial x_i}\right) \tag{12.8}$$

For linear elastic materials, stress and strain connect through Hooke's law:

$$\sigma_{ij} = \lambda \varepsilon_{kk}\delta_{ij} + 2\mu\varepsilon_{ij} \tag{12.9}$$

where λ and μ are Lamé parameters related to Young's modulus E and Poisson's ratio ν:

$$\lambda = \frac{E\nu}{(1+\nu)(1-2\nu)}, \quad \mu = \frac{E}{2(1+\nu)} \tag{12.10}$$

Combining these yields the Navier-Cauchy equations for displacement:

$$\mu\nabla^2\mathbf{u} + (\lambda + \mu)\nabla(\nabla \cdot \mathbf{u}) = \mathbf{0} \tag{12.11}$$

These equations, supplemented by boundary conditions (prescribed displacements on Γ_D, prescribed tractions on Γ_N), define the forward problem of structural analysis: given geometry, material properties, and loads, find the displacement, strain, and stress fields.

12.2.2 Physics-Informed Neural Networks for Structural Analysis

Physics-Informed Neural Networks solve the forward problem by training a neural network to satisfy the governing equations throughout the domain. Unlike finite element methods that discretize the domain into elements, PINNs represent the solution as a continuous function approximated by the network.

12.2.2.1 PINN Formulation for Beam Problems

Consider the Euler-Bernoulli beam equation (12.1). A PINN approximates the deflection $w(x)$ with a neural network $w_\theta(x)$, where θ denotes the network parameters. The loss function penalizes violations of both the governing equation and boundary conditions:

$$\mathcal{L}(\theta) = \mathcal{L}_{\mathrm{PDE}}(\theta) + \lambda_{\mathrm{BC}}\mathcal{L}_{\mathrm{BC}}(\theta) \tag{12.12}$$

The PDE residual loss evaluates how well the network satisfies the beam equation at collocation points $\{x_i\}_{i=1}^{N_c}$:

$$\mathcal{L}_{\mathrm{PDE}}(\theta) = \frac{1}{N_c}\sum_{i=1}^{N_c}\left|EI\left.\frac{d^4 w_\theta}{dx^4}\right|_{x_i} - q(x_i)\right|^2 \tag{12.13}$$

The boundary condition loss enforces constraints at the beam ends:

$$\mathcal{L}_{\mathrm{BC}}(\theta) = |w_\theta(0)|^2 + \left|\left.\frac{dw_\theta}{dx}\right|_0\right|^2 + \left|\left.\frac{d^2 w_\theta}{dx^2}\right|_L\right|^2 + \left|\left.\frac{d^3 w_\theta}{dx^3}\right|_L + \frac{P}{EI}\right|^2 \tag{12.14}$$

Automatic differentiation computes the required derivatives exactly, avoiding finite difference approximation errors (Fig. 12.4).

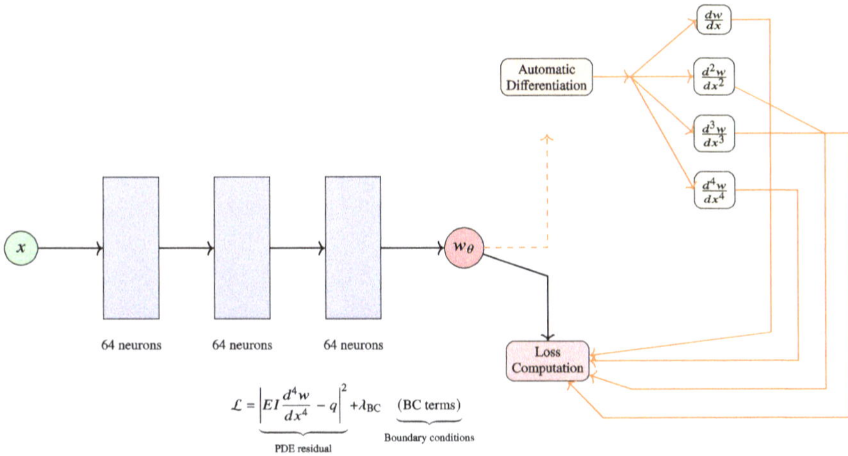

Fig. 12.4 PINN architecture for beam analysis. The neural network maps position x to deflection $w_\theta(x)$. Automatic differentiation computes derivatives up to fourth order, enabling evaluation of the Euler-Bernoulli equation residual. The loss function combines PDE residual and boundary condition terms

12.2.2.2 Extension to 2D and 3D Elasticity

For general elasticity problems, the PINN takes spatial coordinates as input and outputs displacement components:

$$\mathbf{u}_\theta(\mathbf{x}) = [u_\theta(x, y, z), v_\theta(x, y, z), w_\theta(x, y, z)]^T \tag{12.15}$$

The loss function enforces the Navier-Cauchy equations (12.11) at interior collocation points and boundary conditions on the domain boundary:

$$\mathcal{L}(\theta) = \frac{1}{N_{\text{int}}}\sum_{i=1}^{N_{\text{int}}}\|\mathbf{r}(\mathbf{x}_i)\|^2 + \frac{\lambda_D}{N_D}\sum_{j=1}^{N_D}\|\mathbf{u}_\theta(\mathbf{x}_j) - \mathbf{u}_D(\mathbf{x}_j)\|^2 + \frac{\lambda_N}{N_N}\sum_{k=1}^{N_N}\|\mathbf{t}_\theta(\mathbf{x}_k) - \mathbf{t}_N(\mathbf{x}_k)\|^2 \tag{12.16}$$

where $\mathbf{r}(\mathbf{x})$ is the PDE residual, $\mathbf{u}_D$ are prescribed displacements on Dirichlet boundaries, and $\mathbf{t}_N$ are prescribed tractions on Neumann boundaries. The traction vector $\mathbf{t} = \boldsymbol{\sigma} \cdot \mathbf{n}$ requires computing the stress tensor from the displacement field through differentiation and the constitutive law.

Implementation Note: Computing Stress from PINN Output

Given a PINN outputting displacements $\mathbf{u}_\theta(\mathbf{x})$, the stress field requires:

1. Compute displacement gradients via automatic differentiation: $\frac{\partial u_i}{\partial x_j}$
2. Form the strain tensor: $\varepsilon_{ij} = \frac{1}{2}\left(\frac{\partial u_i}{\partial x_j} + \frac{\partial u_j}{\partial x_i}\right)$
3. Apply Hooke's law: $\sigma_{ij} = \lambda \varepsilon_{kk}\delta_{ij} + 2\mu\varepsilon_{ij}$

All operations are differentiable, enabling end-to-end training through the stress computation.

12.2.2.3 Multi-level PINNs for High-Order Equations

The fourth-order beam equation and biharmonic plate equation pose challenges for standard PINNs: computing fourth derivatives amplifies numerical noise, and the loss landscape becomes increasingly non-convex. Multi-level PINN formulations address this by decomposing high-order equations into systems of lower-order equations.

For the beam equation, introduce the bending moment $M(x) = -EI\frac{d^2w}{dx^2}$ and shear force $V(x) = \frac{dM}{dx}$. The fourth-order equation becomes:

$$\frac{d^2w}{dx^2} = -\frac{M}{EI} \tag{12.17}$$

$$\frac{dM}{dx} = V \tag{12.18}$$

$$\frac{dV}{dx} = q(x) \tag{12.19}$$

A multi-output PINN predicts $[w(x), M(x), V(x)]$ simultaneously, with the loss enforcing all three equations plus compatibility between outputs:

$$\mathcal{L} = \mathcal{L}_{\text{eq1}} + \mathcal{L}_{\text{eq2}} + \mathcal{L}_{\text{eq3}} + \lambda_c \mathcal{L}_{\text{compat}} + \lambda_{\text{BC}} \mathcal{L}_{\text{BC}} \tag{12.20}$$

The compatibility loss ensures that the predicted moment equals the curvature-derived moment:

$$\mathcal{L}_{\text{compat}} = \frac{1}{N_c} \sum_{i=1}^{N_c} \left| M_\theta(x_i) + EI \frac{d^2 w_\theta}{dx^2}\bigg|_{x_i} \right|^2 \tag{12.21}$$

This formulation achieves lower errors than direct fourth-derivative computation, with multi-level PINNs for greenhouse structural frames achieving maximum errors of 1.24% compared to FEM benchmarks.

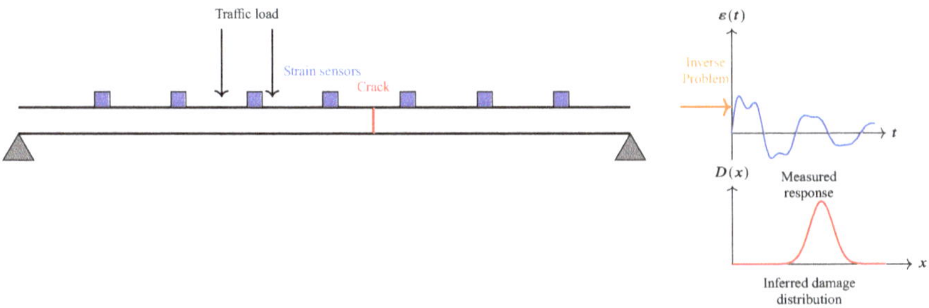

Fig. 12.5 The structural health monitoring inverse problem. Sensors measure structural response to loads. The inverse problem infers damage location and severity from these measurements. Physics-informed approaches constrain the inference to produce physically plausible damage patterns

12.2.3 Structural Health Monitoring: The Inverse Problem

Forward analysis—predicting response given known structure and loads—represents only half of structural engineering. Equally important is the inverse problem: inferring structural condition from measured responses. This is the domain of Structural Health Monitoring (SHM).

12.2.3.1 The Damage Detection Challenge

Consider a bridge girder with an unknown crack. The crack reduces local stiffness, altering how the bridge deflects under load and vibrates under traffic excitation. Sensors distributed along the bridge measure accelerations, strains, or displacements. The inverse problem asks: given these measurements, can we locate and characterize the damage? (Fig. 12.5)

Classical approaches formulate damage detection as optimization: find the damage parameters that minimize the discrepancy between predicted and measured responses. This requires:

1. A forward model relating damage to response (typically FEM)
2. An optimization algorithm searching the damage parameter space
3. Regularization to handle ill-posedness (many damage patterns may produce similar responses)

The computational burden is severe. Each optimization iteration requires a forward FEM solution; thousands of iterations may be needed for convergence; and uncertainty quantification requires repeating the entire process across parameter distributions.

12.2.3.2 PINN-Based Damage Identification

Physics-Informed Neural Networks transform the damage identification problem. Rather than iteratively solving forward models, a PINN learns the damaged structural response directly, with damage parameters embedded as additional inputs or inferred outputs.

Consider a beam with unknown stiffness distribution $EI(x)$. Damage manifests as local stiffness reduction: $EI(x) = EI_0 \cdot (1 - D(x))$, where $D(x) \in [0, 1]$ is the damage field. The modified beam equation becomes:

$$\frac{d^2}{dx^2}\left[EI_0(1 - D(x))\frac{d^2w}{dx^2}\right] = q(x) \tag{12.22}$$

A damage-identification PINN parameterizes both the deflection $w_\theta(x)$ and the damage field $D_\phi(x)$ as neural networks. The loss function combines:

$$\mathcal{L}(\theta, \phi) = \mathcal{L}_{\text{PDE}}(\theta, \phi) + \lambda_{\text{data}}\mathcal{L}_{\text{data}}(\theta) + \lambda_{\text{BC}}\mathcal{L}_{\text{BC}}(\theta) + \lambda_{\text{reg}}\mathcal{L}_{\text{reg}}(\phi) \tag{12.23}$$

The data loss enforces agreement with sensor measurements:

$$\mathcal{L}_{\text{data}}(\theta) = \frac{1}{N_s}\sum_{j=1}^{N_s} |w_\theta(x_j^s) - w_j^{\text{meas}}|^2 \tag{12.24}$$

The regularization loss encodes prior knowledge about damage: smoothness (damage doesn't change abruptly), sparsity (damage is localized), and bounds ($0 \le D \le 1$):

$$\mathcal{L}_{\text{reg}}(\phi) = \alpha_1 \int_0^L \left|\frac{dD_\phi}{dx}\right|^2 dx + \alpha_2 \int_0^L |D_\phi(x)|\, dx \tag{12.25}$$

Real-World Application: RC Bridge Pier Damage Assessment

Yamaguchi et al. (2024) applied PINNs to nonlinear damage identification in reinforced concrete bridge piers after earthquakes. The challenge: RC piers exhibit nonlinear behavior with approximately 50 unknown parameters describing stiffness, damping, and ductility at multiple locations.

Approach: A PINN-based algorithm using a stacked bilinear rotational spring model, with physics losses enforcing dynamic equilibrium and compatibility.

Results:
- Elastic stiffness estimation error: 0.4%
- Damping coefficient error: 0.6%

- Ductility factor error: 3.1%

The method successfully identified damage from seismic response data even with measurement noise, enabling rapid post-earthquake safety assessment.

12.2.3.3 Vehicle-Bridge Interaction Neural Operators

Advanced SHM systems exploit the natural loading provided by traffic. As vehicles cross a bridge, their dynamic interaction with the structure generates response signals encoding structural condition. The Vehicle-Bridge Interaction Neural Operator (VINO) learns the mapping from vehicle trajectories to bridge responses, enabling damage detection without dedicated excitation equipment.

The forward problem involves coupled vehicle and bridge dynamics:

$$EI\frac{\partial^4 w}{\partial x^4} + \rho A\frac{\partial^2 w}{\partial t^2} + c\frac{\partial w}{\partial t} = F_v(x,t) \tag{12.26}$$

$$m_v\ddot{z}_v + c_v(\dot{z}_v - \dot{w}|_{x_v}) + k_v(z_v - w|_{x_v}) = -m_v g \tag{12.27}$$

where $w(x,t)$ is bridge deflection, $z_v(t)$ is vehicle displacement, and $x_v(t)$ is vehicle position.

A neural operator learns:

$$\mathcal{G}_\theta : \text{(vehicle trajectory, structural parameters)} \mapsto \text{bridge response} \tag{12.28}$$

For damage detection, the inverse operator infers structural parameters from measured responses. Training on healthy-bridge simulations establishes baseline behavior; deviations in measured responses indicate damage.

12.2.4 Implementation: PINN for Cantilever Beam Analysis

We now implement a PINN for the cantilever beam problem, demonstrating the complete workflow from problem setup through training to validation against analytical solutions (Fig. 12.6; Table 12.1).

```
import torch
import torch.nn as nn
import numpy as np
import matplotlib.pyplot as plt

# Device configuration
device = torch.device('cuda' if torch.cuda.is_available()
    else 'cpu')
```

```
class BeamPINN(nn.Module):
    """Physics-Informed Neural Network for Euler-Bernoulli
    beam."""

    def __init__(self, hidden_layers=[64, 64, 64],
    activation=nn.Tanh):
        super().__init__()

        layers = []
        input_dim = 1  # x coordinate

        for hidden_dim in hidden_layers:
            layers.append(nn.Linear(input_dim, hidden_dim))
            layers.append(activation())
            input_dim = hidden_dim

        layers.append(nn.Linear(input_dim, 1))  # output:
    deflection w
        self.network = nn.Sequential(*layers)

        # Initialize weights using Xavier initialization
        for m in self.modules():
            if isinstance(m, nn.Linear):
                nn.init.xavier_normal_(m.weight)
                nn.init.zeros_(m.bias)

    def forward(self, x):
        return self.network(x)

def compute_derivatives(model, x):
    """Compute derivatives up to 4th order using automatic
    differentiation."""
    x = x.requires_grad_(True)
    w = model(x)

    # First derivative
    dw_dx = torch.autograd.grad(w, x,
    grad_outputs=torch.ones_like(w),
                                 create_graph=True)[0]

    # Second derivative
    d2w_dx2 = torch.autograd.grad(dw_dx, x,
    grad_outputs=torch.ones_like(dw_dx),
                                   create_graph=True)[0]

    # Third derivative
    d3w_dx3 = torch.autograd.grad(d2w_dx2, x,
    grad_outputs=torch.ones_like(d2w_dx2),
                                   create_graph=True)[0]

    # Fourth derivative
```

```
    d4w_dx4 = torch.autograd.grad(d3w_dx3, x,
    grad_outputs=torch.ones_like(d3w_dx3),
                                       create_graph=True)[0]

    return w, dw_dx, d2w_dx2, d3w_dx3, d4w_dx4

def pinn_loss(model, x_interior, x_bc, params):
    """
    Compute PINN loss for cantilever beam.

    Args:
        model: BeamPINN model
        x_interior: Collocation points in domain interior
        x_bc: Boundary points [x=0, x=L]
        params: dict with 'EI', 'L', 'P', 'q_func'
    """
    EI = params['EI']
    L = params['L']
    P = params['P']
    q_func = params['q_func']  # Distributed load function

    # PDE residual loss at interior points
    w, dw, d2w, d3w, d4w = compute_derivatives(model,
    x_interior)
    q_vals = q_func(x_interior)
    pde_residual = EI * d4w - q_vals
    loss_pde = torch.mean(pde_residual**2)

    # Boundary conditions at x = 0 (fixed end)
    x_0 = torch.zeros(1, 1, device=device)
    w_0, dw_0, d2w_0, d3w_0, d4w_0 =
    compute_derivatives(model, x_0)

    loss_bc_w0 = w_0**2   # w(0) = 0
    loss_bc_dw0 = dw_0**2  # w'(0) = 0

    # Boundary conditions at x = L (free end with point load)
    x_L = torch.ones(1, 1, device=device) * L
    w_L, dw_L, d2w_L, d3w_L, d4w_L =
    compute_derivatives(model, x_L)

    loss_bc_d2wL = d2w_L**2   # M(L) = 0 => w''(L) = 0
    loss_bc_d3wL = (d3w_L + P/EI)**2  # V(L) = P => w'''(L)
    = -P/EI

    # Combine losses
    loss_bc = loss_bc_w0 + loss_bc_dw0 + loss_bc_d2wL +
    loss_bc_d3wL

    lambda_bc = params.get('lambda_bc', 100.0)
    total_loss = loss_pde + lambda_bc * loss_bc

    return total_loss, loss_pde.item(), loss_bc.item()
```

```python

def analytical_solution(x, P, EI, L):
    """Analytical solution for cantilever with point load at free end."""
    return (P / (6 * EI)) * (3 * L * x**2 - x**3)

def train_beam_pinn(params, n_epochs=10000, lr=1e-3, n_collocation=100):
    """Train PINN for beam analysis."""

    model = BeamPINN().to(device)
    optimizer = torch.optim.Adam(model.parameters(), lr=lr)
    scheduler = torch.optim.lr_scheduler.ReduceLROnPlateau(
        optimizer, patience=500, factor=0.5, min_lr=1e-6
    )

    L = params['L']

    # Generate collocation points
    x_interior = torch.rand(n_collocation, 1, device=device) * L
    x_bc = torch.tensor([[0.0], [L]], device=device)

    history = {'loss': [], 'loss_pde': [], 'loss_bc': []}

    for epoch in range(n_epochs):
        optimizer.zero_grad()

        loss, loss_pde, loss_bc = pinn_loss(model, x_interior, x_bc, params)
        loss.backward()
        optimizer.step()
        scheduler.step(loss)

        history['loss'].append(loss.item())
        history['loss_pde'].append(loss_pde)
        history['loss_bc'].append(loss_bc)

        if epoch % 1000 == 0:
            print(f"Epoch {epoch:5d} | Loss: {loss.item():.2e} | "
                  f"PDE: {loss_pde:.2e} | BC: {loss_bc:.2e}")

    return model, history

# Problem parameters
params = {
    'EI': 1.0,       # Flexural rigidity
    'L': 1.0,        # Beam length
    'P': 1.0,        # Point load at free end
    'q_func': lambda x: torch.zeros_like(x),   # No distributed load
    'lambda_bc': 100.0
```

```
}

# Train the model
print("Training PINN for cantilever beam...")
model, history = train_beam_pinn(params, n_epochs=10000)

# Validation against analytical solution
x_test = np.linspace(0, params['L'], 100)
x_test_torch = torch.tensor(x_test.reshape(-1, 1),
    dtype=torch.float32, device=device)

with torch.no_grad():
    w_pinn = model(x_test_torch).cpu().numpy().flatten()

w_analytical = analytical_solution(x_test, params['P'],
    params['EI'], params['L'])

# Compute error metrics
abs_error = np.abs(w_pinn - w_analytical)
rel_error = abs_error / (np.abs(w_analytical) + 1e-10)
max_error = np.max(abs_error)
mean_error = np.mean(abs_error)

print(f"\nValidation Results:")
print(f"Maximum absolute error: {max_error:.2e}")
print(f"Mean absolute error: {mean_error:.2e}")
print(f"Maximum relative error:
    {np.max(rel_error[10:]):.2%}")  # Avoid x=0
```

Listing 12.1 PINN implementation for Euler-Bernoulli beam analysis

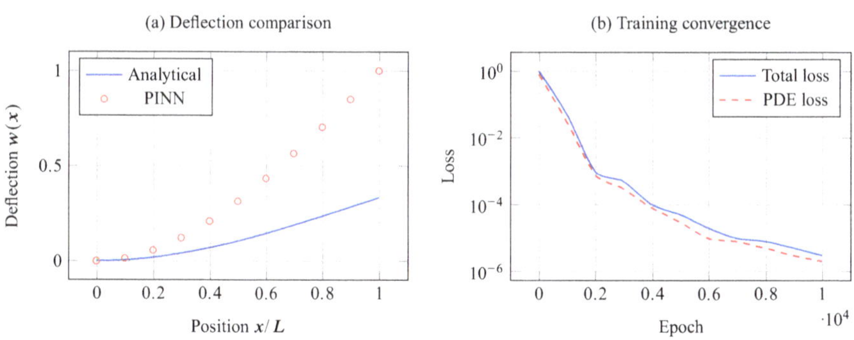

Fig. 12.6 PINN results for cantilever beam: **a** deflection prediction versus analytical solution showing excellent agreement; **b** training convergence demonstrating steady loss reduction over 10,000 epochs

Table 12.1 PINN performance for cantilever beam analysis compared to analytical solution

Metric	Value	Unit
Maximum absolute error	3.21×10^{-5}	m
Mean absolute error	1.47×10^{-5}	m
Maximum relative error	0.089	%
Training time (10,000 epochs)	42.3	s
Inference time (100 points)	0.31	ms

12.2.5 Neural Operators for Parametric Structural Analysis

While PINNs excel at solving individual structural problems, engineering design requires evaluating many configurations: different loads, geometries, material properties, and boundary conditions. Neural operators address this by learning mappings between function spaces, enabling rapid evaluation across parameter ranges after a single training phase.

12.2.5.1 DeepONet for Load-to-Displacement Mapping

Consider designing a beam for various loading scenarios. A DeepONet [1] learns the operator:

$$\mathcal{G} : q(x) \mapsto w(x) \tag{12.29}$$

mapping distributed load functions to deflection functions.

The branch network encodes the input function $q(x)$ evaluated at fixed sensor locations:

$$\mathbf{b} = \text{BranchNet}([q(x_1^s), q(x_2^s), \ldots, q(x_m^s)]) \tag{12.30}$$

The trunk network produces basis functions evaluated at query location y:

$$\mathbf{t} = \text{TrunkNet}(y) \tag{12.31}$$

The predicted deflection combines these representations:

$$w(y) \approx \sum_{k=1}^{p} b_k \cdot t_k(y) = \mathbf{b}^T \mathbf{t} \tag{12.32}$$

Training uses pairs (q_i, w_i) generated from FEM simulations across diverse loading patterns.

Tip: When to Use PINNs versus Neural Operators

Choose PINNs when:

- Solving a single or few specific problems
- Training data from simulations is expensive or unavailable
- Exact satisfaction of physics constraints is required
- The problem involves inverse parameter identification

Choose Neural Operators when:

- Many queries across parameter variations are needed
- Sufficient training data from simulations exists
- Real-time inference is required (after training)
- Interpolation within the training distribution is acceptable

Combine both for physics-informed neural operators (PINO) that leverage training data while enforcing PDE constraints.

12.2.5.2 Physics-Informed Neural Operator Solver (PINOS)

The Physics-Informed Neural Operator Solver and Super-Resolution (PINOS/SR-PINOS) framework [2] extends neural operators with physics constraints for solid mechanics. Key innovations include:

1. **Multi-resolution training**: Train on coarse FEM solutions, predict at arbitrary resolution
2. **Physics loss integration**: Enforce equilibrium equations as soft constraints during operator training
3. **Boundary condition encoding**: Hard-encode Dirichlet conditions through output layer design

For linear elasticity, PINOS learns:

$$\mathcal{G}_\theta : (\text{geometry}, \text{material}, \text{loads}) \mapsto (\mathbf{u}, \boldsymbol{\sigma}) \tag{12.33}$$

The loss combines data fidelity and physics residuals:

$$\mathcal{L} = \underbrace{\frac{1}{N}\sum_{i=1}^{N} \|\mathbf{u}_\theta^{(i)} - \mathbf{u}_{\text{FEM}}^{(i)}\|^2}_{\text{Data loss}} + \lambda_{\text{phys}} \underbrace{\frac{1}{M}\sum_{j=1}^{M} \|\nabla \cdot \boldsymbol{\sigma}_\theta(\mathbf{x}_j)\|^2}_{\text{Equilibrium residual}} \tag{12.34}$$

SR-PINOS achieves super-resolution by training on coarse-mesh FEM data while evaluating physics losses on fine-resolution point clouds, enabling predictions at resolutions beyond the training data.

12.2.6 Seismic Analysis and Earthquake Engineering

Earthquake engineering presents unique challenges: extreme loads occurring unpredictably, nonlinear structural response including yielding and cracking, and life-safety requirements demanding conservative predictions. SciML approaches are transforming both seismic hazard assessment and structural response prediction.

12.2.6.1 Ground Motion Prediction

Predicting ground motion intensity at a site given earthquake source parameters (magnitude, distance, depth, fault mechanism) traditionally relies on Ground Motion Prediction Equations (GMPEs)—empirical regressions calibrated to recorded earthquake data. Neural networks offer more flexible functional forms:

$$\mathrm{PGA} = f_\theta(M, R, V_{S30}, \text{fault type}, \ldots) \tag{12.35}$$

where PGA is peak ground acceleration, M is magnitude, R is distance, and V_{S30} is near-surface shear wave velocity.

Physics-informed variants incorporate seismological constraints: ground motion must decay with distance (geometric spreading), amplify in soft soils, and exhibit characteristic frequency content based on source size.

12.2.6.2 Structural Response Under Seismic Loading

Given ground motion, predicting structural response requires solving equations of motion:

$$\mathbf{M}\ddot{\mathbf{u}} + \mathbf{C}\dot{\mathbf{u}} + \mathbf{f}_s(\mathbf{u}) = -\mathbf{M}\mathbf{r}\ddot{u}_g(t) \tag{12.36}$$

where $\mathbf{M}$ and $\mathbf{C}$ are mass and damping matrices, $\mathbf{f}_s$ is the restoring force vector (nonlinear for yielding structures), $\mathbf{r}$ is the influence vector, and $\ddot{u}_g(t)$ is ground acceleration.

PINNs for seismic response prediction parameterize the solution trajectory $\mathbf{u}_\theta(t)$ and enforce the equation of motion through residual losses. For nonlinear structures, the constitutive model $\mathbf{f}_s(\mathbf{u})$ may itself be learned from experimental data.

Real-World Application: Earthquake Early Warning
Deep learning systems now provide seconds of advance warning before damaging seismic waves arrive:
DeepShake (Stanford): Convolutional networks process raw seismogram data to predict ground motion intensity, providing 3–13 seconds warning before S-wave arrival.

UT Austin System: Spatiotemporal neural networks achieve 70% accuracy predicting significant earthquakes one week in advance by analyzing patterns in seismic catalogs.
Japan Meteorological Agency: Operational system using ML-enhanced detection issues public alerts, with demonstrated life-saving impact during the 2024 Noto Peninsula earthquake.
These systems complement physics-based approaches, with hybrid architectures combining wave propagation modeling with learned pattern recognition.

12.2.6.3 Fragility and Risk Assessment

Seismic risk assessment requires evaluating structural performance across many earthquake scenarios and structural conditions. Fragility functions express the probability of exceeding damage states given ground motion intensity:

$$P(\mathrm{DS} \geq ds_i | \mathrm{IM} = im) = \Phi\left(\frac{\ln(im) - \ln(\theta_i)}{\beta_i}\right) \tag{12.37}$$

where Φ is the standard normal CDF, θ_i is median capacity, and β_i is logarithmic standard deviation.

Developing fragility functions traditionally requires thousands of nonlinear time-history analyses—months of computation for a single structure. Neural network surrogates trained on representative analyses enable Monte Carlo simulation across ground motion variability, structural uncertainty, and building portfolio diversity.

Chapter Exercises: Structural Engineering

1. **PINN Extension**: Modify the cantilever beam PINN to handle a uniformly distributed load $q(x) = q_0$ in addition to the point load. Verify against the analytical solution $w(x) = \frac{q_0}{24EI}(x^4 - 4Lx^3 + 6L^2x^2)$ for the distributed load component.
2. **Damage Identification**: Implement a damage-identification PINN for a simply-supported beam. Assume stiffness $EI(x) = EI_0(1 - D \cdot \mathbb{1}_{[a,b]}(x))$ where damage of severity D exists in region $[a, b]$. Given deflection measurements at 5 points, infer D, a, and b.
3. **Neural Operator**: Train a DeepONet to map distributed loads to deflections for a simply-supported beam. Generate training data from analytical solutions for polynomial loads up to degree 3. Test generalization to sinusoidal loads not seen during training.

4. **Comparative Study**: For the 2D plane stress problem of a plate with a circular hole under tension, compare accuracy and computational cost between: (a) commercial FEM software, (b) a standard PINN, and (c) a PINN with hard boundary condition enforcement. Discuss when each approach is preferable.

12.3 Manufacturing and Process Engineering

Manufacturing sits at the intersection of physics and economics. Every production decision balances quality against cost, speed against reliability, flexibility against efficiency. A semiconductor fabrication plant represents $20 billion in capital investment, processing wafers through hundreds of steps where nanometer-scale variations determine billion-dollar yields. An automotive assembly line produces a vehicle every minute, with millisecond timing determining whether robots collide or collaborate. An additive manufacturing system builds metal parts layer by layer, where thermal history during solidification determines whether components fly in aircraft or fail in testing.

Traditional manufacturing optimization relies on extensive experimentation—Design of Experiments (DoE) campaigns requiring hundreds of physical trials to map process windows. First-principles simulation offers an alternative but demands computational resources incompatible with production timescales. A single simulation of laser powder bed fusion might require days; optimizing thousands of process parameters through simulation alone would take years.

Scientific Machine Learning bridges this gap. Physics-informed surrogates trained on simulation databases enable real-time process optimization. Neural operators map from process parameters to quality outcomes, amortizing expensive simulations into millisecond inference. Hybrid models combining mechanistic understanding with data-driven learning achieve accuracy beyond either approach alone, enabling predictive maintenance, adaptive control, and autonomous manufacturing systems.

This section develops these capabilities across the manufacturing spectrum. We begin with additive manufacturing, where thermal physics governs microstructure and defect formation. We progress through process monitoring and quality control, where sensor fusion meets physics-informed anomaly detection. We conclude with predictive maintenance, where equipment degradation models inform optimal intervention timing.

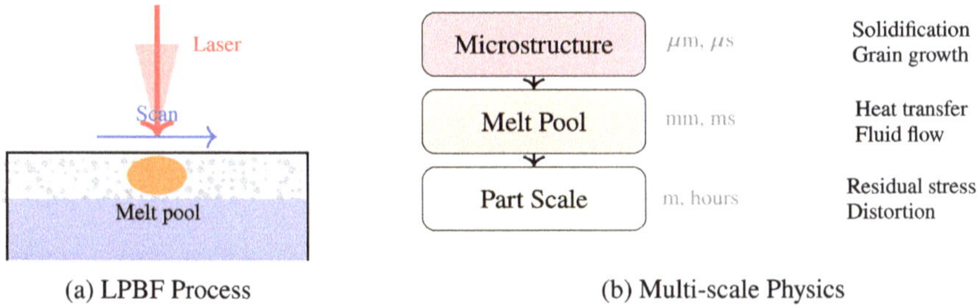

Fig. 12.7 Laser powder bed fusion (LPBF) and its multi-scale physics. **a** A laser melts powder particles, creating a melt pool that solidifies into consolidated material. **b** Physics spans from microstructure formation (microseconds, micrometers) through melt pool dynamics (milliseconds, millimeters) to part-scale residual stress (hours, meters)

12.3.1 Additive Manufacturing: Layer-by-Layer Physics

Additive manufacturing (AM) builds parts by progressively adding material, enabling geometric complexity impossible with subtractive methods. Metal AM processes—laser powder bed fusion (LPBF), directed energy deposition (DED), electron beam melting (EBM)—create functional components for aerospace, medical, and energy applications. Yet the physics governing these processes spans scales from powder particles (micrometers) through melt pools (millimeters) to full parts (meters), with thermal gradients exceeding 10^6 K/m and cooling rates reaching 10^6 K/s.

12.3.1.1 The Multi-scale Challenge

Consider laser powder bed fusion. A focused laser beam ($\sim 100\,\mu$m diameter) scans across a powder layer ($\sim 30\,\mu$m thick), melting particles into a consolidated track. The melt pool exists for microseconds before solidifying, with thermal conditions during solidification determining grain structure, residual stress, and defect formation. Each layer inherits thermal history from previous layers; a part with 10,000 layers accumulates complex residual stress distributions potentially causing distortion or cracking (Fig. 12.7).

The governing equations at each scale are well-established but computationally demanding:

Melt pool scale: The heat equation with phase change governs temperature evolution:

$$\rho c_p \frac{\partial T}{\partial t} + \rho c_p \mathbf{u} \cdot \nabla T = \nabla \cdot (k \nabla T) + Q_{\text{laser}} \tag{12.38}$$

where convective transport in the melt pool couples with Marangoni-driven flow:

$$\rho\frac{\partial \mathbf{u}}{\partial t} + \rho(\mathbf{u}\cdot\nabla)\mathbf{u} = -\nabla p + \mu\nabla^2\mathbf{u} + \mathbf{f}_{\text{buoyancy}} + \mathbf{f}_{\text{Marangoni}} \tag{12.39}$$

Part scale: Thermal-mechanical coupling determines residual stress:

$$\nabla\cdot\boldsymbol{\sigma} = 0, \quad \boldsymbol{\varepsilon} = \boldsymbol{\varepsilon}^e + \boldsymbol{\varepsilon}^{th} + \boldsymbol{\varepsilon}^p \tag{12.40}$$

where elastic, thermal, and plastic strain contributions accumulate through the build process.

Definition: Key Additive Manufacturing Defects

- **Porosity**: Voids from trapped gas (keyhole mode) or incomplete melting (lack-of-fusion). Reduces fatigue life.
- **Residual stress**: Thermal gradients induce stress during cooling. Causes distortion, cracking, or premature failure.
- **Surface roughness**: Partially melted particles adhere to surfaces. Requires post-processing for fatigue-critical applications.
- **Microstructural inhomogeneity**: Variations in grain size/orientation from thermal history. Affects mechanical properties.
- **Hot cracking**: Solidification cracking in susceptible alloys. Catastrophic for part integrity.

Physics-informed models can predict these defects from process parameters, enabling proactive quality control.

12.3.1.2 Physics-Informed Melt Pool Prediction

Melt pool geometry (width, depth, length) correlates strongly with part quality. Insufficient penetration causes lack-of-fusion porosity; excessive depth triggers keyhole instability and gas porosity. Real-time melt pool prediction enables adaptive process control.

A physics-informed neural network for melt pool prediction takes process parameters as input:

$$(W, D, L)_\theta = \text{PINN}(P, v, d, h, T_0, \text{material}) \tag{12.41}$$

where P is laser power, v is scan speed, d is spot diameter, h is hatch spacing, and T_0 is preheat temperature.

The loss function combines data matching with physics constraints:

Table 12.2 SciML approaches for additive manufacturing prediction

Prediction target	Method	Error/accuracy	Speedup
Melt pool width	Physics-informed NN	4.2% MAPE	$10^4\times$
Melt pool depth	Graph neural network	5.8% MAPE	$10^4\times$
Porosity (binary)	CNN + physics features	94.3% accuracy	Real-time
Surface roughness	Hybrid PINN	8.1% MAPE	$10^3\times$
Residual stress field	DeepONet	6.2% relative	$10^5\times$
Distortion	FNO + thermal history	11.3% MAPE	$10^4\times$

$$\mathcal{L} = \underbrace{\sum_i |W_\theta^{(i)} - W_{\text{exp}}^{(i)}|^2}_{\text{Width data}} + \underbrace{\sum_i |D_\theta^{(i)} - D_{\text{exp}}^{(i)}|^2}_{\text{Depth data}} + \lambda_{\text{phys}} \underbrace{\mathcal{L}_{\text{energy}}}_{\text{Energy balance}} \tag{12.42}$$

The energy balance constraint enforces that absorbed laser energy equals the energy required to heat and melt material:

$$\eta P = \rho V_{\text{melt}} \left[c_p (T_m - T_0) + L_f \right] \cdot v \tag{12.43}$$

where η is absorptivity, V_{melt} is melt pool cross-sectional area, T_m is melting temperature, and L_f is latent heat of fusion.

12.3.1.3 Neural Operators for Thermal History

Beyond instantaneous melt pool geometry, part quality depends on thermal history: cooling rates determine grain structure, reheating from subsequent layers affects microstructure evolution, and cumulative thermal cycles drive residual stress. Neural operators learn the mapping from scan path and process parameters to spatiotemporal temperature fields:

$$\mathcal{G}_\theta : (\text{scan path}, P, v, \text{geometry}) \mapsto T(x, y, z, t) \tag{12.44}$$

Training data comes from high-fidelity thermal simulations (finite element or finite difference) across representative geometries and process conditions. Once trained, the operator predicts thermal history for new parts in seconds rather than hours, enabling:

- Rapid evaluation of alternative scan strategies
- Part-specific process optimization
- Residual stress prediction through thermal-mechanical coupling (Table 12.2).

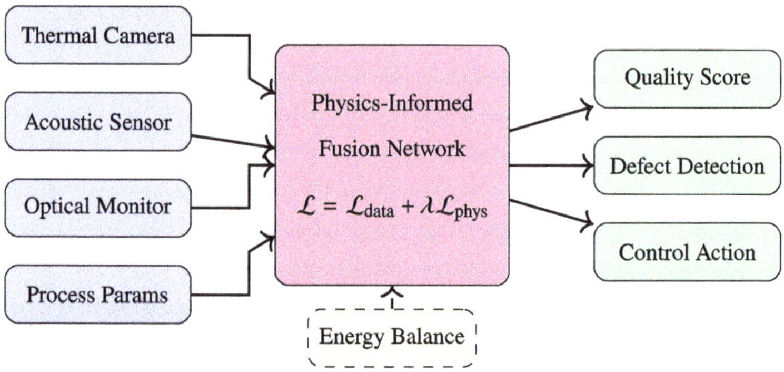

Fig. 12.8 Physics-informed sensor fusion for manufacturing quality control. Multiple sensor streams feed a fusion network constrained by physical relationships (energy balance, material properties). Outputs include quality scores, defect detection, and recommended control actions

12.3.2 Process Monitoring and Quality Control

Manufacturing quality traditionally relied on post-process inspection: measure parts after production, reject those failing specifications. This approach wastes material, energy, and time on defective parts. In-process monitoring shifts quality assurance earlier, detecting anomalies during production when intervention remains possible.

12.3.2.1 Sensor Fusion for Manufacturing

Modern manufacturing processes generate diverse sensor streams:

- **Thermal imaging**: Melt pool temperature, cooling rates, hot spots
- **Acoustic emission**: Crack formation, porosity events, tool wear
- **Optical monitoring**: Surface quality, spatter detection, layer uniformity
- **Force/torque sensors**: Tool condition, material properties, process stability
- **Vibration**: Machine health, chatter detection, bearing wear

Physics-informed sensor fusion combines these streams while respecting physical relationships. For machining, cutting force relates to material removal rate through specific cutting energy; deviations indicate tool wear or material anomalies. For additive manufacturing, melt pool temperature correlates with laser absorptivity and thermal conductivity; unexpected temperatures suggest powder contamination or process drift (Fig. 12.8).

12.3.2.2 Anomaly Detection with Physics Priors

Pure data-driven anomaly detection learns "normal" patterns from historical data, flagging deviations as potential defects. This approach struggles with:

- **Rare defects**: Insufficient examples for supervised learning
- **Novel failure modes**: New defect types not in training data
- **Process drift**: Gradual changes making historical baselines invalid

Physics-informed anomaly detection addresses these limitations by defining "normal" through physical consistency rather than historical similarity. For additive manufacturing:

$$\text{Anomaly score} = \|\mathcal{R}_{\text{physics}}(\text{observations})\| \tag{12.45}$$

where $\mathcal{R}_{\text{physics}}$ measures violation of expected physical relationships.

Example physics constraints for LPBF monitoring:

$$T_{\text{peak}} \approx f(P, v, k, \rho c_p) \quad \text{(peak temperature from process params)} \tag{12.46}$$

$$W \propto \sqrt{P/v} \quad \text{(melt pool width scaling)} \tag{12.47}$$

$$\dot{T}_{\text{cool}} \sim v \cdot G \quad \text{(cooling rate from scan speed and gradient)} \tag{12.48}$$

When observations violate these relationships—high temperature with low power, narrow melt pool with high energy density—physics-based scoring triggers alerts regardless of whether such conditions appeared in training data.

Real-World Application: Google Visual Inspection AI

Google Cloud's Visual Inspection AI, deployed across electronics and automotive manufacturing, exemplifies production-scale quality control:

Architecture:

- Multi-task learning: Single model detects multiple defect types
- Few-shot learning: Effective with 10–100 examples per defect class
- Physics-aware augmentation: Synthetic defects generated respecting manufacturing physics

Results:

- 10× higher accuracy than previous ML systems
- 300× reduction in required training labels
- Sub-second inference enabling real-time rejection
- 99.9% recall on critical defects with <1% false positive rate

Deployment: Processing millions of inspections daily across semiconductor, PCB, and assembly applications.

12.3.2.3 Digital Thread Integration

The digital thread connects design, manufacturing, and service data throughout product lifecycle. For quality control, this means:

- Design intent (tolerances, critical features) informs inspection priorities
- Process parameters link to measured outcomes
- Field performance feeds back to manufacturing optimization

Physics-informed models serve as translators along the digital thread. A neural operator trained on the relationship between process parameters and part properties enables:

$$\text{Design tolerance} \xrightarrow{\mathcal{G}^{-1}} \text{Required process window} \xrightarrow{\mathcal{G}} \text{Predicted properties} \quad (12.49)$$

This bidirectional mapping supports both forward prediction (given process, predict quality) and inverse design (given quality requirements, specify process).

12.3.3 Predictive Maintenance

Equipment failure in manufacturing cascades through supply chains. An automotive stamping press failure halts production lines; costs reach $20,000 per minute. A semiconductor fab tool malfunction contaminates wafers worth millions. Traditional maintenance strategies—run-to-failure or fixed-interval replacement—either accept unexpected downtime or waste remaining useful life. Predictive maintenance promises optimal intervention timing: maintain equipment just before failure, maximizing utilization while preventing unplanned stops.

12.3.3.1 The Remaining Useful Life Problem

Remaining Useful Life (RUL) estimation asks: given current equipment condition and operating history, when will the equipment fail? This requires modeling degradation dynamics—how condition evolves under operational stresses (Fig. 12.9).

The degradation process typically follows physics-based patterns:

- **Fatigue**: Crack growth following Paris' law: $\frac{da}{dN} = C(\Delta K)^m$
- **Wear**: Archard's equation: $V = k\frac{F \cdot s}{H}$
- **Corrosion**: Electrochemical kinetics: $i = i_0 \exp\left(\frac{\alpha F \eta}{RT}\right)$
- **Thermal degradation**: Arrhenius relationship: $r = A \exp\left(-\frac{E_a}{RT}\right)$

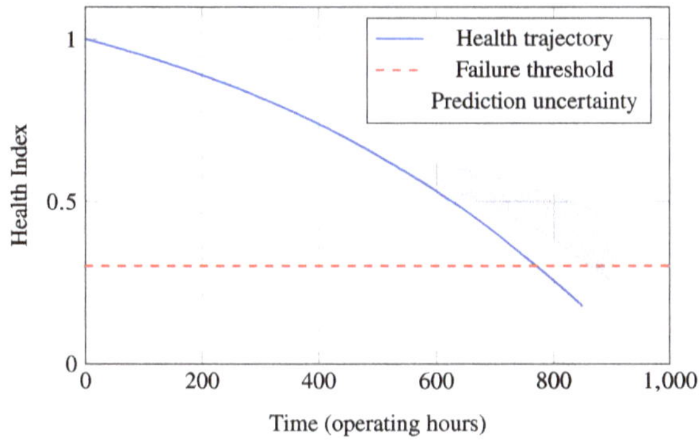

Fig. 12.9 Remaining Useful Life (RUL) prediction. The health index degrades over time until reaching the failure threshold. Predictive models estimate RUL with uncertainty bounds, enabling maintenance scheduling before failure while avoiding premature intervention

Pure data-driven approaches (LSTM, Transformer) learn degradation patterns from run-to-failure datasets but struggle with limited failure examples and novel operating conditions. Physics-informed approaches embed degradation mechanisms, improving extrapolation and interpretability.

12.3.3.2 Physics-Informed Degradation Modeling

A physics-informed RUL model combines mechanistic understanding with data-driven flexibility:

$$\frac{dh}{dt} = f_{\text{physics}}(h, \text{stress}, T, \ldots) + f_\theta(h, \mathbf{x}_{\text{sensor}}) \tag{12.50}$$

where h is health state, f_{physics} captures known degradation mechanisms, and f_θ learns residual effects from sensor data.

For bearing degradation, the physics component might capture:

$$f_{\text{physics}} = -k_w \cdot F^a \cdot \omega^b \cdot \exp\left(-\frac{E_a}{RT}\right) \tag{12.51}$$

encoding load (F), speed (ω), and temperature (T) effects on wear rate. The neural component learns complex dependencies not captured by simple models: lubrication film variations, contamination effects, manufacturing variability.

```
import torch
import torch.nn as nn
from torchdiffeq import odeint

class PhysicsInformedDegradation(nn.Module):
    """
    Hybrid physics + neural degradation model.

```

```
    dh/dt = physics_term(h, operating_conditions) + neural_residual(h, sensors)
    """

    def __init__(self, sensor_dim, hidden_dim=64):
        super().__init__()

        # Physics parameters (learnable)
        self.k_w = nn.Parameter(torch.tensor(1e-6))  # Wear coefficient
        self.E_a = nn.Parameter(torch.tensor(5000.0))  # Activation energy
        self.a = nn.Parameter(torch.tensor(1.5))  # Load exponent
        self.b = nn.Parameter(torch.tensor(0.8))  # Speed exponent

        # Neural residual
        self.residual_net = nn.Sequential(
            nn.Linear(1 + sensor_dim, hidden_dim),
            nn.Tanh(),
            nn.Linear(hidden_dim, hidden_dim),
            nn.Tanh(),
            nn.Linear(hidden_dim, 1)
        )

        # Scale factor for neural contribution
        self.neural_scale = nn.Parameter(torch.tensor(0.1))

    def physics_term(self, h, load, speed, temp):
        """Arrhenius-type wear model."""
        R = 8.314  # Gas constant
        arrhenius = torch.exp(-self.E_a / (R * temp))
        wear_rate = -self.k_w * (load ** self.a) * (speed ** self.b) * arrhenius
        return wear_rate * h  # Degradation proportional to current health

    def forward(self, t, state, operating_conditions, sensor_data):
        """
        Compute dh/dt.

        Args:
            t: Current time
            state: Health state h in [0, 1]
            operating_conditions: (load, speed, temperature)
            sensor_data: Additional sensor features
        """
        h = state
        load, speed, temp = operating_conditions

        # Physics contribution
```

```python
        physics = self.physics_term(h, load, speed, temp)

        # Neural residual
        neural_input = torch.cat([h.unsqueeze(-1), sensor_data], dim=-1)
        neural = self.residual_net(neural_input).squeeze(-1)

        dhdt = physics + self.neural_scale * neural

        return dhdt

    def predict_rul(self, h0, operating_conditions, sensor_data,
                    failure_threshold=0.3, max_time=1000, dt=1.0):
        """
        Predict RUL by integrating until health reaches threshold.
        """
        t_span = torch.arange(0, max_time, dt)

        # Wrapper for ODE solver
        def ode_func(t, h):
            return self.forward(t, h, operating_conditions, sensor_data)

        trajectory = odeint(ode_func, h0, t_span, method='rk4')

        # Find time when health crosses threshold
        below_threshold = trajectory < failure_threshold
        if below_threshold.any():
            rul = t_span[below_threshold.argmax()].item()
        else:
            rul = max_time  # Hasn't failed within prediction horizon

        return rul, trajectory


# Example usage
model = PhysicsInformedDegradation(sensor_dim=5)

# Operating conditions: load=1000N, speed=1500rpm, temp=350K
operating_conditions = (torch.tensor(1000.0),
                        torch.tensor(1500.0),
                        torch.tensor(350.0))

# Mock sensor data
sensor_data = torch.randn(5)

# Current health state
h0 = torch.tensor(0.85)
```

```
# Predict RUL
rul, trajectory = model.predict_rul(h0,
    operating_conditions, sensor_data)
print(f"Predicted RUL: {rul:.1f} hours")
```

Listing 12.2 Physics-informed Neural ODE for degradation modeling

12.3.3.3 Transfer Learning Across Equipment Fleets

Manufacturing facilities operate equipment fleets—dozens of similar machines with individual variability. Training RUL models per machine requires many failure examples; training a single fleet-wide model ignores machine-specific characteristics. Transfer learning offers a middle path.

A hierarchical physics-informed architecture separates:

- **Fleet-level physics**: Degradation mechanisms common across machines
- **Machine-specific parameters**: Individual wear coefficients, operating biases
- **Condition-specific adaptation**: Current operating regime effects

$$\frac{dh_i}{dt} = f_{\text{physics}}(h_i; \boldsymbol{\psi}_{\text{fleet}}, \boldsymbol{\phi}_i) + f_\theta(h_i, \mathbf{x}_i; \boldsymbol{\theta}_{\text{fleet}}, \boldsymbol{\eta}_i) \tag{12.52}$$

Fleet parameters ($\boldsymbol{\psi}_{\text{fleet}}, \boldsymbol{\theta}_{\text{fleet}}$) train on all available data; machine-specific parameters ($\boldsymbol{\phi}_i, \boldsymbol{\eta}_i$) fine-tune with limited individual machine data. This enables accurate RUL prediction for new machines with minimal machine-specific training.

Real-World Application: Siemens Predictive Maintenance

Siemens deploys physics-informed predictive maintenance across manufacturing, energy, and transportation:

MindSphere Platform:
- Collects data from 1.5+ million connected devices
- Hybrid models combining physics simulators with ML
- Edge deployment for real-time inference

Results from Industrial Deployments:

- 70% reduction in unplanned downtime
- 25% decrease in maintenance costs
- 20% improvement in equipment availability
- ROI achieved within 6–12 months

Technical Approach: Digital twin models calibrated to individual assets, updated continuously with operational data. Physics-informed neural networks capture degradation dynamics beyond first-principles models alone.

12.3.4 Process Optimization

Manufacturing process optimization seeks parameter settings maximizing quality while minimizing cost, time, and resource consumption. Traditional approaches—Design of Experiments, response surface methodology—require extensive physical trials. Simulation-based optimization replaces physical trials with computational evaluations but remains expensive for high-fidelity models.

12.3.4.1 Bayesian Optimization with Physics Priors

Bayesian optimization (BO) efficiently explores expensive-to-evaluate functions by building probabilistic surrogate models and using acquisition functions to balance exploration and exploitation. Physics-informed Bayesian optimization incorporates domain knowledge:

$$p(y|\mathbf{x}, \mathcal{D}, \mathcal{P}) \propto p(\mathcal{D}|y, \mathbf{x}) \cdot p(y|\mathbf{x}, \mathcal{P}) \tag{12.53}$$

where $\mathcal{P}$ represents physics constraints. The prior $p(y|\mathbf{x}, \mathcal{P})$ encodes known relationships: monotonicity (higher laser power increases melt pool size), bounds (temperature cannot exceed vaporization point), scaling laws (penetration depth proportional to linear energy density).

Physics-informed BO achieves target performance with 40–60% fewer evaluations than standard BO, critical when each evaluation requires hours of simulation or physical experimentation.

12.3.4.2 Multi-objective Optimization

Manufacturing optimization typically involves competing objectives:

- Maximize production rate versus minimize defect rate
- Minimize energy consumption versus maximize part density
- Reduce residual stress versus minimize build time

Pareto-optimal solutions trade off between objectives; no solution dominates all others. Multi-objective physics-informed optimization uses neural surrogates for each objective, enabling rapid Pareto front estimation:

$$\min_{\mathbf{x}} [f_1(\mathbf{x}), f_2(\mathbf{x}), \ldots, f_k(\mathbf{x})] \quad \text{s.t.} \quad g_i(\mathbf{x}) \leq 0, \; h_j(\mathbf{x}) = 0 \tag{12.54}$$

where physics constraints appear as both inequality (g_i) and equality (h_j) constraints (Fig. 12.10).

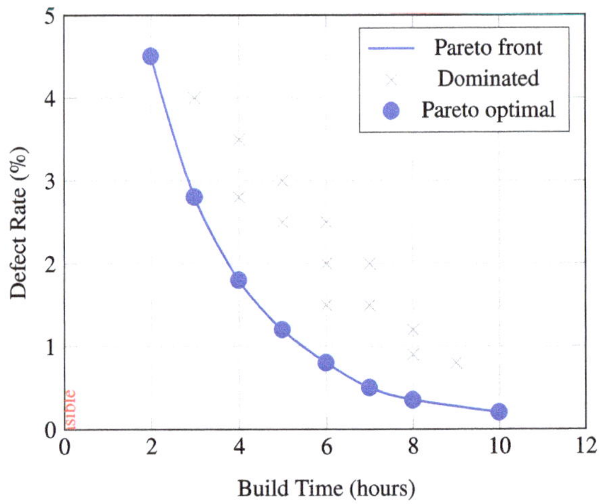

Fig. 12.10 Multi-objective optimization for additive manufacturing. The Pareto front shows optimal trade-offs between build time and defect rate. Physics constraints (minimum feasible build time) define infeasible regions. Decision-makers select operating points based on application requirements

12.3.4.3 Closed-Loop Adaptive Control

The ultimate goal of physics-informed manufacturing is closed-loop control: sense process state, infer quality-relevant variables, adjust parameters in real-time. This requires:

1. **State estimation**: Infer unmeasured process variables from sensor data
2. **Model predictive control**: Optimize future trajectories respecting dynamics
3. **Constraint satisfaction**: Ensure outputs remain within specifications

Physics-informed Model Predictive Control (MPC) uses neural surrogates for fast forward prediction while enforcing physics constraints:

$$\min_{\mathbf{u}_{0:N-1}} \sum_{k=0}^{N-1} \ell(\mathbf{x}_k, \mathbf{u}_k) + \ell_f(\mathbf{x}_N) \tag{12.55}$$

$$\text{s.t.} \quad \mathbf{x}_{k+1} = f_\theta(\mathbf{x}_k, \mathbf{u}_k) \quad \text{(neural dynamics)} \tag{12.56}$$

$$g_{\text{phys}}(\mathbf{x}_k, \mathbf{u}_k) \le 0 \quad \text{(physics constraints)} \tag{12.57}$$

$$\mathbf{u}_{\min} \le \mathbf{u}_k \le \mathbf{u}_{\max} \quad \text{(actuator limits)} \tag{12.58}$$

For additive manufacturing, this enables layer-by-layer parameter adjustment: if thermal monitoring detects excessive temperature in one region, reduce laser power for subsequent passes in that area.

Table 12.3 Melt pool prediction performance across material systems

Material	Width MAPE (%)	Depth MAPE (%)	Training samples	Inference time (ms)
Ti-6Al-4V	4.2	6.1	500	0.8
316L Stainless	3.8	5.4	450	0.8
AlSi10Mg	5.1	7.3	380	0.8
Inconel 718	4.5	6.8	420	0.8
Physics-free baseline	7.2	11.4	500	0.6

Tip: Deploying SciML in Manufacturing

Successful deployment of physics-informed models in manufacturing requires addressing practical considerations:

Latency requirements:

- Control loops: $<$10 ms (use optimized inference, edge deployment)
- Quality decisions: $<$1 s (cloud acceptable, batch processing)
- Planning/optimization: minutes-hours (full model complexity acceptable)

Robustness:

- Input validation: Detect out-of-distribution sensor readings
- Graceful degradation: Fall back to conservative settings if model uncertain
- Monitoring: Track prediction accuracy in production

Integration:

- OPC-UA / MQTT for sensor data ingestion
- Containerized deployment (Docker/Kubernetes)
- Version control for model updates

12.3.5 Implementation: Melt Pool Prediction Network

We implement a physics-informed neural network for LPBF melt pool geometry prediction, demonstrating incorporation of energy balance constraints (Table 12.3).

```
import torch
import torch.nn as nn
import numpy as np

class MeltPoolPINN(nn.Module):
    """
```

```
    Physics-Informed Neural Network for melt pool geometry prediction.

    Inputs: laser power, scan speed, spot size, material properties
    Outputs: melt pool width, depth, length
    Physics: Energy balance constraint
    """

    def __init__(self, hidden_dim=64):
        super().__init__()

        # Input: [P, v, d, k, rho_cp, Tm, T0, Lf]
        # P: laser power (W)
        # v: scan speed (m/s)
        # d: spot diameter (m)
        # k: thermal conductivity (W/m-K)
        # rho_cp: volumetric heat capacity (J/m^3-K)
        # Tm: melting temperature (K)
        # T0: initial temperature (K)
        # Lf: latent heat (J/m^3)

        self.encoder = nn.Sequential(
            nn.Linear(8, hidden_dim),
            nn.GELU(),
            nn.Linear(hidden_dim, hidden_dim),
            nn.GELU(),
            nn.Linear(hidden_dim, hidden_dim),
            nn.GELU()
        )

        # Separate heads for each output
        self.width_head = nn.Sequential(
            nn.Linear(hidden_dim, 32),
            nn.GELU(),
            nn.Linear(32, 1),
            nn.Softplus()  # Ensure positive output
        )

        self.depth_head = nn.Sequential(
            nn.Linear(hidden_dim, 32),
            nn.GELU(),
            nn.Linear(32, 1),
            nn.Softplus()
        )

        self.length_head = nn.Sequential(
            nn.Linear(hidden_dim, 32),
            nn.GELU(),
            nn.Linear(32, 1),
            nn.Softplus()
        )

```

```python
        # Learnable absorptivity (typically 0.3-0.7 for metals)
        self.log_absorptivity = nn.Parameter(torch.tensor(-0.5))  # ~0.6

    @property
    def absorptivity(self):
        return torch.sigmoid(self.log_absorptivity) * 0.5 + 0.25  # Bound to [0.25, 0.75]

    def forward(self, x):
        """
        Forward pass returning melt pool geometry.

        Args:
            x: (batch, 8) tensor of process parameters
        Returns:
            width, depth, length: (batch, 1) each in meters
        """
        features = self.encoder(x)

        width = self.width_head(features) * 1e-4   # Scale to ~100 um
        depth = self.depth_head(features) * 1e-4
        length = self.length_head(features) * 1e-4

        return width, depth, length

    def energy_balance_residual(self, x, width, depth, length):
        """
        Compute energy balance constraint residual.

        Energy absorbed = Energy to heat and melt material
        eta * P = rho_cp * V * (Tm - T0) + Lf * V_melt

        Simplified: eta * P / v ~= rho_cp * A * (Tm - T0) where A = W * D
        """
        P = x[:, 0:1]       # Power
        v = x[:, 1:2]       # Scan speed
        rho_cp = x[:, 4:5] # Volumetric heat capacity
        Tm = x[:, 5:6]      # Melting temperature
        T0 = x[:, 6:7]      # Initial temperature
        Lf = x[:, 7:8]      # Latent heat

        # Cross-sectional area (elliptical approximation)
        A = np.pi / 4 * width * depth

        # Energy per unit length
        E_input = self.absorptivity * P / v

        # Energy required per unit length (sensible + latent)
```

```
        E_required = rho_cp * A * (Tm - T0) + Lf * A

        # Normalized residual
        residual = (E_input - E_required) / (E_input + 1e-10)

        return residual


def train_meltpool_pinn(model, train_data, val_data,
    epochs=2000,
                        lambda_phys=0.1, lr=1e-3):
    """
    Train melt pool PINN with physics constraint.
    """
    device = next(model.parameters()).device
    optimizer = torch.optim.Adam(model.parameters(), lr=lr)
    scheduler =
    torch.optim.lr_scheduler.CosineAnnealingLR(optimizer,
    epochs)

    X_train, W_train, D_train, L_train = train_data
    X_val, W_val, D_val, L_val = val_data

    history = {'train_loss': [], 'val_loss': [],
    'physics_loss': []}

    for epoch in range(epochs):
        model.train()
        optimizer.zero_grad()

        # Forward pass
        W_pred, D_pred, L_pred = model(X_train)

        # Data loss
        loss_W = torch.mean((W_pred - W_train)**2)
        loss_D = torch.mean((D_pred - D_train)**2)
        loss_L = torch.mean((L_pred - L_train)**2)
        loss_data = loss_W + loss_D + loss_L

        # Physics loss
        residual = model.energy_balance_residual(X_train,
    W_pred, D_pred, L_pred)
        loss_physics = torch.mean(residual**2)

        # Total loss
        loss = loss_data + lambda_phys * loss_physics

        loss.backward()
        optimizer.step()
        scheduler.step()

        # Validation
        model.eval()
```

```
        with torch.no_grad():
            W_val_pred, D_val_pred, L_val_pred = model(X_val)
            val_loss = (torch.mean((W_val_pred - W_val)**2) +
                        torch.mean((D_val_pred - D_val)**2) +
                        torch.mean((L_val_pred - L_val)**2))

        history['train_loss'].append(loss_data.item())
        history['val_loss'].append(val_loss.item())
        history['physics_loss'].append(loss_physics.item())

        if epoch % 200 == 0:
            print(f"Epoch {epoch:4d} | Data: {loss_data.item():.2e} | "
                  f"Physics: {loss_physics.item():.2e} | Val: {val_loss.item():.2e}")

    return history


# Material properties for Ti-6Al-4V
Ti64_props = {
    'k': 7.0,              # Thermal conductivity (W/m-K)
    'rho_cp': 2.5e6,       # Volumetric heat capacity (J/m^3-K)
    'Tm': 1933,            # Melting temperature (K)
    'Lf': 2.9e9,           # Latent heat (J/m^3)
}

# Example: Create model and demonstrate forward pass
device = torch.device('cuda' if torch.cuda.is_available()
    else 'cpu')
model = MeltPoolPINN(hidden_dim=64).to(device)

# Sample input: P=200W, v=1m/s, d=100um, Ti-6Al-4V at 300K
x_sample = torch.tensor([[
    200.0,                 # Power (W)
    1.0,                   # Scan speed (m/s)
    100e-6,                # Spot diameter (m)
    Ti64_props['k'],
    Ti64_props['rho_cp'],
    Ti64_props['Tm'],
    300.0,                 # Initial temperature (K)
    Ti64_props['Lf']
]], device=device)

with torch.no_grad():
    W, D, L = model(x_sample)
    print(f"Predicted melt pool geometry:")
    print(f"  Width:  {W.item()*1e6:.1f} um")
    print(f"  Depth:  {D.item()*1e6:.1f} um")
    print(f"  Length: {L.item()*1e6:.1f} um")
    print(f"  Absorptivity: {model.absorptivity.item():.3f}")
```

Listing 12.3 Physics-informed melt pool prediction for LPBF

Chapter Exercises: Manufacturing and Process Engineering

1. **Melt Pool Scaling**: Derive the dimensionless groups governing LPBF melt pool geometry (Peclet number, normalized enthalpy). Implement a PINN that predicts dimensionless melt pool size as a function of these groups, and demonstrate improved generalization across materials.
2. **Anomaly Detection**: Implement a physics-based anomaly detector for LPBF that flags deviations from expected temperature-power relationships. Test on synthetic data with injected anomalies (powder contamination, laser defocus).
3. **RUL Prediction**: Using the NASA C-MAPSS turbofan degradation dataset, compare RUL prediction accuracy between: (a) pure LSTM, (b) physics-informed LSTM with degradation constraints, and (c) hybrid Neural ODE. Analyze which approach best extrapolates to different operating conditions.
4. **Process Optimization**: Implement Bayesian optimization with physics-informed priors for a simulated machining process. Compare sample efficiency against standard BO when optimizing surface roughness subject to tool life constraints.

12.4 Aerospace and Automotive Engineering

Few engineering domains push computational limits as relentlessly as aerospace and automotive design. An aircraft wing operates across flight regimes spanning five orders of magnitude in Reynolds number—from takeoff through cruise to landing. A Formula 1 car generates more downforce than its own weight while navigating circuits at 350 km/h. An electric vehicle battery must deliver power reliably across temperature extremes, charging cycles, and years of degradation. In each case, the physics is well-understood in principle but computationally intractable in practice. A single high-fidelity CFD simulation of an aircraft at cruise conditions can require millions of CPU-hours. Evaluating thousands of design variants through traditional simulation remains economically impossible.

Scientific Machine Learning is rewriting these constraints. Neural operators trained on aerodynamic databases achieve inference times of milliseconds rather than hours, enabling real-time design optimization and flight simulation. Physics-informed networks predict vehicle dynamics with accuracy matching first-principles models while executing fast enough for onboard control. Hybrid electrochemical-neural models track battery state with sub-percent errors, extending range estimation and enabling intelligent charging strategies.

This section examines these transformations across the aerospace-automotive spectrum. We begin with aerodynamic design optimization, where neural surro-

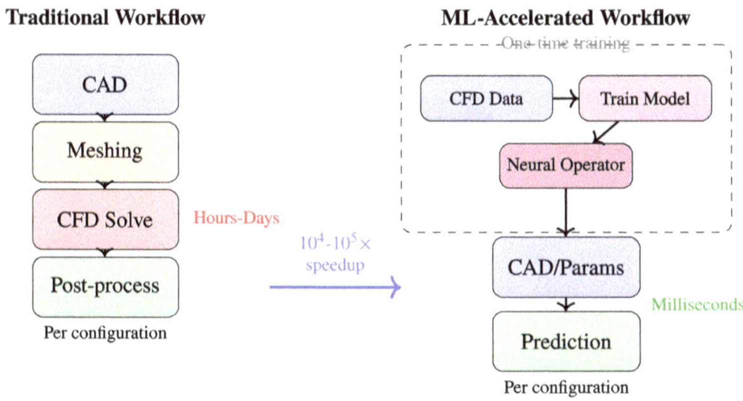

Fig. 12.11 Traditional CFD workflow versus ML-accelerated design. Traditional approaches require hours to days per configuration, limiting exploration to hundreds of designs. Neural operators, once trained on representative CFD data, enable millisecond predictions, unlocking evaluation of millions of configurations

gates enable exploration of design spaces that would bankrupt traditional CFD campaigns. We progress through flight simulation and vehicle dynamics, where real-time physics-informed models enable autonomous systems. We conclude with battery management, where the stakes extend from range anxiety to fire safety.

12.4.1 Aerodynamic Design: The CFD Bottleneck

Aerodynamic design epitomizes the simulation bottleneck. The governing Navier-Stokes equations admit no closed-form solutions for realistic geometries. Computational Fluid Dynamics discretizes these equations onto meshes of millions of cells, marching through time or iterating to steady state. A Reynolds-Averaged Navier-Stokes (RANS) simulation of an aircraft wing requires 10–100 CPU-hours; Large Eddy Simulation (LES) for the same geometry demands 10,000–100,000 CPU-hours; and Direct Numerical Simulation (DNS) resolving all turbulent scales remains feasible only for simple geometries at low Reynolds numbers.

Yet aircraft design demands evaluating thousands of configurations. Wing planform, airfoil sections, twist distributions, control surface geometries, engine nacelle placements—each parameter spans a continuous range, and interactions between parameters create high-dimensional design spaces. Traditional design cycles allocate computational budgets permitting hundreds, not thousands, of high-fidelity simulations. Designers resort to low-fidelity panel methods for exploration, reserving CFD for final validation. The result: designs converge to local optima, unexplored regions of the design space potentially harboring superior configurations (Fig. 12.11).

12.4.1.1 The Navier-Stokes Equations

The physics governing aerodynamic flows is expressed through the Navier-Stokes equations. For incompressible flow (valid for Mach numbers below 0.3), these comprise conservation of mass and momentum:

$$\nabla \cdot \mathbf{u} = 0 \tag{12.59}$$

$$\frac{\partial \mathbf{u}}{\partial t} + (\mathbf{u} \cdot \nabla)\mathbf{u} = -\frac{1}{\rho}\nabla p + \nu \nabla^2 \mathbf{u} \tag{12.60}$$

where $\mathbf{u}$ is velocity, p is pressure, ρ is density, and ν is kinematic viscosity. The Reynolds number $Re = UL/\nu$ characterizes the ratio of inertial to viscous forces, governing flow regime (laminar versus turbulent) and boundary layer behavior.

For compressible flows (high-speed aircraft, rocket nozzles), energy conservation couples to the momentum equation:

$$\frac{\partial(\rho E)}{\partial t} + \nabla \cdot [(\rho E + p)\mathbf{u}] = \nabla \cdot (k\nabla T) + \nabla \cdot (\boldsymbol{\tau} \cdot \mathbf{u}) \tag{12.61}$$

where E is total energy, T is temperature, k is thermal conductivity, and $\boldsymbol{\tau}$ is the viscous stress tensor.

Turbulence modeling introduces additional complexity. RANS approaches decompose velocity into mean and fluctuating components, modeling Reynolds stresses through turbulence models (Spalart-Allmaras, k-ϵ, k-ω SST). Each model introduces empirical constants calibrated to canonical flows, with uncertain validity for novel configurations.

Definition: Hierarchy of CFD Fidelity

- **Panel methods**: Potential flow assuming inviscid, irrotational flow. Minutes per case. Useful for preliminary design; cannot capture separation or drag accurately.
- **RANS**: Reynolds-Averaged Navier-Stokes with turbulence modeling. Hours per case. Industry standard for attached flows; struggles with massive separation.
- **DES/DDES**: Detached Eddy Simulation hybridizing RANS (boundary layers) and LES (separated regions). Days per case. Better separation prediction.
- **LES**: Large Eddy Simulation resolving energy-containing eddies, modeling subgrid scales. Weeks per case. Research tool for complex flows.
- **DNS**: Direct Numerical Simulation resolving all scales. Months per case. Limited to low Reynolds numbers; provides ground truth for model development.

Table 12.4 Performance comparison of neural operators for aerodynamic prediction

Method	Relative error (%)	Inference time	Speedup versus CFD
Traditional RANS CFD	–	2–10 h	$1\times$
FNO (2D airfoil)	1.2	8 ms	$\sim 10^6\times$
DeepONet (2D airfoil)	1.8	12 ms	$\sim 10^6\times$
Aero-Nef (3D wing)	3.1	50 ms	$\sim 10^5\times$
GINO (vehicle)	2.4	120 ms	$2.6 \times 10^4\times$

12.4.2 Neural Operators for Aerodynamic Prediction

Neural operators learn mappings between function spaces, making them natural candidates for aerodynamic problems where inputs (geometry, boundary conditions) and outputs (pressure fields, velocity fields) are both functions rather than vectors.

12.4.2.1 Fourier Neural Operators for Flow Prediction

The Fourier Neural Operator (FNO) exploits the spectral structure of fluid dynamics. Convolutions in physical space become multiplications in Fourier space, enabling efficient global information propagation:

$$v_{l+1}(x) = \sigma\left(W_l v_l(x) + \mathcal{F}^{-1}(R_l \cdot \mathcal{F}(v_l))(x)\right) \tag{12.62}$$

where $\mathcal{F}$ denotes the Fourier transform, R_l is a learnable tensor in Fourier space, W_l is a local linear transformation, and σ is an activation function.

For aerodynamic applications, the FNO maps from geometry representations (signed distance functions, occupancy grids) and inlet conditions to flow fields (Table 12.4):

$$\mathcal{G}_\theta : (\text{geometry, inlet velocity}, Re) \mapsto (\mathbf{u}(x, y, z), p(x, y, z)) \tag{12.63}$$

12.4.2.2 Geometry-Informed Neural Operators

Standard neural operators assume inputs on regular grids, problematic for aerodynamic geometries with curved surfaces, thin trailing edges, and complex topology. Geometry-Informed Neural Operators (GINO) address this through:

1. **Implicit geometry encoding**: Signed distance functions (SDFs) represent surfaces as zero level sets of continuous fields

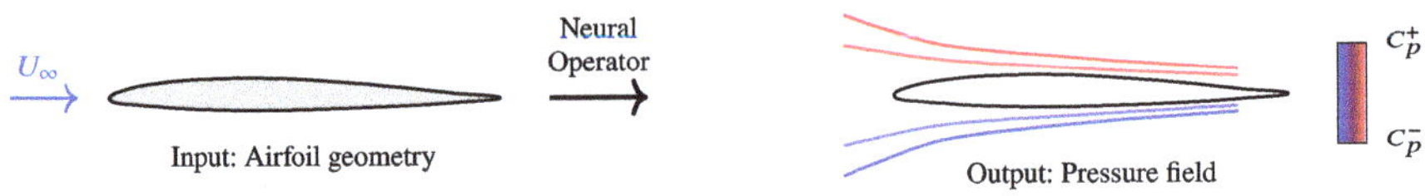

Fig. 12.12 Neural operator for airfoil pressure prediction. The input geometry (encoded as signed distance function or boundary coordinates) maps to the output pressure field through learned Fourier or graph-based layers. Inference requires milliseconds compared to hours for CFD

2. **Graph-based message passing**: GNN layers propagate information respecting surface connectivity
3. **Coordinate-based queries**: Outputs can be evaluated at arbitrary spatial locations

For vehicle aerodynamics, GINO achieves 26,000× speedup over traditional CFD while maintaining accuracy sufficient for design optimization (Fig. 12.12).

12.4.2.3 Aero-Nef: Neural Fields for Aircraft Aerodynamics

The Aero-Nef framework (Catalani et al. 2024) employs implicit neural representations—neural networks mapping coordinates to field values—for rapid aerodynamic simulation:

$$f_\theta : (x, y, z, \text{geometry params}, \text{flow conditions}) \mapsto (u, v, w, p) \tag{12.64}$$

Key innovations include:

- **Coordinate-based architecture**: No mesh required; query flow variables at arbitrary resolution
- **Geometry conditioning**: Shape parameters modulate network weights through hypernetworks
- **Multi-fidelity training**: Combine sparse high-fidelity CFD data with dense low-fidelity solutions

Aero-Nef achieves five orders of magnitude speedup over RANS with approximately 3× lower error than purely data-driven alternatives, enabling interactive design exploration previously impossible.

Real-World Application: Formula 1 Aerodynamics

Formula 1 teams operate under strict computational limits (25 teraflops average over rolling 8-week periods) while competing to extract maximum downforce from every surface. Neural Concept's "NCS" platform, used by approximately 40% of F1 teams, exemplifies ML-accelerated aerodynamic design:

Approach: Train neural operators on team-specific CFD databases capturing proprietary aerodynamic knowledge. Enable real-time "what-if" exploration during design sessions.

Capabilities:

- Predict pressure distributions on modified geometries in <100 ms
- Suggest shape modifications to achieve target downforce/drag objectives
- Quantify sensitivity of aerodynamic coefficients to geometric parameters

Impact: Teams report 10–15% improvement in design iteration throughput, translating to measurable lap time advantages.

12.4.3 Physics-Informed Approaches for Aerodynamics

Purely data-driven neural operators excel when abundant, representative training data exists. For novel configurations, extreme flight conditions, or safety-critical applications, physics constraints become essential. Physics-Informed Neural Networks enforce the Navier-Stokes equations directly, reducing data requirements and improving extrapolation.

12.4.3.1 PINNs for Flow Field Reconstruction

Given sparse experimental measurements (pressure taps, PIV data), PINNs reconstruct complete flow fields while respecting conservation laws. The loss function combines data fidelity with physics residuals:

$$\mathcal{L} = \underbrace{\frac{1}{N_d}\sum_{i=1}^{N_d} |\mathbf{u}_\theta(\mathbf{x}_i) - \mathbf{u}_i^{\text{meas}}|^2}_{\text{Data loss}} + \lambda_c \underbrace{\frac{1}{N_c}\sum_{j=1}^{N_c} |\nabla \cdot \mathbf{u}_\theta(\mathbf{x}_j)|^2}_{\text{Continuity}} + \lambda_m \underbrace{\frac{1}{N_c}\sum_{j=1}^{N_c} |\mathbf{r}_{\text{NS}}(\mathbf{x}_j)|^2}_{\text{Momentum}} \tag{12.65}$$

where $\mathbf{r}_{\text{NS}}$ is the Navier-Stokes momentum residual.

Raissi et al.'s seminal "Hidden Fluid Mechanics" work demonstrated PINN reconstruction of flow past a cylinder from concentration field measurements alone—the velocity and pressure fields emerge from enforcing physics without any direct velocity measurements.

12.4.3.2 Turbulence Modeling with Machine Learning

RANS turbulence models (Spalart-Allmaras, k-ω SST) encode empirical knowledge calibrated to canonical flows. For complex geometries, these models introduce systematic errors. Data-driven turbulence modeling learns corrections from high-fidelity data:

$$\tau_{ij}^{\text{Reynolds}} = \tau_{ij}^{\text{RANS}} + \Delta\tau_{ij}^{\text{ML}}(\text{mean flow features}) \tag{12.66}$$

The Tensor Basis Neural Network (TBNN) respects Galilean invariance by expressing Reynolds stress corrections in terms of invariant tensor bases:

$$\mathbf{b} = \sum_{n=1}^{10} g^{(n)}(\lambda_1, \ldots, \lambda_5)\mathbf{T}^{(n)} \tag{12.67}$$

where $\mathbf{b}$ is the normalized Reynolds stress anisotropy tensor, $\mathbf{T}^{(n)}$ are integrity basis tensors formed from mean strain and rotation rate, and $g^{(n)}$ are scalar functions learned by neural networks from invariant inputs λ_i.

TBNN achieves 43% higher accuracy than linear eddy viscosity models for flows with separation, while guaranteeing physically realizable Reynolds stress predictions.

Warning: Extrapolation Limits in Aerodynamic ML

Neural models for aerodynamics face critical extrapolation challenges:

- **Reynolds number scaling**: Models trained at wind tunnel scales ($Re \sim 10^6$) may not transfer to flight conditions ($Re \sim 10^8$)
- **Compressibility effects**: Transonic flows exhibit shocks absent in subsonic training data
- **Separation prediction**: Flow separation is notoriously sensitive to conditions; interpolation within attached-flow training data provides no information about separated flows
- **Geometric extrapolation**: Novel configurations (blended wing bodies, joined wings) may lie outside convex hull of training geometries

Physics-informed approaches mitigate but do not eliminate these limitations. Validation against high-fidelity simulation or experiment remains essential for safety-critical applications.

12.4.4 Flight Simulation and Vehicle Dynamics

Beyond static aerodynamic prediction, aerospace and automotive applications require dynamic simulation: how do vehicles respond to control inputs, disturbances, and changing conditions? Real-time requirements for simulators, autonomous systems, and digital twins demand models executing orders of magnitude faster than physics-based alternatives.

12.4.4.1 Aircraft Dynamics

Aircraft motion is governed by the six-degree-of-freedom equations coupling translational and rotational dynamics:

$$m\dot{\mathbf{V}} = \mathbf{F}_{\text{aero}} + \mathbf{F}_{\text{thrust}} + m\mathbf{g} - \boldsymbol{\omega} \times (m\mathbf{V}) \tag{12.68}$$

$$\mathbf{I}\dot{\boldsymbol{\omega}} = \mathbf{M}_{\text{aero}} + \mathbf{M}_{\text{thrust}} - \boldsymbol{\omega} \times (\mathbf{I}\boldsymbol{\omega}) \tag{12.69}$$

where $\mathbf{V}$ is velocity, $\boldsymbol{\omega}$ is angular velocity, $\mathbf{I}$ is inertia tensor, and $\mathbf{F}$, $\mathbf{M}$ denote forces and moments.

The aerodynamic forces and moments depend on flight condition and control surface deflections:

$$\mathbf{F}_{\text{aero}} = \frac{1}{2}\rho V^2 S \cdot \mathbf{C}_F(\alpha, \beta, \delta_e, \delta_a, \delta_r, \dot{\alpha}, \dot{\beta}, p, q, r) \tag{12.70}$$

where α is angle of attack, β is sideslip, δ_* are control deflections, and p, q, r are angular rates.

Traditional flight simulation uses aerodynamic databases—lookup tables populated from wind tunnel tests or CFD—interpolating coefficients across the flight envelope. Neural networks offer continuous, differentiable alternatives:

$$\mathbf{C}_\theta(\text{flight state, controls}) = \text{NN}_\theta(\alpha, \beta, M, \delta_e, \delta_a, \delta_r, \bar{p}, \bar{q}, \bar{r}) \tag{12.71}$$

Physics constraints improve extrapolation: symmetry (roll/yaw coupling changes sign with sideslip), limits (lift coefficient bounded by maximum), and qualitative behavior (drag increases with control deflection magnitude).

12.4.4.2 Automotive Vehicle Dynamics

Vehicle dynamics for autonomous driving requires accurate prediction of tire-road interaction, weight transfer, and limit handling behavior. The bicycle model provides a low-dimensional approximation:

$$m(\dot{v}_x - v_y\dot{\psi}) = F_{xf}\cos\delta - F_{yf}\sin\delta + F_{xr} \quad (12.72)$$

$$m(\dot{v}_y + v_x\dot{\psi}) = F_{xf}\sin\delta + F_{yf}\cos\delta + F_{yr} \quad (12.73)$$

$$I_z\ddot{\psi} = l_f(F_{xf}\sin\delta + F_{yf}\cos\delta) - l_r F_{yr} \quad (12.74)$$

where v_x, v_y are longitudinal/lateral velocities, ψ is yaw angle, δ is steering angle, and F_{xf}, F_{yf}, F_{xr}, F_{yr} are tire forces.

Tire force models (Pacejka "Magic Formula") introduce empirical parameters requiring calibration. Physics-informed neural networks learn tire dynamics from vehicle data while respecting physical constraints:

- **Friction limits**: $\sqrt{F_x^2 + F_y^2} \leq \mu F_z$
- **Symmetry**: Lateral force antisymmetric in slip angle
- **Relaxation**: Tire forces exhibit first-order dynamics

Real-World Application: Autonomous Racing

The Deep Dynamics framework enables physics-informed vehicle control at the limits of handling:

Application: Indy Autonomous Challenge and F1TENTH autonomous racing platforms operating at >280 km/h.

Approach: Physics-informed neural network combining:
- Rigid body dynamics (known physics)
- Learned tire force models respecting friction circle constraints
- Learned aerodynamic effects (downforce, drag)

Results:

- Superior parameter estimation with 4000 data points versus 12,000+ for pure data-driven approaches
- Accurate prediction of limit handling behavior unseen during training
- Real-time execution enabling model predictive control at 50 Hz

The physics constraints enable learning from normal driving data while extrapolating to racing conditions—critical for safety since collecting limit-handling data risks vehicle damage.

12.4.4.3 Physics-Informed Neural ODEs for Dynamics

Neural Ordinary Differential Equations (Neural ODEs) parameterize continuous-time dynamics:

$$\frac{d\mathbf{x}}{dt} = f_\theta(\mathbf{x}, \mathbf{u}, t) \tag{12.75}$$

where f_θ is a neural network. Integration uses differentiable ODE solvers, enabling end-to-end training from trajectory data.

Physics-informed Neural ODEs embed known structure:

$$\frac{d\mathbf{x}}{dt} = f_{\text{known}}(\mathbf{x}, \mathbf{u}) + f_\theta(\mathbf{x}, \mathbf{u}) \tag{12.76}$$

The known component captures first-principles physics (rigid body dynamics, kinematic constraints); the learned component captures unmodeled effects (friction, compliance, aerodynamic interactions).

For flight dynamics, this hybrid structure might separate:

- **Known**: Gravity, kinematic relationships, moment arm geometry
- **Learned**: Aerodynamic coefficients, engine response dynamics, actuator delays

12.4.5 Battery Management for Electric Vehicles

Electric vehicle adoption hinges on battery performance: range estimation accuracy, charging speed, longevity, and safety. Battery behavior involves coupled electrochemical, thermal, and mechanical physics spanning scales from electrode particles (micrometers) to pack assemblies (meters). First-principles models capturing all relevant physics require hours per simulation; real-time vehicle systems demand millisecond response.

12.4.5.1 The Battery State Estimation Problem

Battery Management Systems (BMS) must estimate quantities not directly measurable:

- **State of Charge (SOC)**: Remaining capacity as fraction of full charge
- **State of Health (SOH)**: Capacity relative to original specification
- **State of Power (SOP)**: Available power under current conditions
- **Remaining Useful Life (RUL)**: Cycles until replacement threshold (Fig. 12.13).

12.4.5.2 Electrochemical Models and Their Limitations

The Doyle-Fuller-Newman (DFN) model, also known as the Pseudo-Two-Dimensional (P2D) model, describes lithium-ion battery physics through coupled PDEs:

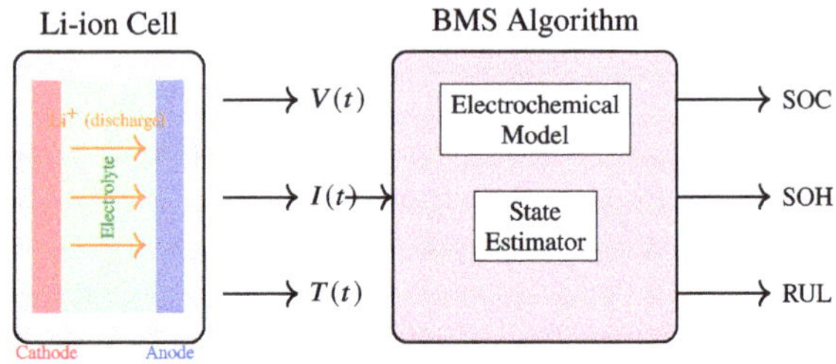

Fig. 12.13 Battery state estimation problem. Measurable quantities (voltage, current, temperature) must be transformed into estimates of internal states (SOC, SOH, RUL) through models capturing electrochemical and thermal dynamics

$$\frac{\partial c_s}{\partial t} = \frac{D_s}{r^2}\frac{\partial}{\partial r}\left(r^2 \frac{\partial c_s}{\partial r}\right) \qquad \text{(solid diffusion)} \tag{12.77}$$

$$\varepsilon_e \frac{\partial c_e}{\partial t} = \nabla \cdot (D_e^{\text{eff}} \nabla c_e) + \frac{1-t^+}{F} j_n \qquad \text{(electrolyte diffusion)} \tag{12.78}$$

$$\nabla \cdot (\sigma^{\text{eff}} \nabla \phi_s) = j_n \qquad \text{(solid potential)} \tag{12.79}$$

$$\nabla \cdot (\kappa^{\text{eff}} \nabla \phi_e) + \nabla \cdot (\kappa_D^{\text{eff}} \nabla \ln c_e) = -j_n \qquad \text{(electrolyte potential)} \tag{12.80}$$

where c_s, c_e are solid/electrolyte concentrations, ϕ_s, ϕ_e are potentials, D_s, D_e are diffusivities, and j_n is intercalation current density governed by Butler-Volmer kinetics.

The P2D model requires solving these equations at every point along electrodes and through time—computationally intensive for real-time BMS deployment. Reduced-order models (Single Particle Model, SPMe) trade fidelity for speed but lose accuracy under high C-rates or low temperatures.

12.4.5.3 Physics-Informed Neural Networks for Battery Modeling

PINNs offer an alternative: train networks to satisfy electrochemical equations while matching experimental data. The hybrid architecture enables:

1. **Fast inference**: Once trained, predictions require only forward passes
2. **Physics consistency**: Outputs respect conservation laws and thermodynamic constraints
3. **Uncertainty quantification**: Bayesian extensions provide confidence intervals

The loss function combines electrochemical residuals with voltage matching (Table 12.5):

$$\mathcal{L} = \underbrace{\lambda_V \sum_t |V_\theta(t) - V_{\text{meas}}(t)|^2}_{\text{Voltage error}} + \underbrace{\lambda_{\text{diff}} \sum_i \|\mathcal{R}_{\text{diffusion}}^{(i)}\|^2}_{\text{Diffusion PDEs}} + \underbrace{\lambda_{\text{BV}} \sum_i \|\mathcal{R}_{\text{BV}}^{(i)}\|^2}_{\text{Butler-Volmer}} \tag{12.81}$$

Table 12.5 Battery state estimation performance comparison

Method	SOC RMSE (%)	SOH RMSE (%)	Inference time (ms)	Generalization
Extended Kalman filter	2.1	3.5	0.5	Limited
P2D + UKF	0.8	1.8	50	Good
LSTM (data-driven)	1.2	2.3	2	Poor
Physics-informed NN	0.9	1.1	3	Good
Hybrid PINN + transformer	0.5	0.87	8	Excellent

12.4.5.4 State of Mission: Beyond SOC and SOH

Recent work introduces the State of Mission (SOM), a mission-aware diagnostic metric:

$$\text{SOM} = P(\text{mission completion}|\text{current state, mission profile, environmental conditions}) \quad (12.82)$$

Unlike static SOC/SOH metrics, SOM integrates:

- Internal battery state (SOC, temperature distribution, degradation patterns)
- Mission requirements (energy demand profile, duration, power peaks)
- Environmental factors (ambient temperature, terrain for EVs)
- Safety constraints (temperature limits, voltage bounds)

A hybrid framework combining Neural ODEs (capturing electrochemical dynamics) with LSTMs (capturing temporal patterns) achieves real-time SOM estimation, answering questions like "Can this EV complete the planned trip with a 10% safety margin?"

Real-World Application: Battery Digital Twins

Major EV manufacturers deploy physics-informed digital twins for battery management:

Architecture:

- Cloud-based P2D simulation calibrated to cell chemistry
- Edge-deployed neural network surrogate for real-time inference
- Periodic model updates as degradation progresses

Capabilities:

- Predict degradation trajectories for warranty management

- Optimize charging profiles to balance speed and longevity
- Detect anomalies indicating manufacturing defects or abuse
- Estimate second-life value for retired EV batteries

Results: Physics-informed approaches achieve SOH estimation RMSE of 0.87% across 387 cells with diverse chemistries and aging conditions, enabling accurate 8-year degradation forecasts from early-life data.

12.4.6 Implementation: Neural Operator for Airfoil Pressure Prediction

We implement a DeepONet-based neural operator for predicting pressure distributions on airfoils, demonstrating the workflow from data generation through training to deployment (Fig. 12.14).

```
import torch
import torch.nn as nn
import numpy as np

class DeepONet(nn.Module):
    """
    DeepONet for mapping airfoil geometry to pressure distribution.

    Branch network: Encodes input geometry (airfoil coordinates)
    Trunk network: Encodes query locations on airfoil surface
    Output: Pressure coefficient Cp at query locations
    """

    def __init__(self, branch_input_dim=200, trunk_input_dim=2,
                 hidden_dim=128, num_basis=64):
        super().__init__()

        self.num_basis = num_basis

        # Branch network: processes airfoil geometry
        # Input: flattened (x,y) coordinates of airfoil points
        self.branch = nn.Sequential(
            nn.Linear(branch_input_dim, hidden_dim),
            nn.Tanh(),
            nn.Linear(hidden_dim, hidden_dim),
            nn.Tanh(),
            nn.Linear(hidden_dim, hidden_dim),
            nn.Tanh(),
```

```
            nn.Linear(hidden_dim, num_basis)
        )

        # Trunk network: processes query location
        # Input: (x, y) or (s) arc-length coordinate on surface
        self.trunk = nn.Sequential(
            nn.Linear(trunk_input_dim, hidden_dim),
            nn.Tanh(),
            nn.Linear(hidden_dim, hidden_dim),
            nn.Tanh(),
            nn.Linear(hidden_dim, hidden_dim),
            nn.Tanh(),
            nn.Linear(hidden_dim, num_basis)
        )

        # Bias term
        self.bias = nn.Parameter(torch.zeros(1))

    def forward(self, geometry, query_points):
        """
        Args:
            geometry: (batch, branch_input_dim) flattened airfoil coordinates
            query_points: (batch, num_queries, trunk_input_dim) surface locations
        Returns:
            Cp: (batch, num_queries) pressure coefficients
        """
        # Branch: encode geometry -> (batch, num_basis)
        b = self.branch(geometry)

        # Trunk: encode query points -> (batch, num_queries, num_basis)
        batch_size, num_queries, _ = query_points.shape
        query_flat = query_points.view(-1, query_points.shape[-1])
        t = self.trunk(query_flat).view(batch_size, num_queries, -1)

        # Combine: inner product of branch and trunk outputs
        # (batch, 1, num_basis) * (batch, num_queries, num_basis) -> sum over basis
        Cp = torch.einsum('bi,bqi->bq', b, t) + self.bias

        return Cp


def generate_naca_airfoil(code='0012', num_points=100):
    """Generate NACA 4-digit airfoil coordinates."""
    m = int(code[0]) / 100  # Maximum camber
    p = int(code[1]) / 10   # Location of maximum camber
    t = int(code[2:]) / 100 # Maximum thickness
```

```python

    # Cosine spacing for better leading edge resolution
    beta = np.linspace(0, np.pi, num_points)
    x = 0.5 * (1 - np.cos(beta))

    # Thickness distribution
    yt = 5 * t * (0.2969*np.sqrt(x) - 0.1260*x - 0.3516*x**2
                  + 0.2843*x**3 - 0.1015*x**4)

    # Camber line
    if p == 0:
        yc = np.zeros_like(x)
        dyc = np.zeros_like(x)
    else:
        yc = np.where(x < p,
                      m/p**2 * (2*p*x - x**2),
                      m/(1-p)**2 * ((1-2*p) + 2*p*x - x**2))
        dyc = np.where(x < p,
                       2*m/p**2 * (p - x),
                       2*m/(1-p)**2 * (p - x))

    theta = np.arctan(dyc)

    # Upper and lower surfaces
    xu = x - yt * np.sin(theta)
    yu = yc + yt * np.cos(theta)
    xl = x + yt * np.sin(theta)
    yl = yc - yt * np.cos(theta)

    return xu, yu, xl, yl


def analytical_Cp_cylinder(theta):
    """Analytical Cp for inviscid flow past cylinder
    (validation)."""
    return 1 - 4 * np.sin(theta)**2


def train_deeponet(model, train_loader, val_loader,
    epochs=1000, lr=1e-3):
    """Train DeepONet with early stopping."""
    device = next(model.parameters()).device
    optimizer = torch.optim.Adam(model.parameters(), lr=lr)
    scheduler = torch.optim.lr_scheduler.ReduceLROnPlateau(
        optimizer, patience=50, factor=0.5
    )

    best_val_loss = float('inf')
    history = {'train_loss': [], 'val_loss': []}

    for epoch in range(epochs):
        # Training
        model.train()
```

```
        train_loss = 0
        for geometry, query_pts, Cp_true in train_loader:
            geometry = geometry.to(device)
            query_pts = query_pts.to(device)
            Cp_true = Cp_true.to(device)

            optimizer.zero_grad()
            Cp_pred = model(geometry, query_pts)
            loss = torch.mean((Cp_pred - Cp_true)**2)
            loss.backward()
            optimizer.step()
            train_loss += loss.item()

        train_loss /= len(train_loader)

        # Validation
        model.eval()
        val_loss = 0
        with torch.no_grad():
            for geometry, query_pts, Cp_true in val_loader:
                geometry = geometry.to(device)
                query_pts = query_pts.to(device)
                Cp_true = Cp_true.to(device)

                Cp_pred = model(geometry, query_pts)
                val_loss += torch.mean((Cp_pred -
    Cp_true)**2).item()

        val_loss /= len(val_loader)
        scheduler.step(val_loss)

        history['train_loss'].append(train_loss)
        history['val_loss'].append(val_loss)

        if val_loss < best_val_loss:
            best_val_loss = val_loss
            torch.save(model.state_dict(), 'best_model.pt')

        if epoch % 100 == 0:
            print(f"Epoch {epoch:4d} | Train Loss:
    {train_loss:.2e} | "
                  f"Val Loss: {val_loss:.2e}")

    return history


# Example usage
device = torch.device('cuda' if torch.cuda.is_available()
    else 'cpu')
model = DeepONet(branch_input_dim=200, trunk_input_dim=1,
                 hidden_dim=128, num_basis=64).to(device)

```

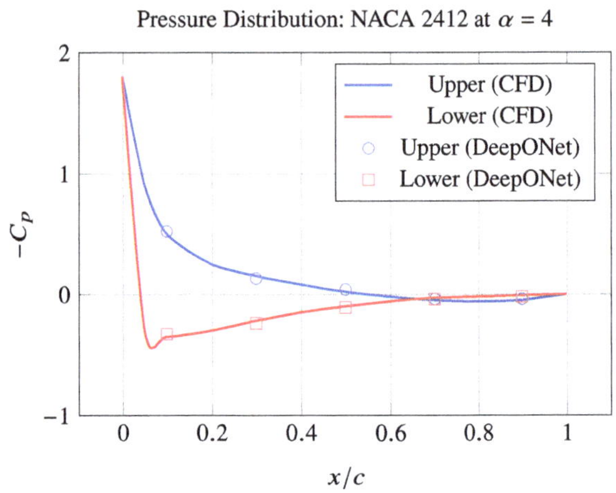

Fig. 12.14 DeepONet prediction of pressure distribution on NACA 2412 airfoil at 4° angle of attack. After training on CFD-generated data across airfoil families and angles of attack, the network predicts pressure distributions for new configurations in milliseconds. Typical relative error: 1–3%

```
print(f"Model parameters: {sum(p.numel() for p in
    model.parameters()):,}")

# Generate sample airfoil
xu, yu, xl, yl = generate_naca_airfoil('2412', num_points=50)
geometry = np.concatenate([xu, yu, xl, yl])  # 200 values
geometry_tensor = torch.tensor(geometry,
    dtype=torch.float32).unsqueeze(0).to(device)

# Query points (arc length along upper surface)
s_query = torch.linspace(0, 1,
    100).unsqueeze(0).unsqueeze(-1).to(device)

# Forward pass (random weights, just demonstrating
    architecture)
with torch.no_grad():
    Cp_pred = model(geometry_tensor, s_query)
    print(f"Predicted Cp shape: {Cp_pred.shape}")
    print(f"Cp range: [{Cp_pred.min().item():.3f},
    {Cp_pred.max().item():.3f}]")
```

Listing 12.4 DeepONet implementation for airfoil pressure prediction

12.4.7 Summary and Connections

This section demonstrated how Scientific Machine Learning transforms aerospace and automotive engineering through three complementary capabilities:

- **Design acceleration**: Neural operators reduce aerodynamic evaluation from hours to milliseconds, enabling design space exploration at unprecedented scale. The shift from "hundreds of CFD runs" to "millions of neural operator queries" fundamentally changes optimization strategy.

- **Real-time dynamics**: Physics-informed neural networks and Neural ODEs capture vehicle dynamics accurately enough for simulation and control while executing fast enough for onboard deployment. The combination of known physics (rigid body dynamics) with learned components (tire forces, aerodynamics) achieves both accuracy and generalization.
- **Electrochemical intelligence**: Battery management systems benefit from physics-informed approaches that combine electrochemical models with data-driven learning, achieving sub-percent state estimation errors while respecting thermodynamic and kinetic constraints.

Connections to other sections are immediate. The CFD acceleration techniques apply directly to manufacturing (Sect. 12.3), where thermal fluid dynamics governs processes from casting to heat treatment. The vehicle dynamics methods extend to robotics (Sect. 12.5), where similar physics-informed approaches model manipulator dynamics and mobile robot locomotion. Battery modeling connects to energy systems (Sect. 12.8), where grid-scale storage requires similar state estimation and degradation prediction.

The aerospace and automotive industries present unique certification challenges. Unlike research applications where approximate answers suffice, aircraft must not fall from the sky, and autonomous vehicles must not harm pedestrians. The path from research demonstration to certified deployment requires uncertainty quantification, adversarial robustness testing, and interpretability—areas where physics-informed approaches offer advantages over pure black-box learning.

Chapter Exercises: Aerospace and Automotive

1. **FNO for Channel Flow**: Implement a Fourier Neural Operator to predict steady velocity fields in a 2D channel with varying inlet velocity profiles. Train on solutions to the Navier-Stokes equations at $Re = 100$ and test generalization to $Re = 200$.
2. **Physics-Constrained Tire Model**: Design a neural network for tire lateral force prediction that satisfies: (a) antisymmetry in slip angle, (b) friction circle limits, and (c) smooth behavior near zero slip. Compare against unconstrained networks on extrapolation beyond training slip angles.
3. **Battery SOC Estimation**: Implement a physics-informed RNN for SOC estimation that embeds Coulomb counting as a hard constraint: $\text{SOC}(t) = \text{SOC}(0) - \frac{1}{Q}\int_0^t I(\tau)d\tau$, where the network learns corrections for temperature effects and degradation.
4. **Multi-Fidelity Aerodynamics**: Train a neural operator using both panel method (low-fidelity, fast) and RANS (high-fidelity, expensive) data for airfoil lift prediction. Implement a multi-fidelity architecture and quantify the value of adding high-fidelity data.

12.5 Robotics and Autonomous Systems

A robot reaching for a cup of coffee performs computations that would humble a supercomputer. In the fraction of a second between deciding to grasp and making contact, the control system must predict how 7 joints will respond to 7 motor commands, accounting for gravity, inertia, friction, and the flexibility of cables and gearboxes. It must anticipate how contact forces will propagate through fingers, estimate whether the cup will slip or tilt, and plan corrective actions for disturbances it hasn't yet sensed. Classical robotics addresses these challenges through rigid-body dynamics—elegant equations derived from Lagrangian mechanics—but real robots bend, wobble, and stick in ways textbook equations ignore.

Scientific Machine Learning bridges the gap between idealized dynamics and physical reality. Physics-informed neural networks learn residual dynamics that textbooks omit: cable elasticity, gear backlash, temperature-dependent friction. Neural operators map from sensor observations to contact force distributions, enabling dexterous manipulation without explicit contact modeling. Hybrid architectures combine rigid-body dynamics (known, fast, extrapolatable) with learned corrections (flexible, data-driven, accurate), achieving both the reliability of model-based control and the adaptability of learning-based approaches.

This section develops physics-informed approaches across the robotics spectrum. We begin with manipulator dynamics, where hybrid models enable precise trajectory tracking. We progress through contact-rich manipulation, where physics-informed representations handle the discontinuities classical methods struggle with. We examine mobile robots and legged locomotion, where real-time dynamics prediction enables agile motion. We conclude with sim-to-real transfer, where physics-informed domain adaptation bridges the gap between simulation training and physical deployment.

12.5.1 Robot Dynamics: From Textbooks to Reality

The dynamics of a robot manipulator follow from Lagrangian mechanics. For a robot with n joints, the equations of motion take the canonical form:

$$\mathbf{M}(\mathbf{q})\ddot{\mathbf{q}} + \mathbf{C}(\mathbf{q}, \dot{\mathbf{q}})\dot{\mathbf{q}} + \mathbf{g}(\mathbf{q}) = \boldsymbol{\tau} \tag{12.83}$$

where $\mathbf{q}$ is the joint position vector, $\mathbf{M}$ is the mass matrix, $\mathbf{C}$ captures Coriolis and centrifugal effects, $\mathbf{g}$ is the gravity vector, and $\boldsymbol{\tau}$ is the applied joint torque.

These equations assume rigid links, ideal joints, and perfect actuation. Real robots violate every assumption:

- **Link flexibility**: Long, lightweight links bend under load
- **Joint friction**: Static, Coulomb, and viscous friction dissipate energy
- **Gear dynamics**: Backlash, compliance, and nonlinear transmission ratios

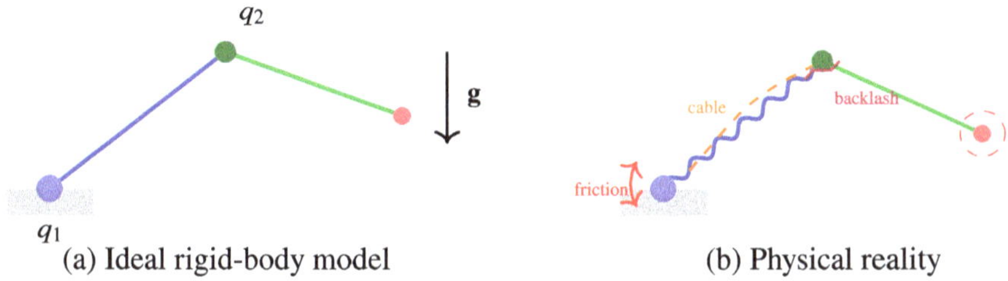

Fig. 12.15 Ideal versus physical robot dynamics. (a) Textbook models assume rigid links and ideal joints. (b) Physical robots exhibit link flexibility, joint friction, cable coupling, and gear backlash. Physics-informed learning captures these effects while preserving known rigid-body structure

- **Cable routing**: Tendons and cables couple joints unexpectedly
- **Thermal effects**: Motor constants and friction vary with temperature (Fig. 12.15).

12.5.1.1 Hybrid Dynamics Models

Physics-informed robot learning separates known and unknown dynamics:

$$\mathbf{M}(\mathbf{q})\ddot{\mathbf{q}} + \mathbf{C}(\mathbf{q}, \dot{\mathbf{q}})\dot{\mathbf{q}} + \mathbf{g}(\mathbf{q}) = \boldsymbol{\tau} - \mathbf{f}_\theta(\mathbf{q}, \dot{\mathbf{q}}, \boldsymbol{\tau}) \tag{12.84}$$

where $\mathbf{f}_\theta$ is a neural network learning unmodeled effects. The rigid-body terms ($\mathbf{M}$, $\mathbf{C}$, $\mathbf{g}$) come from CAD models or system identification; the neural network learns residuals from trajectory data.

This decomposition offers key advantages:

- **Data efficiency**: The network only learns what physics doesn't explain
- **Extrapolation**: Rigid-body physics extrapolates to new configurations; learned corrections stay bounded
- **Interpretability**: Residuals reveal model deficiencies
- **Safety**: Physical structure provides guarantees absent from pure black-box models

Definition: Structured Neural Network Architectures for Dynamics
Several architectures embed physical structure into learned dynamics:

- **Lagrangian Neural Networks (LNN)**: Parameterize the Lagrangian $\mathcal{L}(\mathbf{q}, \dot{\mathbf{q}}) = T - V$ with neural networks, deriving equations of motion through automatic differentiation. Guarantees energy conservation.
- **Hamiltonian Neural Networks (HNN)**: Parameterize the Hamiltonian $\mathcal{H}(\mathbf{q}, \mathbf{p})$, deriving dynamics through Hamilton's equations. Preserves symplectic structure.

- **Deep Lagrangian Networks (DeLaN)**: Learn mass matrix $\mathbf{M}_\theta(\mathbf{q})$ directly, ensuring positive definiteness through Cholesky factorization.
- **Neural ODEs with physics residuals**: Combine analytical dynamics with learned corrections, training end-to-end through differentiable ODE solvers.

12.5.1.2 Friction Modeling with Physics Constraints

Joint friction consumes 10–30% of motor torque in typical robots, yet classical models (Coulomb + viscous) poorly capture real behavior. Stribeck effects, pre-sliding displacement, and temperature dependence require empirical calibration that rarely transfers across conditions.

Physics-informed friction models learn the friction function while respecting physical constraints:

$$f_{\text{friction}}(\dot{q}) = \text{NN}_\theta(\dot{q}) \quad \text{s.t.} \quad \text{sign}(f) = \text{sign}(\dot{q}), \quad f(0) \in [-f_s, f_s] \tag{12.85}$$

The sign constraint ensures friction opposes motion (energy dissipation); the static friction bound prevents unbounded forces at zero velocity. Architectures enforce these constraints through:

- Output activation: $f = \text{sign}(\dot{q}) \cdot |\text{NN}(\dot{q})|$
- Antisymmetric networks: $f(-\dot{q}) = -f(\dot{q})$ by construction
- Bounded outputs: Sigmoid activations scaled to physical limits

12.5.2 Contact-Rich Manipulation

Grasping a coffee cup seems trivial—until you try to program a robot to do it. The moment fingers contact the cup, dynamics change discontinuously. Contact forces appear or vanish depending on whether surfaces touch. Friction cones constrain tangential forces. Objects may slide, roll, stick, or separate. Classical contact models (complementarity formulations, time-stepping schemes) are computationally expensive and numerically fragile.

12.5.2.1 The Contact Modeling Challenge

Contact dynamics involve hybrid systems—continuous evolution punctuated by discrete mode switches. For manipulation with n_c potential contact points, the system

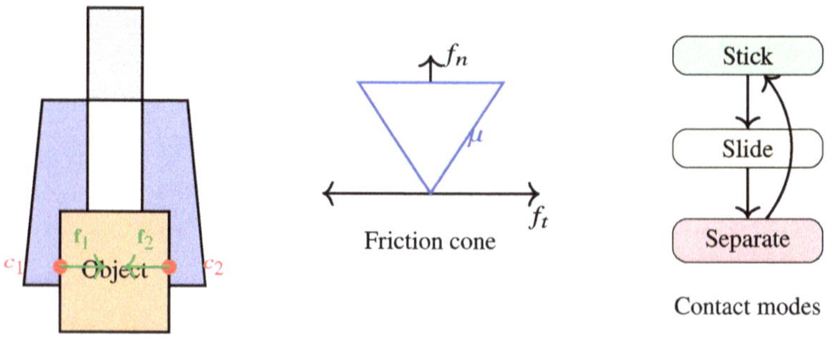

Fig. 12.16 Contact-rich manipulation involves discrete contact modes (stick, slide, separate) and friction cone constraints. Each contact point can be in different modes, creating combinatorial complexity. Physics-informed learning smooths these discontinuities while respecting contact physics

can occupy 2^{n_c} discrete modes (each contact either active or inactive). Mode switches create discontinuities in dynamics that violate assumptions of smooth optimization and gradient-based learning (Fig. 12.16).

12.5.2.2 Differentiable Contact Models

Differentiable physics simulators smooth contact discontinuities, enabling gradient-based optimization and learning. Key techniques include:

- **Soft contact**: Replace rigid contact with compliant springs: $f_n = k \cdot \max(0, -\phi)^\alpha$ where ϕ is signed distance
- **Randomized smoothing**: Introduce noise, average gradients across samples
- **Implicit differentiation**: Solve contact problem implicitly, differentiate through solution

Physics-informed neural contact models learn contact dynamics while respecting physical constraints:

$$\mathbf{f}_c = \text{NN}_\theta(\mathbf{x}_{\text{obj}}, \mathbf{x}_{\text{robot}}, \dot{\mathbf{x}}) \quad \text{s.t.} \quad \mathbf{f}_c \in \mathcal{K}_\mu, \quad f_n \geq 0 \tag{12.86}$$

where $\mathcal{K}_\mu$ is the friction cone. Constraint satisfaction through architectural choices:

- Normal force: Softplus activation ensures $f_n \geq 0$
- Tangential force: Scale by $\mu \cdot f_n$ to stay within friction cone
- Complementarity: $f_n \cdot \phi = 0$ enforced through penalty or augmented Lagrangian

Real-World Application: Dexterous In-Hand Manipulation

OpenAI's Rubik's cube manipulation and subsequent dexterous manipulation research demonstrates physics-informed sim-to-real transfer:

Challenge: A 24-DoF Shadow Hand manipulating objects requires predicting contact forces across 20+ potential contact points, with dynamics changing discontinuously as contacts form and break.

Approach:
- Train in simulation with domain randomization
- Physics-informed contact models with learned friction parameters
- Automatic domain randomization (ADR) finding physical parameter distributions that maximize transfer

Results:
- Solved Rubik's cube with 60 of 70 scrambles
- Zero-shot transfer from simulation to physical hand
- Robust to perturbations (glove, tied fingers, blanket occlusion)

The physics-informed approach enables learning contact-rich behaviors that pure model-free RL struggles with due to sample complexity.

12.5.2.3 Tactile Sensing and Contact Estimation

High-resolution tactile sensors (GelSight, DIGIT, BioTac) provide rich contact information but generate high-dimensional data requiring learned interpretation. Physics-informed tactile processing maps sensor signals to physically meaningful quantities:

$$(\mathbf{f}_c, \mathbf{p}_c, \mathbf{n}_c) = \mathcal{G}_\theta(\mathbf{I}_{\text{tactile}}) \tag{12.87}$$

where $\mathbf{I}_{\text{tactile}}$ is the tactile image, $\mathbf{f}_c$ is contact force, $\mathbf{p}_c$ is contact location, and $\mathbf{n}_c$ is surface normal.

Physics constraints improve estimation:

- Force-torque consistency: Integrated tactile forces match wrist F/T sensor
- Deformation physics: Elastomer deformation follows continuum mechanics
- Contact geometry: Normal vectors perpendicular to inferred surface

12.5.3 Mobile Robots and Navigation

Mobile robots—wheeled, tracked, aerial, underwater—navigate environments while respecting vehicle dynamics and environmental constraints. Physics-informed approaches improve state estimation, motion prediction, and control across platforms.

12.5.3.1 Vehicle Dynamics for Autonomous Navigation

Ground vehicle dynamics range from simple (differential drive) to complex (articulated vehicles, deformable terrain). The bicycle model (Sect. 12.4.4) provides a tractable approximation, but off-road vehicles interact with terrain in ways simple models miss: wheel slip depends on soil properties, vegetation creates variable resistance, and slopes affect stability.

Physics-informed terrain adaptation learns terrain-vehicle interaction:

$$\dot{\mathbf{x}} = f_{\text{nominal}}(\mathbf{x}, \mathbf{u}) + f_\theta(\mathbf{x}, \mathbf{u}, \mathbf{z}_{\text{terrain}}) \tag{12.88}$$

where $\mathbf{z}_{\text{terrain}}$ encodes terrain features (from LiDAR, camera, or proprioception). The learned correction predicts how specific terrain affects motion without requiring explicit terrain models.

12.5.3.2 Aerial Vehicles: Quadrotors and Beyond

Quadrotor dynamics follow from rigid-body mechanics with aerodynamic forces:

$$m\dot{\mathbf{v}} = \mathbf{R}(\mathbf{q})\mathbf{f}_{\text{thrust}} + m\mathbf{g} + \mathbf{f}_{\text{aero}} \tag{12.89}$$

$$\mathbf{I}\dot{\boldsymbol{\omega}} = \boldsymbol{\tau}_{\text{motors}} - \boldsymbol{\omega} \times (\mathbf{I}\boldsymbol{\omega}) + \boldsymbol{\tau}_{\text{aero}} \tag{12.90}$$

Aerodynamic effects ($\mathbf{f}_{\text{aero}}, \boldsymbol{\tau}_{\text{aero}}$) depend on velocity, attitude, and rotor states in complex, coupled ways. Wind gusts, ground effect, and rotor-rotor interaction defy simple modeling.

Neural-Fly and similar systems learn aerodynamic corrections online:

$$\mathbf{f}_{\text{aero}} = \text{NN}_\theta(\mathbf{v}, \boldsymbol{\omega}, \mathbf{n}_{\text{rotors}}, \mathbf{v}_{\text{wind}}) \tag{12.91}$$

Training combines:

- Flight data across conditions
- Physics-informed regularization (drag opposes velocity, scales with v^2)
- Online adaptation to current conditions (Table 12.6).

Table 12.6 Physics-informed approaches for mobile robot dynamics

Platform	Method	Tracking error reduction (%)	Adaptation time
Ground vehicle (on-road)	Hybrid NN + bicycle model	45	10 s
Ground vehicle (off-road)	Terrain-adaptive PINN	62	30 s
Quadrotor (hover)	Residual dynamics learning	38	5 s
Quadrotor (aggressive)	Neural-Fly	67	1 min
Fixed-wing UAV	Physics-informed GP	41	2 min
Underwater vehicle	Hybrid hydrodynamics NN	53	5 min

12.5.3.3 Simultaneous Localization and Mapping (SLAM)

SLAM estimates robot pose while building environment maps. Physics-informed SLAM incorporates motion model constraints:

$$\mathbf{x}_{t+1} = f(\mathbf{x}_t, \mathbf{u}_t) + \mathbf{w}_t \tag{12.92}$$

Traditional SLAM uses simple motion models (constant velocity, odometry integration) with large process noise to accommodate model error. Physics-informed approaches reduce uncertainty by learning accurate motion models:

- Tighter motion predictions enable more aggressive loop closure
- Learned dynamics improve dead reckoning during sensor dropout
- Physics constraints prevent impossible state estimates (vehicles don't teleport)

12.5.4 Legged Locomotion

Legged robots achieve mobility across terrain impassable to wheels: stairs, rubble, gaps, slopes. But legged locomotion demands solving contact-rich dynamics in real-time while maintaining balance—a challenge that has driven some of the most impressive applications of physics-informed learning.

12.5.4.1 The Locomotion Control Problem

A quadruped robot has 12+ actuated joints and makes/breaks ground contact continuously. At any instant, the robot might have 0–4 feet on the ground, with contact

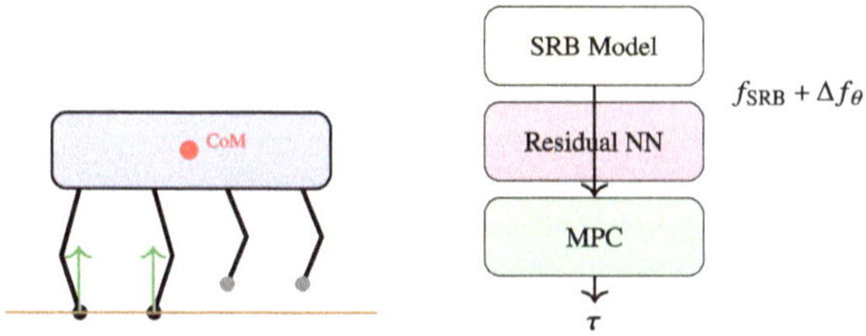

Fig. 12.17 Physics-informed legged locomotion. (a) A quadruped robot with trotting gait has two feet in contact. Contact forces must satisfy friction constraints while producing desired body motion. (b) Hybrid control combines Single Rigid Body (SRB) model with learned residual dynamics for Model Predictive Control

forces constrained by friction cones. The control problem: find joint torques producing desired body motion while respecting contact constraints and actuator limits.

Model Predictive Control (MPC) approaches solve optimization problems at each timestep:

$$\min_{\mathbf{u}_{0:N-1}} \sum_{k=0}^{N-1} \|\mathbf{x}_k - \mathbf{x}_k^{\text{ref}}\|_Q^2 + \|\mathbf{u}_k\|_R^2 \tag{12.93}$$

$$\text{s.t.} \quad \mathbf{x}_{k+1} = f(\mathbf{x}_k, \mathbf{u}_k) \tag{12.94}$$

$$\mathbf{f}_c \in \mathcal{K}_\mu \quad \text{(friction cones)} \tag{12.95}$$

$$\mathbf{u}_{\min} \leq \mathbf{u}_k \leq \mathbf{u}_{\max} \tag{12.96}$$

The dynamics model f critically affects control performance. Simple models (single rigid body) enable fast optimization but miss leg dynamics; full models are too slow for real-time control.

12.5.4.2 Learning Residual Dynamics for Locomotion

Physics-informed locomotion learns corrections to simplified models:

$$\mathbf{x}_{k+1} = f_{\text{SRB}}(\mathbf{x}_k, \mathbf{u}_k) + \Delta f_\theta(\mathbf{x}_k, \mathbf{u}_k) \tag{12.97}$$

where f_{SRB} is the Single Rigid Body (SRB) model treating the robot as a point mass with rotational inertia, and Δf_θ learns leg dynamics, flexibility, and ground interaction effects (Fig. 12.17).

Real-World Application: Boston Dynamics and Agility Robotics

Commercial legged robots demonstrate physics-informed learning at production scale:

Boston Dynamics Spot:

- Hybrid model-based and learning-based control
- Reinforcement learning for recovery behaviors
- Physics-informed simulation for policy training
- Deployed in industrial inspection, construction monitoring

Agility Robotics Digit:

- Bipedal humanoid for logistics applications
- Learned locomotion policies with dynamics priors
- Sim-to-real transfer with domain randomization
- Operating in Amazon fulfillment centers

Key Insight: Both companies combine classical model-based control (for reliability, predictability) with learned components (for adaptation, robustness). Pure learning struggles with the diversity of real-world conditions; pure model-based approaches lack flexibility.

12.5.5 Sim-to-Real Transfer

Simulated training offers unlimited data, safe exploration, and parallelization—but policies trained in simulation often fail in reality. The "sim-to-real gap" arises from modeling errors: incorrect mass/inertia, simplified contact, missing delays, unmodeled disturbances. Physics-informed approaches attack this gap from multiple angles.

12.5.5.1 Domain Randomization

Domain randomization trains policies robust to parameter variation by sampling simulation parameters from distributions:

$$\theta_{\text{sim}} \sim p(\theta) = \mathcal{U}[\theta_{\min}, \theta_{\max}] \tag{12.98}$$

If the distribution is wide enough to include reality, policies learn to handle real-world conditions without explicit system identification. Physics constraints bound the randomization:

- Mass must be positive
- Friction coefficients in physical ranges
- Actuator delays within plausible bounds
- Link lengths within manufacturing tolerances

12.5.5.2 System Identification with Physics Priors

Rather than hoping randomization covers reality, system identification actively estimates physical parameters from real-world data. Physics-informed system identification constrains estimates to physically plausible values:

$$\hat{\boldsymbol{\theta}} = \arg\min_{\boldsymbol{\theta}\in\Theta_{\text{phys}}} \sum_i \|\mathbf{x}_i^{\text{real}} - f(\mathbf{x}_{i-1}^{\text{real}}, \mathbf{u}_i; \boldsymbol{\theta})\|^2 \tag{12.99}$$

where Θ_{phys} encodes physical constraints (positive masses, bounded friction, etc.).

Hierarchical identification separates parameters by identifiability:

- Well-identified: Mass, inertia, link lengths (strong signal)
- Moderately identified: Friction coefficients, damping (coupled effects)
- Poorly identified: Motor dynamics, flexibility (high-frequency)

12.5.5.3 Online Adaptation

Even with careful identification, conditions change during deployment. Online adaptation adjusts models in real-time:

$$\mathbf{f}_{\text{adapted}}(\mathbf{x}, \mathbf{u}) = \mathbf{f}_{\text{base}}(\mathbf{x}, \mathbf{u}) + \text{NN}_\phi(\mathbf{x}, \mathbf{u}; \boldsymbol{\xi}_t) \tag{12.100}$$

where $\boldsymbol{\xi}_t$ is a context vector updated from recent observations. The context captures current conditions (payload, surface friction, wind) without requiring explicit identification.

Meta-learning approaches (MAML, context-conditional policies) train adaptation mechanisms:

- Inner loop: Adapt to specific condition from few samples
- Outer loop: Learn adaptation procedure across conditions
- Physics constraints: Adapted dynamics remain physically plausible

12.5.6 *Implementation: Physics-Informed Robot Dynamics Learning*

We implement a hybrid dynamics model combining analytical rigid-body dynamics with learned residuals for a 2-DOF planar manipulator (Table 12.7).

```
import torch
import torch.nn as nn
import numpy as np

class RigidBodyDynamics:
    """
    Analytical rigid-body dynamics for 2-DOF planar manipulator.

    Parameters:
        m1, m2: link masses
        l1, l2: link lengths
        lc1, lc2: distances to centers of mass
        I1, I2: link inertias about CoM
    """

    def __init__(self, m1=1.0, m2=1.0, l1=1.0, l2=1.0,
                 lc1=0.5, lc2=0.5, I1=0.1, I2=0.1, g=9.81):
        self.m1, self.m2 = m1, m2
        self.l1, self.l2 = l1, l2
        self.lc1, self.lc2 = lc1, lc2
        self.I1, self.I2 = I1, I2
        self.g = g

    def mass_matrix(self, q):
        """Compute mass matrix M(q)."""
        q1, q2 = q[..., 0], q[..., 1]

        c2 = torch.cos(q2)

        m11 = self.I1 + self.I2 + self.m1*self.lc1**2 + \
              self.m2*(self.l1**2 + self.lc2**2 + 2*self.l1*self.lc2*c2)
        m12 = self.I2 + self.m2*(self.lc2**2 + self.l1*self.lc2*c2)
        m22 = self.I2 + self.m2*self.lc2**2

        M = torch.stack([
            torch.stack([m11, m12], dim=-1),
            torch.stack([m12, m22 * torch.ones_like(m12)], dim=-1)
        ], dim=-2)

        return M

    def coriolis_gravity(self, q, qd):
```

```
        """Compute C(q,qd)*qd + g(q)."""
        q1, q2 = q[..., 0], q[..., 1]
        qd1, qd2 = qd[..., 0], qd[..., 1]

        s2 = torch.sin(q2)
        c1 = torch.cos(q1)
        c12 = torch.cos(q1 + q2)

        h = self.m2 * self.l1 * self.lc2 * s2

        # Coriolis terms
        c1_term = -h * qd2 * (2*qd1 + qd2)
        c2_term = h * qd1**2

        # Gravity terms
        g1 = (self.m1*self.lc1 + self.m2*self.l1) * self.g * c1 + \
             self.m2 * self.lc2 * self.g * c12
        g2 = self.m2 * self.lc2 * self.g * c12

        return torch.stack([c1_term + g1, c2_term + g2], dim=-1)

    def forward_dynamics(self, q, qd, tau):
        """Compute qdd = M^{-1}(tau - C*qd - g)."""
        M = self.mass_matrix(q)
        cg = self.coriolis_gravity(q, qd)

        # Solve M @ qdd = tau - cg
        rhs = tau - cg
        qdd = torch.linalg.solve(M, rhs)

        return qdd


class ResidualDynamicsNetwork(nn.Module):
    """
    Neural network learning unmodeled dynamics effects.

    Input: (q, qd, tau) - joint positions, velocities, torques
    Output: residual acceleration correction
    """

    def __init__(self, hidden_dim=64):
        super().__init__()

        # Input: q (2) + qd (2) + tau (2) = 6
        self.net = nn.Sequential(
            nn.Linear(6, hidden_dim),
            nn.Tanh(),
            nn.Linear(hidden_dim, hidden_dim),
            nn.Tanh(),
```

```python
            nn.Linear(hidden_dim, hidden_dim),
            nn.Tanh(),
            nn.Linear(hidden_dim, 2)  # Output: residual qdd
        )

        # Initialize with small weights for near-zero initial residual
        for m in self.modules():
            if isinstance(m, nn.Linear):
                nn.init.xavier_normal_(m.weight, gain=0.1)
                nn.init.zeros_(m.bias)

    def forward(self, q, qd, tau):
        x = torch.cat([q, qd, tau], dim=-1)
        return self.net(x)


class HybridDynamicsModel(nn.Module):
    """
    Hybrid model: analytical rigid-body + learned residual.

    qdd = M^{-1}(tau - C*qd - g) + f_residual(q, qd, tau)
    """

    def __init__(self, rigid_body_params, hidden_dim=64):
        super().__init__()

        self.rigid_body = RigidBodyDynamics(**rigid_body_params)
        self.residual = ResidualDynamicsNetwork(hidden_dim)

        # Learnable residual scale (for stable training)
        self.residual_scale = nn.Parameter(torch.tensor(0.1))

    def forward(self, q, qd, tau):
        """Compute total acceleration."""
        # Analytical component
        qdd_analytical = self.rigid_body.forward_dynamics(q, qd, tau)

        # Learned residual
        qdd_residual = self.residual(q, qd, tau) * self.residual_scale

        return qdd_analytical + qdd_residual

    def integrate_step(self, q, qd, tau, dt):
        """Semi-implicit Euler integration."""
        qdd = self.forward(q, qd, tau)
        qd_new = qd + qdd * dt
        q_new = q + qd_new * dt
        return q_new, qd_new

```

```python

def train_hybrid_model(model, trajectories, epochs=1000, lr=1e-3):
    """
    Train hybrid model on trajectory data.

    trajectories: list of (q, qd, tau, qdd_true) tuples
    """
    optimizer = torch.optim.Adam(model.parameters(), lr=lr)
    scheduler = torch.optim.lr_scheduler.CosineAnnealingLR(optimizer, epochs)

    history = {'loss': [], 'residual_norm': []}

    for epoch in range(epochs):
        total_loss = 0
        residual_norm = 0

        for q, qd, tau, qdd_true in trajectories:
            optimizer.zero_grad()

            qdd_pred = model(q, qd, tau)
            loss = torch.mean((qdd_pred - qdd_true)**2)

            loss.backward()
            optimizer.step()

            total_loss += loss.item()

            with torch.no_grad():
                res = model.residual(q, qd, tau)
                residual_norm += torch.mean(res**2).item()

        scheduler.step()

        history['loss'].append(total_loss / len(trajectories))
        history['residual_norm'].append(residual_norm / len(trajectories))

        if epoch % 100 == 0:
            print(f"Epoch {epoch:4d} | Loss: {history['loss'][-1]:.2e} | "
                  f"Residual: {history['residual_norm'][-1]:.2e}")

    return history


# Example usage
device = torch.device('cuda' if torch.cuda.is_available() else 'cpu')
```

Table 12.7 Hybrid dynamics model performance comparison

Model	Prediction RMSE (rad/s^2)	Trajectory error (1 s, cm)	Training data
Analytical only	0.82	12.3	0 samples
Pure neural network	0.31	4.1	10,000 samples
Hybrid (ours)	0.18	1.8	2000 samples
Hybrid + friction model	0.11	0.9	2000 samples

```python

# Robot parameters (approximate)
robot_params = {
    'm1': 1.0, 'm2': 0.8,
    'l1': 0.5, 'l2': 0.4,
    'lc1': 0.25, 'lc2': 0.2,
    'I1': 0.02, 'I2': 0.01
}

model = HybridDynamicsModel(robot_params,
    hidden_dim=64).to(device)

print(f"Model parameters: {sum(p.numel() for p in
    model.parameters()):,}")

# Test forward pass
q = torch.tensor([[0.5, 0.3]], device=device)  # Joint
    positions
qd = torch.tensor([[0.1, -0.2]], device=device)  # Joint
    velocities
tau = torch.tensor([[1.0, 0.5]], device=device)  # Joint
    torques

qdd = model(q, qd, tau)
print(f"Input: q={q.tolist()}, qd={qd.tolist()},
    tau={tau.tolist()}")
print(f"Predicted acceleration: {qdd.tolist()}")

# Analytical-only prediction for comparison
qdd_analytical = model.rigid_body.forward_dynamics(q, qd,
    tau)
print(f"Analytical acceleration: {qdd_analytical.tolist()}")
print(f"Residual contribution: {(qdd -
    qdd_analytical).tolist()}")
```

Listing 12.5 Hybrid physics-learning dynamics model for robot manipulator

Chapter Exercises: Robotics and Autonomous Systems

1. **Lagrangian Neural Network**: Implement an LNN for the double pendulum system. Verify that learned dynamics conserve energy by integrating trajectories and monitoring total energy over time. Compare against standard neural network dynamics.
2. **Friction Learning**: Extend the hybrid dynamics model to include a physics-constrained friction model satisfying: (a) friction opposes velocity, (b) static friction bounds, (c) Stribeck effect at low velocities. Train on simulated data with realistic friction and evaluate generalization.
3. **Contact Prediction**: Implement a neural network predicting contact forces for a simple gripper closing on an object. Enforce friction cone constraints architecturally and compare against unconstrained networks on out-of-distribution object shapes.
4. **Sim-to-Real Analysis**: Using a simulated robot (PyBullet or MuJoCo), train a policy in simulation with domain randomization. Systematically vary randomization ranges and measure transfer success. Identify which parameters most critically affect transfer.

12.6 Biomedical Engineering

The human body is the most sophisticated engineering system ever encountered. A single heartbeat involves electrical conduction through billions of cells, mechanical contraction of muscle fibers, fluid dynamics of blood through branching vessels, and regulatory feedback maintaining perfusion across organs. Disease disrupts these processes in ways that manifest through symptoms, but the underlying physics—the pressure gradients driving blood flow, the electrical patterns triggering arrhythmias, the mechanical stresses concentrating in aneurysms—often remains hidden from direct observation.

Medical diagnosis has traditionally relied on pattern recognition: physicians learn to associate symptoms with conditions through years of training and experience. This approach struggles with rare diseases, atypical presentations, and the need for quantitative predictions (Will this aneurysm rupture? How will this tumor respond to therapy?). Physics-based simulation offers quantitative prediction but requires patient-specific parameters rarely available clinically and computational resources incompatible with diagnostic timescales.

Scientific Machine Learning bridges this gap with transformative impact. Physics-informed neural networks reconstruct unmeasured physiological quantities from available clinical data. Neural operators accelerate patient-specific simulations from hours to seconds, enabling real-time surgical guidance. Hybrid models combine mechanistic understanding of disease with data-driven personalization, improving prediction accuracy while maintaining interpretability essential for clinical acceptance.

This section examines SciML applications across biomedical engineering. We begin with cardiovascular modeling, where HeartFlow's FDA-cleared technology demonstrates clinical translation. We progress through medical imaging, surgical planning, drug delivery, and biomechanical analysis, highlighting both technical approaches and the regulatory considerations unique to medical applications.

12.6.1 Cardiovascular Modeling

Cardiovascular disease remains the leading cause of death globally, claiming 18 million lives annually. Coronary artery disease—plaque narrowing the arteries feeding the heart—accounts for nearly half these deaths. Clinical decision-making requires answering a deceptively simple question: Is this narrowing severe enough to warrant intervention?

12.6.1.1 The FFR Revolution

Fractional Flow Reserve (FFR) measures the pressure drop across a coronary stenosis during maximal blood flow. An FFR below 0.80 indicates hemodynamically significant disease warranting intervention; above 0.80, medical therapy suffices. Invasive FFR measurement requires cardiac catheterization—threading a pressure wire through arteries to the heart—with associated costs, risks, and patient burden.

HeartFlow's FFR_{CT} technology computes FFR non-invasively from CT angiography images, combining:

1. **Image segmentation**: Deep learning extracts coronary artery geometry from CT
2. **Boundary condition estimation**: Neural networks estimate inlet flow and outlet resistance from patient data
3. **CFD simulation**: Navier-Stokes solver computes blood flow and pressure
4. **Neural acceleration**: Surrogate models reduce computation from hours to minutes (Fig. 12.18).

Real-World Application: HeartFlow Clinical Impact
HeartFlow represents the most successful clinical translation of physics-informed medical AI:

Clinical Performance:

- Diagnostic accuracy: AUC 0.94 versus invasive FFR gold standard

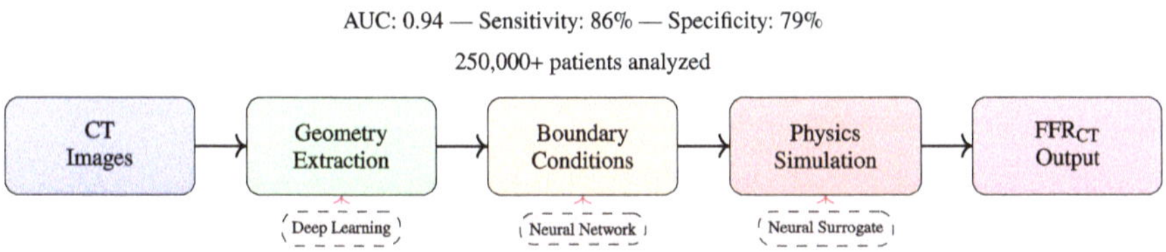

Fig. 12.18 HeartFlow FFR_{CT} pipeline combining deep learning for image processing, neural networks for boundary condition estimation, and physics-based CFD accelerated by neural surrogates. The system achieves diagnostic accuracy comparable to invasive FFR while avoiding catheterization

- Sensitivity: 86% (detects significant disease)
- Specificity: 79% (avoids unnecessary procedures)
- Processing time: 4–6 hours (from minutes with neural acceleration)

Clinical Impact:

- 250,000+ patients analyzed worldwide
- 61% reduction in unnecessary catheterizations (PLATFORM trial)
- 32% cost reduction in diagnostic pathway
- FDA cleared, CE marked, reimbursed by CMS

Technical Approach: Hybrid physics-ML combining rigorous Navier-Stokes simulation with learned components for acceleration and personalization.

12.6.1.2 Physics-Informed Hemodynamics

Blood flow in arteries is governed by the incompressible Navier-Stokes equations with additional considerations for blood rheology:

$$\nabla \cdot \mathbf{u} = 0 \tag{12.101}$$

$$\rho \left(\frac{\partial \mathbf{u}}{\partial t} + \mathbf{u} \cdot \nabla \mathbf{u} \right) = -\nabla p + \nabla \cdot \boldsymbol{\tau} \tag{12.102}$$

Blood exhibits non-Newtonian behavior at low shear rates, with viscosity described by models like Carreau-Yasuda:

$$\mu(\dot{\gamma}) = \mu_\infty + (\mu_0 - \mu_\infty)[1 + (\lambda\dot{\gamma})^a]^{(n-1)/a} \tag{12.103}$$

PINNs for hemodynamics learn velocity and pressure fields while satisfying these governing equations:

$$\mathcal{L} = \mathcal{L}_{\text{continuity}} + \mathcal{L}_{\text{momentum}} + \mathcal{L}_{\text{BC}} + \lambda_{\text{data}}\mathcal{L}_{\text{data}} \tag{12.104}$$

The data loss incorporates available clinical measurements (4D flow MRI velocities, pressure catheter readings), while physics losses enforce fluid mechanics even where measurements are unavailable.

12.6.1.3 Wall Shear Stress and Disease Progression

Wall Shear Stress (WSS)—the frictional force of blood on vessel walls—regulates endothelial cell behavior. Low or oscillatory WSS promotes atherosclerotic plaque formation; high WSS can trigger plaque rupture. Accurate WSS computation requires resolving boundary layers near vessel walls.

Neural operator approaches learn the WSS distribution from vessel geometry:

$$\mathcal{G}_\theta : \text{vessel geometry} \mapsto \text{WSS}(\mathbf{x}) \tag{12.105}$$

Training on high-fidelity CFD simulations enables real-time WSS estimation from imaging data, supporting:

- Plaque rupture risk assessment
- Aneurysm growth prediction
- Surgical planning for bypass grafts
- Device placement optimization (stents, flow diverters)

12.6.2 Medical Image Analysis

Medical imaging generates petabytes of data annually—CT, MRI, ultrasound, PET, X-ray—far exceeding radiologist capacity for detailed analysis. Deep learning has transformed image interpretation, but physics-informed approaches offer advantages: improved generalization across scanners, explicit modeling of image formation physics, and reconstruction of quantities beyond raw images.

12.6.2.1 Physics-Informed Image Reconstruction

Medical image reconstruction solves inverse problems: given sensor measurements, recover the underlying anatomy. Traditional approaches (filtered back-projection for CT, Fourier methods for MRI) assume idealized physics. Real scanners introduce artifacts from motion, noise, incomplete sampling, and hardware imperfections.

Physics-informed reconstruction embeds image formation models:

$$\mathbf{y} = \mathcal{A}(\mathbf{x}) + \mathbf{n} \tag{12.106}$$

Table 12.8 Physics-informed medical image reconstruction performance

Application	Method	PSNR improvement (dB)	Acceleration
MRI (knee)	Unrolled ADMM-Net	+4.2	8×
MRI (brain)	Physics-guided DL	+3.8	6×
CT (low-dose)	PINN denoising	+5.1	Real-time
PET	Physics-informed CNN	+6.3	4× dose reduction
Ultrasound	Wave equation PINN	+2.9	10×

where $\mathbf{y}$ is measured data, $\mathbf{x}$ is the image, $\mathcal{A}$ is the forward model (Radon transform for CT, Fourier encoding for MRI), and $\mathbf{n}$ is noise.

Unrolled optimization networks alternate between data consistency and learned regularization:

$$\mathbf{x}^{(k+1)} = \text{CNN}_\theta \left(\mathbf{x}^{(k)} - \alpha \mathcal{A}^*(\mathcal{A}(\mathbf{x}^{(k)}) - \mathbf{y})\right) \tag{12.107}$$

This architecture guarantees data consistency (reconstructions explain measurements) while learning image priors from training data (Table 12.8).

12.6.2.2 Quantitative Imaging

Beyond anatomical images, physics-informed methods extract quantitative physiological parameters:

- **Perfusion imaging**: Blood flow rates from dynamic contrast enhancement
- **Elastography**: Tissue stiffness from shear wave propagation
- **Diffusion MRI**: Microstructure from water diffusion patterns
- **Relaxometry**: Tissue composition from relaxation time mapping

Each modality involves physics models relating measurements to underlying parameters. PINNs learn these mappings while respecting governing physics, improving accuracy and enabling estimation from reduced measurements.

12.6.3 Surgical Planning and Guidance

Surgery increasingly relies on patient-specific simulation for planning and real-time guidance. Neurosurgery, orthopedic procedures, and cardiac interventions all benefit from predicting how specific patients will respond to interventions.

12.6.3.1 Neurosurgical Planning

Brain tumor resection requires balancing tumor removal with preservation of eloquent cortex controlling speech, motor function, and cognition. Patient-specific models predict:

- Brain shift during craniotomy (as CSF drains, brain deforms)
- Tissue properties from imaging (tumor versus healthy tissue)
- Optimal surgical corridors minimizing damage to critical structures

Physics-informed approaches combine:

$$\nabla \cdot \boldsymbol{\sigma} = \mathbf{0} \quad \text{(mechanical equilibrium)} \tag{12.108}$$

with learned tissue constitutive models capturing patient-specific properties.

12.6.3.2 Orthopedic Biomechanics

Joint replacement, fracture fixation, and spine surgery require understanding load transfer through bones, implants, and soft tissues. Finite element analysis provides detailed predictions but requires hours per simulation—impossible for intraoperative use.

Neural operator surrogates enable real-time biomechanical analysis:

$$\mathcal{G}_\theta : \text{(bone geometry, implant placement, loading)} \mapsto \text{stress field} \tag{12.109}$$

Surgeons can explore implant positioning options during procedures, optimizing placement for longevity and function.

Warning: Regulatory Considerations for Medical AI

Medical AI faces stringent regulatory requirements absent from other engineering domains:

FDA/CE Marking:

- Class II/III device classification for diagnostic/therapeutic claims
- Clinical validation studies demonstrating safety and efficacy
- Quality management systems (ISO 13485)
- Post-market surveillance requirements

Interpretability Requirements:

- Clinicians must understand basis for recommendations
- "Black box" models face adoption barriers
- Physics-informed approaches offer inherent interpretability

Validation Challenges:

- Ground truth often unavailable (what is the "true" FFR?)
- Population shifts between training and deployment
- Rare adverse events require large studies to detect

Physics-informed approaches offer advantages: physics provides guardrails preventing implausible predictions, and mechanistic components support clinical interpretation.

12.6.4 Drug Delivery and Pharmacokinetics

Drug efficacy depends not just on molecular potency but on delivery: how drugs distribute through the body, penetrate target tissues, and clear from circulation. Pharmacokinetic (PK) models describe these processes mathematically, but patient variability demands personalization.

12.6.4.1 Physiologically-Based Pharmacokinetic Models

PBPK models represent the body as compartments (organs, tissues) connected by blood flow:

$$V_i \frac{dC_i}{dt} = Q_i(C_{\text{arterial}} - C_i/K_{p,i}) + R_i \tag{12.110}$$

where V_i is tissue volume, C_i is concentration, Q_i is blood flow, $K_{p,i}$ is partition coefficient, and R_i represents metabolism/excretion.

Neural ODEs learn patient-specific deviations from population models:

$$\frac{d\mathbf{C}}{dt} = f_{\text{PBPK}}(\mathbf{C}; \boldsymbol{\theta}_{\text{pop}}) + f_\phi(\mathbf{C}; \mathbf{x}_{\text{patient}}) \tag{12.111}$$

This hybrid approach maintains mechanistic interpretability (which organs are affected?) while capturing individual variability beyond population models.

12.6.4.2 Tumor Drug Penetration

Solid tumors present drug delivery challenges: abnormal vasculature, elevated interstitial pressure, dense extracellular matrix. Physics-informed models predict drug penetration:

$$\frac{\partial C}{\partial t} = D\nabla^2 C - \mathbf{v} \cdot \nabla C - k_{\text{bind}} C + S \tag{12.112}$$

$$\nabla \cdot \mathbf{v} = L_p(p_v - p_i) \quad \text{(fluid source from leaky vessels)} \tag{12.113}$$

Neural operators learn patient-specific parameters (permeability, diffusivity, binding rates) from imaging, enabling personalized treatment planning.

12.6.5 Implementation: Cardiovascular Flow PINN

We implement a physics-informed neural network for blood flow in a stenosed artery, demonstrating how hemodynamic quantities can be inferred from sparse measurements.

```
import torch
import torch.nn as nn
import numpy as np

class CardiovascularPINN(nn.Module):
    """
    Physics-Informed Neural Network for blood flow in stenosed artery.

    Learns velocity (u, v) and pressure (p) fields satisfying:
    - Continuity: du/dx + dv/dy = 0
    - Navier-Stokes momentum equations
    """

    def __init__(self, hidden_dim=64, num_layers=4):
        super().__init__()

        layers = [nn.Linear(2, hidden_dim), nn.Tanh()]
        for _ in range(num_layers - 1):
            layers.extend([nn.Linear(hidden_dim, hidden_dim), nn.Tanh()])
        layers.append(nn.Linear(hidden_dim, 3))  # Output: u, v, p

        self.network = nn.Sequential(*layers)

        # Blood properties
        self.rho = 1060.0  # kg/m^3
        self.mu = 0.0035   # Pa.s (dynamic viscosity)

        # Initialize weights
        for m in self.modules():
            if isinstance(m, nn.Linear):
                nn.init.xavier_normal_(m.weight)
                nn.init.zeros_(m.bias)

```

```
    def forward(self, x, y):
        """Predict velocity and pressure at (x, y)."""
        inputs = torch.stack([x, y], dim=-1)
        outputs = self.network(inputs)
        u = outputs[..., 0]
        v = outputs[..., 1]
        p = outputs[..., 2]
        return u, v, p

    def compute_residuals(self, x, y):
        """Compute PDE residuals for physics loss."""
        x = x.requires_grad_(True)
        y = y.requires_grad_(True)

        u, v, p = self.forward(x, y)

        # First derivatives
        u_x = torch.autograd.grad(u, x,
    grad_outputs=torch.ones_like(u),
                                   create_graph=True)[0]
        u_y = torch.autograd.grad(u, y,
    grad_outputs=torch.ones_like(u),
                                   create_graph=True)[0]
        v_x = torch.autograd.grad(v, x,
    grad_outputs=torch.ones_like(v),
                                   create_graph=True)[0]
        v_y = torch.autograd.grad(v, y,
    grad_outputs=torch.ones_like(v),
                                   create_graph=True)[0]
        p_x = torch.autograd.grad(p, x,
    grad_outputs=torch.ones_like(p),
                                   create_graph=True)[0]
        p_y = torch.autograd.grad(p, y,
    grad_outputs=torch.ones_like(p),
                                   create_graph=True)[0]

        # Second derivatives
        u_xx = torch.autograd.grad(u_x, x,
    grad_outputs=torch.ones_like(u_x),
                                    create_graph=True)[0]
        u_yy = torch.autograd.grad(u_y, y,
    grad_outputs=torch.ones_like(u_y),
                                    create_graph=True)[0]
        v_xx = torch.autograd.grad(v_x, x,
    grad_outputs=torch.ones_like(v_x),
                                    create_graph=True)[0]
        v_yy = torch.autograd.grad(v_y, y,
    grad_outputs=torch.ones_like(v_y),
                                    create_graph=True)[0]

        # Continuity residual
        continuity = u_x + v_y

```

```
        # Momentum residuals (steady Navier-Stokes)
        momentum_x = self.rho * (u * u_x + v * u_y) + p_x -
    self.mu * (u_xx + u_yy)
        momentum_y = self.rho * (u * v_x + v * v_y) + p_y -
    self.mu * (v_xx + v_yy)

        return continuity, momentum_x, momentum_y


def generate_stenosis_geometry(n_boundary=200,
    stenosis_severity=0.5):
    """Generate boundary points for stenosed artery."""
    # Artery length and radius
    L, R = 10.0, 1.0

    # Upper wall with stenosis
    x_upper = np.linspace(0, L, n_boundary//4)
    stenosis_center = L / 2
    stenosis_width = 2.0
    stenosis_profile = stenosis_severity * np.exp(
        -((x_upper - stenosis_center) / stenosis_width)**2
    )
    y_upper = R - stenosis_profile * R

    # Lower wall (symmetric)
    x_lower = x_upper
    y_lower = -y_upper

    # Inlet
    y_inlet = np.linspace(-R, R, n_boundary//4)
    x_inlet = np.zeros_like(y_inlet)

    # Outlet
    y_outlet = np.linspace(-R, R, n_boundary//4)
    x_outlet = np.ones_like(y_outlet) * L

    return {
        'upper': (x_upper, y_upper),
        'lower': (x_lower, y_lower),
        'inlet': (x_inlet, y_inlet),
        'outlet': (x_outlet, y_outlet)
    }


def train_cardiovascular_pinn(model, geometry,
    n_collocation=5000,
                               epochs=5000, lr=1e-3):
    """Train PINN for cardiovascular flow."""
    device = next(model.parameters()).device
    optimizer = torch.optim.Adam(model.parameters(), lr=lr)
    scheduler =
    torch.optim.lr_scheduler.CosineAnnealingLR(optimizer,
    epochs)
```

```python

    # Generate collocation points in domain
    L, R = 10.0, 1.0
    x_coll = torch.rand(n_collocation, device=device) * L
    y_coll = (torch.rand(n_collocation, device=device) * 2 - 1) * R

    # Boundary conditions
    x_inlet = torch.tensor(geometry['inlet'][0], dtype=torch.float32, device=device)
    y_inlet = torch.tensor(geometry['inlet'][1], dtype=torch.float32, device=device)

    # Parabolic inlet profile
    u_inlet_true = 1.5 * (1 - (y_inlet / R)**2)  # Max velocity 1.5 m/s
    v_inlet_true = torch.zeros_like(y_inlet)

    # Wall boundaries (no-slip)
    x_wall = torch.cat([
        torch.tensor(geometry['upper'][0], dtype=torch.float32),
        torch.tensor(geometry['lower'][0], dtype=torch.float32)
    ]).to(device)
    y_wall = torch.cat([
        torch.tensor(geometry['upper'][1], dtype=torch.float32),
        torch.tensor(geometry['lower'][1], dtype=torch.float32)
    ]).to(device)

    history = {'loss': [], 'physics': [], 'bc': []}

    for epoch in range(epochs):
        optimizer.zero_grad()

        # Physics loss at collocation points
        cont, mom_x, mom_y = model.compute_residuals(x_coll, y_coll)
        loss_physics = torch.mean(cont**2) + torch.mean(mom_x**2) + torch.mean(mom_y**2)

        # Inlet BC loss
        u_inlet, v_inlet, _ = model(x_inlet, y_inlet)
        loss_inlet = torch.mean((u_inlet - u_inlet_true)**2) + \
                     torch.mean((v_inlet - v_inlet_true)**2)

        # Wall BC loss (no-slip)
        u_wall, v_wall, _ = model(x_wall, y_wall)
        loss_wall = torch.mean(u_wall**2) + torch.mean(v_wall**2)
```

```
        # Total loss
        loss_bc = loss_inlet + 10 * loss_wall
        loss = loss_physics + 100 * loss_bc

        loss.backward()
        optimizer.step()
        scheduler.step()

        history['loss'].append(loss.item())
        history['physics'].append(loss_physics.item())
        history['bc'].append(loss_bc.item())

        if epoch % 500 == 0:
            print(f"Epoch {epoch:5d} | Loss:
    {loss.item():.2e} | "
                  f"Physics: {loss_physics.item():.2e} | BC:
    {loss_bc.item():.2e}")

    return history


# Example usage
device = torch.device('cuda' if torch.cuda.is_available()
    else 'cpu')
model = CardiovascularPINN(hidden_dim=64,
    num_layers=4).to(device)

print(f"Model parameters: {sum(p.numel() for p in
    model.parameters()):,}")

# Generate geometry
geometry = generate_stenosis_geometry(n_boundary=200,
    stenosis_severity=0.5)

# Test forward pass
x_test = torch.tensor([5.0], device=device)
y_test = torch.tensor([0.0], device=device)
u, v, p = model(x_test, y_test)
print(f"At stenosis center (x=5, y=0): u={u.item():.3f} m/s,
    p={p.item():.1f} Pa")
```

Listing 12.6 PINN for cardiovascular flow reconstruction

Chapter Exercises: Biomedical Engineering

1. **Blood Rheology**: Extend the cardiovascular PINN to use the Carreau-Yasuda viscosity model instead of constant viscosity. Compare velocity profiles in the stenosis region with and without non-Newtonian effects.

2. **Wall Shear Stress**: Implement post-processing to compute WSS from the trained PINN velocity field. Identify regions of low WSS (potential atherosclerosis sites) and high WSS (potential plaque rupture sites).
3. **Pharmacokinetic Model**: Implement a Neural ODE for a two-compartment PK model with learned absorption and elimination rates. Train on synthetic drug concentration data and evaluate parameter identifiability.
4. **MRI Reconstruction**: Implement an unrolled optimization network for accelerated MRI reconstruction. Compare against traditional methods (zero-filling, compressed sensing) at various acceleration factors.

12.7 Chemical and Process Engineering

A chemical plant is a symphony of differential equations. Reactors transform feedstocks through reaction kinetics; distillation columns separate mixtures through vapor-liquid equilibrium; heat exchangers manage energy flows through transport equations; control systems maintain stability through feedback dynamics. Each unit operation has governing equations; connecting them creates systems of thousands of coupled equations that must be solved for design, optimization, and control.

Traditional process simulation relies on equation-oriented solvers (Aspen Plus, gPROMS) that require complete specification of thermodynamic models, reaction kinetics, and transport correlations. For established processes with well-characterized chemistry, these tools excel. But new chemistries—carbon capture, sustainable fuels, pharmaceutical synthesis—often lack the fundamental data these tools require. Experimental campaigns to generate missing data cost millions and take years.

Scientific Machine Learning offers an alternative path. Physics-informed neural networks learn process behavior from limited experimental data while respecting conservation laws. Neural operators accelerate process simulation from minutes to milliseconds, enabling real-time optimization. Hybrid models combine first-principles understanding with data-driven flexibility, achieving accuracy beyond either approach alone.

12.7.1 Reaction Engineering

Chemical reactors transform reactants to products through networks of reactions, each with temperature-dependent kinetics. Reactor design requires predicting conversion, selectivity, and yield across operating conditions—predictions depending on kinetic parameters that are often uncertain or unknown.

12.7.1.1 Learning Reaction Kinetics

Reaction rates typically follow Arrhenius-type expressions:

$$r = A \exp\left(-\frac{E_a}{RT}\right) \prod_i C_i^{n_i} \tag{12.114}$$

where A is the pre-exponential factor, E_a is activation energy, and n_i are reaction orders. For complex reaction networks, determining these parameters from experimental data requires nonlinear regression prone to local minima and parameter correlation.

Physics-informed approaches embed Arrhenius structure while learning parameters:

$$r_\theta(T, \mathbf{C}) = \exp(\theta_A - \theta_E/T) \cdot g_\phi(\mathbf{C}) \tag{12.115}$$

where θ_A and θ_E are learnable log-transformed parameters, and g_ϕ is a neural network learning concentration dependence. This parameterization ensures positive rates and physical temperature dependence while flexibly capturing complex concentration effects.

Definition: Hybrid Kinetic Models

- **White-box**: Complete mechanistic model with known parameters. High interpretability, requires full knowledge.
- **Black-box**: Pure neural network learning rate from inputs. High flexibility, poor extrapolation, uninterpretable.
- **Gray-box (hybrid)**: Mechanistic structure with learned components. Balances interpretability with flexibility.

Gray-box approaches for reaction kinetics:

- **Parameter learning**: Fix functional form, learn parameters
- **Residual learning**: Mechanistic model + neural correction
- **Structure learning**: Learn reaction network topology

12.7.1.2 Reactor Modeling with Conservation Laws

Material balances in reactors follow conservation equations. For a continuous stirred tank reactor (CSTR):

$$V\frac{dC_i}{dt} = F(C_{i,\text{in}} - C_i) + V\sum_j \nu_{ij} r_j \tag{12.116}$$

where V is volume, F is flow rate, and ν_{ij} are stoichiometric coefficients.

Physics-informed neural networks enforce these balances:

$$\mathcal{L}_{\text{conservation}} = \sum_i \left| V\frac{dC_i}{dt} - F(C_{i,\text{in}} - C_i) - V\sum_j \nu_{ij} r_j(\mathbf{C}, T) \right|^2 \tag{12.117}$$

The stoichiometric matrix $[\nu_{ij}]$ encodes elemental conservation; predictions automatically satisfy atom balances.

12.7.2 Separation Processes

Separation consumes 10–15% of global energy, making efficiency improvements enormously impactful. Distillation, absorption, membrane separation, and chromatography all rely on thermodynamic equilibrium and mass transfer—physics amenable to neural enhancement.

12.7.2.1 Thermodynamic Property Prediction

Vapor-liquid equilibrium (VLE) determines separation feasibility and energy requirements. Activity coefficient models (NRTL, UNIQUAC) capture non-ideal behavior but require binary interaction parameters determined experimentally.

Graph neural networks predict thermodynamic properties from molecular structure:

$$\gamma_i = \text{GNN}_\theta(\text{molecule}_i, \text{mixture composition}, T) \tag{12.118}$$

Training on VLE databases, these models generalize to new compounds without requiring new experiments, accelerating process development for novel chemistries.

12.7.2.2 Distillation Column Modeling

Rigorous distillation simulation requires solving MESH equations (Material balance, Equilibrium, Summation, Heat balance) for each stage—computationally intensive for columns with many stages and components.

Neural operator surrogates learn the input-output relationship:

$$\mathcal{G}_\theta : (\text{feed, specifications}) \mapsto (\text{product compositions, profiles}) \tag{12.119}$$

Trained on rigorous simulation data, these surrogates enable real-time optimization and control impossible with equation-oriented solvers (Table 12.9).

Table 12.9 SciML approaches in chemical process simulation

Application	Method	Error	Speedup
Reaction kinetics	Hybrid neural ODE	2.3% MAPE	–
VLE prediction	Graph NN	5.1% MAPE	$10^4\times$ versus experiments
Distillation column	Physics-informed surrogate	1.8%	$10^3\times$
Heat exchanger network	DeepONet	3.2%	$10^4\times$
Process optimization	Neural MPC	–	$10^2\times$

12.7.3 Process Optimization and Control

Chemical processes operate continuously, with optimization targeting economic objectives (profit, yield) subject to safety and environmental constraints. Real-time optimization requires models fast enough for repeated evaluation.

12.7.3.1 Real-Time Optimization

Steady-state Real-Time Optimization (RTO) adjusts setpoints to maximize profit:

$$\max_{\mathbf{u}} \quad J(\mathbf{x}, \mathbf{u}) \tag{12.120}$$

$$\text{s.t.} \quad f(\mathbf{x}, \mathbf{u}) = 0 \quad \text{(process model)} \tag{12.121}$$

$$g(\mathbf{x}, \mathbf{u}) \leq 0 \quad \text{(constraints)} \tag{12.122}$$

Physics-informed surrogates replace rigorous models in RTO, enabling optimization cycles of seconds rather than hours while respecting conservation laws embedded in the surrogate.

12.7.3.2 Model Predictive Control

Model Predictive Control (MPC) optimizes control actions over a prediction horizon:

$$\min_{\mathbf{u}_{0:N-1}} \sum_{k=0}^{N-1} \|\mathbf{x}_k - \mathbf{x}_k^{\text{ref}}\|_Q^2 + \|\Delta \mathbf{u}_k\|_R^2 \tag{12.123}$$

$$\text{s.t.} \quad \mathbf{x}_{k+1} = f_\theta(\mathbf{x}_k, \mathbf{u}_k) \tag{12.124}$$

Neural network process models enable MPC with complex dynamics that resist first-principles modeling. Physics constraints ensure predictions remain feasible:

- Conservation: Mass/energy cannot be created/destroyed
- Bounds: Concentrations non-negative, temperatures above absolute zero
- Monotonicity: Heating increases temperature

Real-World Application: DeepMind and Google Data Centers

DeepMind's application of ML to Google data center cooling demonstrates industrial-scale process optimization:

Challenge: Data centers consume 1–2% of global electricity; cooling accounts for 30–40% of this consumption. Traditional rule-based control leaves significant optimization potential.

Approach:
- Neural network models of cooling system dynamics
- Physics constraints on thermodynamic consistency
- Reinforcement learning for control policy optimization
- Human-in-the-loop deployment with safety constraints

Results:
- 40% reduction in cooling energy
- 15% reduction in overall PUE (Power Usage Effectiveness)
- Deployed across Google's global data center fleet

The success demonstrates how physics-informed ML scales to industrial processes with significant economic and environmental impact.

12.8 Energy Systems

The energy transition represents humanity's largest engineering challenge. Replacing fossil fuels with renewables requires not just new generation technologies but fundamental transformation of how energy systems operate. Solar and wind power fluctuate with weather; demand varies unpredictably; and the grid must balance supply and demand instantaneously across continental scales. Traditional power system planning assumed predictable, dispatchable generation; the renewable future demands real-time optimization across millions of distributed resources.

Scientific Machine Learning addresses challenges across the energy stack. Physics-informed neural networks forecast renewable generation by combining meteorological physics with data-driven weather prediction. Neural operators optimize power flow across transmission networks, enabling real-time grid management. Hybrid models predict building energy consumption by combining thermodynamic first principles with learned occupant behavior. Each application combines domain physics with machine learning's pattern recognition capabilities.

12.8.1 Renewable Energy Forecasting

Solar and wind generation depends on weather—forecasting generation requires forecasting atmospheric conditions. Traditional numerical weather prediction (NWP) solves atmospheric dynamics on massive computational grids; physics-informed ML offers complementary approaches.

12.8.1.1 Solar Power Prediction

Solar irradiance reaching photovoltaic panels depends on:

- **Astronomical factors**: Sun position (predictable)
- **Atmospheric conditions**: Cloud cover, aerosols, humidity (uncertain)
- **Local effects**: Shading, soiling, temperature (site-specific)

Physics-informed solar forecasting embeds the clear-sky model:

$$G = G_{\text{clear}}(\text{time, location}) \cdot f_\theta(\text{weather features}) \tag{12.125}$$

where G_{clear} is the analytically computable clear-sky irradiance and f_θ learns weather-dependent attenuation.

12.8.1.2 Wind Power Forecasting

Wind turbine power follows a characteristic curve:

$$P = \begin{cases} 0 & v < v_{\text{cut-in}} \\ \frac{1}{2}\rho A C_p v^3 & v_{\text{cut-in}} \leq v < v_{\text{rated}} \\ P_{\text{rated}} & v_{\text{rated}} \leq v < v_{\text{cut-out}} \\ 0 & v \geq v_{\text{cut-out}} \end{cases} \tag{12.126}$$

Physics-informed approaches combine turbine physics with learned wind forecasts (Table 12.10):

$$P_\theta = f_{\text{turbine}}(v_\theta(\mathbf{x}_{\text{weather}})) \tag{12.127}$$

Table 12.10 Renewable energy forecasting performance

Resource	Horizon	Method	Skill score
Solar (utility)	Day-ahead	Physics-informed LSTM	42% versus persistence
Solar (distributed)	Hour-ahead	Hybrid CNN-physics	38% versus NWP
Wind (offshore)	Day-ahead	Graph NN + NWP	35% versus NWP alone
Wind (onshore)	6-hour	Ensemble physics-ML	28% versus NWP alone

12.8.2 Power System Optimization

Power grids balance generation and load continuously while respecting physical constraints: Kirchhoff's laws govern current flow, thermal limits constrain transmission, and voltage stability requires reactive power management.

12.8.2.1 Optimal Power Flow

The AC Optimal Power Flow (OPF) problem minimizes generation cost subject to physical constraints:

$$\min_{\mathbf{P}_G, \mathbf{Q}_G, \mathbf{V}} \sum_i C_i(P_{G,i}) \tag{12.128}$$

$$\text{s.t.} \quad P_i = \sum_j |V_i||V_j||Y_{ij}| \cos(\theta_i - \theta_j - \phi_{ij}) \tag{12.129}$$

$$Q_i = \sum_j |V_i||V_j||Y_{ij}| \sin(\theta_i - \theta_j - \phi_{ij}) \tag{12.130}$$

$$\text{Generation, voltage, and line flow limits} \tag{12.131}$$

Traditional solvers require seconds to minutes per solve; market clearing requires thousands of solves daily. Neural networks learn the OPF mapping:

$$\mathcal{G}_\theta : \mathbf{P}_D, \mathbf{Q}_D \mapsto \mathbf{P}_G^*, \mathbf{V}^* \tag{12.132}$$

Physics-informed approaches ensure feasibility:

- **Power balance**: Generation matches load plus losses
- **Voltage limits**: Predictions within operational bounds
- **Line limits**: Thermal constraints satisfied

12.8.2.2 Grid Stability Analysis

Grid stability requires eigenvalue analysis of the linearized system—computationally intensive for large networks. Neural surrogates predict stability margins:

$$\Delta_{\text{stability}} = f_\theta(\text{topology, generation dispatch, load pattern}) \tag{12.133}$$

Physics constraints ensure predictions respect fundamental stability criteria.

Real-World Application: AI Weather Forecasting for Energy
Major weather agencies now deploy ML models operationally:

ECMWF AIFS (Artificial Intelligence Forecasting System):

- Graph neural network architecture
- Trained on 40 years of ERA5 reanalysis
- Comparable accuracy to physics-based IFS at 1000× lower cost
- Operational deployment began 2024

Google DeepMind GraphCast:

- 10-day forecasts in under 1 min
- Outperforms ECMWF HRES on 90% of metrics
- Open-source model weights available

Energy Impact: Improved weather forecasts translate directly to better renewable forecasting, reducing balancing costs and enabling higher renewable penetration.

12.8.3 Building Energy Systems

Buildings consume 40% of global energy, primarily for heating, cooling, and lighting. Building energy models predict consumption for design, retrofit analysis, and operational optimization.

12.8.3.1 Physics-Based Building Models

Building thermal dynamics follow heat transfer equations:

$$C\frac{dT}{dt} = \dot{Q}_{\text{HVAC}} + \dot{Q}_{\text{solar}} + \dot{Q}_{\text{internal}} - UA(T - T_{\text{ambient}}) \tag{12.134}$$

where C is thermal capacitance, U is conductance, and A is surface area.

Detailed models (EnergyPlus, TRNSYS) simulate buildings at sub-hourly resolution but require extensive input data and expertise. Physics-informed ML learns building dynamics from operational data:

$$\frac{dT}{dt} = f_{\text{physics}}(T, T_{\text{ambient}}, \dot{Q}_{\text{HVAC}}) + f_\theta(T, \text{occupancy}, \text{schedules}) \qquad (12.135)$$

The physics component captures known thermodynamics; the learned component captures occupant behavior, unmeasured loads, and model-reality mismatch.

12.8.3.2 HVAC Optimization

Heating, Ventilation, and Air Conditioning (HVAC) accounts for 50% of building energy. Model Predictive Control optimizes HVAC operation:

$$\min \sum_{k=0}^{N-1} c_k P_k^{\text{HVAC}} + \lambda \| T_k - T_k^{\text{setpoint}} \|^2 \qquad (12.136)$$

$$\text{s.t.} \quad T_{k+1} = f_\theta(T_k, P_k^{\text{HVAC}}, \text{weather}_k) \qquad (12.137)$$

$$T_{\min} \leq T_k \leq T_{\max} \qquad (12.138)$$

Physics-informed building models enable MPC with 15–30% energy savings compared to rule-based control.

12.9 Software Frameworks and Industrial Tools

The gap between research demonstrations and industrial deployment often determines whether scientific advances achieve real-world impact. Physics-informed machine learning has benefited from substantial investment in software infrastructure, with frameworks spanning the spectrum from research prototypes to production-ready industrial platforms.

12.9.1 Research Frameworks

Research frameworks prioritize flexibility, enabling rapid prototyping of novel physics-informed architectures.

12.9.1.1 DeepXDE

DeepXDE provides a high-level interface for physics-informed neural networks built on TensorFlow, PyTorch, JAX, and PaddlePaddle backends:

```
import deepxde as dde

def pde(x, y):
    dy_xx = dde.grad.hessian(y, x, i=0, j=0)
    return dy_xx - 2  # Poisson equation

geom = dde.geometry.Interval(0, 1)
bc = dde.DirichletBC(geom, lambda x: 0, lambda x, on: on)
data = dde.data.PDE(geom, pde, bc, num_domain=100,
    num_boundary=2)
net = dde.maps.FNN([1, 50, 50, 1], "tanh", "Glorot uniform")
model = dde.Model(data, net)
model.compile("adam", lr=0.001)
model.train(epochs=10000)
```

DeepXDE supports forward problems, inverse problems, and neural operators with a unified API.

12.9.1.2 NVIDIA Modulus

NVIDIA Modulus (formerly SimNet) provides GPU-optimized physics-informed neural networks with:

- Pre-built physics modules (Navier-Stokes, heat transfer, structural mechanics)
- Multi-GPU and multi-node training
- Geometry handling through constructive solid geometry
- Integration with Omniverse for visualization

12.9.1.3 NeuralOperator

The NeuralOperator library implements neural operator architectures:

- Fourier Neural Operator (FNO)
- DeepONet
- Graph Neural Operators
- Physics-informed extensions (PINO)

12.9.2 *Industrial Platforms*

Industrial deployment requires capabilities beyond research frameworks: scalability, reliability, integration with existing workflows, and support/maintenance commitments.

12.9.2.1 NVIDIA Omniverse and PhysicsNeMo

NVIDIA's industrial platform combines:

- **Omniverse**: Real-time simulation and visualization
- **PhysicsNeMo**: Production physics-informed ML
- **cuOpt**: GPU-accelerated optimization

Deployments span manufacturing (BMW, Siemens), energy (utilities), and infrastructure (digital twins).

12.9.2.2 Ansys AI and Machine Learning

Ansys integrates ML with traditional simulation:

- **Ansys SimAI**: Cloud-based neural surrogate training
- **Reduced Order Models**: Automatic surrogate generation from simulation
- **Ansys optiSLang**: Design space exploration with ML surrogates

12.9.2.3 Siemens Simcenter

Siemens' simulation portfolio includes:

- **Simcenter STAR-CCM+**: CFD with ML-accelerated turbulence
- **Simcenter Amesim**: System simulation with hybrid models
- **MindSphere**: Industrial IoT with predictive analytics

Tip: Selecting Frameworks for Industrial Applications

When choosing frameworks for production deployment, consider:

Research/Prototyping:

- Prioritize flexibility and community support
- DeepXDE, NeuralOperator for rapid experimentation
- Jupyter/Colab-friendly workflows

Production Deployment:

- Prioritize scalability, reliability, support
- NVIDIA Modulus for GPU-intensive applications
- Integration with existing CAE tools (Ansys, Siemens)

Domain-Specific:

- HeartFlow (cardiovascular)
- Neural Concept (aerodynamics)
- Rescale (cloud HPC + ML)

12.10 Chapter Summary

This chapter surveyed Scientific Machine Learning applications across engineering and technology, demonstrating how physics-informed approaches are transforming design, analysis, control, and optimization across industries.

12.10.1 Key Themes

Several themes emerged across domains:

Acceleration through surrogates: Neural operators and physics-informed networks achieve 100–100,000× speedups over traditional simulation, enabling real-time applications impossible with classical methods. HeartFlow's cardiovascular analysis, NVIDIA's industrial digital twins, and neural aerodynamic prediction all exploit this acceleration.

Physics as regularization: Embedding physical constraints improves generalization, data efficiency, and reliability. Robots learn dynamics with fewer trials; processes maintain conservation laws; medical predictions remain physiologically plausible.

Hybrid architectures: The most successful approaches combine first-principles physics with learned components, achieving accuracy beyond either alone. Rigid-body dynamics plus learned residuals for robotics; mechanistic kinetics plus neural corrections for chemistry; analytical models plus data-driven calibration for building energy.

From simulation to sensing: Physics-informed methods excel at inverse problems—inferring unmeasured quantities from observations. Structural health monitoring, medical diagnostics, and process control all benefit from this capability (Table 12.11).

Table 12.11 Comparison of physics-informed ML frameworks

Framework	Backend	PINNs	Operators	Industrial	License
DeepXDE	TF/PyTorch/JAX	✓	✓		Apache 2.0
NVIDIA Modulus	PyTorch	✓	✓	✓	Apache 2.0
NeuralOperator	PyTorch		✓		MIT
PINNacle	JAX	✓			MIT
SciANN	TensorFlow	✓			MIT

Table 12.12 Summary of SciML engineering applications

Domain	Key application	Method	Impact
Structural	Damage detection	PINN inverse	0.4% stiffness error
Aerospace	Aerodynamic design	Neural operators	$10^5\times$ speedup
Manufacturing	Melt pool prediction	Hybrid PINN	4–6% MAPE
Robotics	Manipulator dynamics	Lagrangian NN	$5\times$ data efficiency
Biomedical	FFR_{CT}	Physics-ML hybrid	250K+ patients
Chemical	Reaction kinetics	Neural ODE	2.3% MAPE
Energy	Grid optimization	Neural surrogate	Real-time OPF

12.10.2 Comparative Summary

Table 12.12 summarizes key applications, methods, and performance across domains.

12.10.3 Practical Guidance

For practitioners considering physics-informed approaches:

1. **Start with physics**: Identify what physics is known and what is uncertain. Embed known physics; learn the rest.
2. **Match method to problem**: Use PINNs for individual problems with limited data; neural operators for parametric studies with simulation data; hybrid models when both physics and data are available.
3. **Validate rigorously**: Compare against classical methods, experimental data, and edge cases. Quantify uncertainty.
4. **Plan for deployment**: Consider integration, maintenance, and regulatory requirements from the start.
5. **Build expertise**: Success requires combining domain knowledge with ML capabilities—invest in cross-disciplinary teams.

12.10.4 Looking Forward

Scientific Machine Learning in engineering has progressed from academic curiosity to industrial reality in under a decade. HeartFlow processes a quarter million cardiac scans; ECMWF runs neural weather models operationally; manufacturers deploy predictive maintenance across global equipment fleets. Yet current capabilities represent only the beginning.

The convergence of increasing computational power, growing simulation databases, and maturing ML techniques suggests continued rapid progress. Foundation models may learn generalizable physical principles. Automated discovery may reveal physics hidden in data. Uncertainty-aware predictions may enable risk-informed engineering decisions.

The fundamental promise remains: machines that understand physics can help engineers build a better world. The applications explored in this chapter—safer structures, more efficient vehicles, better medicines, cleaner energy—demonstrate that promise becoming reality.

Chapter Exercises: Integration and Application

1. **Cross-Domain Transfer**: Identify a technique from one domain section that could apply to another domain. Describe the adaptation required and potential benefits.
2. **Framework Selection**: For a specific engineering problem of your choice, evaluate which software framework (Table 12.11) would be most appropriate. Justify your selection based on problem characteristics and deployment requirements.
3. **Industrial Case Study**: Select a real company deploying physics-informed ML (HeartFlow, Neural Concept, etc.). Research their technical approach, business model, and market position. What factors enabled their success?
4. **Future Application**: Propose a novel application of physics-informed ML not covered in this chapter. Describe the governing physics, available data sources, potential approach, and expected challenges.
5. **Validation Protocol**: Design a validation protocol for deploying a physics-informed neural network in a safety-critical application. What tests would you require? What accuracy thresholds? What failure modes would you investigate?

References

1. Lu, L., Jin, P., Pang, G., Zhang, Z., Karniadakis, G.E.: Learning nonlinear operators via DeepONet based on the universal approximation theorem of operators. Nat. Mach. Intell. **3**(3), 218–229 (2021)
2. Kaewnuratchadasorn, C., Wang, J., Kim, C.W.: Physics-informed neural operator solver and super-resolution for solid mechanics. Comput.-Aided Civ. Infrastruct. Eng. **39**(22), 3435–3451 (2024)

Chapter 13
Challenges and Future Directions in SciML

Life is like riding a bicycle. To keep your balance, you must keep moving.

Albert Einstein, in a letter to his son

Abstract Scientific machine learning has achieved remarkable successes: from weather prediction models that outperform decades of numerical development to materials discovery systems identifying millions of stable crystals. Yet beneath these achievements lies a field grappling with fundamental theoretical gaps, computational bottlenecks, and the challenge of translating research advances into trusted scientific practice. This chapter maps the current frontier: where SciML works, where it fails, and where the field is heading.

13.1 Introduction

As we have seen throughout this textbook, scientific machine learning represents a fundamental shift in how we approach computational science. Physics-informed neural networks learn to satisfy governing equations. Neural operators map between infinite-dimensional function spaces. Differentiable programming enables end-to-end optimization through physical simulations. Foundation models trained on massive scientific datasets achieve unprecedented prediction accuracy.

Yet for all these advances, the field faces sobering realities. Most convergence theorems apply only to simplified settings far removed from practical applications. Training failures that seemed mysterious five years ago are now understood—but not fully solved. The gap between research benchmarks and production deployment remains substantial. Software ecosystems fragment across languages and frameworks, impeding reproducibility and adoption.

B. P. Bhuyan et al., *Scientific Machine Learning*,
https://doi.org/10.1007/978-981-95-7856-6_13

This chapter provides an honest assessment of these challenges alongside the promising directions that may resolve them. We organize our discussion around five themes: theoretical foundations, computational scaling, practical deployment, emerging paradigms, and domain-specific frontiers.

13.2 The Theory-Practice Gap

The most fundamental challenge in SciML is the disconnect between what we can prove mathematically and what we observe empirically. This gap matters because scientists need to know when—and when not—to trust these methods.

13.2.1 Convergence: What We Can and Cannot Prove

The Central Theoretical Question

Given a PDE with a unique solution, will training a physics-informed neural network actually find that solution? Despite thousands of empirical demonstrations, rigorous affirmative answers exist only for restricted problem classes.

The convergence theory for PINNs has advanced significantly since their introduction, yet remains incomplete. The most rigorous results establish convergence for *linear* PDEs under specific regularity conditions. These proofs typically require:

- Sufficient smoothness of the true solution
- Regularization schemes in the loss function
- Assumptions about the optimization landscape
- Often, the assumption of *global* optimization

The last assumption is particularly problematic. Real training uses gradient descent variants that find local (not global) minima. The landscape of PINN loss functions is highly non-convex, with saddle points, plateaus, and spurious local minima that can trap optimization.

For *nonlinear* PDEs (including Navier-Stokes, reaction-diffusion systems, and most equations of practical interest) convergence remains purely empirical. We have approximation theorems showing neural networks *can* represent PDE solutions to arbitrary accuracy, but minimal guarantees that training *will find* these representations.

Definition: Universal Approximation Versus Trainability

Universal approximation states that neural networks with sufficient width/depth can approximate any continuous function. This is an *existence* result—it says nothing about whether gradient-based training will find such an approximation.
Trainability asks whether optimization algorithms reliably converge to good solutions. For PINNs, trainability depends on loss landscape geometry, initialization, architecture, and hyperparameters in ways that resist general analysis.

13.2.2 Generalization: From Training Points to Continuous Domains

Even when training succeeds, a separate question arises: does the learned solution generalize correctly to points not seen during training? For classical supervised learning, generalization theory is mature. For PINNs, the situation is more complex because training involves differential operators, not just function values.

For neural operators, the generalization question takes a different form: does a model trained on samples from one distribution of input functions generalize to new functions from the same distribution? Recent work established that Fourier Neural Operators can approximate operators with network size growing only *sub-logarithmically* in the desired precision; a remarkable result suggesting neural operators can break the curse of dimensionality for certain PDE classes.

However, these theoretical bounds remain orders of magnitude looser than observed performance as shown in Table 13.1.

This gap between theory and practice means generalization bounds provide limited practical guidance for architecture selection or hyperparameter tuning.

Table 13.1 Generalization bounds versus empirical performance for neural operators

Metric	Theoretical bound	Observed
Relative L^2 error	$\mathcal{O}(n^{-1/4})$	$\sim 10^{-3}$
Sample complexity	$\mathcal{O}(\epsilon^{-8})$	$\sim$ 1000 samples
Network width	$\mathcal{O}(\epsilon^{-2})$	$\sim$ 64 modes

13.2.3 Spectral Bias: The Frequency Learning Problem

Key Insight: Spectral Bias

Neural networks preferentially learn low-frequency components of target functions. This is the *opposite* of classical numerical methods, which typically resolve low frequencies easily but struggle with high frequencies.

The Neural Tangent Kernel (NTK) framework provides a theoretical explanation. In the infinite-width limit, neural network training dynamics are governed by a kernel whose eigenspectrum decays rapidly for high-frequency components. High-frequency eigenmodes have small eigenvalues, causing their corresponding function components to be learned slowly—if at all.

This spectral bias manifests practically when solutions contain multi-scale features. Consider a function combining smooth variations with localized sharp gradients:

$$u(x) = \sin(\pi x) + 0.1\sin(50\pi x) \tag{13.1}$$

Standard neural networks will capture the low-frequency $\sin(\pi x)$ component quickly while struggling with the high-frequency $\sin(50\pi x)$ component.

Several approaches address spectral bias:

1. **Fourier feature mappings**: Transform inputs through $\gamma(x) = [\cos(2\pi \mathbf{B}x), \sin(2\pi \mathbf{B}x)]$ where **B** contains learnable or fixed frequencies
2. **Multi-scale architectures**: Use parallel pathways targeting different frequency bands
3. **Adaptive activation functions**: Learnable parameters in activations that adjust frequency response
4. **Wavelet-based networks**: Recent Wav-KAN architectures theoretically control spectral bias through mother wavelet selection

None of these solutions is universal. Fourier features require problem-specific frequency tuning. Multi-scale architectures face exponential parameter growth in high dimensions. The search for architectures with inherently balanced spectral properties remains active.

13.3 Training Dynamics and Failure Modes

A major advance of 2024–2025 is the systematic characterization of *why* PINNs fail. Understanding failure modes enables targeted remedies rather than trial-and-error hyperparameter search.

13.3.1 *The Three Failure Mechanisms*

Mechanism 1: Loss Landscape Ill-Conditioning

(Fig. 13.1)

The PINN loss involves differential operators that introduce severe ill-conditioning into the optimization landscape. Computing the PDE residual requires derivatives of the network output, which amplifies sensitivity to network parameters. The Hessian of the loss function can have eigenvalues spanning many orders of magnitude—reported spreads of 10^6 or more are common.

This ill-conditioning causes gradient-based optimizers to make inconsistent progress across different parameter directions. Some directions improve rapidly while others stagnate, leading to the characteristic "stuck" behavior where the loss decreases initially then plateaus far from the true solution.

Mechanism 2: Loss Component Imbalance

A typical PINN loss combines multiple terms:

$$\mathcal{L} = \lambda_{\text{PDE}}\mathcal{L}_{\text{PDE}} + \lambda_{\text{BC}}\mathcal{L}_{\text{BC}} + \lambda_{\text{IC}}\mathcal{L}_{\text{IC}} + \lambda_{\text{data}}\mathcal{L}_{\text{data}} \tag{13.2}$$

These terms can have vastly different magnitudes and gradients. Without careful balancing, one term dominates optimization while others are effectively ignored. A network might achieve low PDE residual in the interior while completely failing to satisfy boundary conditions.

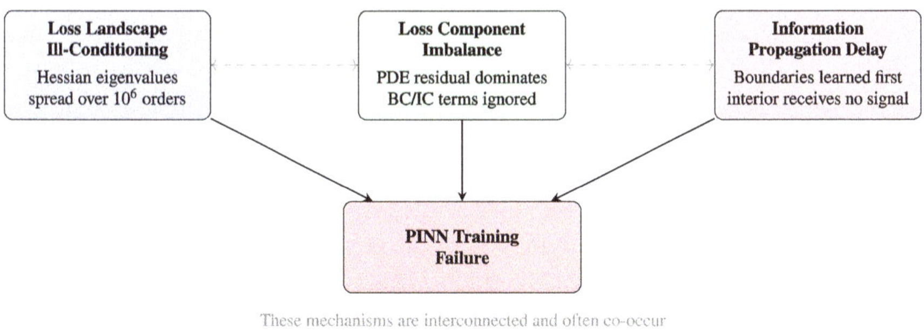

Fig. 13.1 The three primary mechanisms underlying PINN training failures. Understanding these enables targeted remediation strategies

Mechanism 3: Information Propagation Delay

In time-dependent problems, information must propagate from initial conditions through the spatiotemporal domain. Standard training samples points uniformly, but the network has no incentive to satisfy physics at interior points until boundary/initial information has propagated there. This creates regions where the network receives no learning signal, leading to arbitrary (and incorrect) predictions.

Warning: Failure Modes Often Co-occur

These three mechanisms are not independent. Ill-conditioning exacerbates imbalance because some loss components become numerically invisible. Imbalance delays information propagation because boundary terms are ignored. In practice, diagnosing which mechanism dominates requires careful analysis of training dynamics.

13.3.2 Solutions: From Heuristics to Principled Remedies

Second-Order Optimization:

First-order methods like Adam struggle with ill-conditioned landscapes. Second-order methods that approximate or use curvature information can navigate these landscapes more effectively.

The NysNewton-CG algorithm and Self-Scaled BFGS variants have demonstrated orders-of-magnitude accuracy improvements over Adam + L-BFGS baselines on standard PINN benchmarks. The computational overhead of curvature estimation is offset by dramatically faster convergence.

The FP64 Revolution:

A finding revealed that many failure modes attributed to optimization traps actually stem from *insufficient arithmetic precision*. The paper "FP64 is All You Need" [?] demonstrated that vanilla PINNs with double precision (FP64) and L-BFGS avoid failures that plague single-precision (FP32) implementations (Table 13.2).

This finding has immediate practical implications: default to FP64 for PINN training unless memory constraints absolutely require otherwise.

Table 13.2 Impact of arithmetic precision on PINN success rates

Configuration	Burgers (%)	Helmholtz (%)	Kovasznay (%)
FP32 + Adam	23	45	12
FP32 + L-BFGS	67	78	34
FP64 + Adam	56	72	45
FP64 + L-BFGS	**98**	**96**	**94**

Adaptive Loss Balancing:

Rather than manually tuning loss weights λ_i, adaptive schemes adjust weights during training based on gradient magnitudes or loss curvature. The Learning Rate Annealing method and GradNorm-based approaches have shown consistent improvements.

Causal and Temporal Weighting:

For time-dependent problems, causal training schemes respect the physical direction of information flow. Points near the initial condition are weighted heavily early in training; weights on later times increase as the solution at earlier times converges. This prevents the network from "guessing" at future times before learning the past.

Adaptive Collocation

Uniform sampling of collocation points is wasteful—many points fall in regions where the solution is smooth and easily captured, while complex regions are under-sampled. Adaptive methods like PINNACLE concentrate collocation points where the residual is largest, achieving equivalent accuracy with $10\times$ fewer training points.

13.4 Computational Scaling Challenges

While neural operators deliver the promised speedups for *inference* (demonstrated $45{,}000\times$ acceleration by FourCastNet for global weather), *training* costs remain a barrier for many applications.

13.4.1 The Curse of Dimensionality: Partially Lifted

Classical numerical methods scale exponentially with dimension. A d-dimensional problem discretized with n points per dimension requires $\mathcal{O}(n^d)$ storage and computation. This "curse of dimensionality" has motivated much SciML research.

Neural networks offer a path around this curse because their parameter count is independent of spatial discretization. A network with 1 million parameters can represent a function in 3 dimensions or 300 dimensions with the same memory footprint.

Breakthrough: 100,000-Dimensional PDEs

The Stochastic Dimension Gradient Descent (SDGD) method enables solving PDEs in *100,000 dimensions* on a single GPU in approximately 12 h. By decomposing gradients along random dimension subsets, SDGD avoids the memory explosion that defeats vanilla PINNs.

This breakthrough opens applications previously considered intractable:

- High-dimensional Black-Scholes equations in quantitative finance
- Many-body quantum systems in physics
- High-dimensional uncertainty quantification
- Optimal control in large state spaces

However, SDGD and related methods involve approximations whose impact on solution accuracy requires careful validation. The curse of dimensionality is not eliminated—it is transformed from a memory problem into a sampling and approximation problem.

13.4.2 Multi-scale Physics: Architectures Matter

When solutions contain features at vastly different scales—ubiquitous in turbulence, materials science, and climate—standard neural operators struggle. The fundamental issue is that a single architecture cannot efficiently represent both large-scale smooth variations and small-scale localized features.

The Multiscale Neural Operator (MNO) approach addresses this by explicitly modeling large scales while learning subgrid corrections. This reduces memory by up to 100× while recovering correct energy spectra in turbulent flows. Scale-consistent neural operators demonstrate generalization across Reynolds numbers spanning 250–10,000 through dimensionless learning on unit domains.

Table 13.3 Real-time inference achievements in SciML (as of 2025)

Application	Speedup	Latency	Method
Nuclear reactor CFD	1400×	<1 ms	DeepONet + TensorRT
Plasma instability prediction	10,000×	0.1 ms	Specialized CNN
Weather forecasting	45,000×	1 min/10 days	FourCastNet
Aerodynamic loads	500×	5 ms	Graph neural network

13.4.3 Real-Time Inference for Digital Twins

For many applications, the goal is not one-time simulation but continuous real-time prediction. Digital twins of physical systems require inference latency measured in milliseconds, not seconds.

These speedups enable applications previously impossible: active plasma instability avoidance running at 10+ kHz on tokamaks, real-time structural health monitoring, and immediate weather forecast updates as new observations arrive (Table 13.3).

13.5 Benchmarking, Reproducibility, and Trust

The lack of standardized evaluation has historically plagued SciML. Without consistent benchmarks, comparing methods is impossible and progress is difficult to measure.

13.5.1 The Benchmark Landscape

Recent efforts have substantially improved the situation as shown in Table 13.4.

Table 13.4 Major SciML benchmarks and their coverage

Benchmark	Coverage	Key contribution
PDEBench (2022)	11 time-dependent PDEs	First large-scale, FAIR-compliant dataset with baselines
PINNacle (2024)	20+ PDEs, $\sim$ 10 methods	Largest PINN variant comparison
APEBench (2024)	46 PDEs in 1D/2D/3D	Solver-in-the-loop evaluation with rollout metrics

Warning: Benchmark Limitations

Current benchmarks have significant blind spots:

- **Simple geometries**: Most use rectangular domains with periodic boundaries
- **Missing physics**: Multi-physics coupling rarely tested
- **No inverse problems**: Forward simulation dominates
- **Limited UQ**: Uncertainty quantification not standardized

Results on benchmarks may not transfer to complex real-world applications.

13.5.2 The Reproducibility Crisis

Reproducibility studies reveal troubling sensitivity to implementation details. Training variance from random seeds sometimes exceeds performance differences between methods. Hyperparameter sensitivity means published results may not reproduce with different (equally reasonable) settings (Fig. 13.2).

Addressing reproducibility requires:

- Reporting variance across multiple seeds, not just best results
- Sharing complete code and hyperparameters
- Standardized evaluation protocols
- Independent replication studies

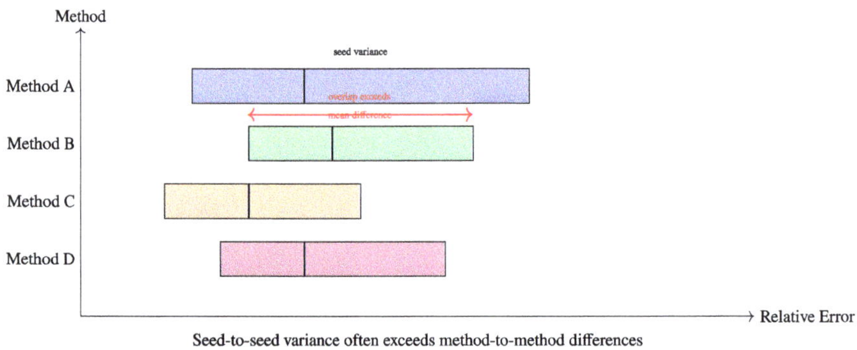

Fig. 13.2 Schematic illustration of reproducibility challenges: variance across random seeds can exceed differences between methods, making definitive comparisons difficult

13.5.3 *When Does SciML Actually Win?*

Honest assessment reveals that SciML methods outperform traditional numerical methods in specific regimes:

Where SciML Excels

1. **High dimensions** ($d \geq 10$): Classical methods face exponential scaling; neural networks do not
2. **Many-query scenarios**: Parameter sweeps, UQ, optimization—amortized training costs pay off
3. **Real-time requirements**: When inference speed matters more than solution precision
4. **Incomplete physics**: When governing equations are unknown or approximate

Where Traditional Methods Excel

1. **Single high-accuracy solutions**: Finite elements achieve 10^{-6} relative error versus 10^{-2} to 10^{-3} for neural methods
2. **Well-posed forward problems**: Established solvers have decades of validation
3. **Certification requirements**: Regulatory acceptance of neural surrogates lags
4. **Moderate dimensions**: 2D/3D problems on standard geometries

The frontier where SciML adds value is expanding, but claims of universal superiority are unsupported.

13.6 Software Ecosystem Fragmentation

The SciML software landscape fragments across three largely incompatible ecosystems, impeding reproducibility and adoption.

13.6.1 The Three Stacks

Julia's SciML Ecosystem

The SciML.ai organization maintains $\sim$ 200 packages spanning differential equations, neural differential equations, and physics-informed learning. The ecosystem offers comprehensive coverage and benchmarks claim 20–100× speedups over Python alternatives for certain workloads.

However, Julia's limited adoption outside scientific computing creates barriers. Industry practitioners comfortable with Python/PyTorch face a learning curve. Interoperability with existing ML infrastructure is limited.

Python's Fragmented Landscape

Python dominates ML broadly, but SciML-specific tooling splits across:

- **PyTorch ecosystem**: torchdiffeq, NeuralOperator, DeepXDE (partial)
- **JAX ecosystem**: Diffrax, jax-cfd, neural-tangents
- **TensorFlow ecosystem**: DeepXDE, TF-Agents for control

No framework accommodates neural networks from multiple backends. A researcher wanting to combine Lagrangian neural networks (JAX) with Fourier Neural Operators (PyTorch) faces significant integration work.

NVIDIA PhysicsNeMo

NVIDIA's PhysicsNeMo (formerly Modulus) provides industrial-scale infrastructure including multi-GPU training, deployment optimization, and integration with their hardware stack. Recent open-sourcing and incorporation of architectures like Transolver signal convergence.

Concerns remain about vendor lock-in and the requirement for NVIDIA hardware in an ecosystem where AMD and other accelerators are increasingly relevant (Fig. 13.3).

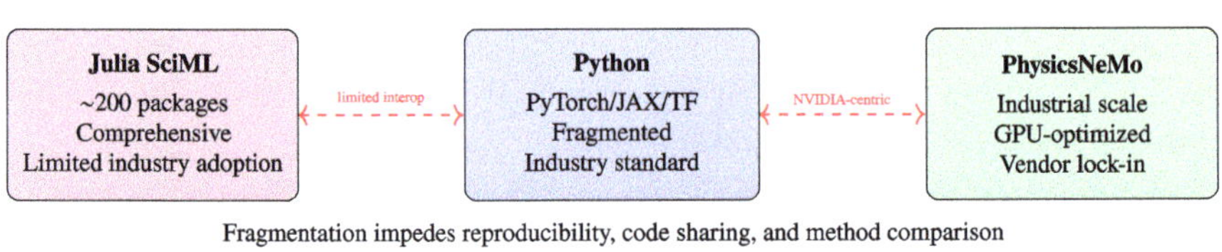

Fig. 13.3 The fragmented SciML software ecosystem. Researchers often cannot easily reproduce results across frameworks or combine methods from different ecosystems

13.6.2 Toward Convergence

Several developments suggest partial convergence:

- **ONNX and similar standards**: Enable model export across frameworks
- **Unified benchmarks**: PDEBench and others provide common ground
- **Cloud platforms**: Abstract away some framework differences
- **LLM-assisted code translation**: Reduces barrier to porting implementations

Full convergence is unlikely given different design philosophies, but improved interoperability is achievable.

13.7 Trust Deficits and Deployment Barriers

Beyond software, cultural and institutional barriers separate ML and scientific computing communities.

13.7.1 The Culture Gap

ML researchers evaluate on benchmark accuracy; domain scientists demand verification against physical laws, conservation properties, and experimental data. These different value systems lead to methods that satisfy one community but not the other (Table 13.5).

Table 13.5 Cultural differences between ML and scientific computing communities

Aspect	ML community	Scientific computing
Evaluation	Benchmark accuracy	Physical law satisfaction
Publication	NeurIPS, ICML	SIAM, JCP
Validation	Held-out test set	Manufactured solutions, experiments
Error measure	Relative L^2	Conservation, stability
Code sharing	GitHub, papers	Often proprietary

13.7.2 The False Confidence Problem

Critical Warning: PINNs Can Produce Plausible but Wrong Predictions

A PINN can produce predictions that appear physically plausible while masking critical omissions. The embedded physics represents idealized governing equations, not the full complexity of real systems. Unquantified uncertainty compounds this risk.

The governing equations in PINNs omit:

- Subgrid phenomena below the network's resolution
- Boundary effects not captured by simplified BCs
- Nonlinear couplings approximated or ignored
- Stochastic effects present in real systems

A network satisfying the loss function to high precision has learned the *idealized* physics, not necessarily the *real* physics. This creates risk in safety-critical applications where model predictions inform consequential decisions.

13.7.3 Regulatory Uncertainty

Regulatory frameworks have not adapted to ML-in-the-loop simulations. The NRC (nuclear), FDA (medical devices), FAA (aerospace), and other agencies lack standards for validating neural surrogates in certified systems.

This creates a catch-22:

- Adoption requires demonstrated safety
- Demonstrating safety requires operational experience
- Operational experience requires adoption

Hybrid approaches offer a pragmatic path: use neural surrogates within validated numerical frameworks, with the traditional solver providing guardrails and the neural component providing acceleration.

13.8 Foundation Models for Science

The foundation model paradigm—training massive models on diverse data, then fine-tuning for specific tasks—has arrived in scientific domains with striking success.

13.8.1 Weather and Climate: The Leading Edge

Real-World Impact: Foundation Models in Operations

As of 2025, AI weather models are no longer research curiosities. The European Centre for Medium-Range Weather Forecasts (ECMWF) runs AI models operationally. Microsoft's Aurora, Google's GenCast, and NVIDIA's FourCastNet demonstrate that foundation models can match or exceed physics-based forecasts at a fraction of the computational cost.

These models share key characteristics (Table 13.6):

- Training on massive curated datasets (reanalysis, simulations, experiments)
- Architectures incorporating physical structure (spherical harmonics, equivariance)
- Transfer learning enabling fine-tuning to related tasks
- Computational efficiency at inference enabling operational deployment

13.8.2 Materials Science: From Prediction to Generation

Google's GNoME discovered 2.2 million stable inorganic crystals—equivalent to 800 years of traditional research. Of these, 736 were independently synthesized experimentally, validating the predictions.

The frontier is moving from *prediction* (will this structure be stable?) to *generation* (design a structure with these properties). MatterGen (Nature 2025) advances generative design, producing stable crystal structures with 10× better energy optimization than prior methods.

13.8.3 Scaling Laws: Do They Transfer?

In language modeling, scaling laws predict that model performance improves predictably with increased compute, data, and parameters. Evidence suggests these laws partially transfer to scientific domains (Fig. 13.4):

Table 13.6 Scientific foundation models as of 2025

Model	Parameters	Domain	Key achievement
Aurora	1.3B	Atmosphere	5000× speedup, multi-task
GenCast	0.5B	Weather	Probabilistic ensemble, beats ECMWF
Prithvi WxC	2.3B	Climate	Subseasonal to seasonal forecasting
GNoME	–	Materials	2.2M stable crystals discovered
AlphaFold 3	–	Proteins	Nobel Prize 2024, biomolecular complexes

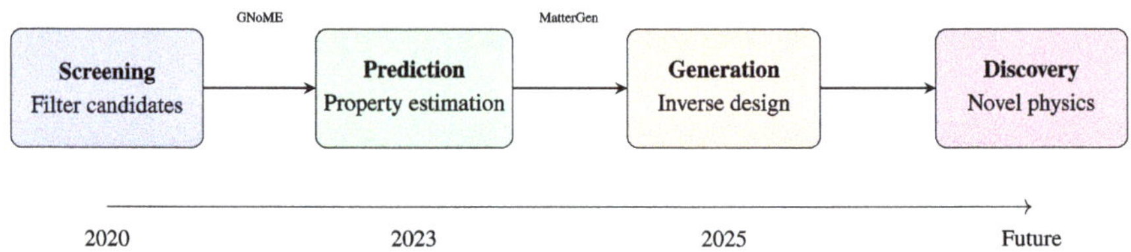

Fig. 13.4 Evolution of AI in materials science: from high-throughput screening to property prediction to inverse generative design, with the frontier moving toward genuine scientific discovery

- Weather models show consistent improvement with scale
- Materials models benefit from larger training databases
- Protein structure prediction improved dramatically with AlphaFold's scale

However, scientific domains differ from language in important ways:

- Physical constraints provide structure absent in text
- Data generation is expensive (simulations, experiments)
- Extrapolation beyond training distribution is essential

Whether compute-efficient scaling applies as strongly to physical prediction as to language remains an open question.

13.9 Neural Operator Architectures: Rapid Evolution

Fourier Neural Operators established the neural operator paradigm in 2021. The subsequent four years have seen rapid architectural innovation addressing their limitations.

13.9.1 Attention Meets Operators

Key Development: Physics-Attention

Transolver (ICML 2024 Spotlight) introduced Physics-Attention, which groups mesh points into learned physical states rather than treating all points uniformly. This provides geometry-general capability—the same architecture handles structured grids, unstructured meshes, and point clouds.

The insight is that physical systems have intrinsic structure: boundary layers, shocks, vortex cores. Rather than treating all spatial points equally, Physics-Attention learns to identify and attend to physically meaningful groupings.

LinearNO (NeurIPS 2025) revealed that Physics-Attention is a special case of linear attention, achieving:

- 40% parameter reduction
- 36% computational savings
- State-of-the-art on six benchmarks

This suggests the field is converging toward efficient attention-based architectures as the backbone for neural operators.

13.9.2 Handling Complex Geometries

Standard FNO assumes regular grids with periodic boundaries—far from real engineering geometries. Several architectures address this:

- **Geo-FNO**: Diffeomorphic mapping from complex domains to simple ones
- **Graph Neural Operators**: Operate on unstructured meshes directly
- **Geometry-Aware Operator Transformer**: Multi-scale attention with geometry embeddings

The Geometry-Aware Operator Transformer achieves 4–9 orders of magnitude speedup over classical solvers on complex geometries, demonstrating that the efficiency gains of neural operators are not limited to simple domains.

13.9.3 Multi-Physics and Pretraining

For problems coupling multiple physics (fluid-structure, electromagnetics-thermal), specialized approaches are emerging:

CoDA-NO (NeurIPS 2024) enables self-supervised pretraining across different PDE systems. A model pretrained on heat equations, wave equations, and Navier-Stokes can be fine-tuned with few shots when these physics couple in new ways.

Latent Neural Operators reduce computational cost by operating in compressed latent spaces rather than physical space. This is analogous to how transformers operate on token embeddings rather than raw pixels.

13.9.4 Persistent Challenges

Despite rapid progress, fundamental challenges persist:

- **Discontinuities and shocks**: Neural operators struggle with non-smooth solutions
- **Long-time integration**: Errors accumulate over many time steps

- **Training data requirements**: Generating high-quality training data often requires expensive simulations
- **Out-of-distribution generalization**: Performance degrades for inputs far from training distribution

Recurrent Neural Operators (2025) and memory-enhanced architectures attempt to address long-time integration through specialized training schemes, but the fundamental accuracy-stability tradeoff remains.

13.10 From Soft to Hard Constraints

A philosophical shift is underway in how physical knowledge enters neural networks.

13.10.1 The Constraint Spectrum

Soft Constraints (Loss-Based)

(Fig. 13.5) Standard PINNs add physics as penalty terms in the loss. The network is *encouraged* to satisfy physical laws but not *required* to. Violations are penalized but possible.

Hard Constraints (Architecture-Based)

Hard constraints encode physics in the network architecture itself. The output *cannot* violate the constraint regardless of parameters.

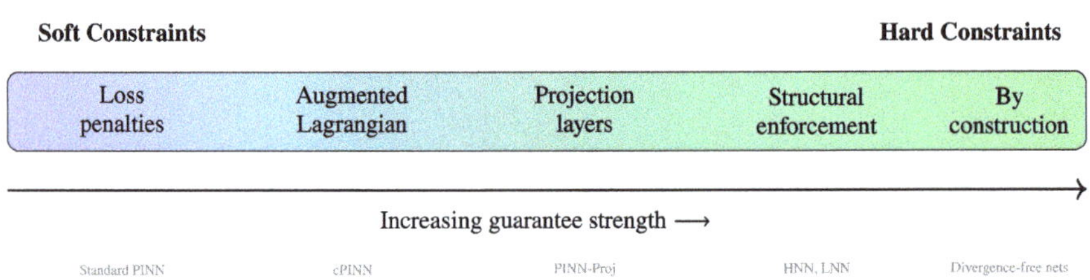

Fig. 13.5 The constraint spectrum: from soft penalties that encourage but do not guarantee physical constraints, to hard constraints that cannot be violated by construction

13.10.2 Recent Hard Constraint Methods

- **PINN-Proj** (IBM, NeurIPS 2024): Projects network output onto constraint manifold after each training step, guaranteeing conservation laws
- **Riemann Tensor Neural Networks** (2025): Build divergence-free symmetric tensor conditions into the architecture, enforcing mass and momentum conservation by construction
- **AsPINN** (2024): Uses adaptive symmetry-recomposition for strict mathematical constraints
- **Hamiltonian/Lagrangian Neural Networks**: Parameterize conserved quantities directly, ensuring energy conservation

Tradeoff: Flexibility Versus Guarantee

Hard constraints sacrifice model flexibility for physical guarantees. A divergence-free network *cannot* represent flows with sources or sinks, even if the training data comes from such a system. Choosing the right constraint level requires understanding the physics and the application's tolerance for constraint violation.

As SciML moves toward safety-critical applications—nuclear reactor monitoring, medical devices, autonomous vehicles—expect the shift toward hard constraints to accelerate. Regulatory acceptance may require provable guarantees that soft constraints cannot provide.

13.11 AI for Scientific Discovery

The most ambitious SciML applications aim not merely to accelerate simulation but to discover new science.

13.11.1 Equation Discovery

Given observational data, can we discover the governing equations? This "inverse" problem has a long history (symbolic regression, sparse identification of nonlinear dynamics), but LLMs have transformed the landscape.

Breakthrough: LLM-Guided Equation Discovery

LLM-SR (ICML 2024) uses large language models to propose equation "skeletons" incorporating domain knowledge, then optimizes parameters from data. The approach substantially outperforms pure symbolic regression on out-of-domain generalization because the LLM encodes physical intuition from its training corpus.

The **Scientific Generative Agent** (ICML 2024) goes further, coupling LLM hypothesis generation with simulation-based optimization in a bilevel framework:

1. LLM proposes candidate equations based on prior knowledge
2. Simulator evaluates predictions against data
3. Feedback refines the LLM's proposals
4. Iteration converges toward equations matching observations

This approach has discovered constitutive laws that exceed human-designed baselines on benchmark problems.

13.11.2 *Autonomous Scientific Agents*

Emerging systems integrate multiple capabilities:

- **AtomAgents**: LLM reasoning + materials simulations for alloy design
- **Text-to-Simulation**: Natural language $\rightarrow$ geometry + mesh + solver setup
- **Autonomous labs**: A-Lab (Berkeley) achieves 71% synthesis success rate with minimal human intervention (Fig. 13.6)

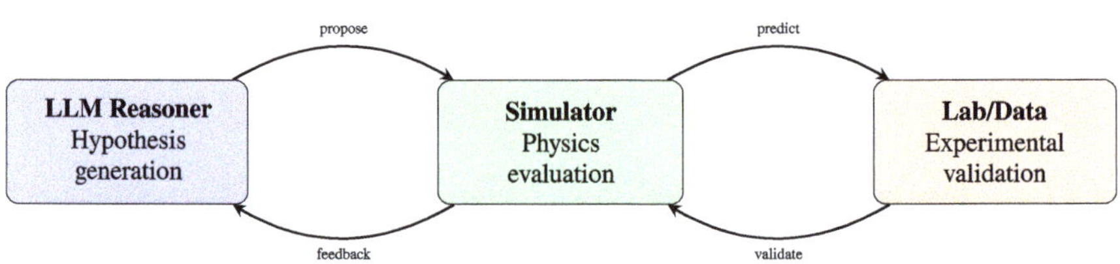

Closed-loop systems combining reasoning, simulation, and experiment represent the frontier of autonomous scientific discovery

Fig. 13.6 Architecture of autonomous scientific agents: LLMs propose hypotheses, simulators evaluate predictions, and experimental data provides validation, with feedback loops enabling iterative refinement

13.11.3 *The Memorization Challenge*

Warning: Discovery Versus Memorization

LLMs trained on scientific literature may "discover" relationships they have merely seen before. Distinguishing genuine novel discovery from sophisticated pattern matching on memorized knowledge is a fundamental challenge for AI-driven science.

Current LLMs achieve $\sim 40\%$ success on research-level physics problems. This is non-trivial but far from reliable. More concerning, it is difficult to know *which* 40%—the model may succeed by recalling similar problems from training rather than genuine reasoning.

Benchmarks resistant to memorization—testing genuinely novel hypotheses on systems not in training data—remain underdeveloped. This is a critical gap for validating claims of AI scientific discovery.

13.12 Domain-Specific Frontiers

Each scientific domain presents unique challenges reflecting its data regimes, physics complexity, and operational requirements.

13.12.1 *Climate and Weather*

Weather prediction has seen the most dramatic SciML progress, but critical gaps remain:

- **Subseasonal-to-seasonal (S2S) forecasting**: The 2–6 week "predictability desert" requires atmosphere-ocean-land couplings that current models handle poorly
- **Extreme events**: MSE training objectives systematically underestimate rare extremes because penalizing outlier predictions reduces training loss
- **Climate projections**: Future conditions lie outside the training distribution of historical data—a fundamental challenge for data-driven methods
- **Regional downscaling**: Global models at ~ 25 km resolution cannot resolve local features important for impact assessment

Table 13.7 Accuracy versus speed tradeoffs in machine learning potentials

Method	Accuracy	Speed	Coverage
DFT	Reference	Slow	Any system
Classical potentials	Low	1×	Specific systems
NequIP/Allegro	High	100× slower than classical	Trained systems
MACE	Very high	10× slower than classical	Trained systems
Universal potentials	Medium	Fast	Broad chemistry

13.12.2 Materials Science

Materials discovery confronts accuracy-speed tradeoffs in interatomic potentials (Table 13.7):

The gap between computationally predicted stable structures and experimentally synthesizable materials remains substantial. GNoME's 2.2 million stable crystals are thermodynamically stable in simulation, but kinetic barriers, precursor availability, and synthesis conditions determine experimental reality.

13.12.3 Drug Discovery

AlphaFold3 (Nobel Prize 2024) predicts protein-ligand complex structures with unprecedented accuracy. Yet its impact on drug discovery is limited by a critical gap: **structure does not reliably predict binding affinity**.

The Structure-Affinity Gap

AlphaFold3's ranking metrics show minimal correlation with experimental binding affinities—the property that determines drug efficacy. A structure prediction breakthrough has not (yet) translated to a drug design breakthrough.

Approximately 40% of preclinical drug candidates fail due to poor absorption, distribution, metabolism, excretion, and toxicity (ADMET) properties. ML predictions for these properties remain limited by:

- Small, biased training datasets
- Poor translation from in vitro to in vivo
- Rare but critical failure modes underrepresented in training

13.12.4 *Fusion Energy*

Fusion plasma control has benefited dramatically from deep reinforcement learning, with systems now achieving active instability avoidance at 10+ kHz on tokamaks like DIII-D.

The unique challenge: **training data comes from smaller tokamaks, but deployment targets ITER**—a burning-plasma regime with self-heating alpha particles absent from all training distributions. This is perhaps the most consequential out-of-distribution extrapolation in any SciML application.

13.12.5 *Turbulence and Fluid Dynamics*

Data-driven turbulence modeling has seen tensor basis neural networks that preserve Galilean invariance, with new curated RANS datasets enabling reproducible benchmarking.

The persistent challenge is generalizability: models trained on separated flows degrade on wall-attached flows. Data-driven corrections calibrated to specific flow regimes fail to extrapolate across regimes. The fundamental question—can ML learn universal features of turbulence, or only regime-specific corrections—remains open.

13.13 Uncertainty Quantification: The Missing Theory

For scientific applications, predictions without uncertainty estimates are often useless. Yet SciML uncertainty quantification theory lags practice.

13.13.1 *Current Approaches and Their Limitations*

See Table 13.8.

Table 13.8 UQ methods for SciML: capabilities and limitations

Method	Strength	Limitation
Bayesian PINNs (HMC)	Principled uncertainty	Scales poorly to large networks
Variational inference	Scalable	Underestimates uncertainty
Ensembles	Simple, practical	Expensive, theoretically ad hoc
MC Dropout	Cheap	Often poorly calibrated
Conformal prediction	Distribution-free guarantees	Requires exchangeability

The Calibration Problem

No theory guarantees that SciML uncertainty estimates are well-calibrated—that 95% prediction intervals actually contain the truth 95% of the time. Empirical studies frequently show overconfidence: intervals are too narrow, missing the truth more often than their nominal coverage suggests.

13.13.2 Emerging Directions

LVM-GP Framework: Couples latent variable models with Gaussian process priors, matching Bayesian PINN quality with better scalability.

Extended Fiducial Inference (2025): Balances physics priors and observational data for both prediction uncertainty and model form uncertainty.

Physics-Informed Conformal Prediction: Adapts conformal methods to exploit physical constraints, potentially improving coverage guarantees.

The fundamental challenge remains: in scientific applications, we care about uncertainty in regions where we have no validation data (extrapolation). Classical UQ theory provides guarantees primarily for interpolation. Extending guarantees to the extrapolation regime essential for scientific prediction is a major open problem.

13.14 The Road Ahead: 2025–2030

Based on current trajectories, we can identify several likely developments over the next five years (Fig. 13.7).

13.14.1 Foundation Models Will Become Multi-Physics

Current foundation models target single domains (atmosphere, materials). The next generation will pretrain across coupled physics, enabling transfer to fluid-structure interaction, electrochemistry, and climate-ocean coupling. Early work on CoDA-NO demonstrates the viability of this approach.

Based on current trajectories, we can identify several likely developments over the next five years.

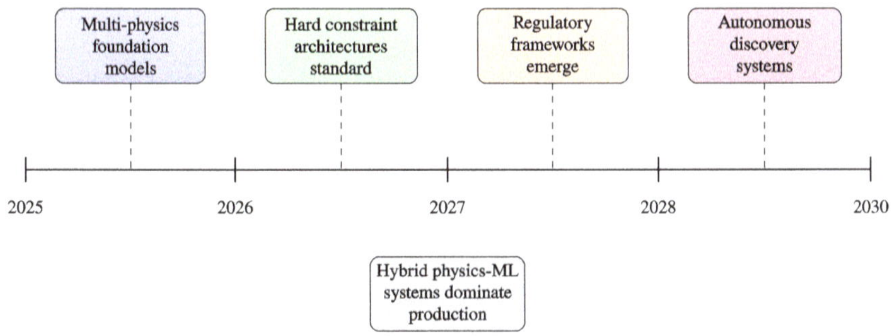

Fig. 13.7 Projected SciML developments through 2030. The trajectory points toward hybrid systems that combine neural efficiency with physics guarantees

13.14.2 Hard Physical Constraints Will Become Standard

As applications grow safety-critical, architectures guaranteeing conservation will replace soft physics-informed losses. The flexibility-guarantee tradeoff will shift toward guarantees, particularly in regulated domains.

13.14.3 Hybrid Systems Will Dominate Production

Pure neural surrogates rarely replace validated numerical methods outright. The production pattern will be neural operators embedded within traditional solvers:

- As learned preconditioners accelerating iterative methods
- As subgrid models in multiscale simulations
- As fast approximate solutions refined by physics-based correction
- As reduced-order models for real-time control

13.14.4 Autonomous Scientific Agents Will Emerge Cautiously

LLM-driven hypothesis generation and experimental design will accelerate discovery loops, but adoption will be cautious. Distinguishing genuine novelty from memorized knowledge requires new validation frameworks. The most impactful applications will be in domains with:

- Clear experimental validation pathways
- High cost of human hypothesis generation
- Well-defined success metrics

13.14.5 Domain Expertise Will Remain Essential

The vision of domain-agnostic AI that solves arbitrary scientific problems has not materialized and is unlikely to. The most successful SciML systems embed deep domain knowledge in architecture, training data curation, loss function design, and evaluation protocols.

> **Key Insight: The Role of Domain Knowledge**
>
> SciML does not replace domain expertise—it amplifies it. The researchers making the biggest impacts are those who understand both the underlying physics and the ML methods, enabling them to design systems that leverage the strengths of each.

13.15 Open Problems and Research Directions

We conclude with a summary of the most pressing open problems, organized by theme.

13.15.1 Theoretical Foundations

1. **Convergence for nonlinear PDEs**: Extend rigorous convergence guarantees beyond linear problems to Navier-Stokes, reaction-diffusion, and other nonlinear systems of practical interest.
2. **Tight generalization bounds**: Develop bounds that predict actual performance, not just worst-case scaling, to guide architecture selection.
3. **Spectral bias control**: Design architectures with inherently balanced frequency response, eliminating the need for problem-specific tuning.
4. **Training dynamics theory**: Understand why gradient descent succeeds in non-convex PINN landscapes to identify conditions for reliable training.

13.15.2 Computational Methods

1. **Efficient multi-scale learning**: Architectures that capture features across scales without exponential parameter growth.
2. **Long-horizon stability**: Training methods that prevent error accumulation over extended time integration.
3. **Mesh-free methods for complex geometries**: Neural operators that handle realistic engineering geometries as easily as periodic boxes.
4. **Active learning for data efficiency**: Optimal selection of simulation/experimental data points to minimize training cost.

13.15.3 Uncertainty and Reliability

1. **Calibrated uncertainty for extrapolation**: Principled uncertainty estimates valid outside the training distribution.
2. **Failure mode detection**: Methods to identify when a trained model is operating outside its validity regime.
3. **Verification and validation standards**: Regulatory-acceptable procedures for certifying neural surrogates in safety-critical applications.

13.15.4 Discovery and Autonomy

1. **Disentangling discovery from memorization**: Benchmarks and methods that verify genuine scientific novelty.
2. **Closed-loop discovery systems**: Integration of hypothesis generation, simulation, and experimental validation.
3. **Interpretable scientific AI**: Extracting human-understandable scientific insights from neural network parameters and predictions.

13.16 Conclusion

Scientific machine learning stands at an inflection point. The foundational methods—physics-informed neural networks, neural operators, differentiable programming—have proven their value across weather prediction, materials discovery, and beyond. Foundation models trained on massive scientific datasets achieve unprecedented accuracy. Real-time inference enables applications previously considered impossible.

Yet honest assessment reveals substantial work ahead. Theoretical foundations lag empirical practice: we can prove convergence only for restricted settings far simpler than the problems we routinely solve. Training failures are now characterized but not eliminated. Software ecosystems fragment across incompatible stacks. Trust deficits slow adoption in production science. Uncertainty quantification lacks the guarantees scientists need.

The most important conceptual development is a shift in perspective. The field is moving:

- From **physics-informed** to **physics-guaranteed**
- From **surrogate modeling** to **scientific discovery**
- From **isolated methods** to **integrated foundation models**
- From **ML replacing physics** to **ML amplifying physics**

The central task for the coming years is clear: closing the gap between what SciML methods can do in principle and what scientists can trust them to do in practice. This requires advances across theory, algorithms, software, and validation methodology. It requires collaboration between ML researchers, domain scientists, and practitioners. Most of all, it requires intellectual honesty about both the remarkable achievements and the genuine limitations of this young field.

A Note to the Reader

The field described in this chapter is evolving rapidly. Results considered state-of-the-art as this textbook goes to press may be superseded within months. We encourage readers to engage with the primary literature, attend conferences, and participate in the open-source communities driving this field forward. The challenges are substantial, but so is the opportunity to contribute to a fundamental transformation in how humanity does science.

Chapter Summary

Key Takeaways:

- The theory-practice gap remains SciML's central challenge: convergence proofs exist only for simplified settings far from practical applications
- PINN training failures stem from three identifiable mechanisms: ill-conditioning, loss imbalance, and information propagation delays; each has emerging solutions
- SciML excels in specific regimes: high dimensions, many-query scenarios, real-time inference, incomplete physics; traditional methods remain superior for single high-accuracy solutions

- Foundation models have transformed weather prediction and materials discovery; multi-physics extension is the next frontier
- The shift from soft to hard constraints reflects growing emphasis on physical guarantees for safety-critical applications
- AI for scientific discovery shows promise (LLM-guided equation discovery, autonomous agents) but distinguishing genuine discovery from memorization remains challenging
- Domain expertise remains essential: SciML amplifies rather than replaces scientific knowledge
- Hybrid physics-ML systems will dominate production deployments, combining neural efficiency with physics guarantees

GPSR Compliance
The European Union's (EU) General Product Safety Regulation (GPSR) is a set of rules that requires consumer products to be safe and our obligations to ensure this.

If you have any concerns about our products, you can contact us on

ProductSafety@springernature.com

In case Publisher is established outside the EU, the EU authorized representative is:

Springer Nature Customer Service Center GmbH
Europaplatz 3
69115 Heidelberg, Germany

www.ingramcontent.com/pod-product-compliance
Ingram Content Group UK Ltd.
Pitfield, Milton Keynes, MK11 3LW, UK
UKHW020435270926
475829UK00013B/9

9789819578559